Stochastic Population Processes

Analysis, Approximations, Simulations

Eric Renshaw

OXFORD
UNIVERSITY PRESS

OXFORD
UNIVERSITY PRESS

Great Clarendon Street, Oxford, OX2 6DP,
United Kingdom

Oxford University Press is a department of the University of Oxford.
It furthers the University's objective of excellence in research, scholarship,
and education by publishing worldwide. Oxford is a registered trade mark of
Oxford University Press in the UK and in certain other countries

© Eric Renshaw 2011

The moral rights of the author have been asserted

First published 2011
First published in paperback 2015

Impression: 1

All rights reserved. No part of this publication may be reproduced, stored in
a retrieval system, or transmitted, in any form or by any means, without the
prior permission in writing of Oxford University Press, or as expressly permitted
by law, by licence or under terms agreed with the appropriate reprographics
rights organization. Enquiries concerning reproduction outside the scope of the
above should be sent to the Rights Department, Oxford University Press, at the
address above

You must not circulate this work in any other form
and you must impose this same condition on any acquirer

Published in the United States of America by Oxford University Press
198 Madison Avenue, New York, NY 10016, United States of America

British Library Cataloguing in Publication Data
Data available

Library of Congress Cataloging in Publication Data
Data available

ISBN 978-0-19-957531-2 (Hbk.)
ISBN 978-0-19-873906-7 (Pbk.)

Printed and bound in Great Britain by
Clays Ltd, St Ives plc

Links to third party websites are provided by Oxford in good faith and
for information only. Oxford disclaims any responsibility for the materials
contained in any third party website referenced in this work.

Preface

In the real world the vast majority of random processes have no memory. That is, the next step in their development depends only on the current state of the system and not on its previous history. Stochastic realizations are then defined in terms of the sequence of event–time pairs, whence the equations for the probability that a system is of size n at time t are straightforward to write down. So from a purely applied perspective the mathematical challenge lies in determining their solution in order to enhance understanding of the particular system under study. Now preceding the Second World War deep and insightful theoretical developments, pioneered by A.A. Markov and A.N. Kolmogorov, enabled many pure mathematically oriented results to be constructed. Whilst these have considerable intrinsic beauty, which stimulated great interest amongst pure probabilists, applied probabilists seeking solutions to systems oriented problems gained relatively little from them. Serious development over a rapidly widening spectrum of applications only really started to grow following a series of thought-provoking and innovative papers read to the Royal Statistical Society in 1949. Although studies in this new field swiftly prospered, with new stochastic results constantly being applied to fresh areas of application, as the associated level of mathematical difficulty increased, more and more researchers started to switch their attention back onto purer probability problems. For these are more likely to lead to intellectually appealing closed-form solutions. Unfortunately, success usually depended on making unrealistic, and hence detrimental, assumptions about the way in which particular scenarios develop. In recent years, however, there has been a resurgence of interest in constructing approximate theoretical solutions and simulation analyses. The latter has prospered greatly by the availability of fast computing power on inexpensive personal computers, which enables quite complex problems to be studied within a reasonable time frame.

This text therefore has three aims: (1) to introduce a variety of stochastic processes that possess relatively simple closed-form solutions; (2) to present a range of approximation techniques which show that in more complex problems considerable behavioural information may still be extracted even when the exact probability structure is mathematically intractable to direct solution; and, (3) to illustrate how to construct basic simulation algorithms which not only yield insight into the way that particular systems develop, but which also highlight intriguing, and previously unknown, properties that may then be studied in their own right. The content is wholly based around stochastic population processes, and is driven almost entirely by the underlying Kolmogorov probability equations for population size. Thus it is not intended as a text for pure-minded mathematicians who require a deep theoretical understanding, but for researchers who want answers to real questions; there are plenty of theoretically oriented books on Markov processes already on the market. What makes this book different is that it concentrates on *practical application*. That is, we start with a problem, create a realistic model which encapsulates its main properties, and then construct and apply appropriate stochastic techniques in order to solve it. This is in direct contrast to 'traditional' books on stochastic processes, which are generally heavy in mathematical theory and attempt to force problems into a rigid framework of closed-form solutions. Because the text provides a personal tour which details the author's experiences in applying stochastic processes over a wide range of different applied arenas, the examples are slanted towards ecological and physical applications. However, only a little imagination is required in order to extend the processes discussed to problems generic to engineering and chemistry. Indeed, the book provides a rich source of ideas and reference not only for researchers in applied probability, but also for any scientists or engineers working with random processes who are prepared to take a flexible approach to the problems they are trying to solve.

Little of the material is covered at a deep mathematical level, so the text should be readily accessible to a wide variety of practitioners. Moreover, considerable effort has been made to ensure that it ties together as a unified whole, with connections being highlighted between all the component processes and issues wherever possible. Many outstanding problems are identified, and signposts are provided en route that point to further reading for readers who wish to acquire a deeper analytic knowledge. One of the main objectives is to use the Kolmogorov equations to expose the high degree of linkage which exists between apparently unconnected processes. For in this way, results and techniques for one may be carried over directly to the other without us having constantly to reinvent the wheel. So in this vein many of the specific processes presented can be thought of as being examples of general application. With regards to teaching, the varied nature of the material means that it is ideal for constructing both honours level undergraduate and graduate courses. Moreover, whilst many readers may wish to progress through each chapter in sequence, it is also easy to treat the book as a toolbox that one can dip into in order to select specific analytic and computational techniques.

Given that a completely definitive treatment would require a huge number of volumes, material has had to be chosen selectively. Examples have been kept sufficiently simple to make the underlying ideas easy to understand, yet sufficiently structured to convince readers that these techniques and approaches may be applied to any problem, no matter how complex. In the first half of the book basic principles and ideas are introduced through the study of single-species populations; in the second these are extended to cover a variety of different scenarios involving bivariate populations and processes that develop in both space and time. Chapter 2 explains the basic principles which underlie the concept of stochastic population dynamics by considering simple forms of model structure for which the resulting mathematical analyses are sufficiently transparent to enable useful conclusions to be drawn from them. This provides the opportunity to explain: the relationship between moments and probabilities; the role undertaken by various forms of generating function; and, a general simulation approach based on event–time pairs. Since any Markov population that either increases or decreases by one at each event can be expressed in terms of a general birth–death process, this topic is introduced at the start of Chapter 3. The examples presented demonstrate models with single and multiple equilibria, and highlight the role played by moment closure and the saddlepoint approximation in the derivation of approximate stochastic solutions. Diffusion and perturbation techniques are also introduced at this point. Attention now switches from continuous to discrete time, with random walks and Markov chains being treated in Chapters 4 and 5. Consideration of path analyses for random walks (Chapter 4) leads on to first passage and return probabilities and the Arc Sine Law. Both absorbing and reflecting barriers are considered in some depth. Since the assumption of independent steps is not universally true, a classic case being the movement of share prices driven by market sentiment, we introduce the correlated random walk. Further generalizations include the extension to Markov chains and branching processes (Chapter 5). Having dealt with processes in discrete space, and discrete and continuous time, Chapter 6 investigates Markov processes in continuous space and time. The material follows a natural progression through the Wiener process, the Fokker–Planck diffusion equation and the Ornstein–Uhlenbeck process, and tracks the relationships between them. When barriers are introduced theoretical development is not straightforward, so care needs to be taken when constructing solutions.

Individuals often do not exist in isolation but instead co-exist with individuals from many other species, and thereby exhibit between-species interaction. Unfortunately, not all the progress made with the analysis of univariate populations carries through to the multivariate case, since the resulting stochastic equations are usually nonlinear and intractable. So developing an understanding of such processes involves greater reliance on simulation and approximation techniques. Chapter 7 examines the general bivariate process, and illustrates the basic approaches involved by first developing a simple process for which the preceding methods of solution do carry across. The univariate saddlepoint approximation is then extended to cover multivariate processes, with cumulant truncation being covered in some detail. A bivariate process of considerable practical importance involves employing total counts as a second variable, especially since this can generate intriguing effects in which the structure of the occupation probabilities depends on whether the population size is odd or even. Various examples are presented, including a family of processes that generates different probability distributions which have the same moment structure. Moreover, although complex systems often exhibit extremely rich dynamic behaviour, gaining a direct understanding of the underlying structure may not be possible if the system remains hidden and observations can only be made on external counts. We show that a surprisingly high level of analytic information can still be gained from the counts alone, and demonstrate how to employ Markov chain Monte Carlo techniques in such situations. Chapter 8 derives specific results for the bivariate logistic process in the contexts of competition, predator–prey and epidemic processes. Here minor modifications to the transition rates cause substantial changes in the structure of stochastic realizations. Earlier discussions on cumulative size and power-law processes are woven into the discussion, and strong emphasis is placed on the development of computer simulation algorithms.

Since many processes develop through both space and time, Chapter 9 introduces a spatial dimension by presuming that individuals now develop over a number of interacting sites. The level of mathematical tractability swiftly decreases as the level of spatial interaction increases, so a variety of exact and approximate approaches are introduced for constructing moments and probability distributions. The techniques covered include slightly connected processes, generating functions, integral equations and Riccati matrix representations. The removal of boundary effects eases analytic progress, with wavefront profiles and their velocities of propagation producing interesting results. For their construction not only ties in with the earlier saddlepoint work, but for non-exponentially bounded contact distributions the velocities explode. Moreover, moving from single to multi-type systems generates Turing processes in which inherently unstable stochastic systems can be made stable by injecting spatial structure. Whilst the concept of a 'stochastic dynamic' enables highly volatile systems which exhibit frequent local extinctions to persist indefinitely. Further insight is gained by replacing Markov processes by approximating Markov chains, thereby giving rise to stochastic cellular automata; two simple examples highlight the spread and control of forest fires and foot-and-mouth disease. Although the exact event–time pairs algorithm is easy to construct, and hence is ideal for simulating fairly simple dynamical systems, it is typically too computationally expensive to be of use in more complex situations. A suite of alternative approximate strategies is therefore presented based on time increments, tau-leaping, Langevin-leaping and stochastic differential equations.

Finally, in Chapter 10 we consider two spatial–temporal extensions. First, large-scale spatial interaction allows the general contact distribution to possess a power-law structure. Spectral results are derived which intertwine with those obtained via a parallel fractional integration route. The solution to the corresponding inverse problem is particularly useful since it enables us to construct processes which possess any observed spectral structure; examples show the generation of sea waves and optical caustic surfaces. Second, employing marked point processes allows us to replace the discrete (usually lattice) site-structure by a continuous space domain. To enable fast compute times, and to draw parallels with Gibbs processes, analysis proceeds within a deterministic space–time framework with stochasticity being induced through random arrivals and departures; interactive death occurs when the mark size decays to zero. As the resulting pattern structures depend heavily on the chosen growth and spatial interaction functions, a wide variety of different pattern types may be produced, many of which have direct practical application to packed systems in physics, forestry, geology, chemistry and ecology. Specific issues covered include convergence, the generation of fractal structure, and the construction of densely packed systems for particles moving under interaction pressure.

The potential for extending many of the single- and multi-species model structures considered in this text is virtually without limit. Not only may we have different types of species, or marks, involving competition, attraction, predator–prey, infection, chemical reaction, etc., but severe edge-effect problems may arise when these processes are placed in a finite bounded domain. The aim is therefore to ensure that readers are sufficiently enthused to explore both theory and application in this ever-widening arena of applied stochastic processes, which, for far too long, has been ignored in favour of developing appealing results in pure probability theory that often have scant practical relevance to the real world. The development of genuinely applied stochastic temporal and spatial–temporal analyses generates an exciting field of study which provides great mathematical, statistical and computational challenges.

Contents

1	**Introduction**	1
	1.1 Some simple stochastic processes	3
	1.2 Single-species population dynamics	7
	1.3 Bivariate populations	15
	1.4 Spatial–temporal processes	21
2	**Simple Markov population processes**	31
	2.1 Simple Poisson process	31
	2.1.1 Time to subsequent events	31
	2.1.2 Number of events	34
	2.1.3 General moments	37
	2.1.4 Simulation	40
	2.1.5 Generalizations	43
	2.1.6 Do moments uniquely define a distribution?	47
	2.2 Pure death process	50
	2.2.1 Stochastic model	50
	2.2.2 Reverse transition probabilities	53
	2.2.3 Bridge probabilities	54
	2.3 Pure birth process	55
	2.3.1 Stochastic model	56
	2.3.2 Laplace transform solution	57
	2.3.3 Solution via the p.g.f.	59
	2.3.4 Time to a given state	60
	2.3.5 Generalized birth process	63
	2.4 Simple birth–death process	70
	2.4.1 Stochastic model	70
	2.4.2 Extinction	74
	2.4.3 A simple mixture representation	76
	2.4.4 The backward equations	77
	2.4.5 Time to a given state	79
	2.4.6 Reverse transition probabilities	80
	2.4.7 Simulating the simple birth–death process	81
	2.4.8 The dominant leader process	85
	2.5 Simple immigration–birth–death process	86
	2.5.1 Stochastic model ($\lambda = 0$)	87
	2.5.2 Stochastic model ($\lambda \geq 0$)	88
	2.5.3 Equilibrium probabilities	90
	2.5.4 Perfect simulation	91

	2.6	Simple immigration–emigration process	95
		2.6.1 Equilibrium distribution	96
		2.6.2 Time-dependent solutions	97
	2.7	Batch events	101
		2.7.1 Birth–mass annihilation and immigration process	101
		2.7.2 Mass immigration–death process	103
		2.7.3 Regenerative phenomena	104
3	**General Markov population processes**		**107**
	3.1	Classification of states	108
	3.2	Equilibrium solutions	117
		3.2.1 Logistic population growth	123
		3.2.2 Simulated realizations	126
		3.2.3 Multiple equilibria	129
		3.2.4 Power-law processes	132
	3.3	Time-dependent solutions	141
		3.3.1 Approximate solutions	141
		3.3.2 Probability of ultimate extinction	143
		3.3.3 Mean time to ultimate extinction	145
	3.4	Moment closure	148
		3.4.1 Cumulant equations	149
	3.5	Avoiding the Kolmogorov equations	153
		3.5.1 Generating cumulant equations directly	153
		3.5.2 Local approximations	156
		3.5.3 Application to Africanized honey bees	159
	3.6	Diffusion approximations	164
		3.6.1 Equilibrium probabilities	165
		3.6.2 Perturbation methods	168
		3.6.3 Additive mass-immigration process	175
	3.7	The saddlepoint approximation	181
		3.7.1 Basic derivation	183
		3.7.2 Examples	185
		3.7.3 Relationship with the Method of Steepest Descents	188
		3.7.4 The truncated saddlepoint approximation	190
		3.7.5 Final comments	197
4	**The random walk**		**199**
	4.1	The simple unrestricted random walk	200
		4.1.1 Normal approximation	201
		4.1.2 Laws of large numbers	202
		4.1.3 Using the 'reflection' principle	203
		4.1.4 First passage and return probabilities	206
		4.1.5 Probability of long leads: the First Arc Sine Law	208
		4.1.6 Simulated illustration	212
	4.2	Absorbing barriers	214
		4.2.1 Probability of absorption at time n	215

		4.2.2	One absorbing barrier	221
		4.2.3	Number of steps to absorption	223
		4.2.4	Further aspects of the unrestricted random walk	225
	4.3	Reflecting barriers	228
		4.3.1	General equilibrium probability distribution	228
		4.3.2	Relation between reflecting and absorbing barriers	232
		4.3.3	Time-dependent probability distribution	234
	4.4	The correlated random walk	238
		4.4.1	Occupation probabilities: direct solution	239
		4.4.2	Occupation probabilities: p.g.f. solution	241
		4.4.3	Application to share trading	246
5	**Markov chains**		**252**
	5.1	Two-state Markov chain	256
		5.1.1	Occupation probabilities	257
		5.1.2	Matrix solution 1	258
		5.1.3	Matrix solution 2	260
		5.1.4	The Discrete Telegraph Wave	261
		5.1.5	Relation to the continuous-time process	262
	5.2	Examples of m-state Markov chains	265
		5.2.1	The Ehrenfest model	265
		5.2.2	The Perron–Frobenius Theorem	272
	5.3	First return and passage probabilities	274
		5.3.1	Classification of states	274
		5.3.2	Relating first return and passage probabilities	276
		5.3.3	Closed sets of states	280
		5.3.4	Irreducible chains	281
	5.4	Branching processes	282
		5.4.1	Population size moments	283
		5.4.2	Probability of extinction	286
	5.5	A brief note on martingales	290
6	**Markov processes in continuous time and space**	**295**
	6.1	The basic Wiener process	296
		6.1.1	Diffusion equations for the Wiener process	298
		6.1.2	Wiener process with reflecting barriers	300
		6.1.3	Wiener process with absorbing barriers	304
	6.2	The Fokker–Planck diffusion equation	309
		6.2.1	Simulation of the simple immigration–death diffusion process	312
		6.2.2	Equilibrium probability solution	317
		6.2.3	Boundary conditions	321
		6.2.4	Time-dependent probability solutions	322
		6.2.5	The associated stochastic differential equation	323
	6.3	The Ornstein–Uhlenbeck process	325
		6.3.1	The OU process as a time-transformed Wiener process	329
		6.3.2	Rapid oscillations of the Wiener and OU processes	329

7 Modelling bivariate processes — 331
- 7.1 Simple immigration–death–switch process — 331
  - 7.1.1 Generating moments — 334
- 7.2 Count-dependent growth — 337
  - 7.2.1 Stochastic representation — 337
- 7.3 Bivariate saddlepoint approximation — 339
  - 7.3.1 Simple illustrations — 340
  - 7.3.2 Cumulant truncation — 342
  - 7.3.3 A cautionary tale! — 346
  - 7.3.4 A spatial example — 347
- 7.4 Counting processes — 349
  - 7.4.1 Paired-immigration–death process — 350
  - 7.4.2 Single-paired-immigration–death counting process — 353
  - 7.4.3 Batch-immigration–death counting process — 357
  - 7.4.4 Summary and further developments — 364
- 7.5 Applying MCMC to hidden event times — 367
  - 7.5.1 Introducing Markov chain Monte Carlo — 367
  - 7.5.2 Fitting the simple immigration–death process to incomplete observations — 369
  - 7.5.3 Extension to the single/paired-immigration–death process — 378
  - 7.5.4 A comparison of Metropolis Q- and direct P-matrix strategies — 381

8 Two-species interaction processes — 389
- 8.1 Competition processes — 389
  - 8.1.1 Deterministic analysis — 390
  - 8.1.2 Stability — 393
  - 8.1.3 Stochastic behaviour — 398
  - 8.1.4 Moment equations — 408
- 8.2 Predator–prey processes — 413
  - 8.2.1 The Lotka–Volterra process — 414
  - 8.2.2 The Volterra process — 419
  - 8.2.3 A model for prey cover — 420
  - 8.2.4 Sustained deterministic and stochastic limit cycles — 428
- 8.3 Epidemic processes — 434
  - 8.3.1 Simple epidemic — 435
  - 8.3.2 General epidemic — 445
  - 8.3.3 Recurrent epidemics — 454
- 8.4 Cumulative size processes — 464
  - 8.4.1 Deterministic models — 465
  - 8.4.2 Stochastic simulation — 466
  - 8.4.3 Probability solutions — 468
  - 8.4.4 Power-law processes — 471

9 Spatial processes — 474
- 9.1 General results — 474
- 9.2 Two-site models — 477

			Contents	xi

	9.2.1	Moments	478
	9.2.2	Exact probabilities	481
	9.2.3	An approximate stochastic solution	484
	9.2.4	Slightly connected processes	488
	9.2.5	Sequences of integral equations	490
	9.2.6	Riccati representations	493
9.3	Stepping-stone processes		496
	9.3.1	Birth–death–migration processes on the infinite line	498
	9.3.2	Birth–death–migration processes on the finite line	508
	9.3.3	Basic simulation algorithms	514
	9.3.4	Tau-leaping and other extensions	519
9.4	Velocities of propagation		524
	9.4.1	Wave profiles for two-way migration	526
	9.4.2	Wave profiles for non-nearest-neighbour migration	533
	9.4.3	Travelling waves	541
9.5	Turing's model for morphogenesis		549
	9.5.1	Solution of the linearized equations	551
	9.5.2	An example of wave formation	553
	9.5.3	Stability and the 'Stochastic Dynamic'	559
9.6	Markov chain approach		561
	9.6.1	A more refined approximating process	563
	9.6.2	Simulating the Markov chain representation	565
	9.6.3	Stochastic cellular automata	567

10 Spatial–temporal extensions 575
10.1 Power-law lattice processes 575
  10.1.1 First- and second-order moments 576
  10.1.2 The spectrum 578
  10.1.3 General power-law spectra 584
  10.1.4 The inverse problem 585
  10.1.5 Simulated realizations 589
  10.1.6 An application to sea waves 592
10.2 Space–time marked point processes 596
  10.2.1 The general model 598
  10.2.2 Choosing growth and interaction functions 599
  10.2.3 Parameter selection 603
  10.2.4 Simulation algorithm 604
  10.2.5 Convergence issues 608
  10.2.6 Application to forestry 610
  10.2.7 Application to tightly packed particle systems 613
  10.2.8 Other stochastic strategies 619
  10.2.9 Final comments 624

References 628

Subject Index 647

1
Introduction

The nineteenth century saw the birth of many truly great Russian mathematicians whose brilliance and dedication has been central to the foundation and subsequent development of modern-day Mathematics. In particular, the field of Probability Theory owes a tremendous debt of gratitude to Andrei Andreyevich Markov (1856–1922). His deep insight that progress would be greatly facilitated by assuming that the next step in a random process depends only on the current state, and not on any previous history, led to massive strides forward. For the entire development of a stochastic realization is then defined in terms of the sequence of event–time pairs, which means that in a *Markov process* the equations for the probability that the population is of size n at time t, namely $p_n(t)$, are straightforward to write down. The mathematical challenge then lies purely in determining their solution. Whilst this was wonderful news for pure mathematicians, their more applied colleagues proved to be remarkably slow in appreciating the great practical potential that such theoretical advances offered. Indeed, it was not until after the conclusion of the Second World War that serious development over a widening spectrum of applications really grew, primarily spurred on by the thought-provoking and innovative papers read to the Royal Statistical Society Symposium on Stochastic Processes in 1949. Interest in this new field of Applied Probability then started to accelerate, and by the early 1960s had attracted the attention of several outstanding statisticians. At a purely personal level, there are four key milestones relating to this period. First, there was Maurice Bartlett's influential monograph on Stochastic Population Models, published in 1960. Second, in 1965 David Cox and Hilton Miller published one of the best and most inspiring early texts on the Theory of Stochastic Processes. Third, was William Feller's (1968, 1971) two-volume masterly treatment of a general, and appealingly holistic, introduction to Probability Theory. Whilst fourth was the establishment of the Applied Probability Trust in 1964 whose *Journal of Applied Probability*, and the later *Advances in Applied Probability*, not only provided a forum for researchers to bring their work to the attention of like-minded colleagues around the world, but also ensured that current developments could easily be tracked. For in those halcyon days these two journals provided the main outlet for publication.

Through them studies in this field swiftly prospered, as increasing numbers of talented PhD students applied new stochastic developments to fresh areas of application. However, as the associated level of mathematical difficulty gradually increased, with solutions becoming ever more challenging and tricky to construct, younger researchers started to switch attention back onto more theoretically oriented problems since these were more likely to lead to intellectually appealing results based on 'nice' closed-form solutions. Unfortunately, success usually depended on making unrealistic assumptions about the way in which particular stochastic scenarios developed, with the result that the ensuing solutions became increasingly less applicable to real-life problems. In recent years, however, there has been a resurgence of interest in developing approximate theoretical solutions and simulation analyses. The latter prospered as soon as the speeds offered by inexpensive personal computers were sufficient to enable quite complex problems to be studied within a reasonable time frame. This text therefore has three aims. First, to set the ball rolling, we introduce a variety of stochastic processes that possess relatively simple closed-form solutions; these often form the basis of undergraduate level courses. Second, a range of approximation techniques are presented in order to show that in more complex problems considerable behavioural information may still be extracted even when their exact probability structure is mathematically intractable to direct solution. If required, these results may be cemented within a deeper analytic framework by making a comparative study with the convergence and characterization results developed in Ethier and Kurtz (1986). Third, we illustrate how computer simulation can not only yield insight into the way that particular systems develop, but that it may also highlight intriguing, and previously unknown, properties which may then be studied theoretically. For different realizations of a random process may vary enormously, and since the probabilities $p_n(t)$ relate to the distributional properties of a large ensemble of such realizations, they do not provide explicit information on the structure of individual time-histories.

Given that a completely definitive treatment of this subject would require a huge number of volumes, choice of material has had to be made on a highly selective basis. Moreover, examples have been kept sufficiently simple to make the underlying ideas easy to understand, yet sufficiently structured to convince readers that these techniques and approaches may be applied to any problem, no matter how complex. In the first half of the book basic principles and ideas are introduced through the study of single-species populations; in the second half these are extended to cover a variety of different situations involving bivariate populations and processes that develop in both space and time. Note that attention has been deliberately restricted to the consideration of random events and times of occurrence. For there are many other mechanisms for inducing randomness, such as letting parameters be random, incorporating feedback mechanisms, or allowing for external random inputs to the system, and trying to encompass all of them within a single text would be futile.

Readers who are relatively new to Stochastic Processes may wish to move directly on to Chapter 2 and begin their tour through the field; the material gradually unfolds in a natural progression. For more experienced readers we now present a concise summary of the topics covered which not only provides a road map to specific content but also highlights the strong relationships which exist between the various processes under study.

1.1 Some simple stochastic processes

Over the years interest in modelling population dynamics has polarized, with Applied Mathematicians generally adopting a purely deterministic approach in which random variation is ignored and investigations are reduced to solving a set of differential equations. Mathematical expertise then relates to using novel theoretical approximation or numerical analysis techniques in order to extract behavioural properties. In contrast, statisticians and probabilists generally model the probability structure of population size from the outset, and whilst at a first glance this appears to be a superior approach it has the potential downside that the resulting equations may well be intractable to mathematical solution. The ideal route is to consider the deterministic and stochastic representations of a particular system together, since both have important roles to play. If it transpires that random variation plays only a minor role in the development of the population under study then a deterministic analysis may well offer the most pragmatic way forward. However, unless a stochastic analysis is made first we will not know whether this is a realistic presumption. Conversely, if the deterministic path is totally avoided then a relatively simple, and hence readily understood, description of population development may be missed. Pursuing both approaches simultaneously ensures that we do not become trapped either by deterministic fantasy or unnecessary mathematical detail.

The basic principles which underlie the concept of stochastic population dynamics are easily explained by considering simple forms of model structure in which individual members of a population are assumed to develop *independently* from each other. For then the resulting mathematical analyses are sufficiently transparent to enable useful conclusions to be drawn. So in Chapter 2 we examine the effect of assuming independence in various simple situations involving birth, death, immigration and emigration, before going on to develop the ideas of density-dependent population growth in the ensuing chapters.

The simplest construct is to assume that individuals enter a population at a constant rate α, and then neither reproduce, die nor leave thereafter. This gives rise to the Poisson process, named after the French mathematician Siméon-Denis Poisson (1781–1840), and it possesses two key properties. The time between successive events, i.e. arrivals, follows the exponential distribution with parameter α; and, the probability that n individuals arrive during the time interval $(0, t)$ is given by $p_n(t) = (\alpha t)^n e^{-\alpha t}/n!$ for $n = 0, 1, 2, \ldots$. Since here individuals neither reproduce nor die, $\{p_n(t)\}$ is completely determined by the times between successive immigrations. Not only are these easy to construct, but their method of solution is generic to the study of far more complex problems, and thereby illustrates a core element in the probabilist's armoury. Indeed, the sheer simplicity of the Poisson process not only provides an excellent way of introducing theoretical methods for constructing probability density functions (p.d.f.'s) and moments, together with the general relationships which exist between them, but it is also ideal for introducing basic simulation procedures. Uniform random variables, $U \sim (0, 1)$, form the essential building block, since any other random variable can be generated from them provided its cumulative distribution function can be inverted. This procedure is trivial for exponential variables and remarkably easy even for Poisson ones. However, care must be taken to choose a reliable generator for U; the mixed congruential generator is our preferred option, though correct parameter choice is vital.

Although we shall be considering basic types of stochastic process, real-life often complicates matters. So before starting our progression through an increasingly complex array of model structures we take the opportunity to mention four embellishments, namely time-dependent parameters, multi-rate processes, simultaneous occurrences and change of state space. All of which can be used to generate extensions to the later material. The last of these is particularly interesting, for by defining the Poisson distribution over $n = 0, 1, 2, 4, 8, \ldots$, rather than $n = 0, 1, 2, \ldots$, it is possible to produce an infinite number of probability distributions that possess the same moment structure. This highlights an often forgotten truth, namely that moments do not uniquely define a distribution.

Allowing individuals to die independently of each other at a constant rate μ also leads to simple results for the population size $X(t)$ at time t; not only for the transition probabilities $p_{ij}(t) = \Pr(X(t) = j | X(0) = i)$, but also for the reverse transition probabilities $\overleftarrow{p}_{ij}(t) = \Pr(X(0) = i | X(t) = j)$ and the bridge probabilities $p_{inj}(t) = \Pr(X(s) = n | X(0) = i, X(t) = j)$. Whilst in the pure immigration process $X(t)$ increases indefinitely, the pure death process follows the progress of the original i individuals through to their ultimate extinction. In contrast, replacing death by birth at rate λ produces a process whose expected population size increases exponentially fast (the Poisson population increases linearly). Since the pure birth process is multiplicative, rather than additive as for pure immigration, the associated theory is slightly more convoluted and provides an excellent opportunity to demonstrate two general methods of solution via probability generating functions (p.g.f.'s) and Laplace transforms. Moreover, it also provides an ideal platform from which to introduce two important stochastic process benchmarks, namely: the time to a given state; and, how to analyse processes whose individuals do not develop independently. That is, for $X(t) = n$ the population birth rate λn is replaced by the general function λ_n; if λ_n increases with n faster than λn then $X(t)$ can literally explode to infinity in a finite time. Nonlinear transition rates are central to subsequent chapters. Sadly, although the birth process was first proposed by Yule in 1925 to describe the rate of evolution of a new species within a genus, it took a further 30 years before its fundamental role in the study of stochastic processes was truly appreciated.

Combining birth and death into a single process results in a more complex structure, since population size is no longer restricted to be either increasing or decreasing over time. The implicit assumptions that individual birth and death rates remain constant though time, and that they are independent of both age and population size, will clearly be violated in virtually all natural populations involving population growth. However, whilst the *construction* of realistic stochastic models in astronomy, physics, finance, etc. is essentially unconstrained by theoretical considerations, their subsequent *mathematical analysis* most certainly is, since direct theoretical solution will be intractable in all but the simplest of cases. So models like the simple (i.e. linear) birth–death process should not be envisaged as providing a literal representation of reality, but should be treated instead as a means of obtaining greater understanding of the behavioural characteristics of the processes which they attempt to mimic. Derivation of the population size probabilities follows as for the pure birth process, but we now also have the opportunity of determining the probability of extinction. If $\lambda \leq \mu$ then ultimate extinction is certain, but if $\lambda > \mu$ it has probability $p_0(\infty) = (\mu/\lambda)^{n_0} < 1$. So now two types of vastly different behaviour are generated by the same process: either the population eventually dies out completely with a probability that decreases geometrically with increasing initial population size n_0; or else it becomes infinite. Comparison with the corresponding (Malthusian) deterministic exponential growth highlights the danger in blindly following a deterministic approach. The usual way of constructing the equation for $p_{ij}(t)$ involves taking an initial population of size i at $t = 0$ followed by two time points t and $t + dt$ later on that lie very close together, and this is called the *forward* approach. The converse procedure is to take two initial time points close together, 0 and dt, followed by a single time point $t + dt$ later on. This generates the *backward* equations. The general characteristic of the forward equations is that the same initial state is involved throughout, whereas in the corresponding backward equations the same final state enters in all terms. So the forward equations are likely to be more useful if there is a single initial state of particular interest and we require the population size probabilities at some later fixed time t for various final states. Whereas the backward equations are appropriate if there is a single final state of particular interest, and we require the probabilities of reaching this at time t from various initial states. As these dual processes produce the same solution, this means that we usually have two different approaches open to us. The exception involves dishonest processes that reach an infinite population size in a finite time with positive probability; for here only the backward equation route is possible. Simulation requires an added step to the basic algorithm used for the pure immigration, death and birth processes, since these only require the determination of the time to the next event. For we now also need to determine whether the next event is a birth or a death. Not only is this relatively trivial to do, but it yields a totally general construct for any continuous-time Markov process whose variable changes by either $+1$ or -1 at the occurrence of each event.

Note that discrete-time processes are especially easy to simulate, since they only require computation of the event type. In a rather extreme example, a nominated 'dominant leader' has probability p of giving birth to twins at each time step $t = 0, 1, 2, \ldots$, with all remaining members of the population producing twins only if she does. This exposes a huge difference between deterministic and stochastic solutions. For the deterministic population size increases geometrically fast in spite of the (stochastic) population becoming extinct geometrically fast. So even at this early stage in the text the danger in taking deterministic analyses at their face value becomes apparent.

In the simple birth–death process the population either becomes extinct or else it becomes indefinitely large. However, provided $\lambda < \mu$, combining birth, death and immigration produces a process that persists indefinitely yet remains finite, which means that an equilibrium distribution $\pi_n = \lim_{t \to \infty} p_{in}(t)$ now exists. Such equilibrium probabilities are considerably easier to construct than time-dependent probabilities, not only because in equilibrium we can exploit the fact that the 'probability flow' between neighbouring states must be in balance, but also because they are independent of the initial population size i. So as long as the process has 'burnt-in' by some time T, $\{\pi_n\}$ provides full probability information on the system for all times $t > T$. Any concern about whether burn-in has actually occurred by T can be overcome by employing a version of the 'perfect simulation' approach initiated by Propp and Wilson (1996).

So far we have considered three basic forms for the population transition rates, namely birth at rate λn, death at rate μn, and immigration at rate α which unlike the first two is independent of n. However, there is a fourth basic form, namely emigration at rate β for $n > 0$, which is fundamentally different to the previous three. For if the population is empty, i.e. $n = 0$, then there are no individuals available to emigrate. So whereas the birth–death–immigration equations are valid for all $n = 0, 1, 2, \ldots$, once emigration is introduced the equation for $n = 0$ has a different structure to those for $n > 0$. This $n = 0$ equation is said to be anomalous, and its presence causes a considerable increase in algebraic complexity. Unfortunately, this feature often arises in the form of 'queueing processes': individuals arrive into and depart from a system with various events occurring to them within the system itself. So this increase in complexity cannot be ignored. Although working in terms of the Laplace transform of the occupation probabilities, namely $p_n^*(s) = \mathcal{L}[p_n(t)]$, does yield analytic results, we are still left with a considerable inversion problem if we are to recover the $p_n(t)$. However, the anomalous equation issue disappears if we regard our queueing process as being imbedded in a larger process over $-\infty < n < \infty$ by placing $\alpha p_{-1}(t) = \beta p_0(t)$ and making an appropriate choice for the initial conditions $p_{-1}(0), p_{-2}(0), \ldots$. For then the $p_n(t)$ can be determined directly in terms of modified Bessel functions of the first kind. This technique, which has its origins in electrostatics, is called the *Method of Images*. Although it is seldom used to solve stochastic process problems, it is nevertheless extremely powerful, and its resurrection, together with other early twentieth century approaches, is to be strongly encouraged. In addition to Bessel functions, other likely polynomial candidates for constructing probability solutions include Hermite, Laguerre, Legendre and confluent hypergeometric functions.

Our tour of simple stochastic models concludes by considering an extension to the birth–death–immigration structure which has considerable practical importance and has recently aroused attention in the literature. For suppose that the whole population instantaneously dies, i.e. it suffers a total catastrophe, at rate γ, after which it is restarted by the arrival of a batch of j new immigrants at rate α_j. So now state 0 can be reached either by a death occurring in state 1 or by a catastrophe occurring in any of the states $i \geq 1$. Then although the associated analytic detail is more involved than before, determining the population size probabilities and moments is still perfectly tractable; the solutions are just more cumbersome. Moreover, the model possesses two particularly illuminating features. First, the mass immigration–death and birth–death–immigration processes have identical equilibrium structures. So although individual realizations exhibit large behavioural differences between these two models, it is not possible to differentiate between models by using the equilibrium probability structure alone. This highlights a fundamental weakness in trying to describe a stochastic process purely in terms of the equilibrium population size probabilities. Second, on letting $\sum_{j=1}^{\infty} \alpha_j \to \infty$ the process experiences instantaneous immigration as soon as it reaches state 0. The mathematical implications of this limit are surprisingly far-reaching, and provide a bridge that allows for genuine and useful interaction between pure theoreticians and applied practitioners.

1.2 Single-species population dynamics

In practice the assumption that individuals develop independently from each other will often be violated. If individuals exhibit only a weak association with each other, then simple models like those in Chapter 2 may well provide a useful approximation to reality. However, if the individual birth and death rates exhibit substantial dependence on population size then this approach becomes untenable. In most situations the growth of an expanding population will eventually be limited by a shortage of resources, and a stage is then reached when the demands made on these resources exerts a net downwards pressure on population size. The converse situation of 'mutual enhancement' is much rarer, and involves individuals growing in symbiosis with each other. A classic example of this is the speculative dot-com boom from 1995 to 2000 (the NASDAQ peaked at 5132 on March 10th 2000) in which many investors replaced their usual financial wisdom by mutual greed. Though if one takes the subsequent crash into account, we may well argue that this situation is simply an aggressive version of the first one.

To construct a general model we replace the population birth and death rates, λn and μn, by general non-negative functions λ_n and μ_n. However, whilst the resulting deterministic and stochastic equations are still trivial to construct, their full theoretical solution can be very difficult, or impossible, to derive, even for minor modifications to the simple birth and death rates. Nevertheless, not only are both forms generally amenable to approximation techniques or numerical solution, but we can also construct properties of the general stochastic process. So Chapter 3 starts by first developing an integral representation of the transition probability matrix, $\mathbf{P}(t)$, in terms of the eigenvalues of the transition rate matrix. For this viewpoint highlights an intimate connection between the theory of birth–death processes and orthogonal polynomials, which enables the probabilities $p_n(t)$ to be expressed in terms of the integral of a simple weighted product of orthogonal polynomials. This, in turn, leads to powerful results on underlying properties such as existence and uniqueness, together with general conditions for recurrence, ergodicity and transience. Now although the time-dependent properties of a process are important in the opening stages of development, if the process possesses an equilibrium mode of behaviour then problems of mathematical intractability vanish. For then the equilibrium distribution $\{\pi_n\}$ over $n \geq 0$ is simple to construct. Moreover, even if extinction does eventually occur we can still evaluate the corresponding quasi-equilibrium distribution $\{\tilde{\pi}_n\}$ over $n > 0$ just as easily. To be strictly accurate we should work in terms of the probability size distribution as $t \to \infty$ conditional on extinction not having occurred, i.e. $\pi_n^{(Q)} = \lim_{t \to \infty} p_n(t)/[1 - p_0(t)]$. Though this is generally not a pragmatic option, since not only are the $\{p_n(t)\}$ often inherently difficult to obtain, but also in many scenarios $\{\tilde{\pi}_n\}$ and $\{\pi_n^{(Q)}\}$ will be virtually identical.

As regards specific models, the simplest case to consider is the classic logistic process for which both λ_n and μ_n are quadratic. For this facilitates not only an easy derivation of deterministic and equilibrium solutions, but also a straightforward extension to power-law processes in which $\lambda_n = an(1 - bn^s)$ and $\mu_n = cn(1 + dn^s)$ for positive constants $a, \ldots, d$ and s. This type of model is particularly intriguing, since the higher the value of the power-term s, the more 'aggressively' the population grows. Simulations neatly highlight the way in which the behaviour of individual realizations changes with s. Moreover, extending this power-law structure by letting the transition rates take quotient polynomial forms enables the process to exist in two or more equilibrium states with realizations intermittently flip-flopping between them. For example, combining simple death at rate $\mu_n = n$ with quotient quadratic birth at rate $\lambda_n = (170n^2 + 100000)/(n^2 + 8000)$ gives rise to a process which oscillates around the two local equilibrium points 20 (for $0 \leq n < 50$) and 100 (for $n > 50$), with the intervening locally unstable equilibrium point at 50 acting as a bridge between them. Note that this phenomenum is not displayed by the corresponding deterministic trajectories. Whilst logistic-type processes and multiple equilibrium distributions have major roles to play in biology, in physics interest often centres around more complex systems which not only have the potential for exhibiting extremely rich dynamic behaviour but which also frequently generate large fluctuations that may be described by 'scale-free' probability density functions, typically with power-law structure $p_n \sim 1/n^\nu$. Such random processes play an important role in the study of complex, nonlinear correlated behaviour, for which the class of Lévy-stable densities provides a suite of models. The mass-immigration–death process with $\alpha_1 = \nu^2$ and $\alpha_j = \nu^2(1-\nu)\ldots(j-1-\nu)/j!$ gives a particularly neat illustration of this effect; including birth as well opens up the range of possibilities still further.

However, although surprisingly deep mathematical inroads can be made by exploring the structure of nonlinear processes, time-dependent solutions for their probability distributions, $\{p_n(t)\}$, are rarely analytically tractable. We therefore first develop a basic approximation technique, together with a general study on the probability of ultimate extinction and the associated mean time to ultimate extinction, before proceeding to examine in detail the precise nature of the mathematical intractability which surrounds the Kolmogorov forward equations. The fundamental problem is easily exposed by contructing the differential equation for the moment generating function (m.g.f.) of population size. For not only does this involve differentials of order $s + 1$, which shrieks intractability, but the associated differential equation for the rth-order moment involves moments of higher order. So no matter how many moment equations are generated they will always contain a greater number of variables. This puzzle is a well-known feature in applied mathematics where many partial differential equations (p.d.e.'s) can be reduced to an infinite set of ordinary differential equations (o.d.e.'s) by taking Laplace transforms. The classic solution is to ignore all higher-order Fourier coefficients beyond a certain point (the Galerkin approximation); here we follow a parallel route by ignoring moments (preferably cumulants) beyond a given order. The resulting truncation procedure is a powerful approach, and numerical investigations into its accuracy have been highly encouraging.

An alternative approach is to avoid the Kolmogorov equations completely by adding stochastic noise to the underlying deterministic process. For then approximate quasi-equilibrium moments can be determined explicity. Whilst other local approximation techniques yield useful moment expressions that are algebraically simple and structurally transparent. The accuracy of all these various levels of approximation is examined by studying the rapid colonization by Africanized honey bees of South and Central America (Matis and Kiffe, 2000). This is regarded as being one of the most remarkable biological events of the twentieth century since the range of the population explosion expanded relentlessly every year from 1956 when 26 reproductive swarms escaped in Brazil: during the next 40 years millions of colonies developed and spread over 20×10^6 km^2. This rich scenario generates a whole host of intriguing ecological, and hence stochastic, problems.

Now if we have already determined expressions for the population size moments, then we can insert them into any appropriate standard distribution and thereby obtain an approximation to the probability density function. However, if we wish to generate more detailed mathematical forms for the probabilities, $p_n(t)$, then since analysis of models with a discrete population size $N = n$ is intrinsically more demanding than that required for continuous N, it is sensible to take the variable N over the continuous values $n \geq 0$. One approach is to replace the forward Kolmogorov equation by a diffusion equation, and the simplest option is to employ the Taylor series expansion. This technique yields a particularly neat representation for equilibrium probabilities, and works reasonably well provided the mean is not too small and we keep away from the tails. However, if we are especially interested in the tails then perturbation methods offer a far more reliable way forward. Essentially these work by expanding $\ln[p_n(t)]$ instead of $p_n(t)$. The downside is that it is not possible to analyse specific processes on 'automatic pilot', since each one may have its own particular features that need careful handling. Nevertheless, it does yield a solution for the general birth–death process, and its rather awkward nature presents an interesting future challenge if we are to shape it into a more user-friendly form. In equilibrium the technique operates much more smoothly, though since exact solutions can always be found, benefit is restricted to the relative transparency of the resulting expressions. A return to this old-fashioned perturbation style of mathematics would be potentially rewarding, as demonstrated by an example in which $l(n,t) = \ln[p_n(t)]$ is expressed as a power series, $\sum_{i=0}^{\infty} \epsilon^i l_i(n,t)$, in ϵ.

Whilst the routes taken by the above analytic approaches are dependent on the specific models under study, there is an optimal technique, called the *Saddlepoint Approximation*, which can be applied in exactly the same way to all processes. Instead of constructing approximations $p(x,t)$ ($x \geq 0$) to the probabilities $p_n(t)$ ($n = 0, 1, 2, \ldots$) directly, attention now switches to their derivation via the associated cumulant generating function (c.g.f.). For this enables the construction of a completely general form of solution. As the underlying concepts are quite deep, we first introduce the basic ideas by considering specific Taylor expansion analyses for the mass immigration and immigration–emigration processes. These are then generalized by following Maurice Bartlett's lead in paralleling the above ϵ-power-series approach. Finally, we extract the core analysis that is necessary for the development of a full analytic proof of the saddlepoint approximation; this was first presented in wonderful detail by Henry Daniels in the 1950s, and later extended in the 1960s and 1980s. This technique is optimal in the sense that a good approximation to a p.d.f. can nearly always be obtained via the powerful *Method of Steepest Descents*. The saddlepoint approximation is the dominant term in the associated asymptotic expansion, and for sample size m has relative error $O(m^{-1})$ as opposed to $O(m^{-1/2})$ for the Normal approximation. The key point is that this approximation still works exceedingly well even when $m = 1$.

After presenting several examples to illustrate both its general use and the neatness of the resulting expressions, we return to address the moment closure problem. Suppose we have derived approximate theoretical or numerical values for the first r cumulants $\kappa_1(t), \ldots, \kappa_r(t)$; the simplest way is to solve the set of first-order o.d.e.'s for $\kappa_i(t)$ by placing $\kappa_s(t) \equiv 0$ for all $s > r$. Then replacing the full c.g.f. by the truncated one, namely, $\sum_{i=1}^{r} \kappa_i(t) \theta^i / i!$, yields a set of approximations as r increases through $r = 2, 3, \ldots$. However, cumulants are notoriously fickle to work with, and contrary to intuition precision does not automatically increase with r. So each process needs to be examined separately in order to decide on the optimal choice of r that balances algebraic simplicity and accuracy. Nevertheless, this truncation procedure has immense potential for yielding a general theoretical solution. The challenge is to find it.

Whereas Chapter 2 shows that specific results for population size probabilities, moments, probabilities of extinction, etc. can be constructed relatively easily for low-level (usually linear) stochastic processes, Chapter 3 shows that the construction of time-dependent properties in the general case is far more problematic. The difficulty is that the Kolmogorov equations, being based on event–time pairs, are too difficult to solve exactly, which is why we are often forced back onto developing approximation techniques. However, if instead of considering the probability of moving from say i to j in time t, we consider the probability $p_{ij}^{(n)}$ of moving from i to j in n steps, the the first-order o.d.e.'s for the $p_{ij}(t)$ reduce to a simpler system of first-order difference equations which are intrinsically easier to work with. So in Chapter 4 we switch from continuous to discrete time, starting with an analysis of the simplest possible scenario in which an individual moves over the infinite integer line $i = \ldots, -1, 0, 1, \ldots$, where at each step $\Pr(i \to i + 1) = p$ and $\Pr(i \to i - 1) = 1 - p$. For the position, X_n, after n steps is then just the sum of n Bernoulli random variables, and hence is amenable to direct solution. Moreover, it leads us neatly onto two fundamental results, namely the Weak and Strong Laws of Large Numbers. This simple random walk has two particularly appealing features. First, it is usually a student's first encounter with the reality that theoretical conclusions may seem to run totally counter to intuition and common sense. Second, since the 2^n sample paths and their associated probabilities are easily enumerated, we no longer have to rely on being able to solve sets of equations. The basic ideas stem from a very simple result, namely that if in a ballot candidates C and D score c and $d < c$ votes, then the probability that there are always more votes for C than D is $(c - d)/(c + d)$. This leads naturally to Bertrand's Ballot Theorem and the powerful Reflection Principle, and on to a classic result for the number of paths joining the two endpoints 0 and 0 which are wholly positive, or non-negative, inbetween them. The beauty of such expressions is that since they concern admissible paths, and not probabilities, they relate to *any* birth–death process with general rates λ_n and μ_n. Further insight is gleaned by developing the first passage and return probabilities for the symmetric simple random walk (as $p = 0.5$ all paths are equally likely). One result, for example, runs counter to the intuitive belief that the fraction of time spent purely on one side of the axis is likely to be close to $1/2$; the exact opposite is true. Indeed, the First Arc Sine Law gives the probability for the fraction of time spent on one particular side. Not only are all such results trivial to simulate, but simply observing a series of simulation runs may well produce unexpected behavioural characteristics which can then be analysed theoretically by evaluating the corresponding sample paths.

Such path-based analyses are mathematically appealing, though if the process is not symmetric (i.e. $p \neq 0.5$) then construction of concise solutions requires a different strategy, especially if the walk is restricted by external barriers. A return to generating functions offers the best prospects. Detailed algebraic investigations cover the time to absorption together with the probabilities of being in non-absorbing and absorbing states, both at finite and infinite time n, given two *absorbing* barriers at a and $b < a$. Analysis of the single absorbing barrier first proceeds by letting $b \to -\infty$, then an alternative approach provides a different perspective since it does not rely on deriving the two-barrier solution. Note that important properties of the unrestricted random walk can be deduced by imposing absorbing barriers, with $\max(X_n)$, $\min(X_n)$ and the probability of first return to the origin being particular examples in point. Many processes, however, do not involve physical absorption, but merely have upper and lower limits on population size. Such situations correspond to processes with *reflecting* barriers, and these are guaranteed to continue indefinitely. As such they easily yield to the construction of a general equilibrium solution. There is an intriguing relationship between these two different barrier scenarios, for it transpires that they are mathematically equivalent. Specifically, if we know the probability of eventual absorption at a barrier for an arbitrary start point then we can immediately write down the equilibrium distribution for the corresponding reflecting barrier process. Such holistic viewpoints are unfortunately a rarity, since all too often models are analysed in total isolation from each other. This relationship suggests that the derivation of the time-dependent (reflecting barrier) probability distribution is likely to be just as messy as that for the construction of the absorption probabilities, and this proves to be the case. Nevertheless, the construction of this p.d.f. is useful since it reuses the Method of Images introduced in the previous chapter and highlights the algebraic care that needs to be taken when evaluating the associated mirrored coefficients for the exterior site locations $i < b$ and $i > a$.

Although the assumption of independent steps is valid in many applications, it does not always hold true. For example, if the random walk relates to genuine motion then a particle's momentum ensures that it is more likely to travel in the same direction at the next step than to reverse direction. Similarly, in the financial markets, investor sentiment suggests that an increase (decrease) on day n is more likely to be followed by an increase (decrease) on day $n+1$ than a decrease (increase). In the simplest case the probability that a particle or share price moves in the same direction as at the last step is p, with the converse probability being $1-p$. Though analyses can be undertaken similar to those developed for the independent step random walk with barriers, not to mention extension to higher dimensions, here we just introduce some basic results by considering the pure one-dimensional unrestricted process. The occupation probabilities are determined first via a path analysis approach, and then through a double generating function procedure since this has the added advantage of also yielding moments. Simulation merely involves a slight extension to the simple random walk algorithm, and is particularly appropriate since it highlights the different modes of behaviour depending on whether $p \sim 0$, 0.5 or 1. Since studies of realizations for $p \geq 0.5$ show an overall structure that bears some similarity to time-series of daily share prices, this correlated random walk process offers considerable scope for use as a trading strategy. The problem for private investors is that p needs to be high enough to counteract the inherent transaction costs associated with broker charges, stamp duty and buy–sell spread prices. Institutional transactions are better placed since they suffer far smaller relative costs, and can deal on a millisecond time-scale. In both cases successful implementation requires careful analysis of each stock being considered. Considerable progress can be made through the construction of theoretical results, but for practical implementation fast simulation studies are most likely to offer the best way forward.

The simple random walk yields to analytic solution primarily because the underlying transition probability matrix $\mathbf{P} = \{p_{ij}^{(n)}\}$ is tri-diagonal, and hence amenable to algebraic decomposition. However, if the step-size increment is no longer restricted to the three values -1, 0 and 1, then single jumps from state i to many other states j are now possible, which means that $\mathbf{P}$ may well be far more difficult to work with. Indeed, if the number of states $m > 4$, then the associated eigenvalue equation no longer has a general closed-form solution. Such processes, called Markov chains, are considered in Chapter 5 and are best introduced through a matrix-based approach. Since the two-state process, on say 0 and 1, is analytically tractable (the characteristic equation is quadratic) it yields easily to a complete analysis, and thereby provides an excellent illustration of the matrix solution for the occupation probabilities $(p_0(t), p_1(t))$. That discrete-time Markov chains have an inherently more complex structure than their continuous-time Markov process counterparts is easily seen by comparing model structures for the birth–death process. In both the simple and general representations, analysed in Chapters 2 and 3, respectively, only three types of event can occur during the infinitesimal time increment $(t, t+dt)$, namely a birth, death or no-change. The same holds true in the random walk paradigm (Chapter 4), where the 'time' n relates either to the nth event (birth or death) or to clock-time ndt (birth, death or no-change). However, suppose we now examine the development of every individual member of the population separately between the integer times t and $t+1$. As each one either gives birth, dies or remains unchanged, $\Pr(X_{t+1} = j | X_t = i)$ is the sum of trinomial probabilities for which $j - i = r - s$ where r and s denote the total number of births and deaths, respectively. Although this enables us to write down the full joint probabilities over all times $1, 2, \ldots, n$, together with the occupation probabilities $\mathbf{p}^{(n)}$ at time n, the resulting expression is totally opaque. Initial progress is best made by first decomposing the $m \times m$ transition matrix $\mathbf{P}$ into the form $\mathbf{Q}\mathbf{\Lambda}\mathbf{Q}^{-1}$

where the columns of $\mathbf{Q}$ are the column eigenvectors and $\mathbf{\Lambda}$ is the diagonal matrix of eigenvalues, since this yields the vector-matrix solution $\mathbf{p}^{(n)} = \mathbf{p}^{(0)}\mathbf{Q}\mathbf{\Lambda}^n\mathbf{Q}^{-1}$. So the construction of user-friendly expressions in specific cases depends heavily on being able to determine neat forms for the eigenvalues, eigenvectors and matrix inverse. An equivalent representation avoids the inverse issue by involving both row and column eigenvectors.

Not only do these general matrix results yield the two-state solution for $(p_0^{(n)}, p_1^{(n)})$ directly, but other features can also be determined such as the distributions of the time to recurrence of state 0 (or 1) and the length of spells spent in state 0. Moreover, a simple limiting argument highlights the relation between the discrete- and continuous-time processes. In principle, the extension of these ideas from 2 to m states is straightforward provided $\mathbf{P}$ is sufficiently highly structured to enable algebraically amenable expressions to be determined for the eigenvalues and eigenvectors. For example, in the Ehrenfest model, which describes the exchange of molecules between two connected containers, $\mathbf{P}$ is tri-diagonal. Although this leads to a simple expression for the *stationary* distribution $\boldsymbol{\pi}$, i.e. the solution to the equation $\boldsymbol{\pi} = \boldsymbol{\pi}\mathbf{P}$, the algebra involved in the determination of the time-dependent probabilities $\mathbf{p}^{(n)}$ is non-trivial even though this is a very basic process. Fortunately, considerable progress can be made in the general case by considering specific properties. First, it is worthwhile constructing the conditions under which there will be a genuine approach to equilibrium. In the case of the Ehrenfest model, for example, the process always moves away from its current state which means that the limiting solution $\mathbf{p}^{(\infty)}$ continually oscillates and so cannot be an *equilibrium* distribution. The key results are enshrined in the Perron–Frobenius Theorem for *irreducible* (i.e. all states connect) and non-negative square matrices. In practical situations questions often centre around the number of steps for the process to make its first return to a given state j (the recurrence time), and to enter state k from j for the first time (the first passage time). These are best treated through a generating function approach, and this has the advantage of providing a particularly neat relation between the two solutions. Moreover, it also leads directly on to a proof of a classic Markov chain limit theorem which provides conditions under which chains are transient or recurrent; that is, ultimate return to j is uncertain or certain. Systems that are *ergodic*, i.e. are recurrent and aperiodic, form a particularly important class; they not only have unique equilibrium distributions but these are also unique stationary distributions. Lastly, we just touch upon Branching Processes. For this is a huge field in its own right, and readers interested in pursuing this particular path already have a large literature to delve into. In essence, instead of an individual being able to produce only 0 or 1 individuals at the next generation, it now has probability $g_0, g_1, g_2, \ldots$ of producing $k = 0, 1, 2, \ldots$ offspring with associated p.g.f. $G(z)$. If every individual in a generation fails to reproduce then the population becomes extinct. On defining $F_n(z)$ to be the p.g.f. of the size of the nth generation, this gives rise to the functional form $F_n(z) = F_{n-1}(G(z))$, and the inherent algebraic complexity enshrined within this seducingly appealing expression is clearly immense. Specific analyses are therefore restricted to the study of moments and the probability of extinction.

Having dealt with Markov processes in discrete space and continuous time, and Markov chains in discrete space and time, Chapter 6 completes the trilogy through a study of Markov processes in continuous time and space. Situations in which this third option are particularly appropriate often involve populations which have the potential for containing huge numbers of individuals. For then it makes more sense to study change in population *density* rather than integer *numbers*; modelling marine organisms in the oceans is an obvious example. So as well as visualizing continuous time as the limit as $dt \to 0$ of the discrete time space $0, dt, 2dt, \ldots$, we now also envisage continuous population density or size as being the limit as $dx \to 0$ of the discrete state space $0, dx, 2dx, \ldots$. This incremental *diffusion* approach was first proposed by Robert Brown in 1827 to describe the motion of particles suspended in a fluid that move under the rapid, successive impacts of neighbouring particles. However, it was not until 1905 that Albert Einstein first advanced an explanation, which was famously later made rigorous by Norbert Wiener.

Although the Wiener process is most easily understood by considering the Normal approximation to the binomial distribution, a more formal procedure is to take the Taylor series expansion of the discrete-space forward (or backward) Kolmogorov equation for appropriately small dx and dt. This generates the well-known partial differential equation for heat conduction through a solid, and the Normal form for the solution is easily derived by taking moment generating functions. This link between probability and heat highlights the strong, and often forgotten, connection between stochastic processes and applied problems in physics and engineering. The reflecting barrier case is easily solved by considering the earlier simple random walk solution with position i replaced by x/dx and upper barrier a by a/dx, and then letting $dx \to 0$. However, considerable care is needed not to push this limiting paradigm too far. For whereas in the simple random walk the process spends an amount of time $\pi_0 > 0$ at the lower barrier 0, the proportion of time spent there by the Wiener process is $\pi_0 dx \to 0$ as $dx \to 0$. So the barrier condition has to be changed if we are to inject an atom of probability, π_0, there. That the associated trajectories are radically different in the discrete- and continuous-time scenarios may be seen by using the random walk simulation algorithm with an appropriately small time increment dt. Time-dependent solutions may be generated by once again employing the Method of Images, and this approach also carries over to the absorbing barrier case which is discussed in some detail. An alternative method is to base the solution on the Method of Separation of Variables. Prior experience of working with partial differential equations, and coping with their various eccentricities, is clearly an advantage in this setting.

The Wiener equation is extended to cover general homogeneous time-dependent random walks in which we can move from state k to states $k+1$ and $k-1$ with probabilities $\theta_k(t)$ and $\phi_k(t)$; a similar limiting argument to the above yields the corresponding forward and backward Fokker–Planck equations. The backward version is particularly useful, since unlike the forward equation its form is not affected by the type of boundary condition that may be imposed on the process. A consequence of this difference is that the forward and backward equations are no longer dual processes. Whether either can be solved depends heavily on the structure of the particular model being examined. Fortunately, when θ_k and ϕ_k are not dependent on t the general equilibrium equation yields a fairly neat form of solution. Moreover, this enables us to make a precise analysis of the effect of taking the diffusion limit in specific situations; when this effect is small the diffusion solution is often to be preferred since it is likely to be more combinatorially transparent. Simulation plays a key role here in exploratory time-dependent studies, and as with the Wiener process only minor modifications need to be made to the random walk algorithm. Though careful choice of dx and dt may be required to ensure that $X(t)$ neither explodes nor remains constant in a finite time period when the step size changes from ± 1 to $\pm dx$. Examination of the simple immigration–death diffusion process, together with several variations to the model structure, illustrates the main issues involved. As the analysis of the general Focker–Planck equation with boundaries is even more awkward than that for the simple Wiener process, we just point out possible lines of enquiry. A parallel way of visualizing a diffusion process is to construct the corresponding stochastic differential equation (s.d.e.). Though as this research field has generated a huge, and easy explorable, literature in its own right, we simply note the correspondence between the population, diffusion and s.d.e. approaches.

Even though the velocity, $U(t)$, of a particle becomes infinite as $dt \to 0$, the Wiener and Focker–Planck processes still provide an excellent representation for random independent movement. However, if velocity plays an important role in the underlying dynamics, as happens for the correlated random walk, then it sensible to model this, rather than the displacement of a process. The classic approach developed by Uhlenbeck and Ornstein (1930) generalizes the correlated random walk procedure by allowing $U(t)$ to take any real value, and not just ± 1, by employing the s.d.e. representation $dU(t) = -\beta U(t) + dY(t)$ where $Y(t)$ is a Wiener process. This leads to a Normal form for the solution, so $U(\infty)$ is in accord with Maxwell's law for the velocity of particles in equilibrium. Analysis of $X(t)$ proceeds by considering the integral $X(t) = X(0) + \int_0^t U(s)\,ds$. Although the Wiener and Ornstein–Uhlenbeck processes have different moment structures, since both have Normal solutions it is not too surprising that by making an appropriate choice of time-scale we can transform one into the other. Moreover, both possess the interesting property that when $X(t)$ passes through x_0 at time t_0, then $X(t) = x$ infinitely often in the interval $(t_0 < t < t_0 + \epsilon)$ no matter how small $\epsilon > 0$ may be. In consequence, constructing a model with 'realistic' barrier properties demands considerable creativity.

1.3 Bivariate populations

So far we have examined single-species population dynamics, yet individuals often do not exist in isolation but instead coexist with those from many other species, and thereby exhibit between-species interaction. Unfortunately, not all the progress made with the analysis of univariate populations carries through to the multivariate case since the resulting stochastic equations are usually nonlinear and intractable. This places far greater reliance on simulation and approximation techniques than we have seen so far. Whilst specific scenarios are studied in later chapters, Chapter 7 sets the scene by examining the general bivariate model.

First, we develop a simple immigration–death–switching process for which the preceding methods of solution do carry across. Two types of individual now immigrate into a population with immigration rates α_i and death rates μ_i, with type i individuals changing into type j at rate ν_i $(i,j=1,2)$. The bivariate p.g.f., $G(z_1, z_2; t)$, for the population size $(X_1(t), X_2(t))$ is linear, and so can be solved using the same approach as for the univariate case, though its solution is understandably longer. Moment equations can be constructed directly, either by differentiating G or by transforming the forward equations, and for linear processes these may be solved via a standard matrix procedure. However, in more complex situations the equation for G is highly likely to be nonlinear, and its structure, never mind the probabilities $p_{mn}(t)$ (assuming we can extract them), may reveal little insight into the characteristics of the process. A simple count-dependent growth example, with the deterministic representation $dX_1(t)/dt = \lambda X_1(t) - \mu X_1(t) X_2(t)$ where $X_2(t) = \int_0^t X_1(s)\, ds$, highlights the general underlying problem. For G is now given by a second-order bivariate p.d.e. which is not amenable to direct solution. Even exploring the associated moment equations is not without difficulty. For not only are the equations for the means (called the stochastic update equations) different from their deterministic counterparts, but they also contain a second-order component; similarly the equations for the rth-order moments contain terms of higher order. This therefore gives rise to moment closure issues, and as we have already analysed univariate processes successfully by using the saddlepoint approximation this approach is extended to the multivariate case. Recall that it offers several key advantages: no distributional assumptions are required; it works regardless of the selected moment order; and, it is guaranteed to provide an algebraic form for the associated p.d.f.

Several examples are given to illustrate how this straightforward extension operates when G is known, before we move on to consider the vast raft of situations in which G is either very difficult, or impossible, to determine. We truncate the associated c.g.f. $K(\theta_1, \theta_2; t) \equiv G(e^{\theta_1}, e^{\theta_2}; t)$ by equating all but low-order cumulants, κ_{rs}, to zero, and then use this to develop an algorithm for computing numerical values for the κ_{rs}. This procedure is covered in some detail, for successful implementation requires care and good judgement, especially if the proposed cumulant structure exhibits awkward characteristics. Though not dealt with here, parallel algebraic solutions can also be constructed by using algebraic software such as *Mathematica*. The beauty of this powerful procedure is that it provides an excellent way of deriving approximate probability solutions to otherwise intractable stochastic processes.

Analytic problems tend to occur as soon as birth enters the frame, since events are then multiplicative rather than additive. However, in many situations it is still possible to gain a considerable understanding of a process by excluding it. One area of application that has been particularly successful in this regard is quantum optics. An especially interesting problem concerns the stochastic evolution of populations of photons within optical cavities which are monitored through *external* counts of the number of photons leaving the population. For such an approach not only provides insight into the quantum formulation of the process, but it also enables models to be developed that generate *non-classical*, i.e. sub- or super-Poissonian distributed, light. The paired-immigration–death process is particularly appealing in this regard, for its analytic tractability enables the probability and moment structure to be completely determined. Particularly illuminating is the emergence of a 'saw-tooth' odd–even effect, in which the form of the resulting expressions, and hence their solution, depends on whether the population size, n, is odd or even. Moreover, this and other structural characteristics of the process can be inferred purely by observing the exit-times of emigrants escaping the system. This attribute has immense benefit across a wide range of different fields since it strongly suggests that properties of a 'hidden' stochastic process can be inferred directly from the conditional probability and moment properties of the counting variable.

Although algebraically tractable extensions are presented that involve batch immigration, Poisson–Schoenberg immigration, and birth, trying to use counting measures in order to make statistical inferences in all but the simplest of model structures is extremely difficult. For computation of the associated counting likelihood is normally plagued by the necessity of having to integrate the joint likelihood with respect to the missing data. However, great progress has been made in tackling problems involving partial information numerically through reversible jump Markov chain Monte Carlo (MCMC) techniques. Here we simply explain the basic ideas involved, and illustrate them on single- and paired-immigration–death processes, but the algorithm developed is straightforward to extend to far more complex models. A personal view is that it is far better to construct algorithms oneself than to use professionally developed software, no matter how user-friendly this might appear to be. For the user then knows exactly what is being calculated. For example, posterior marginal distributions with long tails require very careful handling, as do multi-parameter systems, and given that applying MCMC to stochastic processes is still a black-art it is vital to know precisely what the program does at each stage in the computation.

Chapter 8 concentrates on deriving specific results for the bivariate logistic process. Although each of the six parameters in our general model may be either positive (enhancement), negative (impedance) or zero (no effect), which gives rise to $3^6 = 729$ possible model configurations, few of these have any real practical relevance. Three, however, are of fundamental importance, and involve competition, predation and infection. Each one is examined in detail since they exhibit remarkably different behavioural characteristics. Competition is considered first, for this provides a natural extension to the single-species logistic process; all individuals impede the growth of each other, regardless of type. Here the key issue is to determine the conditions under which both species can coexist in harmony, rather than one destroying the other with the surviving species then developing as a univariate logistic process. The deterministic model yields a particularly neat, and easily interpretable, solution provided the inhibitory effects are the same for both populations. Moreover, this special case confirms the well-known observation that having a higher reproductive ability is no guarantee of eventual success. The general quadratic model is less algebraically amenable, but yields quite easily to a more informal examination and shows the existence of four possible types of behaviour: (i) X_1 always wins; (ii) X_2 always wins; (iii) stable equilibrium; and, (iv) unstable equilibrium with the winning species depending on the start position. The simple techniques used include solving the o.d.e.'s numerically, constructing stability diagrams, and using linearization in order to investigate local stability.

Global stability is considered by constructing the force-field at all points in the (X_1, X_2)-plane.

These deterministic approaches provide valuable qualitative information on process development, though care must be taken not to place too much reliance on them since random effects can cause radically different outcomes to be produced from identical start conditions. The easiest way to demonstrate stochastic behaviour is through simulation, especially since the algorithm for general bivariate interaction processes involves a straightforward extension to that for the general univariate birth–death process. The only change is that the two events $X \to X \pm 1$ now become four, namely $(X_1, X_2) \to (X_1 \pm 1, X_2 \pm 1)$. Simulation runs confirm that the stochastic competition process is far less intuitively predictable than its deterministic counterpart. Generating repeated simulations based on different start seeds enables us to construct empirical values for the occupation probabilities $p_{ij}(t)$, the extinction probabilities $p_{i0}(t)$ and $p_{0j}(t)$ $(i, j > 0)$, and the probability of total extinction $p_{00}(t)$. Direct numerical integration of the respective differential and difference equations may also be used, though this does carry the complication of having to truncate the unbounded population sizes. Only a slight change is required to convert this procedure into one that determines the expected number of steps, and time, to extinction. Derivation of the associated moment structure is arguably meaningless for this process, since it relates to a combined measure across four different distributions, namely (a) $p_{ij}(t)$ $(i, j > 0)$, (b) $p_{i0}(t)$ $(i > 0)$, (c) $p_{0j}(t)$ $(j > 0)$ and (d) $p_{00}(t)$. Nevertheless, it is still useful to show how the bivariate raw moment equations may be developed either directly from the probability equations or by differentiating the forward equation for the p.g.f. In principle these may be solved via moment closure either numerically, or algebraically by using *Mathematica* and then employing the saddlepoint approximation. However, the quadmodal nature of the true population size p.d.f. means that any unimodal approximating p.d.f. is likely to be suspect, and a surer way to proceed is to work in terms of the three conditional moment structures defined through (a)–(c) above.

Predation is fundamentally different to competition in that whilst (in the simplest model) the prey population grows logistically with birth rate $r_1 X_1$ and death rate $b_1 X_1 X_2$, the signs in the predator equation are switched. So X_2 dies at rate $r_2 X_2$ and grows at rate $b_2 X_1 X_2$. Thus whereas under competition both species can exist quite happily in the absence of the other, now the predators die out in the absence of prey. Linearizing the deterministic representation shows that this early Lotka–Volterra process (developed in 1925 and 1926) generates neutrally stable population cycles, and the crude similarity between this structure and many two-species interaction cycles observed in the field meant that too much importance was initially awarded to it. For corresponding stochastic simulations exhibit boom-and-bust dynamics, with trajectories quickly spiralling out to produce early prey extinction, often during the first cycle. Partial insight into why this occurs may be gleaned by constructing the 'expected update' equations, for these enable us to make a quick assessment of the stochastic stability of variations to the Lotka–Volterra theme. The first of these involves making the prey equation logistic in both X_1 and X_2, which guarantees to keep the prey population under control. This produces moderately smooth cyclic behaviour, which can be characterized by constructing the linearized deterministic solution and the linearized second-order moments. Though such results produce more of a qualitative than a quantitative perspective, they are nevertheless extremely useful in portraying the likely behavioural characteristics of a process, which can then be studied further through a stochastic simulation exercise. An approximation to the mean time to extinction provides a measure of global stability. In another variation we include prey cover, since then trajectories automatically avoid the prey axis.

The strong differences between deterministic and stochastic behaviour for all three models highlight a serious issue, namely the potential for polarizing viewpoints between deterministic and stochastic analysts. A more pragmatic modelling approach is to strike a balance between key information about the process under study and the overall transparency of the resulting model. An obvious way forward is to construct a process that generates sustained stable limit cycles under both the deterministic and stochastic regimes, and this is easily achieved if the logistic components involve linear fractional terms in X_1 and X_2. The beauty of this approach is that the deterministic and stochastic analyses now blend together. The former provides detailed theoretical insight into behavioural characteristics, whilst the latter enables the assessment of stochastic variability, mainly through change to the amplitude near the origin. For there the deterministic trajectories are packed sufficiently tightly together for the stochastic process to jump between them with relative ease.

An apparently minor change to the Lotka–Volterra process opens up a totally different field of study, namely epidemic modelling. For suppose we envisage prey and predators as corresponding to the number of individuals susceptible to, and infected by, a disease. Then in the simplest model a single infection results in $(X_1, X_2) \to (X_1 - 1, X_2 + 1)$ at rate $\beta X_1 X_2$, and a single death or removal of an infective results in $(X_1, X_2) \to (X_1, X_2 - 1)$ at rate γX_2. Whence the associated deterministic equations, namely $dx_1/dt = -\beta x_1 x_2$ and $dx_2/dt = \beta x_1 x_2 - \gamma x_2$, correspond to the Lotka–Volterra process with a zero birth rate. Here, however, the similarity ends. For in the predator–prey scenario prey-death and predator-birth are separate events, with $(X_1, X_2) \to (X_1 - 1, X_2)$ and $(X_1, X_2) \to (X_1, X_2 + 1)$ occurring independently at rates $\beta_1 X_1 X_2$ and $\beta_2 X_1 X_2$. So it is highly likely that stochastic trajectories will develop in quite different ways. One pleasant effect induced by this switch in outlook, which reduces the number of model parameters from four to two, is to render the stochastic equations tractable to analytic solution. Nevertheless, the apparently simple nature of the epidemic process proves to be surprisingly deceptive. For although the forward equations can be solved sequentially in terms of series expansions, this does yield combinatorial problems; whilst using Laplace transforms from the outset involves working with very inelegant expressions. The inherent difficulties essentially stem from the integer population sizes. However, making a problem initially more complicated may result in an easier solution, and in our case we apply an ϵ-perturbation to the infection rate, derive the solution, and then let $\epsilon \to 0$. Moment expressions can be obtained from this analysis directly, though a more elegant result can be developed by considering the p.g.f. of the ϵ-modified process. Of some concern is the potentially strong disparity between the deterministic mean and the stochastic expectation for the number of infectives. For the deterministic equations are equivalent to first-order truncation of the stochastic moment equations, and here the second-order moments are relatively large. So dissimilarity between deterministic and expected behaviour is not due to the presence of small population sizes, a common misconception, but to the variance being non-negligible relative to the mean.

Constructing the infective (or susceptible) generating function enables us to derive approximate moments through moment closure, and hence approximated p.d.f.'s through the saddlepoint approximation. Whilst exact, and algebraically amenable, results for the duration time can be developed by exploiting the similarity between simple (linear) birth and nonlinear (quadratic) infection in the opening stages of the process. Moreover, although the simple epidemic process lacks a non-trivial equilibrium solution since every susceptible must eventually become infected, this is easily rectified by assuming that each infected individual recovers and returns to the susceptible state (rather than becoming immune or dying). This assumption is perfectly reasonable for many mild diseases.

In order to obtain a genuinely bivariate epidemic process we need to remove the direct link between infectives and susceptibles. One common way is to allow infectives to be removed from circulation either through isolation or death, which gives rise to the so-called 'general epidemic process'. The deterministic equations readily yield an approximate expression for the number of removals at time t, which leads to the celebrated *Deterministic Threshold Theorem* of Kermack and McKendrick (1927). Unfortunately, this remarkable early result languished virtually unnoticed for three decades; if it had not then current epidemic theory might be in a far more advanced state. Now in this deterministic analysis an outbreak of infection either does, or does not, occur as the number of susceptibles switches from being just above to just below a threshold value. This is clearly unreasonable; far more likely is that the probability that an outbreak occurs will change. Whilst some exact analytic progress can be made by trying to solve the forward equations, the opening stages of the process relate quite neatly to a simple birth–death process, and Whittle (1955) took full advantage of this to produce his famous *Stochastic Threshold Theorem*. Below the threshold there is zero probability of a major outbreak, whilst above it the probability rises geometrically with the initial number of susceptibles. Further exploitation of the birth–death approximation leads to other useful stochastic results; whilst the final configuration of the epidemic system yields to a path-based analysis.

Given that the basic infection-removal process corresponds to the simplest possible model structure, it is clear why even slightly more complex models, like the general epidemic process, do not readily lend themselves to a full exact analytic treatment. Nevertheless, when handled in the right way they can still advance understanding quite considerably, which is of prime importance when we consider the huge devastation caused by the wide variety of epidemic diseases currently afflicting the human population. Cross-over infection from birds and animals to humans, coupled with mutating viruses, means that the field of epidemic modelling has to evolve constantly in order to keep up. The earliest extension, initially applied to measles and chicken-pox, allows for the investigation of recurrent epidemics by replacing prey birth in the Lotka–Volterra process by prey immigration; for then both infectives and susceptibles can oscillate around equilibrium values. A deterministic analysis provides considerable insight into the population dynamics, though the implied damping of infective cycles towards a steady state contradicts what happens in practice. However, our earlier experience with predator–prey models, in which deterministic converging spirals switch into sustained stochastic cycles, suggests that the same might well apply here. Not only do simulated stochastic realizations bear this out, but a linearized second-order moment analysis quantifies the nature of the resulting cycles. This success through the use of an outside agency, namely immigration from an external source, raises the question as to whether similar behaviour may be obtained by purely internal means? One likely solution is to replace prey immigration by logistic prey birth, since this returns the deterministic Volterra equations, and simulations of the corresponding epidemic process do indeed exhibit the right form of cyclic behaviour.

Whilst the predator–prey, competition and epidemic processes discussed in Chapter 8 play crucial roles in helping us to understand important mechanisms, and hence our ability to predict likely behaviour, there is clearly a huge potential for developing other important fields of application. We therefore conclude this chapter by presenting a different kind of interaction effect which has so far received much less attention. Suppose that the development of a population depends not only on the current population size but also on either: (i) the (discrete) cumulative number of individuals, $C(t)$, born by time t; or, (ii) the (continuous) integrated population size, $F(t)$. The example used relates to the growth of aphids. Then although the second variable is now the sum or integral of the first, the process still provides transparent deterministic solutions. Moreover, both of the associated stochastic simulation algorithms still only involve a slight modification to the general two-species algorithm used so far. Not only is there a considerable difference between deterministic and stochastic realizations, but stochastic realizations based on $C(t)$ and $F(t)$ exhibit substantially different structures even though they possess the same underlying deterministic process. This highlights the importance of comparing deterministic and stochastic realizations for all possible processes before making a final decision on model choice. Further care is required when developing the moment structure via the p.d.e. for the bivariate process $\{X(t), C(t)\}$. For second-order cumulant truncation approximates the means well but the variances poorly, and including higher-order cumulants offers little improvement. So here we have an example for which even the saddlepoint approach is unlikely to produce a good approximation to the probability structure. Extension to the power-law logistic process promotes even more complex issues which require further theoretical development.

1.4 Spatial–temporal processes

So far we have assumed that individuals mix homogeneously at a single site. Now whilst this assumption facilitates analytic tractability, there are unfortunately many situations in which it is violated. For not only may a population be spatially distributed across several sites, but the chance of two individuals interacting may well depend on the inter-site distance. As the addition of a spatial dimension clearly hinders analytic development, our aim is to expose the underlying theoretical difficulties, highlight the directions in which some degree of progress can be made, and show that the introduction of space generates the whole new concept of a *Stochastic Dynamic*.

Let $X_i(t)$ denote the number of individuals in site $i = 1, \ldots, n$, and suppose that an individual may experience three types of event, namely arrival, departure and transfer to another site. Then if the associated rates depend only on the numbers in the sites affected by the transition the process is described as simple, and is considerably more amenable to mathematical analysis than if it is not. Note that the close paradigm between 'site' and 'individual type' means that the term spatial has a much wider interpretation than geographic location. If the process is reversible then general expressions may be derived for the stationary distribution (provided it exists), so solutions to many special cases are immediately available. Sadly, this is often not fully appreciated.

Since the simple birth–death–immigration models investigated in Chapter 2 are far more mathematically tractable than the nonlinear processes considered in subsequent chapters, we initially explore the two-site birth–death–migration process in order to illustrate the general issues involved. For as the population size p.g.f. is the solution of a first-order o.d.e., moment equations constructed from it can be solved exactly without the need for moment closure. Moreover, this model provides an excellent example of the way in which the level of mathematical tractability involved in determining the population size probabilities, $\{p_{ij}(t)\}$, swiftly decreases as the number of parameters increases. Although the full p.g.f. equation is likely to evade a closed-form solution for some time yet, approximate probabilities can be developed by: (a) slightly modifying the birth mechanism; and, (b) developing a power-series solution for the backward equations when sites are 'slightly connected', i.e. all the migration rates lie close to zero. Whilst these two approaches are certainly useful, they do not present a universal way forward and their deficiencies highlight the issues to be faced when considering general interconnected processes. So as an alternative to constructing approximate solutions, a third procedure is presented which involves two sequences of integral equations. One converges monotonically downwards and the other upwards, and the true, unknown, solution is sandwiched between them. The reason why adding a spatial dimension induces intractability is best seen by examining the nature of the first-order quadratic o.d.e.'s for the backward p.g.f.'s. For these are Riccati in nature, and it is the presence of a single specific term in the matrix representation that causes all the problems.

Given that the linear two-site birth–death–migration process raises such serious analytic issues, it is apparent that there are few mathematical tools available to enable us to conduct useful stochastic analyses even in the linear n-site case, never mind general spatial situations involving nonlinear interactions such as competition, predation and infection. All we can realistically hope to determine is the first few moments, which can then be placed within the saddlepoint approximation. For linear processes, general solutions for the means and variance-covariances are straightforward to write down in vector-matrix form. However, being able to use these to construct solutions in specific cases hinges on having a decomposable matrix structure. A fairly straightforward way of effecting this is to presume that the migration rate matrix is tri-diagonal; so we have migration between nearest-neighbour sites only. Suppose that the process is homogeneous and that it develops over the full integer axis; the one-dimensional nature of the process is crucial here. Then the forward p.d.e. for the population size p.g.f., $G(\mathbf{z};t)$, may be written down directly using the 'random variable' technique (Bailey, 1964). This equation is then easily differentiated to yield the moment equations, which are solved by exploiting their association with modified Bessel functions. Moreover, the probability structure can also be determined if birth is replaced by immigration. It transpires that here each site behaves as though it is independent of all other sites, with the size of colony i being distributed as a Poisson variable whose parameter depends on both i and t. Considerable care needs to be taken when using this result; for although the sites are independent *in probability* they are not independent *in realization*. As for the two-site process, if birth is present then approximate probabilities may still be developed by modifying the birth mechanism.

Whilst allowing the process to develop on the infinite line is mathematically convenient, since this avoids edge-effects, in practice the number of separate colony locations will often be finite. Considerable algebraic effort is required even to derive expressions for the expected colony sizes using the Method of Images. Moreover, experimenting with different boundary conditions shows that edge-effects can permeate right through a large dynamical system, so it is vital to ensure that the correct condition is chosen if we are to avoid the tail wagging the dog. Given that the probability equations for the basic two-site process are intractable to direct solution, stochastic investigations for this more complex spatial structure have little choice but to follow the simulation route. Provided the number of sites, n, is not too large we can still use the exact multi-species algorithm based on 'event–time pairs', simply by calling species sites. That is, events are chosen by comparing each of the $4n-2$ possible birth, death and migration rates against a $U(0,1)$ random number. Though this exact procedure may take too long to run if either n or the time interval are big, and in such cases alternative strategies may well need to be employed. A general approach that works well if the sum of the transition rates is not too large is to assume that all sites develop independently during each small time increment $(t, t+dt)$. Potential inefficiencies can be overcome by allowing dt to change dynamically. Whilst if n substantially exceeds the total number of individuals, then an individual-based approach may be more computationally efficient. All three techniques cover general nonlinear processes with interaction between every site. However, very large and complex systems as found, for example, in chemical reaction systems, would still take too long to analyse, and this provides the motivation for generating multi-scale algorithms that maximize speed and efficiency by operating at the most coarse-grained level possible. In essence, suppose that a time interval of length τ is small enough to ensure that the rates change relatively little over $(t, t+\tau)$. Then during this time period we may assume that the numbers of births, etc., are distributed as Poisson random variables. Whence using these in place of the Bernoulli variables in the 'dt' algorithm generates a popular alternative approach called tau-leaping. The degree of rate variability experienced clearly depends on τ, and this, in turn, relates to the 'stiffness' of the system of associated o.d.e.'s. Probabilists wishing to use this technique are advised to study recent developments in this field before adopting it as a standard procedure.

Although these four algorithms should be sufficient for most population modelling studies, other approaches may also be considered. For example, replacing the Poisson tau-leaping variables by Normal ones produces Langevin-leaping. Whilst taking the reverse limit $\tau \to dt$ returns the underlying stochastic differential (a.k.a. chemical Langevin) equation, about which huge amounts are known. Finally, ignoring the stochastic component yields the classic deterministic representation (a.k.a. the reaction rate equations). Virtually all of the above procedures have been developed in splendid isolation from each other, and much work needs to be done not only to compare their respective properties, but also to draw them far closer together in a unified whole.

Two key measures of spatial population growth are: the total range of spread, defined by the positions of the outmost non-empty sites; and, the velocity of propagation of these end sites. Now although the deterministic analysis is much simpler than its stochastic counterpart, it does appear to suffer a serious drawback when used to construct these measures. For suppose that the initial population comprises a single individual in site $i = 0$ at time $t = 0$. Then the mean population size *instantaneously* becomes positive at *every* site, i.e. $m_i(t) > 0$ for all $-\infty < i < \infty$ and infinitesimally small $t > 0$, which defies the whole concept of population spread. The way round this is to define the position, i^*, of the wavefront at time $t > 0$ through the condition $m_{i^*}(t) = K$ for say $K = 1$. This yields velocities, c, which are in close accord not only with simulated values, but also with those derived through the saddlepoint approximation. The latter approach also yields a general form for the wave profile. Construction of the parallel diffusion process shows that unless the net growth rate is small in comparison to the migration rate, the discrete and continuous state space cases can give rise to substantially different velocities. So care needs to be taken when employing continuous approximations in spatial–temporal scenarios.

When operating in a finite spatial domain, once the process has reached all available sites interest mainly lies with the change in spatial population density through time. A good illustration is provided by an experiment conducted by Neyman, Park and Scott (1956) to determine the distribution of flour beetles within a cubic container. Our one-dimensional deterministic analysis extends neatly to two and three dimensions, and highlights the marked difference in expected population numbers between the interior, sides and corners. However, this example still involves migration to neighbouring sites, so what happens when we consider general migration rates over the full infinite line where an individual in site i migrates to site $i + j$ ($-\infty < i, j < \infty$) at rate νh_j? Here $\{h_j\}$ is called the 'contact distribution'. The generating function for $\{m_i(t)\}$ is easy to construct, and in the case of geometrically distributed migration the $m_i(t)$ may be extracted from it. Though the velocity of propagation and wave profile are best derived via the saddlepoint approximation. Life becomes much more interesting though when the contact distribution is non-exponentially bounded. A good (extreme) choice is to take the Cauchy form $h_j \propto [\prod_{u=0}^{r}(j+u)]^{-1}$, since then only the first $r - 2$ moments exist and the asymptotic velocity of the right-hand wavefront is infinite. When $r = 3$ the mean exists, but the variance does not, so the stochastic wavefronts progress in wilder and wilder leaps forward. Whilst when $r = 4$ both the mean and variance exist, which results in a mixture of steady progress and great leaps forward that is unlikely to be mirrored by any deterministic approximation. Nevertheless, the deterministic process is partly tractable, and gives rise to expressions involving

Gauss hypergeometric series and Stirling numbers. Unfortunately, trying to find a family of velocities for $r = 2, 3, \ldots$ remains an open problem, and even reverting to the saddlepoint approximation exposes awkward analytic issues.

A comparison of simulated stochastic realizations for nearest-neighbour, geometric and Cauchy ($r = 4$) migration is most easily made via an individual-based algorithm, and highlights the increasing volatility of the associated population spread. Inspection shows that whilst the deterministic analyses summarize overall behaviour quite well in the first two cases, they provide only a very rough mirror of reality in the third, extreme, case. Further stochastic analyses clearly need to be developed if we are to gain a greater insight into such spatial–temporal structures.

Multivariate situations are even more intriguing. For whereas all the individuals in a birth–death–migration process develop independently of each other, in nonlinear systems, such as those involving predator–prey, competition and infection interactions, this is no longer the case since population size and spatial distribution are closely linked. Thus whereas in a (non-spatial) Lotka–Volterra process the prey usually die out before the end of the first cycle, provided prey travel faster than predators both species may propagate outwards indefinitely in a travelling wave. Series expansions may be derived for the respective deterministic population sizes, $x(s;t)$ and $y(s;t)$, near the wavefront at $i^*(t) = ct$ in terms of the relative displacement $s = i - ct$; provided, of course, that the asymptotic velocity c is finite (which rules out using Cauchy-type contact distributions). The more terms we employ in these series, the greater the level of accuracy that is achieved. Whilst this 'wavefront technique' is totally general in terms of the models it may be applied to, it provides no information within the body of the wave itself. However, if an equilibrium situation exists then the deterministic equations may be linearized and hence solved, thereby yielding the period of oscillation and rate of attenuation. Whether or not the imposition of such spatial structure stabilizes the stochastic system in this way, i.e. population cycles persist within it over long periods of time, is best investigated through simulation.

The general question of whether an inherently unstable stochastic system can be made stable by inducing spatial structure is an old one, and was first examined in a wonderfully elegant paper by Turing (1952) in a study of chemical reaction structures. Here two variables $X_i(t)$ and $Y_i(t)$ interact within each site $i = 1, \ldots, n$ and also 'migrate' to neighbouring sites. Linearizing the deterministic equations and taking Fourier transforms yields the general linearized solution for any system possessing an equilibrium structure, and this highlights two particularly striking situations. In one, a series of 'morphologically stable' waves develops as t increases; whilst in the other the waves oscillate through time. The first of these, which involves stationary waves of finite length, stimulated tremendous interest in the pattern-generating community, since it offered a highly credible explanation both for the growth of symmetric tentacled-species such as the sea-anemone, and the development of skin-patterns on animals, reptiles and plants, through the diffusion of morphogens. Biological verification of this theory followed several decades later! Now whilst properties of the linearized deterministic system can clearly be used to determine parameter values appropriate to those interaction functions which lead to the replication of observed patterns, a key question remains as to whether such strong similarities carry across to the corresponding stochastic system. Particular care needs to be exercised here, since (as seen previously) there is no unique way of forming the associated birth and death rates from the deterministic equations. Thus each interaction function has several potential birth and death functions contained within it, all of which need to be examined on their own merit. As the single-species algorithms considered earlier extend in a straightforward manner to multi-species processes, simulation experiments are easy to conduct. The fact that many exhibit a considerable discrepancy between stochastic and deterministic dynamics offers great scope for future research in this area. This is especially pertinent given the close similarity between deterministic prediction and real-life observation.

The deterministic–stochastic divide becomes even more acute for highly volatile systems which exhibit frequent local extinctions yet persist virtually indefinitely. Consider, for example, a large connected system of Lotka–Volterra processes. Then within each site the prey and predator populations will quickly boom-and-bust. However, provided a prey can migrate to a neighbouring empty colony, followed by a predator, a new cycle will be established there, which then induces yet another cycle at a neighbouring site, *ad infinitum*. So taken as a whole the process may persist almost indefinitely, in spite of the inevitable swift collapse of population cycles at individual sites. This gives rise to the concept of a *Stochastic Population Dynamic*, and to date no theory exists either for the conditions for long-term persistence or the time to extinction itself. Issues become even more involved when nearest-neighbour transitions are replaced with other forms of connected spatial structure. Examples considered include: bicycle wheels with and without hubs; general connectivity; distance-dependent rates; and, Small World effects.

One way of making stochastic progress is to aim for a qualitative solution rather than a fully quantitative one. For example, if our main interest lies in determining the probability of extinction we may define the population size at a given site as being either empty or not empty. Thus a predator–prey system of n connected sites now has 4^n total states, which may well be analytically and computationally manageable provided n is not too large. Extending the number of qualitative labels to four, namely empty, few, increasing and many, produces an even more credible system. Indeed, by employing some careful basic algebra the resulting matrix representation enables us to make reasonably detailed studies of how different types of network connectivity affect the quasi-equilibrium structure. However, this approach can only provide a broad outline of likely outcomes, and so the development of precise statements concerning occupation and extinction probabilities has to follow the simulation route. Fortunately, it is relatively straightforward to construct a general algorithm which parallels the earlier multi-site algorithm based on event–time pairs. Although the events themselves have to be handled with some care, time is no longer an issue since we are now operating within a Markov chain. This means that extremely quick simulation experiments can be performed which provide a good feel for how the within- and between-colony transition rules interact with spatial structure to produce long-term persistence and interesting patterns of population size. The latter catapulted to prominence in the 1970s when Conway's "Game of Life" appeared in the popular scientific press. A large number of fascinating patterns generated both by this algorithm and its many variations have been produced since the original discovery, and these raise wide-ranging and important questions across the mathematical spectrum. Here, we are primarily concerned with determining what happens when deterministic rules are replaced by stochastic ones. To illustrate how such patterns develop, we present two simple stochastic cellular automata for the spread (and hence potential control) of forest fires and foot-and-mouth disease. Both of these scenarios pose interesting threshold questions concerning how pattern structure and time to extinction change with the underlying probability transition rates. Given the obvious link with percolation theory, which presents many fascinating mathematical challenges in its own right, the stochastic cellular automata arena offers great scope for further development.

To enable some degree of mathematical tractability we have concentrated our attention on spatial interaction between adjacent sites which are located on a lattice structure. In practice, however, not only might interaction occur across much larger spatial levels, but there is no reason why sites have to lie at integer locations. In Chapter 10 we therefore conclude by presenting two extensions involving long-range dependence and process development over real space. Taking the former case first, the construction of reasonably transparent solutions requires the adoption of a pragmatic modelling approach. The sensible option is to superimpose white noise on a linear growth-interaction process, since this allows us to use a set of general interaction weights $\{a_r\}$ for variables located a distance r apart; the extension to multi-dimensional lattices is immediate. Although employing a standard generating procedure easily yields expressions for the first- and second-order moments, the resulting autocovariance structure may be substantially more involved than the process itself. Fortunately, any related difficulties can be avoided by working with the associated spectral density function instead. The expression for the spectrum is surprisingly concise and informative, and shows that this process is 'Ω-stationary' in the sense that it is stationary only within certain scales of pattern; outwith these scales it diverges. Sadly, although the analytic power of the spectrum has been recognized by physicists and electrical engineers for many years, statisticians and probabilists have generally shunned its appeal. It would be nice if this ceased to be the case. Five specific regimes are presented to show how key properties of the underlying system may be constructed. (1) Nearest-neighbour interaction produces the universally encountered inverse square law; it also allows us to reflect on the difference between our spatial–temporal model and non-temporal simultaneous and conditional schemes. (2) Geometric interaction allows all sites to interact with each other, and, being exponentially bounded, gives rise to a similar spectrum together with a well-defined threshold frequency. (3) Logarithmic interaction is non-exponentially bounded and decays much more slowly; this also yields straightforward expressions for the threshold frequency and spectrum, and the effect induced by the heavy tail of the logarithmic interaction is clearly evident. (4) Cauchy interaction has a fundamentally different structure to the last three forms, since variables that are far apart are now highly correlated with each other. That is, they exhibit 'long-range memory dependence'. Although the corresponding spectrum is more convoluted, it nevertheless exposes a strong change in spectral nature, since at small frequencies it corresponds to 'pure $1/\omega$-noise'. Whilst (5) binomial interaction produces more general ω^{-d} spectral structure.

The inverse question is equally important, namely how to determine the interaction rates which give rise to any specific target spectrum? This problem is solved by considering the parallel, albeit purely spatial, fractional integration process, since this points us towards the properties that the interaction structure should possess. The two approaches are duals of each other in that a surface with given spectral structure can either be generated statically or grown through time. Pure power-law examples are provided to illustrate these techniques. Simulating such processes presents its own problems since computers cannot handle the infinite number of locations required for long-range interaction. Fortunately, only high frequencies are affected, and as the (finite) number of locations employed rises so does the upper cut-off value. Simulations are presented both to show the strong agreement exhibited between theoretical and simulated spectra within the admissible inner and outer scales of pattern, and to illustrate the computational difficulties involved. We conclude with two two-dimensional examples. First, we examine a power-law application to sea waves (of major importance to the offshore engineering industry), and show how to construct spatial interaction regimes which give rise to fully developed and developing seas. Second, we invoke the property that fractals may be characterized by the inverse-power-law form of their spectrum in order to gain a deeper insight into the structure seen by the scattering of light from a choppy illuminated swimming pool. Currently, within the stochastic process domain the fields of spatial and spatial-temporal ray propagation lie virtually untouched.

The assumption that sites are located over integer arrays has been crucial to the development of our theoretical results pertaining to wavefronts, velocities of spread, Turing phenomena, stochastic cellular automata, and spectral and ray propagation issues. Moving from this discrete, to a continuous, spatial domain means that mathematical theory becomes far harder to construct, which places heavy reliance on simulation-based approaches. In a 'pure point process' each individual can live anywhere in $[0,1]^2$ (say), so a 'site', x_i, corresponds to the location of the single individual i. Whilst in a 'marked point process' each point i also carries an attribute, or 'mark', $m_i(t)$. Now whereas in lattice space the separation of neighbouring individuals is governed by the known inter lattice point distance, so marks play a more important role than points, in continuous space both marks and points are equally significant. Indeed, their own spatial structures can be highly intertwined and the problem of disentangling them can be challenging.

Various modelling strategies are open to us, but the main one presented here involves each mark $m_i(t)$ developing deterministically through a growth function $f(m_i(t))$ and a spatial interaction function $h(m_i(t), m_j(t); \|x_i - x_j\|)$ taken over all points j that interact with i. If $m_i(t) \leq 0$ then point i dies 'interactively' and is deleted, as happens for 'natural' death. Whence coupling this deterministic growth–interaction process with basic stochastic arrival and departure mechanisms, such as the simple immigration–death process, results in a powerful, yet highly flexible and computationally fast, generator of space–time structure that bears similarities to Gibbs processes. Although the choice of plausible growth and interaction functions is huge, careful selection of four growth functions, and two interaction functions, combine to produce a rich range of pattern structure. The former centre around variants of the power-law logistic and polynomial forms, whilst the latter involve simple hard-core and soft-core interaction. This construct is sufficient for handling biologically oriented scenarios involving plant and tree growth. Though in 'maximum packing' situations, observed in say materials science, we also need to let points move under interaction pressure in order for the system to converge to the high packing intensities often observed in practice.

The most pragmatic way of simulating marked point processes is to update the n marks $i = 1, \ldots, n$ at the time points $t = dt, 2dt, \ldots$ for appropriately small dt. Random arrivals occur at rate αdt (whence $n \to n+1$) with uniformly distributed locations. Whilst each individual i dies naturally at rate μdt and interactively if $m_i(t+dt) \leq 0$ (whence $n \to n-1$). However, if such stochastic events are not present, as in the packing of homogenous materials, then we have to use instead an augmented version of the exact event–time algorithm mentioned earlier. To effect this we first need to decompose the growth and interaction functions into general stochastic birth and death components. So the individual growth and interaction rates $f_i \to f_i^+ - f_i^-$ and $h_{ij} \to h_{ij}^+ - h_{ij}^-$, where f_i^+ and f_i^- denote pure birth and death, whilst h_{ij}^+ and h_{ij}^- denote spatial enhancement and inhibition. Write $\lambda_i = f_i^+ + \sum_{j \neq i} h_{ij}^+$ and $\mu_i = f_i^- + \sum_{j \neq i} h_{ij}^-$. Then on assuming that in the time interval $(t, t+dt)$ each mark i may change size from m_i to $m_i + \delta$ or $m_i - \delta$ at rate λ_i and μ_i, respectively, simulation parallels that for the n-site single-species population process. Extensions, such as allowing particles to move, are easily developed from this basic framework. Now in the absence of particles randomly entering or leaving the system the problem of determining when the process has burnt in is not really an issue. However, if random arrivals and departures do occur then the problem is less clear-cut. Note that using 'perfect simulation' (based on Propp and Wilson, 1996) is no real help here, since this procedure generally involves far too long a compute time. If the exact algorithm causes compute time problems then we can modify it in accord with the approximate leaping techniques discussed earlier, namely: dt, tau, Langevin and s.d.e. In each case we have to assess the trade-off between the accuracy of pattern structure and computational efficiency. In the absence of natural death care needs to be exercised if new marks arrive with small size ϵ. For under some (f, h)-regimes new marks that arrive in empty space not only grow but they also reduce the size of neighbouring touching marks, thereby releasing more empty space for future immigrants to arrive and establish themselves. So for $\epsilon, dt \sim 0$ the process may become fractal as $t \to \infty$. Related convergence issues are discussed in some detail.

Finally, two examples are presented which illustrate how this modelling approach may be applied in practice. Thinning strategies are a prime factor in generating commercially efficient spatial patterns in managed forests, since they exert a strong effect on stand development and hence product yields. Moreover, trees have long life spans relative to the usual length of research projects, which means that forest modelling is a key tool in the formulation and development of optimal management strategies. Fortunately, our highly flexible algorithms are easily adapted to enable comparative studies to be made of different thinning regimes. Two specific ones are analysed which highlight the improvement that can be achieved in respect of two signatures of forest output, namely stand basal area and quadratic mean diameter. Our second example, which is of considerable importance in chemistry and engineering, relates to the study of porous and granular material through the modelling and statistical analysis of random systems of hard particles. For the way in which individual particles pack together, and the resulting packing intensity, are crucial when defining the properties of the material. Although several simulation strategies are already well-established, notably ones based on sequential packing under gravity, collective rearrangement and force-biased, the advantage of using the growth-interaction approach lies in the ability of marks to exploit constantly the available space local to them by changing their size and position in response to the changing interaction pressure induced by their neighbours. The change in packing intensity and spatial pattern caused by allowing movement is particularly dramatic. Current interest centres on the determination of optimal ways of selecting the growth and interaction functions, $f(\cdot)$ and $h(\cdot)$, together with their associated parameters, in order to obtain the best 'fit' to observed two- and three-dimensional patterns. Ideally, this should proceed through the construction of efficient recursive simulation experiments in which functions and parameters are tweaked until both visual and summary statistical measures are optimized. However, virtually no theoretical studies have so far been made on the underpinning mathematical and statistical properties of either these univariate spatial structures, or their natural multivariate extensions. The development of general results for spatial–temporal marked point processes that would enable us to enhance our analytic understanding provides a great challenge for the future.

2
Simple Markov population processes

To introduce the basic principles which underlie the concept of stochastic population dynamics, let us consider some simple types of population structure. Suppose that individuals develop completely independently from each other, and may either enter a population (i.e. immigrate) randomly at rate $\alpha > 0$, depart (i.e. emigrate) at rate β, reproduce at rate λ or die at rate μ. Then combining these different types of behaviour generates a variety of interesting processes, each with its own characteristics, and we shall examine six specific combinations in roughly increasing order of mathematical complexity.

2.1 Simple Poisson process

The simplest construct is to assume that individuals enter a population randomly at constant rate α, and then neither reproduce, die nor leave thereafter. Denote $N(t, t+dt)$ to be the number of arrivals in the small time interval $(t, t+dt)$. Then the probability transition rates are given by

(i) $\Pr\{N(t, t+dt) = 0\} = 1 - \alpha dt + \mathrm{o}(dt)$,
(ii) $\Pr\{N(t, t+dt) = 1\} = \alpha dt + \mathrm{o}(dt)$ and
(iii) $\Pr\{N(t, t+dt) > 1\} = \mathrm{o}(dt)$, where
(iv) $N(t, t+dt)$ is independent of all arrival events in $(0, t)$,

and these define the simple Poisson process. Here $\mathrm{o}(dt)$ denotes all terms that are of smaller order than dt, i.e. $\mathrm{o}(dt)/dt \to 0$ as $dt \to 0$. Thus dt^2, dt^3, etc., are all of order $\mathrm{o}(dt)$. Key questions relate to the time to the next event, and to the number of events in a given time period.

The sheer simplicity of this process provides an excellent opportunity to introduce theoretical methods for constructing probability density functions and moments, together with the relationship between them. In our later studies of more complex processes we shall see that such exact analyses become mathematically intractable, at which point progress relies on the construction of approximation techniques and simulated stochastic realizations.

2.1.1 Time to subsequent events

Suppose we fix the time origin at t_0. Then condition (iv) ensures that whatever happens to the process after time t_0 is totally independent of what happens at or

32 Simple Markov population processes

Figure 2.1 Poisson process of arrivals starting at time t_0.

before t_0. So if $t_0 + Z$ is the time of the first event after t_0, then the random variable Z is independent of whether an event occurs at t_0 and of occurrences before t_0. To determine the probability distribution of Z we could, in theory, work directly with the distribution function $F(t) = \Pr(Z \leq t)$ $(t \geq 0)$. However, for any given $Z \in (0,t)$ there are an infinite number of possible event–time pairs in the remaining interval (Z,t), and we would have to integrate out over all such possibilities in order to recover $F(t)$. In contrast, the converse function $G(t) = 1 - F(t) = \Pr(Z > t)$ involves only the single null event in $(0,t)$, and so is much easier to work with. This highlights a common approach in stochastic processes, namely that it may well be better to answer a much simpler related question, and then to work back from this towards the original problem.

Consider the small time increment $dt > 0$. Then on splitting the events that occur in $(t_0, t_0 + t + dt)$ into those that occur in the initial large interval $(t_0, t_0 + t)$ followed by those in the subsequent incremental interval $(t_0 + t, t_0 + t + dt)$ (as illustrated in Figure 2.1), we see that

$$G(t+dt) = \Pr\{Z > t + dt\}$$
$$= \Pr\{Z > t \text{ and no event occurs in } (t_0 + t, t_0 + t + dt)\}$$
$$= \Pr\{Z > t\} \Pr\{\text{no event occurs in } (t_0 + t, t_0 + t + dt) \mid Z > t\},$$

which, on using condition (iv), reduces to

$$G(t+dt) = G(t) \Pr\{\text{no event occurs in } (t_0 + t, t_0 + t + dt)\}.$$

Use of condition (i) then yields

$$G(t+dt) = G(t)(1 - \alpha dt + o(dt)).$$

Whence letting $dt \to 0$ produces

$$dG(t)/dt = -\alpha G(t), \tag{2.1.1}$$

which integrates directly to give

$$G(t) = G(0)e^{-\alpha t}. \tag{2.1.2}$$

Since $G(0) = \Pr(Z > 0) = 1$, it follows that the cumulative distribution function (c.d.f.) of Z is

$$F(t) = \Pr(Z \leq t) = 1 - G(t) = 1 - e^{-\alpha t}. \tag{2.1.3}$$

So the corresponding probability density function is

$$f(t) = F'(t) = \alpha e^{-\alpha t} \qquad (\alpha > 0), \tag{2.1.4}$$

namely the exponential distribution with parameter α.

Without any loss of generality, we may assume that the process starts at time $t_0 = 0$. Then on letting Z_n denote the time interval between the $(n-1)$th and nth events, the events themselves occur at times $\{Z_1, Z_1 + Z_2, \ldots\}$. Clearly Z_1 has p.d.f. (2.1.4), and on repeating this argument with $t_0 = Z_1$ we see that Z_2 has the same p.d.f. independently of Z_1, and so on. Thus $\{Z_1, Z_2, \ldots\}$ is a sequence of independent and identically distributed (i.i.d.) random variables with the exponential p.d.f. (2.1.4). This property characterizes the Poisson process.

To find the p.d.f. of the time to the rth event, namely $T_r = Z_1 + Z_2 + \cdots + Z_r$, we first introduce the moment generating function (m.g.f.)

$$M(\theta) \equiv \mathrm{E}[e^{-\theta Z_i}] = \int_0^\infty e^{-\theta t} f(t)\, dt. \tag{2.1.5}$$

Here

$$M(\theta) = \int_0^\infty e^{-\theta t} \alpha e^{-\alpha t}\, dt = \frac{\alpha}{\alpha + \theta}. \tag{2.1.6}$$

So as the $\{Z_i\}$ are independent, the m.g.f. of T_r is

$$\mathrm{E}[e^{-\theta(Z_1 + \cdots + Z_r)}] = \mathrm{E}[e^{-\theta Z_1}] \ldots \mathrm{E}[e^{-\theta Z_r}] = \left(\frac{\alpha}{\alpha + \theta}\right)^r. \tag{2.1.7}$$

This inverts to yield the p.d.f.

$$\frac{\alpha(\alpha t)^{r-1} e^{-\alpha t}}{(r-1)!} \qquad (t > 0), \tag{2.1.8}$$

which is the gamma distribution with parameter α. To verify this result we simply note that (2.1.8) has the associated m.g.f.

$$\int_0^\infty \frac{\alpha(\alpha t)^{r-1} e^{-(\alpha+\theta)t}\, dt}{(r-1)!} = \left(\frac{\alpha}{\alpha+\theta}\right)^r \int_0^\infty \frac{(\alpha+\theta)[(\alpha+\theta)t]^{r-1} e^{-(\alpha+\theta)t}\, dt}{(r-1)!}$$

$$= \left(\frac{\alpha}{\alpha+\theta}\right)^r$$

in agreement with (2.1.7), since the integrand is a gamma p.d.f. with parameter $(\alpha + \theta)$ and so integrates to 1.

Now the mean and variance of the exponential p.d.f. (2.1.4) are given by $\mathrm{E}(Z) = 1/\alpha$ and $\mathrm{Var}(Z) = 1/\alpha^2$, respectively. So as successive intervals are independent, the mean and variance of the gamma random variable T_r are

$$\mathrm{E}(T_r) = r/\alpha \qquad \text{and} \qquad \mathrm{Var}(T_r) = r/\alpha^2. \tag{2.1.9}$$

Hence as r increases, use of the Central Limit Theorem shows that T_r is asymptotically Normally distributed as a $N(r/\alpha, r/\alpha^2)$ random variable. If required, plots of the p.d.f.

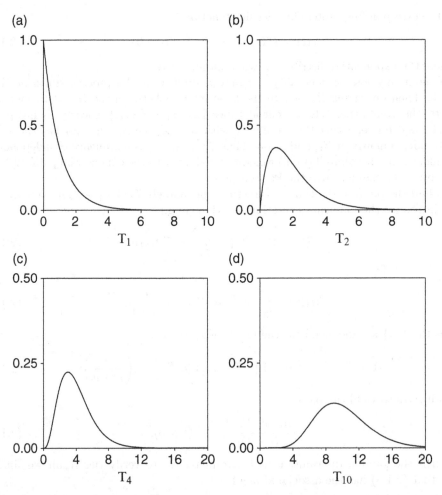

Figure 2.2 Gamma p.d.f.'s for the time, T_r, to the rth event in a Poisson process of rate $\alpha = 1$: (a) $r = 1$, (b) $r = 2$, (c) $r = 4$ and (d) $r = 10$.

(2.1.8) are easily obtained through standard packages such as *Minitab*, *Matlab*, *Splus* and *SigmaPlot*. Figure 2.2 shows the form of this gamma p.d.f. with $\alpha = 1$ as r increases through 1, 2, 4 and 10, and highlights the shift first from J-shaped to skew-right, and then to Normal.

2.1.2 Number of events

Having considered the distribution of the time between two events, let us now examine the dual situation in which we are interested in determining the number of events $N(t)$ in an interval $(0, t)$ of fixed length. Recall that the numbers of events in non-overlapping intervals are mutually independent because of the defining properties of the Poisson process. We write $p_i(t) = \Pr\{N(t) = i\}$ and construct a series of differential equations for the probabilities $\{p_i(t)\}$. This procedure is a simple illustration of a general technique for analysing discrete state space processes in continuous time, and is one of the main pillars of stochastic analysis.

On splitting the time interval $(0, t + dt)$ into a large interval $(0, t)$ and the small adjacent interval $(t, t + dt)$ (i.e. Figure 2.1 with $t_0 = 0$), with $N(t)$ and $N(t, t + dt)$ denoting the associated number of events in each, we have

$$p_i(t + dt) = \Pr\{N(t + dt) = i\}$$
$$= \Pr\{N(t) = i \text{ and } N(t, t + dt) = 0$$
$$\text{or } N(t) = i - 1 \text{ and } N(t, t + dt) = 1$$
$$\text{or } N(t) = i - r \text{ and } N(t, t + dt) = r \text{ for } r = 2, 3, \ldots, i\}$$
$$= \Pr\{N(t) = i\} \Pr\{N(t, t + dt) = 0 | N(t) = i\}$$
$$+ \Pr\{N(t) = i - 1\} \Pr\{N(t, t + dt) = 1 | N(t) = i - 1\} \quad (2.1.10)$$
$$+ \sum_{r=2}^{i} \Pr\{N(t) = i - r\} \Pr\{N(t, t + dt) = r | N(t) = i - r\}.$$

Note that the general nature of equation (2.1.10) means that it holds for any process in which transitions never lead to a decrease in population size. On using the Poisson properties (i)–(iv) this becomes

$$p_i(t + dt) = p_i(t)(1 - \alpha dt) + p_{i-1}(t)\alpha dt + \sum_{r=2}^{i} p_{i-r}(t) \times o(dt). \quad (2.1.11)$$

Letting $dt \to 0$, as before, then leads to the differential equations

$$dp_0(t)/dt = -\alpha p_0(t) \quad (2.1.12)$$
$$dp_i(t)/dt = -\alpha p_i(t) + \alpha p_{i-1}(t) \quad (i = 1, 2, \ldots). \quad (2.1.13)$$

Since no events can occur in a time interval of length zero, the initial conditions are

$$p_0(0) = 1 \quad \text{and} \quad p_i(0) = 0 \quad (i = 1, 2, \ldots). \quad (2.1.14)$$

There are several methods available for solving equations of this type. The simplest (which in general is seldom applicable) is to solve recursively. Here (2.1.12) gives $p_0(t) = e^{-\alpha t}$, whence successive substitution into (2.1.13) leads to $p_1(t) = (\alpha t)e^{-\alpha t}$, $p_2(t) = (\alpha t)^2 e^{-\alpha t}/2, \ldots$, and hence (via an induction argument) to

$$p_i(t) = \frac{(\alpha t)^i e^{-\alpha t}}{i!} \quad (i = 0, 1, 2, \ldots). \quad (2.1.15)$$

An alternative approach, which is neater and mathematically more appealing, is to convert the infinite set of equations for $\{p_i(t)\}$ into a single equation for the probability generating function (p.g.f.)

36 Simple Markov population processes

$$G(z;t) \equiv \sum_{i=0}^{\infty} p_i(t) z^i. \tag{2.1.16}$$

Thus $p_i(t)$ is given by the coefficient of z^i in the power series expansion of $G(z;t)$. On multiplying equations (2.1.12)–(2.1.13) by z^0, z^1, z^2, ... and summing, we obtain

$$\frac{d}{dt} \sum_{i=0}^{\infty} p_i(t) z^i = -\alpha \sum_{i=0}^{\infty} p_i(t) z^i + \alpha z \sum_{i=1}^{\infty} p_{i-1}(t) z^{i-1}$$

(note that we have taken z outside the third summation to enable the subscript and superscript to match). So

$$\frac{dG(z;t)}{dt} = -\alpha G(z;t) + \alpha z G(z;t). \tag{2.1.17}$$

Whilst from the initial condition (2.1.14) we see that

$$G(z;0) = 1. \tag{2.1.18}$$

For any fixed z, (2.1.17) is an ordinary linear differential equation in t with the solution

$$G(z;t) = A(z) e^{-\alpha t(1-z)},$$

where the arbitrary 'constant' of integration, $A(z)$, is a function of z. Taking $t = 0$ and using (2.1.18) then gives

$$G(z;0) \equiv A(z) = 1,$$

whence

$$G(z;t) = e^{-\alpha t(1-z)}. \tag{2.1.19}$$

Expanding $G(z;t)$ in the form

$$G(z;t) = e^{-\alpha t} \sum_{i=0}^{\infty} \frac{z^i (\alpha t)^i}{i!},$$

and isolating the coefficient of z^i, then recovers the Poisson distribution (2.1.15).

Moments of this process are easily derived from the p.g.f. solution (2.1.19). For differentiating this once with respect to z, and placing $z = 1$, yields the mean

$$\mu(t) \equiv \sum_{i=0}^{\infty} i p_i(t) = \left. \frac{\partial G(z;t)}{\partial z} \right|_{z=1} = \alpha t; \tag{2.1.20}$$

whilst differentiating once more yields the variance

$$\text{Var}(t) \equiv \sum_{i=0}^{\infty} i^2 p_i(t) - [\mu(t)]^2 = \sum_{i=0}^{\infty} i(i-1) p_i(t) + \sum_{i=0}^{\infty} i p_i(t) - [\mu(t)]^2 \tag{2.1.21}$$

$$= \left. \frac{\partial^2 G(z;t)}{\partial z^2} \right|_{z=1} + \mu(t) - [\mu(t)]^2 = (\alpha t)^2 + (\alpha t) - (\alpha t)^2 = \alpha t. \tag{2.1.22}$$

Thus as t increases, application of the Central Limit Theorem shows that $N(t)$ becomes Normally distributed with asymptotic mean and variance both equal to αt.

2.1.3 General moments

At this point it is worthwhile introducing four types of moments that we shall use throughout this text, together with three associated types of generating function. First, result (2.1.20) is easily extended to yield the rth-order factorial moments

$$\mu_{(r)}(t) \equiv \mathrm{E}[N(N-1)\ldots(N-r+1)]$$

$$= \sum_{i=r}^{\infty} i(i-1)\ldots(i-r+1)p_i(t) = \left.\frac{\partial^r G(z;t)}{\partial z^r}\right|_{z=1} ; \qquad (2.1.23)$$

note that the first r probabilities $p_0(t),\ldots,p_{r-1}(t)$ do not feature in (2.1.23). So for the Poisson distribution (2.1.15)

$$\mu_{(r)}(t) = (\alpha t)^r e^{-\alpha t(1-z)}\Big|_{z=1} = (\alpha t)^r. \qquad (2.1.24)$$

Alternatively, we can replace z by $1+y$ to obtain the factorial moment generating function (f.m.g.f.)

$$H(y;t) \equiv G(1+y;t) = \sum_{i=0}^{\infty} p_i(t)(1+y)^i \qquad (2.1.25)$$

$$= \sum_{i=0}^{\infty} p_i(t) \sum_{j=0}^{i} \binom{i}{j} y^j = \sum_{j=0}^{\infty} y^j \sum_{i=j}^{\infty} \binom{i}{j} p_i(t)$$

$$= \sum_{j=0}^{\infty} \frac{y^j}{j!} \sum_{i=j}^{\infty} i(i-1)\ldots(i-j+1)p_i(t) = \sum_{j=0}^{\infty} \frac{y^j}{j!}\mu_{(j)}(t). \qquad (2.1.26)$$

So for the Poisson distribution we obtain the extremely simple form

$$H(y;t) = e^{\alpha y t} = \sum_{j=0}^{\infty} \frac{y^j}{j!}(\alpha t)^j, \qquad (2.1.27)$$

from which the coefficient of $y^j/j!$ yields $\mu_{(j)}(t) = (\alpha t)^j$ without any need for differentiation.

Although this is certainly a neat representation, the structure of the expectation $\mathrm{E}[N(N-1)\ldots(N-r+1)]$ is hardly transparent, and the rth raw moment (or moment about the origin)

$$\mu'_r(t) \equiv \mathrm{E}[N^r] \equiv \sum_{i=1}^{\infty} i^r p_i(t) \qquad (2.1.28)$$

is far easier to interpret. One way of evaluating $\mu'_r(t)$ is to place $z = e^\theta$ in the p.g.f. $G(z;t)$, thereby obtaining the moment generating function

$$M(\theta;t) \equiv \sum_{i=0}^{\infty} e^{i\theta} p_i(t) = \sum_{r=0}^{\infty} \frac{\theta^r}{r!} \sum_{i=0}^{\infty} i^r p_i(t) = \sum_{r=0}^{\infty} \frac{\mu'_r(t)\theta^r}{r!}. \qquad (2.1.29)$$

Thus $\mu'_r(t)$ is the coefficient of $\theta^r/r!$ in the expansion of $M(\theta;t)$; equivalently, we can differentiate $M(\theta;t)$ to obtain

$$\mu'_r(t) = \left.\frac{\partial^r M(\theta;t)}{\partial \theta^r}\right|_{\theta=0}. \qquad (2.1.30)$$

Since the Poisson p.g.f. (2.1.19) leads to

$$M(\theta;t) = \exp\{-\alpha t(1 - e^\theta)\}, \qquad (2.1.31)$$

which does not possess a neat form, either for its expansion in powers of θ or its rth derivative, the μ'_r clearly have a more awkward structure than the simple representation (2.1.24) for the factorial moments $\mu_{(r)}$. The first few values are best obtained by substituting for $\mu_{(r)} = (\alpha t)^r$ in the general relations

$$\mu'_1 = \mu_{(1)} = \mu$$

$$\mu'_2 = \sum_{i=0}^{\infty} [i(i-1) + i] p_i(t) = \mu_{(2)} + \mu_{(1)} \qquad (2.1.32)$$

$$\mu'_3 = \sum_{i=0}^{\infty} [i(i-1)(i-2) + 3i(i-1) + i] p_i(t) = \mu_{(3)} + 3\mu_{(2)} + \mu_{(1)}$$

with

$$\mu'_4 = \mu_{(4)} + 6\mu_{(3)} + 7\mu_{(2)} + \mu_{(1)}, \quad \text{etc.}$$

The rth-order relation between μ'_r and $\mu_{(1)}, \ldots, \mu_{(r)}$, which involves Bernoulli polynomials, is given in Kendall and Stuart (1977).

Because $E[N^r]$ depends heavily on the mean μ, it has little intuitive appeal. It is far more instructive to consider the rth-order central moment (i.e. moment about the mean)

$$\mu_r \equiv E[(N-\mu)^r] = \sum_{i=0}^{\infty} (i-\mu)^r p_i(t) = \sum_{s=0}^{r} \binom{r}{s} (-\mu)^{r-s} \mu'_s, \qquad (2.1.33)$$

as this is unaffected by the value of μ. If we have already determined the raw moments, μ'_r, then (2.1.33) may be used to determine the central moments μ_r; the first four relationships are

$$\mu_1 = 0$$
$$\mu_2 = \mu_2' - \mu^2$$
$$\mu_3 = \mu_3' - 3\mu\mu_2' + 2\mu^3 \qquad (2.1.34)$$
$$\mu_4 = \mu_4' - 4\mu\mu_3' + 6\mu^2\mu_2' - 3\mu^4.$$

In practice, theoretical moments of order higher than the fourth are rarely required; whilst estimators of μ_r for $r > 4$ are so sensitive to sampling fluctuations that values computed from moderate numbers of observations are subject to a large margin of error. So specification of μ, μ_2, μ_3 and μ_4 is usually sufficient for most practical purposes.

A more direct way of obtaining central moments is to work in terms of the cumulant generating function (c.g.f.)

$$K(\theta;t) \equiv \sum_{i=1}^{\infty} \frac{\kappa_i \theta^i}{i!} = \ln[G(e^\theta;t)] = \ln[M(\theta;t)]. \qquad (2.1.35)$$

For on substituting for $M(\theta;t)$ from (2.1.29) we have

$$\frac{\kappa_1 \theta}{1!} + \frac{\kappa_2 \theta^2}{2!} + \cdots = \ln[1 + \frac{\mu_1' \theta}{1!} + \frac{\mu_2' \theta^2}{2!} + \cdots],$$

whence expanding the logarithmic term and picking out the coefficient of $\theta^r/r!$ ($r = 1, 2, \ldots$) yields

$$\kappa_1 = \mu_1'$$
$$\kappa_2 = \mu_2' - \mu_1'^2$$
$$\kappa_3 = \mu_3' - 3\mu_2'\mu_1' + 2\mu_1'^3 \qquad (2.1.36)$$
$$\kappa_4 = \mu_4' - 4\mu_3'\mu_1' - 3\mu_2'^2 + 12\mu_2'\mu_1'^2 - 6\mu_1'^4, \quad \text{etc.}$$

Whilst replacing the raw moments μ_r' by the central moments μ_r gives

$$\kappa_1 = \mu$$
$$\kappa_2 = \mu_2$$
$$\kappa_3 = \mu_3 \qquad (2.1.37)$$
$$\kappa_4 = \mu_4 - 3\mu_2^2, \quad \text{etc.}$$

Thus the first three cumulants provide the mean, variance and skewness, whilst the kurtosis, μ_4, is simply $\kappa_4 + 3\mu_2^2$. For higher-order relations up to μ_{10}', μ_{10} and κ_{10} see Kendall and Stuart (1977).

Note that inserting the Poisson m.g.f. (2.1.31) into definition (2.1.35) yields

$$K(\theta;t) = -\alpha t(1 - e^\theta) = \alpha t \sum_{r=1}^{\infty} \frac{\theta^r}{r!},$$

whence $\kappa_r \equiv \alpha t$ for all $r \geq 1$. So all the cumulants of the Poisson distribution are equal, which provides a useful touchstone for 'randomness'.

2.1.4 Simulation

Although we know the probability distribution of population size through (2.1.15), the probabilities $\{p_i(t)\}$ do not explicitly tell us what a *particular realization* (i.e. an individual time history) of the process might look like, but provide instead the distributional properties of a *large ensemble* of such realizations. Moreover, mean behaviour, represented for the Poisson process by $\mu(t) = \alpha t$, may be very different from the behaviour of individual realizations. A further complication, examined later, is that deterministic and mean behaviour will generally differ for nonlinear processes, i.e. those for which individual members of the population do not develop independently of one another.

Fortunately, information on the shape of individual realizations may be readily obtained by simulating them. Not only does this technique give considerable insight into the underlying process, but behavioural features may well emerge that had not previously been anticipated. Furthermore, it also enables the construction of empirical probability distributions which are either theoretically intractable to direct solution or which involve massive computational effort for numerical solution. Since here each event comprises the population size increasing by one, all we have to do is to construct the inter-event times.

Suppose we wish to simulate a continuous random variable X with cumulative distribution function $F(x) = \Pr(X \leq x)$ for which the inverse function $F^{-1}(x)$ is well-defined over $0 \leq x \leq 1$. Let U be a continuous uniform random variable on the interval $[0, 1]$, so that $\Pr(U \leq u) = u$. Then on placing $X = F^{-1}(U)$ we have

$$\Pr(X \leq x) = \Pr(F^{-1}(U) \leq x) = \Pr(U \leq F(x)) = F(x). \tag{2.1.38}$$

So in general, to simulate values of X we simply choose independent values from the $U(0, 1)$ distribution and evaluate $F^{-1}(U)$.

Now we have already shown in (2.1.3) and (2.1.4) that for the Poisson process the inter-event times $\{Z\}$ follow the exponential p.d.f. $f(z) = \alpha e^{-\alpha z}$, with c.d.f. $F(z) = 1 - e^{-\alpha z}$. So to use (2.1.38) in order to simulate realizations from this exponential p.d.f. we place

$$F(Z) = 1 - e^{-\alpha Z} = U,$$

thereby obtaining

$$Z = -(1/\alpha) \ln(U).$$

Since $1 - U$ is also a $U(0, 1)$ variable, this result simplifies to give the inter-event times

$$Z = -(1/\alpha) \ln(U). \tag{2.1.39}$$

Thus to simulate a Poisson process with rate α we:
(i) construct a set of independent pseudo-random $U(0, 1)$ random variables $\{U_n\}$;
(ii) evaluate the inter-event times $Z_n = -(1/\alpha) \ln(U_n)$;
(iii) output the event-times $t_n = Z_1 + \cdots + Z_n$ for $n = 1, 2, \ldots$.

To demonstrate this procedure for the Poisson process with rate 2, we select $U(0,1)$-values from a table of uniform random numbers and evaluate

n	1	2	3	4	5	...
U_n	0.8398	0.2215	0.4420	0.7356	0.8376	...
$Z_n = -[\ln(U_n)]/2$	0.0873	0.7537	0.4082	0.1535	0.0886	...
t_n	0.0873	0.8410	1.2492	1.4027	1.4913	...

Figure 2.3 shows two such independent realizations for $0 \leq t \leq 25$. The severe skewness of the exponentially distributed inter-event times (see Figure 2.2a) explains why the plots exhibit apparent clustering followed by relatively large gaps. This behaviour typifies randomness, and highlights the fallibility of human intuition which is often tempted to associate randomness with regularity. Realizations 1 and 2 contain 50 and 42 points, respectively, in line with the expected value $\alpha t = 50$. Since the variance also equals αt, approximate 95% confidence limits for the number of points contained in this time interval are 50 ± 7, i.e. 36 to 64.

Whilst to simulate the number, n, of Poisson events in the interval $[0, T]$, we simply note that n satisfies the relation $t_n \leq T < t_{n+1}$, i.e.

$$(-1/\alpha) \sum_{i=1}^{n} \ln(U_i) \leq T < (-1/\alpha) \sum_{i=1}^{n+1} \ln(U_i).$$

Thus n is the largest integer such that

$$\prod_{i=1}^{n} U_i \leq e^{-\alpha T}, \qquad (2.1.40)$$

which is both trivial and quick to compute.

If simulation is being used purely to illustrate various aspects of dynamic behaviour, then the quality of the pseudo-random number generator which produces the $\{U_i\}$ is of limited concern. However, we shall shortly be meeting processes for which the derivation of exact theoretical results is currently not possible, and so we will be forced into using approximation or simulation techniques in order to make progress. In this latter case, employing a reliable generator can be of paramount importance; for a

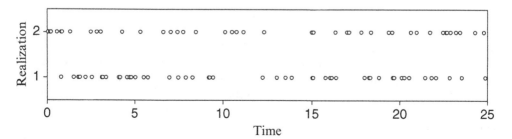

Figure 2.3 Two simulations of a Poisson process with $\alpha = 2$ over $0 \leq t \leq 25$.

discussion of various approaches and examples see (for example) Morgan (1984) and Ripley (1987).

Most popular generators are based on the mixed congruential form

$$X_i = (aX_{i-1} + c) \bmod(M) \qquad (2.1.41)$$

for appropriate integer values of a, c and M, and initial seed X_0. This generates all M possible values of $\{X_i\}$, namely $0, 1, \ldots, M-1$ (i.e. the generator has full period), provided:

(a) the greatest common divisor (g.c.d.) of $(c, M) = 1$;
(b) $a \equiv 1 \bmod(p)$ for each prime factor p of M; and,
(c) $a \equiv 1 \bmod(4)$ if 4 divides M.

If $c = 0$ the generator is called multiplicative, and for $M \geq 16$ a power of 2, has maximal period $M/4$ provided $a \bmod(8) = 3$ or 5. Working with integer arithmetic avoids the problem of round-off error, whence the $\{U_i\}$ are easily determined via $U_i = X_i/M$ where the integer M is replaced by its real double-precision value.

Provided M is taken to be reasonably large, full and maximal period generators will yield $\{U_i\}$ that are nearly uniformly distributed on the interval $[0, 1)$. They may, however, be completely predictable; for example, taking $M = 32$, $a = 1$ and $c = 1$ produces the successive values $(1/32)[0, 1, \ldots, 31, 0, \ldots]$. Whilst on taking $M = 2048$, $a = 1365$ and $c = 1$, a plot of the pairs $\{U_i, U_{i+1}\}$ yields three equi-spaced lines, so given U_i the next U_{i+1} can take only one of three possible values. Although this structure is easily seen from the plot, the once very popular generator RANDU with $M = 2^{31}$, $a = 2^{16} + 3$ and $c = 0$ provides a far less obvious example. For with integer $c_1, \ldots, c_4$,

$$\begin{aligned}
X_{i+2} &= (2^{16} + 3)X_{i+1} \bmod(2^{31}) \\
&= (2^{16} + 3)X_{i+1} + c_1 2^{31} \\
&= (2^{16} + 3)[(2^{16} + 3)X_i + c_2 2^{31}] + c_1 2^{31} \\
&= (2^{16} + 3)^2 X_i + c_2 2^{31}(2^{16} + 3) + c_1 2^{31} \\
&= (6.2^{16} + 9)X_i + 2^{31}[2X_i + c_2(2^{16} + 3) + c_1] \\
&= 6(2^{16} + 3)X_i - 9X_i + c_3 2^{31} \\
&= 6X_{i+1} - 9X_i + c_4 2^{31}.
\end{aligned}$$

So $U_{i+2} - 6U_{i+1} + 9U_i$ is integer, and it may be shown that (U_i, U_{i+1}, U_{i+2}) lies on one of 15 planes in the unit cube. Thus if the pair (U_{i-2}, U_{i-1}) is known to even limited accuracy, then U_i is quite predictable. Figure 2.4b shows a three-dimensional scatter plot of 10000 U_i-triples, suitably rotated to highlight the 15-plane lattice structure. Care must be taken to spin the plot in order to see this feature, as it may well not show up under default settings (Figure 2.4a); the software package *MathCad* is particularly good for this purpose. This extremely undesirable property of RANDU caused considerable grief before it was eventually exposed.

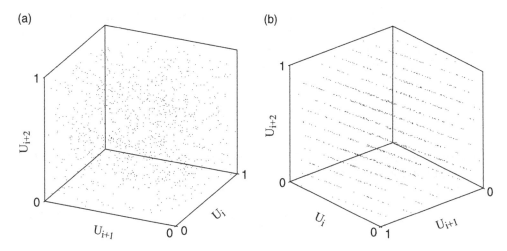

Figure 2.4 Three-dimensional scatter plot of 10000 (U_i, U_{i+1}, U_{i+2})-triples generated by RANDU: (a) default and (b) rotated viewpoints.

The main objective when selecting a pseudo-random number generator is to avoid this type of breakdown. Though as pointed out by Marsaglia (1968), and highlighted in Theorem 2.5 of Ripley (1987), the k-tuples $(U_1, \ldots, U_{i+k-1})$ will always lie on a number of hyperplanes in $[0, 1)^k$. Choice of c merely shifts the lattice (which means that for mixed congruential forms we can always take $c = 1$); M should be chosen large enough to ensure that the $\{U_i\}$ are finely spread; whilst the associated value of a should be selected to provide 'good lattice structure'. Whence for all practical purposes the $\{U_i\}$ can be regarded as being independent; see Morgan (1984) and Ripley (1987) for statistical test procedures. A highly recommended and popular choice is the multiplicative routine G05CAF of the NAG Fortran Library, for which $M = 2^{59}$, $a = 13^{13}$ and $c = 0$, i.e.

$$X_i = (13^{13} X_{i-1}) \bmod (2^{59}). \qquad (2.1.42)$$

The seed X_0 is initially set to $123456789(2^{32} + 1)$, but this can easily be changed.

2.1.5 Generalizations

We shall soon discover that although the Poisson process is highly amenable to theoretical analysis, it takes little embellishment to produce processes that are intractable to exact mathematical solution. Before we proceed along this route, let us briefly see how to handle four generalizations that occur in real-life situations. For our simple Poisson process allows us to highlight the ideas involved without introducing opaque mathematical detail.

Time-dependent parameter

In practice the arrival rate α may not be constant, but may depend on time (e.g. through a daily or seasonal effect). So suppose that α is replaced by the time-dependent

44 Simple Markov population processes

function $\alpha(t)$. Since this does not alter the basic property that the number of events in $(t, t+dt)$ is independent of the number, $N(t)$, in $(0,t)$, equations (2.1.12) and (2.1.13) remain unchanged. Thus the p.g.f. equation (2.1.17) with boundary condition (2.1.18) is also unchanged, and takes the solution

$$G(z;t) = \exp\{(z-1)\int_0^t \alpha(u)\,du\}. \tag{2.1.43}$$

Extracting the coefficient of z^i in (2.1.43) shows that $N(t)$ follows a time-dependent Poisson distribution with mean

$$\tau(t) = \int_0^t \alpha(u)\,du. \tag{2.1.44}$$

We may view $\tau(t)$ as being a nonlinear transformation of the time-scale. For as $dt/d\tau = 1/\alpha$, equations (2.1.12) and (2.1.13) transform into

$$dp_0(\tau)/d\tau = -p_0(\tau) \quad \text{and} \quad dp_i(\tau)/d\tau = -p_i(\tau) + p_{i-1}(\tau), \tag{2.1.45}$$

which conforms to a Poisson process of constant unit rate. So this device clearly makes a heterogeneous Poisson process homogeneous. However, its effective use in more general multi-parameter problems depends on a single transformation of the time (or space) scale being able to remove all the temporal (or spatial) dependencies. This will not usually be possible.

Multi-rate processes

A further type of generalization involves viewing a multi-type process as a superposition of k elementary processes which operate according to *independent* Poisson processes with rates $\alpha_1, \alpha_2, \ldots, \alpha_k$. For example, when studying the times to failure of a computer we can examine individual components such as circuit boards, disc drives, screen, keyboard, etc., separately. Let $N(t, t+dt)$ denote the number of events of any type in the small time interval $(t, t+dt)$, with $N_i(t, t+dt)$ denoting the number corresponding to the ith process. Then as the k processes are assumed to operate independently of each other,

$$\Pr\{N(t,t+dt) = 0\} = \prod_{i=1}^k \Pr\{N_i(t,t+dt) = 0\}$$

$$= \prod_{i=1}^k \{1 - \alpha_i dt + o(dt)\} = 1 - \alpha dt + o(dt) \tag{2.1.46}$$

where $\alpha = \sum_{i=1}^{k} \alpha_i$. Similarly,

$$\Pr\{N(t, t+dt) = 1\} = \sum_{i=1}^{k} \Pr\{N_i(t, t+dt) = 1\} \times \Pr\{N_j(t, t+dt) = 0 \text{ for all } j \neq i\}$$

$$= \sum_{i=1}^{k} \{\alpha_i dt + o(dt)\} \prod_{j \neq i} \{1 - \alpha_j dt + o(dt)\} = \alpha dt + o(dt).$$
(2.1.47)

Whence on combining (2.1.46) and (2.1.47) we see that $\Pr\{N(t, t+dt) > 1\} = o(dt)$. Finally, the number of events in $(t, t+dt)$ is independent of occurrences in all k processes at or before time t. Thus the combined series of elementary events follows a Poisson process with rate $\alpha = \sum_{i=1}^{k} \alpha_i$.

It follows from (2.1.4) that the time $Z_i = t$ taken by process i to reach its first event takes the exponential p.d.f. $\alpha_i e^{-\alpha_i t}$ $(i = 1, \ldots, k)$, whilst the time $Z = \min(Z_1, \ldots, Z_k)$ to the first event in the combined process has p.d.f. $\alpha e^{-\alpha t}$. So given that $Z = t$, the conditional probability that this event is of type i is

$$\Pr\{Z_i = t, Z_j > t \ (j \neq i) \mid Z = t\}$$
$$= \Pr\{Z_i = t, Z_j > t \ (j \neq i)\} / \Pr\{Z = t\} \quad (2.1.48)$$
$$= (\alpha_i e^{-\alpha_i t} dt) \times (e^{-\alpha_1 t} \ldots e^{-\alpha_{i-1} t}) \times (e^{-\alpha_{i+1} t} \ldots e^{-\alpha_k t}) / (\alpha e^{-\alpha t} dt) = \alpha_i / \alpha.$$

Since this expression does not involve t, it follows that the set of independent Poisson processes can be described as follows. Intervals of length z between successive events are independently distributed with p.d.f. $\alpha e^{-\alpha z}$; events are assigned at random to the individual types $1, \ldots, k$ with constant probabilities $\alpha_1/\alpha, \ldots, \alpha_k/\alpha$, respectively.

For example, suppose that a Xerox machine may fail from three different types of paper jam according to independent Poisson processes with rates 0.10, 0.15 and 0.25. Then the time to the next failure is exponentially distributed with parameter 0.50, whilst the three types of paper jam occur independently of each other with probabilities $\{0.2, 0.3, 0.5\}$. This decomposition into event–time pairs plays a central role in the simulation of stochastic processes and we shall constantly return to it.

Simultaneous occurrences

Our initial condition (iii), namely that $\Pr\{N(t, t+dt) > 1\} = o(dt)$, fails to hold as soon as multiple occurrences, such as mass immigration or the simultaneous birth of twins or triplets, are allowed. For illustration, suppose that aircraft arrive at an airport according to a Poisson process of rate α, and that the ith aircraft contains V_i passengers, where the $\{V_i\}$ are independently and identically distributed. Let the p.d.f.

$$\Pr(V_i = j) = w_j \quad (j = 0, 1, 2, \ldots) \quad (2.1.49)$$

46 *Simple Markov population processes*

have the associated p.g.f.

$$\Pi(z) \equiv \sum_{j=0}^{\infty} w_j z^j. \qquad (2.1.50)$$

Then to determine the probability that $N(t)$ passengers arrive in the fixed time interval $[0, t]$ we need the following two general results.

RESULT 2.1 *Let the random variables X_i ($i = 1, \ldots, p$) be independently distributed with p.g.f.'s $G_i(z)$. Then $X_1 + \cdots + X_p$ has p.g.f.*

$$\prod_{i=1}^{p} G_i(z). \qquad (2.1.51)$$

Proof For convenience write $U = X_1$ and $V = X_2$. Then $U + V$ has p.g.f.

$$G_{U+V}(z) = \sum_{i=0}^{\infty} (u_0 v_i + u_1 v_{i-1} + \cdots + u_i v_0) z^i$$

$$= u_0(v_0 + v_1 z + \cdots) + u_1 z(v_0 + v_1 z + \cdots) + \cdots$$

$$= (u_0 + u_1 z + u_2 z^2 + \cdots)(v_0 + v_1 z + v_2 z^2 + \cdots)$$

$$= G_U(z) G_V(z).$$

Similarly, with $W = X_3$

$$G_{U+V+W}(z) = G_{(U+V)+W}(z) = G_{U+V}(z) G_W(z) = G_U(z) G_V(z) G_W(z),$$

whence the result follows via a simple induction argument. □

RESULT 2.2 *It follows that if the $\{X_i\}$ are independent and identically distributed with p.g.f. $\Pi(z)$, then $X_1 + \cdots + X_p$ has p.g.f.*

$$\Pi_p(z) \equiv [\Pi(z)]^p. \qquad (2.1.52)$$

Note that the cumulant generating function approach is particularly powerful in this general situation. For we see from (2.1.35) that the products in (2.1.51) and (2.1.52) are replaced by sums, and so the jth cumulant of $X_1 + \cdots + X_p$ is simply the sum of the jth cumulants of $X_1, \ldots, X_p$.

Returning to our example, conditional on the number of aircraft arrivals in $[0, t]$ being $M(t) = m$, $N(t)$ is the sum of m independent Poisson processes. Whence applying Result 2.2 shows that $N(t)$ has p.g.f. $[\Pi(z)]^m$. So as $M(t)$ follows a Poisson process with rate α, the unconditional p.g.f. of $N(t)$ is given by

$$G_N(z; t) \equiv \sum_{m=0}^{\infty} \Pr(M(t) = m) \sum_{i=0}^{\infty} \Pr\{N(t) = i | M(t) = m\} z^i$$

$$= \sum_{m=0}^{\infty} \frac{(\alpha t)^m e^{-\alpha t}}{m!} [\Pi(z)]^m = \exp\{\alpha t [\Pi(z) - 1]\}. \qquad (2.1.53)$$

Extracting the coefficient of z^i then yields

$$\Pr(N(t) = i) = (\alpha t)^i e^{-\alpha t} \sum_{j_0+j_1+\cdots=i} \frac{w_0^{j_0} w_1^{j_1} \ldots w_k^{j_k} \ldots}{j_0! j_1! \ldots j_k! \ldots}. \qquad (2.1.54)$$

When $w_1 = 1$, and hence $w_j = 0$ for $j \neq 1$, we recover the simple Poisson process. The associated cumulant structure is easily determined from (2.1.53) since

$$K_N(\theta; t) \equiv \ln[G_N(e^\theta; t)] = \alpha t \sum_{j=1}^{\infty} w_j e^{j\theta}, \qquad (2.1.55)$$

whereupon extracting the coefficient of $\theta^r/r!$ yields

$$\kappa_r(t) = \alpha t \sum_{j=0}^{\infty} w_j j^r. \qquad (2.1.56)$$

2.1.6 Do moments uniquely define a distribution?

Any probability distribution $\{p_r\}$ possesses its own uniquely defined set of moments. For example, the summation defining the rth raw moment, $\mu_r' = \sum_{i=0}^{\infty} i^r p_r$, can clearly take only one value. So it is interesting to ask whether the converse also holds true. That is, does a given set of moments, say $\{\mu_r'\}$, uniquely define a probability distribution $\{p_r\}$? To answer this question let us consider our fourth generalization by defining the Poisson distribution not over the non-negative integers but over powers of two, viz:

$$p_{2^i} = 2^i e^{-2}/i! \qquad \text{for} \quad i = 0, 1, 2, 3, \ldots \qquad (2.1.57)$$

and $p_j = 0$ otherwise, i.e. $j \neq 1, 2, 4, 8, \ldots$ (Schoenberg, 1983). Then the associated moments

$$\mu_r' = \sum_{i=0}^{\infty} 2^{ri} 2^i e^{-2}/i! = \exp\{2(2^r - 1)\}. \qquad (2.1.58)$$

To construct a family of distributions which possess the same $\{\mu_r'\}$-values (2.1.58), consider the function

$$h(z) \equiv \prod_{k=1}^{\infty} (1 - z/q^k) \qquad \text{for} \quad q = 2, 3, \ldots, \qquad (2.1.59)$$

and denote its Taylor series expansion around zero by

$$h(z) \equiv \sum_{m=0}^{\infty} c_m z^m. \qquad (2.1.60)$$

Now we see from (2.1.59) that

$$h(qz) = (1 - qz/q) \prod_{k=2}^{\infty} (1 - z/q^{k-1}) = (1 - z)h(z). \qquad (2.1.61)$$

Whence combining (2.1.60) and (2.1.61) gives

$$\sum_{m=0}^{\infty} c_m q^m z^m = (1-z) \sum_{m=0}^{\infty} c_m z^m, \quad \text{i.e.} \quad \sum_{m=0}^{\infty} c_m(1-q^m) z^m = \sum_{m=0}^{\infty} c_m z^{m+1}. \tag{2.1.62}$$

Comparing (2.1.59) and (2.1.60) at $z = 0$ shows that $c_0 = 1$. Whilst taking the coefficient of z^m in (2.1.62) for $m = 1, 2, \ldots$ yields $c_m(1-q^m) = c_{m-1}$, i.e. $c_m = c_{m-1}/(1-q^m)$, from which we obtain

$$c_m = [(1-q)(1-q^2)\ldots(1-q^m)]^{-1}. \tag{2.1.63}$$

Whence setting $a_m = m! c_m$ leads to

$$|a_m| = \left(\frac{1}{q-1}\right)\left(\frac{2}{q^2-1}\right)\ldots\left(\frac{m}{q^m-1}\right) \le 1 \tag{2.1.64}$$

for all $m = 0, 1, 2, \ldots$ and $q = 2, 3, \ldots$. Define

$$p_{2m}^\theta = p_{2m}(1 + \theta a_m) = e^{-2} 2^m (1 + \theta a_m)/m! \tag{2.1.65}$$

for $-1 \le \theta \le 1$. Then since $|a_m \theta| \le 1$ we see that $p_{2m} \ge 0$, whilst from (2.1.57), (2.1.59) with $q = 2$ and (2.1.60) we have

$$\sum_{m=0}^{\infty} \theta a_m p_{2m} = \theta e^{-2} \sum_{m=0}^{\infty} a_m 2^m/m! = \theta e^{-2} h(2) = \theta e^{-2} \prod_{k=1}^{\infty}(1 - 2/2^k) = 0.$$

Hence

$$\sum_{m=0}^{\infty} p_{2m}^\theta = \sum_{m=0}^{\infty} p_{2m}(1 + \theta a_m) = 1,$$

and so $\{p_{2m}^\theta\}$ is a proper probability distribution. Before developing the associated raw moments $\{\mu'_{r\theta}\}$, we first note that when $q = 2$

$$\sum_{m=0}^{\infty} 2^{m(r+1)} a_m/m! = h(2^{r+1}) = \prod_{k=1}^{\infty}(1 - 2^{r+1}/2^k) = 0,$$

since $1 - 2^{r+1}/2^k = 0$ at $k = r+1$. So

$$\mu'_{r\theta} = \sum_{m=0}^{\infty}(2^m)^r \times 2^m e^{-2}(1 + \theta a_m)/m!$$

$$= \sum_{m=0}^{\infty} 2^{m(r+1)} e^{-2}/m! + \theta e^{-2} \sum_{m=0}^{\infty} 2^{m(r+1)} a_m/m!$$

$$= \exp\{2(2^r - 1)\} + 0 = \mu'_r, \tag{2.1.66}$$

using (2.1.58). Thus $\{p_{2m}^\theta\}$ is a set of probability distributions over $-1 \le \theta \le 1$ which have distinct probability structures, but identical sets of moments.

To see how the p.d.f. $\{p_{2m}^\theta\}$ varies as θ ranges between -1 and $+1$, let us compare the two 'end' distributions

$$p_{2^m}^{-1} = (1-a_m)p_{2^m} \quad \text{and} \quad p_{2^m}^{+1} = (1+a_m)p_{2^m} \qquad (2.1.67)$$

against the pure Poisson 'middle' distribution

$$p_{2^m}^0 = p_{2^m} = 2^m e^{-2}/m!. \qquad (2.1.68)$$

The first four probabilities take the form

m	$i = 2^m$	a_m	p_i^{-1}		p_i		p_i^{+1}
0	1	1	0	<	e^{-2}	<	$2e^{-2}$
1	2	-1	$4e^{-2}$	>	$2e^{-2}$	>	0
2	4	2/3	$(2/3)e^{-2}$	<	$2e^{-2}$	<	$(10/3)e^{-2}$
3	8	$-2/7$	$(12/7)e^{-2}$	>	$(4/3)e^{-2}$	>	$(20/21)e^{-2}$

and the dramatic *sawtooth* nature of p_i^{-1} and p_i^{+1} relative to p_i is clearly evident. However, inspection of (2.1.64) shows that a_m tends rapidly to zero as m increases, and so the three distributions effectively have identical tail structures. For example, even with m as small as 8, i.e. $i = 256$, we have $a_8 = 0.20238 \times 10^{-5}$, which leads to the three virtually equal values

$$p_{256}^{-1} = 0.0008592699004 < p_{256} = 0.0008592716393 < p_{256}^{+1} = 0.0008592733784.$$

We shall return to such sawtooth structure in Chapter 7, but here our example serves to highlight the danger in arbitrarily choosing an 'appropriate' distribution based solely on known moments. Figure 2.5 shows p_i^{-1}, p_i and p_i^{+1} for $i = 1, 2, 4, \ldots, 256$ (remember that all probabilities are zero for $i \neq 2^0, 2^1, 2^2, \ldots$), and there is clearly a

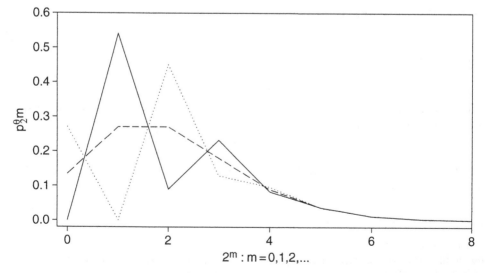

Figure 2.5 Three distributions $\{p_{2^m}^\theta\}$ having identical moment structures: $\theta = 0$ (Poisson) (- - -); $\theta = -1$ (——) and $\theta = +1$ ($\cdots$).

50 Simple Markov population processes

massive difference between the three probability structures at low values of i, in spite of virtually exact agreement once i reaches 2^8.

The family of distributions $\{p_{2m}^\theta\}$ is just one of a whole class of distributions which have equal moments of all orders, but different probability structures, and we refer readers to Stoyanov (1988) for a comprehensive review of this field.

2.2 Pure death process

The longevity of individuals varies greatly, from 0.4×10^{-23} seconds for rho-mesons, to a few hours for healthy bacteria, and to several millenia for the bristlecone pine *Pinus longaeva*. A question of considerable interest is to determine how the chance of death relates to the length of time an individual has been alive. So suppose we wish to follow the life histories of n_0 specific individuals. Then given that we are not interested in the production of new members, the simplest assumptions to make are that: (i) individuals develop completely independently from each other; and, (ii) the death rate μ is the same for all individuals and does not change with time. This last assumption is equivalent to saying that individuals do not age; in biological situations, for example, this means that the death rate is usually determined by the species liability to accidental causes of death rather than to natural causes.

To gain an initial insight into this process let us first consider the deterministic approach. Suppose that in a small time interval of length dt the decrease in population size due to all $N(t)$ individuals present at time t is $\mu \times dt \times N(t)$. Thus

$$N(t+dt) = N(t) - \mu dt N(t), \tag{2.2.1}$$

which, on dividing both sides by dt, gives

$$[N(t+dt) - N(t)]/dt = \mu N(t).$$

Letting $dt \to 0$ then yields the differential equation

$$dN(t)/dt = -\mu N(t) \tag{2.2.2}$$

which integrates to give

$$N(t) = N(0)e^{-\mu t} \quad (t \geq 0), \tag{2.2.3}$$

where $N(0)$ denotes the initial population size at time $t=0$. Note that taking logarithms gives the linear relationship

$$\ln[N(t)] = \ln[N(0)] - \mu t, \tag{2.2.4}$$

which enables a quick graphical check to be made on whether a given data set exhibits exponential decay.

2.2.1 Stochastic model

For any given individual the time to death is clearly identical to the time to the first event in a Poisson process of rate μ. Hence an argument identical to that leading to (2.1.2) shows that

Pure death process

$$G(t) = \Pr(\text{death occurs after time } t) = e^{-\mu t}, \tag{2.2.5}$$

whence

$$F(t) = \Pr(\text{death occurs before time } t) = 1 - G(t) = 1 - e^{-\mu t}. \tag{2.2.6}$$

So given that at time $t = 0$ a population comprises $N(0) = n_0$ individuals, all of which behave independently of each other, the number $N(t)$ that are still alive at time $t > 0$ must follow the binomial distribution with probabilities

$$p_N(t) = \binom{n_0}{N} e^{-N\mu t} (1 - e^{-\mu t})^{n_0 - N} \quad (N = 0, 1, \ldots, n_0). \tag{2.2.7}$$

The associated mean and variance are therefore given by the standard results

$$m(t) = n_0 e^{-\mu t} \quad \text{and} \quad V(t) = n_0 e^{-\mu t}(1 - e^{-\mu t}). \tag{2.2.8}$$

Once all n_0 individuals have died, the population is *extinct*, and we see from (2.2.7) that the time to extinction, T_0, has the distribution function

$$\Pr(T_0 \leq t) = p_0(t) = (1 - e^{-\mu t})^{n_0} \quad (t \geq 0). \tag{2.2.9}$$

Differentiating (2.2.9) then yields the p.d.f. of the random variable T_0, namely

$$f_0(t) = n_0 \mu e^{-\mu t}(1 - e^{-\mu t})^{n_0 - 1}. \tag{2.2.10}$$

Though means, variances and higher-order moments may be derived directly from (2.2.10), it is instructive to develop an alternative approach which does not rely on first constructing $f_0(t)$.

Let Z_N be the length of time for which the population is of size N. Then given that individuals develop independently, we see from (2.2.5) and (2.2.6) that

$$\Pr(Z_N \leq t) = 1 - \Pr(\text{all } N \text{ are alive at time } t) = 1 - e^{-N\mu t}, \tag{2.2.11}$$

and so Z_N follows the exponential distribution with parameter $N\mu$. As all the successive times Z_N for $N = n_0, n_0 - 1, \ldots, 1$ are independent, with

$$T_0 = Z_{n_0} + Z_{n_0 - 1} + \cdots + Z_1, \tag{2.2.12}$$

it follows that the expected value of T_0 is

$$\mathrm{E}(T_0) = \sum_{N=n_0}^{1} \mathrm{E}(Z_N) = \sum_{N=1}^{n_0} (1/N\mu). \tag{2.2.13}$$

Although this summation is easy to compute numerically, a useful algebraic approximation may be derived by noting that as $n_0 \to \infty$

$$1 + \frac{1}{2} + \frac{1}{3} + \cdots + \frac{1}{s} \to \gamma + \ln(s) \tag{2.2.14}$$

where $\gamma = 0.577216\ldots$ denotes Euler's constant. Thus for large n_0

$$\mathrm{E}(T_0) \simeq (1/\mu)[\gamma + \ln(n_0)]. \tag{2.2.15}$$

52 *Simple Markov population processes*

Similarly,

$$\text{Var}(T_0) = \sum_{N=n_0}^{1} \text{Var}(Z_N) = \sum_{N=1}^{n_0} (1/N\mu)^2. \qquad (2.2.16)$$

Whence the result

$$\sum_{s=1}^{\infty} 1/s^2 = \pi^2/6 \qquad (2.2.17)$$

leads to

$$\text{Var}(T_0) \sim \pi^2/6\mu^2. \qquad (2.2.18)$$

Thus whilst the variance of the time to extinction is bounded above by $\pi^2/6\mu^2$, irrespective of the value of n_0, the expected time to extinction increases with $\ln(n_0)$.

The fact that $\text{Var}(T_0)$ is bounded suggests that the process effectively develops deterministically until N becomes fairly small. Now comparing (2.2.3) and (2.2.15) shows that the deterministic population size which corresponds to stochastic extinction is given by

$$N_{ext} = n_0 e^{-\mu T_0} \simeq n_0 \exp\{-\mu E(T_0)\} = n_0 \exp\{-\gamma - \ln(n_0)\}$$
$$= n_0 e^{-\gamma} n_0^{-1} = e^{-\gamma} \simeq 0.562 \approx 0.5.$$

So on taking a deterministic viewpoint, we may interpret extinction as having occurred once $N(t)$ drops to around 0.5. This makes perfect sense since the integer nature of the stochastic population size, $X(t)$, means that $0 \leq N(t) \leq 0.5$ is effectively equivalent to $X(t) = 0$.

The times to death for each of the n_0 individuals initially alive are easy to simulate since these are independent Exponential(μ) random variables. Hence it follows from (2.1.39) that we just select n_0 $U(0,1)$ uniform random variables and form

$$-(1/\mu)\ln(U_i) \qquad (i = 1, \ldots, n_0). \qquad (2.2.19)$$

Moreover, to simulate the decay in population size, starting at time $t = 0$, we observe that the times of population change from n_0 to $n_0 - 1, n_0 - 2, \ldots$ are given by $t_1 = \tau_0$, $t_2 = \tau_0 + \tau_1, \ldots, t_i = \tau_0 + \cdots + \tau_{i-1}, \ldots$ where τ_i are Exponential$[(n_0 - i)\mu]$ random variables. Thus we evaluate

$$\tau_i = -[(n_0 - i)\mu]^{-1} \ln(U_i) \qquad (i = 0, \ldots, n_0 - 1) \qquad (2.2.20)$$

and plot $n_0 - i$ against t_i. Figure 2.6 shows three simulated realizations for $\mu = 1$ and $n_0 = 100$; these correspond to the fastest, median and slowest time to extinction from nine independent runs, and illustrate that the level of between-run variability is small until N drops down close to 1.

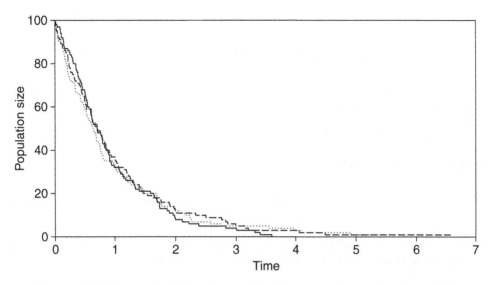

Figure 2.6 Three realizations of a pure death process with $\mu = 1$ and $n_0 = 100$.

2.2.2 Reverse transition probabilities

On denoting $n_0 = i$ and $N = j$, we can express the population size probabilities (2.2.7) as the transition probabilities

$$p_{ij}(t) \equiv \Pr(X(t) = j | X(0) = i) = \binom{i}{j} e^{-\mu j t}(1 - e^{-\mu t})^{i-j} \qquad (j = 0, 1, \ldots, i).$$

(2.2.21)

Suppose we now view this result in reverse, by taking j as fixed and focussing on the values i that lead to it. Then this binomial p.d.f. over $j = 0, 1, \ldots, i$ translates into a negative binomial form over $i = j, j+1, \ldots$ for the *reverse* transition probabilities $\{\overleftarrow{p}_{ij}(t)\}$. Intuitively, we might therefore anticipate that, in the absence of any prior knowledge of $\{p_i(0)\}$,

$$\overleftarrow{p}_{ij}(t) \equiv \Pr(X(0) = i | X(t) = j) = k(t) \binom{i}{j} e^{-\mu j t}(1 - e^{-\mu t})^{i-j} \qquad (i = j, j+1, \ldots)$$

(2.2.22)

for some appropriate normalizing function $k(t)$. Now on writing $x = i + 1$, $j = m + 1$ and $p = e^{-\mu t}$, the standard negative binomial p.d.f.

$$\binom{x-1}{m-1} p^m (1-p)^{x-m} \qquad (x = m, m+1, \ldots)$$

(2.2.23)

takes the form

$$\binom{i}{j} e^{-\mu(j+1)t}(1-e^{-\mu t})^{i-j},$$

whence direct comparison with (2.2.22) yields $k(t) = e^{-\mu t}$.

To verify this negative binomial assertion we first note that since

$$\Pr(X(t) = j | X(0) = i)\Pr(X(0) = i) = \Pr(X(0) = i | X(t) = j)\Pr(X(t) = j),$$

we have the general result

$$\overleftarrow{p}_{ij}(t) = p_i(0)p_{ij}(t) / \sum_{i=0}^{\infty} p_i(0)p_{ij}(t); \tag{2.2.24}$$

where for our simple death process $p_{ij}(t) \equiv 0$ for $i = 0, 1, \ldots, j-1$. Suppose the initial population at time $t = 0$ is geometrically distributed with

$$p_i(0) = \theta^i(1-\theta) \qquad (i = 0, 1, \ldots). \tag{2.2.25}$$

Then under this prior regime (2.2.24) leads to

$$\overleftarrow{p}_{ij}(t) = \binom{i}{j}\theta^{i-j}(1-e^{-\mu t})^{i-j}[1-\{\theta(1-e^{-\mu t})\}]^{j+1}. \tag{2.2.26}$$

Now we may argue that, as $\theta \to 1$, $\{p_i(0)\}$ approaches a uniform distribution on the non-negative integers, which corresponds to having no prior knowledge on the initial states $\{i\}$. Taking this limit in (2.2.26) then recovers the above negative binomial form, i.e.

$$\overleftarrow{p}_{ij}(t) = \binom{i}{j}e^{-\mu(j+1)t}(1-e^{-\mu t})^{i-j} = e^{-\mu t}p_{ij}(t). \tag{2.2.27}$$

2.2.3 Bridge probabilities

As well as conditioning on $X(0) = i$ or $X(t) = j$, we can also condition on both of these events simultaneously. For on writing

$$p_{inj}(s;t) \equiv \Pr(X(s) = n | X(0) = i, X(t) = j) \qquad (0 < s < t) \tag{2.2.28}$$

we have

$$p_{inj}(s;t) = \frac{\Pr(\{X(s) = n | X(0) = i\} \bigcup \{X(t) = j | X(t-s) = n\})}{\Pr(X(t) = j | X(0) = i)},$$

which, as $(0, s)$ and (s, t) are non-overlapping time intervals, yields the general result

$$p_{inj}(s;t) = p_{in}(s)p_{nj}(t-s)/p_{ij}(t) \qquad (n = 0, 1, \ldots). \tag{2.2.29}$$

Whence substituting the binomial probabilities (2.2.7) for $p_{n_0,N}(t)$ into (2.2.29) with $(n_0, N; t)$ replaced by $(i, n; s)$, $(n, j; t-s)$ and $(i, j; t)$ produces the *bridge* probabilities

$$p_{inj}(s;t) = \binom{i-j}{i-n}\left(\frac{1-e^{-\mu s}}{1-e^{-\mu t}}\right)^{i-n}\left(\frac{e^{-\mu s}-e^{-\mu t}}{1-e^{-\mu t}}\right)^{n-j} \qquad (n = j, j+1, \ldots, i).$$
(2.2.30)

That is, a set of Binomial$[i-j, i-n, (e^{-\mu s} - e^{-\mu t})/(1-e^{-\mu t})]$ probabilities which depend only on the differences $i-j$ and $i-n$, and not on the actual values of i and j themselves.

Note that although expressions (2.2.27) and (2.2.30) for the reverse and bridge probabilities specifically relate to the pure death process, the general results (2.2.24) and (2.2.29) from which they are derived may be applied to any stochastic process for which the transition probability structure $\{p_{ij}(t)\}$ is known.

2.3 Pure birth process

The converse of death is birth, and since this generates new individuals the resulting stochastic process involves a fundamental increase in complexity. For in the pure death process the total number of individuals, i.e. alive plus dead, remains constant. Suppose that a population of individuals develops over a short period of time (relative to their lifespan) in crowd-free conditions and with unlimited resource for development. Then we may make the following assumptions: (i) individuals do not die; (ii) they develop without interacting with each other; and, (iii) the birth rate λ is the same for all individuals, regardless of their age, and does not change with time. This last assumption is particularly appropriate to single-celled organisms that reproduce by dividing.

The deterministic pure death process equation (2.2.2) now becomes

$$dN(t)/dt = \lambda N(t), \qquad (2.3.1)$$

which integrates to give

$$N(t) = N(0)e^{\lambda t} \qquad (t \geq 0), \qquad (2.3.2)$$

i.e. simple exponential growth. This result is purely deterministic since it assumes that each organism reproduces on a completely predictable basis at constant rate. In real life, however, population growth is random. For example, given a population of cells which grow by division we cannot say that a particular cell *will* divide in a specific time interval, only that there is a certain *probability* that it will do so. Moreover, whilst observed population sizes can only take the integer values $1, 2, 3, \ldots$, expression (2.3.2) gives rise to all real numbers from $N(0)$ upwards, and this can be particularly unsatisfactory when $N(0)$ is small. Both of these objections may be overcome by considering the stochastic form of the pure birth process.

2.3.1 Stochastic model

Suppose that in the short time interval $(t, t + dt)$ the probability that a particular individual gives birth is $\lambda dt + o(dt)$. Then for a population of size i at time t, the next birth is governed by the superposition of i independent Poisson processes each of rate λ. So

$$\Pr(\text{birth in } (t, t+dt)|N(t) = i) = \lambda i dt + o(dt), \qquad (2.3.3)$$

whence

$$\Pr(\text{no birth in } (t, t+dt)|N(t) = i) = 1 - \lambda i dt + o(dt). \qquad (2.3.4)$$

On paralleling results (2.1.10) and (2.1.11) for the pure death process, we therefore have the Kolmogorov forward equations

$$p_i(t + dt) = \Pr\{N(t) = i \text{ and no birth occurs in } (t, t+dt)\}$$
$$+ \Pr\{N(t) = i - 1 \text{ and 1 birth occurs in } (t, t+dt)\}$$
$$+ \Pr\{N(t) = i - r \text{ and } r \text{ births occur in } (t, t+dt); r = 2, 3, \ldots, i\}$$
$$= p_i(t)(1 - \lambda i dt) + p_{i-1}(t)\lambda(i-1)dt + o(dt). \qquad (2.3.5)$$

Thus

$$[p_i(t + dt) - p_i(t)]/dt = -\lambda i p_i(t) + \lambda(i-1)p_{i-1}(t) + o(1),$$

whence letting $dt \to 0$ yields

$$dp_i(t)/dt = -\lambda i p_i(t) + \lambda(i-1)p_{i-1}(t) \qquad (i = n_0, n_0 + 1, \ldots). \qquad (2.3.6)$$

Here $N(0) = n_0$, so the initial condition is $p_{n_0}(0) = 1$ with $p_i(0) = 0$ $(i > n_0)$.

We are able to construct equations (2.1.12) and (2.1.13) for the Poisson process, and also equation (2.3.6) for the simple birth process, purely because we assume that the probability of an event in $(t, t+dt)$ depends solely on the state of the process at time t and not on any previous history of the process at times $s < t$. This key assumption is called the *Markov property*, and is of fundamental importance in allowing the construction of mathematically tractable sets of equations for the probabilities $\{p_i(t)\}$. Although this appears to place a strong restriction on the types of process that can be developed, in principle the underlying state space can be extended to allow for dependence on a fixed number of previous times, time-delays, etc. There is, of course, no guarantee that such extensions will yield equations that are amenable to solution.

The simple recurrent structure of equations (2.3.6) means that we can obtain a direct sequential solution. Since $p_{n_0-1}(t) \equiv 0$, at $i = n_0$ we have

$$dp_{n_0}(t)/dt = -\lambda n_0 p_{n_0}(t), \qquad (2.3.7)$$

which, with our initial condition, gives

$$p_{n_0}(t) = e^{-\lambda n_0 t}. \qquad (2.3.8)$$

Whence writing (2.3.6) in the form

$$\frac{d}{dt}[e^{\lambda it}p_i(t)] = \lambda(i-1)e^{\lambda it}p_{i-1}(t), \quad \text{i.e.}$$

$$p_i(t) = \lambda(i-1)\int_0^t e^{\lambda i(s-t)}p_{i-1}(s)\,ds,$$

and then integrating sequentially to form $p_{n_0+1}(t), p_{n_0+2}(t), \ldots$, yields the negative binomial distribution

$$p_i(t) = \binom{i-1}{n_0-1} e^{-\lambda n_0 t}(1-e^{-\lambda t})^{i-n_0} \quad (i = n_0, n_0+1, \ldots). \tag{2.3.9}$$

When $n_0 = 1$ this reduces to the geometric form

$$p_i(t) = e^{-\lambda t}(1-e^{-\lambda t})^{i-1} \quad (i = 1, 2, \ldots), \tag{2.3.10}$$

and because $1 - e^{-\lambda t}$ lies between 0 and 1 for all positive values of λt this distribution is permanently J-shaped. In contrast, the negative binomial form (2.3.9) with $n_0 > 1$ is J-shaped only when t is small; as t increases it soon switches into a skew unimodal form with the severity of the skewness decreasing as time develops. The associated mean and variance are given by the standard results

$$m(t) = n_0 e^{\lambda t} \quad \text{and} \quad \text{Var}(t) = n_0 e^{\lambda t}(e^{\lambda t} - 1). \tag{2.3.11}$$

Even in this simple model of population growth, direct sequential solution for the probabilities $\{p_i(t)\}$ is rather unwieldy, and only a slight increase in model complexity, such as including a death component, makes the resulting derivation unattractive. Throughout this text we shall therefore introduce a variety of different techniques in order to maximize the opportunities for finding an amenable method of solution in any given situation.

2.3.2 Laplace transform solution

Denote the Laplace transform of the probabilities $p_i(t)$ by

$$\mathcal{L}[p_i(t)] \equiv p_i^*(s) = \int_0^\infty e^{-st}p_i(t)\,dt \quad (s > 0). \tag{2.3.12}$$

Then we may effectively remove the differential component in equation (2.3.6) by noting that

$$\mathcal{L}[dp_i(t)/dt] = \int_0^\infty e^{-st}\frac{dp_i(t)}{dt}\,dt = [e^{-st}p_i(t)]_0^\infty - \int_0^\infty p_i(t)(-s)e^{-st}\,dt$$

$$= s\mathcal{L}[p_i(t)] - p_i(0). \tag{2.3.13}$$

For on applying this result to (2.3.6) we obtain

$$sp_{n_0}^*(s) - 1 = -\lambda n_0 p_{n_0}^*(s) \tag{2.3.14}$$

$$sp_i^*(s) = -\lambda i p_i^*(s) + \lambda(i-1)p_{i-1}^*(s) \quad (i = n_0+1, n_0+2, \ldots). \tag{2.3.15}$$

Thus
$$p_{n_0}^*(s) = \left(\frac{1}{s+\lambda n_0}\right) \tag{2.3.16}$$

and
$$p_i^*(s) = \left(\frac{\lambda(i-1)}{s+\lambda i}\right) p_{i-1}^*(s), \tag{2.3.17}$$

which solve recursively to give
$$p_{n_0+r}^*(s) = \frac{\lambda^r (n_0+r-1)!}{(n_0-1)!} \prod_{j=n_0}^{n_0+r} \left(\frac{1}{s+\lambda j}\right) \quad (r=0,1,\ldots). \tag{2.3.18}$$

Now this product may be written as the partial fraction expansion
$$\prod_{j=n_0}^{n_0+r} \left(\frac{1}{s+\lambda j}\right) \equiv \sum_{j=n_0}^{n_0+r} \frac{c_j}{s+\lambda j} \tag{2.3.19}$$

where
$$c_j = \left(\frac{1}{-\lambda j + \lambda n_0}\right) \cdots \left(\frac{1}{-\lambda j + \lambda(j-1)}\right)\left(\frac{1}{-\lambda j + \lambda(j+1)}\right) \cdots \left(\frac{1}{-\lambda j + \lambda(n_0+r)}\right)$$
$$= \left(\frac{1}{\lambda}\right)^r \frac{(-1)^{j-n_0}}{(j-n_0)!(n_0+r-j)!}. \tag{2.3.20}$$

So as
$$\mathcal{L}(e^{-\lambda jt}) = \int_0^\infty e^{-st} e^{-\lambda jt}\,dt = \frac{1}{s+\lambda j}, \tag{2.3.21}$$

we have
$$\mathcal{L}^{-1}\left(\frac{1}{s+\lambda j}\right) = e^{-\lambda jt}. \tag{2.3.22}$$

Hence the Laplace transform (2.3.18) inverts to give
$$p_{n_0+r}(t) = \lambda^r \binom{n_0+r-1}{r} \sum_{j=n_0}^{n_0+r} \lambda^{-r} \frac{(-1)^{j-n_0} e^{-\lambda jt} r!}{(j-n_0)!(n_0+r-j)!}$$
$$= \binom{n_0+r-1}{r} e^{-\lambda n_0 t} \sum_{j=n_0}^{n_0+r} (-1)^{j-n_0} e^{-\lambda(j-n_0)t} \binom{r}{j-n_0},$$

which on putting $m = j - n_0$ yields
$$p_{n_0+r}(t) = \binom{n_0+r-1}{r} e^{-\lambda n_0 t} \sum_{m=0}^{r} (-1)^m e^{-\lambda m t} \binom{r}{m} = \binom{n_0+r-1}{r} e^{-\lambda n_0 t} (1-e^{-\lambda t})^r. \tag{2.3.23}$$

Placing $i = n_0 + r$ shows that this is in exact agreement with (2.3.9).

Although for this simple case the Laplace transform approach is clearly less straightforward than the above recursive method, for more complex processes it can prove to be a powerful analytic tool. In general, if we can solve the difference equations for $p_i^*(s)$, then application of standard Laplace transform inversion formulae (such as (2.3.22)) should yield $p_i(t)$ directly.

2.3.3 Solution via the p.g.f.

For our third method of solution we parallel Section 2.1.2 for the derivation of the Poisson distribution by introducing the p.g.f. $G(z;t)$. Multiplying the Kolmogorov forward equations (2.3.6) by z^i, summing over $i = 0, 1, \ldots$, and noting that $p_i(t) \equiv 0$ for $i < n_0$, gives

$$\sum_{i=0}^{\infty} \frac{dp_i(t)}{dt} z^i = -\lambda \sum_{i=0}^{\infty} i p_i(t) z^i + \lambda \sum_{i=0}^{\infty} (i-1) p_{i-1}(t) z^i$$

$$= -\lambda z \sum_{i=0}^{\infty} i p_i(t) z^{i-1} + \lambda z^2 \sum_{i=1}^{\infty} (i-1) p_{i-1}(t) z^{i-2}.$$

Since

$$\frac{\partial G(z;t)}{\partial t} = \sum_{i=0}^{\infty} \frac{dp_i(t)}{dt} z^i \quad \text{and} \quad \frac{\partial G(z;t)}{\partial z} = \sum_{i=0}^{\infty} i p_i(t) z^{i-1}, \qquad (2.3.24)$$

we therefore have

$$\frac{\partial G(z;t)}{\partial t} = -\lambda z \frac{\partial G(z;t)}{\partial z} + \lambda z^2 \frac{\partial G(z;t)}{\partial z}.$$

Thus

$$\frac{\partial G(z;t)}{\partial t} + \lambda z(1-z) \frac{\partial G(z;t)}{\partial z} = 0 \qquad (2.3.25)$$

with

$$G(z;0) = \sum_{i=0}^{\infty} p_i(0) z^i = z^{n_0}. \qquad (2.3.26)$$

One way of solving equation (2.3.25) would be to take the Laplace transform

$$\mathcal{L}[G(z;t)] \equiv G^*(z;s) = \int_0^{\infty} e^{-st} G(z;t)\, dt, \qquad (2.3.27)$$

since invoking (2.3.13) leads to the ordinary differential equation

$$sG^*(z;s) - z^{n_0} + \lambda z(1-z) \frac{\partial G^*(z;s)}{\partial z} = 0 \qquad (2.3.28)$$

which can be solved directly. Taking the inverse Laplace transform of $G^*(z;s)$ then gives $G(z;t)$, the expansion of which yields $p_i(t)$ as the coefficient of z^i.

However, bearing in mind the relatively simple structure of the negative binomial solution (2.3.9), this approach is clearly heavy-handed, and a more popular technique is

60 *Simple Markov population processes*

to treat (2.3.25) as a Lagrange equation (see, for example, Piaggio, 1962). On writing the right-hand side as $0 \times \partial G/\partial G$, we use the differential coefficients to form the auxiliary equations

$$\frac{dt}{1} = \frac{dz}{\lambda z(1-z)} = \frac{dG}{0}. \qquad (2.3.29)$$

Two independent solutions are $G = a$ and $ze^{-\lambda t}/(1-z) = b$, where a and b are 'arbitrary constants'. Since an 'arbitrary function' f of an arbitrary constant must also be an arbitrary constant, we can write $a = f(b)$, i.e.

$$G(z;t) = f(ze^{-\lambda t}/(1-z)). \qquad (2.3.30)$$

On inserting the initial condition (2.3.26) we have

$$z^{n_0} = f(z/(1-z)),$$

whence replacing $z/(1-z)$ by ξ, i.e. z by $\xi/(1+\xi)$, yields

$$f(\xi) = \{\xi/(1+\xi)\}^{n_0}.$$

We now put $\xi = ze^{-\lambda t}/(1-z)$ to obtain the solution

$$G(z;t) = \left(\frac{ze^{-\lambda t}}{1-z+ze^{-\lambda t}}\right)^{n_0}. \qquad (2.3.31)$$

On expanding this p.g.f. in the form

$$G(z;t) = z^{n_0} e^{-\lambda n_0 t} [1 - z(1 - e^{-\lambda t})]^{-n_0}$$

$$= z^{n_0} e^{-\lambda n_0 t} \sum_{r=0}^{\infty} \binom{n_0 + r - 1}{r} z^r (1 - e^{-\lambda t})^r,$$

we see that the coefficient of z^{n_0+r} is precisely our solution (2.3.23).

2.3.4 Time to a given state

It follows from expressions (2.3.11) for the mean and variance that, as t increases, $\text{Var}(t)/n_0$ increases virtually as the square of $m(t)/n_0$, which can create the misleading impression that the process wanders further and further way from the mean value $m(t)$. A much better guide to what really happens is provided by the coefficient of variation

$$\sqrt{\{\text{Var}(t)\}}/m(t) = \sqrt{\{n_0 e^{\lambda t}(e^{\lambda t} - 1)\}}/n_0 e^{\lambda t} \sim 1/\sqrt{(n_0)} \qquad (2.3.32)$$

for large t. For this approaches zero as the initial population size n_0 increases, indicating not only that large variation about $m(t)$ is associated with small values of n_0, but also that the variance of the time taken to reach any given large population size should have a finite upper bound. To prove this last assertion let T_a denote the time at which state $a > n_0$ is first occupied. Then if $T_a \leq t$, at time t the population size $N(t)$ must be at least a, whence we have the fundamental result that

$$\Pr(T_a \leq t) = \Pr(N(t) \geq a). \qquad (2.3.33)$$

For example, in the study of microbial infection (Morgan and Watts, 1980) a host is initially infected with a number of bacteria, n_0, and exhibits symptoms of disease only when the number of bacteria, $N(t)$, has grown to some threshold value a. The time T_a from initial infection to first showing symptoms is called the incubation period.

In the case of the simple birth process determination of the first passage time T_a is in principle straightforward because $N(t)$ cannot decrease. Indeed, we can immediately insert the probabilities $\{p_i(t)\}$ from (2.3.9) into (2.3.33) to obtain the explicit result that

$$\Pr(T_a \leq t) = \sum_{N=a}^{\infty} \binom{N-1}{n_0-1} e^{-\lambda n_0 t}(1 - e^{-\lambda t})^{N-n_0}. \tag{2.3.34}$$

In particular, when $n_0 = 1$ the distribution function (2.3.34) reduces to

$$\Pr(T_a \leq t) = \sum_{N=a}^{\infty} e^{-\lambda t}(1 - e^{-\lambda t})^{N-1} = (1 - e^{-\lambda t})^{a-1}, \tag{2.3.35}$$

whence the random variable T_a takes the p.d.f.

$$g(t) = \lambda(a-1)e^{-\lambda t}(1 - e^{-\lambda t})^{a-2}. \tag{2.3.36}$$

Although means, variances and higher-order moments may be derived directly from (2.3.36), it is instructive to develop an alternative approach (for general n_0) which does not rely on first constructing $g(t)$.

Exactly as for the pure death process (Section 2.2.1) let Z_n be the length of time for which the population is of size n. Then on paralleling result (2.2.13) we see that

$$E(T_a) = \sum_{N=n_0}^{a-1} E(Z_n) = \sum_{N=n_0}^{a-1} (1/\lambda N). \tag{2.3.37}$$

Whence use of (2.2.14) gives

$$E(T_a) = \sum_{N=1}^{a-1}(1/\lambda N) - \sum_{N=1}^{n_0-1}(1/\lambda N)$$
$$\simeq (1/\lambda)[(\gamma + \ln(a-1)) - (\gamma + \ln(n_0-1))] \simeq (1/\lambda)\ln(a/n_0). \tag{2.3.38}$$

Note that this asymptotic approximation agrees with the deterministic prediction (2.3.2), since $N(T_a) = a = n_0 \exp(\lambda T_a)$.

Similarly, as the variance of Z_n is $(1/\lambda N)^2$, we have

$$\text{Var}(T_a) = \text{Var}(Z_{n_0} + \cdots + Z_{a-1}) = \sum_{N=n_0}^{a-1}(1/\lambda N)^2. \tag{2.3.39}$$

62 Simple Markov population processes

For large a this approximates to

$$\operatorname{Var}(T_a) \simeq \sum_{N=1}^{\infty}(1/\lambda N)^2 - \sum_{N=1}^{n_0-1}(1/\lambda N)^2,$$

and using (2.2.17) leads to

$$\operatorname{Var}(T_a) \simeq (\pi^2/6\lambda^2) - \sum_{N=1}^{n_0-1}(1/\lambda N)^2. \tag{2.3.40}$$

Thus in spite of the considerable variation between different realizations, the variance of T_a is always less than $\pi^2/6\lambda^2$ no matter how large a is. Indeed, as n_0 increases, the variance of T_a tends to zero for all $a > n_0$, which confirms that the variation between different realizations develops early on in the process when population size is relatively small.

Since $\operatorname{Var}(T_a)$ is bounded, the distribution of $T_a - \operatorname{E}(T_a) \simeq T_a - (1/\lambda)\ln(a/n_0)$ must tend to a limiting form as $a \to \infty$. Suppose for simplicity that $n_0 = 1$, and introduce the standardized random variable

$$U_a = \lambda T_a - \ln(a). \tag{2.3.41}$$

Now $T_a = Z_1 + \cdots + Z_{a-1}$, where the Z_i are independent and each takes the exponential p.d.f. $\lambda_i \exp(-\lambda_i t)$. So on writing the m.g.f. of Z_i in the form

$$\operatorname{E}(e^{-\theta Z_i}) = \int_0^{\infty} e^{-\theta t}\lambda_i e^{-\lambda_i t}\, dt = \left(\frac{\lambda i}{\theta + \lambda i}\right), \tag{2.3.42}$$

it follows that the m.g.f. of T_a is

$$\operatorname{E}(e^{-\theta T_a}) = \prod_{i=1}^{a-1}\left(\frac{\lambda i}{\theta + \lambda i}\right). \tag{2.3.43}$$

Thus the m.g.f. of U_a is

$$\operatorname{E}(e^{-\theta U_a}) = \operatorname{E}(e^{-\theta[\lambda T_a - \ln(a)]}) = a^{\theta}\operatorname{E}(e^{-(\theta\lambda)T_a}) = a^{\theta}\prod_{i=1}^{a-1}\left(\frac{i}{\theta + i}\right). \tag{2.3.44}$$

On using Euler's product form of the gamma function, namely

$$\Gamma(1+\theta) = \lim_{a \to \infty}\frac{(a-1)!\,a^{\theta}}{(\theta+1)(\theta+2)\ldots(\theta+a-1)}, \tag{2.3.45}$$

we then have that for large a

$$\operatorname{E}(e^{-\theta U_a}) \sim \Gamma(1+\theta). \tag{2.3.46}$$

Now the corresponding inverse function, when considered as a two-sided Laplace transform, is the extreme-value distribution

$$f_U(u) = \exp\{-u - e^{-u}\} \qquad (-\infty < u < \infty). \tag{2.3.47}$$

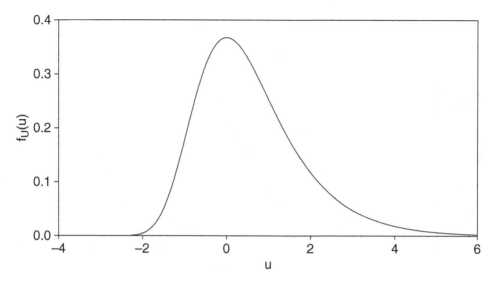

Figure 2.7 Extreme value p.d.f. $f_U(u) = \exp\{-u - e^{-u}\}$ over $-4 < u < 6$.

At $u = 0$, i.e. $T_a \simeq \mathrm{E}(T_a)$, $f_U(0) = e^{-1}$. Whilst as $u \to \infty$, $f_U(u) \sim e^{-u}$ is approximately Exponential(1), which contrasts with $f_U(u) \sim e^{-u}\exp\{-e^{-u}\}$ as $u \to -\infty$ since this decays much faster than exponential. Figure 2.7 illustrates this strong asymmetry, and shows that virtually all the probability mass lies in the range $-2 < u < 6$; the substantial positive skewness means that reaching a 'early' is much less likely than reaching a 'late'. Note that since here $n_0 = 1$, combining (2.3.41) with (2.3.37) and (2.3.40) yields $\mathrm{E}(U_a) \simeq \gamma$ and $\mathrm{Var}(U_a) \simeq \pi^2/6$, respectively; γ (Euler's constant) is given in (2.2.14).

The development of the stochastic simulation algorithm parallels that for the pure death process, except that now a population of size N proceeds for a time distributed as Exponential(λN), whereupon it increases to $N+1$. Thus we evaluate

$$\tau_i = -\ln(U_i)/[(n_0 + i)\lambda] \qquad (i = 0, 1, 2, \ldots), \tag{2.3.48}$$

plot $n_0 + i$ against $t_i = \tau_0 + \cdots + \tau_i$, and terminate the simulation run when either an appropriate population size or time has been reached. Figure 2.8 shows three simulated realizations for $\lambda = 1$ and $n_0 = 1$, corresponding to the fastest, median and slowest time to reach $N = a = 500$ from nine independent runs. The variability between them is clearly associated with small N, i.e. with the time taken to 'get going', and although only nine runs are involved, the substantial shift of the right-most curve away from the expected value hints at the extreme nature of $U_{500} - \ln(500)$ discussed above.

2.3.5 Generalized birth process

There are numerous ways in which the forward equation (2.3.6) may be generalized; here we consider two that retain the Markov property. The first, and simplest, is to

64 Simple Markov population processes

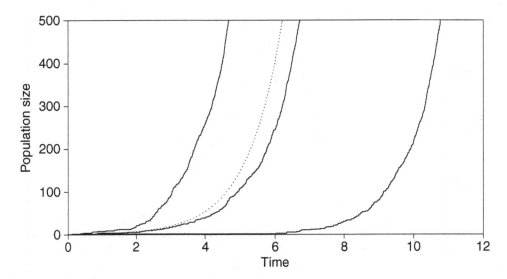

Figure 2.8 Three realizations (——) of a pure birth process for $\lambda = 1$, $n_0 = 1$ and $a = 500$; together with the expected value $m(t) = e^t$ ($\cdots$).

allow the birth rate λ to be a function of time, namely $\lambda(t)$. This generalization exactly parallels that made for the Poisson process in Section 2.1.5, since on transforming the time-scale to

$$\tau(t) = \int_0^t \lambda(u)\,du, \qquad (2.3.49)$$

we have $d\tau/dt = \lambda(t)$. Whence equation (2.3.6) (with λ replaced by $\lambda(N,t)$) reduces to

$$dp_i(\tau)/d\tau = -ip_i(\tau) + (i-1)p_{i-1}(\tau), \qquad (2.3.50)$$

i.e. the forward equation of a simple birth process with $\lambda = 1$. Note that this is far easier than allowing λ to depend on an individual's age, since instead of looking at total population size we would then have to track each individual separately.

Second, we can replace the population birth rate λi by the general function λ_i. This is of major importance since it allows for mutual interaction between individual population members. If λ_i/i decreases as i increases, then we may envisage the process as involving 'competition for resource', whilst conversely we have 'mutual enhancement'. General examples of such processes are considered later on in Chapter 3, but for the moment let us stay within the confines of the birth process, whence equation (2.3.6) becomes

$$dp_i(t)/dt = -\lambda_i p_i(t) + \lambda_{i-1} p_{i-1}(t) \qquad (i = n_0, n_0+1, \ldots). \qquad (2.3.51)$$

The Laplace transform results (2.3.16) and (2.3.17) now take the form

$$p^*_{n_0}(s) = \left(\frac{1}{s+\lambda_{n_0}}\right) \quad \text{with} \quad p^*_i(s) = \left(\frac{\lambda_{i-1}}{s+\lambda_i}\right) p^*_{i-1}(s) \quad (i > n_0), \qquad (2.3.52)$$

which immediately yield the solution

$$p^*_{n_0+r}(s) = \left(\frac{1}{s+\lambda_{n_0}}\right)\left(\frac{\lambda_{n_0}}{s+\lambda_{n_0+1}}\right)\cdots\left(\frac{\lambda_{n_0+r-1}}{s+\lambda_{n_0+r}}\right). \qquad (2.3.53)$$

On generalizing (2.3.19) and (2.3.20), we can write this product as the partial fraction expansion

$$\prod_{j=n_0}^{n_0+r}\left(\frac{1}{s+\lambda_j}\right) \equiv \sum_{j=n_0}^{n_0+r}\frac{c_j}{s+\lambda_j}, \qquad (2.3.54)$$

where

$$c_j = \left(\frac{1}{-\lambda_j+\lambda_{n_0}}\right)\left(\frac{1}{-\lambda_j+\lambda_{n_0+1}}\right)\cdots\left(\frac{1}{-\lambda_j+\lambda_{j-1}}\right)$$
$$\times \left(\frac{1}{-\lambda_j+\lambda_{j+1}}\right)\cdots\left(\frac{1}{-\lambda_j+\lambda_{n_0+r}}\right) \qquad (2.3.55)$$

(provided the λ_j are all different). Whence the transform (2.3.53) inverts to give the general solution

$$p_{n_0+r}(t) = (\lambda_{n_0}\ldots\lambda_{n_0+r-1})\sum_{j=n_0}^{n_0+r} c_j e^{-\lambda_j t}. \qquad (2.3.56)$$

Other, more mathematically challenging, generalizations include allowing for multiple births and introducing different types of individual.

The general birth process (2.3.51) provides an excellent illustration of a phenomenon called 'dishonesty' that is of considerable importance in the study of Markov processes in continuous time. On paralleling the simple birth result (2.3.43) for the m.g.f. of T_a, the time taken for the population to increase from 1 individual to a individuals, we have

$$\mathrm{E}(e^{-\theta T_a}) = \prod_{i=1}^{a-1}\frac{1}{(1+\theta/\lambda_i)}. \qquad (2.3.57)$$

Now this product tends to a finite non-zero limit as $a \to \infty$ if and only if $\sum_{i=0}^{\infty} 1/\lambda_i < \infty$ (see, for example, Titchmarsh, 1939), in which case (2.3.57) is the Laplace transform of a well-defined probability distribution of a potentially infinite random variable. That is, there is a non-zero probability that the population size will become infinite in a finite time. Conversely, if $\sum_{i=0}^{\infty} 1/\lambda_i = \infty$ then the product converges to zero for all $s \neq 0$, and the population size will always be finite.

66 Simple Markov population processes

A similar conclusion is reached if we examine the equivalent deterministic equation, namely

$$dn(t)/dt = \lambda(n(t)), \qquad (2.3.58)$$

for general growth rate function $\lambda(n(t))$. For the time taken for the population to grow from size 1 to n is

$$\int_1^n \frac{dx}{\lambda(x)}, \qquad (2.3.59)$$

and if this integral converges as $n \to \infty$ then the population explodes to infinity in a finite time.

Bearing in mind that $\sum_{i=1}^{\infty} p_i(t) = \lim_{a \to \infty} \sum_{i=1}^{a} p_i(t)$ is the probability that the population is finite at time t, the solution of equation (2.3.51) has

$$\sum_{i=1}^{\infty} p_i(t) = 1 \quad \text{for all } t \qquad \text{if} \quad \sum_{i=0}^{\infty} 1/\lambda_i = \infty \qquad (2.3.60)$$

$$\sum_{i=1}^{\infty} p_i(t) < 1 \quad \text{for some } t \qquad \text{if} \quad \sum_{i=0}^{\infty} 1/\lambda_i < \infty. \qquad (2.3.61)$$

Processes satisfying (2.3.60) are called *honest*, whilst those satisfying (2.3.61) are called *dishonest*. In the latter case the missing probability $1 - \sum_{i=1}^{\infty} p_i(t)$ is the probability of having escaped to infinity by time t.

The simple birth process is clearly honest, since as $\lambda_i = i\lambda$ we have

$$\sum_{i=1}^{\infty} \frac{1}{\lambda_i} = \frac{1}{\lambda} \sum_{i=1}^{\infty} \frac{1}{i} = \infty.$$

However, if the probability of an individual giving birth in a population of size i increases very slightly with i, say from λi to $\lambda i^{1+\delta}$ ($\delta > 0$), then explosion to infinity within a finite time now becomes possible since

$$\sum_{i=1}^{\infty} \frac{1}{i^{1+\delta}} < \infty \qquad \text{for all } \delta > 0. \qquad (2.3.62)$$

Thus this linear model, with its assumption of independent births (i.e. $\lambda_i = \lambda i$), is essentially a watershed, since even a tiny amount of mutual self-enhancement renders the process dishonest.

The Laplace transform of the p.d.f. $l(t)$ of the time taken to reach infinite population size is given by the m.g.f. (2.3.57) with $a = \infty$, i.e.

$$l^*(s) = \prod_{i=1}^{\infty} \frac{1}{(1 + s/\lambda_i)}. \qquad (2.3.63)$$

Whence the Laplace transform, $L^*(s)$, of the distribution function, $L(t)$, may be obtained by writing

$$l^*(s) = \int_0^\infty e^{-st} l(t)\, dt = \int_0^\infty e^{-st} L'(t)\, dt = sL^*(s) - L(0)$$

i.e., as $L(0) = 0$,

$$L^*(s) = \frac{1}{s} \prod_{i=1}^\infty \frac{1}{(1 + s/\lambda_i)}. \tag{2.3.64}$$

On expanding this infinite product into partial fractions we therefore have

$$L^*(s) = \frac{d_0}{s} + \sum_{i=1}^\infty \frac{d_i}{s + \lambda_i}, \tag{2.3.65}$$

where the constants

$$d_0 = 1 \quad \text{and} \quad d_i = \lim_{s \to -\lambda_i} \{(s + \lambda_i) L^*(s)\} \quad (i \geq 1). \tag{2.3.66}$$

The simplest case to consider is $\lambda_i = i^2$; that is the individual birth rate equals the current population size. Here

$$d_i = \frac{(-1)^i}{i^2} \times \frac{1^2}{(i-1)(i+1)} \times \frac{2^2}{(i-2)(i+2)} \times \cdots \times \frac{(i-1)^2}{1(2i)} \times i^2 \times \frac{(i+1)^2}{1(2i+1)}$$

$$\times \frac{(i+2)^2}{2(2i+1)} \times \cdots,$$

and since all terms greater than $2i$ in the numerator also appear in the denominator we can cancel these to obtain

$$d_i = \frac{(-1)^i}{i^2} \times \frac{(2i)!(2i)!}{(i-1)!(i+1)\ldots(2i-1)(2i)!} = 2(-1)^i.$$

Thus (2.3.65) becomes

$$L^*(s) = \frac{1}{s} + 2 \sum_{i=1}^\infty \frac{(-1)^i}{s + i^2},$$

which inverts to give the c.d.f.

$$L(t) = 1 + 2 \sum_{i=1}^\infty (-1)^i e^{-ti^2}. \tag{2.3.67}$$

This theta function rapidly approaches 1 as t increases: for example, at

t	0.1	0.5	1.0	2.0	5.0	10.0
$L(t)$	2×10^{-10}	0.0361	0.3006	0.7300	0.9865	0.99991

68 Simple Markov population processes

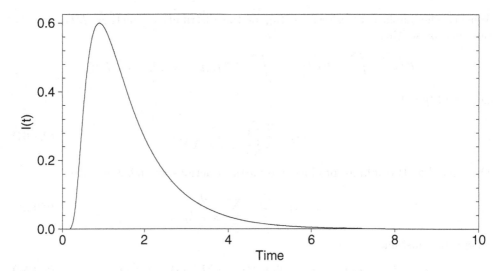

Figure 2.9 P.d.f. $l(t)$ for the time taken to reach infinite population size.

The associated p.d.f.

$$l(t) = 2\sum_{i=1}^{\infty}(-1)^{i+1}i^2 e^{-ti^2}, \qquad (2.3.68)$$

shown in Figure 2.9, highlights the initial very slow departure from zero, followed by a swift rise to the maximum and a slower decline back towards zero thereafter.

Moreover, since the time taken for the population size to explode to infinity, T_∞, is the sum of independent Exponential(i^2) random variables, each of which has mean $1/i^2$ and variance $1/i^4$, it follows that

$$\mathrm{E}(T_\infty) = \sum_{i=1}^{\infty}\frac{1}{i^2} = \pi^2/6 \simeq 1.6449 \qquad (2.3.69)$$

and

$$\mathrm{Var}(T_\infty) = \sum_{i=1}^{\infty}\frac{1}{i^4} = \pi^4/90 \simeq 1.0823. \qquad (2.3.70)$$

This swift approach to infinite population size shows that even minor changes to model structure may lead to extreme types of behaviour.

For the more general case of $\lambda_i = i^r$ ($r = 2, 3, \ldots$), $\mathrm{E}(T_\infty)$ is given by the Riemann zeta function

$$\zeta(r) = \sum_{i=1}^{\infty}\frac{1}{i^r} \qquad (2.3.71)$$

whose properties are described, for example, in Section 23.2 of Abramowitz and Stegun (1970). Whilst $\mathrm{Var}(T_\infty) = \zeta(2r)$. Indeed, the provision of tabulated values for $\zeta(r)$ (e.g. Jolley, 1961 for $r = 1, \ldots, 27$) enables higher-order moments to be constructed, not only for our simple $\lambda_i = i^2$ example, but also for even more explosive processes with $r > 2$.

Simulation runs may be constructed as for the pure birth process, except that (when $r = 2$) the inter-event times (2.3.48) are modified to

$$\tau_i = -\ln(U_i)/[\lambda(n_0 + i)^2]. \qquad (2.3.72)$$

For example, one run with $n_0 = \lambda = 1$ gave $t_\infty = 4.3577577713\ldots$, with the population size n at each decimal point gain in precision being given by

t	n	t	n
0	1	4.357	1,295
3.835	2	4.3577	17,359
3.973	3	4.35775	128,171
4.080	4	4.357757	1,269,833
4.3	14	4.3577577	11,457,274
4.35	128	4.35775777	70,584,429

i.e. a roughly 10-fold increase at each stage. Figure 2.10 shows a realization for $\lambda = 1$ and $n_0 = 1$, plotted as $\log_{10}(n(t))$ against $-\log_{10}(t_\infty - t)$ for $n = 1, \ldots, 240$ and $\log_{10}(n) = 2.38, 2.39, \ldots, 8.00$ thereafter. Once the initial 'time to get going' period is past, the process rises in line with the deterministic prediction derived from (2.3.58), namely $n(t) = 1/(t_\infty - t)$. Provided, that is, we replace the deterministic value of

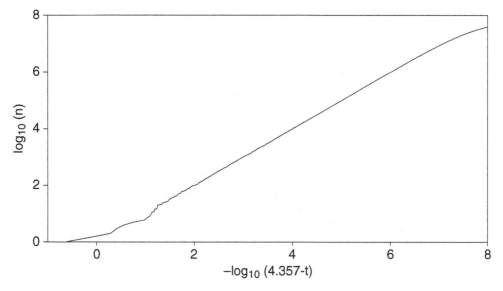

Figure 2.10 Simulation of a quadratic (explosive) birth process with $\lambda = 1$ and $n_0 = 1$.

70 *Simple Markov population processes*

$t_\infty = 1$ obtained from (2.3.59) by the stochastic value, namely $t_\infty = 4.357\ldots$ for this particular realization.

2.4 Simple birth–death process

Combining birth and death into a single process results in a more complex structure, since population size is no longer restricted to be either purely increasing or decreasing over time. Moreover, our implicit assumptions that individual birth and death rates remain constant through time, and that they are independent of both age and population size, will clearly be violated in virtually all natural situations involving population growth. However, whilst the construction of realistic stochastic models in research areas such as astronomy, chemistry, ecology, physics, etc., is essentially unconstrained by theoretical considerations, their subsequent mathematical analysis is usually intractable in all but the simplest of cases. This necessitates the use of approximation procedures, either through simplifying the underlying model, or by replacing the underlying equations by simpler (often linear) relationships. Models such as the simple (i.e. linear) birth–death process should therefore not be looked upon as providing a literal representation of reality, but should be treated instead as a means of obtaining greater understanding of the behavioural characteristics of the processes which they attempt to mimic. Linearity implies that there is no interaction between individuals: more realistic (i.e. nonlinear) birth and death rates can, of course, be tried, but the price paid in terms of added mathematical complexity is usually high (Renshaw, 1991).

The deterministic approach proceeds in exactly the same way as for the pure death and pure birth processes, with their respective deterministic equations (2.2.2) and (2.3.1) combining to produce

$$dN(t)/dt = (\lambda - \mu)N(t). \tag{2.4.1}$$

The solution

$$N(t) = N(0)\exp\{(\lambda - \mu)t\} \tag{2.4.2}$$

is formally the same as (2.2.3) and (2.3.2), though the net rate of increase $(\lambda - \mu)$ may now be either positive or negative.

2.4.1 Stochastic model

Analysis of stochastic behaviour follows along exactly the same lines as that already developed for the pure birth process, except that in the short time interval $(t, t+dt)$ there is now a probability $\lambda dt + o(dt)$ that a particular individual gives birth and a probability $\mu dt + o(dt)$ that it dies. For a population of size i at time t, the probability that no event occurs in $(t, t+dt)$ is $1 - (\lambda + \mu)idt + o(dt)$, whence equation (2.3.6) extends to give

$$dp_i(t)/dt = \lambda(i-1)p_{i-1}(t) - (\lambda + \mu)ip_i(t) + \mu(i+1)p_{i+1}(t) \tag{2.4.3}$$

Simple birth–death process

over $i = 0, 1, 2, \ldots$ and $t \geq 0$. As before, the initial condition corresponding to $N(0) = n_0$ is $p_{n_0}(0) = 1$ with $p_i(0) = 0$ $(i \neq n_0)$; whilst for $i = 0$ in (2.4.3) we define $p_{-1}(t) \equiv 0$.

Since population size may now both decrease and increase, direct sequential solution for the probabilities $p_i(t)$ or their Laplace transforms $p_i^*(s)$, as given by (2.3.9) and (2.3.18) for the pure birth process, is no longer possible. For when $i = 0$ equation (2.4.3) becomes

$$dp_0(t)/dt = \mu p_1(t) \tag{2.4.4}$$

with Laplace transform

$$sp_0^*(s) = \mu p_1^*(s). \tag{2.4.5}$$

So a sequential solution has to be expressed in terms of the unknown $p_0(t)$ or $p_0^*(s)$, which then have to be found through the relationship

$$\sum_{i=0}^{\infty} p_i(t) = 1, \quad \text{i.e.} \quad \sum_{i=0}^{\infty} p_i^*(s) = 1/s. \tag{2.4.6}$$

The simplest approach is to evaluate the p.g.f. $G(z;t)$, since this involves only a minor amendment to the pure birth solution (2.3.31). On paralleling the derivation of (2.3.25), equation (2.4.3) gives rise to the p.d.e.

$$\frac{\partial G(z;t)}{\partial t} - (\lambda z - \mu)(z - 1)\frac{\partial G(z;t)}{\partial z} = 0, \tag{2.4.7}$$

whence the auxiliary equations (2.3.29) now become

$$\frac{dt}{1} = \frac{-dz}{(\lambda z - \mu)(z - 1)} = \frac{dG}{0}. \tag{2.4.8}$$

One independent solution is $G = a$ (as before), whilst integrating the first part of (2.4.8) gives the second, namely

$$t + \text{constant} = \left(\frac{1}{\lambda - \mu}\right) \ln\left(\frac{\lambda z - \mu}{z - 1}\right), \quad \text{i.e.}$$

$$\left(\frac{\lambda z - \mu}{z - 1}\right) e^{-(\lambda - \mu)t} = b. \tag{2.4.9}$$

Thus the general 'solution' $a = f(b)$, for arbitrary function f and arbitrary constants a and b, is

$$G(z;t) = f\left[\left(\frac{\lambda z - \mu}{z - 1}\right) e^{-(\lambda - \mu)t}\right]. \tag{2.4.10}$$

On inserting the initial condition (2.3.26), we see that at $t = 0$

$$z^{n_0} = f[(\lambda z - \mu)/(z - 1)].$$

Whence replacing $(\lambda z - \mu)/(z - 1)$ by ξ, i.e. z by $(\xi - \mu)/(\xi - \lambda)$, yields

$$f(\xi) = [(\xi - \mu)/(\xi - \lambda)]^{n_0}.$$

We then put $\xi = [(\lambda z - \mu)/(z-1)]e^{-(\lambda-\mu)t}$ to obtain

$$G(z;t) = \left\{\frac{\mu(1-z) - (\mu - \lambda z)e^{-(\lambda-\mu)t}}{\lambda(1-z) - (\mu - \lambda z)e^{-(\lambda-\mu)t}}\right\}^{n_0}. \tag{2.4.11}$$

The probabilities $\{p_N(t)\}$ may now be recovered by expanding (2.4.11) as a power series in z and extracting the coefficient of z^N (for the moment we write $p_N(t)$ rather than $p_i(t)$). First we combine z-terms by placing

$$\beta(t) = (\lambda - \lambda e^{-(\lambda-\mu)t})/(\lambda - \mu e^{-(\lambda-\mu)t}) \quad \text{with} \quad \alpha(t) = \mu\beta(t)/\lambda, \tag{2.4.12}$$

so that

$$G(z;t) = [\alpha(t) + z(1 - \alpha(t) - \beta(t))]^{n_0}[1 - \beta(t)z]^{-n_0}. \tag{2.4.13}$$

For when $n_0 = 1$ this yields

$$p_0(t) = \alpha(t) \tag{2.4.14}$$

$$p_N(t) = [1 - \alpha(t)][1 - \beta(t)][\beta(t)]^{N-1} \quad (N = 1, 2, \ldots), \tag{2.4.15}$$

which takes the form of a geometric distribution over $N > 0$. Note that if we remove the probability of extinction $\alpha(t)$ by considering

$$\tilde{p}_N(t) = \Pr\{N(t) = N | N \neq 0\}, \tag{2.4.16}$$

then we obtain the pure geometric distribution

$$\tilde{p}_N(t) = [1 - \beta(t)][\beta(t)]^{N-1} \tag{2.4.17}$$

over $N = 1, 2, \ldots$. In general, on writing

$$G(z;t) = \sum_{i=0}^{n_0} \binom{n_0}{i} z^i (1 - \alpha - \beta)^i \alpha^{n_0-i} \sum_{j=0}^{\infty} \binom{n_0 + j - 1}{j} z^j \beta^j,$$

we see that extracting the coefficient of z^N (i.e. putting $i + j = N$) yields

$$p_N(t) = \alpha^{n_0}\beta^N \sum_{i=0}^{N} \binom{n_0}{i}\binom{n_0 + N - i - 1}{N - i}[(1 - \alpha - \beta)/\alpha\beta]^i \tag{2.4.18}$$

where we denote $\binom{n}{m} \equiv 0$ for $m > n$. So the basic functional form of the solution as given by the p.g.f. (2.4.13), namely the convolution of a binomial and a negative binomial, is much less easy to discern when we examine the probabilities themselves. Indeed, it is apparent that adding only a slight degree of complexity to the model would render the solution $\{p_N(t)\}$ virtually opaque to informed interpretation.

Since individuals are assumed to develop independently of each other, we may consider the development of a population of initial size n_0 as being equivalent to the development of n_0 separate populations each of initial size 1. Thus in order to understand the stochastic behaviour of the process it is sufficient just to consider $n_0 = 1$. Fortunately, in this case the population size $N(t)$ takes the simple geometric form (2.4.14)–(2.4.15). Whence on exploiting the fact that each of the n_0 populations grows independently, we have the mean

$$m(t) = n_0 \sum_{N=0}^{\infty} N p_N(t) = n_0(1-\alpha)(1-\beta) \sum_{N=1}^{\infty} N \beta^{N-1}$$
$$= n_0(1-\alpha)(1-\beta)(1-\beta)^{-2} = n_0 e^{(\lambda-\mu)t}, \qquad (2.4.19)$$

which is the same as the deterministic value (2.4.2); whilst the variance

$$V(t) = n_0 \left[\sum_{N=0}^{\infty} N^2 p_N(t) - \left\{ \sum_{N=0}^{\infty} N p_N(t) \right\}^2 \right]$$
$$= n_0(1-\alpha)(1-\beta) \sum_{N=1}^{\infty} N^2 \beta^{N-1} - n_0 e^{2(\lambda-\mu)t}$$
$$= n_0(1-\alpha)(1-\beta)(1+\beta)(1-\beta)^{-3} - n_0 e^{2(\lambda-\mu)t}$$
$$= n_0(1-\alpha)(1+\beta)(1-\beta)^{-2} - n_0 e^{2(\lambda-\mu)t},$$

which reduces to

$$V(t) = \frac{n_0(\lambda+\mu)}{\lambda-\mu} e^{(\lambda-\mu)t}(e^{(\lambda-\mu)t} - 1). \qquad (2.4.20)$$

Note that unlike $m(t)$, not only does $V(t)$ depend on the difference between the birth and death rates, but it also increases directly with $\lambda + \mu$: thus predictions about the future size of a population will be less precise if births and deaths occur in rapid succession than if they occur only occasionally. When births and deaths are in balance (i.e. $\lambda = \mu$) we see that $m(t)$ takes the constant value n_0, whilst $V(t)$ increases linearly as $2n_0\mu t$.

General moments may be obtained by following any of the procedures described in Section 2.1.3. For example, to derive the rth-order factorial moments, $\mu_{(r)}$, we can either follow: (2.1.23) by differentiating the p.g.f. $G(z;t)$ r times with respect to z, and then placing $z = 1$; or, (2.1.26) by expanding $H(y) = G(1+y)$ as a power series in y and forming the coefficient of $y^r/r!$. To derive the raw moments, μ'_r, we differentiate the m.g.f. $G(e^\theta;t)$ r times with respect to θ and then place $\theta = 0$, as in (2.1.29)–(2.1.30); whilst the cumulants κ_r may be obtained from $\ln[G(e^\theta;t)]$ either by determining the coefficient of $\theta^r/r!$ in its power-series expansion, or, equivalently, by evaluating the rth-derivative with respect to θ at $\theta = 0$. Here the factorial moments are especially easy to obtain, since (2.4.13) yields

$$H(y;t) \equiv G(1+y;t) = \sum_{i=0}^{n_0} \binom{n_0}{i} y^i \left(\frac{1-\alpha-\beta}{1-\beta} \right)^i \sum_{j=0}^{\infty} \binom{n_0+j-1}{j} y^j \left(\frac{\beta}{1-\beta} \right)^j,$$

$$(2.4.21)$$

whence placing $i + j = N$ and extracting the coefficient of $y^N/N!$ gives

$$\mu_{(r)} = N! \left(\frac{\beta}{1-\beta}\right)^N \sum_{i=0}^{N} \binom{n_0}{i}\binom{n_0 + N - i - 1}{N - i}\left(\frac{1 - \alpha - \beta}{\beta}\right)^i. \qquad (2.4.22)$$

If required, raw and central moments can be constructed from these $\mu_{(r)}$ through the relationships (2.1.32) and (2.1.34).

2.4.2 Extinction

The behaviour of individual realizations clearly depends on the relative sizes of λ and μ. If $\lambda < \mu$ then it is intuitively reasonable to suppose that since deaths predominate, the population will eventually die out, and as there are no arrivals from outside the system (i.e. immigrants) this implies *extinction*. If $\lambda > \mu$ then births predominate, and either an initial downward surge causes the population to become extinct or else $N(t)$ avoids becoming zero early on and the population grows indefinitely.

That such differences in behaviour do occur can be seen directly from the p.g.f. (2.4.11), since the probability of extinction by time t is

$$p_0(t) \equiv G(0; t) = \left\{\frac{\mu - \mu e^{-(\lambda-\mu)t}}{\lambda - \mu e^{-(\lambda-\mu)t}}\right\}^{n_0}. \qquad (2.4.23)$$

Whence on letting $t \to \infty$, it follows that the probability of *ultimate extinction*

$$p_0(\infty) = (\mu/\lambda)^{n_0} \qquad \text{if } \lambda > \mu. \qquad (2.4.24)$$

Thus although extinction can occur if $\lambda > \mu$, the probability that it does so decreases geometrically fast at rate μ/λ with increasing n_0. Conversely, if $\lambda < \mu$ then

$$p_0(t) \to (\mu/\mu)^{n_0} = 1. \qquad (2.4.25)$$

Whilst if $\lambda = \mu$, on placing $\lambda = \mu + \epsilon$ in (2.4.11) we see that

$$G(z;t) = \left\{\frac{\mu(1-z) - (\mu - \mu z - \epsilon z)(1 - \epsilon t + O(\epsilon^2))}{(\mu+\epsilon)(1-z) - (\mu - \mu z - \epsilon z)(1 - \epsilon t + O(\epsilon^2))}\right\}^{n_0}.$$

So in the limit as $\epsilon \to 0$,

$$G(z;t) = \left\{\frac{z + \mu t(1-z)}{1 + \mu t(1-z)}\right\}^{n_0}. \qquad (2.4.26)$$

Whence placing $z = 0$ gives

$$p_0(t) = [\mu t/(1+\mu t)]^{n_0}. \qquad (2.4.27)$$

Thus as t increases, $p_0(t) \to 1$, and so ultimate extinction is certain even though the birth and death rates are equal. So although the expected population size $m(t)$ remains permanently fixed at n_0, stochastic fluctuations about n_0 will inevitably lead to extinction no matter how large n_0 is. At first sight this result appears to be counter-intuitive, but it simply reflects the behavioural variability enjoyed by individual stochastic realizations.

For the pure death process with large initial population size n_0, we see from (2.2.15) that the expected time to extinction, namely $E[T_0(n_0)]$, is (asymptotically) equal to $(1/\mu)[\gamma + \ln(n_0)]$. Since $\gamma = 0.577216\ldots$ (Euler's constant) is relatively small, it follows that $E[T_0(n_0)]$ is effectively equal to $(1/\mu)\ln(n_0)$, which is the time required for the deterministic population size $N(t)$ to decrease in size from n_0 to 1. This suggests that an approximation to the mean time to extinction for the simple birth–death process with $\lambda < \mu$ may be obtained from (2.4.2) by writing

$$1 \simeq n_0 \exp\{(\lambda - \mu)E[T_0(n_0)]\},$$

which gives

$$E[T_0(n_0)] \simeq (\mu - \lambda)^{-1} \ln(n_0). \qquad (2.4.28)$$

To determine the corresponding p.d.f., we first parallel results (2.2.9) and (2.2.10) for the pure death process by using (2.4.23) to form the c.d.f.

$$\Pr(T_0 \leq t) = p_0(t) = \left\{\frac{\mu - \mu e^{-(\lambda-\mu)t}}{\lambda - \mu e^{-(\lambda-\mu)t}}\right\}^{n_0}. \qquad (2.4.29)$$

Whence the p.d.f. of T_0 is given by

$$f_0(t) = p_0'(t) = n_0 \mu(\lambda - \mu)^2 e^{-(\lambda-\mu)t} \frac{\{\mu - \mu e^{-(\lambda-\mu)t}\}^{n_0 - 1}}{\{\lambda - \mu e^{-(\lambda-\mu)t}\}^{n_0 + 1}}. \qquad (2.4.30)$$

Thus as $t \to \infty$ we have

$$f_0(t) \sim \begin{cases} n_0(1/\mu)(\lambda - \mu)^2 e^{(\lambda-\mu)t} & (\lambda < \mu) \\ n_0(1/\lambda)(\lambda - \mu)^2 (\mu/\lambda)^{n_0} e^{-(\lambda-\mu)t} & (\lambda > \mu). \end{cases} \qquad (2.4.31)$$

So whilst exponential decay at rate $|\lambda - \mu|t$ occurs in both cases, when $\lambda > \mu$ the p.d.f. $f_0(t)$ contains the additional geometrically decreasing term (2.4.24) in n_0.

Note that $p_0(t)$ plays a dual role, representing both (i) $\Pr(X(t) = 0)$ and (ii) $\Pr(T_0 \leq t)$. From (ii) it follows that

$$E(T_0) = \int_0^\infty t p_0'(t)\, dt. \qquad (2.4.32)$$

Now in many situations $p_0(\infty) > 0$, i.e. state 0 may be occupied at time $t = \infty$, and in such cases straight integration of (2.4.32) by parts is clearly inadmissible. So let us first write (2.4.32) in the equivalent form

$$E(T_0) = -\int_0^\infty t \frac{d}{dt}\{p_0(\infty) - p_0(t)\}\, dt, \qquad (2.4.33)$$

which can now be integrated by parts to give

$$-[t\{p_0(\infty) - p_0(t)\}]_0^\infty + \int_0^\infty \{p_0(\infty) - p_0(t)\}\, dt.$$

76 Simple Markov population processes

Since the first term vanishes in all but the severest of pathological cases, we may therefore conclude that

$$E[T_0(n_0)] = \int_0^\infty \{p_0(\infty) - p_0(t)\}\, dt. \tag{2.4.34}$$

For $n_0 = 1$, use of the extinction probability (2.4.23) then easily yields

$$E[T_0(1)] = \begin{cases} -(1/\lambda)\ln[1 - \lambda/\mu] & (\lambda < \mu) \\ -(1/\lambda)\ln[1 - \mu/\lambda] & (\lambda > \mu). \end{cases} \tag{2.4.35}$$

Whilst for $n_0 > 1$, we have the less transparent result that, for $\lambda < \mu$, on taking $i + j \neq 0$

$$E[T_0(n_0)] = \left(\frac{1}{\lambda - \mu}\right) \sum_{i=0}^{n_0} \sum_{j=0}^{\infty} \binom{n_0}{i}\binom{n_0 + j - 1}{j} \frac{(-1)^i (\lambda/\mu)^j}{i + j}, \tag{2.4.36}$$

with a similar expression holding for $\lambda > \mu$. If required, expressions for the rth-order raw moments of $T_0(n_0)$ are straightforward to determine; we simply expand the binomial $\{\mu - \mu e^{-(\lambda-\mu)t}\}^{n_0-1}$ and negative binomial $\{\lambda - \mu e^{-(\lambda-\mu)t}\}^{-n_0-1}$ components of (2.4.30) in powers of $e^{-|\lambda-\mu|t}$ and note that all the resulting terms are gamma integrals.

2.4.3 A simple mixture representation

Given that the birth–death probabilities (2.4.18) take the rather awkward form of a convolution of binomial and negative binomial probabilities, one may reasonably ask whether there is a simpler, albeit approximate, representation which encapsulates the main features of the process. For illustration, suppose we wish to retain the probability of extinction, $p_0(t)$, and the mean, $m(t)$, given by (2.4.23) and (2.4.19), respectively. Then on writing $q_0(t) = p_0(t)$, and choosing the simple Poisson form

$$q_N(t) = \frac{k(t)[\theta(t)]^N e^{-\theta(t)}}{N!} \qquad (N > 0) \tag{2.4.37}$$

for some suitable $\theta(t)$, we see that the normalizing function $k(t)$ must satisfy

$$p_0(t) + k(t)(1 - e^{-\theta(t)}) = 1. \tag{2.4.38}$$

Now equating the two mean values gives

$$k(t)\theta(t) = m(t), \tag{2.4.39}$$

thereby yielding the transcendental form

$$[1 - e^{-\theta(t)}]/\theta(t) = [1 - p_0(t)]/m(t) \qquad \text{with} \qquad k(t) = m(t)/\theta(t). \tag{2.4.40}$$

So although we have a neat 'functional' solution, it is unfortunately not possible to express these approximating probabilities $q_N(t)$ in terms of a simple closed expression involving $p_0(t)$ and $m(t)$. Though if this is not an issue, replacing the Poisson form by (for example) a negative binomial expression would enable the variance $V(t)$ to be accommodated as well.

2.4.4 The backward equations

To generate the Kolmogorov *forward* equations (2.4.3), we took an initial population size $i = n_0$ at time $t = 0$ followed by two time points t and $t + dt$ later on that lie very close together. Conversely, if we take two initial points close together, say 0 and dt, followed by a single time point, $t + dt$, later on, then we generate the Kolmogorov *backward* equations. The general characteristic of the forward equations is that the same initial state is involved throughout, whereas in the corresponding backward equations the same final state enters in all terms. So the forward equations are likely to be more useful if there is a single initial state of particular interest and we require the population size probabilities at some later fixed time t for various final states. Whereas the backward equations are appropriate if there is a single final state of particular interest, and we require the probabilities of reaching this at time t from various initial states.

Denote

$$p_{ki}(t) \equiv p_{ki}(0, t) = \Pr(N(t) = i | N(0) = k). \qquad (2.4.41)$$

Then since (i) $p_{ki}(t + dt)$ is the probability that $N(t+dt) = i$ and either a birth, death or no event takes place in the small initial time interval $(0, dt)$, and (ii) $p_{ki}(dt, t+dt) = p_{ki}(t)$, we have

$$p_{ki}(t + dt) = (k\lambda dt)p_{k+1,i}(t) + (k\mu dt)p_{k-1,i}(t) + [1 - k(\lambda + \mu)dt]p_{ki}(t) + o(dt).$$

Whence letting $dt \to 0$ gives

$$dp_{ki}(t)/dt = \lambda k p_{k+1,i}(t) - (\lambda + \mu) k p_{ki}(t) + \mu k p_{k-1,i}(t). \qquad (2.4.42)$$

Note that unlike the corresponding forward equation (2.4.3), i.e.

$$dp_{ki}(t)/dt = \lambda(i-1)p_{k,i-1}(t) - (\lambda + \mu)i p_{ki}(t) + \mu(i+1)p_{k,i+1}(t), \qquad (2.4.43)$$

the coefficients on the right-hand side of (2.4.42) relate to a constant population size (k).

Given that both equations concern the same set of probabilities, $\{p_{ki}(t)\}$, it is intuitively reasonable to presume that in spite of their different form they will yield the same solution. This is easily demonstrated for the pure death process (i.e. $\lambda = 0$), since we see from (2.2.7) that the forward solution takes the binomial form

$$p_{ki}(t) = \binom{k}{i} e^{-i\mu t}(1 - e^{-\mu t})^{k-i} \qquad (i = 0, 1, \ldots, k), \qquad (2.4.44)$$

and it is a trivial exercise to verify that (2.4.44) is also a solution to the backward equation (2.4.43). To show that this equivalence is a general feature, let us replace the (simple) transition rates λk and μk by the general rates λ_k and μ_k, whence the forward and backward equations become

$$dp_{ki}(t)/dt = \lambda_{i-1}p_{k,i-1}(t) - (\lambda_i + \mu_i)p_{ki}(t) + \mu_{i+1}p_{k,i+1}(t) \quad \text{and} \qquad (2.4.45)$$

$$dp_{ki}(t)/dt = \lambda_k p_{k+1,i}(t) - (\lambda_k + \mu_k)p_{ki}(t) + \mu_k p_{k-1,i}(t), \qquad (2.4.46)$$

respectively. Now define the matrix $\mathbf{P}(t)$ as having $p_{ki}(t)$ as its (k,i)th element, and the transition matrix

$$\mathbf{Q} = \begin{pmatrix} -(\lambda_0 + \mu_0) & \lambda_0 & 0 & 0 & \cdots \\ \mu_1 & -(\lambda_1 + \mu_1) & \lambda_1 & 0 & \cdots \\ 0 & \mu_2 & -(\lambda_2 + \mu_2) & \lambda_2 & \cdots \\ 0 & 0 & \mu_3 & -(\lambda_3 + \mu_3) & \cdots \\ \vdots & \vdots & \vdots & \vdots & \end{pmatrix}. \qquad (2.4.47)$$

Then in this matrix representation, the forward and backward equations take the respective forms

$$\mathbf{P}'(t) = \mathbf{P}(t)\mathbf{Q} \quad \text{and} \quad \mathbf{P}'(t) = \mathbf{Q}\mathbf{P}(t). \qquad (2.4.48)$$

Subject to the initial condition $\mathbf{P}(0) = \mathbf{I}$, where $\mathbf{I}$ is the unit diagonal matrix, both of these equations have the same formal solution

$$\mathbf{P}(t) = \exp(\mathbf{Q}t) \equiv \sum_{r=0}^{\infty} \mathbf{Q}^r \frac{t^r}{r!}. \qquad (2.4.49)$$

We shall return to this general form later on in Section 3.1, but the key point is that this argument shows that for 'reasonable processes' the forward and backward solutions are identical. When $\mathbf{Q}$ is a finite matrix, that is the number of states of the process is finite (as is the case for the pure death process), then the summation (2.4.49) is convergent and $\mathbf{P}(t)$ is the unique solution of both the forward and backward equations. However, when $\mathbf{Q}$ is an infinite matrix the series may not converge and careful treatment of such processes may be necessary.

To see how the algebraic construction of the backward solution differs from the forward version, denote the backward p.g.f.

$$g_k(z;t) \equiv \sum_{k=0}^{\infty} p_{ki}(t) z^k \qquad (2.4.50)$$

and regard $p_{-1,i}(t) \equiv 0$. Then it follows from (2.4.42) with $k = 1$ that

$$dg_1/dt = \lambda g_2 - (\lambda + \mu)g_1 + \mu g_0. \qquad (2.4.51)$$

Since individuals develop independently of each other, $g_2(z;t) = [g_1(z;t)]^2$ (compare with (2.4.11) for the forward p.g.f. $G(z;t)$); whilst $g_0(z;t) \equiv 1$ since an empty population remains empty. So (2.4.51) simplifies to

$$dg_1/dt = \lambda g_1^2 - (\lambda + \mu)g_1 + \mu. \qquad (2.4.52)$$

Note that being an ordinary differential equation, this backward version is simpler in structure than the corresponding partial differential equation (2.4.7) for the forward p.g.f. $G(z;t)$. Using partial fractions, (2.4.52) gives

$$\int \left[\frac{\lambda}{\lambda g_1 - \mu} - \frac{1}{g_1 - 1}\right] dg_1 = -(\lambda - \mu)t + \text{constant},$$

which integrates to

$$(\lambda g_1 - \mu)/(g_1 - 1) = k(z) e^{-(\lambda-\mu)t}$$

where $k(z) = (\lambda z - \mu)/(z - 1)$ since $g_1(z; 0) = z$. Thus

$$g_1(z; t) = \frac{\mu(z - 1) - (\lambda z - \mu) e^{-(\lambda-\mu)t}}{\lambda(z - 1) - (\lambda z - \mu) e^{-(\lambda-\mu)t}}, \qquad (2.4.53)$$

in exact agreement with (2.4.11) when $n_0 = 1$.

Equation (2.4.52) for $g_1(z; t)$ can also be derived without constructing the backward probability equation (2.4.42). For on denoting $\theta = \lambda + \mu$, the time to the first event is distributed as an Exponential(θ) random variable, with the probability that this first event is a birth being λ/θ and a death being μ/θ. So it follows that

$$g_1(z; t) = z \times \Pr\{\text{no event in } (0, t)\}$$

$$+ \int_0^t \Pr\{\text{first event at } \tau\} \Pr\{\text{event is a birth}\} g_1^2(z; t - \tau) \, d\tau$$

$$+ \int_0^t \Pr\{\text{first event at } \tau\} \Pr\{\text{event is a death}\} z^0 \, d\tau.$$

Since a birth at time τ results in two independently developing populations whose total population size after a further time $t - \tau$ has p.g.f. $g_1^2(z; t - \tau)$, whilst a death results in extinction which has p.g.f. z^0, i.e. 1. Thus

$$g_1(z; t) = z e^{-\theta t} + \int_0^t \theta e^{-\theta \tau} [(\lambda/\theta) g_1^2(z; t - \tau) + (\mu/\theta)] \, d\tau,$$

whence putting $t - \tau = s$ and multiplying both sides by $e^{\theta t}$ gives

$$g_1(z; t) e^{\theta t} = z + \int_0^t e^{\theta s} [\lambda g_1^2(z; s) + \mu] \, ds.$$

Differentiating both sides with respect to t then recovers (2.4.52).

2.4.5 Time to a given state

As well as investigating the time to extinction $T_0(n_0)$ (Section 2.4.2), we can also parallel Section 2.3.4 for the simple birth process by considering the time T_a at which state a is first occupied. Though we now have to modify the fundamental result (2.3.33) to

$$\Pr(T_a \leq t) = \sum_{N=a}^{\infty} p_N(t) \quad \text{if } a > n_0 \qquad \text{and} \qquad \Pr(T_a \leq t) = \sum_{N=0}^{a} p_N(t) \quad \text{if } a < n_0,$$

$$(2.4.54)$$

where the $\{p_N(t)\}$ are given by (2.4.14) and (2.4.15). The resulting expressions are fairly opaque for large n_0, but they do simplify considerably for small n_0. For example, when $n_0 = 1$ we see that for $a > 1$

$$\Pr(T_a \le t) = [1 - \alpha(t)][\beta(t)]^{a-1} \tag{2.4.55}$$

which decreases geometrically fast in $\beta(t)$.

In general, the associated p.d.f. $g(t)$ of the random variable T_a will be even more opaque than the c.d.f. (2.4.54), and so we might wish to develop an expression for the p.g.f. (from which we could obtain moments). To achieve this consider a new process in which we make a an 'absorbing' state, i.e. once there we cannot leave. So now $p_{aa}(t) \equiv 1$, and $p_{ka}(t)$ is the distribution function of the first passage time from state k to the absorbing barrier at a. This is a situation in which the backward equations are more convenient to work with than the forward equations. Their solution is relatively simple for small values of a, though the general solution is quite complicated (see Saaty, 1961).

2.4.6 Reverse transition probabilities

Since the probabilities $\{p_N(t)\}$ in (2.4.18) contain $N+1$ terms, the associated bridge probabilities $\{p_{inj}(s;t)\}$ given by (2.2.29) will not have a simple form, such as (2.2.30) for the pure death process. Though if $i = 1$ use of (2.4.15) does yield

$$p_{1nj}(s;t) = \left[\frac{(1-\alpha(s))(1-\beta(s))[\beta(s)]^{n-1}}{(1-\alpha(t))(1-\beta(t))[\beta(t)]^{n-1}}\right] p_{nj}(t-s), \tag{2.4.56}$$

i.e. a geometric weighting of the 'second-stage' probabilities $\{p_{nj}(t-s)\}$. However, the reverse transition probabilities developed in Section 2.2.2 for the death process do carry through equally well for the birth-death process, though it is now easier to work via the p.g.f. $G(z;t)$.

As before, let $\{p_i(0)\}$ take the geometric form (2.2.25), and write the p.g.f. (2.4.11) as $G_i(z;t)$ for $n_0 = i$. Then

$$\sum_{i=0}^{\infty} p_i(0) G_i(z;t) = (1-\theta) \sum_{i=0}^{\infty} \theta^i [G_1(z;t)]^i = \frac{1-\theta}{1-\theta G_1(z;t)}.$$

So

$$\sum_{i=0}^{\infty} p_i(0) p_{ij}(t) = (1-\theta) \times \text{coeff. } z^j \text{ in } [1 - \theta G_1(z;t)]^{-1},$$

which reduces to

$$(1-\theta)\{e^{-(\lambda-\mu)t} + O(1-\theta)\}.$$

Thus (2.2.24) becomes

$$\overleftarrow{p}_{ij}(t) = \lim_{\theta \to 1} \left\{ \frac{(1-\theta)\theta^i p_{ij}(t)}{(1-\theta)e^{-(\lambda-\mu)t} + o(1-\theta)} \right\} = e^{(\lambda-\mu)t} p_{ij}(t), \qquad (2.4.57)$$

which takes the same form as the pure death result (2.2.27).

2.4.7 Simulating the simple birth–death process

Whereas simulating the pure death (Figure 2.6) and pure birth (Figure 2.8) processes just requires the construction of the time to the next death and birth, respectively, we now also need to determine what the next event is. In the small time interval $(t, t+dt)$ the probabilities of a birth or death, given that the population is of size N at time t, are $\lambda N dt$ and $\mu N dt$, respectively. So as the probability of some event (i.e. either a birth or a death) occurring in $(t, t+dt)$ is $(\lambda+\mu)N dt$, we have

$$\Pr(\text{birth} \mid \text{event in } (t, t+dt)) = \frac{\lambda N dt}{(\lambda+\mu)N dt} = \frac{\lambda}{\lambda+\mu}. \qquad (2.4.58)$$

Similarly,

$$\Pr(\text{death} \mid \text{event in } (t, t+dt)) = \frac{\mu}{\lambda+\mu}. \qquad (2.4.59)$$

Whilst the time τ to the next event is exponentially distributed with parameter $(\lambda+\mu)N$. So for a sequence $\{U\}$ of independent uniform $U(0,1)$ random variables, $\tau = -\ln(U)/[(\lambda+\mu)N]$; and this event is a birth if, for the next U in the sequence, $\lambda/(\lambda+\mu) \leq U$, else it is a death. This leads to the following simulation procedure.

Exact simple birth–death algorithm (A2.1)
    (i) input λ and μ
    (ii) set $N = n_0$ and $t = 0$
    (iii) print N and t
    (iv) select $U \sim U(0,1)$
    (v) update t to $t - \ln(U)/[(\lambda+\mu)N]$
    (vi) select $U \sim U(0,1)$
    (vii) if $U \leq \lambda/(\lambda+\mu)$ then $N \to N+1$ else $N \to N-1$
    (viii) if $N = 0$ stop
    (ix) return to (iii)

For example, suppose we take $\lambda = 2$, $\mu = 1$ and $n_0 = 1$. Then $\tau = -\ln(U)/(3N)$ and $\lambda/(\lambda+\mu) = 1/3$. Whence for $\{U\} = \{0.6237, 0.3281, \ldots\}$ the first three cycles generate the event–time sequence:

step	n	t	U	τ	t(new)	U	event	n(new)
1	1	0.0000	0.6237	0.1574	0.1574	$0.3281 \leq 1/3$	birth	2
2	2	0.1574	0.4735	0.1508	0.3082	$0.0642 \leq 1/3$	birth	3
3	3	0.3082	0.3967	0.1207	0.4109	$0.8028 > 1/3$	death	2
4	2	0.4109	etc.					

Figure 2.11 shows five typical realizations for $\mu = 1$ and $n_0 = 3$ with $\lambda = 1.1$, 1 and 0.9. Taking $\lambda = 1.1$ first, we see from (2.4.19) that at time $t = 30$ the mean population size $m(t) = n_0 e^{(\lambda-\mu)t} = 60.26$, whilst from (2.4.20) the corresponding standard deviation $\sqrt{\{V(t)\}} = \sqrt{\{n_0[(\lambda+\mu)/(\lambda-\mu)]e^{(\lambda-\mu)t}(e^{(\lambda-\mu)t}-1)\}} = 155.40$. Since a crude 95% confidence interval of $m(30) \pm 2\sqrt{\{V(30)\}}$ covers the range from 0 to over 370, there is clearly a large variety of possible trajectory types. Figure 2.11a shows one realization that remains fairly close to its initial value during $0 \le t \le 30$, and another that becomes extinct by $t = 9.75$. Note from (2.4.24) that because $p_0(\infty) = (\mu/\lambda)^{n_0} = 0.7513$, only a quarter of all realizations escape the ultimate fate of extinction. Figure 2.11b shows one such trajectory; by $t = 30$ the population size N has reached 243 from where the probability of extinction has reduced to $(1/1.1)^{243} = 8.75 \times 10^{-11}$. When $\lambda > \mu$ and a realization has moved sufficiently far away from zero, the probability of extinction thereafter is clearly negligible. In contrast, if $\lambda = \mu$ then the mean population size $m(t)$ remains fixed at n_0 in spite of the certainty of ultimate extinction, i.e. $\Pr(N(\infty) = 0) = 1$. This apparent paradox is studied in some depth later on in Section 4.1; we discover that $N(t)$ may become indefinitely large before it eventually sinks back down to extinction. Figure 2.11c shows a realization for $\lambda = \mu = 1$ which exhibits substantial positive growth before the onset of extinction just before time $t = 40$; together with an example of early decay to extinction for $\lambda = 0.9 < \mu = 1$. In this latter case result (2.4.28) provides a rough value for the mean time to extinction of $T_0(3) \simeq 10 \ln(3) = 11$; if required, the exact value may be computed from (2.4.36).

This procedure extends directly to multi-species situations in which a large number of different types of event may occur. Moreover, although it was occasionally used in the 1960s to simulate simple stochastic biological models, a modified version made a considerable impact on the relatively complex stochastic formulation of chemical kinetics in the 1970s (e.g. Gillespie, 1977), and this is sometimes referred to as "Gillespie's method". For instead of computing the time to the next event, we can select a time increment of size dt which is sufficiently small to ensure that for all practical purposes the probability of a multiple event during the time interval $(t, t+dt)$ is negligible. Whence on breaking the interval $(0,t)$ into a large number of neighbouring small intervals of length dt, we may dispense with determining the inter-event times, and hence all the U's associated with them, and simply modify rule (2.4.58)–(2.4.59) for determining the next event to: birth if $U \le N\lambda dt$, death if $N\lambda dt < U \le N(\lambda + \mu)dt$, else no change. On rearranging these 'if'-checks in order to improve efficiency, algorithm A2.1 becomes:

Approximate simple birth–death algorithm (A2.2)
  (i) input λ, μ and dt
  (ii) set $N = n_0$ and $t = 0$
  (iii) print N and t
  (iv) if $N = 0$ stop
  (v) select $U \sim U(0,1)$
  (vi) if $U \le 1 - N(\lambda + \mu)dt$ then $N \to N$

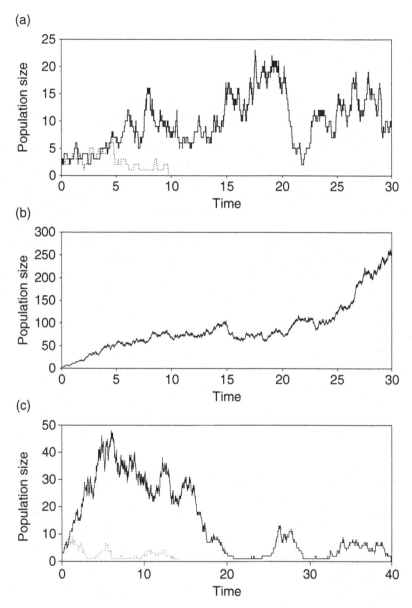

Figure 2.11 Five realizations of the simple birth–death process for $\mu = 1$ and $n_0 = 3$ showing: (a) slow growth (—) and early extinction ($\cdots$) for $\lambda = 1.1$; (b) fast growth for $\lambda = 1.1$; (c) initial population expansion (—) for $\lambda = 1$, and early extinction ($\cdots$) for $\lambda = 0.9$.

else if $U \leq 1 - N\mu dt$ then $N \to N+1$
else $N \to N-1$
(vii) update t to $t + dt$
(viii) return to (iii)

Variations on this theme are discussed later on in Sections 9.3.3 and 9.3.4.

Note that even if N remains fairly constant, this procedure is still highly wasteful of U's since step (vi) will often result in no change. So given that in the current context this algorithm is an approximation to a simple exact procedure, there is little merit in adopting it here. Also, if N exhibits considerable variation, then choosing dt sufficiently small to ensure that $\max\{N(\lambda + \mu)dt\} \ll 1$ means that for $N \ll \max(N)$ the vast proportion of simulated events will be 'no change'. Though this problem can be eased by changing dt with N, for example by placing say $dt = 0.01/[N(\lambda + \mu)]$. Where the procedure does merit serious consideration is when the population comprises a large number of sub-populations linked, for example, by migrating individuals (see Chapter 9). Moreover, if the process is too large to be simulated on a single computer, then by linking each individual sub-population to a processor on a supercomputer, using constant time-steps enables the individual processors to maintain synchrony with each other (Smith and Renshaw, 1994). Whereas employing the exact algorithm A2.1 on each sub-processor would involve far too much time being spent in 'internal housekeeping'.

An alternative approach (seldom used) may be developed if uniform random numbers are either in short supply (unlikely in an office environment, but quite possible in a field situation), or computationally expensive to generate. For whereas in algorithm A2.1 two $U(0,1)$ random variables are required to generate each event–time pair (and potentially many more to generate each genuine event change in algorithm A2.2), we can save half of the required U's by dispensing with the need to determine the type of the next event. To do this, we associate one uniform random variable with the time to the next birth and another to the next death. Whichever leads to the shorter time then determines the next event type. We therefore generate successive times between births and deaths through two sequences of i.i.d. $U(0,1)$ random variables, $\{U_i\}$ and $\{V_i\}$. To demonstrate this procedure, let the population be of initial size $N = n_0$ at time $t_0 = 0$. Then the times to the first birth and death are $\theta_1 = -\ln(U_1)/\lambda N$ and $\phi_1 = -\ln(V_1)/\mu N$, respectively. So if $\theta_1 < \phi_1$ then the first event is a birth at time $t_1 = t_0 + \theta_1$, at which point the population changes in size from N to $N+1$. Thus the time to the second birth is $\theta_2 = -\ln(U_2)/\lambda(N+1)$; whilst the time to the first death must now be modified to take account of the change in population size, i.e. $\phi_1' = t_1 + (\phi_1 - t_1)[\mu N/\mu(N+1)]$. Suppose that the second event is also a birth, at time $t_2 = t_1 + \theta_2 > \phi_1'$. Then the time to the third birth is $\theta_3 = -\ln(U_3)/\lambda(N+2)$; whilst the time of the first death must be further modified to $\phi_1'' = t_2 + (\phi_1' - t_2)[\mu(N+1)/\mu(N+2)]$. So if the third event is a death, then this must therefore occur at time $t_3 = \phi_1'' < t_2 + \theta_3$, and we must now modify the time to the third birth, etc. In tabular form we therefore have the construction:

i	birth	death	event time	pop. size	modified θ or ϕ
0			$t_0 = 0$	N	
1	θ_1	$< \phi_1$	$t_1 = t_0 + \theta_1$	$N+1$	$\phi_1' = t_1 + (\phi_1 - t_1)[N/(N+1)]$
2	$t_1 + \theta_2$	$< \phi_1'$	$t_2 = t_1 + \theta_2$	$N+2$	$\phi_1'' = t_2 + (\phi_1' - t_2)$ $\times[(N+1)/(N+2)]$
3	$t_2 + \theta_3$	$> \phi_1''$	$t_3 = t_0 + \phi_1''$	$N+1$	$\theta_3' = (t_3 - t_2) + [\theta_3 - (t_3 - t_2)]$ $\times[(N+2)/(N+1)]$
4	$t_2 + \theta_3'$	$< t_3 + \phi_2$	$t_4 = t_2 + \theta_3'$	$N+2$	$\phi_2' = (t_4 - t_3) + [\phi_2 - (t_4 - t_3)]$ $\times[(N+1)/(N+2)]$
5	etc.	etc.			

Although the running modifications to $\{\theta_i\}$ and $\{\phi_i\}$ mean that this procedure is less transparent than A2.1 and A2.2, it does, nevertheless, provide an interesting way of simulating successive events. Indeed, it may even be computationally faster should the specific algorithm employed to generate the underlying $U(0,1)$ random variables be slow relative to the speed of employing 'if'-checks and modifying the inter-birth and inter-death times.

2.4.8 The dominant leader process

We have seen that although the deterministic and expected population sizes, namely (2.4.2) and (2.4.19), both have the same form $N(t) = m(t) = n_0 e^{(\lambda - \mu)t}$, the deterministic and stochastic processes are themselves quite different in nature. For example, if $\lambda = \mu$ then whilst $N(t) \equiv n_0$ for all $t \geq 0$ it follows from (2.4.27) that $p_0(\infty) = 1$ and so ultimate extinction is certain. So the deterministic population size $N(t)$ remains permanently fixed at n_0, whilst stochastic fluctuations about n_0 inevitably lead to extinction no matter how large n_0 is. We shall see later on that even $N(t)$ and $m(t)$ generally no longer agree when the linear birth and death rates, λi and μi, are replaced with nonlinear forms.

The conflict that these differences generate is exacerbated by the polarizing standpoints often taken by deterministic and stochastic analysts. For the mathematical techniques which underlie these two approaches, namely the solution of ordinary and partial differential equations for the former, and the development of stochastic approximation techniques and existence and uniqueness properties for the latter, belong to quite different research fields, and any resistance to master both is perfectly understandable. Simulation provides a way forward here, since (at least in principle) it is relatively easy to perform, and so enables direct comparison to be made between the deterministic and stochastic viewpoints. If there appears to be little structural difference in the resulting simulated trajectories, then on purely pragmatic grounds the deterministic approach is optimal. However, if substantial differences do exist between them, then the stochastic approach should be undertaken; for deterministic results in such circumstances are at best questionable, and at worst worthless.

To expose the huge differences that can occur, let us conclude this section by considering the rather extreme 'dominant leader' process. For simplicity, suppose we replace continuous time $t \geq 0$ by the discrete time steps $t = 0, 1, 2, \ldots$ (considered in detail in Chapters 4 and 5). At each time point an individual either gives birth to twins with probability p, or does not with probability $1 - p$, and then dies. However, unlike the simple birth–death process, individuals do not behave independently of each other. For there is a dominant leader who nominates one of her twins as the next leader before she dies. All remaining members of the population copy the dominant leader, namely producing twins only if she does. So the stochastic population doubles at each time step with probability p. With $N(0) = n_0$, we therefore have

$$\Pr(N(t) = 2^t n_0) = p^t \quad \text{with} \quad \Pr(N(t) = 0) = 1 - p^t. \tag{2.4.60}$$

Thus the expected population size at time t is given by $m(t) = n_0(2p)^t \to \infty$ for $p > 1/2$, even though $p_0(t) \to 1$ irrespective of how large n_0 is. This effect is clearly far more severe than that of the simple birth–death scenario with $\lambda > \mu$, for which $p_0(\infty) \to 0$ with increasing n_0. To parallel the development of the deterministic equation (2.4.1) we take a *purely individual based* approach. For as far as each individual is concerned, it splits into two at rate p and dies at rate $1 - p$, and so becomes $2p$ individuals at the next time point. Thus $N(t) = n_0(2p)^t = m(t)$, though unlike simple birth–death we can no longer argue that the two processes are compatible when n_0 becomes large.

This example illustrates the danger of taking a purely deterministic stance. For it leads to the conclusion that the population size increases geometrically fast, which is in direct contradiction to the population becoming extinct geometrically fast. Whilst a stochastic analysis does indeed yield the same geometric growth for the *expected* population size, construction of the extinction probability, $p_0(t)$, clarifies what is actually happening. Note that even without performing this analysis, simulating the deterministic and stochastic processes would highlight the gross difference which exists between them.

2.5 Simple immigration–birth–death process

For the simple birth–death process we have seen that if $\lambda \leq \mu$ then $p_0(\infty) = 1$ and so the population is doomed to extinction, whilst if $\lambda > \mu$ then the population either becomes extinct with probability $(\mu/\lambda)^{n_0}$ or else it becomes infinitely large with probability $1 - (\mu/\lambda)^{n_0}$. To obtain a process that persists indefinitely yet remains finite, suppose we take $\lambda < \mu$ and allow new individuals to enter the population at rate α independently of all other individuals. Though the process with $\lambda > \mu$ is indeed worthy of consideration, that with $\lambda < \mu$ is fundamentally more interesting. For as the net growth rate $\alpha + (\lambda - \mu)N$ is negative for all $N > \alpha/(\mu - \lambda)$, we may anticipate the existence of an *equilibrium* solution in which, as $t \to \infty$, the probabilities $\{p_N(t)\}$ approach a steady state which is independent of t.

The deterministic analysis points to such a possibility, since

$$dN(t)/dt = (\lambda - \mu)N(t) + \alpha \tag{2.5.1}$$

leads to

$$N(t) = [N(0) + \alpha/(\lambda - \mu)]e^{(\lambda-\mu)t} - \alpha/(\lambda - \mu), \qquad (2.5.2)$$

namely an exponential approach to the equilibrium value $\alpha/(\mu - \lambda)$ (provided $\lambda < \mu$).

2.5.1 Stochastic model ($\lambda = 0$)

For algebraic simplicity let us first consider the case $\lambda = 0$. This pure immigration–death process then relates to a population of individuals which randomly arrive in a region of space and then either die or leave; early examples include the movement of colloidal particles in suspension (Chandrasekhar, 1943) and spermatozoa (Rothschild, 1953). Since each individual currently in the system leaves it at constant rate μ, when i individuals are present the total death (i.e. departure) rate is $i\mu$. Note, however, that if individuals interact, say because of crowding, then the basic underlying assumption of independence is violated and the model no longer holds. We shall consider this issue in the next chapter

The birth–death forward equation (2.4.3) is easily extended to give

$$dp_i(t)/dt = [\alpha + \lambda(i-1)]p_{i-1}(t) - [\alpha + (\lambda + \mu)i]p_i(t) + \mu(i+1)p_{i+1}(t). \qquad (2.5.3)$$

Whence paralleling the construction of the p.d.e. (2.4.7) for the birth–death p.g.f. $G(z;t)$ leads to

$$\frac{\partial G(z;t)}{\partial t} + \mu(z-1)\frac{\partial G(z;t)}{\partial z} = \alpha(z-1)G(z;t); \qquad (2.5.4)$$

note the additional term $\alpha(z-1)G(z;t)$ on the right-hand side of the equation. On writing this as $\alpha(z-1)G\,\partial G/\partial G$ we then have the auxiliary equations

$$\frac{dt}{1} = \frac{dz}{\mu(z-1)} = \frac{dG}{\alpha(z-1)G}. \qquad (2.5.5)$$

Two independent solutions are

$$(z-1)e^{-\mu t} = a \quad \text{and} \quad Ge^{-\alpha z/\mu} = b,$$

for arbitrary constants a and b, so the general solution is

$$G(z;t) = e^{\alpha z/\mu}f[(z-1)e^{-\mu t}] \qquad (2.5.6)$$

for some arbitrary function f. On inserting the initial condition $N(0) = n_0$ we have

$$z^{n_0} = e^{\alpha z/\mu}f(z-1).$$

Whence replacing z by $1+\xi$ gives

$$f(\xi) = (1+\xi)^{n_0}e^{-\alpha(1+\xi)/\mu}.$$

Finally, we place $\xi = (z-1)e^{-\mu t}$, which yields

$$G(z;t) = [1 + (z-1)e^{-\mu t}]^{n_0}\exp\{(\alpha/\mu)(z-1)(1-e^{-\mu t})\}. \qquad (2.5.7)$$

88 Simple Markov population processes

The probability $p_i(t)$ is easily derived from this p.g.f. by forming the coefficient of z^i from the combined binomial and exponential expansions. Moreover, independently of n_0,

$$\lim_{t \to \infty} G(z;t) = \exp\{(\alpha/\mu)(z-1)\}, \tag{2.5.8}$$

which is the p.g.f. of a Poisson process with mean α/μ (see expression (2.1.19) and sequel). Thus the equilibrium probabilities

$$p_i(\infty) = \frac{(\alpha/\mu)^i e^{-\alpha/\mu}}{i!}. \tag{2.5.9}$$

Since individuals develop independently from each other, the Poisson mean α/μ is also the deterministic equilibrium value. For when $\lambda = 0$ the deterministic solution (2.5.2) approaches α/μ exponentially fast at rate μ.

We shall discuss the general structure of equilibrium solutions in Chapter 3. They are particularly important since we are often more interested in the long-term, i.e. 'persistent' behaviour of a process, than in its initial, i.e. 'transient', behaviour.

2.5.2 Stochastic model ($\lambda \geq 0$)

If we now reintroduce birth by letting $\lambda \geq 0$, then on comparing the p.g.f. equations (2.4.7) and (2.5.4) it is apparent that $G(z;t)$ satisfies the equation

$$\frac{\partial G(z;t)}{\partial t} - (\lambda z - \mu)(z-1)\frac{\partial G(z;t)}{\partial z} = \alpha(z-1)G(z;t). \tag{2.5.10}$$

This results in a slight modification to the auxiliary equations (2.5.5), namely

$$\frac{dt}{1} = \frac{-dz}{(\lambda z - \mu)(z-1)} = \frac{dG}{\alpha(z-1)G}, \tag{2.5.11}$$

for which two independent solutions are

$$\left(\frac{z-1}{\lambda z - \mu}\right) e^{(\lambda - \mu)t} = a \quad \text{(as in (2.4.9))} \quad \text{and} \quad G(\lambda z - \mu)^{\alpha/\lambda} = b.$$

On paralleling the development of (2.5.7) we therefore have (e.g. Bailey, 1964)

$$G(z;t) = [\Theta(z;t)]^{n_0} \Phi(z;t) \tag{2.5.12}$$

where

$$\Theta(z;t) = \frac{(\lambda z - \mu) - \mu(z-1)e^{(\lambda-\mu)t}}{(\lambda z - \mu) - \lambda(z-1)e^{(\lambda-\mu)t}} \quad \text{and} \quad \Phi(z;t) = \left(\frac{(\lambda z - \mu) - \lambda(z-1)e^{(\lambda-\mu)t}}{\lambda - \mu}\right)^{-\alpha/\lambda}. \tag{2.5.13}$$

In particular, the probability of extinction

$$p_0(t) = G(0;t) = \left[\frac{\mu e^{(\lambda-\mu)t} - \mu}{\lambda e^{(\lambda-\mu)t} - \mu}\right]^{n_0} \left[\frac{\lambda e^{(\lambda-\mu)t} - \mu}{\lambda - \mu}\right]^{-\alpha/\mu}. \tag{2.5.14}$$

Direct comparison with (2.4.11) shows that the associated probabilities $\{p_i(t)\}$ are formed by the convolution of the birth–death process which operates on the initial n_0 individuals (having p.g.f. $[\Theta(z;t)]^{n_0}$), and that which operates on each new arriving immigrant. The p.g.f. of the latter process, $\Phi(z;t)$, corresponds to the negative binomial probabilities

$$p_i(t) = \binom{i-1+(\alpha/\lambda)}{i} \left(\frac{\lambda(1-s)}{\mu-\lambda s}\right)^i \left(\frac{\mu-\lambda}{\mu-\lambda s}\right)^{\alpha/\lambda} \quad (i=0,1,2,\ldots) \quad (2.5.15)$$

where $s = e^{(\lambda-\mu)t}$. Thus if the population is initially empty, i.e. $n_0 = 0$, then $G(z;t) \equiv \Phi(z;t)$ and the probabilities (2.5.15) relate to the full process. On writing $r = \alpha/\lambda$, $p = (\mu-\lambda)/(\mu-\lambda s)$ and $q = 1-p = \lambda(1-s)/(\mu-\lambda s)$, use of standard negative binomial results shows that the mean and variance of population size are respectively

$$rq/p = \alpha(1-s)/(\mu-\lambda) \quad \text{and} \quad rq/p^2 = \alpha(1-s)(\mu-\lambda s)/(\mu-\lambda)^2. \quad (2.5.16)$$

Following on from the generalizations developed in Section 2.3.5 for the birth process, an obvious further extension is to replace the population growth rate λN by the general function $\lambda(N,t)$. Consider, for example, the case of the classic Pólya process with $\lambda(N,t) = \lambda(1+\gamma N)/(1+\gamma\lambda t)$. For small t, $\lambda(N,t) \simeq \lambda + (\lambda\gamma)N$ which corresponds to immigration at rate λ and simple birth at rate $\lambda\gamma$. Whilst when t becomes large, these rates switch into the time-dependent forms $1/\gamma t$ and $1/t$, respectively. As the general birth equations (2.3.6) now take the form

$$dp_i(t)/dt = \lambda[-(1+\gamma i)p_i(t) + (1+\gamma(i-1))p_{i-1}(t)]/(1+\gamma\lambda t), \quad (2.5.17)$$

let us write

$$\tau(t) = \int_0^t \frac{du}{1+\lambda\gamma u} = (1/\lambda\gamma)[\ln(1+\gamma\lambda t)]. \quad (2.5.18)$$

Then $d\tau/dt = (1+\gamma\lambda t)$, whence (2.5.17) becomes

$$dp_i(\tau)/d\tau = -\lambda(1+\gamma i)p_i(\tau) + \lambda(1+\gamma(i-1))p_{i-1}(\tau), \quad (2.5.19)$$

which, in this new τ-scale, corresponds to immigration at rate λ and birth at rate $\gamma\lambda$. So results (2.5.14) and (2.5.15) apply directly.

The simple nature of the negative binomial solution (2.5.15) disappears when $n_0 > 0$, since the p.g.f. $G(z;t)$ is then a product of a binomial and two negative binomial expansions. So although the derivation of $\{p_i(t)\}$ is still technically trivial, its structural form conveys little visual information. Yet this is still an extremely basic stochastic process; in general, exact expressions for the $\{p_i(t)\}$ will usually be totally opaque. The 'best solution' then involves deriving approximate probability expressions (as discussed later on in Section 3.7); and differentiating or expanding the p.g.f. to obtain moments of the process (as in Section 2.1.3).

2.5.3 Equilibrium probabilities

Such algebraic difficulties disappear if we examine the equilibrium structure of the process. For denote the equilibrium probabilities by

$$\pi_i = \lim_{t \to \infty} p_i(t). \qquad (2.5.20)$$

Then we see from (2.5.13) that, for $\lambda < \mu$,

$$\Theta(z; \infty) = 1 \quad \text{and} \quad \Phi(z; \infty) = [(\mu - \lambda z)/(\mu - \lambda)]^{-\alpha/\lambda}, \qquad (2.5.21)$$

whence (2.5.12) yields the negative binomial probabilities

$$\pi_i = \binom{i - 1 + \alpha/\lambda}{i} \left(\frac{\lambda}{\mu}\right)^i \left(1 - \frac{\lambda}{\mu}\right)^{\alpha/\lambda} \quad (i = 0, 1, 2, \ldots). \qquad (2.5.22)$$

Note that this result may be derived directly without recourse to the time-dependent p.g.f. $G(z; t)$. For on letting $t \to \infty$ in the Kolmogorov forward equations (2.5.3) we have

$$0 = [\alpha + \lambda(i - 1)]\pi_{i-1} - [\alpha + (\lambda + \mu)i]\pi_i + \mu(i + 1)\pi_{i+1} \qquad (2.5.23)$$

since $dp_i(\infty)/dt = d\pi_i/dt = 0$. To solve this difference equation we first write it in the balanced form

$$\alpha\pi_0 - \mu\pi_1 = 0 \qquad (i = 0) \qquad (2.5.24)$$

$$[\alpha + \lambda(i - 1)]\pi_{i-1} - \mu i \pi_i = (\alpha + \lambda i)\pi_i - \mu(i + 1)\pi_{i+1} \quad (i = 1, 2, \ldots). \qquad (2.5.25)$$

Now the two sides of equation (2.5.25) are identical, apart from $i - 1$ on the left being replaced by i on the right, and so

$$[\alpha + \lambda(i - 1)]\pi_{i-1} - \mu i \pi_i = \text{constant}$$

(independent of $i = 1, 2, \ldots$). Since (2.5.24) gives

$$\alpha\pi_0 - \mu\pi_1 = \text{constant} - 0,$$

it follows that

$$\mu i \pi_i = [\alpha + \lambda(i - 1)]\pi_{i-1} \quad (i = 1, 2, \ldots). \qquad (2.5.26)$$

Whence solving equation (2.5.26) successively for $i = 1, 2, \ldots$ then gives

$$\pi_i = \pi_0 (\lambda/\mu)^i [(i - 1) + \alpha/\lambda][(i - 2) + \alpha/\lambda] \ldots [\alpha/\lambda]/(i!), \qquad (2.5.27)$$

which is identical in form to (2.5.22). The multiplicative constant $\pi_0 = (1 - \lambda/\mu)^{\alpha/\lambda}$ is determined from

$$\sum_{i=0}^{\infty} \pi_i = 1. \qquad (2.5.28)$$

We see via (2.5.16) that the equilibrium mean and variance are $\alpha/(\mu - \lambda)$ and $\alpha\mu/(\mu - \lambda)^2$, respectively.

Equation (2.5.26) may also be written down directly on intuitive grounds, since in equilibrium the 'probability flow' between neighbouring states must be 'in balance'. That is, the probability of having population size i followed by a death, i.e. $(\mu i)\pi_i$, is the same as the probability of having $i-1$ followed by an immigration or birth, i.e. $[\alpha + \lambda(i-1)]\pi_{i-1}$. This simple concept of *reversibility* is particularly useful if we wish to make a quick study of different transition regimes. For example, if immigration occurs only when the population is empty, then we may replace (2.5.26) by the recursive equations

$$\alpha\pi_0 = \mu\pi_1 \qquad (2.5.29)$$
$$\mu i \pi_i = \lambda(i-1)\pi_{i-1} \qquad (i = 1, 2, \ldots),$$

and immediately obtain

$$\pi_i = (\lambda/\mu)^{i-1}(\alpha\pi_0/\mu i) \quad (i > 0) \qquad \text{where} \qquad \pi_0 = \{1 - (\alpha/\lambda)\ln[1 - (\lambda/\mu)]\}^{-1}. \qquad (2.5.30)$$

2.5.4 Perfect simulation

Although in theory equilibrium is only truly attained at $t = \infty$, in practice we may often assume that the process has *burnt-in* by some finite time, say T_{eq}, in that by this time the effect of the initial population size has effectively vanished. Since both the deterministic population size (2.5.2) and the stochastic p.g.f. (2.5.12) involve t only through $e^{(\lambda-\mu)t}$, we may reasonably place

$$e^{(\lambda-\mu)T_{eq}} = \epsilon, \qquad \text{i.e.} \quad T_{eq} = \ln(\epsilon)/(\mu - \lambda),$$

for some appropriately small choice of $\epsilon > 0$ (e.g. $\epsilon = 10^{-6}$). The process should then be in both deterministic and stochastic equilibrium for all $t \geq T_{eq}$.

Nevertheless, start positions, $N(0)$, far away from the deterministic equilibrium value $\alpha/(\mu - \lambda)$ (for $\lambda < \mu$) will clearly require a longer burn-in period than ones lying nearby, and we see from (2.5.2) that $|N(t) - \alpha/(\mu - \lambda)| < \epsilon$ provided

$$|N(0) - \alpha/(\mu - \lambda)|e^{(\lambda-\mu)t} < \epsilon,$$

i.e. provided

$$t \geq \tilde{T}_{eq} = (\mu - \lambda)^{-1}[-\ln(\epsilon) + \ln|N(0) - \alpha/(\mu - \lambda)|]. \qquad (2.5.31)$$

Thus from a deterministic viewpoint, $\tilde{T}_{eq}$ varies logarithmically both with ϵ and $N(0)$. For example, if $\alpha = 1$, $\lambda = 0.15$ and $\mu = 0.25$, then $N(\infty) = 10$; whence with $\epsilon = 10^{-6}$ we require $t \geq 138.16 + 10\ln|N_0 - 10|$. However, whilst such ball-park estimates are useful in providing an order of magnitude for the burn-in period, as more sensitive indicators they fail on two counts. First, for a given simulated run they may overestimate the burn-in period; second, irrespective of how large we take the burn-in period to be, there is no absolute guarantee that equilibrium will have been achieved.

92 Simple Markov population processes

Both of these objections can be overcome by using the following pragmatic (i.e. non-rigorous) version of the *perfect simulation* approach initiated by Propp and Wilson (1996).

(a) Select a start seed S for the pseudo-random number generator.
(b) Choose a maximum initial population size L.
(c) Using the same seed S simulate the process for $N(0) = 0$ and L, and plot the respective population sizes $N_i^{(0)}$ and $N_i^{(L)}$ against i for events $i = 0, 1, \ldots$. Record the value $i = \tilde{i}$ where the two trajectories converge.

Once two event trajectories coalesce they must clearly remain inseparable thereafter, since they are governed by the same pseudo-random number sequence. Thus for $i \geq \tilde{i}$ the process must be in true equilibrium for all start values $N(0) = 0, 1, \ldots, L$, since $\{N_i\}$ is then independent of the initial population size $N(0) = 0, 1, \ldots, L$. Whilst for $i < \tilde{i}$, $N_i^{(0)}$ and $N_i^{(L)}$ form the lower and upper envelope of the process, as trajectories cannot cross.

Figure 2.12a shows the development of a birth–death–immigration process with $\alpha = 1$, $\lambda = 0.15$ and $\mu = 0.25$ over the first 500 events, for $N(0) = 0, 10, \ldots, 50$. The fact that the probability that the next event is a decrease is $\mu N/(\alpha + \mu N + \lambda N) \sim \mu/(\mu + \lambda)$ for large N, explains why the size trajectories look remarkably similar. Note that placing $s = 0$ in (2.5.16) yields the equilibrium mean $\alpha/(\mu - \lambda) = 10$ and variance $\alpha\mu/(\mu - \lambda)^2 = 25$, so $N(0) = 50$ corresponds to eight standard deviations above the mean. To form the event–time trajectories (Figure 2.12b) we simply note the times taken for the six trajectories to reach count 314, where full coalescence occurs. Denoting these by $s^{(0)}, s^{(10)}, \ldots, s^{(50)}$, and observing that $s^{(0)}$ is the largest, we then delay the start of the remaining runs to times $s^{(0)} - s^{(10)}, \ldots, s^{(0)} - s^{(50)}$. Here full coalescence (on the $N(0) = 0$ trajectory time-scale) occurs at time $t = 89.28$. Moreover, for $N(0) = 0, 10, \ldots, 100$ (i.e. up to a massive 18 standard deviations above the mean), the number of events to neighbouring coalescence, together with the time, trajectory time-delay, and times to coalescence with all lower trajectories are given by:

$N(0)$	events to coalescence	with	at time	initial time delay	time to 'convergence'
10	208	0	53.61	18.01	35.60
20	173	10	42.55	23.42	30.19
30	215	20	55.02	30.96	24.06
40	223	30	57.43	37.29	20.14
50	314	40	89.28	56.21	33.07
60	298	50	85.62	63.41	25.87
70	363	60	99.38	71.39	27.99
80	353	70	97.19	75.33	21.86
90	413	80	109.28	82.14	27.14
100	450	90	119.71	86.85	32.86

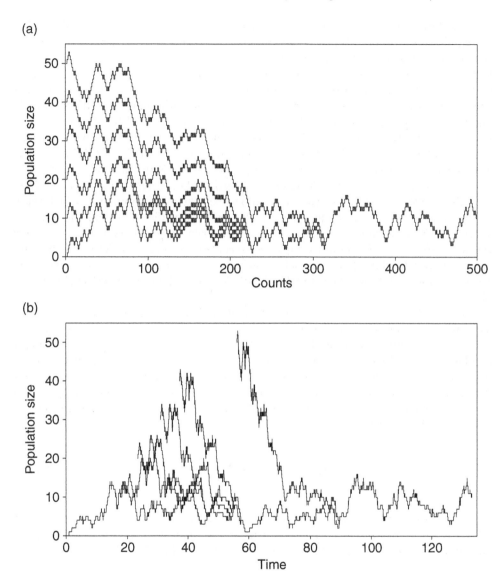

Figure 2.12 Perfect simulation of a birth–death–immigration process with $\alpha = 1$, $\lambda = 0.15$, $\mu = 0.25$ and $N(0) = 0, 10, \ldots, 50$, showing convergence of the six population size trajectories against (a) counts and (b) time.

Thus here equilibrium has effectively been attained by time $t = 120$, which is slightly less than the deterministic value $T_{eq} = 138$ for $\epsilon = 10^{-6}$. Indeed, the coalescent time 120 corresponds to taking $\epsilon = 0.0005$, and it would be interesting to see whether this ϵ-value represents a general ball-park figure for inferring stochastic equilibrium from deterministic population sizes.

So perfect simulation enables us to generate realizations which lie wholly within the equilibrium regime. Ideally, we should look *backwards* in time, usually by starting the process at some time $-T$ and observing whether coalescence has occurred by time zero; if not, we start at time $-2T$, then $-4T$, etc., until coalescence at zero occurs. The state occupied is then a 'perfectly simulated' value, whilst the realization from then on is in genuine equilibrium. The problem with going forwards in time is highlighted by an extreme three-state example on $\{0, 1, 2\}$ (cited by Möller, 2000) in which the transitions 00, 01, 10, 12, 21 and 22 each occur with probability $1/3$. This simple random walk takes the uniform equilibrium distribution $(1/3)\{1, 1, 1\}$, yet it is easy to show that coalescence can occur only on the states 0 and 2. To circumvent this underlying bias, Wilson (2000) develops a forward read-once algorithm (also described in Möller, 2000). A reasonable compromise between the biased approach, which takes the *random* point of coalescence itself as the onset of equilibrium, and the rigorous Wilson (2000) procedure, is to examine whether coalescence occurs at (say) the *fixed* times $t = 10, 20, \ldots$. So in our example, since coalescence based on $N(0) = 0$ and 100 occurs at time $t = 119.71$ we may safely take the realization from $t = 120$ onwards as reflecting true equilibrium. See Dimakos (2001) for a useful review of perfect simulation ideas.

Since all the trajectories have coupled events (Figure 2.12a), it follows that their corresponding start times relative to $N(0) = 0$, namely $s^{(0)} - s^{(i)}$ for $i = 10, \ldots, 50$, must differ (Figure 2.12b). Indeed, specific values (column 5 of the above table) highlight a general increase with $N(0)$, as one would expect since the larger the initial population size the faster the overall event rate. So the question arises as to where the trajectories are likely to have been at the earlier time $t = 0$? On employing the reverse transition probability approach introduced for the simple death process in Section 2.2.2, we see from (2.2.24) that

$$\overleftarrow{p}_{ij}(t) = p_i(0) p_{ij}(t) / \sum_{i=0}^{\infty} p_i(0) p_{ij}(t) \qquad (2.5.32)$$

where in this context i and j denote population size at $t = 0$ and at the time of full coalescence, respectively. Whence on using the limiting geometric prior (2.2.25), namely $p_i(0) = \theta^i (1 - \theta)$ $(i = 0, 1, \ldots)$ with $\theta \to 1$, it follows from (2.5.12) that

$$\overleftarrow{p}_{ij}(t) = (1 - \theta) \theta^i p_{ij}(t) / \{(1 - \theta) \times \text{coeff. of } z^j \text{ in } \sum_{i=0}^{\infty} [\theta \Theta(z;t)]^i \Phi(z;t)\}$$

$$= \theta^i p_{ij}(t) / \{\text{coeff. of } z^j \text{ in } \frac{\Phi(z;t)}{1 - \theta \Theta(z;t)}\}.$$

On letting $\theta \to 1$

$$\overleftarrow{p}_{ij}(t) = p_{ij}(t) / \{\text{coeff. of } z^j \text{ in } \frac{\Phi(z;t)}{1 - \Theta(z;t)}\}, \qquad (2.5.33)$$

which simplifies to

$$\overleftarrow{p}_{ij}(t) = p_{ij}(t)/\{\text{coeff. of } z^j \text{ in } \frac{\Phi(z;t)}{(\lambda-\mu)s} \times [(\lambda s - \mu) + (\lambda - \mu)(z + z^2 + \cdots)]\}.$$

(2.5.34)

However, in terms of the much simpler deterministic representation (2.5.2), we see that for our example the initial population size required for the population to reach size 100 in time 86.85 (last line of above table) is a fairly massive 506272. So in practical terms, extending the simulated realizations of Figure 2.12b backwards to time $t = 0$ has limited appeal.

2.6 Simple immigration–emigration process

So far we have considered three basic forms for the transition rates, namely:

- λi simple linear birth at rate λ per individual;
- μi simple linear death at rate μ per individual; and,
- α immigration at rate α independent of the population size $i \geq 0$.

However, there is a fourth basic form, namely

- β emigration at rate β independent of $i > 0$,

which is fundamentally different to the previous three since its structure changes at $i = 0$. For if the population is empty then no one can emigrate. So whereas the birth–death–immigration equations (2.5.3) are valid for all $i = 0, 1, 2, \ldots$, once emigration is introduced the equation for $i = 0$ is fundamentally different from those for $i > 0$. This $i = 0$ equation is said to be *anomalous*, and its presence causes a considerable increase in algebraic complexity. Although the solution of even the simplest such process, with $\lambda = \mu = 0$, is an order of difficulty beyond what we have seen so far, it is instructive to consider it here since it illustrates two neat 'tricks' that lead to the solution for the time-dependent probabilities $\{p_i(t)\}$.

This model is best described in terms of a simple queueing situation in which customers arrive at a service area according to a Poisson process with rate α. Customers are served one at a time and queue in the order of their arrival. The length of time taken to serve a customer is exponentially distributed with mean $1/\beta$. Clearly, if the rate of arrivals (α) exceeds the rate of departure (β), then we can expect the queue of unserved customers to increase indefinitely. In practice, the laws governing this process, which is called an M/M/1 queue, would change once the queue became too large; for example, by new customers being deterred from joining it, or by employing additional servers. Such scenarios are intrinsic to the highly applicable field of Queueing Theory.

In contrast, if $\alpha < \beta$ then it is entirely reasonable to expect that the system will settle down to a situation of stable statistical equilibrium. In this case, on denoting $\{p_i(t)\}$ to be the probability that the queue contains i customers at time t (including

96 Simple Markov population processes

the one being served), we have $p_i(t) \to \pi_i$ as $t \to \infty$. Examples of properties of direct interest include:

(a) the equilibrium distribution $\{\pi_i\}$ of the number of customers in the queue;
(b) the distribution of the customers' queueing time;
(c) the probability distribution of the length of a server's busy period, and, conversely, the distribution of the time he is 'idle';
(d) the distribution of the time taken for the system to become completely empty for the first time, given that n_0 customers were already queueing at the commencement of service.

2.6.1 Equilibrium distribution

Constructing the forward equations, as before, we obtain

$$dp_0(t)/dt = -\alpha p_0(t) + \beta p_1(t) \tag{2.6.1}$$

$$dp_i(t)/dt = \alpha p_{i-1}(t) - (\alpha + \beta) p_i(t) + \beta p_{i+1}(t) \quad (i > 0). \tag{2.6.2}$$

The anomalous form (2.6.1) arises because the emigration rate β is zero when the queue is empty. Before we introduce techniques to handle this anomaly, let us first consider the equilibrium solution.

On taking $\alpha < \beta$ and letting $t \to \infty$, equations (2.6.1) and (2.6.2) become

$$0 = -\alpha \pi_0 + \beta \pi_1 \tag{2.6.3}$$

$$0 = \alpha \pi_{i-1} - (\alpha + \beta) \pi_i + \beta \pi_{i+1}. \tag{2.6.4}$$

Whence on writing (2.6.4) in the balanced form

$$\alpha \pi_{i-1} - \beta \pi_i = \alpha \pi_i - \beta \pi_{i+1} \quad (i > 0),$$

we see that

$$\alpha \pi_{i-1} - \beta \pi_i = \text{constant}.$$

Now comparison with (2.6.3) at $i = 0$ shows that

$$\alpha \pi_0 - \beta \pi_1 = \text{constant} = 0.$$

So

$$\pi_i = (\alpha/\beta) \pi_{i-1}, \tag{2.6.5}$$

which, on denoting the *traffic intensity* $\rho = \alpha/\beta$, yields

$$\pi_i = \rho^i \pi_0$$

for $\rho < 1$. Use of the normalizing condition

$$1 = \sum_{i=0}^{\infty} \pi_i = \pi_0 \sum_{i=0}^{\infty} \rho^i = \pi_0/(1-\rho)$$

then gives the solution
$$\pi_i = (1-\rho)\rho^i \quad (i = 0, 1, 2, \ldots). \tag{2.6.6}$$
Thus the equilibrium population size distribution is geometric with mean $\rho/(1-\rho)$.

Further properties of the system can be deduced from this equilibrium result. For example, the probability that the server is idle is $\pi_0 = (1-\rho)$, and this tends to zero as $\rho \uparrow 1$ (a situation called *heavy traffic*). Moreover, let W denote the total time that a given individual spends in the queue. Then W is the sum of the service times of the i customers ahead of him plus his own service time; the time already spent by the customer currently being served when our individual arrives may be discounted because of the 'lack of memory' property of the exponential distribution. Thus conditional on i, the total time W is the sum of $i+1$ random variables exponentially distributed with parameter β. As each of these has Laplace transform (see (2.3.12))
$$\int_0^\infty e^{-st}\beta e^{-\beta t}\, dt = \beta/(\beta+s), \tag{2.6.7}$$
the conditional Laplace transform of W is $\{\beta/(\beta+s)\}^{i+1}$. So the unconditional Laplace transform of W is
$$\sum_{i=0}^\infty \left(\frac{\beta}{\beta+s}\right)^{i+1} \rho^i(1-\rho) = \frac{\beta-\alpha}{\beta-\alpha+s}, \tag{2.6.8}$$
whence it follows from (2.6.7) that the p.d.f. of W is exponential with parameter $\beta - \alpha$.

2.6.2 Time-dependent solutions

Whilst equation (2.6.2) holds for $i = 1, 2, \ldots$, equation (2.6.1) for the boundary state $i = 0$ has a different structure. So trying to parallel our solution for the simple birth–death process, by introducing the p.g.f. $G(z;t) \equiv \sum_{i=0}^\infty p_i(t)z^i$ and forming a p.d.e. from (2.6.1) and (2.6.2), results in a more difficult equation than we have seen before, namely
$$\frac{\partial G(z;t)}{\partial t} = \left(-\alpha - \beta + \alpha z + \frac{\beta}{z}\right)G(z;t) + \beta\left(1 - \frac{1}{z}\right)p_0(t). \tag{2.6.9}$$
For unlike equation (2.4.7) for the birth–death process, or (2.5.10) for the immigration–death process, (2.6.9) involves $p_0(t) = G(0;t)$ which is unknown. So it has the general character of a boundary value problem.

A Laplace transform approach

To examine the nature of this equation, let us remove the differential component by introducing the Laplace transforms
$$G^*(z;s) = \int_0^\infty G(z;t)e^{-st}\, dt \quad \text{and} \quad p_0^*(s) = \int_0^\infty p_0(t)e^{-st}\, dt \tag{2.6.10}$$

(as in Cox and Miller, 1965). Then on assuming that at time $t = 0$ the queue contains n_0 individuals, so $G(z;0) = z^{n_0}$, use of result (2.3.13), that is

$$\mathcal{L}[\partial G(z;t)/\partial t] = s\mathcal{L}[G(z;t)] - G(z;0), \qquad (2.6.11)$$

leads to

$$sG^*(z;s) - z^{n_0} = (-\alpha - \beta + \alpha z + \beta/z)G^*(z;s) + \beta(1 - 1/z)p_0^*(s).$$

Whence

$$G^*(z;s) = \frac{z^{n_0+1} + \beta(z-1)p_0^*(s)}{z(s + \alpha + \beta) - \alpha z^2 - \beta}. \qquad (2.6.12)$$

One might hope that we could put $z = 0$ in (2.6.12) and then solve for $G^*(0;s) = p_0^*(s)$. Unfortunately this approach does not work since it just produces the identity $p_0^*(s) = p_0^*(s)$. We therefore need a new form of argument, and one method of attack is to examine the analytic nature of (2.6.12) (Bailey, 1954). Since $\sum_{i=0}^{\infty} p_i(t) = 1$ we have $G(1;t) \equiv 1$, and so $G(z;t)$ must converge in and on the unit circle $|z| = 1$ for all $t \geq 0$. Hence provided $\mathcal{R}(z) \geq 0$, $G^*(z;s)$ must converge in and on $|z| = 1$. Thus if the denominator of (2.6.12), when considered as a function of z, has zeros inside $|z| = 1$, then these must also be zeros of the numerator. Now solving for

$$\alpha z^2 - (s + \alpha + \beta)z + \beta = 0$$

yields the two roots

$$z_1(s), z_0(s) = (1/2\alpha)[(s + \alpha + \beta) \pm \sqrt{\{(s + \alpha + \beta)^2 - 4\alpha\beta\}}], \qquad (2.6.13)$$

and it is easy to show that $z_0(s)$ is the single root inside $|z| = 1$. Thus the numerator of (2.6.12) vanishes at $z = z_0(s)$, whence it follows that

$$p_0^*(s) = \frac{\{z_0(s)\}^{n_0+1}}{\beta\{1 - z_0(s)\}}. \qquad (2.6.14)$$

If we assume that the queue is initially empty, i.e. $n_0 = 0$, then this result simplifies to

$$p_0^*(s) = \frac{z_0(s)}{\beta\{1 - z_0(s)\}} = \frac{(s + \alpha + \beta) - \sqrt{\{(s + \alpha + \beta)^2 - 4\alpha\beta\}}}{\beta[2\alpha - (s + \alpha + \beta) + \sqrt{\{(s + \alpha + \beta)^2 - 4\alpha\beta\}}]}. \qquad (2.6.15)$$

Indeed, on noting that if $\mathcal{L}[g(t)] = h(s)$ then $\mathcal{L}[e^{at}g(t)] = h(s-a)$, we have the even simpler Laplace transform result

$$\beta\mathcal{L}[e^{(\alpha+\beta)t}p_0(t)] = \frac{2\alpha}{2\alpha - s + \sqrt{(s^2 - 4\alpha\beta)}} - 1. \qquad (2.6.16)$$

As substituting (2.6.14) into (2.6.12) yields the complete Laplace transform solution for $G^*(z;s)$, in principle the probabilities $p_i(t)$ are now fully determined. In practice, however, there is still the non-trivial problem of inverting $G^*(z;s)$ by expanding (2.6.12) in terms of suitable functions of s in order to use known Laplace transform results, and hence deriving a series solution in terms of standard functions. Since we

An extended state-space approach

The potentially awkward algebraic inversion of $G^*(s;t)$ may be avoided by working with $G(z;t)$ directly. We first make the anomalous equation (2.6.1) take the general form (2.6.2) by extending the state space from $N = 0, 1, 2, \ldots$ to $N = 0, \pm 1, \pm 2, \ldots$. To achieve this we simply note that placing $i = 0$ in (2.6.2) gives

$$dp_0(t)/dt = \alpha p_{-1}(t) - (\alpha + \beta)p_0(t) + \beta p_1(t), \tag{2.6.17}$$

and that this agrees with (2.6.1) provided

$$\alpha p_{-1}(t) - \beta p_0(t) = 0. \tag{2.6.18}$$

The initial conditions are now taken to be $p_{n_0}(0) = 1$ and $p_i(0) = 0$ ($i = 0, 1, \ldots, n_0 - 1, n_0 + 1, \ldots$) with $p_i(0) = q_i$ ($i = -1, -2, \ldots$) for some unknown constants q_i. On replacing $G(z;t)$ by $H(z;t) \equiv \sum_{i=-\infty}^{\infty} p_i(t) z^i$, equation (2.6.2) yields

$$\frac{\partial H(z;t)}{\partial t} = (\alpha z - \alpha - \beta + \beta/z) H(z;t), \tag{2.6.19}$$

and unlike (2.6.9) this equation does not contain the nuisance term $p_0(t)$. So we can integrate it using our initial conditions, namely

$$H(z;0) = z^{n_0} + \sum_{i=-\infty}^{-1} q_i z^i, \tag{2.6.20}$$

to obtain the solution

$$H(z;t) = \left(z^{n_0} + \sum_{i=-\infty}^{-1} q_i z^i\right) \exp\{t(\alpha z - \alpha - \beta + \beta/z)\}. \tag{2.6.21}$$

It now remains to extract the coefficients of z^0 and z^{-1} from (2.6.21) so that we can ensure the solution satisfies the boundary equation (2.6.18) by making an appropriate choice for $\{q_i\}$. Now (2.6.21) hints heavily of the modified Bessel function

$$I_n(t) = \sum_{r=0}^{\infty} \frac{(t/2)^{n+2r}}{r!(n+r)!} \tag{2.6.22}$$

since

$$\exp\{(t/2)(z + z^{-1})\} = \sum_{n=-\infty}^{\infty} I_n(t) z^n \quad (z \neq 0) \tag{2.6.23}$$

(results §9.6.10 and §9.6.33 of Abramowitz and Stegun, 1970). Indeed, this generating function takes the form of (2.6.21) once we write it as

$$\exp\{(t\sqrt{\alpha\beta})(z\sqrt{\alpha/\beta} + z^{-1}\sqrt{\beta/\alpha})\} = \sum_{n=-\infty}^{\infty} I_n(2t\sqrt{\alpha\beta})(z\sqrt{\alpha/\beta})^n. \tag{2.6.24}$$

100 Simple Markov population processes

So on abbreviating $I_n(2t\sqrt{\alpha\beta})$ to I_n, we can write (2.6.21), with $\rho = \alpha/\beta$, as

$$H(z;t) = e^{-(\alpha+\beta)t}(z^{n_0} + \sum_{n=-\infty}^{-1} q_n z^n) \sum_{n=-\infty}^{\infty} I_n(\rho^{1/2}z)^n. \qquad (2.6.25)$$

Whence extracting the coefficients of z^0 and z^{-1} yields

$$p_0(t) = e^{-(\alpha+\beta)t}[I_{-n_0}\rho^{-n_0/2} + \sum_{n=1}^{\infty} q_{-n}I_n\rho^{n/2}] \qquad (2.6.26)$$

$$p_{-1}(t) = e^{-(\alpha+\beta)t}[I_{-n_0-1}\rho^{-(n_0+1)/2} + \sum_{n=1}^{\infty} q_{-n}I_{n-1}\rho^{(n-1)/2}]. \qquad (2.6.27)$$

The coefficients q_n are then determined uniquely by equating to zero the coefficients of each order of Bessel function in (2.6.18), i.e. $p_0(t) = \rho p_{-1}(t)$. On recalling that $I_{-n} \equiv I_n$, we have $q_n = 0$ $(n = -1, -2, \ldots, -n_0)$, $q_{-n_0-1} = \rho^{-n_0-1}$, $q_{-n_0-2} = \rho^{-n_0-2}(1-\rho)$ and $q_{-n-1} = \rho^{-1}q_{-n}$ $(n = -n_0 - 2, -n_0 - 3, \ldots)$. Whence the final Method of Images solution for $H(z;t)$ becomes

$$H(z;t) = e^{-(\alpha+\beta)t}\{z^{n_0} + \rho^{-n_0-1}z^{-n_0-1}$$

$$+ (1-\rho)\sum_{j=n_0+2}^{\infty}(\rho z)^{-j}\} \sum_{n=-\infty}^{\infty} I_n(2t\sqrt{\alpha\beta})(\rho^{1/2}z)^n. \qquad (2.6.28)$$

Since this expression just involves the product of two power series in z, any required $p_i(t)$ can be derived from $H(z;t)$ in terms of a series of Bessel functions. In particular, when $n_0 = 0$ we have

$$p_0(t) = e^{-(\alpha+\beta)t}\{I_0(2t\sqrt{\alpha\beta}) + \rho^{-1/2}I_1(2t\sqrt{\alpha\beta}) + (1-\rho)\sum_{n=2}^{\infty} \rho^{-n/2}I_n(2t\sqrt{\alpha\beta})\}.$$
$$(2.6.29)$$

So, if required, inversion of the previous Laplace transform solution (2.6.16) would be best tackled through standard Laplace transform results for modified Bessel functions. Though the implicit level of difficulty involved in even this fairly simple example can be seen by noting that in order to apply the standard form

$$\mathcal{L}[a^n I_n(at)] = \frac{(s - \sqrt{(s^2 - a^2)})}{\sqrt{(s^2 - a^2)}} \qquad (n > -1) \qquad (2.6.30)$$

(result §29.3.59 of Abramowitz and Stegun, 1970) to (2.6.16), we would first have to rearrange and then expand (2.6.16) in an appropriately amenable form. Nevertheless, modified Bessel functions often appear in applied probability results, and are a natural first choice for trial solutions. In general, other likely candidates include Hermite, Laguerre, Legendre and confluent hypergeometric functions. Note that further insight into the process can be gained by employing a third approach, namely the spectral representation (Karlin and McGregor, 1958), and we shall shortly be introducing this technique in Section 3.1.

2.7 Batch events

So far we have allowed only single events to occur, namely $i \to i+1$ at rate $\alpha + \lambda i$ and $i \to i-1$ at rate $\beta + \mu i$. Whilst for many situations this is fine, occasionally we do have to deal with genuinely multiple events. For example, a population of aphids on a bush may all be killed simultaneously by chemical spraying, a coach may suddenly download its content of passengers at an air terminal, or an individual may give birth to twins, triplets, etc. This last scenario is a multiplicative process, and as such involves more awkward algebra than we have so far experienced (discussed later in Chapter 5.4 in terms of branching processes). Fortunately, however, the first two scenarios are much more amenable to mathematical analysis.

To conclude this chapter, let us therefore briefly mention an extension of the simple birth process in which the whole population can instantaneously die (i.e. a suffer a catastrophe) at rate γ, after which the population is restarted by the arrival of a batch of j new immigrants at rate α_j. Note that adding this mass annihilation and immigration structure to a simple birth–death process substantially increases the underlying complexity, since state 0 can then be reached either by a death occurring in state 1 or a catastrophe occurring in any of the states $i \geq 1$.

2.7.1 Birth–mass annihilation and immigration process

The corresponding forward Kolmogorov equations for the population size probabilities $\{p_i(t)\}$ take the form

$$dp_0(t)/dt = \gamma(1 - p_0(t)) - (\sum_{j=1}^{\infty} \alpha_j) p_0(t) \qquad (2.7.1)$$

$$dp_i(t)/dt = \alpha_i p_0(t) + \lambda(i-1) p_{i-1}(t) - (\gamma + \lambda i) p_i(t) \qquad (i \geq 1), \qquad (2.7.2)$$

which is well-suited to using a Laplace transform approach. For on taking the process to be empty at time $t=0$, i.e. $p_0(0) = 1$, applying (2.3.12) yields, for $a = \sum_{j=1}^{\infty} \alpha_j$, the first-order recurrence equations

$$sp_0^*(s) - 1 = (\gamma/s) - \gamma p_0^*(s) - ap_0^*(s) \qquad (2.7.3)$$

$$sp_i^*(s) = \alpha_i p_0^*(s) + \lambda(i-1) p_{i-1}^*(s) - (\gamma + i\lambda) p_i^*(s) \qquad (i \geq 1). \qquad (2.7.4)$$

Equation (2.7.3) solves directly to give

$$p_0^*(s) = [1 + (\gamma/s)]/(s + \gamma + a), \qquad (2.7.5)$$

whence repeated application of (2.7.4) yields

$$p_i^*(s) = \sum_{j=0}^{i-1} \frac{(i-1)!\lambda^j}{(i-1-j)!(s+\gamma+i\lambda)\dots(s+\gamma+(i-j)\lambda)} \alpha_{i-j} p_0^*(s). \qquad (2.7.6)$$

Inversion of (2.7.5) (or direct solution of (2.7.1)) recovers

$$p_0(t) = (\gamma + a \exp\{-(\gamma+a)t\})/(\gamma+a); \qquad (2.7.7)$$

whilst for $i \geq 1$ splitting the denominator of (2.7.6) into partial fractions and then inverting each individual component leads to the solution (Renshaw and Chen, 1997)

$$p_i(t) = \sum_{j=1}^{i} \alpha_j \binom{i-1}{j-1} \left\{ \sum_{k=j}^{i} \frac{(-1)^{k-j}\binom{i-j}{i-k}(k\lambda)\exp\{-(\gamma+k\lambda)t\}}{(a-k\lambda)(\gamma+k\lambda)} \right.$$

$$\left. + \frac{a(i-j)!\lambda^{i-j}\exp\{-(a+\gamma)t\}}{(i\lambda-a)\dots(j\lambda-a)(a+\gamma)} + \frac{\gamma(i-j)!\lambda^{i-j}}{(\gamma+i\lambda)\dots(\gamma+j\lambda)(a+\gamma)} \right\}. \qquad (2.7.8)$$

Although this expression contains a considerable number of terms, it does, nevertheless, highlight the way in which $p_i(t)$ changes with i and t.

Intuitively, it seems reasonable to presume that an equilibrium situation will develop as $t \to \infty$. For on moving from state 0 to state j (at rate α_j) the process grows exponentially at rate $\lambda > 0$ but then, regardless of how large j is, returns to state 0 after an Exponential(γ) length of time. Such exponential boundedness suggests that the process is *recurrent*, and we see from (2.7.7) and (2.7.8) that the equilibrium probabilities are given by

$$\pi_0 = \gamma/(\gamma+a) \qquad (2.7.9)$$

$$\pi_i = \pi_0 \sum_{j=1}^{i} \frac{\alpha_j(i-1)!\lambda^{i-j}}{(j-1)!(\gamma+i\lambda)\dots(\gamma+j\lambda)} \qquad (i \geq 1). \qquad (2.7.10)$$

Moreover, for $i \geq 1$ we have $p_i(t) \to \pi_i$ at rate $\exp\{-[\gamma+\min(a,\lambda)]t\}$ as t increases. We shall return to make a formal study of *recurrence* in Chapter 3.

As far as determining moments of the process is concerned, there is no mathematical hardship in including death at rate $\mu > 0$. Applying the p.g.f.'s $G(z;t) \equiv \sum_{i=0}^{\infty} p_i(t) z^i$ and $H(z) \equiv \sum_{j=1}^{\infty} \alpha_j z^j$ to the (new) forward equations (2.7.1) and (2.7.2) yields

$$\frac{\partial G(z;t)}{\partial t} = (\lambda z - \mu)(z-1)\frac{\partial G(z;t)}{\partial z} + \gamma[1 - G(z;t)] + p_0(t)[H(z) - H(1)]. \qquad (2.7.11)$$

Recalling relation (2.1.23), denote the rth factorial moment

$$\mu_{(r)} = \left.\frac{\partial^r G(z;t)}{\partial z^r}\right|_{z=1} \quad \text{with} \quad a_{(r)} = \left.\frac{\partial^r H(z)}{\partial z^r}\right|_{z=1}. \qquad (2.7.12)$$

Then on differentiating equation (2.7.11) r times and placing $z = 1$ it may be shown that, for $a_{(r)} < \infty$,

$$\mu_{(r)} = \int_0^t [r(r-1)\lambda\mu_{(r-1)}(s) + p_0(s)a_{(r)}]\exp\{[\gamma - r(\lambda - \mu)](s-t)\}\,ds. \quad (2.7.13)$$

So once $p_0(t)$ is known (as in the above $\mu = 0$ case), the factorial moments $\{\mu_{(r)}(t)\}$ may be obtained by successive integration. In particular, when $\mu = 0$ and $a_{(1)} = \sum_{j=1}^{\infty} j\alpha_j < \infty$, the mean

$$\mu_{(1)} = \begin{cases} \left(\frac{a_{(1)}}{\gamma+a}\right)\left[\gamma t + \left(\frac{a}{\lambda+a}\right)\{1 - e^{-(\gamma+a)t}\}\right] & \text{if } \lambda = \gamma \\ \left(\frac{a_{(1)}}{\gamma+a}\right)\left[\left(\frac{\gamma}{\lambda-\gamma}\right)\{e^{(\lambda-\beta)t} - 1\} + \left(\frac{a}{\lambda+a}\right)\{e^{(\lambda-\gamma)t} - e^{-(\gamma+a)t}\}\right] & \text{if } \lambda \neq \gamma. \end{cases}$$

(2.7.14)

Hence if $\lambda > \gamma$ then the mean increases as $O(e^{(\lambda-\gamma)t})$, if $\lambda = \gamma$ it still increases but at the much slower rate of $O(t)$, whilst if $\lambda < \gamma$ then it converges to the equilibrium value

$$\mu_{(1)}(\infty) = a_{(1)}\gamma/[(\gamma + a)(\gamma - \lambda)]. \quad (2.7.15)$$

Development of the probability structure of the general process for $\lambda, \mu > 0$ is more demanding. For the presence of the $p_0(t)$ term in equation (2.7.11) means that use of Laplace transform techniques (for example) for solving this equation will give rise to an algebraically complex result. Though such difficulties are not insurmountable, Renshaw and Chen (1997) present a much neater approach based on Chen and Renshaw (1990, 1993b), which interested readers may wish to pursue.

2.7.2 Mass immigration–death process

No such algebraic problems exist if we return to the basic immigration–death process and just replace single arrivals at rate α by j arrivals at rate α_j. For on denoting $p_i(t) \equiv 0$ for $i < 0$, the forward equations take the form

$$dp_i(t)/dt = \sum_{j=1}^{i-1} \alpha_j p_{i-j}(t) - (a + i\mu)p_i(t) + (i+1)\mu p_{i+1}(t), \quad (2.7.16)$$

and the associated Lagrangian equation for the p.g.f. $G(z;t)$, namely

$$\frac{\partial G(z;t)}{\partial t} + \mu(z-1)\frac{\partial G(z;t)}{\partial z} = [H(z) - a]G(z;t), \quad (2.7.17)$$

clearly has a much simpler structure than (2.7.11). Indeed, solution of the equilibrium equation is trivial. For as $t \to \infty$, (2.7.17) reduces to

$$\mu\frac{dG(z;\infty)}{dz} = \sum_{j=1}^{\infty} \alpha_j \left(\frac{1-z^j}{1-z}\right) G(z;\infty) = \sum_{j=1}^{\infty} \alpha_j(1 + z + \cdots + z^{j-1})G(z;\infty),$$

which integrates to yield

$$G(z;\infty) = \exp\{(1/\mu) \sum_{j=1}^{\infty} \alpha_j \sum_{r=1}^{j} (1/r)(z^r - 1)\} \qquad (2.7.18)$$

since $G(1;\infty) = 1$.

As an illustration, consider the geometric case

$$\alpha_j = \alpha \theta^{j-1}(1-\theta) \qquad (0 < \theta < 1; j = 1, 2, \ldots). \qquad (2.7.19)$$

Inserting (2.7.19) into (2.7.18) and rearranging the θ^j and z^r terms gives

$$G(z;\infty) = \exp\{(\alpha/\theta\mu) \sum_{r=1}^{\infty} \left[\frac{(\theta z)^r}{r} - \frac{\theta^r}{r}\right]\}$$

$$= \exp\{(\alpha/\theta\mu)[-\ln(1-\theta z) + \ln(1-\theta)]\} = \left(\frac{1-\theta z}{1-\theta}\right)^{-\alpha/\theta\mu}. \qquad (2.7.20)$$

Now letting $t \to \infty$ in result (2.5.12) shows that, for the simple birth–death–immigration process with rates λ, μ and α, the equilibrium p.g.f. takes the form

$$G(z;\infty) = \left(\frac{\mu - \lambda z}{\mu - \lambda}\right)^{-\alpha/\lambda}. \qquad (2.7.21)$$

Whence comparing (2.7.20) and (2.7.21) with $\lambda = \mu\theta$ shows that the mass immigration–death process and the birth–death–immigration process have identical equilibrium structures. So although individual realizations will exhibit large behavioural differences between these two processes, in terms of the equilibrium probability structure it is not possible to differentiate between them. This highlights a fundamental weakness in trying to describe a stochastic process purely in terms of the equilibrium population size probabilities.

2.7.3 Regenerative phenomena

Throughout this text we concern ourselves with the development of a stochastic process $\{X(t)\}$ over $0 \leq t < \infty$. We have seen that for the simple Poisson process (Section 2.1) growth is roughly linear with $E[X(t)] = \text{Var}[X(t)] = \alpha t$, whilst for the simple birth process (Section 2.3) growth is essentially exponential with $E[X(t)] = n_0 e^{\lambda t}$ and $\text{Var}[X(t)] = n_0 e^{\lambda t}(e^{\lambda t} - 1)$. In neither case does the population size ever decrease. However, when death is introduced (Section 2.4) this is no longer the case, and the process may now become extinct. Results (2.4.24)–(2.4.27) show that if $\lambda \leq \mu$ then extinction is certain, but if $\lambda > \mu$ then it is not. Moreover, since $p_\infty = (\mu/\lambda)^{n_0}$ if $\lambda > \mu$, the probability of extinction decreases geometrically as the initial population size, n_0, increases. In either case, if the process becomes extinct at time T_0 then $X(t) \equiv 0$ for all $t \geq T_0$ thereafter.

If this birth–death process is to become positive again then we clearly need to introduce some kind of re-starting mechanism. One solution would be to switch $X(T_0) = 0$ into $X(T_0) = n_0$, so that a fresh batch of n_0 individuals arrives in the system immediately the previous batch dies out. However, this not only changes the fundamental nature of the process, but it also induces considerable mathematical complexity into the corresponding analysis. Far simpler is to allow immigration to occur at rate α (Sections 2.5 and 2.6). For suppose that the population is initially empty (i.e. $n_0 = 0$ at $t = t_0 = 0$), it increases to $X(t) = 1$ at $t = t' > t_0$, and we record the *renewal* time $t_1 > t'$ at which $\{X(t)\}$ next returns to zero. Then if we now regard t_1 as the new start time, we can record the renewal time t_2 to the second return to zero; similarly we can denote $t_3, t_4, \ldots$ to be the times to the third, fourth, etc. return to zero. As the stochastic structures of the sectionalized processes $\{X(t) : 0 \le t \le t_1\}$, $\{X(t) : t_1 < t \le t_2\}$, $\{X(t) : t_2 < t \le t_3\}$, etc. are identical, the event 'return to state zero' is called a *regenerative phenomenon*. Questions of obvious interest now relate, for example, to the distributions and associated moments of the inter-event times $s_1 = t_1 - t_0, s_2 = t_2 - t_1, \ldots$, the number of events and the maximum population size within a specific inter-event interval (t_r, t_{r+1}), and the number of renewals within a given time period $(0, t)$. Moreover, there is no reason why the regenerative event has to relate to zero population size. It could just as easily refer to the population size $X(t)$ reaching a specific size $M > 0$, or, as discussed in Section 2.3.5, a dishonest process reaching infinity. Though in this last case $X(t)$ would have to be reset to say n_0 immediately this 'explosion' occurred. Nor do we have to restrict attention to simple birth–death–immigration structures. For in Section 2.7 we have already seen how to cope both with a general mass arrival distribution $\{\alpha_j\}$ and catastrophes at rate γ, whilst in Chapter 3 we shall allow for general arrival and departure rates, λ_N and μ_N, which depend on the population size N. The regenerative structure of such processes raises highly interesting and non-trivial issues whose analysis leads to important insights into analytical problems connected with this type of Markov process. Although their theoretical development lies well beyond the scope of this book, readers interested in delving further into this field are advised to cut their teeth on Kingman's (1972) excellent monograph and the Markov Oscillation Problem discussed in Dai and Renshaw (2000).

Fortunately, if we move away from asking questions of a deep theoretical nature concerning regenerative structure, onto more basic applied probability questions which revolve around the probability density function, $f(s)$, of the renewal times $s_1, s_2, \ldots$, then life becomes much simpler. For we may regard the 'lifetime' of our population process, $\{X(t)\}$, as being a direct paradigm of the time to failure (and subsequent replacement) of a manufactured component such as an electric light bulb. This area of study swiftly became an important feature in the fast developing field of operations research, and readers who are interested in problems which relate to regenerative stochastic population processes should be aware of this industrial paradigm in order to avoid reinventing the wheel. Feller (1968) and Cox (1962) provide excellent introductions to the analysis of probability problems in renewal theory in discrete and continuous time, respectively; whilst Cox and Lewis (1966) provide a

106 Simple Markov population processes

survey of some of the related statistical issues that stem from the analysis of renewal data.

Although renewal theory does not play a central role in this text, it does, nevertheless, make several appearances in one guise or another, generally in terms of the time of first return to a given state. Discrete results are discussed at some length in Chapters 4 and 5. Whilst continuous-time results for specific models may be deduced from the probability relations pertinent to them. For example, differentiating expression (2.4.23) yields the p.d.f., $f_0(t)$, of the time to extinction for a simple birth–death process which starts with n_0 individuals at time $t = 0$. So the associated renewal event is the batch immigration of n_0 new individuals immediately the population dies out. Whence the distribution and moments of the number of renewals in a fixed time interval may now be deduced by inserting $f_0(t)$ into standard renewal theory results.

3
General Markov population processes

So far we have considered processes which develop under the assumption that the probabilities that a given individual will reproduce or die remain constant and are independent of population size. Clearly, this can only be true if there is no interference amongst individual population members. However, in most real-life situations the growth of an expanding population will eventually be limited by either a shortage of resources or space in which to develop further. A stage is then reached when the demands made on these resources either exert a net downwards pressure on population size, or, in extreme situations, preclude further growth altogether. This stage is often called the *carrying capacity* of the system.

At first glance it might appear that such implied generality means that a large number of separate theoretical models have to be analysed, each one corresponding to a different type of interaction between individuals. Fortunately this is not the case, since the total number of individuals, $N(t)$, can change for only four reasons, namely:

(a) birth   – rate depends on N
(b) death $(N > 0)$ – rate depends on N
(c) immigration   – rate independent of N
(d) emigration $(N > 0)$ – rate independent of N.

Moreover, if we combine (a) and (c) to form a general birth rate λ_N, and (b) and (d) to form a general death rate μ_N, then between them λ_N and μ_N encompass any modelling situation in which the population size N changes by one at each event. For example, in the simple immigration–birth–death process (Section 2.5) $\lambda_N = \alpha + \lambda N$ and $\mu_N = \mu N$. The wide variety of population equations to be found in the literature simple reflects the large number of special cases available. Such processes, based on λ_N and μ_N, represent the most elementary formulation of probabilistic models describing population growth, since at each event N either increases or decreases by just one individual. Nevertheless, they play a fundamentally important role, not least because they lend themselves to formal analysis, and much analytic progress has been made since the stimulating pioneering presentation by Feller (1950). Readers interested in a lucid survey of this early work should consult Kendall (1952) and the extensive bibliography contained therein.

Whilst the associated deterministic equation takes the deceptively simple form

$$dN/dt = \lambda_N - \mu_N, \tag{3.0.1}$$

deriving exact theoretical solutions to (3.0.1) can often be tricky, though numerical solutions can always be obtained. Similar considerations hold for the Kolmogorov forward and backward equations (2.4.45) and (2.4.46) for $\Pr(N(t) = i|N(0) = k)$, namely

$$dp_{ki}(t)/dt = \lambda_{i-1}p_{k,i-1}(t) - (\lambda_i + \mu_i)p_{ki}(t) + \mu_{i+1}p_{k,i+1}(t) \quad \text{and} \tag{3.0.2}$$

$$dp_{ki}(t)/dt = \lambda_k p_{k+1,i}(t) - (\lambda_k + \mu_k)p_{ki}(t) + \mu_k p_{k-1,i}(t), \tag{3.0.3}$$

respectively, where we define $\lambda_{-1} = \mu_0 = 0$. Here full theoretical solutions can be very difficult, or impossible to construct, even for minor modifications to the simple birth and death rates $\lambda_N = \lambda N$ and $\mu_N = \mu N$.

3.1 Classification of states

Write the matrix of transition rates as

$$\mathbf{A} = \begin{pmatrix} -(\lambda_0 + \mu_0) & \lambda_0 & 0 & 0 & \cdots \\ \mu_1 & -(\lambda_1 + \mu_1) & \lambda_1 & 0 & \cdots \\ 0 & \mu_2 & -(\lambda_2 + \mu_2) & \lambda_2 & \cdots \\ \vdots & \vdots & \vdots & \vdots & \end{pmatrix}, \tag{3.1.1}$$

and denote the corresponding matrix of transition probabilities by

$$\mathbf{P}(t) = (p_{ij}(t)) \qquad (i,j = 0, 1, 2, \ldots). \tag{3.1.2}$$

Then as we have already noted in Section 2.4.4, the forward and backward equations (3.0.2) and (3.0.3) take the simple matrix forms

$$d\mathbf{P}(t)/dt = \mathbf{P}(t)\mathbf{A} \quad \text{and} \quad d\mathbf{P}(t)/dt = \mathbf{A}\mathbf{P}(t). \tag{3.1.3}$$

For a general matrix $\mathbf{A}$ these two equations define a Markov chain in continuous time (e.g. Anderson, 1991).

Karlin and McGregor (1955, 1957ab) provide a definitive treatment of the existence, uniqueness and properties of the matrices $\mathbf{P}(t)$ which satisfy (3.1.3) for the initial identity matrix condition

$$\mathbf{P}(0) = \mathbf{I}. \tag{3.1.4}$$

As there are always infinitely many matrices which satisfy these relations, additional properties are required if we are to select those matrices $\mathbf{P}(t)$ which may serve as transition probability matrices. Two such properties are

$$p_{ij}(t) \geq 0 \quad \text{and} \quad \sum_{j=0}^{\infty} p_{ij}(t) \leq 1. \tag{3.1.5}$$

The second inequality in (3.1.5) allows for the possibilities that: the population size reaches infinity in finite time t with probability $1 - \sum_{j=0}^{\infty} p_{ij}(t)$ (as for the dishonest birth process of Section 2.3.5); or, if $\mu_0 > 0$, that the process disappears. Note that this latter feature is different from an extinct population which retains a conceptual presence at zero (e.g. the simple birth–death process with $\mu > \lambda$). An alternative way of visualizing this construct is to consider population size as a diffusing particle over the states $i = 0, 1, 2, \ldots$. So dishonesty corresponds to $i = \infty$, whilst removal corresponds to absorption from state $i = 0$ into the state $i = -1$ from which there is no return.

Another property, called the *semi-group property*, is expressed by the Chapman–Kolmogorov equation

$$p_{ij}(t+s) = \sum_{k=0}^{\infty} p_{ik}(t) p_{kj}(s), \quad (3.1.6)$$

namely that the probability of moving from i to j in time $t + s$ is the probability of first moving from i to an intermediate state k in time t followed by an onwards move to j in time s, summed over all possible states k. Feller (1952) notes that the forward equation is in general considerably more complicated than the basic form in (3.1.3), since a process might return from $i = \infty$ to the finite states with positive probability. Although this feature generally lies beyond the realms of practical population modelling, it is worth noting that there are interesting families of such processes for which the forward equation, when the state at infinity is disregarded, is exactly (3.1.3). For these special processes the inequality

$$p_{ij}(t+s) \leq \sum_{k=0}^{\infty} p_{ik}(t) p_{kj}(s) \quad (0 \leq i, j < \infty) \quad (3.1.7)$$

is satisfied.

The approach adopted by Karlin and McGregor is to look for an integral representation of the transition probability matrix $\mathbf{P}(t)$ in terms of the eigenvalues of the transition rate matrix $\mathbf{A}$. This point of view leads to the revelation of an intimate connection between the theory of birth–death processes and an associated system of orthogonal polynomials. Whilst Reuter and Lederman (1953) present an independent discovery of the integral representations of birth–death processes. Not only is the time-dependence of the transition probabilities displayed in a particularly simple and lucid manner in the integral representation, but the deep understanding that already underpins the theory and application of orthogonal polynomials can immediately be transferred across to the study of birth–death processes. However, the detailed algebra involved operates at a substantially higher level of complexity than can comfortably be presented in this text, so here we shall restrict ourselves to quoting the main results and refer readers to the cited papers for a full analytic discussion.

For $N(t) = n$, assume that λ_n ($n \geq 0$) and μ_n ($n > 0$) are *strictly positive* (which excludes all processes having a finite population range) and that $\mu_0 \geq 0$. Let $Q_0(x), Q_1(x), \ldots$ denote a sequence of polynomials in x, and the column vector $\mathbf{Q}(x) = (Q_0(x), Q_1(x), \ldots)^T$. Then the backward recurrence relations

$$-xQ_0(x) = -(\lambda_0 + \mu_0)Q_0(x) + \lambda_0 Q_1(x) \tag{3.1.8}$$

$$-xQ_n(x) = \mu_n Q_{n-1}(x) - (\lambda_n + \mu_n)Q_n(x) + \lambda_n Q_{n+1}(x) \quad (n \geq 1), \tag{3.1.9}$$

or more compactly

$$-x\mathbf{Q}(x) = \mathbf{A}\mathbf{Q}(x), \tag{3.1.10}$$

together with the normalizing condition

$$Q_0(x) = 1, \tag{3.1.11}$$

determine a sequence $\{Q_n(x)\}$ of orthogonal polynomials with respect to some measure $\psi(x)$. That is, there exists a function $\psi(x)$ such that

$$\int_0^\infty Q_i(x)Q_j(x)\,d\psi(x) = 0 \quad (i \neq j) \quad \text{and} \quad \int_0^\infty Q_j^2(x)\,d\psi(x) = 1/\omega_j, \tag{3.1.12}$$

where we denote

$$\omega_0 = 1 \quad \text{and} \quad \omega_n = \frac{\lambda_0 \lambda_1 \ldots \lambda_{n-1}}{\mu_1 \mu_2 \ldots \mu_n} \quad (n > 0). \tag{3.1.13}$$

Now consider the sequence of functions

$$f_i(x,t) \equiv \sum_{j=0}^\infty p_{ij}(t)Q_j(x), \tag{3.1.14}$$

i.e. in vector-matrix form

$$\mathbf{f}(x,t) = \mathbf{P}(t)\mathbf{Q}(x). \tag{3.1.15}$$

Then it follows from (3.1.10) and (3.1.15) that this vector satisfies the equation

$$\frac{\partial \mathbf{f}(x,t)}{\partial t} = \mathbf{P}'(t)\mathbf{Q}(x) = \mathbf{P}(t)\mathbf{A}\mathbf{Q}(x) = -x\mathbf{P}(t)\mathbf{Q}(x) = -x\mathbf{f}(x,t), \tag{3.1.16}$$

together with the initial condition

$$\mathbf{f}(x,0) = \mathbf{P}(0)\mathbf{Q}(x) = \mathbf{Q}(x) \tag{3.1.17}$$

(since $\mathbf{P}(0) = \mathbf{I}$). Hence

$$\mathbf{f}(x,t) = e^{-xt}\mathbf{f}(x,0) = e^{-xt}\mathbf{Q}(x). \tag{3.1.18}$$

So on extracting the ith element we have

$$f_i(x,t) \equiv \sum_{j=0}^\infty p_{ij}(t)Q_j(x) = e^{-xt}Q_i(x,t). \tag{3.1.19}$$

Multiplying both sides of this expression by $Q_j(x)$ and integrating with respect to $\psi(x)$ yields

$$\sum_{j=0}^\infty p_{ij}(t) \int_0^\infty Q_j(x)Q_j(x)\,d\psi(x) = \int_0^\infty e^{-xt}Q_i(x)Q_j(x)\,d\psi(x). \tag{3.1.20}$$

Whence use of the orthogonality relation (3.1.12) gives

$$p_{ij}(t) = \omega_j \int_0^\infty e^{-xt} Q_i(x) Q_j(x) \, d\psi(x). \tag{3.1.21}$$

Thus $p_{ij}(t)$ is simply the jth Fourier coefficient of $f_i(x,t)$. The value of this integral representation lies in the fact that not only is the time dependence contained entirely in the simple monotonic factor e^{-xt} in the integrand, but also that the dependence on i and j is transparently clear. Indeed, (3.1.21) exposes the remarkable property that

$$p_{ij}(t)/p_{ji}(t) = \pi_j/\pi_i. \tag{3.1.22}$$

The beauty of this approach is that it leads to powerful results on the underlying properties of the process, such as existence, uniqueness and recurrence. The downside is that determining the measure $\psi(x)$ and the associated probabilities (3.1.21) is complicated, even for simple linear immigration–birth–death processes for which the underlying orthogonal structure involves classical Meixner and Laguerre polynomials (see Karlin and McGregor, 1958); these in turn are related to continued fractions (Guillemin and Pinchon, 1999). A worthwhile and potentially rewarding challenge would be to develop a simplified user-friendly version of this technique which would work over a wide range of stochastic processes.

When $\mu_0 = 0$ there is one and only one matrix $\mathbf{P}(t)$ which satisfies (3.1.1)–(3.1.6) if and only if

$$\sum_{n=0}^\infty \left(\omega_n + \frac{1}{\lambda_n \omega_n} \right) = \infty. \tag{3.1.23}$$

Since we know of no application in which (3.1.23) is not satisfied, we shall always assume that $\mathbf{P}(t)$ is unique. If $\mu_0 > 0$ then considerably more care is required since (3.1.23) becomes

$$\sum_{n=0}^\infty \omega_n \left(1 + \mu_0 \sum_{k=0}^{n-1} \frac{1}{\lambda_k \omega_k} \right)^2 = \infty; \tag{3.1.24}$$

it may be shown that a sufficient condition for (3.1.24) to hold is that

$$\sum_{n=1}^\infty \frac{1}{\mu_n} = \infty. \tag{3.1.25}$$

Note that the linear death rate $\mu_n = \mu n$ just satisfies this last condition: if $\mu_n = \mu n^{1+\delta}$, then no matter how small $\delta > 0$ is we have

$$\frac{1}{\mu} \sum_{n=1}^\infty \frac{1}{n^{1+\delta}} = \frac{1}{\mu} \zeta(1+\delta) < \infty \quad \text{(for } \mu_0 > 0\text{)}, \tag{3.1.26}$$

where $\zeta(1+\delta)$ denotes the Riemann zeta function (2.3.71). Fortunately, real-life situations rarely involve processes that disappear, and so we may safely assume that $\mu_0 = 0$; the case of $\mu_0 > 0$ is best left to mathematical abstraction.

Two features of particular interest are the time of first return to state i, and the time of first reaching state j from state i. Let us therefore consider the *recurrence time distribution*

$$F_{ii}(t) = \Pr\{X(\tau_1) \neq i, X(\tau_2) = i \text{ for some } 0 < \tau_1 < \tau_2 \leq t \mid X(0) = i)\}, \quad (3.1.27)$$

and for $j \neq i$ the *first passage time distribution*

$$F_{ij}(t) = \Pr\{X(\tau) = j \text{ for some } 0 < \tau \leq t \mid X(0) = i\}. \quad (3.1.28)$$

Thus $F_{ii}(t)$ is the probability that the particle, having started at i, leaves i and returns to i before time t; whilst $F_{ij}(t)$ is the probability that the particle, having started at i, visits j before time t. For a rigorous discussion of these two quantities see Chung (1956).

Now we have already shown in (2.4.24), (2.4.25) and (2.4.27) that for the simple birth–death process $p_{i0}(\infty) = (\mu/\lambda)^i < 1$ if $\lambda > \mu$ and $p_{i0}(\infty) = 1$ if $\lambda \leq \mu$. This suggests that in general, on denoting $f_{ii}(t)$ and $f_{ij}(t)$ to be the probability density functions corresponding to the distribution functions $F_{ii}(t)$ and $F_{ij}(t)$, it may happen that

$$F_{ii}(\infty) = \int_0^\infty f_{ii}(u)\,du < 1 \quad \text{and} \quad F_{ij}(\infty) = \int_0^\infty f_{ij}(u)\,du < 1, \quad (3.1.29)$$

i.e. the distributions are dishonest. The integral $\int_0^\infty f_{ii}(u)\,du$ is the probability that the particle starts at i, leaves i, and then returns to i in a finite time. The ith state is called *recurrent* if $\int_0^\infty f_{ii}(u)\,du = 1$, i.e. eventual return to i is certain, and *transient* otherwise. A recurrent state i is called *positive recurrent* or *null recurrent* according as its expected recurrence time $\int_0^\infty u f_{ii}(u)\,du$ is finite or infinite. If all states of a process connect, a process is called positive recurrent, null recurrent or transient if every one of its states has the corresponding property.

To determine the relationship between $F_{ii}(t)$, $F_{ij}(t)$ and $p_{ii}(t)$, $p_{ij}(t)$, we first note that since

$$\Pr\{X(\tau) = i \text{ for all } 0 \leq \tau \leq t \mid X(0) = i\} = \exp\{-(\lambda_i + \mu_i)t\}, \quad (3.1.30)$$

we have

$$p_{ii}(t) = \Pr\{(\text{never leaves } i \text{ in } (0,t)) \bigcup (\text{returns to } i \text{ for the first time in}$$
$$(u, u+du) \text{ for } 0 \leq u < t \text{ and is at } i \text{ at time } t)\}$$
$$= e^{-(\lambda_i + \mu_i)t} + \int_0^t f_{ii}(u) p_{ii}(t-u)\,du. \quad (3.1.31)$$

Likewise for $j \neq i$

$$p_{ij}(t) = \int_0^t f_{ij}(u) p_{jj}(t-u)\,du. \quad (3.1.32)$$

On introducing the Laplace transforms

$$p_{ij}^*(s) = \int_0^\infty e^{-st} p_{ij}(t)\, dt \quad \text{and} \quad F_{ij}^*(s) = \int_0^\infty e^{-st} f_{ij}(t)\, dt, \qquad (3.1.33)$$

we then see from (3.1.31) and (3.1.32) that

$$p_{ii}^*(s) = \frac{1}{(s+\lambda_i+\mu_i)} + p_{ii}^*(s) F_{ii}^*(s), \qquad \text{i.e.}$$

$$p_{ii}^*(s) = \frac{1}{(s+\lambda_i+\mu_i)(1-F_{ii}^*(s))}, \qquad (3.1.34)$$

and

$$p_{ij}^*(s) = p_{jj}^*(s) F_{ij}^*(s) \qquad (j \neq i). \qquad (3.1.35)$$

Since

$$F_{ii}^*(s) \to \int_0^\infty f_{ii}(t)\, dt \leq 1$$

as $s \to 0$, it follows from (3.1.34) that the recurrence distributions are honest if and only if $p_{ii}^*(1) = \infty$, i.e. $\int_0^\infty p_{ii}(t)\, dt = \infty$. On integrating (3.1.21) with $i = j$, this last condition reduces to

$$\int_0^\infty p_{ii}(t)\, dt = \omega_i \lim_{t\to\infty} \int_0^\infty \frac{e^{-xt}}{-x} Q_i^2(x)\, d\psi(x) = \infty,$$

i.e.

$$\int_0^\infty \frac{Q_i^2(x)}{x}\, d\psi(x) = \infty. \qquad (3.1.36)$$

Because $Q_i(0) = 1$ for all i, (3.1.36) clearly holds for any i if and only if

$$\int_0^\infty \frac{d\psi(x)}{x} = \infty. \qquad (3.1.37)$$

The classical result that either all states or no states are recurrent is now apparent. Readers interested in the extensive algebraic detail that develops from this analysis should consult Karlin and McGregor (1957b), but the key result is that because the process is recurrent if $\int_0^\infty f_{ii}(s)\, ds = 1$ (i.e. there is zero probability that the individual remains at infinity), and since (from their result (9.9))

$$\int_0^\infty \frac{d\psi(x)}{x} = \sum_{n=0}^\infty \frac{1}{\lambda_n \omega_n}, \qquad (3.1.38)$$

we obtain:

THEOREM 3.1 *The process is recurrent if and only if*

$$\sum_{n=0}^\infty \frac{1}{\lambda_n \omega_n} = \infty. \qquad (3.1.39)$$

114 General Markov population processes

Since a birth–death process is defined in terms of the birth and death rates, λ_n and μ_n, this condition should be easy to check.

THEOREM 3.2 (a) *The process is positive recurrent if and only if $\sum_{n=0}^{\infty} w_n < \infty$ and $\sum_{n=0}^{\infty} 1/(\lambda_n w_n) = \infty$.*
(b) *The process is null recurrent if and only if $\sum_{n=0}^{\infty} w_n = \infty$ and $\sum_{n=0}^{\infty} 1/(\lambda_n w_n) = \infty$.*
(c) *The process is transient if and only if $\sum_{n=0}^{\infty} 1/(\lambda_n w_n) < \infty$.*

Example Consider the simple birth–death process with immigration when empty, i.e. $\lambda_0 = \alpha$, $\lambda_n = n\lambda$ $(n > 0)$ and $\mu_n = n\mu$ $(n \geq 0)$. Then (3.1.13) becomes $w_0 = 1$ and $w_n = (\alpha/n\lambda)(\lambda/\mu)^n$ $(n > 0)$, whence

$$\sum_{n=0}^{\infty} w_n = \begin{cases} 1 - (\alpha/\lambda)\ln[1 - (\lambda/\mu)] & \text{if } \lambda < \mu \\ \infty & \text{if } \lambda \geq \mu \end{cases} \quad (3.1.40)$$

and

$$\sum_{n=0}^{\infty} 1/(\lambda_n w_n) = \begin{cases} (1/\alpha)/[1 - (\mu/\lambda)] & \text{if } \mu < \lambda \\ \infty & \text{if } \mu \geq \lambda. \end{cases} \quad (3.1.41)$$

Hence on using Theorem 3.2 we have:

(a) $\sum_{n=0}^{\infty} w_n < \infty$ if $\lambda < \mu$ and $\sum_{n=0}^{\infty} 1/(\lambda_n w_n) = \infty$ if $\lambda \leq \mu$, so the process is positive recurrent if $\lambda < \mu$;
(b) $\sum_{n=0}^{\infty} w_n = \infty$ if $\lambda \geq \mu$ and $\sum_{n=0}^{\infty} 1/(\lambda_n w_n) = \infty$ if $\lambda \leq \mu$, so the process is null recurrent if $\lambda = \mu$;
(c) $\sum_{n=0}^{\infty} 1/(\lambda_n w_n) < \infty$ if $\mu < \lambda$, so the process is transient if $\lambda > \mu$.

Note that the general relationships (3.1.34) and (3.1.35) between the transition probability transforms $p_{ii}^*(s)$ and $p_{ij}^*(s)$ and the first return and first passage probability transforms $F_{ii}^*(s)$ and $F_{ij}^*(s)$ are of fundamental importance in their own right. For example, suppose that we wish to evaluate the distribution $F_{i0}(t)$ of the time taken for the M/M/1 queue to become empty, given that it contains $i > 0$ customers at time $t = 0$. To determine this result from first principles requires a parallel analysis to Section 2.6 with an absorbing barrier at zero (i.e. once the queue becomes empty no more customers arrive). Whilst to determine the first return distribution given that the queue is initially empty requires the convolution between $F_{10}(t-\tau)$ and the p.d.f. $f_{01}(\tau) = (1/\alpha)e^{-\tau/\alpha}$. In marked contrast, given that we already know from (2.6.14) with $n_0 = i$ that

$$p_{i0}^*(s) = \frac{\{z_0(s)\}^{i+1}}{\beta\{1 - z_0(s)\}}, \quad (3.1.42)$$

where $z_0(s)$ is defined in (2.6.13), use of (3.1.34) with $i = 0$ yields

$$F_{00}^*(s) = 1 - \frac{\beta\{1 - z_0(s)\}}{(s+\alpha)z_0(s)}; \quad (3.1.43)$$

whilst for $i > 0$ (3.1.35) gives

$$F_{i0}^*(s) = p_{i0}^*(s)/p_{00}^*(s) = \{z_0(s)\}^i. \qquad (3.1.44)$$

In principle, the corresponding c.d.f.'s $F_{00}(t)$ and $F_{i0}(t)$ ($i > 0$) may now be determined by taking the inverse Laplace transforms of (3.1.43) and (3.1.44). Though in practice this may well be non-trivial, as indicated in the discussion following (2.6.29).

This approach is not restricted to honest processes. For example, we see from the probability transforms (2.3.52) and (2.3.53) for the general pure birth process that

$$p_{ii}^*(s) = \left(\frac{1}{s+\lambda_i}\right) \quad \text{and} \quad p_{ij}^*(s) = \frac{\lambda_i \lambda_{i+1} \ldots \lambda_{j-1}}{(s+\lambda_i)(s+\lambda_{i+1})\ldots(s+\lambda_j)} \quad (j > i), \qquad (3.1.45)$$

whence $F_{ii}^*(s)$ and $F_{ij}^*(s)$ ($j > i$) follow by applying results (3.1.34) and (3.1.35), respectively. We examined the special case $\lambda_i = i^{1+\delta}$ ($\delta \geq 0$) in Section 2.3.5, and showed that whilst the process drifts towards infinity if $\delta = 0$ (simple birth–death process with $\mu = 0$) it reaches infinity in a finite time if $\delta > 0$. Indeed, if $\delta = 1$ the distribution function of the time taken to reach infinity is given by the theta function (2.3.67), which has mean $\pi^2/6$ and variance $\pi^4/90$.

In general, a transient process will approach infinity with probability one, and the point at infinity can be regarded as a permanent absorbing state which may be reached in finite (e.g. $\delta > 0$) or infinite (e.g. $\delta = 0$) time. Paralleling (3.1.29), we define a transient process to be dishonest if for some finite $t > 0$ and some i we have $\sum_{j=0}^{\infty} p_{ij}(t) < 1$, otherwise it is called honest. Moreover, for transient dishonest processes $\sum_{j=0}^{\infty} p_{ij}(t) \to 0$ exponentially fast as $t \to \infty$; and, for given i and (finite) t,

$$\Pr\{\lim_{\tau \to t-0} X(\tau) = \infty \mid X(0) = i\} = 1 - \sum_{j=0}^{\infty} p_{ij}(t). \qquad (3.1.46)$$

THEOREM 3.3 *A transient process is honest, that is $\sum_{j=0}^{\infty} p_{ij}(t) = 1$ for all $i = 0, 1, 2, \ldots$ and $t > 0$, if and only if*

$$\sum_{n=0}^{\infty} \omega_n \sum_{i=n}^{\infty} \frac{1}{\lambda_i \omega_i} = \infty. \qquad (3.1.47)$$

The next theorem is obvious for dishonest transient processes, but it does provide a useful insight into the path functions of honest transient processes since it shows that they must always 'end-up' at infinity.

THEOREM 3.4 *For any transient process*

$$\Pr\{\lim_{t \to \infty} X(t) = \infty \mid X(0) = i\} = 1. \qquad (3.1.48)$$

116 *General Markov population processes*

Let $\tilde{T}_i$ denote the time at which the process reaches infinity from the initial state $X(0) = i$. Then

$$H_i(t) \equiv \Pr(\tilde{T}_i \leq t) = 1 - \sum_{j=0}^{\infty} p_{ij}(t). \qquad (3.1.49)$$

Now for transient honest processes $\Pr(\tilde{T}_i = \infty) = 1$. However, for transient dishonest processes we have:

THEOREM 3.5   *If* $\Pr(\tilde{T}_i < \infty) > 0$, *then* $\Pr(\tilde{T}_i < \infty) = 1$ *and* $1 - H_i(t)$ *tends exponentially to zero as* $t \to \infty$.

Thus with probability one transient processes which are honest take an infinite time to reach infinite size, whilst dishonest ones not only take a finite time but the probability that the process is of finite size at time t decays exponentially with increasing t.

THEOREM 3.6   *For initial state* $X(0) = 0$, *the raw moments of* $\tilde{T}_0$ *are given by*

$$E(\tilde{T}_0) = \sum_{j=0}^{\infty} \omega_j \sum_{n=j}^{\infty} \frac{1}{\lambda_n \omega_n}, \qquad (3.1.50)$$

$$E(\tilde{T}_0^2) = 2 \sum_{j=0}^{\infty} \omega_j \sum_{m=j}^{\infty} \frac{1}{\lambda_m \omega_m} \sum_{i=0}^{m} \omega_i \sum_{k=i}^{\infty} \frac{1}{\lambda_k \omega_k}, \qquad (3.1.51)$$

and in general

$$E(\tilde{T}_0^n) = n! \sum_{j=0}^{\infty} \omega_j \int_0^{\infty} \frac{Q_j(x)}{x^n} d\psi(x). \qquad (3.1.52)$$

For example, in Section 2.3.5 we consider the dishonest pure birth process with $\lambda_n = n^2$, and show that $E(\tilde{T}_1) = \pi^2/6$ and $\text{Var}(\tilde{T}_1) = \pi^4/90$ (results (2.3.69) and (2.3.70)). Suppose we now generalize this process by superimposing linear death with $\mu_n = n\mu$ $(n \geq 0)$ and $\lambda_0 = \lambda$ (i.e. immigration at rate λ when the process is empty) on $\lambda_n = \lambda n^2$ $(n > 0)$. Then (3.1.13) gives

$$\omega_0 = 1 \quad \text{and} \quad \omega_n = \frac{(\lambda/\mu)^n (n-1)!}{n} \quad (n > 0). \qquad (3.1.53)$$

From Theorem 3.2c we see that as

$$\sum_{n=0}^{\infty} \frac{1}{\lambda_n \omega_n} = \frac{1}{\lambda} + \sum_{n=1}^{\infty} \left[\frac{1}{\lambda n^2} \times \frac{n}{(\lambda/\mu)^n (n-1)!} \right] = (1/\lambda) e^{\mu/\lambda} < \infty, \qquad (3.1.54)$$

the process is clearly transient, as one would anticipate from the explosive nature of the birth rate λn^2 as n increases. Whilst from Theorem 3.6 we have

$$\mathrm{E}(\tilde{T}_0) = \frac{1}{\lambda} + \sum_{n=1}^{\infty}\left[\frac{1}{\lambda n^2} \times \frac{n}{(\lambda/\mu)^n(n-1)!}\right]$$
$$+ \sum_{j=1}^{\infty} \frac{(\lambda/\mu)^j(j-1)!}{j} \sum_{n=j}^{\infty}\left[\frac{1}{\lambda n^2} \times \frac{n}{(\lambda/\mu)^n(n-1)!}\right],$$

which reduces to

$$\mathrm{E}(\tilde{T}_0) = (1/\lambda)[e^{\mu/\lambda} + \sum_{r=0}^{\infty}(\mu/\lambda)^r \sum_{j=1}^{\infty}\frac{(j-1)!}{j(r+j)!}]. \qquad (3.1.55)$$

Although this representation does not appear to have a simple closed form solution, it is nevertheless fairly neat and is easily obtained from the general result (3.1.50). Construction of $\mathrm{Var}(\tilde{T}_0)$ follows from (3.1.51) along parallel lines.

3.2 Equilibrium solutions

Whilst the time-dependent properties of a process are clearly important in the opening (i.e. transient) stages of development, should the process then settle down into a persistent (i.e. stationary) mode of behaviour then problems of mathematical intractability no longer apply. In practice, this means that we have neither extinction nor eventual approach to infinite population size, as happens in the simple birth–death process, nor do we have population explosion, as in the case of the dishonest (i.e. divergent) birth process. Let us therefore now assume that as $t \to \infty$ the population size probabilities $\{p_N(t)\}$ approach the constant values $\{\pi_N\}$, where $\pi_N > 0$ over a set of N-values with $\sum_{N=0}^{\infty} \pi_N = 1$. The Kolmogorov forward equation (3.0.2) then reduces to the much simpler form

$$0 = \mu_{N+1}\pi_{N+1} - (\lambda_N + \mu_N)\pi_N + \lambda_{N-1}\pi_{N-1} \qquad (3.2.1)$$

for $N = 0, 1, 2, \ldots$, since $d\pi_N/dt = 0$.

We have already analysed a particular case of this general situation, namely the immigration–birth–death process (Section 2.5.3), and the solution of equation (3.2.1) proceeds on exactly the same lines. First rewrite it in the form

$$\mu_{N+1}\pi_{N+1} - \lambda_N\pi_N = \mu_N\pi_N - \lambda_{N-1}\pi_{N-1} \qquad (N = 1, 2, \ldots). \qquad (3.2.2)$$

Then since the two sides of this equation are identical, apart from N on the left being replaced by $N-1$ on the right, we may write

$$\mu_N\pi_N - \lambda_{N-1}\pi_{N-1} = \text{constant} \qquad (3.2.3)$$

independent of $N = 1, 2, \ldots$. Take $\mu_0 = 0$ (if there are no individuals present, then none can die), together with $\pi_{-1} = 0$ (by definition, since N cannot be negative). Placing $N = 0$ in (3.2.3) then gives

$$\mu_0\pi_0 - \lambda_{-1}\pi_{-1} = \text{constant} = 0,$$

118 *General Markov population processes*

and hence the balance equation

$$\mu_N \pi_N = \lambda_{N-1} \pi_{N-1}. \tag{3.2.4}$$

That is, the probability flow from state $N-1$ to N matches that from state N to $N-1$ (see the remark preceeding (2.5.29)). Repeated application of this recurrence relation now yields

$$\pi_N = \frac{\lambda_0 \lambda_1 \ldots \lambda_{N-1}}{\mu_1 \mu_2 \ldots \mu_N} \pi_0. \tag{3.2.5}$$

Hence on recalling from (3.1.13) the notation

$$\omega_0 = 1 \quad \text{and} \quad \omega_N = \frac{\lambda_0 \lambda_1 \ldots \lambda_{N-1}}{\mu_1 \mu_2 \ldots \mu_N} \quad (N > 0), \tag{3.2.6}$$

and choosing π_0 to ensure that $\sum_{N=0}^{\infty} \pi_N = 1$ (i.e. the process is recurrent), we obtain the equilibrium solution

$$\pi_N = \omega_N / \sum_{i=0}^{\infty} \omega_i \quad (N = 0, 1, 2, \ldots). \tag{3.2.7}$$

To illustrate this approach consider the birth–death–immigration process of Section 2.5 for which $\lambda_N = \alpha + \lambda N$ and $\mu_N = \mu N$. Direct substitution of these rates into (3.2.6) yields

$$\omega_N = \frac{(\alpha)(\alpha+\lambda)\ldots(\alpha+\lambda(N-1))}{(\mu)(2\mu)\ldots(N\mu)} = \binom{N-1+(\alpha/\lambda)}{N}(\lambda/\mu)^N, \tag{3.2.8}$$

whence

$$\sum_{i=0}^{\infty} \omega_i = [1 - (\lambda/\mu)]^{-\alpha/\lambda} = 1/\pi_0.$$

Thus (3.2.7) yields

$$\pi_N = \pi_0 \binom{N-1+(\alpha/\lambda)}{N}(\lambda/\mu)^N \quad (N = 0, 1, 2, \ldots) \tag{3.2.9}$$

in exact agreement with (2.5.22).

Although the results of Section 3.1 hinge on λ_N ($N \geq 0$) and μ_N ($N > 0$) being strictly positive, from an applied probability perspective this assumption is unnecessarily restrictive. For if for some integers $0 < a < b$ we have $\mu_a = \lambda_b = 0$, then provided $a \leq X(0) \leq b$ the process is forced to lie in the range $a \leq X(t) \leq b$ for all $t \geq 0$; so if all states in $[a, b]$ connect then the process has to be recurrent. For example, suppose we combine the simple birth rate $\lambda_N = \lambda N$ with the severe density-dependent death rate $\mu_N = \mu N(N-1)$. Then since $\mu_1 = 0$ the process is defined over $N = 1, 2, \ldots$. Thus the balance equations (3.2.4) now give

$$\pi_N = \tilde{\omega}_N \pi_1 \quad (N \geq 1)$$

where, in contrast to (3.2.6), we denote

$$\tilde{\omega}_1 = 1 \quad \text{and} \quad \tilde{\omega}_N = \frac{\lambda_1 \lambda_2 \ldots \lambda_{N-1}}{\mu_2 \mu_3 \ldots \mu_N} \quad (N > 1). \tag{3.2.10}$$

So for this specific example

$$\tilde{\omega}_N = \frac{(\lambda/\mu)^{N-1}}{N!}.$$

Whence

$$\sum_{N=1}^{\infty} \tilde{\omega}_N = (\lambda/\mu)^{-1}(e^{\lambda/\mu} - 1),$$

thereby yielding

$$\pi_N = \tilde{\omega}_N / \sum_{i=1}^{\infty} \tilde{\omega}_i = \frac{(\lambda/\mu)^N}{N!}(e^{\lambda/\mu} - 1)^{-1} \tag{3.2.11}$$

which is a censored Poisson distribution over $N = 1, 2, \ldots$.

Over this range the equilibrium distribution (3.2.11) has an identical structure to that of the immigration–death process (expression (2.5.9)) in spite of the considerable dissimilarity between the two sets of birth and death rates. This highlights yet again that the nature of the equilibrium p.d.f. sheds remarkably little light on the structure of the underlying generating mechanism. That must come from direct knowledge of the real-life process under study.

The equivalence between the balance equations (3.2.4), which represent zero net flow between adjacent states N and $N+1$, and the full Kolmogorov forward equilibrium equations (3.2.1), which represent zero net flow into and out of state N, is a direct consequence of single-jump population processes (i.e. N can change only to $N-1$ or $N+1$). For consider the three-state periodic process which cycles through $0 \to 1 \to 2 \to 0$, etc. at unit rate; so now $\lambda_0 = \lambda_1 = \mu_2 = 1$ and $\mu_0 = \mu_1 = \lambda_2 = 0$. Then equations (3.2.1), namely

$$0 = \pi_2 - \pi_0 = \pi_0 - \pi_1 = \pi_1 - \pi_2,$$

yield the uniform probabilities $\pi_0 = \pi_1 = \pi_2 = 1/3$. These correspond to the *stationary* distribution of the process, and they do *not* satisfy the balance equations (3.2.4) since (for example) $\pi_0 \lambda_0 = 1/3 \neq \pi_1 \mu_1 = 0$.

In many situations the finite nature of the resource available for population growth ensures that an isolated population cannot grow without limit. So provided we wait long enough, a downward fluctuation in population size that is sufficiently violent to drive the population extinct is bound to occur. However, if a long time (T_{ext}) has to elapse before the probability of extinction becomes non-negligible, then although a proper equilibrium distribution $\{\pi_N\}$ of population size will not formally exist (we exclude the trivial case $\pi_0 = 1$) there will be a stationary *conditional* distribution $\{\pi_N^{(Q)}\}$ for which

$$\pi_N^{(Q)} = \lim_{M \to \infty} \{(1/M) \times \text{number of realizations containing exactly } N > 0 \text{ individuals}\}. \tag{3.2.12}$$

Here M denotes the number of non-extinct realizations at some large time $T < T_{ext}$. Thus for all practical purposes this *quasi-equilibrium* distribution $\{\pi_N^{(Q)}\}$ is effectively a true equilibrium distribution. Ideally we therefore need to work in terms of the probability size distribution as $t \to \infty$ conditional on extinction not having occurred, by defining

$$\pi_N^{(Q)} = \lim_{t \to \infty} p_N(t)/[1 - p_0(t)]. \tag{3.2.13}$$

On denoting $q_N(t) = p_N(t)/(1 - p_0(t))$ with $p_N'(t) = dp_N(t)/dt$, etc., the forward equations (3.0.2) with $\lambda_{-1} = \mu_0 = \lambda_0 = 0$, namely

$$p_0'(t) = \mu_1 p_1(t) \tag{3.2.14}$$

$$p_N'(t) = \lambda_{N-1} p_{N-1}(t) - (\lambda_N + \mu_N) p_N(t) + \mu_{N+1} p_{N+1}(t) \quad (N = 1, 2, \ldots), \tag{3.2.15}$$

take the form

$$p_0'(t) = \mu_1 (1 - p_0(t)) q_1(t) \tag{3.2.16}$$

$$q_N'(t) - q_N(t) p_0'(t)/(1 - p_0(t)) = \lambda_{N-1} q_{N-1}(t) - (\lambda_N + \mu_N) q_N(t) + \mu_{N+1} q_{N+1}(t). \tag{3.2.17}$$

Whence combining these two equations yields (for $N = 1, 2, \ldots$)

$$q_N'(t) - \mu_1 q_1(t) q_N(t) = \lambda_{N-1} q_{N-1}(t) - (\lambda_N + \mu_N) q_N(t) + \mu_{N+1} q_{N+1}(t). \tag{3.2.18}$$

Letting $t \to \infty$ then gives $\pi_N^{(Q)} = \lim_{t \to \infty} q_N(t)$ as the solution to the system of equations

$$\lambda_{N-1} \pi_{N-1}^{(Q)} - (\lambda_N + \mu_N) \pi_N^{(Q)} + \mu_{N+1} \pi_{N+1}^{(Q)} = -\mu_1 \pi_1^{(Q)} \pi_N^{(Q)}. \tag{3.2.19}$$

If λ_N and μ_N are nonlinear, then apart from a few trivial exceptions this system will not give rise to a closed form solution for $\{\pi_N^{(Q)}\}$. Numerical solutions can, however, easily be achieved by solving (3.2.19) iteratively over an appropriate range $N_{min} \leq N \leq N_{max}$ which covers all non-negligible values of $\pi_N^{(Q)}$. One obvious approach when $N_{min} = 1$ is to employ the:

Exact quasi-equilibrium probability algorithm (A3.1)

(i) place $\pi_1^{(Q)} = 1$
(ii) use this to determine $\pi_N^{(Q)}$ for $N = 2, 3, \ldots, N_{max}$
(iii) form $\gamma = \sum_{N=N_{min}}^{N_{max}} \pi_N^{(Q)}$
(iv) rescale via $\pi_N^{(Q)} \to \pi_N^{(Q)}/\gamma$
(v) return to (ii) with this new value for $\pi_1^{(Q)}$

The iterations cease when an appropriate convergence criterion is met. Provided $\pi_1^{(Q)}$ is not so small that it leads to round-off error, this procedure should be perfectly adequate. If it is not then a considerable improvement can easily be made by first solving the balance equation $\lambda_{N-1}\pi_{N-1} = \mu_N \pi_N$ with $\pi_{N-1} \simeq \pi_N$ in order to determine the modal value, N_{mod}, and then replacing steps (ii) and (v) by:

(ii)' determine $\pi_N^{(Q)}$ over $N = N_{mod} - 1, \ldots, N_{min}$ and $N = N_{mod} + 1, \ldots, N_{max}$
(v)' return to (ii)' with this new value for $\pi_{N_{mod}}^{(Q)}$

However, a far simpler, and more pragmatic, way of determining the nature of $\{\pi_N^{(Q)}\}$ is to use the balance equations (3.2.4) over $N > 0$ to obtain the analytic solution (3.2.10)–(3.2.11), namely

$$\tilde{\pi}_N = \tilde{\omega}_N / \sum_{i=1}^{\infty} \tilde{\omega}_i \quad \text{where} \quad \tilde{\omega}_N = \frac{\lambda_1 \lambda_2 \ldots \lambda_{N-1}}{\mu_2 \mu_3 \ldots \mu_N} \quad (N = 1, 2, \ldots). \quad (3.2.20)$$

For this direct, albeit approximate, approach is clearly much easier to work with than the system (3.2.19). How safely we can use it depends on the degree of influence exerted by the $-\mu_1 \pi_1^{(Q)} \pi_N^{(Q)}$ term on the right-hand side of (3.2.19). Intuitively, provided $\pi_1^{(Q)}$ is tiny compared with $\pi_{N_{mod}}^{(Q)}$ the difference between the exact and approximate distributions should be negligible. In essence, whilst the conditional nature of (3.2.12) and (3.2.13) ensures that transitions from 1 to 0 do not occur, to prevent this happening under (3.2.20) we have imposed the condition that $\mu_1 = 0$, which alters the underlying process. Note that there is no unique way of effecting this, but leaving all the other rates $\mu_2, \mu_3 \ldots$ and $\lambda_1, \lambda_2, \ldots$ unchanged is clearly the least disruptive. One could argue that although (3.2.13) provides the mathematically correct definition of quasi-equilibrium, it is the approximation (3.2.20) to it that is intuitively correct. For it is based purely on two eminently sensible assumptions. First, that the probability flow, $\lambda_{N-1}\tilde{\pi}_{N-1}$, from state $N - 1$ to N is in exact balance with the probability flow, $\mu_N \tilde{\pi}_N$, from N to $N - 1$. Second, that if the process had become extinct then we would not be observing it, so movement from state 1 to 0 is excluded automatically.

An alternative strategy is to change the original process by allowing the presence of an immortal individual. For then extinction is impossible and so the new quasi-equilibrium probabilities, $\check{\pi}_N$, are the true equilibrium probabilities of this modified process. Here each death rate μ_N is replaced by μ_{N-1}, so the balance equations, $\mu_{N-1}\check{\pi}_N = \lambda_{N-1}\check{\pi}_{N-1}$, now yield the solution

$$\check{\pi}_N = \check{\omega}_N / \sum_{i=1}^{\infty} \check{\omega}_i \quad \text{where} \quad \check{\omega}_N = \frac{\lambda_1 \lambda_2 \ldots \lambda_{N-1}}{\mu_1 \mu_3 \ldots \mu_{N-1}} \quad (N = 1, 2, \ldots). \quad (3.2.21)$$

Comparing this expression with (3.2.20) shows that the two processes have a very similar structure, except for $\tilde{\omega}_N$ in the first being replaced by $\check{\omega}_N = (\mu_N/\mu_1)\tilde{\omega}_N$ in the second. There is some evidence to suggest that $\tilde{\pi}_N$ provides a better approximation to $\pi_N^{(Q)}$ when the time to extinction is long, and that $\check{\pi}_N$ does better when the time to extinction is short. Note that replacing step (i) in algorithm A3.1 by (i)′ place $\pi_1^{(Q)} = \tilde{\pi}_1$ or $\check{\pi}_1$

reduces the time to convergence.

Provided the probability of being in state 1 is very small, which is generally the case when quasi-equilibrium is being considered in real-life situations, $\pi_N^{(Q)}$ and $\tilde{\pi}_N$ will often be in close agreement. We just have to be on guard for situations in which this assumption does not hold, generally when the time to extinction is not long. Nåsell (2001) presents an interesting analysis of such quasi-equilibrium issues in which he identifies three parameter regions which exhibit qualitatively different behaviour. In one the quasi-equilibrium distribution is approximately Normal and the time to extinction is long. Another region has a short time to extinction and a quasi-equilibrium distribution that is approximately a truncated geometric. Whilst the third is a transition region between these two in which the time to extinction is moderately long and the quasi-equilibrium distribution has a more complicated structure. With this caveat in mind we shall retain $\tilde{\pi}_N$ as our working definition of quasi-equilibrium.

To illustrate a rather extreme case where the time to extinction is likely to be short, consider a Markov process over just three states 0, 1 and 2 having transition rates $\lambda_0 = \lambda_2 = 0$, $\lambda_1 = \lambda$ and $\mu_1 = \mu_2 = \mu$. Then the population size probabilities $p_N(t)$ satisfy the linear differential equations

$$p_0'(t) = \mu p_1(t)$$
$$p_1'(t) = \mu p_2(t) - (\lambda + \mu) p_1(t) \quad (3.2.22)$$
$$p_2'(t) = \lambda p_1(t) - \mu p_2(t).$$

Eliminating $p_1(t)$ and $p_2(t)$ produces the third-order ordinary differential equation

$$p_0'''(t) + (\lambda + 2\mu) p_0''(t) + \mu^2 p_0'(t) = 0, \quad (3.2.23)$$

which, on writing $\theta_1, \theta_2 = (1/2)[-(\lambda + 2\mu) \pm \sqrt{\{(\lambda^2 + 4\lambda\mu)\}}] < 0$, yields the solution

$$p_0(t) = 1 - Ae^{\theta_1 t} - Be^{\theta_2 t} \quad (3.2.24)$$

for constants of integration A and $B = 1 - A$. Here we avoid the trivial starting state 0 by placing $p_0(0) = 0$; note that $p_0(\infty) = 1$ since 0 is an absorbing state. If we further assume that the process starts in state 2 at time $t = 0$, i.e. $p_2(0) = 1$, then we have $A = 1/[\theta_1(\theta_1 - \theta_2)]$ and $B = 1/[\theta_2(\theta_2 - \theta_1)]$. Substituting for $p_0(t)$ into the first two equations for $p_1(t)$ and $p_2(t)$ yields the conditional probabilities

$$q_1(t) = p_1(t)/[1 - p_0(t)] = -(1/\mu)[A\theta_1 e^{\theta_1 t} + B\theta_2 e^{\theta_2 t}]/[Ae^{\theta_1 t} + Be^{\theta_2 t}] \quad (3.2.25)$$

$$q_2(t) = p_2(t)/[1 - p_0(t)] = -(1/\mu^2)[A\theta_1(\theta_1 + \lambda + \mu)e^{\theta_1 t}$$
$$+ B\theta_2(\theta_2 + \lambda + \mu)e^{\theta_2 t}]/[Ae^{\theta_1 t} + Be^{\theta_2 t}]. \quad (3.2.26)$$

Whence letting $t \to \infty$ gives $\pi_1^{(Q)} = -\theta_1/\mu$ and $\pi_2^{(Q)} = 1 + \theta_1/\mu$. To compare these values with those derived through (3.2.20), we simply note that states 1 and 2 are in balance when $\lambda\tilde{\pi}_1 = \mu\tilde{\pi}_2$. Since in quasi-equilibrium $\tilde{\pi}_1 + \tilde{\pi}_2 = 1$, we therefore have $\tilde{\pi}_1 = \mu/(\mu + \lambda)$ and $\tilde{\pi}_2 = \lambda/(\mu + \lambda)$, which can clearly differ quite markedly from

$\pi_1^{(Q)}$ and $\pi_2^{(Q)}$. For example, if $\lambda = \mu = 1$, then $\tilde{\pi}_1 = 1/2$ as compared with $\pi_1^{(Q)} = (3 - \sqrt{5})/2 = 0.3820$.

3.2.1 Logistic population growth

Historically, development of quasi-equilibrium situations took the following route. Suppose that the net growth rate per individual, say $f(N)$, is a function of the total population size N at time $t \geq 0$. Then the deterministic rate of increase

$$dN/dt = Nf(N). \tag{3.2.27}$$

On assuming that the inhibitory effect on further growth increases with N, it follows that for large N the derivative $df(N)/dt$ must be negative. Now the simplest assumption to make is that $f(N)$ is linear, i.e.

$$f(N) = r[1 - (N/K)] \tag{3.2.28}$$

for some positive constants r and K. Combining (3.2.27) and (3.2.28) then yields

$$dN/dt = rN[1 - (N/K)], \tag{3.2.29}$$

which is the well-known Verhulst–Pearl logistic equation.

Integrating (3.2.28) with $N(0) = n_0$ yields the solution

$$N(t) = K/[1 + \{(K - n_0)/n_0\} \exp(-rt)] \quad (t \geq 0), \tag{3.2.30}$$

though a slightly neater representation is

$$N(t) = K/[1 + \exp\{-r(t - t_0)\}] \quad (t \geq 0) \tag{3.2.31}$$

where t_0 is defined by

$$(K - n_0)/n_0 = \exp(rt_0), \quad \text{i.e.} \quad t_0 = (1/r)\ln[(K/n_0) - 1]. \tag{3.2.32}$$

As $N(t_0) = K/2$, the value t_0 denotes the time taken for the population to reach half its maximum size K (for $n_0 < K/2$). Yet another representation may be obtained by replacing t by $t + 1$ and n_0 by $N(t)$, so forming

$$N(t+1) = K/[1 + \{(K - N(t))/N(t)\}e^{-r}] = e^r N(t)/[1 + N(t)(e^r - 1)/K]. \tag{3.2.33}$$

This expression is not only useful for sketching the shape of $N(t)$, but it may also be regarded as a possible model for a population growing at the discrete time points $t = 0, 1, 2, \ldots$. Though care must be taken if this form is used as a template for constructing discrete-time generalizations, namely $N_{t+1} = F(N_t)$ for $t = 0, 1, 2, \ldots$, in an attempt to mimic the general continuous-time construct (3.2.27). For example, both the quadratic, $N_{t+1} = rN_t(1 - N_t)$, and exponential, $N_{t+1} = N_t \exp\{r(1 - N_t/K)\}$, models produce a remarkable range of behaviour ranging from stable equilibrium points, to stable oscillations, through to chaos with increasing r (see Renshaw 1994c, for a discussion of deterministic and stochastic behaviour).

It is interesting to note that although Verhulst suggested the use of the simple logistic form (3.2.30) as early as 1838 to describe the growth of human populations, his work was virtually ignored until 1920 when Pearl and Reed derived it as an empirical curve which meets the following realistic conditions:

(i) it is asymptotic to the line $N(t) = K$ as $t \to +\infty$;
(ii) $N(t) = 0$ at $t = -\infty$; and,
(iii) it has a point of inflexion at some time t_0.

Shortly after this, Lotka (1925) provided a rational, as distinct from an empirical, derivation; further early references are contained in Pearl (1930).

The deterministic logistic curve (3.2.30) has been applied to a huge variety of different scenarios, meeting with varying degrees of success in summarizing the population under study. Sometimes such summaries can be surprisingly precise. Renshaw (1991), for example, presents a yeast growth situation in which the data points lie almost exactly on the linear rearranged form

$$Y(t) \equiv \ln[(K - N(t))/N(t)] = \ln[(K - n_0)/n_0] - rt. \qquad (3.2.34)$$

As yeast is a simple structure growing by cell division, there are good a priori reasons for believing that the (mathematical) representation (3.2.30) should successfully mimic the (biological) growth of the yeast population. Yet one of the surprising features of the logistic curve is that although for many situations the simple arguments which underpin it cannot hold true, population growth often closely follows this curve even when it does not satisfy the Verhulst–Pearl assumptions. The temptation to accept it as a universal law of population growth must be resisted; its prime importance lies in the parameter estimates which provide useful summary statistics.

This apparent paradigm with 'real life' throws up a number of interesting mathematical challenges. For example, Sang (1950) points out that in Pearl's (1927) laboratory experiments on the growth of *Drosophila melanogaster* the yeast that was the source of food was itself a growing population, yet the logistic still fits the data reasonably well. Is this an artefact of the corresponding two-species stochastic growth process? Moreover, within the confines of a long-running laboratory experiment, mutation is likely to occur, whilst less well-controlled populations may be subject to external influences. So how do such perturbations to the logistic system affect the population dynamics? Such questions clearly need to be answered.

Though not all data sets show good agreement with the deterministic logistic curve, many do exhibit reasonable agreement during the initial sigmoid stage followed by some degree of stabilization around the asymptote K. Two such cases, based on the growth of the sheep populations in Tasmania and South Australia from 1818–1936 and 1838–1936, respectively, are also discussed in Renshaw (1991). In stark contrast to the yeast scenario, these sheep population sizes exhibit a substantial degree of variability around the underlying deterministic asymptote, which highlights the need for a stochastic analysis. Whilst construction of the initial time-dependent probabilities $\{p_N(t)\}$, and hence the associated true quasi-equilibrium probabilities $\{\pi_N^{(Q)}\}$, poses considerable difficulties, that of the pragmatic quasi-equilibrium distribution $\{\tilde{\pi}_N\}$ is extremely simple. For under logistic growth λ_N must decrease and μ_N must increase as N increases. So we may write

$$\lambda_N = N(a_1 - b_1 N) \quad \text{and} \quad \mu_N = N(a_2 + b_2 N) \qquad (3.2.35)$$

for some positive constants a_1, a_2, b_1 and b_2. Note that since the associated deterministic equation

$$dN/dt = \lambda_N - \mu_N = N[(a_1 - a_2) - (b_1 + b_2)N], \qquad (3.2.36)$$

comparison with (3.2.29) shows that we require $a_1 - a_2 = r$ and $b_1 + b_2 = r/K$. Substituting (3.2.35) into (3.2.10) then gives

$$\tilde{\omega}_1 = 1 \quad \text{and} \quad \tilde{\omega}_N = \frac{[a_1 - b_1]\ldots[a_1 - (N-1)b_1]}{[a_2 + 2b_2]\ldots[a_2 + Nb_2]N} \quad (N > 1) \qquad (3.2.37)$$

as the component terms in the quasi-equilibrium distribution (3.2.20) for $\{\tilde{\pi}_N\}$. Because $\lambda_N = N(a_1 - b_1 N)$ is a birth rate, and so cannot be negative, N is now restricted to the range $1 \leq N \leq a_1/b_1$.

The literature contains a variety of interesting discussions concerning different ways of constructing quasi-equilibrium probability distributions. In the epidemiological context, the stochastic logistic model is called the endemic SIS model (see Section 8.3), and was first discussed by Weiss and Dishon (1971). Early work on the stochastic logistic model has been summarized well by Kryscio and Lefèvre (1989), Jacquez and Simon (1993) and Nåsell (1996) (1996, 1999), the latter within an epidemic setting. Whilst Ovaskainen (2001) provides a mathematically rigorous approximation formula for the (scaled) logistic process as $N \to \infty$.

Construction of quasi-equilibrium probabilities for *any* birth–death process with given birth and death rates λ_N and μ_N can clearly be developed just as easily along parallel lines. Consider, for example, the power-law logistic rates (Banks 1994)

$$\lambda_N = \begin{cases} N(a_1 - b_1 N^s) & \text{for} \quad N \leq (a_1/b_1)^{1/s} \\ 0 & \text{otherwise} \end{cases} \qquad (3.2.38)$$

$$\mu_N = N(a_2 + b_2 N^s) \qquad (3.2.39)$$

for integer $s \geq 1$. As for the ordinary logistic process, a_1 and a_2 denote the intrinsic growth rates, i.e. the per-capita birth and death rates in the absence of any population density pressure induced by the crowding coefficients b_1 and b_2. Direct substitution of (3.2.38) and (3.2.39) into (3.2.10) then yields the quasi-equilibrium terms

$$\tilde{\omega}_1 = 1 \quad \text{and} \quad \tilde{\omega}_N = \frac{[a_1 - b_1][a_1 - b_1 2^s]\ldots[a_1 - b_1(N-1)^s]}{[a_2 + b_2 2^s]\ldots[a_2 + b_2 N^s]N} \quad (N > 1). \qquad (3.2.40)$$

Although the same power s is used in both the birth and death rates, there is no reason why this has to hold in general.

This power-law logistic process is of considerable interest in its own right (see, for example, Matis and Kiffe, 2000). The deterministic representation

$$dN/dt = \lambda_N - \mu_N = N[(a_1 - a_2) - (b_1 + b_2)N^s] = aN - bN^{s+1} \qquad (3.2.41)$$

has the solution

$$N(t) = K/[1 + c\exp(-ast)]^{1/s}, \qquad (3.2.42)$$

where the carrying capacity $K = (a/b)^{1/s}$ and $c = (K/n_0)^s - 1$. The point of inflexion (for $n_0 < K$) occurs at $N = N_{infl}$, where

$$0 = \frac{d^2N}{dt^2} = a - b(s+1)N^s, \qquad \text{i.e.} \quad N^s = \frac{a}{b(s+1)} = \frac{K^s}{(s+1)}.$$

So

$$N_{infl} = K/(s+1)^{1/s}, \qquad (3.2.43)$$

and substituting this value into (3.2.42) shows that this is attained at time

$$t_{infl} = (1/as)\ln(c/s). \qquad (3.2.44)$$

For the ordinary logistic process ($s = 1$) we have $N_{infl} = K/2$, whilst as $s \to \infty$ we see that $N_{infl} \to K$. For example, N_{infl} is $K/1.732$ when $s = 2$, $K/1.587$ when $s = 3$ and $K/1.495$ when $s = 4$. This change in N_{infl} with s is an important consideration which may be useful in model discrimination, since there is evidence that for some populations maximum growth occurs substantially in excess of $N = K/2$. In such situations, models with $s > 1$ are clearly of considerable interest. Note that for large s we have $t_{infl} \sim (1/a)\ln(K/n_0)$.

To interpret the role of s, observe that $b_1 + b_2 = (a_1 - a_2)K^{-s}$. So for fixed carrying capacity K and linear coefficients a_1 and a_2, the nonlinear coefficients b_1 and b_2 must approach zero as s increases. Thus the higher the value of s, the nearer N is to K before the nonlinear birth and death components 'kick-in'. Hence we can relate s to the aggression of the individual population members: the larger s, the larger the population size before individual growth is affected by competition. To consider the limit as $s \to \infty$, suppose we take $b_1 = b_2 = b/2 = aK^{-s}/2$. Then

$$\lambda_N - N[a_1 - (1/2)(a_1 - a_2)(N/K)^s] \quad \text{and} \quad \mu_N = N[a_2 + (1/2)(a_1 - a_2)(N/K)^s] \qquad (3.2.45)$$

for

$$0 \leq N \leq (a_1/b_1)^{1/s} = K[2a_1/(a_1 - a_2)]^{1/s} \to K \quad \text{as } s \to \infty. \qquad (3.2.46)$$

Thus on presuming K to be integer, $\lambda_N \to a_1 N$ and $\mu_N \to a_2 N$ (i.e. a simple birth–death process) for $N = 0, 1, \ldots, K-1$, with $\lambda_K = 0$ and $\mu_K = (a_1 + a_2)N/2$.

3.2.2 Simulated realizations

Figure 3.1 shows three simulated realizations of this process for $s = 1, 2$ and 5 with $a_1 = 2.2$, $a_2 = 0.2$ and $n_0 = 1$; taking $b_1 = b_2 = 0.1^s$ ensures that the carrying capacity remains fixed at $K = 10$. We see from (3.2.38) that $\lambda_N > 0$ for $N \leq (a_1/b_1)^{1/s} = 10 \times 2.2^{1/s}$, i.e. provided $N \leq 22$ ($s = 1$), 14.83 ($s = 2$) and 11.71 ($s = 5$), whence the corresponding maximum population sizes are 22, 15 and 12, respectively. In each case

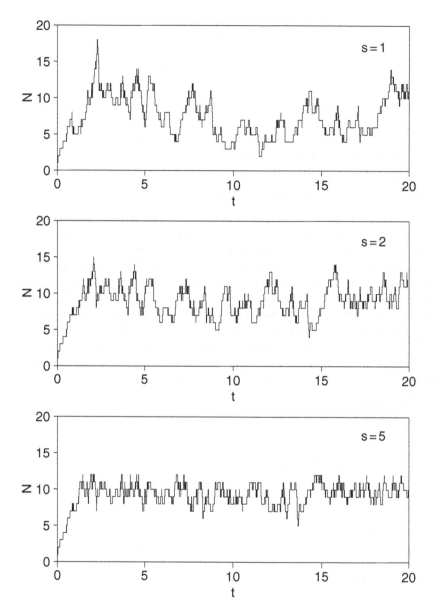

Figure 3.1 Simulated realizations of the power-law logistic process with rates $\lambda_N = N(2.2 - (0.1N)^s)$ and $\mu_N = N(0.2 + (0.1N)^s)$ for $n_0 = 1$.

there is a swift initial rise towards the carrying capacity, with subsequent behaviour around it becoming increasingly constrained as s increases.

The simulation algorithm is a straightforward extension of A2.1 (listed in Section 2.4.7) that we have already developed for the simple birth–death process. For

{U} a sequence of independent uniform $U(0,1)$ random variables, the time τ to the next event is exponentially distributed with parameter $\lambda_N + \mu_N$, so

$$\tau = -\ln(U)/(\lambda_N + \mu_N); \tag{3.2.47}$$

and this next event is a birth if

$$\lambda_N/(\lambda_N + \mu_N) \leq U, \tag{3.2.48}$$

otherwise it is a death. Here $\lambda_N = \max\{0, N(2.2 - (0.1N)^s)\}$ and $\mu_N = N(0.2 + (0.1N)^s)$. Thus steps (v) and (vii) change to

(v)′ update t to $t - \ln(U)/(\lambda_N + \mu_N)$
(vii)′ if $U \leq \lambda_N/(\lambda_N + \mu_N)$ then $N \to N+1$ else $N \to N-1$

Since (3.2.47) and (3.2.48) define a general univariate Markov process, it is worthwhile presenting the associated *Fortran* algorithm. For as it has a highly transparent structure it is easily translated into other popular languages such as *C* and *Pascal*. Note that this is the only place in the text where we provide specific code; if required, all the algorithms developed later on can be programmed by using this as a template.

The program is illustrated for the above power-law logistic process with $s = 2$; just change the code between the asterisks to simulate using other rate functions. The random number generator calls the highly recommended NAG routines G05CAF and G05CBF (see (2.1.42)), but any suitable random number generators may be used. The initial **seed** should be a large odd integer, whilst **maxtime** and **maxevents** denote the maximum duration time and number of events, respectively.

General birth–death Fortran code

```
      real*8 G05CAF, y, B, D, rate, test, time, rexp, s
      real maxtime
      integer count, seed
      seed = 4249
      n0 = 1
      maxtime = 20
      maxevents = 10000
      call G05CBF(seed)
      time = 0.0
      n = n0
      write(02,8001) time, n
      do 10 count=1,maxevents
C ***********************************
      B = N*(2.2-0.01*(N**2))
      if(B.lt.0.0) B=0.0
      D = N*(0.2+0.01*(N**2))
C ***********************************
      y = G05CAF(y)
      rate = B + D
      test = B/rate
```

```
            rexp = -(DLOG(y))/rate
            time = time + rexp
            if(time.GT.maxtime) STOP
            nold = n
            y = G05CAF(y)
            n = n-1
            if(y.LE.test) n=n+2
            write(1,8001) time, nold
            write(1,8001) time, n
   10    continue
 8001    format(f10.4, i10)
         stop
         end
```

3.2.3 Multiple equilibria

For the simple and power-law logistic processes, once the population size reaches the vicinity of the carrying capacity it remains at that level apart from random fluctuations (see Figure 3.1). We shall see later how to determine the variance of such oscillations, but the key point is that there is a single quasi-equilibrium value. Suppose, however, that a population can exist in two (or more) equilibrium states reflecting both moderate and high population densities. How can we model such scenarios, and is it possible to flip to and fro between the two population density levels?

To answer such questions suppose we replace the quadratic deterministic logistic equation (3.2.29) by the quotient form

$$dx/dt = x - x^2/(a^2 + x^2). \qquad (3.2.49)$$

Then on placing $dx/dt = 0$, we see that there are two equilibrium points given by the solutions $x = \theta_1, \theta_2$ of the quadratic equation

$$x^2 - x + a^2 = 0. \qquad (3.2.50)$$

Provided $4a^2 < 1$, the two roots

$$\theta_1, \theta_2 = (1/2)[1 \mp \sqrt{(1 - 4a^2)}]$$

of (3.2.50) are both positive and satisfy $0 < \theta_1 < 1/2 < \theta_2 < 1$. On writing $x = \theta_i \pm \epsilon$ it is easily seen that $dx/dt > 0$ for $0 < x < \theta_1$ and $x > \theta_2$, whilst $dx/dt < 0$ for $\theta_1 < x < \theta_2$. So θ_1 corresponds to a locally stable equilibrium, and θ_2 to globally unstable equilibrium. Thus as long as $0 < x < \theta_2$ the process converges to θ_1, but as soon as $x > \theta_2$ the process explodes exponentially.

Now for a population process we clearly require such equilibrium solutions to be larger than 1, so let us write $N = cx$ for some $c > 1$. Then (3.2.49) becomes

$$dN/dt = \lambda_N - \mu_N = N - N^2c/(N^2 + a^2c^2), \qquad (3.2.51)$$

which corresponds to simple birth at rate $\lambda_N = N$ and a more complex death component at rate $\mu_N = N^2c/(N^2 + a^2c^2)$. So when N is small we have $\mu_N \sim N^2/a^2c$, i.e.

a severe density dependent death rate; this switches as N increases into $\mu_N \sim c$, i.e. a simple emigration process. Suppose we require the locally stable and globally unstable equilibrium points to be at $N = K_1$ and K_2, respectively. Then since $dN/dt = 0$ at

$$N^2 - Nc + a^2c^2 = 0,$$

which has roots

$$K_1, K_2 = (c/2)[1 \mp \sqrt{(1 - 4a^2)}],$$

we see that we need to place

$$c = K_1 + K_2 \quad \text{and} \quad 4a^2 = 1 - [(K_2 - K_1)/(K_2 + K_1)]^2. \tag{3.2.52}$$

So if, for example, $K_1 = 10$ and $K_2 = 20$, then (3.2.52) yields $c = 30$ and $a^2 = 2/9$, which gives $\mu_N = 30N^2/(N^2 + 200)$. Since $\lambda_N > \mu_N$ for $N > K_2$ leads to a likely population explosion if $N > 20$, suppose we now place an upper bound K_3 on the population size by the simple expedient of defining $\lambda_N = 0$ for $N \geq K_3 = 32$ (say). Figure 3.2a shows the resulting stochastic behaviour, together with the three different types of deterministic behaviour for (i) $n_0 = 1, 19 < K_2 = 20$, (ii) $n_0 = 20 = K_2$ and (iii) $n_0 = 21 > K_2$. Whilst deterministic growth is wholly dependent on which of these three starting regimes is chosen, we see that the stochastic simulation not only exhibits short term fluctuations around the moderate (K_1) or high (K_3) values, but there is also a medium term fluctuation between these two states. In particular, the stable deterministic equilibrium value $K_1 = 10$ does not translate into stable stochastic equilibrium, since realizations can easily move from $N = K_1$ into the higher regime $K_2 < N \leq K_3$. Indeed, if we remove the artificial upper barrier by letting $K_3 \to \infty$, the process grows exponentially fast with increasingly high probability as N rises above the unstable equilibrium point K_2. The proportion of time spent in each of the individual states $N = 1, \ldots, 32$ (Figure 3.2b) is given by the quasi-equilibrium probabilities. That is, on using the pragmatic definition (3.2.20), by

$$\pi_N = \tilde{\omega}_N / \sum_{i=1}^{32} \tilde{\omega}_i, \tag{3.2.53}$$

where the $\tilde{\omega}_N$ are easily derived by first placing $\tilde{\omega}_1 = 1$ and then sequentially forming

$$\tilde{\omega}_N = \frac{\lambda_{N-1}}{\mu_N}\tilde{\omega}_{N-1} = \frac{(N^2 + 200)(N - 1)}{30N^2}\tilde{\omega}_{N-1} \quad (N = 2, \ldots, 32). \tag{3.2.54}$$

Although this is just a simple illustration, more complex, and potentially more realistic, multi-equilibrium structures can be generated by replacing the quadratic quotient in equation (3.2.51) by higher-order polynomial quotients. For example, suppose we require $dN/dt = 0$ at $N = 20, 50$ and 100, with the population oscillating around the two local equilibrium points $N_1 = 20$ (for $0 \leq N < 50$) and $N_3 = 100$ (for $N > 50$). So the locally unstable equilibrium point $N_2 = 50$ acts as a bridge between these two locally stable regimes, thereby enabling the process to switch backwards and forwards between them. Then on writing

$$(N - 20)(N - 50)(N - 100) = 0 \tag{3.2.55}$$

(a)

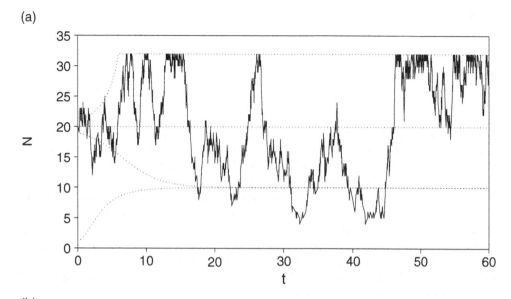

(b)

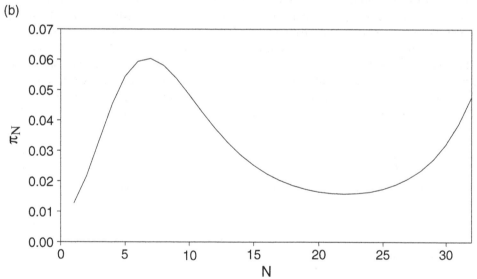

Figure 3.2 (a) Simulated realization of a double equilibrium process with rates $\lambda_N = N$ $(N < 32)$ and $\mu_N = 30N^2/(N^2 + 200)$, and the associated deterministic population sizes $(\cdots)$ for $n_0 = 1$, 19, 20 and 21; (b) corresponding quasi-equilibrium probabilities $\tilde{\pi}_N$ $(N = 1, \ldots, 32)$.

in the form

$$N = \frac{170N^2 + 100000}{N^2 + 8000}, \qquad (3.2.56)$$

we see that placing

$$\mu_N = N \quad \text{and} \quad \lambda_N = \frac{170N^2 + 100000}{N^2 + 8000} \qquad (3.2.57)$$

achieves the desired effect. For $\lambda_N > \mu_N$ when $N < 20$ and $50 < N < 100$, whilst $\lambda_N < \mu_N$ when $20 < N < 50$ and $N > 100$. So the process is attracted towards N_1 for $N < N_2$, and N_3 for $N > N_2$, as required. Note that unlike the previous example, $\lambda_0 = 12.5 > 0$, so not only can extinction not occur, but there is a strong upwards drift away from zero. Moreover, we no longer need the artificial device of placing $\lambda_N = 0$ for N greater than some upper bound in order to prevent population explosion. Figure 3.3a shows the resulting stochastic behaviour, and the intermittent switching between the two locally stable equilibria at $N_1 = 20$ and $N_3 = 100$ is clearly evident. That this phenomenon is not displayed by the corresponding deterministic trajectories (superimposed on this figure), highlights the large difference between stochastic and deterministic behaviour. The mean time spent in each of these two local regimes before switching to the other can be determined in two ways. The easiest, by far, is to construct a single simulated realization starting from say $N_0 = 20$, and then to record the successive times $\tau_1, \tau_1', \tau_2, \tau_2', \ldots$ taken to move between the $N = 20$ and 100 regimes. These can then be used directly to determine the estimated mean passage times and higher-order moments. An alternative, more laborious, approach involves first solving the Kolmogorov forward equations (3.0.2) for $p_{20,i}(t)$ and $p_{100,i}(t)$ numerically, then extracting $p_{20,100}(t)$ and $p_{100,20}(t)$, and finally evaluating the associated raw nth-order moments

$$\mu_{20,100}^{(n)} = \int_0^\infty t^n p_{20,100}(t)\, dt \quad \text{and} \quad \mu_{100,20}^{(n)} = \int_0^\infty t^n p_{100,20}(t)\, dt. \qquad (3.2.58)$$

Since $\lambda_0 > 0$, we can now evaluate the full equilibrium values via

$$\pi_N = \omega_N / \sum_{i=0}^\infty \omega_i, \qquad (3.2.59)$$

where the ω_N are derived by sequentially forming

$$\omega_N = \frac{\lambda_{N-1}}{\mu_N} \omega_{N-1} = \frac{170(N-1)^2 + 100000}{N[(N-1)^2 + 8000]} \omega_{N-1} \quad (N = 1, 2, \ldots) \qquad (3.2.60)$$

with $\omega_0 = 1$.

3.2.4 Power-law processes

Whilst logistic-type processes and multiple equilibrium p.d.f.'s have a major role to play in biology, in physics we are often interested in more complex systems that not only have the potential for exhibiting extremely rich dynamic behaviour, but which frequently generate large fluctuations that may be described by 'scale-free'

(a)

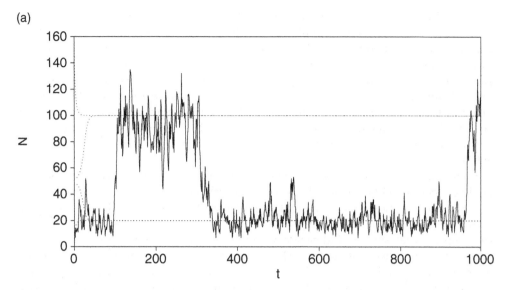

(b)

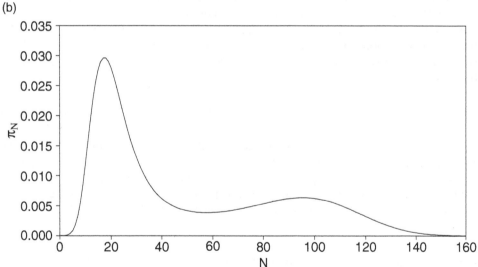

Figure 3.3 (a) Simulated realization of a triple-equilibrium process with rates $\mu_N = N$ and $\lambda_N = (170N^2 + 100000)/(N^2 + 8000)$, and the associated deterministic population sizes ($\cdots$) for $n_0 = 1, 49, 51$ and 160; (b) corresponding equilibrium probabilities π_N ($N \geq 0$).

probability density functions. A wide range of examples exhibiting such scale-free fluctuations that have been observed in discrete random phenomena (see Matthews, Hopcraft and Jakeman, 2003) include systems as diverse as the World Wide Web (Barabási and Albert, 1999), organic metabolisms (Jeong, Tombor, Albert, Oltvai and Barabási, 2000), protein interactions (Jeong, Mason, Barabási and Oltvai, 2001) and social networks (Albert and Barabási, 2000). The associated order-distributions of such systems describe the number, N, of links that connect the nodes, and these are typically scale-free with p.d.f. $p_N \sim 1/N^\nu$. In the classic sandpile paradigm, the distance travelled by grains in an avalanche, called the 'flight length', raises a paradox since it has a power-law density and so the variance of the distance travelled does not exist, thereby implying the need for infinite energy to transport the particles. Detailed analysis, however, reveals that individual flight lengths comprise a sum of steps where the length of each step does have finite variance, and it is because the number of steps has a power-law tail that causes $p^N \sim 1/N^\nu$ (Hopcraft, Jakeman and Tanner, 2001, 2002).

Such scale-free random processes play an important role in the study of complex, nonlinear correlated behaviour. They are frequently described in terms of realizations possessing fluctuations with self-similar characteristics, for which the class of Lévy-stable densities (Feller, 1968) can provide a suite of models. The term 'stable' refers to the property that sums of independent and identically distributed random variables retain the same distribution to within a scaling factor. For example, the Normal distribution is stable, since if $X_i \sim N(0, \sigma^2)$ then so is $Y = (1/n) \sum_{i=1}^n X_i$. The (continuous) Lévy-stable class (Lévy, 1937) forms a broader group, with members sharing the property that their p.d.f.'s have power-law tails with $-3 < \nu < -1$, so the variance does not exist. However, properties of discrete-stable distributions differ from their continuous analogues, which means that they have to be developed afresh (Hopcraft, Jakeman and Matthews, 2004). They are defined through their generating function

$$G(z) \equiv \sum_{N=0}^{\infty} p_N z^N = \exp\{-A(1-z)^\nu\}, \quad (3.2.61)$$

where the constant $A > 0$ acts as a scale factor and $0 < \nu \leq 1$ (see, for example, Matthews, Hopcraft and Jakeman, 2003). For it may be shown that if $N \gg 1$ then $p_N \sim 1/N^{1+\nu}$, thereby giving rise to a power-law asymptote with index in the range -2 to -1. Clearly, even the means of such distributions do not exist. The associated stability property results from replacing $1 - z$ by $(1 - s)/(An)^{1/\nu}$ to yield

$$G(z) \to G_s(s) = \exp\{-(1-s)^\nu/n\},$$

whence the p.g.f. of the sum Y takes the power-law form

$$[G_s(s)]^n = \exp\{-(1-s)^\nu\}. \quad (3.2.62)$$

Since the case $\nu = 1$ corresponds to the Poisson distribution, the Poisson therefore has the same significance for discrete-stable distributions as does the Normal for continuous Lévy-stable densities.

Note that distributions which have a power-law asymptote with index in the range -3 to -2 lie outside the discrete-stable regime of validity; although the mean exists, the variance does not. An 'obvious' power-law p.d.f. to try is

$$p_N = \frac{1}{\zeta(\mu, a)(n+a)^\mu} \quad (3.2.63)$$

where a is real and positive, $\mu > 1$, and the generalized Riemann zeta function

$$\zeta(\mu, a) = \sum_{N=0}^{\infty} \frac{1}{(N+a)^{\mu}} \qquad (3.2.64)$$

(Abramowitz and Stegun, 1970) provides the normalization. The associated p.g.f. can be written in terms of the Lerch Transcendent (Gradshteyn and Ryzhik, 1994)

$$\Phi(z; \mu, a) \equiv \zeta(\mu, a) G(z; \mu, a) = \sum_{N=0}^{\infty} \frac{z^N}{(a+N)^{\mu}}. \qquad (3.2.65)$$

The properties of this function for values of $z \sim 1$ determine the associated asymptotics (for large N), and there are surprisingly few documented results for Φ. Expressions that can be obtained are non-trivial and display subtle behaviour (in so far as the singularities of the function are concerned) as μ passes through integer values (Hopcraft, Jakeman and Matthews, 2004). It is the nature of these singularities that ultimately influences the asymptotic behaviour of the distribution which has p.g.f. (3.2.62).

Although the associated algebraic arguments are too involved to reproduce here, we can easily develop population processes which either possess the equilibrium power-law p.g.f. (3.2.61), or else satisfy the detailed balance equations. Taking the p.g.f. approach first, suppose we return to the simple mass-immigration–death process of Section 2.7.2, in which immigrants arrive randomly in batches of size j at rate α_j ($j = 1, 2, \ldots$), and then die independently at rate μ. Now provided the mean batch size is finite an equilibrium distribution will clearly exist, and on letting $t \to \infty$ in the forward p.g.f. equation (2.7.17) we obtain

$$\mu(z-1)\frac{dG(z; \infty)}{dz} = [H(z) - a]G(z; \infty) \qquad (3.2.66)$$

for $a = \sum_{j=1}^{\infty} \alpha_j$. Whence inserting (3.2.61) into equation (3.2.66) yields the required form

$$H(z) = a - \mu\nu A(1-z)^{\nu} \qquad (0 < \nu \leq 1). \qquad (3.2.67)$$

Finally, we choose $a = \mu\nu A$ to ensure that $H(0) = 0$ (i.e. $\alpha_0 = 0$), which gives rise to the immigration structure

$$\alpha_1 = \mu A \nu^2 \quad \text{and} \quad \alpha_j = \mu A \nu^2 (1-\nu)(2-\nu)\ldots(j-1-\nu)/j! \quad (j > 1). \qquad (3.2.68)$$

Since death occurs at rate μN and immigration at rate $H(1) = \mu\nu A$, it follows that the next event is unlikely to be an immigration if $N \gg \nu A$. So the process essentially comprises sudden large upward surges from small population sizes, immediately followed by fast decay. To simulate this process we therefore need to extend the general event-time procedure (3.2.47)–(3.2.48) in order to allow moves from a given state N to any state $M > N$ and not just to $N+1$. On paralleling (3.2.47), as the overall event rate is $r_N = N\mu + \mu\nu A$, we see that for $\{U\}$ a sequence of independent

$U(0,1)$ random variables the time τ to the next event is exponentially distributed with $\tau = -\ln(U)/r_N$. This next event is a death (i.e. $N \to N-1$) if $U \leq (N\mu)/r_N$, otherwise it is an immigration. Since all jumps of size $s = 1, 2, \ldots$ are possible, and bearing in mind that the partial sum $\sum_{j=1}^{s} \alpha_j$ does not possess a closed form solution, to determine s we need to generalize the simple birth–death algorithm A2.1 by incorporating a sequential search procedure.

Exact mass immigration–death algorithm (A3.2)
  (i) input parameters
  (ii) set $N = n_0$ and $t = 0$
  (iii) print N and t
  (iv) evaluate r_N
  (v) select $U \sim U(0,1)$
  (vi) update t to $t - \ln(U)/r_N$
  (vii) set $\alpha_1 = \mu A \nu^2$ and $sum = N\mu + \alpha_1$
  (viii) select $U \sim U(0,1)$
  (ix) if $U \leq sum/r_N$ then set $s = 1$, else
  (x) cycle over $j = 2, 3, \ldots$
    set $\alpha_j = \alpha_{j-1}[(j-1-\nu)/j]$
    place $sum = sum + \alpha_j$
    if $U \leq sum/r_N$ then set $s = j$, else return to (x)
  (xi) return to (iii)

Figure 3.4a shows a realization over $0 \leq t \leq 600$ for $A = \mu = 1$ and $\nu = 0.5$, starting with $N = n_0 = 5$ at time $t = 0$. The process clearly spends most of the time hovering near zero, with intermittent medium, and occasionally large, upward jumps immediately followed by rapid decay, as predicted above. However, the picture presented by this figure is far too conservative, for at time $t = 606.4615$ there is a *huge* switch from $N = 4$ to $N = 1066433$. That jumps of this size are quite possible can be seen by inspecting the immigration rates (3.2.68) for α_j. For although these rates initially decrease quite quickly, with $\alpha_1 = 0.25$, $\alpha_2 = 0.0625$ and $\alpha_{10} = 0.004637$, thereafter the decrease occurs much more slowly since $\alpha_j/\alpha_{j-1} = 1 - (1+\nu)/j \simeq 1$. Indeed, we soon have almost logarithmic fall-off with

s	α_s	$\sum_{j=1}^{s} \alpha_j$	
10^3	4.462×10^{-6}	0.4911	
10^4	1.411×10^{-7}	0.4972	(3.2.69)
10^5	4.460×10^{-9}	0.49911	
10^6	1.411×10^{-10}	0.49972	etc.

Since $\sum_{j=1}^{\infty} \alpha_j = 0.5$, there is still sufficient probability remaining by $s = 10^6$ to enable even larger jumps to occur during a non-negligible 0.056% of the time.

Figure 3.4b shows the empirical log–log probability distribution for $\{p_N\}$ over $N = 1, 2, \ldots, 10^5$, formed by evaluating the proportion of time state N is occupied during a simulation comprising 10^8 event-changes. This plot exhibits remarkable linearity

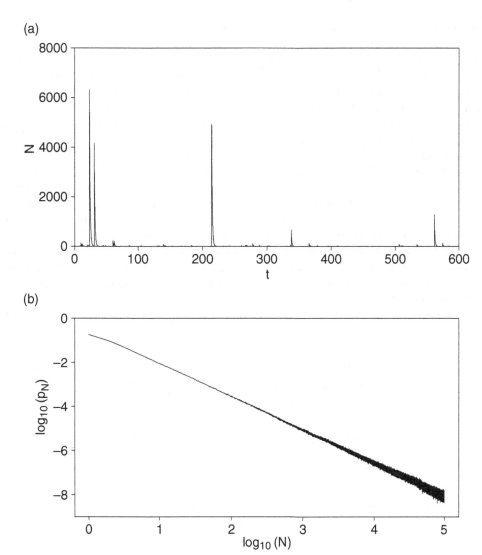

Figure 3.4 Simulated realization of a mass-immigration–death process with power-law parameters $\nu = 0.5$ and $A = 1$, and death rate $\mu = 1$, over 10^8 events starting from $N(0) = 5$ showing: (a) N against time over $0 \leq t \leq 600$; and, (b) empirical log-probabilities $\log_{10}(p_N)$ against $0 \leq \log_{10}(N) \leq 5$.

almost from the outset, in spite of the underpinning result $p_N \sim constant/N^{1+\nu}$, i.e. $\log(p_N) \sim constant - (1+\nu)\log(N)$, being theoretically valid only for large N. Simple regression over these 10^5 points yields the slope estimate -1.521, close to the theoretical power-law value of $-(1+\nu) = -1.5$. Note that increasing the number of event-changes would further improve the precision of the probability estimates, and hence allow us to reach deeper into the tail of the distribution. Although the search routine in algorithm A3.2 for determining the immigration jump size s is potentially time-consuming, it is seldom invoked for large s and so this empirical approach is surprisingly speedy. Moreover, it avoids the computational problems associated with determining the equilibrium solution to the corresponding forward equations. For since $\sum_{j=1}^{s} 1/N^{1+\nu}$ converges to $\mu\nu A$ very slowly as s increases (see (3.2.69)), to ensure 'reasonable' numerical accuracy we would need to determine the solution over a huge state space, say $0 \leq N \leq 10^{12}$, and although the associated probability transition matrix is sparse there are clearly substantial computational issues involved in constructing this. An alternative route would be to determine p_N theoretically, by expanding the p.g.f. (3.2.61) and then obtaining the coefficient of z^N. Though the resulting expression is not exactly neat since it involves the exponential expansion of a fractional binomial power-series.

In contrast, employing the detailed balance equations (3.2.4), namely

$$\lambda_N \pi_N = \mu_{N+1} \pi_{N+1}, \qquad (3.2.70)$$

for the general birth–death process is intrinsically far simpler. For to construct a process which possesses (say) the exact equilibrium Pareto form

$$\pi_N = 1/[\zeta(\nu) N^{1+\nu}] \qquad (N = 1, 2, \ldots; \nu > 0), \qquad (3.2.71)$$

we can use any rates λ_N and μ_N provided they satisfy the relation

$$\frac{\lambda_N}{\mu_{N+1}} = \frac{N^{1+\nu}}{(N+1)^{1+\nu}}. \qquad (3.2.72)$$

Three obvious choices are

model	A	B	C
λ_N	$N^{1+\nu}$	1	$\left(\frac{1}{N+1}\right)^{1+\nu}$
μ_N	$N^{1+\nu}$	$\left(\frac{N}{N-1}\right)^{1+\nu}$	$\left(\frac{1}{N-1}\right)^{1+\nu}$
λ_N/μ_N	1	$\left(\frac{N-1}{N}\right)^{1+\nu}$	$\left(\frac{N-1}{N+1}\right)^{1+\nu}$

(3.2.73)

where in each case we define $\mu_1 = 0$. Whilst all three models produce the required equilibrium Pareto p.d.f. (3.2.71), they clearly do so in different ways. Indeed, for large N we have $\lambda_N, \mu_N \sim N^{1+\nu}$, 1 and $N^{-1-\nu}$, respectively. So before proceeding further let us first examine their behaviour. On recalling Theorem 3.2 (Section 3.1), we see that this is determined by whether the two summations $\sum_{N=1}^{\infty} \omega_N$ and $\sum_{N=1}^{\infty} 1/(\lambda_N \omega_N)$

either converge or diverge where (here)

$$\omega_1 = 1 \quad \text{and} \quad \omega_N = \frac{\lambda_1 \lambda_2 \ldots \lambda_{N-1}}{\mu_1 \mu_2 \ldots \mu_N} \quad (N > 1). \tag{3.2.74}$$

Now on applying the balance equations (3.2.70), this expression takes the form

$$\omega_N = \frac{\pi_2}{\pi_1} \times \frac{\pi_3}{\pi_2} \times \cdots \times \frac{\pi_N}{\pi_{N-1}} = \frac{\pi_N}{\pi_1} = \frac{1}{N^{1+\nu}} \tag{3.2.75}$$

regardless of the model under consideration. Thus $\sum_{N=1}^{\infty} \omega_N < \infty$ for all $\nu > 0$, and so model behaviour is determined solely by whether $\sum_{N=1}^{\infty} 1/(\lambda_N \omega_N) = \infty$ or is $< \infty$. In the former case the process is positive recurrent (Theorem 3.2a), whilst in the latter case it is transient (Theorem 3.2c). On examining (3.2.73) we see that $1/(\lambda_N \omega_N) = 1$, $N^{1+\nu}$ and $[N(N+1)]^{1+\nu}$ under models A, B and C, respectively, so it follows that all three processes must be positive recurrent. However, whilst $\lambda_N < \mu_N$ for all $N \geq 2$ under models B and C, i.e. there is always at least a slight downwards drift irrespective of how large N is, under model A each step change is equally likely to be a birth or a death. Now in Chapter 5 we shall see that if we just consider event changes, i.e. we ignore inter-event times, then this *embedded* process (which is a simple random walk with a reflecting barrier at state 1) is transient. By this we mean that although eventual return to say state 1 is certain, the mean number of steps taken to achieve this is infinite. The reason why including inter-event times turns this transient random walk into a positive recurrent process lies in the swiftly increasing birth and death rates $N^{1+\nu}$. For when N becomes extremely large, the corresponding inter-event times, which are exponentially distributed with mean $1/(\lambda_N + \mu_N) = 0.5 N^{-(1+\nu)}$, become negligibly small, and so a large excursion is brought back to small N in a finite time. In contrast, although model C suffers a permanent downwards drift, the inter-event times now have mean $\sim 0.5 N^{1+\nu}$, which swiftly become extremely large with increasing N. So although a break-away from the zero state is highly unlikely, the proportion of time spent in each large state N is very high.

Figures 3.5a&c respectively show the first 10^5 events with $\nu = 0.5$ for model A starting from $N = 1$ and 1000 at time $t = 0$ using the same initial seed. Whilst the excursions away from zero in Figure 3.5a are modest compared to the those for the mass-immigration process (Figure 3.4), in spite of there being no downwards drift (since $\lambda_N \equiv \mu_N$), the two realizations shown in Figure 3.5c suggest that once the process moves away from zero it wanders quite aimlessly. Indeed, as we shall later see, it may become infinite before it returns to zero again. In contrast, the realization shown in Figure 3.5b for model C exhibits a substantial pull towards zero for small N, and it would take a set of rare events for N to become large enough for λ_N and μ_N to be in effective balance. However, should such an extremely rare excursion occur then the expected time spent in each of these large N-states increases dramatically. Note that to investigate the length of burn-in time required to achieve equilibrium, we could (at least in principle) use the perfect simulation approach of Section 2.5.4 to determine the times to coalescence of trajectories starting at $N = 1$ and say $N = 10, 100, 1000, \ldots$.

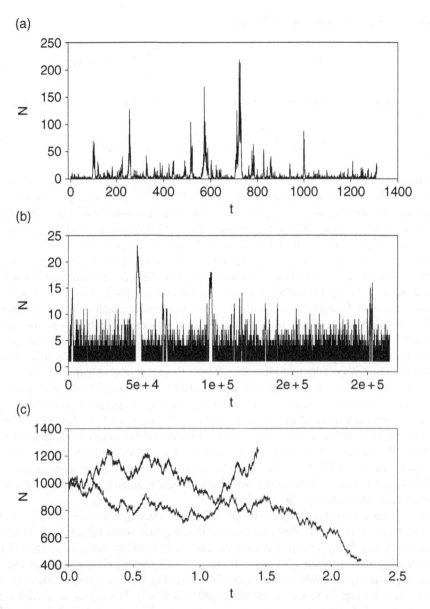

Figure 3.5 10^5 event changes for simulated realizations of the power-law birth–death models (a) A and (b) C with $\nu = 0.5$, starting from $N = 1$ at $t = 0$ and using the same seed; (c) shows the equivalent realization (lower) for model A starting from $N = 1000$ together with a second realization (upper) based on a different seed.

3.3 Time-dependent solutions

Although direct exact solutions to the forward and backward difference-differential equations (3.0.2) and (3.0.3) cannot be obtained for general birth–death rates λ_N and μ_N, progress can be made if we impose additional assumptions. For example, suppose that for the forward equation

$$dp_N(t)/dt = \lambda_{N-1}p_{N-1}(t) - (\lambda_N + \mu_N)p_N(t) + \mu_{N+1}p_{N+1}(t) \qquad (3.3.1)$$

we make some presumption on $p_0(t)$ (e.g. it is exponential or logistic); or on $p_0^*(s)$ in the equivalent Laplace transform representation (see (2.3.12)–(2.3.13))

$$sp_N^*(s) - p_N(0) = \lambda_{N-1}p_{N-1}^*(s) - (\lambda_N + \mu_N)p_N^*(s) + \mu_{N+1}p_{N+1}^*(s). \qquad (3.3.2)$$

Then in principle $\{p_N(t)\}$ and $\{p_N^*(s)\}$ may be obtained recursively for $N = 1, 2, \ldots$ by successive substitution into (3.3.1) and (3.3.2). In general, however, the form of $p_0(t)$ and $p_0^*(s)$ will not be known, and although both equations can still be solved successively in terms of them, the functions $p_0(t)$ and $p_0^*(s)$ themselves cannot be recovered by using the total probability law

$$\sum_{N=0}^{\infty} p_N(t) = 1, \quad \text{i.e.} \quad \sum_{N=0}^{\infty} p_N^*(s) = 1/s. \qquad (3.3.3)$$

Such intractability is in marked contrast to what we have seen for the simple and general birth processes (for which $\mu_N \equiv 0$), where we have already determined the full solutions (2.3.23) and (2.3.56).

3.3.1 Approximate solutions

Nevertheless, if we can deduce an acceptable approximation for $p_0(t)$ at the outset then this sequential approach can produce useful results. The most basic assumption is to let μ_1 be considerably smaller than λ_1. For then, even if the initial population consists of just a single individual (i.e. $n_0 = 1$), the population is unlikely to become extinct in the opening stages of the process (and even less so if $n_0 > 1$). Hence we may reasonably argue that the process settles down to quasi-equilibrium prior to extinction, i.e.

$$p_1(t)/[1 - p_0(t)] \simeq \tilde{\pi}_1. \qquad (3.3.4)$$

Whence on taking $\lambda_0 = \mu_0 = 0$, the forward equation becomes

$$dp_0(t)/dt = \mu_1 p_1(t) \simeq \mu_1 \tilde{\pi}_1 [1 - p_0(t)]. \qquad (3.3.5)$$

This integrates to give

$$p_0(t) \simeq 1 - \exp(-\mu_1 \tilde{\pi}_1 t), \qquad (3.3.6)$$

and so

$$p_N(t) \simeq \tilde{\pi}_N [1 - p_0(t)] \simeq \tilde{\pi}_N \exp(-\mu_1 \tilde{\pi}_1 t) \qquad (N = 1, 2, \ldots). \qquad (3.3.7)$$

One way of extending this approach, in order to take the initial transient behaviour into account, is to combine the exponential approximation (3.3.6) with expression (2.4.23) for the simple birth–death process with $n_0 = 1$, namely

$$p_0(t) = \frac{\mu - \mu e^{-(\lambda-\mu)t}}{\lambda - \mu e^{-(\lambda-\mu)t}}, \qquad (3.3.8)$$

to form

$$p_0(t) \simeq \frac{1 - e^{-\mu_1 \tilde{\pi}_1 t}}{1 - A e^{-\mu_1 \tilde{\pi}_1 t}} \qquad (3.3.9)$$

for some appropriate constant A. Since this implies $p_0(\infty) = 1$, here we are assuming that ultimate extinction is certain; note that $p_0(0) = 0$, as required. So now the equation

$$dp_0(t)/dt = \mu_1 p_1(t) \qquad (3.3.10)$$

leads to

$$p_1(t) = \frac{\tilde{\pi}_1(1-A)e^{-\mu_1 \tilde{\pi}_1 t}}{(1 - A e^{-\mu_1 \tilde{\pi}_1 t})^2}. \qquad (3.3.11)$$

Whence taking (for example) $n_0 = 1$ gives $p_1(0) = 1 = \tilde{\pi}_1(1-A)$, i.e. $A = 1 - \tilde{\pi}_1$. Thus

$$p_0(t) \simeq \frac{1 - e^{-\mu_1 \tilde{\pi}_1 t}}{1 - (1-\tilde{\pi}_1)e^{-\mu_1 \tilde{\pi}_1 t}} \quad \text{and} \quad p_1(t) \simeq \frac{\tilde{\pi}_1^2 e^{-\mu_1 \tilde{\pi}_1 t}}{(1 - (1-\tilde{\pi}_1)e^{-\mu_1 \tilde{\pi}_1 t})^2}. \qquad (3.3.12)$$

So for large t, $p_1(t)/(1 - p_0(t)) \sim \tilde{\pi}_1$, as required.

Whilst in principle $p_N(t)$ for $N \geq 2$ may now be determined recursively via the forward equation (3.3.1), i.e.

$$p_{N+1}(t) \simeq (1/\mu_{N+1})[dp_N(t)/dt - \lambda_{N-1}p_{N-1}(t) + (\lambda_N + \mu_N)p_N(t)], \qquad (3.3.13)$$

considerable care needs to be taken with this approach since errors inherent in $p_0(t), p_1(t), \ldots$ might be magnified at each successive differentiation of $p_1(t), p_2(t), \ldots$. A simple way of following this (rather dubious) route is to write

$$p_N(t) \equiv \sum_{r=0}^{\infty} c_{N,r} e^{-r\mu_1 \tilde{\pi}_1 t}, \qquad (3.3.14)$$

and then to substitute (3.3.14) into (3.3.1) and extract the coefficient of $e^{-r\mu_1 \tilde{\pi}_1 t}$ to form

$$c_{N+1,r} = (1/\mu_{N+1})[(\lambda_N + \mu_N - r\mu_1 \tilde{\pi}_1)c_{N,r} - \lambda_{N-1}c_{N-1,r}]. \qquad (3.3.15)$$

For since (3.3.12) expands to give

$$p_0(t) \simeq (1 - e^{-\mu_1 \tilde{\pi}_1 t})[1 + (1-\tilde{\pi}_1)e^{-\mu_1 \tilde{\pi}_1 t} + (1-\tilde{\pi}_1)^2 e^{-2\mu_1 \tilde{\pi}_1 t} + \cdots],$$

we have

$$c_{0,0} = 1 \quad \text{and} \quad c_{0,r} = -\tilde{\pi}_1(1-\tilde{\pi}_1)^{r-1} \quad (r > 0), \qquad (3.3.16)$$

whence applying the recurrence relation (3.3.15) yields $\{c_{N,r}\}$.

Bearing in mind that expression (3.3.12) for $p_0(t)$ is an approximation, a more pragmatic (and translucent) approach is simply to write

$$p_N(t) \simeq \tilde{\pi}_N(1 - p_0(t))(1 - e^{-\mu_1 \tilde{\pi}_1 t}) \qquad (N \geq 2), \qquad (3.3.17)$$

since this satisfies both the conditions $p_N(0) = 0$ and $p_N(t)/(1 - p_0(t)) \sim \tilde{\pi}_N$. Expression (3.3.17) automatically reduces the right-hand side of the forward equation (3.3.1) to zero; whilst the left-hand side becomes

$$dp_N(t)/dt \simeq \frac{-\mu_1 \tilde{\pi}_N e^{-\mu_1 \tilde{\pi}_1 t}}{(1 - e^{-\mu_1 \tilde{\pi}_1 t})^2} \times \tilde{\pi}_1^2 (1 + O(\tilde{\pi}_1)),$$

which, since $\tilde{\pi}_1$ is presumed small, is itself very near zero. So between them, the simple approximations (3.3.12) and (3.3.17) provide a useful representation for $p_N(t)$ over all $N = 0, 1, 2, \ldots$ once the initial burn-in stage has been passed.

For illustration, suppose we replace the simple birth and death rates $\lambda_N = \lambda N$ and $\mu_N = \mu N$ by the inverse rates $\lambda_N = \lambda/N$ and $\mu_N = \mu/N$ ($N > 0$) with $\lambda_0 = \mu_0 = 0$. So now the expected inter-event times increase, rather than decrease, with N. This suggests that, as N increases, the quasi-equilibrium probabilities decay to zero more slowly than those for the simple birth–death process. The balance equations $\lambda_{N-1} \tilde{\pi}_{N-1} = \mu_N \tilde{\pi}_N$ give

$$\tilde{\pi}_N = \left(\frac{\lambda}{\mu}\right)\left(\frac{N}{N-1}\right)\tilde{\pi}_{N-1},$$

i.e.

$$\tilde{\pi}_N = (\lambda/\mu)^{N-1} N \tilde{\pi}_1. \qquad (3.3.18)$$

Whence using $\sum_{N=1}^{\infty} \tilde{\pi}_N = 1$ yields

$$\tilde{\pi}_1 = (1 - \lambda/\mu)^2, \qquad (3.3.19)$$

and hence the quasi-equilibrium solution

$$\tilde{\pi}_N = (1 - \lambda/\mu)^2 N (\lambda/\mu)^{N-1}. \qquad (3.3.20)$$

Placing $\tilde{\pi}_1$ in expression (3.3.12), with $\mu_1 = \mu$, recovers $p_0(t)$, whence the approximation for $p_N(t)$ follows directly through (3.3.17). Note that (3.3.18) is of order $O(N^2)$ larger than the equivalent result for the simple birth–death process, for which $\tilde{\pi}_N = (\lambda/\mu)^{N-1} N^{-1} \tilde{\pi}_1$.

3.3.2 Probability of ultimate extinction

If the population process does not involve immigration when empty (i.e. $\lambda_0 = 0$) and $\mu_1 > 0$, then the population may become extinct. As the probability of ultimate extinction, $p_0(\infty)$, may depend on the initial population size, let us denote r_i to be the value of $p_0(\infty)$ corresponding to $n_0 = i$. On considering the first event, we have for $i = 1, 2, \ldots$

$$r_i = \Pr\{\text{first event is a birth}\} \times r_{i+1} + \Pr\{\text{first event is a death}\} \times r_{i-1}$$
$$= (\lambda_i r_{i+1} + \mu_i r_{i-1})/(\lambda_i + \mu_i). \tag{3.3.21}$$

Since $\lambda_0 = 0$ we clearly have $r_0 = 1$, whence r_i can be calculated for $i = 2, 3, \ldots$ in terms of r_1 by repeated application of equation (3.3.21). That is,

$$r_2 = r_1 + \frac{\mu_1}{\lambda_1}(r_1 - 1) \tag{3.3.22}$$

$$r_3 = r_2 + \frac{\mu_2}{\lambda_2}(r_2 - r_1) = r_2 + \frac{\mu_1 \mu_2}{\lambda_1 \lambda_2}(r_1 - 1) = r_1 + \left(\frac{\mu_1}{\lambda_1} + \frac{\mu_1 \mu_2}{\lambda_1 \lambda_2}\right)(r_1 - 1) \tag{3.3.23}$$

and, in general,

$$r_n = r_1 + (r_1 - 1)\sum_{i=1}^{n-1} \xi_i \quad \text{where} \quad \xi_i = \frac{\mu_1 \mu_2 \cdots \mu_i}{\lambda_1 \lambda_2 \cdots \lambda_i}. \tag{3.3.24}$$

Although result (3.3.24) is not a complete solution, in that it is expressed in terms of r_1 which is still unknown, it does lead to two useful results.

First, since r_n is a probability we must have $0 \leq r_n \leq 1$ for all n. Whence it follows from (3.3.24) that if $\sum_{i=1}^{n-1} \xi_i$ does not converge to a finite limit as $n \to \infty$ then we must have $r_1 - 1 = 0$. Thus $r_1 = 1$ and hence $r_n = 1$ for all $n = 2, 3, \ldots$. So in a closed population with $\lambda_0 = 0$ ultimate extinction is certain if

$$\sum_{i=1}^{\infty} \xi_i = \sum_{i=1}^{\infty} \frac{\mu_1 \mu_2 \cdots \mu_i}{\lambda_1 \lambda_2 \cdots \lambda_i} = \infty. \tag{3.3.25}$$

Note that this result directly parallels Theorem 3.1 which states that a process (with $\lambda_0 > 0$) is recurrent if and only if

$$\sum_{i=0}^{\infty} \frac{1}{\lambda_i \omega_i} = \sum_{i=0}^{\infty} \frac{\mu_1 \mu_2 \cdots \mu_i}{\lambda_0 \lambda_1 \cdots \lambda_i} = \infty. \tag{3.3.26}$$

For irrespective of the value of $\lambda_0 > 0$, recurrence implies that the process will visit state 1 infinitely often, and each time it is does so there is a probability $\mu_1/(\mu_1 + \lambda_1) > 0$ that it will move to state 0 rather than to state 2 at the next step. Thus if $\lambda_0 = 0$ then the probability of ultimate extinction must be one.

Second, we can evaluate $p_0(\infty)$ for those models in which condition (3.3.25) is not satisfied. For example, if the population can grow indefinitely large we may argue that as n increases, r_n must decay to zero (i.e. $r_\infty = 0$). Formally, this means that once the population size becomes infinite it remains so with probability one. Expression (3.3.24) then yields $0 = r_1 + (r_1 - 1)\sum_{i=1}^{\infty} \xi_i$, so $r_1 = [\sum_{i=1}^{\infty} \xi_i]/[1 + \sum_{i=1}^{\infty} \xi_i]$, whence (3.3.24) reduces to

$$r_n = [\sum_{i=n}^{\infty} \xi_i]/[1 + \sum_{i=1}^{\infty} \xi_i]. \tag{3.3.27}$$

To illustrate how these results may be applied, let us return to the simple birth–death process for which we have already shown in Section 2.4.2 that ultimate extinction is certain if and only if $\lambda \leq \mu$. On substituting $\lambda_i = \lambda i$ and $\mu_i = \mu i$ into (3.3.24) we see that $\xi_i = (\mu/\lambda)^i$, whence (3.3.25) becomes

$$\sum_{i=1}^{\infty} \xi_i = \sum_{i=1}^{\infty} (\mu/\lambda)^i.$$

This summation clearly diverges if and only if $\lambda \leq \mu$, as required. If $\lambda > \mu$, so that indefinitely prolonged growth is possible, then (3.3.27) gives

$$r_n = \sum_{i=n}^{\infty} (\mu/\lambda)^i / [1 + \sum_{i=1}^{\infty} (\mu/\lambda)^i] = (\mu/\lambda)^n, \qquad (3.3.28)$$

in exact agreement with result (2.4.24).

Finally, we note that in many closed populations with a single deterministically stable steady state, say $N(\infty) = K$, the death rate μ_N will exceed the birth rate λ_N for all $N > K$. Result (3.3.25) shows that in such situations ultimate extinction is certain.

3.3.3 Mean time to ultimate extinction

Another quantity of major interest is the mean time to extinction, T_n, of a population that initially consists of n individuals and for which eventual extinction is certain. Specifically, T_n can be used as an index of stability. For we may define a population as being *stable* if it persists for a large number of generations, and *unstable* if it persists for only a few. So as the order of magnitude of T_n represents the degree of stability, we may define the *stability index* corresponding to an initial state n as

$$\zeta_n = \ln(T_n). \qquad (3.3.29)$$

Now on considering the first event in a population of initial size i, we have

$$T_i = \text{mean time to the first event} + \Pr(\text{first event is a birth}) \times T_{i+1}$$
$$+ \Pr(\text{first event is a death}) \times T_{i-1}. \qquad (3.3.30)$$

Since the time to the first event is distributed exponentially with parameter $\lambda_i + \mu_i$, the mean time to the first change in population size is $1/(\lambda_i + \mu_i)$. Thus (3.3.30) becomes

$$T_i = (1 + \lambda_i T_{i+1} + \mu_i T_{i-1})/(\lambda_i + \mu_i) \qquad (3.3.31)$$

for $i = 1, 2, \ldots$. Paralleling the derivation of r_n in (3.3.24), first note that since $T_0 = 0$, when $i = 1$ we have $(\lambda_1 + \mu_1)T_1 = 1 + \lambda_1 T_2$, i.e.

$$T_2 = T_1 + \xi_1(T_1 - q_1), \qquad (3.3.32)$$

where ξ_i $(i = 1, 2, \ldots)$ is given in (3.3.24) and $q_1 = 1/\mu_1$. Define

$$q_i = \frac{\lambda_1 \lambda_2 \ldots \lambda_{i-1}}{\mu_1 \mu_2 \ldots \mu_i} \qquad (i > 1). \qquad (3.3.33)$$

146 General Markov population processes

Then substituting (3.3.32) into (3.3.31) with $i = 2$ yields

$$T_3 = T_2 + \xi_2(T_1 - q_1 - q_2),$$

whence a simple induction proof verifies that

$$T_n = T_{n-1} + \xi_{n-1}(T_1 - \sum_{i=1}^{n-1} q_i). \qquad (3.3.34)$$

Let us now consider the limiting case $n \to \infty$. Since extinction from an initial population of size n must eventually involve a move to $n - 1$ followed by extinction from $n - 1$,

$$T_n - T_{n-1} = \xi_{n-1}(T_1 - \sum_{i=1}^{n-1} q_i) > 0.$$

Hence if $\sum_{i=1}^{\infty} q_i$ is infinite, then T_1 must also be infinite. Moreover, since extinction from n must first involve movement to 1 followed by extinction from 1, we must have $T_n > T_1$ for all $n > 1$. Thus the mean time to extinction from any initial population size $n \geq 1$ is finite if and only if

$$\sum_{i=1}^{\infty} q_i = \frac{1}{\mu_1} + \sum_{i=2}^{\infty} \frac{\lambda_1 \lambda_2 \ldots \lambda_{i-1}}{\mu_1 \mu_2 \ldots \mu_i} < \infty \qquad (3.3.35)$$

(Nisbet and Gurney, 1982). When this condition is satisfied we see from successive use of (3.3.34) that

$$T_2 = T_1 + \xi_1(T_1 - q_1)$$
$$T_3 = T_2 + \xi_2(T_1 - q_1 - q_2) = T_1 + \xi_1(T_1 - q_1) + \xi_2(T_1 - q_1 - q_2)$$

and, in general,

$$T_n = T_1 + \xi_1(T_1 - q_1) + \cdots + \xi_{n-1}(T_1 - q_1 - \cdots - q_{n-1}),$$

which is easily verified on using a simple induction proof. Thus

$$T_n = T_1 + \sum_{m=1}^{n-1} \xi_m(T_1 - \sum_{i=1}^{m} q_i), \qquad (3.3.36)$$

and so we have reduced the problem of calculating the mean extinction time, T_n, to that of determining T_1.

Returning to (3.3.34), we see that

$$T_1 - \sum_{i=1}^{n} q_i = \frac{T_{n+1} - T_n}{\xi_n} = \frac{\lambda_1 \lambda_2 \ldots \lambda_n}{\mu_1 \mu_2 \ldots \mu_n}(T_{n+1} - T_n). \qquad (3.3.37)$$

So bearing in mind that the probability of extinction is one, it is reasonable to assume that $0 < T_{n+1} - T_n < A$ for all $n \geq 1$ and some constant $A > 0$. In which case (3.3.35) gives

$$\sum_{n=2}^{\infty} \frac{\lambda_1 \lambda_2 \ldots \lambda_n}{\mu_1 \mu_2 \ldots \mu_n}(T_{n+1} - T_n) < A \sum_{n=2}^{\infty} \frac{\lambda_1 \lambda_2 \ldots \lambda_n}{\mu_1 \mu_2 \ldots \mu_n} = A(\sum_{i=1}^{\infty} q_i - \frac{1}{\mu_i}) < \infty \qquad (3.3.38)$$

provided $\sum_{i=1}^{\infty} q_i < \infty$. Thus each term in the left-hand summation must tend to zero as $n \to \infty$, i.e.

$$\lim_{n \to \infty} [\frac{\lambda_1 \lambda_2 \ldots \lambda_n}{\mu_1 \mu_2 \ldots \mu_n}(T_{n+1} - T_n)] = 0. \qquad (3.3.39)$$

Whence on letting $n \to \infty$ in (3.3.37) we have the final result that

$$T_1 = \sum_{i=1}^{\infty} q_i, \qquad (3.3.40)$$

with T_n for $n > 1$ being given by (3.3.36).

Now substituting the logistic birth and death rates (3.2.35), namely $\lambda_N = N(a_1 - b_1 N)$ and $\mu_N = N(a_2 + b_2 N)$, into (3.3.40) and then (3.3.36) leads to rather cumbersome expressions. So to gain a more intuitive feel for the mean time to extinction of the logistic process let us consider the simplified rates

$$\lambda_N = rN \qquad \text{and} \qquad \mu_N = rN^2/K \qquad (3.3.41)$$

(i.e. $b_1 = a_2 = 0$). Then (3.3.33) gives

$$q_i = K^i/[ri(i!)] \qquad (i \geq 1), \qquad (3.3.42)$$

whilst from (3.3.24)

$$\xi_i = (i!)/K^i \qquad (i \geq 1). \qquad (3.3.43)$$

Since $\xi_i \to \infty$ as $i \to \infty$ it follows that $\sum_{i=1}^{\infty} \xi_i = \infty$, which implies that extinction is certain (result (3.3.25)). Whence (3.3.40) yields

$$T_1 = \sum_{i=1}^{\infty} q_i = (1/r) \sum_{i=1}^{\infty} K^i/[i(i!)]. \qquad (3.3.44)$$

Though this summation does not have a simple closed form solution, we see that

$$T_1 < (1/r) \sum_{i=1}^{\infty} K^i/i! = (1/r)(e^K - 1).$$

Whilst as q_i reaches a maximum around $i \simeq K$, for large K we have

$$T_1 \simeq (1/r) \sum_{i=1}^{\infty} K^{i-1}/i! = (1/rK)(e^K - 1) \sim (1/rK)e^K. \qquad (3.3.45)$$

To provide analytic justification of this result write

$$T_1 = f(K)e^K \qquad (3.3.46)$$

for some function $f(K)$. Then differentiating (3.3.44) gives

$$\frac{dT_1}{dK} = \frac{df(K)}{dK}e^K + f(K)e^K = \left(\frac{1}{r}\right)\sum_{i=1}^{\infty}\frac{K^{i-1}}{i!}$$

$$= (1/rK)(e^K - 1) \simeq (1/rK)e^K$$

for large K. Thus

$$\frac{df(K)}{dK} + f(K) = \frac{1}{rK}.$$

Whence we may argue that since placing $f(K) = 1/rK$ yields $df(K)/dK = -1/rK^2 \simeq 0$, taking $f(K) \simeq 1/rK$ provides a suitable approximation. Thus (3.3.46) gives

$$T_1 \simeq (1/rK)e^K$$

in agreement with (3.3.45). Inserting this result into (3.3.29) then shows that for large K the stability index

$$\zeta_1 = \ln(T_1) \simeq K - \ln(rK) \simeq K, \tag{3.3.47}$$

so here the carrying capacity K is itself an index of stability.

3.4 Moment closure

To illustrate the nature of the mathematical intractability surrounding the forward equation (3.3.1), namely

$$dp_N(t)/dt = \lambda_{N-1}p_{N-1}(t) - (\lambda_N + \mu_N)p_N(t) + \mu_{N+1}p_{N+1}(t), \tag{3.4.1}$$

let us return to the power-law logistic birth and death rates (3.2.38) and (3.2.39) for integer $s \geq 1$. If $\phi = (a_1/b_1)^{1/s}$ is also integer, then $\lambda_\phi = 0$, which (for $n_0 \leq \phi$) constrains the population size to the finite range $N = 0, 1, \ldots, \phi$. So if we replace the transition rates by the more algebraically amenable form

$$\lambda_N = N(a_1 - b_1 N^s) \quad \text{and} \quad \mu_N = N(a_2 + b_2 N^s) \quad (N = 0, 1, \ldots), \tag{3.4.2}$$

then the solution of (3.4.1) remains unchanged in spite of the birth rates now taking the illegal values $\lambda_N < 0$ for $N > \phi$. Although it seems eminently plausible to assume that the original and modified birth rates will yield highly similar solutions for $p_N(t)$ when ϕ is non-integer, numerical comparison over $0 \leq N \leq \phi_{int}$ for specific parameter values is nevertheless advised if accuracy is paramount. Inserting the rates (3.4.2) into equation (3.4.1) gives

$$dp_N(t)/dt = (N-1)[a_1 - b_1(N-1)^s]p_{N-1}(t) - N[(a_1 + a_2) + (-b_1 + b_2)N^s]p_N(t)$$
$$+ (N+1)[a_2 + b_2(N+1)^s]p_{N+1}(t), \tag{3.4.3}$$

whence multiplying both sides by $e^{N\theta}$ and summing over $N = 0, 1, \ldots$ gives

$$\sum_{N=0}^{\infty} \frac{dp_N(t)}{dt} e^{N\theta} = a_1 e^{\theta} \sum_{N=0}^{\infty} (N-1) p_{N-1}(t) e^{(N-1)\theta} - b_1 e^{\theta} \sum_{N=0}^{\infty} (N-1)^{s+1} p_{N-1}(t) e^{(N-1)\theta}$$

$$- (a_1 + a_2) \sum_{N=0}^{\infty} N p_N(t) e^{N\theta} + (b_1 - b_2) \sum_{N=0}^{\infty} N^{s+1} p_N(t) e^{N\theta}$$

$$+ a_2 e^{-\theta} \sum_{N=0}^{\infty} (N+1) p_{N+1}(t) e^{(N+1)\theta}$$

$$+ b_2 e^{-\theta} \sum_{N=0}^{\infty} (N+1)^{s+1} p_{N+1}(t) e^{(N+1)\theta}. \quad (3.4.4)$$

Now differentiating the m.g.f. $M(\theta; t) \equiv \sum_{N=0}^{\infty} p_N(t) e^{N\theta}$ with respect to θ gives

$$\frac{\partial^r M(\theta; t)}{\partial \theta^r} = \sum_{N=0}^{\infty} p_N(t) N^r e^{N\theta}. \quad (3.4.5)$$

So on taking $p_{-1}(t) \equiv 0$ (as before), it follows that equations (3.4.4) reduce to

$$\frac{\partial M(\theta; t)}{\partial t} = [a_1(e^{\theta} - 1) + a_2(e^{-\theta} - 1)] \frac{\partial M(\theta; t)}{\partial \theta}$$

$$+ [(-b_1)(e^{\theta} - 1) + b_2(e^{-\theta} - 1)] \frac{\partial^{s+1} M(\theta; t)}{\partial \theta^{s+1}}, \quad (3.4.6)$$

and the intractability of this equation for $s > 0$ is self-evident.

3.4.1 Cumulant equations

Although equation (3.4.6) is too complex to yield explicit solutions for the probabilities $\{p_N(t)\}$, it does enable us to obtain equations for the moments which turn out to be more malleable. For as we have already noted in Section 2.1.3, the raw moments, μ'_r, are given by the coefficients of $\theta^r/r!$ in the expansion of $M(\theta; t)$; or, equivalently, from

$$\mu'_r \equiv \left. \frac{\partial^r M(\theta; t)}{\partial \theta^r} \right|_{\theta=0}. \quad (3.4.7)$$

The central moments, μ_r, can then be derived from them via relations (2.1.34). For example, applying (3.4.7) with $r = 1$ to equation (3.4.6) yields

$$d\mu'_1/dt = (a_1 - a_2)\mu'_1 - (b_1 + b_2)\mu'_{s+1}. \quad (3.4.8)$$

Unfortunately, this involves the higher-order moment μ'_{s+1}, which exposes a fundamental problem of working with nonlinear stochastic processes. Namely that the equation for the rth-order moment involves moments of higher-order, and so no matter how many moment equations are generated they will always contain a greater number of variables.

This conundrum is a classic feature in applied mathematics. For example, many partial differential equations can be reduced to an infinite set of ordinary differential equations by taking Fourier transforms. Each such o.d.e. describes the evolution of a single Fourier coefficient, or, from an alternative perspective, the evolution in a given coordinate direction in an infinite-dimensional function space. Since no computer can represent the full infinite set of o.d.e.'s, in this scenario we have, of necessity, to truncate the set to a finite number of equations. Classically, this was achieved by ignoring all higher-order Fourier coefficients beyond a certain point (the *Galerkin approximation*). Note that in probability terms the p.d.e. to o.d.e. procedure is the converse of our generating function approach. For here we start with an infinite set of o.d.e.'s for the $\{p_N(t)\}$ (e.g. (3.4.3)) and from them derive a single p.d.e. (e.g. (3.4.6)) in the hope of being able to obtain an explicit solution for the generating function from which the $p_N(t)$ might be recovered. It is the inability to achieve this goal that forces the re-expansion of the generating function in terms of moments. However, whilst placing high-frequency Fourier terms to zero may indeed be plausible in many physics or engineering situations, setting $\mu'_r = \sum_{N=0}^{\infty} N^r p_N(t) = 0$ for r greater than some large fixed R is clearly a non-starter since $N^r p_N(t) \to \infty$ as $r \to \infty$ for $N > 1$ and $p_N(t) > 0$. Two options are therefore open to us. First, we can choose a different type of moment structure to work with, where this objection is weaker. Second, we could note that an important class of generalizations of the classical Galerkin method has recently appeared, motivated by the theory of *inertial manifolds*. In the differential equation context an inertial manifold M yields a function Ψ that gives each high-order Fourier coefficient as a function of a finite set of low-order coefficients; so *non-Galerkin* methods correspond to using an exact or approximated Ψ. Whilst in theory this approach might eventually prove useful in stochastic process settings (Stark, Ianelli and Baigent, 2001), its potential for providing useful non-numerical moment solutions is weak, not only because of the rapidly increasing complexity of the moment equations themselves as r increases, but also because of the need to develop a rigorous theoretical underpinning of the functional forms to be used. We shall therefore remain within the *moment closure* framework in which high-order moments are replaced by zero.

That cumulants provide an ideal structure in which to operate is highlighted by the ubiquitous Normal distribution. For the m.g.f. is given by

$$M(\theta;t) = \frac{1}{\sqrt{2\pi\sigma^2}} \int_{-\infty}^{\infty} \exp\{-(x-\mu)^2/2\sigma^2\} e^{x\theta} dx$$

$$= \exp\{\mu\theta + \sigma^2\theta^2/2\} \times \frac{1}{\sqrt{2\pi\sigma^2}} \int_{-\infty}^{\infty} \exp\{-[x-(\mu+\sigma^2\theta)]^2/2\sigma^2\} dx$$

$$= \exp\{\mu\theta + \sigma^2\theta^2/2\}, \tag{3.4.9}$$

since the last integral relates to a Normal p.d.f. with mean $\mu + \sigma^2\theta$ and variance σ^2. Thus the corresponding c.g.f. takes the simple quadratic form

$$K(\theta;t) \equiv \ln[M(\theta;t)] = \mu\theta + \sigma^2\theta^2/2. \tag{3.4.10}$$

Because under the Normal distribution all cumulants of third- and higher-order are therefore zero, it follows that truncating cumulants at the second-order will automatically produce the Normal distribution. Our aspiration is that by successively introducing the cumulants $\kappa_3, \kappa_4, \ldots$ we will obtain increasingly better approximations to the true p.d.f. Later on we shall see that the truth of this assertion depends on the particular situation under study.

On replacing $M(\theta;t)$ in (3.4.6) by $\exp\{K(\theta;t)\}$ we have

$$\frac{\partial K(\theta;t)}{\partial t} = [a_1(e^\theta - 1) + a_2(e^{-\theta} - 1)]\frac{\partial K(\theta;t)}{\partial \theta} \qquad (3.4.11)$$

$$+ [(-b_1)(e^\theta - 1) + b_2(e^{-\theta} - 1)]e^{-K(\theta;t)}\frac{\partial^{s+1}\exp\{K(\theta;t)\}}{\partial \theta^{s+1}},$$

and comparing the $(s+1)$th derivative term with that in (3.4.6) shows that, as s increases, the cumulant equations become steadily more awkward to work with than those for the raw moments. Nevertheless, use of symbolic computer packages such as *Mathematica* (Wolfram, 1999) ensures that such problems can be surmounted fairly easily. So although we shall restrict discussion to the standard logistic process, for which $s = 1$, the ideas are readily extended to cover general $s = 1, 2, \ldots$.

Taking $s = 1$ in (3.4.11) gives

$$\frac{\partial K(\theta;t)}{\partial t} = [a_1(e^\theta - 1) + a_2(e^{-\theta} - 1)]\frac{\partial K(\theta;t)}{\partial \theta} \qquad (3.4.12)$$

$$+ [(-b_1)(e^\theta - 1) + b_2(e^{-\theta} - 1)]\left[\frac{\partial^2 K(\theta;t)}{\partial \theta^2} + \left(\frac{\partial K(\theta;t)}{\partial \theta}\right)^2\right].$$

Whence expanding both sides in powers of θ, equating coefficients of θ, θ^2 and θ^3, and denoting $a = a_1 - a_2$, $b = b_1 + b_2$, $c = a_1 + a_2$ and $d = b_1 - b_2$, gives

$$\dot\kappa_1 = (a - b\kappa_1)\kappa_1 - b\kappa_2 \qquad (3.4.13)$$

$$\dot\kappa_2 = (c - d\kappa_1)\kappa_1 + (2a - d - 4b\kappa_1)\kappa_2 - 2b\kappa_3 \qquad (3.4.14)$$

$$\dot\kappa_3 = (a - b\kappa_1)\kappa_1 + (3c - b - 6d\kappa_1 - 6b\kappa_2)\kappa_2 + (3a - 3d - 6b\kappa_1)\kappa_3 - 3b\kappa_4. \qquad (3.4.15)$$

For comparable expressions for $s = 2$ and $s = 3$ see Matis, Kiffe and Parthasarathy (1998) and Matis and Kiffe (2000).

On placing $\kappa_2 = 0$ in (3.4.13) we immediately recover the deterministic equation (3.2.36). So given that the variance function, $\kappa_2(t)$, is inherently positive for $t > 0$, it follows that since $b > 0$ the deterministic solution for $N(t)$ exceeds the analogous mean value function, $\kappa_1(t)$, for all $t > 0$. The implications for processes whose realizations exhibit considerable variability, like those shown in Figure 3.1, is clear. For nonlinear processes, $N(t)$ and $\kappa_1(t)$ should always be regarded as relating to different properties of the system. Whilst the latter refers to mean values, the former is best considered in terms of the *expected update equation*: since for realizations $\{X(t)\}$, we may write the deterministic equation $dN(t)/dt = \lambda_N - \mu_N$ as

$$N(t + dt) - N(t) = (\lambda_N - \mu_N)dt = E[X(t+dt) - X(t)], \qquad (3.4.16)$$

i.e. the expected change in $X(t)$ between times t and $t + dt$.

Just as placing $\kappa_2 = 0$ in (3.4.13) yields the deterministic equation, so placing $\kappa_3 = 0$ in (3.4.14) yields two nonlinear o.d.e.'s for $\kappa_1(t)$ and $\kappa_2(t)$. Similarly, placing $\kappa_4 = 0$ in (3.4.15) yields three equations for $\kappa_1(t)$, $\kappa_2(t)$ and $\kappa_3(t)$. This truncation approach is a powerful procedure, and although the accuracy of the resulting cumulant approximations clearly depends on the specific process and parameter values under consideration, numerical investigations concerning their accuracy have been very encouraging. Algebraic solutions may be derived by using *Mathematica*, whilst numerical estimates can be obtained via standard differential equation solvers in, say, *Matlab* and *Mathcad*. Though be careful to note that there is little point in pursuing the latter approach if the Kolmogorov equations (3.4.1) are themselves amenable to direct numerical solution. For example, the power-law logistic birth rate (3.2.38) is zero for $N \geq \phi_{int} =$ integer part of $(a_1/b_1)^{1/s}$, which means that there are at most $\phi_{int} + 1$ equations to solve. So provided ϕ_{int} is not too large, exact numerical values for $p_N(t)$, and hence for the moments $\mu'_i(t)$, $\mu_i(t)$ and cumulants $\kappa_i(t)$, can be computed directly. Unless high accuracy is required, simply replacing $dp_N(t)/dt$ by $[p_N(t+h) - p_N(t)]/h$ for small h (say 0.01 or 0.001) is often perfectly adequate, thereby enabling the equations to be solved via a few simple lines of program code. Where moment closure comes into its own is when the number of equations to be solved for $\{p_N(t)\}$ becomes computationally unmanageable. For example, with the simple birth–death process with $\lambda > \mu$ and $n_0 = 1$, we see from (2.4.19) and (2.4.20) that the mean $m(t) = e^{(\lambda-\mu)t}$ and variance $V(t) \sim [(\lambda + \mu)/(\lambda - \mu)] e^{2(\lambda-\mu)t}$ (for large t). So if we wish to ensure that the N in equations (3.4.1) cover say $m(t) + 3$ standard deviations, then for $\lambda = 2$ and $\mu = 1$ we require $(1 + 3\sqrt{3})e^t$ equations. With $t = 10$ this implies using more than 136500 equations, whilst with $t = 100$ the number rises to a staggering 1.67×10^{44}. Note that the simplest way of restricting the range of the probability equations to say $N = 0, 1, \ldots, N'$, for some suitable upper bound N', is to place $\lambda_{N'} = 0$ with $n_0 \leq N'$.

The moment closure equations are particularly attractive in the investigation of steady state solutions. Figure 3.1, for example, shows that initial transient growth disappears well before time $t = 5$, whence thereafter we may safely place $\dot{\kappa}_i(t) = 0$. Using (3.4.13) with $\kappa_2 = 0$ then gives the first cumulant closure approximation

$$\kappa_1^{(1)} = a/b = K, \qquad (3.4.17)$$

i.e. the deterministic carrying capacity. This is in stark contrast to the true value $\kappa_1(\infty) = 0$; for extinction must ultimately occur since $N(t)$ is bounded above and $\mu_1 > 0$. So the moment closure estimates relate solely to non-extinct realizations in quasi-equilibrium. To obtain the second cumulant closure approximations, $\kappa_1^{(2)}$ and $\kappa_2^{(2)}$, we place $\kappa_i = 0$ for $i > 2$, whence (3.4.13) and (3.4.14) yield

$$\kappa_2^{(2)} = (a - b\kappa_1^{(2)})\kappa_1^{(2)}/b = (c - d\kappa_1^{(2)})\kappa_1^{(2)}/(4b\kappa_1^{(2)} - 2a + d). \qquad (3.4.18)$$

Thus $\kappa_1^{(2)}$ is the solution near $\kappa_1^{(1)} = K$ of the quadratic equation

$$4b^2 \kappa_1^2 - 6ab\kappa_1 + (bc + 2a^2 - ad) = 0, \qquad (3.4.19)$$

i.e. the positive root

$$\kappa_1^{(2)} = [3a + \sqrt{(a^2 - 4bc + 4ad)}]/4b, \qquad (3.4.20)$$

with $\kappa_2^{(2)}$ following on directly from (3.4.18). Including equation (3.4.15) with $\kappa_4 = 0$ then leads to the third-order approximations $\kappa_1^{(3)}$, $\kappa_2^{(3)}$ and $\kappa_3^{(3)}$, though at this level full analytic solutions are probably best investigated by employing a computer algebra package. If required, $\kappa_1^{(4)}, \ldots, \kappa_4^{(4)}$, etc., can then be determined in a similar manner, though construction of the cumulant equations clearly gets progressively more awkward. Numerical solutions may be obtained iteratively, with start values for $\kappa_1^{(r)}, \ldots, \kappa_r^{(r)}$ being given by the previous set of estimates $\kappa_1^{(r-1)}, \ldots, \kappa_{r-1}^{(r-1)}, 0$.

3.5 Avoiding the Kolmogorov equations

Success in using the above approach depends on being able to form the m.g.f. equation (as in (3.4.6)), which essentially means that λ_N and μ_N involve the sum of terms involving small integer powers of N. So rates which have, for example, fractional linear forms or non-integer powers (3.4.2) are not amenable to the generation of exact cumulant equations such as (3.4.13)–(3.4.15). An alternative compromise technique is to envisage the development of the stochastic process in terms of

$$\text{stochastic} = \text{deterministic} + \text{zero-mean noise},$$

since this dispenses with the need to work in terms of the Kolmogorov equations. We first write the deterministic equation

$$dN(t)/dt = f(N(t))$$

in the form

$$N(t + dt) - N(t) = f(N(t))dt$$

for infinitesimal time increment dt, and then consider the random change in the population size $X(t)$ during the time interval $(t, t + dt)$ to be

$$X(t + dt) - X(t) = f(X(t))dt + dZ(t). \qquad (3.5.1)$$

That is, we form the stochastic update by superimposing a noise component, $dZ(t)$, with $\Pr(dZ = +1) = \lambda_X dt$ and $\Pr(dZ = -1) = \mu_X dt$, on the deterministic update term.

3.5.1 Generating cumulant equations directly

To illustrate how this representation may be used to generate the cumulant equations directly, i.e. without recourse to either the probabilities $\{p_N(t)\}$ or the p.g.f. $G(z;t)$, suppose that a logistic population lies in a state of quasi-equilibrium with variance V and third central moment (i.e. cumulant) W around the mean value m. Then

$$E[X(t)] = m = E[X(t + dt)]. \qquad (3.5.2)$$

Whence taking expectations on both sides of equation (3.5.1), namely

$$X(t+dt) - X(t) = X(t)(a - bX(t))dt + dZ(t), \qquad (3.5.3)$$

yields

$$0 = E[aX(t) - bX^2(t)]dt + E[dZ(t)].$$

On placing $E[dZ(t)] = 0$ (see below) this becomes

$$(a - bm)m - bV = 0, \qquad (3.5.4)$$

which agrees with the cumulant equation (3.4.13) for $\dot{\kappa}_1 = 0$. To introduce third-order terms we first square (3.5.3) to form

$$X^2(t+dt) = X^2(t) + X^2(t)(a - bX(t))^2 dt^2 + dZ^2(t) + 2X^2(t)(a - bX(t))dt$$
$$+ 2X(t)dZ(t) + 2X(t)(a - bX(t))dtdZ(t). \qquad (3.5.5)$$

Now in quasi-equilibrium $E[X^2(t+dt)] = E[X^2(t)]$, whilst taking $\lambda_X = X(a_1 - b_1 X)$ and $\mu_X = X(a_2 + b_2 X)$ gives

$$E[dZ(t)] = (+1)(a_1 - b_1 X(t))X(t)dt + (-1)(a_2 + b_2 X(t))X(t)dt. \qquad (3.5.6)$$

Whence replacing $X(t)$ by $E[X(t)] = m$ and $X^2(t)$ by $E[X^2(t)] = V + m^2$ gives

$$E[dZ(t)] = (a_1 - a_2)mdt - (b_1 + b_2)(V + m^2)dt = [(a - bm)m - bV]dt = 0 \qquad (3.5.7)$$

from (3.5.4), as required. Similarly,

$$E[dZ^2(t)] = (+1)^2(a_1 - b_1 X(t))X(t)dt + (-1)^2(a_2 + b_2 X(t))X(t)dt$$
$$= (a_1 + a_2)mdt - (b_1 - b_2)(V + m^2)dt$$
$$= [cm - d(V + m^2)]dt. \qquad (3.5.8)$$

On neglecting terms of $O(dt^2)$ and taking expectations on both sides of (3.5.5), we therefore obtain

$$0 = [cm - d(V + m^2)]dt + 2aE[X^2(t)]dt - 2bE[X^3(t)]dt + 2E[X(t)dZ(t)]. \qquad (3.5.9)$$

Suppose that for $X(t) \simeq m$ we may reasonably take

$$E[X(t)dZ(t)] \simeq E[X(t)]E[dZ(t)] = 0. \qquad (3.5.10)$$

Then as

$$E[X^3(t)] = E[\{(X(t) - m) + m\}^3] = W + 3mV + m^3, \qquad (3.5.11)$$

equation (3.5.9) becomes

$$0 = [cm - d(V + m^2)] + 2a(V + m^2) - 2b(W + 3mv + m^3) \qquad (3.5.12)$$
$$= [(c - dm)m + (2a - d - 4bm)V - 2bW] + 2m[(a - bm)m - bV].$$

On applying (3.5.4) this simplifies to

$$0 = (c - dm)m + (2a - d - 4bm)V - 2bW, \qquad (3.5.13)$$

which is in exact agreement with the cumulant equation (3.4.14) for $\dot{\kappa}_2 = 0$ with $\kappa_1 = m$, $\kappa_2 = V$ and $\kappa_3 = W$.

Although this approach successfully reproduces the first two quasi-equilibrium cumulant equations for κ_1, κ_2 and κ_3, it yields an approximate third cumulant equation, rather than the exact one (i.e. (3.4.15) with $\dot{\kappa}_3 = 0$). This time we cube (3.5.3) to obtain

$$X^3(t+dt) = X^3(t) + X^3(t)(a - bX(t))^3 dt^3 + dZ^3(t)$$
$$+ 3X^3(t)(a - bX(t))dt + 3X^3(t)(a - bX(t))^2 dt^2 + 3X^2(t)dZ(t)$$
$$+ 3X(t)dZ^2(t) + 3X^2(t)(a - bX(t))^2 dt^2 dZ(t) + 3X(t)(a - bX(t))dt dZ^2(t)$$
$$+ 6X^2(t)(a - bX(t))dt dZ(t). \quad (3.5.14)$$

So in quasi-equilibrium, on taking expectations we have to $O(dt)$

$$0 = \mathrm{E}[dZ^3(t)] + 3a\mathrm{E}[X^3(t)]dt - 3b\mathrm{E}[X^4(t)]dt + 3\mathrm{E}[X^2(t)dZ(t)] + 3\mathrm{E}[X(t)dZ^2(t)]. \quad (3.5.15)$$

Suppose we extend assumption (3.5.10) to include the third-order terms

$$\mathrm{E}[X^2(t)dZ(t)] = \mathrm{E}[X^2(t)]\mathrm{E}[dZ(t)] = 0 \quad (3.5.16)$$

(from (3.5.7)) and

$$\mathrm{E}[X(t)dZ^2(t)] = \mathrm{E}[X(t)]\mathrm{E}[dZ^2(t)] = m[cm - d(V + m^2)]dt \quad (3.5.17)$$

(from (3.5.8)). Then on noting via (3.5.6) that

$$\mathrm{E}[dZ^3(t)] = (+1)^3(a_1 - b_1 X(t))X(t) + (-1)^3(a_2 + b_2 X(t))X(t) = \mathrm{E}[dZ(t)] = 0, \quad (3.5.18)$$

equation (3.5.15) becomes

$$0 = a\mathrm{E}[X^3(t)] - b\mathrm{E}[X^4(t)] + m[cm - d(V + m^2)]. \quad (3.5.19)$$

Using the κ_i notation, we have already shown in (3.5.11) that $\mathrm{E}[X^3(t)] = \kappa_3 + 3\kappa_1\kappa_2 + \kappa_1^3$, whilst on recalling (2.1.37)

$$\mathrm{E}[X^4(t)] = \mathrm{E}[\{(X(t) - \kappa_1) + \kappa_1\}^4] = (\kappa_4 + 3\kappa_2^2) + 4\kappa_3\kappa_1 + 6\kappa_2\kappa_1^2 + 0 + \kappa_1^4. \quad (3.5.20)$$

So equation (3.5.19) may be expressed in the form

$$0 = -b\kappa_4 + (a - 4b\kappa_1)\kappa_3 + (3a\kappa_1 - 3b\kappa_2 - 6b\kappa_1^2)\kappa_2 + (a\kappa_1^2 - b\kappa_1^3 + c\kappa_1 - d\kappa_1^2)\kappa_1. \quad (3.5.21)$$

In spite of (3.5.21) having a similar structure to the third cumulant equation (3.4.15) with $\dot{\kappa}_3 = 0$, it is nevertheless different, even when simplified by application of (3.5.4) and (3.5.13). Further work is clearly needed here to make this approach more accurate, possibly by giving greater thought to suppositions (3.5.16) and (3.5.17).

However, any such investigation would have to be undertaken with considerable care. For example, suppose we replace assumptions (3.5.16) and (3.5.17) by

$$E[X^2(t)dZ(t)] = E[X^2\{(+1)(a_1X - b_1X^2) + (-1)(a_2X + b_2X^2)\}]dt$$
$$= aE(X^3)dt - bE(X^4)dt \quad \text{and} \quad (3.5.22)$$
$$E[X(t)dZ^2(t)] = E[X\{(+1)^2(a_1X - b_1X^2) + (-1)^2(a_2X + b_2X^2)\}]dt$$
$$= cE(X^2)dt - dE(X^3)dt. \quad (3.5.23)$$

Then whilst both approaches yield the same truncated cumulant equations for κ_1 and κ_2 as generated through the full Kolmogorov equation approach, once κ_3 is introduced (by placing $\kappa_4 = 0$ in the third cumulant equation) they produce different algebraic approximations for κ_1, κ_2 and κ_3. So an important question arises as to how robust the approximation procedure is when the nature of the underlying assumptions changes. At present it looks as though such lines of enquiry would have to be conducted on a case by case basis. For whilst numerical solutions for the $\kappa_i(t)$ may well remain relatively unaffected, we shall soon see in Section 3.5.3 that this is by no means guaranteed.

3.5.2 Local approximations

Since under quasi-equilibrium the first two cumulant equations generated by both the c.g.f. and direct approaches are the same, either method produces the same equations for κ_1 and κ_2 when we close them by placing $\kappa_i = 0$ for $i > 2$. Thus if the principal aim of determining κ_1 and κ_2 is to use them to obtain an approximate p.d.f. of population size, then to a first approximation we may replace them by the truncated cumulants $\kappa_1^{(2)}$ and $\kappa_2^{(2)}$ given by (3.4.20) and (3.4.18). Three basic p.d.f.'s worth considering are the Normal (symmetric), binomial (asymmetric, $\kappa_2 < \kappa_1$) and negative binomial (asymmetric, $\kappa_2 > \kappa_1$). For the Normal approximation

$$\pi_N \simeq \Pr(N - 1/2 < X < N + 1/2) \simeq \frac{e^{-(N-\kappa_1)^2/2\kappa_2}}{\sqrt{2\pi\kappa_2}} \quad (N = 1, 2, \ldots) \quad (3.5.24)$$

provided $\sum_{N=-\infty}^{0} \pi_N$ is negligibly small. If it is not, then the $\{\pi_N\}$ could be scaled to ensure that $\sum_{N=1}^{\infty} \pi_N = 1$. For the binomial$(n, p; N)$ case, we equate $\kappa_1 = np$ and $\kappa_2 = npq$ which gives $\kappa_2/\kappa_1 = q$, i.e. $p = 1 - q = 1 - \kappa_2/\kappa_1$, and take n to be the nearest integer to $\kappa_1/(1 - \kappa_2/\kappa_1)$. Thus provided $\pi_0 \simeq 0$ and $\sum_{N=n+1}^{\infty} \pi_N \simeq 0$, we have

$$\pi_N \simeq \binom{n}{N}[1 - \kappa_2/\kappa_1]^N[\kappa_2/\kappa_1]^{n-N} \quad (N = 1, 2, \ldots, n). \quad (3.5.25)$$

Whilst for the negative binomial case we put $\kappa_1 = nq/p$ and $\kappa_2 = nq/p^2$, so $p = \kappa_1/\kappa_2$ with $q = 1 - \kappa_1/\kappa_2$, and n is now the nearest integer to $p\kappa_1/q = \kappa_1^2/(\kappa_2 - \kappa_1)$. Whence for $\pi_0 \simeq 0$ we have

$$\pi_N \simeq \binom{n+N-1}{N}(\kappa_1/\kappa_2)^n(1 - \kappa_1/\kappa_2)^N \quad (N = 1, 2, \ldots). \quad (3.5.26)$$

Although it would be reasonable to use $\kappa_1^{(2)}$ and $\kappa_2^{(2)}$ in the Normal form (3.5.24), this sits less well in the binomial (3.5.25) and negative binomial (3.5.26) cases. For there is a direct contradiction between their asymmetry and the presumption that $\kappa_3 = 0$ in order to ensure second-order cumulant truncation. Yet if we are to involve $\kappa_3 \neq 0$, pragmatism demands that we work with a more user-friendly set of equations than (3.4.13)–(3.4.15) with $\kappa_4 = 0$. Fortunately, the 'stochastic = deterministic + noise' route developed above can be made much simpler by working in terms of the *proportional* variation about the quasi-equilibrium mean value m.

To achieve this we follow the approach of Bartlett, Gower and Leslie (1960) by writing

$$X(t) = m[1 + u(t)] \tag{3.5.27}$$

where the random variable $u(t)$ is presumed sufficiently small for $u^2(t)$ to be negligible in comparison with $u(t)$. This linearization technique is popular in mathematical ecology, though the rather crude nature of the underlying assumption (3.5.27) is sometimes forgotten. That some degree of caution is needed is easily seen from the stochastic logistic simulation shown in Figure 3.1 (with $s = 1$). For over $2 \leq t \leq 20$ (i.e. once quasi-equilibrium is reached) the population size ranges over $2 \leq X(t) \leq 18$. Given that here $m \simeq 10$, this corresponds to $|u(t)| \leq 0.8$. So when $X(t)$ lies near to these outer limits, $u^2(t) \simeq 0.64$ which is clearly not negligible in comparison to $|u(t)| \simeq 0.8$. However, the process seldom reaches these extreme values, and the approach generally performs better than one might anticipate.

With this in mind, substituting for $X(t)$ from (3.5.27) into (3.5.3), and then differentiating (3.5.27) to form

$$dX(t) = mdu(t), \tag{3.5.28}$$

leads to

$$mdu = m(1 + u)(a - bm - bmu)dt + dZ.$$

Since $m \simeq \kappa_1^{(1)} = a/b$, this gives

$$mdu \simeq -bm^2 u(1 + u)dt + dZ$$

which, on ignoring u^2, reduces to

$$du \simeq -bmudt + dZ/m. \tag{3.5.29}$$

Thus

$$u + du \simeq u(1 - adt) + (b/a)dZ, \tag{3.5.30}$$

whence squaring gives

$$(u + du)^2 \simeq u^2(1 - adt)^2 + 2u(b/a)(1 - adt)dZ + (b/a)^2 dZ^2. \tag{3.5.31}$$

Now

$$E(dZ^2) \simeq (+1)^2 \lambda_m dt + (-1)^2 \mu_m dt$$
$$= (a_1 + a_2)mdt - (b_1 - b_2)m^2 dt = m(c - dm)dt \qquad (3.5.32)$$

(the exact result is given by (3.5.8)). So taking expectations of both sides of (3.5.31), and noting that since the process is in quasi-equilibrium $E(u) = 0$ and $E[(u+du)^2] = E(u^2) = \text{Var}(u)$, yields to $O(dt)$

$$\text{Var}(u) \simeq (1 - 2adt)\text{Var}(u) + 0 + (b/a)^2 m(c-dm)dt,$$

i.e. with $m \simeq a/b$

$$\text{Var}(u) \simeq (bc - ad)/2a^2 = (a_1 b_2 + a_2 b_1)/(a_1 - a_2)^2.$$

Thus

$$\text{Var}(X) = m^2 \text{Var}(u) \simeq (a/b)^2 \text{Var}(u) \simeq (a_1 b_2 + a_2 b_1)/(b_1 + b_2)^2 = \hat{V} \qquad \text{(say)}. \qquad (3.5.33)$$

We are now in a position to obtain a better estimate for $\kappa_1 = m \simeq a/b$. For writing (3.5.4) in the form

$$m = a/b - V/m,$$

and then replacing V by $\hat{V}$ and m by a/b, yields the improved approximation

$$\hat{m} = a/b - b\hat{V}/m \simeq (a/b) - (b/a)\hat{V}. \qquad (3.5.34)$$

These approximations for $\kappa_1 \simeq \hat{m}$ and $\kappa_2 \simeq \hat{V}$ are clearly much simpler, and hence far more manageable, than the cumulant closure solutions obtained by solving the Kolmogorov cumulant equations (3.4.13)–(3.4.15) with $\dot{\kappa}_i = 0$. Moreover, an equally simple approximation for the third cumulant, namely

$$\kappa_3 \simeq \hat{\kappa}_3 = V(b_2 - b_1)/(b_1 + b_2), \qquad (3.5.35)$$

can be obtained by considering $E[(X(t) - m)^3]$ (Barlett, Gower and Leslie, 1960; see also Bartlett, 1960). This result implies that κ_3 can range from $-V$ when $b_2 = 0$ (skew left) to $+V$ when $b_1 = 0$ (skew right). Remembering that $\lambda_N = N(a_1 - b_1 N)$ and $\mu_N = N(a_2 + b_2 N)$, it follows that positive skewness is associated with competition affecting the death rate more than the birth rate (i.e. $b_2 > b_1$), and conversely for negative skewness ($b_2 < b_1$).

This (BGL) approximation essentially involves replacing the fourth central moment, $E[(X(t) - \kappa_1(t))^4]$, contained within the raw moment equation (3.5.15), by the Normal approximation $\mu_4 = 3V^2$, and thereby represents an early example of moment closure since this condition is equivalent to placing $\kappa_4 = 0$. Ignoring terms of order $o(dt)$ in the expansion of

$$[X(t+dt) - m]^3 = [(X(t) - m) + (\lambda_X - \mu_X)dt + dZ(t)]^3 \qquad (3.5.36)$$

gives
$$\mathrm{E}[(X-m)(\lambda_X - \mu_X) + (X-m)dZ^2]dt = 0. \qquad (3.5.37)$$

For the logistic process this reduces to

$$\mathrm{E}[(X-m)^2\{(a_1 - a_2)(X-m+m) - (b_1+b_2)(X-m+m)^2\}]$$
$$+ \mathrm{E}[(X-m)\{(a_1+a_2)(X-m+m) - (b_1-b_2)(X-m+m)^2\}] = 0, \quad (3.5.38)$$

that is, to

$$(a_1 - a_2)(\mu_3 + mV) - (b_1+b_2)(\mu_4 + 2m\mu_3 + m^2 V) + (a_1+a_2)(V+0)$$
$$- (b_1 - b_2)(\mu_3 + 2mV + 0) = 0.$$

Whence taking $\sigma^2 \simeq 0$ and supposing $m \gg 1$ to be large in comparison to a_1, b_1, a_2 and b_2, yields

$$-2m\mu_3(b_1+b_2) \simeq (b_1-b_2)2mV, \qquad \text{i.e.}$$
$$\mu_3 \simeq V(b_2 - b_1)/(b_1+b_2). \qquad (3.5.39)$$

Although the level of approximation used in this derivation is clearly generous, numerical studies nevertheless indicate that (3.5.39) can be surprisingly precise.

On a general point, theoretically oriented readers who might harbour concerns about the nature of our approximation devices should consult Ethier and Kurtz (1986); for this text places approximations for a broad range of Markov population processes within a fully rigorous framework by proving related limit theorems.

3.5.3 Application to Africanized honey bees

To illustrate the accuracy of these various levels of approximation it is worthwhile examining the rapid colonization by Africanized honey bees (AHB) of South and Central America (Matis and Kiffe, 2000). This is regarded as being one of the most remarkable biological events of the twentieth century (see Spivak, Fletcher and Breed, 1991) since the range of the AHB population expanded relentlessly every year from 1956 when 26 reproductive swarms escaped in Brazil: during the next 40 years the subsequent population explosion resulted in millions of AHB colonies occupying over 20×10^6 km^2. This rich scenario easily lends itself to a variety of ecological problems such as predicting the leading edge of the invasion (Matis, Rubink and Makela, 1992) and modelling the dynamics of population growth (Otis, 1991), but it also produces a useful non-symmetric equilibrium distribution to work with.

In order to investigate a possible equilibrium distribution for the stochastic model, Matis and Kiffe (1996) propose the parameter values $a_1 = 0.3$, $b_1 = 0.015$, $a_2 = 0.02$ and $b_2 = 0.001$, which yields $K = (a_1 - a_2)/(b_1 + b_2) = 17.5$ as the deterministic carrying capacity. Since this value is only just below the maximum population size $N_{max} = 20$ (so $\lambda_{20} = 0$), we anticipate fairly strong negative skewness. The quasi-equilibrium distribution (3.2.20), namely

$$\tilde{\pi}_N = \tilde{\omega}_N / \sum_{i=1}^{N_{max}} \tilde{\omega}_i \qquad (N = 1, \ldots, 20), \qquad (3.5.40)$$

is easily obtained via (3.2.10) by taking $\tilde{\omega}_1 = 1$ and then successively computing

$$\tilde{\omega}_N = \left(\frac{\lambda_{N-1}}{\mu_N}\right) \tilde{\omega}_{N-1} = \left(\frac{(N-1)[0.3 - 0.015(N-1)]}{N(0.02 + 0.001N)}\right) \tilde{\omega}_{N-1}. \qquad (3.5.41)$$

Figure 3.6 shows how the resulting values of $\tilde{\pi}_N$ and $\log(\tilde{\pi}_N)$ vary with N, and it is clear that virtually all the probability mass is contained in the range $10 \leq N \leq 20$.

At this point it is worth recalling our earlier discussion in Section 3.2 concerning the subtle difference between the pragmatic, $\{\tilde{\pi}_N\}$, and true, $\{\pi_N^{(Q)}\}$, quasi-equilibrium distributions given by (3.5.40) and the solution to equation (3.2.19), respectively. In the former we restrict attention to those realizations which are non-extinct in the time period under study, i.e. we are essentially placing $\mu_1 = 0$ and hence regarding $\{\tilde{\pi}_N\}$ as being a proper equilibrium distribution over $N \geq 1$. So this definition relates to situations in which we follow a time-series of population size that does not lead to extinction over some suitably large, but finite, period of time following the initial burn-in period. In contrast, the latter relates to a conditional probability limit which reflects not only those realizations following burn-in which reach state 1 and then increase again, but also those that continue downwards to 0. Although for many processes $\tilde{\pi}_1$ will be small relative to say $\tilde{\pi}_K$, in which case we would not (intuitively) expect there to be much difference between $\tilde{\pi}_N$ and $\pi_N^{(Q)}$ over the main region of support, there is still a clear need to solve for the $\{\pi_N^{(Q)}\}$ so that we can proceed on a formal analytical basis.

For the AHB example, computation of $\tilde{\pi}_N$ and $\pi_N^{(Q)}$ via (3.5.40) and algorithm A3.1 shows agreement to more than seven significant figures for $N = 9, \ldots, N_{max} = 20$. Since $\pi_9 \simeq 0.00039 \ll \tilde{\pi}_{18} \simeq 0.25$, this clearly implies a very high level of agreement over the full effective range of the distribution. For $i = 1, \ldots, 7$ agreement is to i significant figures, with $\tilde{\pi}_1 \simeq 0.778 \times 10^{-11}$ being just slightly higher than $\pi_1^{(Q)} \simeq 0.720 \times 10^{-11}$. The first three central moments agree exactly to eight significant figures, so in this context both the balance and conditional arguments yield the same values. Where slight discrepancies do arise is in the approximate probability of extinction (3.3.6), i.e. $p_0(t) \simeq 1 - \exp(-\mu_1 \tilde{\pi}_1 t)$. For whilst this gives the time taken for half the realizations to become extinct as $t = (\ln(2))/(\mu_1 \tilde{\pi}_1) \simeq 4.24 \times 10^{12}$, using $\pi_1^{(Q)}$ results in the slightly higher value $t = (\ln(2))/(\mu_1 \hat{\pi}_1) \simeq 4.59 \times 10^{12}$. Though both times are effectively infinite for all practical purposes. To obtain the exact value of $p_0(t)$ we would need to solve the forward equation (3.4.1) numerically over the full range $N = 0, 1, \ldots, N_{max}$.

Given that the quasi-equilibrium probabilities $\{\tilde{\pi}_N\}$ are easy to compute over $N = 1, \ldots, N_{max}$ for any given birth and death rates λ_N and μ_N (determining $\{\pi_N^{(Q)}\}$ is only slightly more complicated), it follows that the raw and central moments,

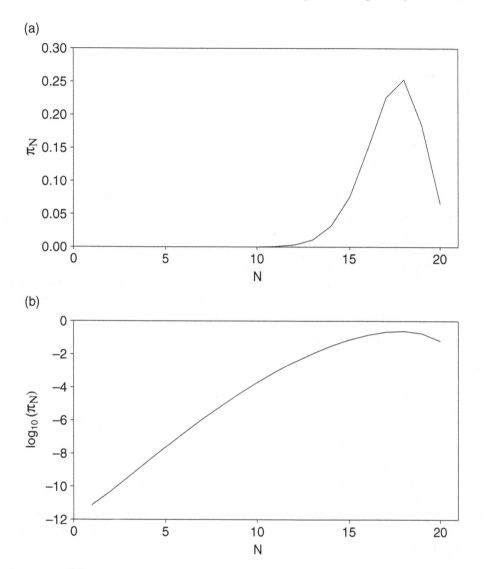

Figure 3.6 (a) Quasi-equilibrium probabilities $\tilde{\pi}_N$ and (b) $\log(\tilde{\pi}_N)$ for the AHB logistic model with rates $\lambda_N = N(0.3 - 0.015N)$ and $\mu_N = N(0.02 + 0.001N)$.

$$\mu'_r = \sum_{N=1}^{N_{max}} N^r \tilde{\pi}_N \quad \text{and} \quad \mu_r = \sum_{N=1}^{N_{max}} (N - \mu'_1)^r \tilde{\pi}_N, \quad (3.5.42)$$

are also easy to evaluate. Though care is required if $\lambda_N > 0$ for all $N \geq 1$, since then N_{max} must be chosen large enough to ensure that neither the probabilities nor the moments are affected by truncation error (within the required level of precision). For the AHB example we have

$$\kappa_1 = 17.35644, \quad \kappa_2 = 2.49170, \quad \kappa_3 = -2.21009 \quad \text{and} \quad \kappa_4 = 1.71048. \qquad (3.5.43)$$

In contrast, the local linearization procedure produces results which are algebraically transparent, but involve accepting some degree of approximation. For the logistic AHB case we have from (3.5.33)–(3.5.35) that

$$\hat{V} = (a_1 b_2 + a_2 b_1)/(b_1 + b_2)^2 = 2.34375 \qquad (5.94\% \text{ too low})$$

$$\hat{m} = (a_1 - a_2)/(b_1 + b_2) - \hat{V}(b_1 + b_2)/(a_1 - a_2)$$
$$= 17.36607 \qquad (0.06\% \text{ too high}) \qquad (3.5.44)$$

$$\hat{\kappa}_3 = \hat{V}(b_2 - b_1)/(b_1 + b_2) = -2.05078 \qquad (7.21\% \text{ too high}),$$

and these moment estimates are surprisingly good (especially $\hat{m}$) given the fairly crude level of approximation used.

The beauty of expressions (3.5.44) lies in their simplicity, and given that we have already seen that the exact numerical values (3.5.43) for the logistic process are so easy to compute, it raises the question as to whether it is worthwhile deriving equilibrium moment closure equations such as (3.4.13)–(3.4.15) with $\dot{\kappa}_r = 0$. Considering their level of precision first, we see from (3.4.17) that

$$\kappa_1^{(1)} = a/b = (a_1 - a_2)/(b_1 + b_2) = 17.5 \qquad (0.83\% \text{ too high}); \qquad (3.5.45)$$

whilst from (3.4.20) and (3.4.18)

$$\kappa_1^{(2)} = [3a + \sqrt{(a^2 - 4bc + 4ad)}]/4b = 17.36396 \qquad (0.04\% \text{ too high})$$

$$\kappa_2^{(2)} = (a - b\kappa_1^{(2)})\kappa_1^{(2)}/b = 2.36226 \qquad (5.20\% \text{ too low}), \qquad (3.5.46)$$

which offer a slight increase in accuracy over the Bartlett estimators (3.5.44) at the expense of a slight loss in transparency. Given that $\kappa_3 = -2.21$ implies substantial skewness (see Figure 3.6a), it is not surprising that $\kappa_2^{(2)}$ is not particularly accurate since its derivation presumes that $\kappa_3 = 0$. On returning to the third-order cumulant equation (3.4.15) with $\dot{\kappa}_3 = \dot{\kappa}_4 = 0$, although we now have three algebraically non-trivial equations to solve for $\kappa_1^{(3)}$, $\kappa_2^{(3)}$ and $\kappa_3^{(3)}$, their numerical solution follows easily, via say *Maple*, yielding

$$\kappa_1^{(3)} = 17.35679 \qquad (0.002\% \text{ too high})$$

$$\kappa_2^{(3)} = 2.48576 \qquad (0.24\% \text{ too low}) \qquad (3.5.47)$$

$$\kappa_3^{(3)} = -2.10886 \qquad (4.58\% \text{ too high}).$$

So $\kappa_1^{(3)}$ and $\kappa_3^{(2)}$ lie close to their true values, though $\kappa_3^{(3)}$ is only marginally better than the crude Bartlett estimate $\hat{\kappa}_3$. Both $\kappa_3^{(3)}$ and $\hat{\kappa}_3$ are affected by the replacement of the true value $\kappa_4 = 1.71$ by zero. Thus whilst these procedures work well for the mean and variance, further theoretical work is required if we are to develop reliable estimators for higher-order moments.

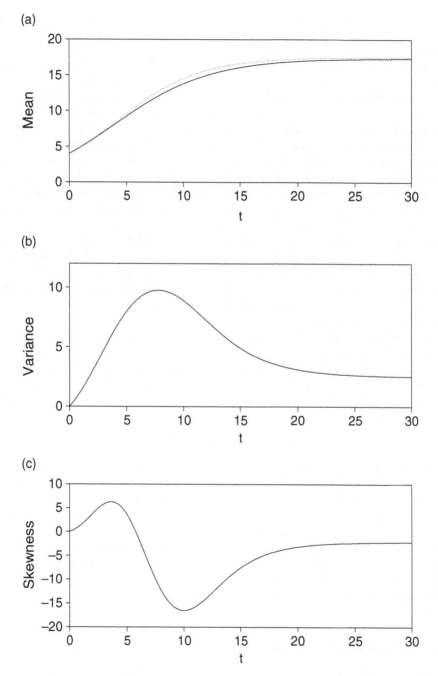

Figure 3.7 Approximate cumulant functions for the AHB logistic model: (a) $\kappa_1(t)$ (—) and $N(t)$ ($\cdots$), (b) $\kappa_2(t)$ and (c) $\kappa_3(t)$.

In summary, the local linearization approach produces useful moment estimators without the need for constructing the Kolmogorov forward equations and the associated differential equation for the c.g.f. Moreover, the resulting expressions are algebraically simple and structurally transparent. In contrast, moment closure generally necessitates the construction of the c.g.f. equation in order to obtain the cumulants, but the pay-back is that the resulting mean and variance estimators are excellent under third-order truncation, whilst the skewness estimate is an improvement on the BGL value. Should algebraic solutions not be required, then in most situations direct numerical computation of the quasi-equilibrium probabilities, and hence determination of exact values for the corresponding moments, will be computationally trivial. However, the exact cumulant equations do carry a considerable advantage when time-dependent moments are required. For example, Matis, Kiffe and Parthasarathy (1998) investigate the way in which $\kappa_1(t)$, $\kappa_2(t)$ and $\kappa_3(t)$, defined through the c.g.f. equation (3.4.11), change with t for the power-law logistic process with rates (3.4.2) over a range of s-values. Curiously, although the difficulty of obtaining approximate cumulant functions increases with s, in the examples investigated the error rates of the approximations decrease with s. In such studies it is more expedient to work directly with the moment closure equations than to construct cumulants via numerical computation of the vector time-series $\{p_N(t); N = 0, 1, 2, \ldots\}$, provided the small resulting errors can be tolerated.

Figure 3.7 shows (a) $\kappa_1(t)$, (b) $\kappa_2(t)$ and (c) $\kappa_3(t)$ generated from (3.4.13)–(3.4.15) for the AHB process over $0 \leq t \leq 30$ with $n_0 = 4$. Although the mean, $\kappa_1(t)$, has the classic sigmoid shape of the deterministic trajectory $N(t)$ given by (3.2.30), it always lies slightly below it (see Figure 3.7a). The reason becomes apparent on comparing the deterministic equation (3.2.36) with the mean equation (3.4.13). For replacing κ_1 by N shows that

$$d\kappa_1/dt = \kappa_1(a - b\kappa_1) - b\kappa_2 < N(a - bN) = dN/dt \qquad (t > 0);$$

since $\kappa_2(t) > 0$ for all $t > 0$, whilst $b = b_1 + b_2 > 0$ by definition. Specifically, we see from (3.5.43) that the asymptotic mean value $\kappa_1(\infty) = 17.35644$ is 0.82% lower than the deterministic carrying capacity $N(\infty) = 17.5$. The variance $\kappa_2(t)$ is substantially greater in the opening (i.e. transient) stages of the process than in the quasi-equilibrium phase; whilst the skewness $\kappa_3(t)$ is initially positive (maximum of 6.28) before dropping to a local minimum (-16.53) which is considerably below the quasi-equilibrium value of -2.21. Note that the structure of $\kappa_i(t)$ depends heavily on the initial value n_0. For example, when $n_0 = 1$, $\kappa_3(t)$ rises to the much larger value 98.82 before dropping to the now higher local minimum of -2.96.

3.6 Diffusion approximations

If we know the algebraic or numerical values for the time-dependent cumulants $\kappa_i(t)$, or equilibrium cumulants $\kappa_i(\infty)$, then we can insert them into any appropriate standard distribution to obtain approximate forms for the probability distributions $\{p_N(t)\}$ and $\{\pi_N\}$. Simple examples previously considered include the Normal, binomial and negative binomial distributions (3.5.24)–(3.5.26). If we wish to generate more complex mathematical forms to describe these probabilities, then since mathematical analysis of models with a discrete population size $N = 0, 1, 2, \ldots$ is intrinsically more demanding than that required for continuous N, it is sensible to envisage N as a continuous variable.

One approach is to replace the forward differential equation (3.3.1), viz.

$$dp_N(t)/dt = \lambda_{N-1} p_{N-1}(t) - (\lambda_N + \mu_N) p_N(t) + \mu_{N+1} p_{N+1}(t) \qquad (N = 0, 1, 2, \ldots) \tag{3.6.1}$$

by a *diffusion equation* in which N is allowed to take any non-negative value. Since there are a whole variety of related procedures and limiting arguments that may be considered, we shall defer discussion of this wide-ranging topic until Chapter 6 and here just use a crude approximation procedure. On taking the Taylor series expansion of $p_{N-1}(t)$, $p_{N+1}(t)$, λ_{N-1} and μ_{N+1}, and denoting $\partial(\cdot)/\partial N$ by $(\cdot)'$, equation (3.6.1) yields

$$\partial p_N(t)/\partial t = (\lambda_N - \lambda'_N + \lambda''_N/2! - \cdots)(p_N(t) - p'_N(t)$$
$$+ p''_N(t)/2! + \cdots) - (\lambda_N + \mu_N) p_N(t)$$
$$+ (\mu_N + \mu'_N + \mu''_N/2! + \cdots)(p_N(t) + p'_N(t) + p''_N(t)/2! + \cdots).$$

Whence ignoring terms with a combined order greater than two gives

$$\partial p_N(t)/\partial t = -[p_N(t)(\lambda'_N - \mu'_N) + p'_N(t)(\lambda_N - \mu_N)]$$
$$+ (1/2)[p_N(t)(\lambda''_N + \mu''_N) + 2p'_N(t)(\lambda'_N + \mu'_N) + p''_N(t)(\lambda_N + \mu_N)],$$

i.e.
$$\frac{\partial p_N(t)}{\partial t} = -\frac{\partial}{\partial N}\{(\lambda_N - \mu_N) p_N(t)\} + \frac{1}{2} \frac{\partial^2}{\partial N^2}\{(\lambda_N + \mu_N) p_N(t)\} \qquad (N, t \geq 0). \tag{3.6.2}$$

Care must be taken when using this equation since there are various pitfalls that can beset the unwary, so for the moment we shall restrict our attention to examining the associated quasi-equilibrium structure. Although a parallel diffusion equation can also be developed for the conditional equations for $\tilde{\pi}_N(t) = p_N(t)/(1 - p_0(t))$, it has a rather nasty nonlinear form which is intractable analytically and more difficult to solve numerically than the discrete equations which it purports to approximate (Nisbet and Gurney, 1982).

3.6.1 Equilibrium probabilities

Before we investigate how the diffusion equation (3.6.2) can be used to construct an approximation to the quasi-equilibrium probabilities $\{\tilde{\pi}_N\}$, let us first consider processes which possess a true equilibrium structure. That is, they are subject to neither extinction nor population explosion. So $p_N(t) \to \pi_N$ as $t \to \infty$, whence we may set $\partial p_N(t)/\partial t = 0$. On writing $f(N) = \lambda_N - \mu_N$ and $g(N) = \lambda_N + \mu_N$, for large t equation (3.6.2) becomes

166   General Markov population processes

$$\frac{1}{2}\frac{\partial^2}{\partial N^2}[g(N)\pi_N] - \frac{\partial}{\partial N}[f(N)\pi_N] = 0, \qquad (3.6.3)$$

which integrates to give

$$\frac{1}{2}\frac{\partial}{\partial N}[g(N)\pi_N] - f(N)\pi_N = C \qquad (3.6.4)$$

for some constant C. We shall provide a more formal examination of this equation in Section 6.2.2, but for the moment let us argue that when N is very large, i.e. well in excess of its carrying capacity, then both π_N and $\partial \pi_N/\partial N$ must be very small. Thus $C = 0$, and so we may write (3.6.4) in the standard form

$$\frac{\partial}{\partial N}[g(N)\pi_N] - 2\left[\frac{f(N)}{g(N)}\right][g(N)\pi_N] = 0,$$

which integrates directly to yield

$$\pi_N \simeq A[g(N)]^{-1} \exp\{2\int \frac{f(N)}{g(N)} dN\} \qquad (0 \leq N \leq N_{max}). \qquad (3.6.5)$$

Here the normalizing constant A is chosen to ensure that

$$\int_0^{N_{max}} \pi_N \, dN = 1, \qquad (3.6.6)$$

where N_{max} denotes the maximum permissible population size.

Because the p.d.f. $\{\pi_N\}$ now relates to a continuous distribution over $N \geq 0$, the exact integer population size probabilities $\{\pi_n\}$ may be deduced by taking (for example)

$$\pi_n \simeq \int_{n-1/2}^{n+1/2} \pi_N \, dN \quad (n > 0) \quad \text{with} \quad \pi_0 \simeq \int_0^{1/2} \pi_N \, dN. \qquad (3.6.7)$$

Furthermore, if ultimate extinction is inevitable, then result (3.6.5) provides the quasi-equilibrium probabilities provided we change (3.6.6) to (say)

$$\int_{1/2}^{N_{max}} \pi_N \, dN = 1. \qquad (3.6.8)$$

A possible drawback to solution (3.6.5) is that it necessitates evaluating $\int [f(N)/g(N)] dN$. However, this presents no problem for the logistic quadratic birth and death forms $\lambda_N = N(a_1 - b_1 N)$ and $\mu_N = N(a_2 + b_2 N)$. For example, when $b_1 = b_2$ we have

$$\pi_N \simeq \frac{A}{N(a_1 + a_2)} \exp\left\{-\left(\frac{2b_1}{a_1 + a_2}\right)\left[N - \left(\frac{a_1 - a_2}{2b_1}\right)\right]^2\right\} \qquad (1 \leq N \leq a_1/b_1). \qquad (3.6.9)$$

So provided the population size N remains close to the deterministic carrying capacity $K = (a_1 - a_2)/2b_1$, and π_N is negligibly small for $N < 1$ and $N > a_1/b_1$, we may regard N as being approximately Normally distributed with mean K (in agreement with (3.4.17) for $\kappa_1^{(1)}$) and variance $(a_1 + a_2)/4b_1$ (in agreement with (3.5.33) for $\hat{V}$). To obtain a similar Normal form for the case $b_1 \neq b_2$, we first write

$$\int \frac{f(N)}{g(N)} dN = \int \frac{(a_1 - a_2) - N(b_1 + b_2)}{(a_1 + a_2) - N(b_1 - b_2)} dN,$$

and then replace N by K in the denominator. For this also yields a Normal distribution with mean K and variance $\hat{V}$. Note that if we replace N by K *before* cancelling N top and bottom in the integrand, then we obtain the non-Normal form

$$\pi_N \propto \exp\{[(a_1 - a_2)N^2 - (2/3)(b_1 + b_2)N^3]/g(K)\}. \qquad (3.6.10)$$

For general λ_N and μ_N, on writing $N = K + \epsilon$ (for appropriately small ϵ) we have

$$f(N) \simeq f(K) + \epsilon \frac{\partial f(N)}{\partial N}\bigg|_{\epsilon=0} = 0 + (N - K) \frac{\partial f(N)}{\partial N}\bigg|_{N=K}$$

since $f(K) = 0$. Whilst placing $g(N) \simeq g(K)$ yields

$$2\int \frac{f(N)}{g(N)} dN \simeq \frac{(N - K)^2}{g(K)} \frac{\partial f(N)}{\partial N}\bigg|_{N=K}.$$

Thus

$$\pi_N \propto \exp\left\{\frac{(N - K)^2}{g(K)} \frac{\partial f(N)}{\partial N}\bigg|_{N=K}\right\}, \qquad (3.6.11)$$

whence it follows that N is approximately Normally distributed with mean K and variance

$$V_{Normal} = -g(K) / \left[2 \frac{\partial f(N)}{\partial N}\bigg|_{N=K}\right]. \qquad (3.6.12)$$

For example, the simple immigration–death process has rates $\lambda_N = \alpha$ and $\mu_N = N\mu$, whence $K = V_{Normal} = \alpha/\mu$, as required for the underlying Poisson(α/μ) process. Whilst for the standard logistic process with $\lambda_N = N(a_1 - b_1 N)$ and $\mu_N = N(a_2 + b_2 N)$ we have $K = (a_1 - a_2)/(b_1 + b_2)$ and $V_{Normal} = (a_1 b_2 + a_2 b_1)/(b_1 + b_2)^2$, in agreement with $\hat{V}$ in (3.5.33). Extending this result to the power-law logistic process with rates (3.4.2), i.e. $\lambda_N = N(a_1 - b_1 N^s)$ and $\mu_N = N(a_2 + b_2 N^s)$, gives

$$K = [(a_1 - a_2)/(b_1 + b_2)]^{1/s} \quad \text{and} \quad V_{Normal} = \frac{(a_1 - a_2)^{1/s-1}}{(b_1 + b_2)^{1/s+1}}(a_1 b_2 + a_2 b_1).$$

$$(3.6.13)$$

Employing local linearization both to verify underlying Normality and to construct approximate values for the mean and variance, irrespective of the complexity of λ_N and μ_N, is clearly an extremely useful procedure. Whilst formulations such as (3.6.10) are sufficiently simple to provide useful indications of departures from Normality.

In general, however, the integral $\int [f(N)/g(N)] dN$ in the diffusion solution (3.6.5) may not be sufficiently transparent to provide any real advantage for using the approximate diffusion solution instead of the exact equilibrium probabilities (3.2.7) or the quasi-equilibrium probabilities (3.2.20). It is therefore advisable to construct and compare both the discrete and the diffusion solutions. For example, the simple immigration–death rates $\lambda_N = \alpha$ and $\mu_N = \mu N$ have $f(N) = \alpha - \mu N$ and $g(N) = \alpha + \mu N$. So

$$2 \int \frac{f(N)}{g(N)} dN = -2N + (4\alpha/\mu) \ln(\alpha + \mu N),$$

which yields

$$\pi_N \simeq A e^{-2N} (\alpha + \mu N)^{(4\alpha/\mu)-1}. \tag{3.6.14}$$

This is marginally less transparent than the exact Poisson(α/μ) solution (2.5.9), since (3.6.14) implies that $N + \alpha/\mu$ follows a Gamma distribution. Whilst a parallel calculation for the basic logistic process with rates $\lambda_N = N(a_1 - b_1 N)$ and $\mu_N = N(a_2 + b_2 N)$ yields (for $a = a_1 - a_2$, $b = b_1 + b_2$, $c = a_1 + a_2$ and $d = b_1 - b_2$)

$$\pi_N \simeq (A/N) e^{2bN/d} (c - dN)^{[2(bc-ad)/d^2]-1}. \tag{3.6.15}$$

Being $1/N$ times a Gamma p.d.f., this is (arguably) less informative than the exact result (3.2.37) which expresses the quasi-equilibrium probabilities in terms of a ratio of linear products. We shall return to this issue in Section 6.2.2.

3.6.2 Perturbation methods

The Normal form given by the solution to the approximating diffusion equation usually works reasonably well provided that the mean is not too small and we keep away from the tails of the distribution. However, if we are interested in tail probabilities then since it is the *relative*, rather than the *absolute*, error of the approximation that concerns us, it would seem more satisfactory to seek an approximation to $\ln[p_N(t)]$ rather than $p_N(t)$. At this point tribute must be paid to Henry Daniels, who argued strongly in favour of using perturbation methods to generate good approximate solutions to problems in stochastic processes. Though the concept of 'approximation' is (oddly) anathema to some applied probabilists, there is no point in shunning approximate solutions when exact solutions are mathematically intractable. Readers interested in discovering more about this powerful, albeit partially forgotten, field of mathematics should consult Bellman (1964) or Simmonds and Mann (1986).

To illustrate the approach, consider the simple immigration process for which the Kolmogorov forward equation is given by

$$dp_N(t)/dt = \alpha p_{N-1}(t) - \alpha p_N(t) \tag{3.6.16}$$

with $p_{-1}(t) \equiv 0$. This gives rise to a Poisson process (discussed extensively in Section 2.1) with exact probabilities

$$p_N(t) = (\alpha t)^N e^{-\alpha t}/N! \qquad (N = 0, 1, 2, \ldots), \tag{3.6.17}$$

and a Normal $N(\alpha t, \alpha t)$ diffusion approximation. Suppose we now write

$$l(N, t) = \ln[p_N(t)], \qquad \text{i.e.} \qquad p_N(t) = e^{l(N,t)}. \tag{3.6.18}$$

Then equation (3.6.16) becomes

$$dl(N,t)/dt = \alpha e^{l(N-1,t)-l(N,t)} - \alpha. \tag{3.6.19}$$

Whence taking the first-order Taylor series expansion of $l(N-1,t)$ about N (in contrast to the second-order expansion used in (3.6.2)) leads to the first-order nonlinear differential equation

$$\frac{\partial l(N,t)}{\partial t} \simeq \alpha e^{-\partial l(N,t)/\partial N} - \alpha. \tag{3.6.20}$$

A standard method of solution (e.g. Forsyth, 1928) is to put

$$\frac{\partial l(N,t)}{\partial N} = -a \tag{3.6.21}$$

for some constant a. Then

$$\frac{\partial l(N,t)}{\partial t} = \alpha(e^a - 1), \tag{3.6.22}$$

and so the combined solution is given by

$$l(N,t) = \alpha t(e^a - 1) - aN + c \tag{3.6.23}$$

for some constant c. Suppose that the population is initially empty at time $t = 0$, i.e. $p_0(0) = 1$ and $p_N(0) = 0$ for $N > 0$. Then $l(0,0) = 0$ and $l(N,0) = -\infty$ for $N > 0$, whence it follows that the expansion procedure breaks down near $t = 0$. Nevertheless, let us examine what happens when the planes (3.6.21) and (3.6.22) are made to pass through $l(0,0) = 0$. We see from (3.6.23) that the required condition is $c = 0$, and so the envelope is

$$l(N,t) = \alpha t(e^a - 1) - aN. \tag{3.6.24}$$

If we now argue that $l(N,t)$ should not change with a, then on placing

$$\frac{\partial l}{\partial a} = \alpha t e^a - N = 0$$

we have

$$e^a = N/\alpha t,$$

whence (3.6.24) becomes

$$l(N,t) = (N - \alpha t) - N \ln(N/\alpha t).$$

So

$$p_N(t) \simeq e^{l(N,t)} = \frac{(\alpha t)^N e^{-\alpha t}}{N^N e^{-N}}. \tag{3.6.25}$$

Hence on using the basic form of Stirling's approximation, namely

$$N! \sim \sqrt{2\pi} N^{N+1/2} e^{-N}, \tag{3.6.26}$$

we have

$$p_N(t) \simeq \sqrt{2\pi N} \frac{(\alpha t)^N e^{-\alpha t}}{N!}. \qquad (3.6.27)$$

That is, we recover the exact Poisson probabilities (3.6.17) provided we replace the pre-multiplier N by its expected value αt, and then normalize (for fixed t) the resulting $p_N(t)$ to ensure that they sum to unity over the (now) discrete values $N = 0, 1, 2, \ldots$. Note that since (3.6.27) yields real values of $p_N(t)$ only in the range $N > 0$, negative values of N are automatically excluded.

Clearly, such a normalization would be greatly facilitated if the approximation retained the arbitrary constant c. This would have appeared if the planes (3.6.23) had been made to pass through $l(0,0) = c$, which amounts to requiring that as $t \to 0$ the distribution becomes concentrated within a negligibly small neighbourhood of $N = 0$. Evidently no more precise condition is allowable near $t = 0$ where the expansion of $l(N,t)$ becomes invalid. Since it is precisely at small t that the assumptions underlying the procedure break down, a more worthwhile approach would be to ignore this initial condition and to determine c through $\sum_{N=0}^{\infty} p_N(t) = 1$ (having first replaced $\sqrt{N}$ by $\sqrt{\alpha t}$). There is no point in trying to circumvent the initial value difficulty by introducing the second-order expansion in N, since a similar problem arises. We shall therefore shortly consider a more satisfactory approach.

Note that we obtain exactly the same result (3.6.25) if, instead of putting $\partial l/\partial N = -a$ as in (3.6.21), we first place

$$\frac{\partial l(N,t)}{\partial t} = a. \qquad (3.6.28)$$

For then (3.6.20) becomes

$$\frac{\partial l(N,t)}{\partial N} = \ln\left(\frac{\alpha}{a+\alpha}\right), \qquad (3.6.29)$$

and the combined solution now takes the form

$$l(N,t) = at + N\ln[\alpha/(a+\alpha)] + c \qquad (3.6.30)$$

for some constant c. Taking c to be independent of a, as before, and forming the envelope with respect to a by placing

$$\frac{\partial l(N,t)}{\partial a} = t - \frac{N}{a+\alpha} = 0, \qquad (3.6.31)$$

then yields $a = N/t - \alpha$. So the first-order approximation is

$$l(N,t) = N - \alpha t + N\ln(\alpha t/N) + c. \qquad (3.6.32)$$

Thus

$$p_N(t) \simeq e^{l(N,t)} = A(\alpha t/N)^N e^{N-\alpha t} \qquad (3.6.33)$$

in agreement with (3.6.25), for normalizing constant $A = e^c$.

In principle, this approach can be extended to more complex stochastic processes, though the algebraic detail may soon become messy. Consider, for example, the general birth process with rate λ_N which, for convenience, we write as $\lambda(N)$. Then the forward equations (2.3.51), namely

$$dp_N(t)/dt = \lambda(N-1)p_{N-1}(t) - \lambda(N)p_N(t), \tag{3.6.34}$$

become

$$dl(N,t)/dt = \lambda(N-1)e^{l(N-1,t)-l(N,t)} - \lambda(N). \tag{3.6.35}$$

So (3.6.19) is the special case with $\lambda(N) = \alpha$. Following Daniels (1960), suppose we assume that $\lambda(N)$ varies smoothly with N and that $\lambda'(N)$ is small compared with $\lambda(N)$, at least when N is not small. Then the first-order Taylor series expansion of (3.6.35) gives

$$\frac{\partial l(N,t)}{\partial t} \simeq \lambda(N-1)\exp\{l(N,t) - \frac{\partial l(N,t)}{\partial N} - l(N,t)\} - \lambda(N)$$

i.e.
$$\frac{\partial l(N,t)}{\partial t} \simeq \lambda(N)[e^{-\partial l(N,t)/\partial N} - 1]. \tag{3.6.36}$$

Unlike (3.6.21) above, we cannot now put $-\partial l(N,t)/\partial N = -a$, since this still leaves a function of N on the right-hand side of (3.6.36); note that the simple immigration process with $\lambda(N) \equiv a$ is the only exceptional case. So instead, let us follow the second approach (3.6.28), i.e. we take $\partial l(N,t)/\partial t = a$, whence (3.6.36) yields

$$\frac{\partial l(N,t)}{\partial N} = \ln\left(\frac{\lambda(N)}{\lambda(N)+a}\right). \tag{3.6.37}$$

Then the complete solution of (3.6.36) for $N(0) = \xi$ is

$$l(N,t) = at + \int_\xi^N \ln\left(\frac{\lambda(w)}{\lambda(w)+a}\right) dw + c. \tag{3.6.38}$$

Whence on writing the constant c as $\ln(A)$, we have the general result

$$p_N(t) \simeq e^{l(N,t)} = A\exp\{at + \int_\xi^N \ln\left(\frac{\lambda(w)}{\lambda(w)+a}\right)\} dw. \tag{3.6.39}$$

Here we obtain a by paralleling (3.6.31), i.e. we take c to be independent of a and form

$$\frac{\partial l(N,t)}{\partial a} = t - \int_\xi^N \frac{dw}{\lambda(w)+a} = 0. \tag{3.6.40}$$

Solving (3.6.40) for a as a function of N, and substituting the result into (3.6.39), then yields an approximation to the general birth process which provides a completely different perspective on the structure of the probabilities $p_N(t)$ to the exact partial fraction expansion (2.3.55)–(2.3.56).

However, as a means of examining specific processes this approach is clearly of limited use. For example, on taking the simple birth rate $\lambda(N,t) = \lambda_N = \lambda N$, (3.6.40) integrates to

$$a = \lambda(\xi e^{\lambda t} - N)/(1 - e^{\lambda t}), \tag{3.6.41}$$

whence (3.6.39) yields

$$p_N(t) \simeq A\exp\left\{\frac{\lambda t(\xi e^{\lambda t} - N)}{1 - e^{\lambda t}}\right\}[(N-\xi)\ln(1-e^{-\lambda t}) + \int_\xi^N \ln\left(\frac{w}{w-\xi}\right)dw], \tag{3.6.42}$$

which is considerably more complex than the exact negative binomial solution (2.3.9) (with $n_0 = \xi$) it purports to approximate. Whilst for the quadratic birth process with $\lambda_N = N^2$ (see Section 2.3.5), on integrating (3.6.40) we see that a is the solution to the transcendental equation

$$\sqrt{a}t = \tan^{-1}(N/\sqrt{a}) - \tan^{-1}(\xi/\sqrt{a}). \tag{3.6.43}$$

So there is not even a closed form expression for a which can inserted into the integrand in (3.6.39).

This approach becomes even more convoluted when we allow both births and deaths. For placing $l(N,t) = \ln[p_N(t)]$ in the forward equation (3.6.1) yields

$$dl(N,t)/dt = \lambda_{N-1}e^{l(N-1,t)-l(N,t)} - (\lambda_N + \mu_N) + \mu_{N+1}e^{l(N+1,t)-l(N,t)}. \tag{3.6.44}$$

Whence paralleling the construction of (3.6.36) by taking first-order Taylor series expansions in the exponents, and regarding λ'_N and μ'_N small in comparison to λ_N and μ_N, leads to

$$\partial l(N,t)/\partial t \simeq \lambda_N[e^{-\partial l(N,t)/\partial N} - 1] + \mu_N[e^{\partial l(N,t)/\partial N} - 1]. \tag{3.6.45}$$

On placing $\partial(N,t)/\partial t = a$, we now obtain the quadratic equation

$$\mu_N e^{2\partial(N,t)/\partial N} - (a + \lambda_N + \mu_N)e^{\partial l(N,t)/\partial N} + \lambda_N = 0$$

which solves to give

$$\frac{\partial l(N,t)}{\partial N} = \ln\left\{\frac{(a+\lambda_N+\mu_N) \pm \sqrt{\{(a+\lambda_N+\mu_N)^2 - 4\lambda_N\mu_N\}}}{2\mu_N}\right\}. \tag{3.6.46}$$

So the complete solution of (3.6.45) for $N(0) = \xi$ and

$$I(w) = [(a + \lambda(w) + \mu(w)) \pm \sqrt{\{(a + \lambda(w) + \mu(w))^2 - 4\lambda(w)\mu(w)\}}/(2\mu(w))]$$

is

$$l(N,t) = at + \int_\xi^N \ln[I(w)]\,dw + c. \tag{3.6.47}$$

Which, on writing $A = e^c$, yields the general result that

$$p_N(t) \simeq e^{l(N,t)} = A\exp\{at + \int_\xi^N \ln[I(w)]\,dw\}. \tag{3.6.48}$$

This expression is substantially more messy than the special birth-only result (3.6.39). For not only is the parallel equation to (3.6.40) formed by placing $\partial l(N,t)/\partial a = 0$ much more awkward in nature, but we also have to separate out the two branches of the quadratic solution (3.6.46). Currently, this approximate solution to the general birth–death process is clearly too complicated to be of much practical use in all but the simplest of special cases. However, it does offer a completely general representation and, as such, must surely have considerable potential if it could be shaped into a more user-friendly form. This open problem offers an interesting challenge.

In equilibrium the technique operates much more smoothly, since as the balance equations $\lambda_N \pi_N = \mu_{N+1} \pi_{N+1}$ are first-order, we bypass the quadratic bifurcation problem. On replacing π_N by $e^{l(N)}$ and taking $\mu_{N+1} \simeq \mu_N$, as above, we now have

$$\lambda_N e^{l(N)} = \mu_{N+1} e^{l(N+1)} \simeq \mu_N e^{l(N) + dl(N)/dN}$$

which yields

$$\frac{\partial l(N)}{\partial N} \simeq \ln(\lambda_N / \mu_N). \tag{3.6.49}$$

Thus

$$l(N) \simeq \text{constant} + \int \ln(\lambda_N / \mu_N) \, dN, \tag{3.6.50}$$

i.e. for appropriate constant A,

$$\pi_N = e^{l(N)} \simeq A \exp\{ \int \ln(\lambda_N / \mu_N) \, dN \}. \tag{3.6.51}$$

For example, in the case of the simple immigration–death process with $\lambda_N = \alpha$ and $\mu_N = \mu N$, we have

$$\pi_N \simeq A (\alpha/\mu)^N N^{-N} e^N. \tag{3.6.52}$$

Whence on assuming that the main probability mass lies near to the mean equilibrium value $K = \alpha/\mu$, substituting for

$$N! \sim \sqrt{2\pi} e^{-N} N^{N+1/2} \simeq \sqrt{2\pi K} e^{-N} N^N, \tag{3.6.53}$$

and then choosing A to ensure that $\sum_{N=0}^{\infty} \pi_N = 1$, recovers the exact Poisson probabilities (2.5.9), namely

$$\pi_N = (\alpha/\mu)^N e^{-\alpha/\mu} / N!. \tag{3.6.54}$$

Since the equivalent result (3.6.14) based on the diffusion solution (3.6.5) is nowhere near as successful, one might conjecture that the 'tilted' solution (3.6.51) is generally superior to the diffusion solution, though further investigation is clearly necessary here. The logistic probabilities emerge just as easily, for with $\lambda_N = N(a_1 - b_1 N)$ and $\mu_N = N(a_2 - b_2 N)$ we have

$$\int \ln(\lambda_N / \mu_N) \, dN = [(a_1 b_2 + a_2 b_1)/b_2^2] \ln(a_2 + b_2 N) - (b_1/b_2) N,$$

which gives

$$\pi_N = e^{l(N)} \simeq A(a_2 + b_2 N)^{(a_1 b_2 + a_2 b_1)/b_2^2} e^{-(b_1/b_2)N}. \tag{3.6.55}$$

This has a similar structure to the diffusion result (3.6.15): numerical studies would show whether (3.6.55) follows the Poisson example in providing more accurate probability estimates.

The application and further development of expansion techniques in the stochastic process arena is clearly a potentially rewarding field of study, and to highlight the way in which flexible thinking may be employed let us return to the simple immigration equation (3.6.19), but this time write it in the form

$$dl(N,t)/dt = \alpha \exp\{(1/\epsilon)[l(N-\epsilon,t) - l(N,t)]\} - \alpha. \tag{3.6.56}$$

So when $\epsilon = 0$ we obtain the first-order approximating equation (3.6.20), whilst placing $\epsilon = 1$ recovers the exact equation (3.6.19). Intuitively, it is therefore reasonable to expect that taking $0 < \epsilon < 1$ will yield a more accurate result than working with (3.6.20). Write

$$l(N,t) \equiv \sum_{i=0}^{\infty} \epsilon^i l_i(N,t). \tag{3.6.57}$$

Then on denoting $f' = \partial f/\partial N$ (as before), (3.6.56) expands to give

$$\frac{1}{\alpha}\frac{\partial}{\partial t}(l_0 + \epsilon l_1 + \epsilon^2 l_2 + \cdots)$$
$$= \exp\{-l' + (\epsilon/2!)l'' - (\epsilon^2/3!)l''' + \cdots\} - 1$$
$$= \exp\{-(l_0' + \epsilon l_1' + \epsilon^2 l_2' + \cdots) + (\epsilon/2)(l_0'' + \epsilon l_1'' + \cdots) - (\epsilon^2/6)(l_0''' + \cdots) + \cdots\} - 1$$
$$= \exp\{-l_0' + \epsilon(-l_1' + l_0''/2) + \epsilon^2(-l_2' + l_1''/2 - l_0'''/6) + \cdots\} - 1. \tag{3.6.58}$$

Whence extracting the coefficient of ϵ^0 recovers the original approximating equation (3.6.20), namely

$$(1/\alpha)\partial l_0(N,t)/\partial t = e^{-\partial l_0(N,t)/\partial N} - 1. \tag{3.6.59}$$

Now on noting from (3.6.32) that the solution takes the form

$$l_0(N,t) = N - \alpha t + N \ln(\alpha t) - N \ln(N) + c, \tag{3.6.60}$$

we have

$$\partial l_0(N,t)/\partial N = \ln(\alpha t) - \ln(N) \quad \text{and} \quad \partial^2 l_0(N,t)/\partial N^2 = -1/N.$$

Thus extracting the coefficient of ϵ^1 in (3.6.58), namely

$$\frac{\partial l_1(N,t)}{\partial t} = \alpha e^{-\partial l_0(N,t)/\partial N}\left(-\frac{\partial l_1(N,t)}{\partial N} + \frac{1}{2}\frac{\partial^2 l_0(N,t)}{\partial N^2}\right), \tag{3.6.61}$$

yields

$$t\frac{\partial l_1(N,t)}{\partial t} = -N\frac{\partial l_1(N,t)}{\partial N} - \frac{1}{2}. \quad (3.6.62)$$

This has the solution

$$l_1(N,t) = \ln(t) - (3/2)\ln(N). \quad (3.6.63)$$

So on taking the first two terms in the expansion (3.6.57), and placing $\epsilon = 1$ and $A = e^c$, we have

$$p_N(t) \simeq e^{l_0(N,t)+l_1(N,t)} = (t/N\sqrt{N}) \times A(\alpha t/N)^N e^{N-\alpha t}. \quad (3.6.64)$$

Whence applying Stirling's formula (3.6.53) yields

$$p_N(t) \simeq A'(t/N)(\alpha t)^N e^{-\alpha t},$$

and replacing N by its expected value αt then recovers the exact Poisson(αt) probabilities. Comparison with the derivation of result (3.6.27) shows that the effect of including the second-order term $l_1(N,t)$ is to remove the necessity for the normalization constant to be a function of t.

3.6.3 Additive mass-immigration process

Although the above approach yields the reasonably neat formulation (3.6.39) for the solution of the general birth equation (3.6.34), we have seen through (3.6.48) that the solution becomes considerably more opaque once deaths are allowed. We shall therefore now illustrate how progress can be made by switching attention away from the transition rates onto the cumulative generating function and, in the first instance, by changing from continuous time $t \geq 0$ to discrete time $n = 0, 1, 2, \ldots$. Suppose we parallel the continuous-time mass-immigration process of Section 2.7.2 by assuming that immigrants arrive in batches of size k with probability α_k where $\sum_{k=1}^{\infty} \alpha_k = 1$. Then the probabilities $\{p_N(n)\}$ of being in state N after n transitions satisfy the forward equations

$$p_N(n+1) = \sum_{k=1}^{N} p_{N-k}(n)\alpha_k \quad (n = 1, 2, \ldots). \quad (3.6.65)$$

Given that we shall be considering Taylor series expansions, which are essentially continuous, let us replace N by x and k by y, and rewrite these equations in the general form

$$p(x, n+1) = \int p(x-y, n)\, dG(y). \quad (3.6.66)$$

So in terms of $l(x,n) = \ln[p(x,n)]$, (3.6.66) becomes

$$e^{l(x,n+1)-l(x,n)} = \int_{-\infty}^{\infty} e^{l(x-y,n)-l(x,n)}\, dG(y), \quad (3.6.67)$$

where for our mass-immigration process the p.d.f. $dG(y) \equiv 0$ for $y < 0$. Allowing $dG(y) > 0$ for $y < 0$ means that we can also have mass emigration as well as immigration, but since this implies the possibility of $x < 0$ we no longer have a true population process. Expanding $l(x, n+1)$ and $l(x-y, n)$ to first-order yields the approximating equation

$$\exp\{l(x,n) + \partial l(x,n)/\partial n - l(x,n)\} = \int \exp\{l(x,n) - y\partial l(x,n)/\partial x - l(x,n)\} dG(y)$$

i.e.
$$\exp\{\partial(x,n)/\partial n\} = \int_{-\infty}^{\infty} \exp\{-y\partial l(x,n)/\partial x\} dG(y). \quad (3.6.68)$$

Thus in terms of the c.g.f. of transition probabilities, namely

$$\tilde{K}(\theta) = \ln\left\{\int_{-\infty}^{\infty} e^{\theta y} dG(y)\right\}, \quad (3.6.69)$$

we have
$$\exp\{\partial l(x,n)/\partial n\} = \exp\{\tilde{K}[-\partial l(x,n)/\partial x]\}$$

i.e.
$$\partial l(x,n)/\partial n = \tilde{K}[-\partial l(x,n)/\partial x]. \quad (3.6.70)$$

Paralleling our earlier solution of the simple immigration process, we first invoke (3.6.21), namely

$$\frac{\partial l(x,n)}{\partial x} = -a. \quad (3.6.71)$$

Whence (3.6.70) becomes

$$\frac{\partial l(x,n)}{\partial n} = \tilde{K}(a), \quad (3.6.72)$$

and the combined solution (3.6.23) now takes the form

$$l(x,n) = n\tilde{K}(a) - ax + c. \quad (3.6.73)$$

On taking c to be independent of a, differentiating the envelope (3.6.71) with respect to a yields

$$0 = n\frac{\partial \tilde{K}(a)}{\partial a} - x.$$

So to first order we have the approximation

$$p(x,n) \simeq e^{l(x,n)} = Ae^{n\tilde{K}(a) - ax} \quad (3.6.74)$$

where a is the appropriate root of

$$\tilde{K}'(a) = x/n \quad (3.6.75)$$

for normalizing constant $A = e^c$.

Consider, for example, the simple Bernoulli case corresponding to single and paired immigrants arriving with probability rates $\alpha_1 = (1-\gamma)$ and $\alpha_2 = \gamma$. Then after n

events the population size x lies between n (all single immigrants) and $2n$ (all paired immigrants), with the occupation probabilities taking the binomial form

$$p(r+n, n) = \binom{n}{r} \gamma^r (1-\gamma)^{n-r} \qquad (x = r+n;\ r = 0, 1, \ldots, n). \qquad (3.6.76)$$

Now on denoting $\delta(\cdot)$ to be the Kronecker delta function, we may write the transition rate distribution as

$$dG(y) = (1-\gamma)\delta(1-y) + \gamma\delta(2-y), \qquad (3.6.77)$$

so the c.g.f. (3.6.69) becomes

$$\tilde{K}(\theta) = \ln[(1-\gamma)e^\theta + \gamma e^{2\theta}] = \theta + \ln[(1-\gamma) + \gamma e^\theta]. \qquad (3.6.78)$$

Using (3.6.75) with a replaced by θ then gives

$$\tilde{K}'(\theta) = 1 + \gamma e^\theta / [(1-\gamma) + \gamma e^\theta] = x/n,$$

which has the solution

$$e^\theta = \left(\frac{x-n}{2n-x}\right)\left(\frac{1-\gamma}{\gamma}\right).$$

Since

$$e^{\tilde{K}(\theta)} = e^\theta[(1-\gamma) + \gamma e^\theta],$$

on substituting for e^θ the first-order approximation (3.6.74), namely

$$p(x, n) \simeq A e^{n\tilde{K}(\theta) - \theta x} = A e^{n\theta}[(1-\gamma) + \gamma e^\theta]^n e^{-x\theta},$$

simplifies to

$$p(r+n, n) \simeq \frac{n^n}{r^r(n-r)^{n-r}} A\gamma^r(1-\gamma)^{n-r} \qquad (x = r+n;\ r = 0, 1, \ldots, n). \qquad (3.6.79)$$

This differs from the exact binomial probability (3.6.76) only in that the factorials are replaced by a crude form of Stirling's approximation.

As a second example, suppose that the immigration–emigration distribution is Normal with mean μ and variance σ^2. Note that for a proper population process (i.e. $x \geq 0$), we would essentially require $\mu > 0$ and say $x > 3\sigma$ at $n = 0$ in order to ensure that there is negligible probability of the population size becoming negative over a reasonably long time period. Since here

$$\tilde{K}(\theta) = \mu\theta + \sigma^2\theta^2/2, \qquad (3.6.80)$$

equation (3.6.75), namely

$$\tilde{K}'(\theta) = \mu + \sigma^2\theta = x/n, \qquad (3.6.81)$$

yields

$$\theta = \frac{1}{\sigma^2}\left(\frac{x}{n} - \mu\right). \qquad (3.6.82)$$

178 General Markov population processes

Thus (3.6.74) takes the form

$$p(x,n) \simeq Ae^{n\tilde{K}(\theta)-\theta x} = Ae^{-(x-n\mu)^2/2n\sigma^2}. \qquad (3.6.83)$$

That is, the population size x follows an approximate Normal distribution with mean $n\mu$ and variance $n\sigma^2$. However, since x is the sum of n i.i.d. $N(\mu,\sigma^2)$ random variables it follows that this result is exact. Moreover, differentiating (3.6.81) gives $\tilde{K}''(\theta) = \sigma^2$, so we may write the normalizing constant as $A = 1/\sqrt{(2\pi n\sigma^2)} = 1/\sqrt{(2\pi n\tilde{K}''(\theta))}$. Thus for this additive Normal process the first-order approximation (3.6.74) takes the exact form

$$p(x,n) = e^{n\tilde{K}(\theta)-x\theta}/\sqrt{(2\pi n\tilde{K}''(\theta))}. \qquad (3.6.84)$$

This result can also be expressed in terms of the c.g.f. $K(\theta)$ of the population size x. For since x is the sum of n independent increments, each having c.g.f. $\tilde{K}(\theta)$, we have

$$K(\theta) = n\tilde{K}(\theta), \qquad (3.6.85)$$

whilst equation (3.6.75) becomes

$$K'(\theta) = x.$$

So for this special case (3.6.84) has the equivalent representation

$$p(x,n) = e^{K(\theta)-x\theta}/\sqrt{2\pi K''(\theta)} \quad \text{where} \quad K'(\theta) = x. \qquad (3.6.86)$$

We shall soon see that this *saddlepoint* result, which here has been derived for a simple additive discrete-time Markov process, is also directly applicable to any continuous-time Markov process. Its power lies in the fact that it is often relatively easy to derive an expression for the c.g.f. $K(\theta)$ of total population size N at time t, so once this is known, $p(x,n)$ automatically provides a good approximation to $p_N(t)$.

However, before we provide a general proof of result (3.6.86), it is instructive to indicate how Daniels (1995) generalizes the above approach in order to allow for time-dependence. The continuous-time extension of the discrete-time equations (3.6.66) is

$$\frac{\partial p(x,t)}{\partial t} + \alpha p(x,t) = \alpha \int_{-\infty}^{\infty} p(x-y,t)\,dG(y), \qquad (3.6.87)$$

where α denotes the overall Poisson immigration–emigration event rate. So in terms of $p(x,t) = e^{l(x,t)}$, we have

$$\frac{\partial l(x,t)}{\partial t} + \alpha = \alpha \int_{-\infty}^{\infty} e^{l(x-y,t)-l(x,t)}\,dG(y). \qquad (3.6.88)$$

Daniel's next step is to parallel the perturbation approach based on (3.6.56) by writing (3.6.88) as

$$\frac{\partial l(x,t)}{\partial t} + \alpha = \alpha \int_{-\infty}^{\infty} \exp\{(1/\epsilon)[l(x-\epsilon y,t) - l(x,t)]\}\,dG(y). \qquad (3.6.89)$$

Diffusion approximations

As before, $\epsilon = 0$ corresponds to the first-order approximation, and $\epsilon = 1$ to the exact solution. Taking the Taylor series expansion in the exponent gives

$$\frac{\partial l(x,t)}{\partial t} + \alpha = \alpha \int_{-\infty}^{\infty} \exp\{-y[\frac{\partial l(x,t)}{\partial x} - (\epsilon y/2)\frac{\partial^2 l(x,t)}{\partial x^2} + \cdots]\} dG(y),$$

which itself expands to

$$\frac{\partial l(x,t)}{\partial t} + \alpha = \alpha \int_{-\infty}^{\infty} \exp\{-y\frac{\partial l(x,t)}{\partial x}\}[1 + (\epsilon y^2/2)\frac{\partial^2 l(x,t)}{\partial x^2} + \cdots] dG(y). \quad (3.6.90)$$

Denote the m.g.f. of the mass immigration–emigration probabilities by

$$M(\theta) \equiv \int_{-\infty}^{\infty} e^{\theta x} dG(x), \quad \text{so} \quad \frac{d^r M(\theta)}{d\theta^r} = \int_{-\infty}^{\infty} x^r e^{\theta x} dG(x). \quad (3.6.91)$$

Then (3.6.90) may be expressed as

$$(1/\alpha)\frac{\partial l(x,t)}{\partial t} + 1 = M\left(-\frac{\partial l(x,t)}{\partial x}\right) + (\epsilon/2)\frac{\partial^2 l(x,t)}{\partial x^2} M''\left(-\frac{\partial l(x,t)}{\partial x}\right) + \cdots. \quad (3.6.92)$$

We can remove the 1 on the left-hand side by making the trivial transformation $H(\theta) \equiv M(\theta) - 1$, and then noting that $H(0) = 0$ and $H^{(r)}(0) = E(x^r)$ $(r > 0)$. On paralleling (3.6.57) by writing

$$l(x,t) \equiv \sum_{i=0}^{\infty} \epsilon^i l_i(x,t), \quad (3.6.93)$$

equation (3.6.92) expands to give

$$(1/\alpha)\frac{\partial}{\partial t}(l_0 + \epsilon l_1 + \cdots) = H[-\frac{\partial l_0}{\partial x} - \epsilon\frac{\partial l_1}{\partial x} - \cdots] + (\epsilon/2)[\frac{\partial^2 l_0}{\partial x^2} + \epsilon\frac{\partial^2 l_1}{\partial x^2} + \cdots]$$

$$\times H''(-\frac{\partial l_0}{\partial x} - \epsilon\frac{\partial l_1}{\partial x} - \cdots) + \cdots.$$

Equating coefficients of ϵ^0 then yields

$$(1/\alpha)\frac{\partial l_0}{\partial t} = H(-\frac{\partial l_0}{\partial x}). \quad (3.6.94)$$

Whence on following the (now) standard procedure of putting

$$\frac{\partial l_0}{\partial x} = -\theta, \quad (3.6.95)$$

equation (3.6.94) becomes

$$\frac{\partial l_0}{\partial t} = \alpha H(\theta),$$

which leads to

$$l_0 = \alpha t H(\theta) - x\theta + c(\theta) \quad (3.6.96)$$

for some arbitrary function $c(\theta)$. So on taking l_0 to be independent of θ, we have

$$0 = \frac{\partial l_0}{\partial \theta} = \alpha t H'(\theta) - x + c'(\theta). \tag{3.6.97}$$

Similarly, expanding $H(\cdot)$ about $-\partial l_0/\partial x$ and then equating coefficients of ϵ^1 yields

$$\frac{\partial l_1}{\partial t} = -\alpha \frac{\partial l_1}{\partial x} H'(-\frac{\partial l_0}{\partial x}) + \frac{1}{2} \frac{\partial^2 l_0}{\partial x^2} H''(-\frac{\partial l_0}{\partial x}), \tag{3.6.98}$$

which, on using (3.6.95), reduces to

$$\frac{\partial l_1}{\partial t} + \alpha H'(\theta) \frac{\partial l_1}{\partial x} = -\frac{1}{2} \frac{\partial \theta}{\partial x} H''(\theta). \tag{3.6.99}$$

Since we are now working in terms of $\theta = -\partial l_0/\partial x$, let us make the change of variables from (t, x) to (t, θ). On employing the standard notation for partial derivatives we have

$$\left(\frac{\partial l_1}{\partial t}\right)_\theta = \left(\frac{\partial l_1}{\partial t}\right)_x + \left(\frac{\partial l_1}{\partial x}\right)_t \left(\frac{\partial x}{\partial t}\right)_\theta. \tag{3.6.100}$$

Differentiating (3.6.97) with respect to t and then x yields

$$0 = \alpha H'(\theta) - \left(\frac{\partial x}{\partial t}\right)_\theta \quad \text{and} \quad 0 = \alpha t H''(\theta)\left(\frac{\partial \theta}{\partial x}\right)_t - 1 + c''(\theta)\left(\frac{\partial \theta}{\partial x}\right)_t$$

i.e.

$$\left(\frac{\partial x}{\partial t}\right)_\theta = \alpha H'(\theta) \quad \text{and} \quad \left(\frac{\partial \theta}{\partial x}\right)_t = 1/[\alpha t H''(\theta) + c''(\theta)]. \tag{3.6.101}$$

Whence on combining (3.6.99)–(3.6.101) we see that

$$\left(\frac{\partial l_1}{\partial t}\right)_\theta = -\alpha H'(\theta)\left(\frac{\partial l_1}{\partial x}\right)_t - \frac{1}{2}\left(\frac{\partial \theta}{\partial x}\right)_t H''(\theta) + \left(\frac{\partial l_1}{\partial x}\right)_t \alpha H'(\theta),$$

i.e

$$\left(\frac{\partial l_1}{\partial t}\right)_\theta = \frac{-(1/2)H''(\theta)}{tH''(\theta) + c''(\theta)}. \tag{3.6.102}$$

This integrates to give

$$l_1 = -(1/2)\ln[tH''(\theta) + c''(\theta)] + b(\theta) \tag{3.6.103}$$

for some arbitrary function $b(\theta)$.

We now return to the expansion (3.6.93), and, with $\epsilon = 1$, just select the first two terms, i.e.

$$l(x,t) \simeq l_0(x,t) + l_1(x,t)$$
$$= \alpha t H(\theta) - x\theta + c(\theta) - (1/2)\ln[tH''(\theta) + c''(\theta)] + b(\theta). \tag{3.6.104}$$

To determine $b(\theta)$ and $c(\theta)$ we need to refer back to the initial conditions, yet we have already seen in Section 3.6.2 that imposing

$$p(x,0) = \delta(x) \tag{3.6.105}$$

is awkward. Let us therefore consider the Normal form
$$p(x,0) \equiv e^{l(x,0)} = (1/\sqrt{2\pi\sigma^2})e^{-x^2/2\sigma^2}, \qquad (3.6.106)$$
since this essentially reverts to (3.6.105) as $\sigma \to 0$. Comparison with (3.6.104) at $t=0$ shows that
$$-x^2/2\sigma^2 - \ln(\sigma) - \ln(\sqrt{2\pi}) = -x\theta + c(\theta) - (1/2)\ln[c''(\theta)] + b(\theta), \qquad (3.6.107)$$
whilst from (3.6.97) we have
$$0 = -x + c'(\theta). \qquad (3.6.108)$$
Placing $b(\theta) = -\ln(\sqrt{2\pi})$ and $c(\theta) = \sigma^2\theta^2/2$ then enables us to reduce (3.6.107) to
$$-x^2/2\sigma^2 = -x\theta + \sigma^2\theta^2/2, \quad \text{i.e. to} \quad (x - \sigma^2\theta)^2 = 0,$$
which has the single root
$$\theta = x/\sigma^2. \qquad (3.6.109)$$
So all that remains is to substitute for $b(\theta)$ and $c(\theta)$ in (3.6.104) to obtain
$$p(x,t) = e^{l(x,t)} \simeq \frac{\exp\{\alpha t H(\theta) - x^2/2\sigma^2\}}{\sqrt{\{2\pi[tH''(\theta) + \sigma^2]\}}} \quad \text{for} \quad \theta = x/\sigma^2. \qquad (3.6.110)$$

Thus for any discrete-time process (3.6.66) with superimposed Poisson(α) event-times, we simply form $H(\theta)$ and substitute directly into (3.6.110). For example, for the single-paired immigration rate (3.6.77) we have $H(\theta) = M(\theta) - 1 = (1-\gamma)e^\theta + \gamma e^{2\theta} - 1$.

On comparing the second-order perturbation solution (3.6.110) with the saddlepoint result (3.6.86), we see that the latter is potentially simpler. This is particularly true for Normally distributed increments since then $M(\theta)$ takes the exponential form whilst $K(\theta)$ is merely quadratic. Moreover, it turns out that the representation (3.6.86) is fundamentally 'optimal'. So having introduced the basic ideas which underlie perturbation techniques, let us now consider a formal, continuous-time, development of the saddlepoint approximation itself.

3.7 The saddlepoint approximation

The development of the saddlepoint result (3.6.86) is proved in full analytic detail by Daniels (1954), and though here we shall just extract the essential detail pertinent to our discussion of stochastic population processes, readers who acquire a keen interest in this extremely powerful technique are highly recommended to consult Daniels' superb account (see also Daniels 1960, 1983 and 1987). Unfortunately, Daniels' notable contribution was neglected for a quarter of a century until keynote papers such as Barndorff-Nielson and Cox (1979), and the later review of Reid (1988), brought the saddlepoint technique into general statistical use.

Although the integrated orthogonal polynomial representation (3.1.21) provides a general solution for the transition probabilities $p_{ij}(t)$, this rarely gives sufficient insight into the structure of the process itself. For equilibrium and quasi-equilibrium probabilities we have the general results (3.2.7) and (3.2.11), but for time-dependent solutions we usually have to rely on a variety of approximation techniques as developed earlier in this chapter. So apart from a few basic processes involving independent growth (e.g. the simple birth–death process), the essential problem is to approximate the distribution of some statistic whose exact form cannot be conveniently obtained. In situations for which the first few moments are known the classic procedure is to fit a law of Pearson or Edgeworth type which has the same moments to some desired order. Although this method is satisfactory in practice, at least if one is interested only in the central region of the distribution, it suffers the drawback that errors in the tail regions of the distribution may be substantial. Moreover, the Edgeworth expansion is particularly notorious for yielding negative probabilities in the tails. Though approximating a true probability of say 0.00001 by the negative value −0.00001 is arguably much better than approximating it by a larger, albeit positive, value such as +0.001.

The rationale which underlies the usual probability or moment generating function approach is that it reduces the difference-differential Kolmogorov forward (or backward) equations for $p_{ij}(t)$ in j (or i) and t to a differential equation in t alone. So provided this can be solved for $G(z,t)$ or $M(\theta,t)$, the problem then reduces to the analytic one of obtaining an explicit inversion. Daniels (1954) shows that for a statistic such as the mean of a sample of size n, or the ratio of two such means, a satisfactory approximation to its probability density, when it exists, can nearly always be obtained by the Method of Steepest Descents. This gives an asymptotic expansion in powers of n^{-1} whose dominant term, called the *saddlepoint approximation*, has a number of desirable features. The error incurred by its use is $O(n^{-1})$ as against the more usual $O(n^{-1/2})$ associated with the Normal approximation. Moreover, it is shown that in an important class of cases, the *relative* error of the approximation is uniformly $O(n^{-1})$ over the whole admissible range of the variable.

On a historical note, the Method of Steepest Descents was first used systematically by Debye for Bessel functions of large order (see Watson, 1952), and was introduced by Darwin and Fowler (see Fowler, 1936) into statistical mechanics where it has proved to be a tool of considerable importance. Though recognized by Jeffreys (1948) and Cox (1948) has having potential in statistical applications, relatively few later papers consider its statistical development. Daniels (1954) works in terms of the asymptotic expansion for the probability density of the mean $\bar{x}$ of a sample of size n, and he shows how the steepest descents technique is related to an alternative method used by Khinchin (1949) and, in a slightly different context, by Cramer (1938). He establishes general conditions under which the relative error of the saddlepoint approximation is $O(n^{-1})$ uniformly for all admissible $\bar{x}$, together with corresponding results for the asymptotic expansion.

The key point is that although Daniels' classic paper, together with the many others that have subsequently been developed from it, work in terms of $n > 1$, for a stochastic realization $n = 1$ since there is a single population value for each time point t. Yet the technique still works extremely well. So although the forthcoming introductory account is developed in terms of n in order to present a more complete picture, when we subsequently apply the saddlepoint approximation n will effectively disappear.

3.7.1 Basic derivation

Let X be a random variable with density function $f(x)$ and distribution function $F(x)$, and suppose that the associated m.g.f. $M(\theta)$ and c.g.f. $K(\theta)$, i.e.

$$M(\theta) = e^{K(\theta)} = \int_{-\infty}^{\infty} e^{\theta x} f(x)\, dx, \qquad (3.7.1)$$

converge for real θ in some nonvanishing interval $(-c_1, c_2)$ containing the origin. Then $f(x)$ may be recovered by applying the Fourier inversion formula

$$f(x) = \frac{1}{2\pi} \int_{-\infty}^{\infty} M(i\theta) e^{-i\theta x}\, d\theta. \qquad (3.7.2)$$

To consider the mean of n independent variables $X_1, \ldots, X_n$, we first note that the m.g.f. of $S = X_1 + \cdots + X_n$ is $M_S(\theta) = M^n(\theta)$. So the m.g.f. of $\overline{X} = S/n$ is

$$M_{\overline{X}}(\theta) = \mathrm{E}(e^{\theta S/n}) = \mathrm{E}(e^{(\theta/n)S}) = M_S(\theta/n) = M^n(\theta/n).$$

Whence on applying the inversion formula (3.7.2) to $\overline{X}$, instead of X, we see that the p.d.f. of $\overline{X}$ is given by

$$h(\overline{x}) = \frac{1}{2\pi} \int_{-\infty}^{\infty} M^n(i\theta/n) e^{-i\theta \overline{x}}\, d\theta. \qquad (3.7.3)$$

That is, on writing $\psi = \theta/n$, by

$$h(\overline{x}) = \frac{n}{2\pi} \int_{-\infty}^{\infty} M^n(i\psi) e^{-n i \psi \overline{x}}\, d\psi. \qquad (3.7.4)$$

Since it is convenient to work in terms of the imaginary, rather than the real, axis, we now put $\phi = i\psi$, giving

$$h(\overline{x}) = \frac{n}{2\pi i} \int_{-i\infty}^{i\infty} M^n(\phi) e^{-n \phi \overline{x}}\, d\phi = \frac{n}{2\pi i} \int_{-i\infty}^{i\infty} e^{n[K(\phi) - \phi \overline{x}]}\, d\phi. \qquad (3.7.5)$$

When n is large, an approximation to $h(\overline{x})$ can be found by choosing the path of integration to pass through a saddlepoint of the integrand in such a way that the integrand is negligible outside its immediate neighbourhood. The saddlepoints are situated where the exponent in (3.7.5) has zero derivative, i.e. where

$$K'(\phi) = \overline{x}. \qquad (3.7.6)$$

Now Daniels (1954) proves that under general conditions (3.7.6) has a single real root ϕ_0 in $(-c_1, c_2)$ for every value of $\overline{x}$ such that $0 < F_n(\overline{x}) < 1$, and that $K''(\phi_0) > 0$. Let us choose the path of integration to be a straight line through ϕ_0 parallel to the imaginary axis. Since $K(\phi) - \phi x$ has a minimum at ϕ_0 for real ϕ, the modulus of the integrand must have a maximum at ϕ_0 on the chosen path. Now use of a standard argument (see, for example, Wintner, 1947) shows that on any admissible path parallel to the imaginary axis the integrand attains its maximum modulus only when the line crosses the real axis, and that for the particular path chosen only the neighbourhood of ϕ_0 need be considered when n is large.

The argument then proceeds formally as follows. Write $\phi = \phi_0 + iy$. Then on expanding $K(\phi)$ about ϕ_0 we have

$$K(\phi) = K(\phi_0) + iyK'(\phi_0) - (y^2/2)K''(\phi_0) - (iy^3/6)K'''(\phi_0) + (y^4/24)K^{iv}(\phi_0) + \cdots,$$

i.e., as $iy = \phi - \phi_0$ and $K'(\phi_0) = \bar{x}$,

$$K(\phi) - \phi\bar{x} = K(\phi_0) - \phi_0\bar{x} - (y^2/2)K''(\phi_0) - (iy^3/6)K'''(\phi_0) + (y^4/24)K^{iv}(\phi_0) + \cdots.$$

(3.7.7)

On setting $y = v/\sqrt{nK''(\phi_0)}$, noting that $d\phi = idy = idv/\sqrt{nK''(\phi_0)}$, and expanding the integral in (3.7.5), we then obtain

$$h(\bar{x}) = \left(\frac{n}{2\pi i}\right)\left(\frac{i}{\sqrt{nK''(\phi_0)}}\right)e^{n[K(\phi_0)-\phi_0\bar{x}]} \qquad (3.7.8)$$

$$\times \int_{-\infty}^{\infty} \exp\{n[\frac{-v^2 K''(\phi_0)}{2nK''(\phi_0)} - \frac{iv^3 K'''(\phi_0)}{6(nK''(\phi_0))^{3/2}} + \frac{v^4 K^{iv}(\phi_0)}{24(nK''(\phi_0))^2} + \cdots]\} dv.$$

Whence on writing

$$\lambda_j(\phi) = K^{(j)}(\phi)/[K''(\phi)]^{j/2} \qquad (j \geq 3), \qquad (3.7.9)$$

expanding each of the exponents yields

$$h(\bar{x}) = \frac{\sqrt{n}e^{n[K(\phi_0)-\phi_0\bar{x}]}}{2\pi\sqrt{K''(\phi_0)}} \int_{-\infty}^{\infty} e^{-v^2/2}\left[1 + \left(\frac{-iv^3\lambda_3(\phi_0)}{6\sqrt{n}}\right) + \frac{1}{2}\left(\frac{-iv^3\lambda_3(\phi_0)}{6\sqrt{n}}\right)^2 + \cdots\right]$$

$$\times \left[1 + \frac{v^4\lambda_4(\phi_0)}{24n} + \cdots\right] \times \cdots dv. \qquad (3.7.10)$$

Since the Normal distribution is symmetric, the odd powers of v must vanish in the integration, thereby yielding an expansion in powers of n^{-1}, viz:

$$h(\bar{x}) = \frac{\sqrt{n}e^{n[K(\phi_0)-\phi_0\bar{x}]}}{2\pi\sqrt{K''(\phi_0)}} \int_{-\infty}^{\infty} e^{-v^2/2}\left[1 + \frac{1}{n}\left(\frac{-v^6\lambda_3^2(\phi_0)}{72} + \frac{v^4\lambda_4(\phi_0)}{24}\right) + \cdots\right] dv.$$

(3.7.11)

So as the even moments for the $N(0,1)$ distribution are given by

$$\mu_{2r} = (2r)!/(2^r r!), \qquad (3.7.12)$$

$h(\bar{x})$ takes the form of an expansion in powers of n^{-1}, namely

$$h(\bar{x}) = g(\bar{x})\{1 + (1/24n)[3\lambda_4(\phi_0) - 5\lambda_3^2(\phi_0)] + (1/1152n^2)[385\lambda_4(\phi_0) \qquad (3.7.13)$$
$$- 630\lambda_3^2(\phi_0)\lambda_4(\phi_0) + 105\lambda_4^2(\phi_0) - 24\lambda_6(\phi_0) + 168\lambda_3(\phi_0)\lambda_5(\phi_0)] + \cdots\}.$$

Here

$$g(\bar{x}) = \frac{e^{n[K(\phi_0)-\phi_0\bar{x}]}}{\sqrt{(2\pi/n)K''(\phi_0)}} \qquad (3.7.14)$$

is called the *saddlepoint approximation* to $f(\bar{x})$, and, for given $\bar{x}$, the value of ϕ_0 is determined from
$$K'(\phi_0) = \bar{x}. \qquad (3.7.15)$$

3.7.2 Examples

Example 1 Suppose X is a $N(\mu, \sigma^2)$ random variable, so that
$$f(x) = (1/\sqrt{2\pi\sigma^2})e^{-(x-\mu)^2/2\sigma^2} \quad (-\infty < x < \infty). \qquad (3.7.16)$$
Then
$$M(\phi) = \frac{1}{\sqrt{2\pi}} \int_{-\infty}^{\infty} e^{\phi x} e^{-(x-\mu)^2/2\sigma^2} \, dx$$
$$= e^{\mu\phi + \sigma^2 \phi^2/2} \times \frac{1}{\sqrt{2\pi}} \int_{-\infty}^{\infty} e^{-[x-(\mu+\sigma^2\phi)]^2/2\sigma^2} \, dx = e^{\mu\phi + \sigma^2\phi^2/2},$$
since comparison with (3.7.16) shows that the integral is that of a $N(\mu + \sigma^2\phi, \sigma^2)$ random variable over the full range $(-\infty, \infty)$. So
$$K(\phi) = \ln[M(\phi)] = \mu\phi + \sigma^2\phi^2/2, \qquad (3.7.17)$$
whence it follows from (3.7.15) that ϕ_0 is the solution to
$$K'(\phi_0) = \mu + \sigma^2\phi_0 = \bar{x}, \quad \text{i.e.} \quad \phi_0 = (\bar{x} - \mu)/\sigma^2. \qquad (3.7.18)$$
Moreover,
$$K''(\phi_0) = \sigma^2 \quad \text{and} \quad K^{(j)}(\phi_0) = 0 \quad (j > 2),$$
and so the expansion (3.7.8) reduces to
$$h(\bar{x}) = g(\bar{x}) = (1/\sqrt{2\pi\sigma^2/n})\exp\{-(\bar{x}-\mu)^2/(2\sigma^2/n)\}. \qquad (3.7.19)$$
Thus here the saddlepoint approximation yields the exact distribution for $\bar{X}$, namely $N(\mu, \sigma^2/n)$.

Example 2 The approximation also works well with discrete random variables, and to illustrate that taking $n = 1$ is no handicap suppose we consider $X \sim \text{Poisson}(\alpha)$. Then since the Poisson distribution is non-zero only at the integer points $x = 0, 1, 2, \ldots$, where it takes probability mass $\alpha^x e^{-\alpha}/x!$, we have
$$M(\phi) = \int_{-\infty}^{\infty} e^{\phi x} f(x) \, dx = \sum_{x=0}^{\infty} \frac{e^{\phi x} \alpha^x e^{-\alpha}}{x!} = e^{-\alpha} \sum_{x=0}^{\infty} \frac{(\alpha e^\phi)^x}{x!} = e^{\alpha(e^\phi - 1)}.$$
Hence
$$K(\phi) = \alpha(e^\phi - 1), \qquad (3.7.20)$$
with ϕ_0 being the solution to
$$K'(\phi_0) = \alpha e^{\phi_0} = x. \qquad (3.7.21)$$

So
$$K''(\phi_0) = \alpha e^{\phi_0} = x.$$

Whence using the general saddlepoint result (3.7.14) and (3.7.15) with $n = 1$, namely

$$g(x) = \frac{e^{[K(\phi_0) - \phi_0 x]}}{\sqrt{2\pi K''(\phi_0)}} \quad \text{with} \quad K'(\phi_0) = x, \qquad (3.7.22)$$

yields

$$g(x) = \frac{\alpha^x e^{-\alpha}}{\sqrt{2\pi} e^{-x} x^{x+1/2}}. \qquad (3.7.23)$$

This differs from the exact Poisson probability only in that $x!$ is replaced by Stirling's approximation $x! \sim \sqrt{2\pi} e^{-x} x^{x+1/2} \equiv \varsigma(x)$. So whilst the approximation works well for x greater than say 6 ($\varsigma(6) = 710.08$ which is only 1.38% lower than $6! = 720$), for smaller values of x we clearly have to be careful; though as $\varsigma(1) = 0.922$, which is only 8% too low, the saddlepoint approximation still performs quite well in the tail. Note, however, that as $\varsigma(0) = 0$ we have $g(0) = \infty$, and so the approximation is not defined at the end point of its range of support. This, unfortunately, is a general feature of the technique, with values being obtainable only *within* the region of support. The function $g(x)$ can always be multiplied by an appropriate normalizing factor to ensure that the saddlepoint probabilities sum to 1 over the required range.

One of the intriguing delights of the saddlepoint approximation is just how well it works when $n = 1$, which is the value we are naturally faced with when considering a single random variable X. For although result (3.7.13) is an expansion in powers of n^{-1}, the second term still provides a commendable correction factor when $n = 1$. For example, in the above Poisson example we have $K^{(j)}(\phi_0) = \alpha e^{\phi_0} = x$ for $j \geq 1$, whence (3.7.9) yields $\lambda_j(\phi_0) = x^{1-j/2}$, so the $O(n^{-1})$ correction term

$$(1/n)[(1/8)\lambda_4(\phi_0) - (5/24)\lambda_3^2(\phi_0)] = -1/12x.$$

On applying this we have

$$h(x) \simeq g_1(x) = g(x)[1 - 1/(12x)] = \frac{\alpha^x e^{-\alpha}}{\varsigma(x)/[1 - 1/(12x)]}, \qquad (3.7.24)$$

and the denominator equals 720.079 (a mere 0.01% too high) at $x = 6$. Whilst at $x = 1$ it equals 1.00597, and although this is 0.6% too high, it is still impressive given that we are now at the extreme end of the admissible range.

Even greater accuracy can be achieved if we include the $O(n^{-2})$ correction term in (3.7.13). For then our Poisson example approximation (3.7.24) is enhanced to

$$h(x) \simeq g_2(x) = g(x)[1 - 1/(12x) + 1/(288x^2)] = \frac{\alpha^x e^{-\alpha}}{\varsigma(x)/[1 - 1/(12x) + 1/(288x^2)]}, \qquad (3.7.25)$$

and the denominator now equals 720.00886 (a mere 0.0012% too high) at $x = 6$, and 1.00217 (0.22% too high) at the effective end-point $x = 1$. Note that although $g_2(0) = 0$

at the value $x = 0$ (which is illegitimate since it lies outwith the open interval $(0, \infty)$), $g_2(0.071) = 1.001 \simeq 0!$, so the breakdown occurs quite close to the boundary value.

Example 3 Bearing in mind the apparently smooth nature of these general saddlepoint results, it would be unreasonable to expect them to produce good approximations to distributions that are intrinsically unsmooth. Care should therefore be taken not to switch on the automatic pilot too early. To illustrate this danger consider the saw-tooth Poisson(α) distribution given by

$$p_{2x} = \alpha^x e^{-\alpha}/x! \quad \text{and} \quad p_{2x+1} = 0 \quad \text{for} \quad x = 0, 1, 2, \ldots. \tag{3.7.26}$$

Then the straight Poisson c.g.f. (3.7.20) becomes

$$K(\phi) = \alpha(e^{2\phi} - 1), \tag{3.7.27}$$

whence the saddlepoint approximation (3.7.22) changes from (3.7.23) to

$$g(x) = \frac{\alpha^{x/2} e^{-\alpha}}{2\sqrt{2\pi} e^{-x/2} (x/2)^{(x+1)/2}} \sim \frac{(1/2)\alpha^{x/2} e^{-\alpha}}{(x/2)!} \tag{3.7.28}$$

where $(x/2)!$ denotes $\Gamma[(x/2) + 1]$. Now whilst the original distribution (3.7.26) is non-zero only at the even integers $x = 0, 2, 4, \ldots$, $g(x)$ takes its support over $x = 0, 1, 2, \ldots$. So the saw-teeth are smoothed into a Poisson 'curve'. Thus the only way of recovering the original structure is to select $x = 0, 2, 4, \ldots$, and then to replace $g(x)$ by $2g(x)$ in order to ensure that the probabilities sum to unity over these even values.

It is worthwhile remarking that the integral representation (3.7.8) can be used to develop an exact form for $h(\bar{x})$. For

$$h(\bar{x}) = \frac{e^{n[K(\phi_0) - \phi_0 \bar{x}]}}{2\pi \sqrt{K''(\phi_0)/n}} \times \int_{-\infty}^{\infty} e^{-v^2/2} \exp\{\sum_{r=0}^{\infty} \frac{(iv)^r \lambda_r(\phi_0)}{r! n^{r/2-1}}\} \, dv,$$

which, on using (3.7.14) and expanding the exponential terms, gives

$$h(\bar{x}) = g(x) \times \frac{1}{\sqrt{2\pi}} \int_{-\infty}^{\infty} e^{-v^2/2} \prod_{r=3}^{\infty} \sum_{j=0}^{\infty} \left[\frac{(iv)^r \lambda_r(\phi_0)}{r! n^{r/2-1}}\right]^j \left(\frac{1}{j!}\right) dv.$$

Whence on defining $a(s, \phi_0)$ to be the coefficient of v^s in the integrand, we have

$$h(\bar{x}) = g(\bar{x}) \times \frac{1}{\sqrt{2\pi}} \int_{-\infty}^{\infty} e^{-v^2/2} \sum_{s=0}^{\infty} a(s, \phi_0) v^s \, dv.$$

So as all the odd moments disappear, we are left with

$$h(\bar{x}) = g(\bar{x}) \times \frac{1}{\sqrt{2\pi}} \int_{-\infty}^{\infty} e^{-v^2/2} [1 + \sum_{s=1}^{\infty} a(2s, \phi_0) v^{2s}] \, dv,$$

which, on using the moment result (3.7.12), yields

$$h(\bar{x}) = g(\bar{x})[1 + \sum_{s=1}^{\infty} a(2s, \phi_0) \frac{(2s)!}{2^s s!}]. \tag{3.7.29}$$

Although this is a neat theoretical result, the opaque nature of the $a(2s, \phi_0)$ terms means that this formulation is unlikely to be of much practical use. It does, however, highlight the structure of the full saddlepoint expansion.

3.7.3 Relationship with the Method of Steepest Descents

It is not apparent from the above formal development that expression (3.7.11) for $h(\bar{x})$ is a proper asymptotic expansion in which the remainder is of the same order as the last term neglected. The asymptotic nature of an expansion of this type is usually established by the Method of Steepest Descents with the aid of a lemma due to Watson (1952), in which the path of integration is taken to be the curve of steepest descent through ϕ_0, since on this the modulus of the integrand decreases most rapidly. The analysis is simplified by using a "truncated" version of Watson's lemma introduced by Jeffreys and Jeffreys (1950), namely:

LEMMA 3.1 *If $\psi(z)$ is analytic in a neighbourhood of $z = 0$, and bounded for real $z = w$ in an interval $-A \leq w \leq B$ with $A > 0$ and $B > 0$, then*

$$\sqrt{\frac{n}{2\pi}} \int_{-A}^{B} e^{-nw^2/2} \psi(w)\, dw \sim \psi(0) + \frac{\psi''(0)}{2n} + \cdots + \frac{\psi^{(2r)}(0)}{(2n)^r r!} + \cdots$$

(3.7.30)

is an asymptotic expansion in powers of n^{-1}.

To apply this lemma, Daniels (1954) deforms the contour so that for $|\phi - \phi_0| \leq \delta$ the line $\phi = \phi_0 + iy$ is replaced by the curve of steepest descent. This is that branch of $\Im\{K(\phi) - \phi\bar{x}\} = 0$ touching $\phi = \phi_0 + iy$ at ϕ_0, where δ is chosen small enough to exclude possible saddlepoints other than ϕ_0. The contour is thereafter continued along the orthogonal curves of constant $\Re\{K(\phi) - \phi\bar{x}\}$. Now on the curve of steepest descent $K(\phi) - \phi\bar{x}$ is real and decreases steadily on each side of ϕ_0. Whilst on paralleling expansion (3.7.7), with iy being replaced by $\phi - \phi_0$, we have

$$K(\phi) = K(\phi_0) + (\phi - \phi_0)K'(\phi_0) + (\phi - \phi_0)^2 K''(\phi_0)/2! + (\phi - \phi_0)^3 K'''(\phi_0)/3! + \cdots .$$

(3.7.31)

Whence on remembering that $K'(\phi_0) = \bar{x}$, the function

$$-w^2/2 \equiv K(\phi) - \phi\bar{x} - K(\phi_0) + \phi_0\bar{x}$$
$$= (1/2)(\phi - \phi_0)^2 K''(\phi_0) + (1/6)(\phi - \phi_0)^3 K'''(\phi_0) + \cdots . \quad (3.7.32)$$

Moreover, on placing $z = (\phi - \phi_0)[K''(\phi_0)]^{1/2}$, and recalling from (3.7.9) that $\lambda_j(\phi_0) = K^{(j)}(\phi_0)/[K''(\phi_0)]^{j/2}$ ($j \geq 3$), we have

$$(\phi - \phi_0)^j K^{(j)}(\phi_0) = z^j \lambda_j(\phi_0).$$

Thus (3.7.32) becomes

$$-w^2/2 = z^2/2 + \lambda_3(\phi_0)z^3/6 + \lambda_4(\phi_0)z^4/24 + \cdots , \quad (3.7.33)$$

where w is chosen to have the same sign as $\Im(z)$ on the contour. To invert this expression we substitute for $z = aw + bw^2 + cw^3 + \cdots$. Then comparing coefficients of $w^2, w^3, w^4, \ldots$ yields $a = i$, $b = \lambda_3(\phi_0)/6$, $c = i[(\lambda_4(\phi_0)/24) - (5\lambda_3^2(\phi_0)/72)]$, ..., and hence the expansion

$$z = iw + \lambda_3(\phi_0)w^2/6 + \{3\lambda_4(\phi_0) - 5\lambda_3^2(\phi_0)\}iw^3/72 + \cdots \quad (3.7.34)$$

which is convergent in some neighbourhood of $w = 0$. Using (3.7.32), it follows that the contribution to (3.7.5) from the required part C of the contour (see above) is then

$$\frac{n}{2\pi i} \int_C e^{n[K(\phi) - \phi\bar{x}]} d\phi = \frac{n}{2\pi i} e^{n[K(\phi_0) - \phi_0 \bar{x}]} \int_C e^{-nw^2/2} d\phi, \quad (3.7.35)$$

which, as $dz = d\phi[K''(\phi_0)]^{1/2}$, becomes

$$\frac{n}{2\pi i} \frac{e^{n[K(\phi_0) - \phi_0 \bar{x}]}}{\sqrt{K''(\phi_0)}} \int_C e^{-w^2/2} dz = \frac{n}{2\pi i} \frac{e^{n[K(\phi_0) - \phi_0 \bar{x}]}}{\sqrt{K''(\phi_0)}} \int_C e^{-w^2/2} \left(\frac{dz}{dw}\right) dw. \quad (3.7.36)$$

Applying Watson's lemma (3.7.30) to the first derivative of (3.7.34), namely

$$\psi(w) = dz/dw = i + \lambda_3(\phi_0)w/3 + \{3\lambda_4(\phi_0) - 5\lambda_3^2(\phi_0)\}iw^2/24 + \cdots, \quad (3.7.37)$$

then recovers a series which is identical to (3.7.13) (see Daniels, 1954, for details).

Continuing this historical vein, it is worth noting that the work of Cramér (1928, 1937) on the Edgeworth series can also be employed to establish the asymptotic nature of these expansions, using a technique similar to that adopted by Cramér (1938) and Khinchin (1949). For given that on any admissible path of the form $\phi = \tau + iy$ the integrand attains its maximum modulus only at $\phi = \tau$, it follows that (3.7.13) is only one member of a whole family of series for $h(\bar{x})$ which can be derived in a similar way by integrating along $\phi = \tau + iy$, where τ takes any value in $(-c_1, c_2)$. It may be shown that $h(\bar{x})$ can be expressed in terms of the family of asymptotic expansions

$$h(\bar{x}) \sim \frac{\exp\{K(\tau) - \tau\bar{x} - [K'(\tau) - \bar{x}]^2/2K''(\tau)\}}{\sqrt{2\pi K''(\tau)/n}} \{1 + A_1/n^{1/2} + A_2/n + \cdots\} \quad (3.7.38)$$

where

$$A_1 = (1/3!)\lambda_3(\tau)H_3([K'(\tau) - \bar{x}]/\sqrt{K''(\tau)/n}), \quad (3.7.39)$$

$$A_2 = (1/4!)\lambda_4(\tau)H_4([K'(\tau) - \bar{x}]/\sqrt{K''(\tau)/n})$$
$$+ (10/6!)\lambda_3^2(\tau)H_6([K'(\tau) - \bar{x}]/\sqrt{K''(\tau)/n}), \quad \text{etc.},$$

and $H_n(\cdot)$ denotes the Hermite polynomial (e.g. Abramowitz and Stegun, 1970)

$$H_n(x) = n! \sum_{m=0}^{[n/2]} \frac{(-1)^m (2x)^{n-2m}}{m!(n-2m)!}. \quad (3.7.40)$$

For a detailed discussion of this result see Daniels (1954). In particular, when $\tau = 0$ we obtain the Edgeworth series for $h(\bar{x})$ whose asymptotic nature was demonstrated by Cramér (1928). Whilst when $\tau = \phi_0$, so that $K'(\phi_0) = \bar{x}$, all the odd powers of $n^{-1/2}$ vanish and we recover the saddlepoint approximation (3.7.13). Note that the dominant term, $g(\bar{x})$, has the same order of accuracy as the first two terms of the Edgeworth series, and, unlike the latter, can never be negative. Since selecting any other admissible value of τ generates a power series in $n^{-1/2}$, it follows that the saddlepoint expansion is optimal.

3.7.4 The truncated saddlepoint approximation

Whilst this saddlepoint approach is ideal for constructing a good approximation to the p.d.f. of a population process whose c.g.f. $K(\theta, t)$ is known, in general the Kolmogorov equations will not be analytically tractable to exact solution, thereby forcing us back onto developing approximation techniques. As we have already noted, this is purely a time-dependent problem since expressions (3.2.6)–(3.2.7) and (3.2.10)–(3.2.11) provide the equilibrium and quasi-equilibrium solutions, respectively, for general birth and death rates λ_N and μ_N.

As an illustration, consider the power-law logistic process with transition rates (3.2.38) and (3.2.39), namely

$$\lambda_N = \max\{0, N(a_1 - b_1 N^s)\} \quad \text{and} \quad \mu_N = N(a_2 + b_2 N^s). \qquad (3.7.41)$$

These impose a long-term upper bound, N_{max}, on N, and the threshold nature of λ_N makes construction of the associated c.g.f. rather tricky. However, provided $\Pr(N = N_{max})$ is negligible over the time period of interest, we can sensibly replace λ_N by the much more amenable form $\lambda_N = N(a_1 - b_1 N^s)$, whence it is now relatively easy to construct the p.d.e. (3.4.11) for $K(\theta, t)$. Expanding this equation in powers of θ, and equating coefficients of $\theta^1, \theta^2, \ldots$, then yields a sequence of ordinary first-order differential equations for the cumulants $\kappa_i(t)$. Equations (3.4.13)–(3.4.15) show the first three equations for the standard logistic case for which $s = 1$.

We have already seen in Section 3.4 that we may obtain a set of closed equations for $\kappa_1(t), \ldots, \kappa_n(t)$ by placing $\kappa_i(t) = 0$ for all $i > n$, and even if these equations cannot be solved algebraically, numerical solution is always possible. Indeed, given that the Kolmogorov equations for $\{p_N(t)\}$ can also be solved numerically within an appropriate finite space–time domain $0 \leq N \leq M; 0 \leq t \leq T$, the resulting numerical values for $p_N(t)$ can then be used to construct the raw moments $\mu'_r(t)$ and hence the associated cumulants $\kappa_r(t)$. So there is a fundamental question as to whether such cumulant values can be used to find the structure of the underlying process itself? That some degree of success is possible follows from the widespread popularity of the Normal approximation which is based solely on $\kappa_1(t)$ and $\kappa_2(t)$. What is not so obvious is whether including higher-order cumulants $\kappa_3(t), \kappa_4(t), \ldots$ will deliver successive improvements. If they do, then the approach will have an impact on equilibrium, as well as time-dependent, probabilities. For although we have already derived the general forms for the full- and quasi-equilibrium probabilities, namely

$$\pi_N = \frac{\lambda_0 \ldots \lambda_{N-1}}{\mu_1 \ldots \mu_N} \pi_0 \quad (N > 0) \quad \text{and} \quad \tilde{\pi}_N = \frac{\lambda_1 \ldots \lambda_{N-1}}{\mu_2 \ldots \mu_N} \tilde{\pi}_1 \quad (N > 1) \quad (3.7.42)$$

where π_0 and $\tilde{\pi}_1$ are determined from $\sum_{N=0}^{\infty} \pi_N = 1 = \sum_{N=1}^{\infty} \tilde{\pi}_N$, in practice it would be advantageous to obtain good approximations which take a much simpler form. For example, in Section 3.2.1 we use the power-law logistic rates to obtain

$$\tilde{\pi}_N = \frac{(a_1 - b_1)(a_1 - 2^s b_1) \ldots (a_1 - (N-1)^s b_1)}{N(a_2 + 2^s b_2)(a_2 + 3^s b_2) \ldots (a_2 + N^s b_2)} \tilde{\pi}_1 \qquad (N = 2, 3, \ldots, N_{max}), \tag{3.7.43}$$

and although this is a fairly neat algebraic representation it still has too complex a form to allow us to discern its shape and statistical properties from a summary algebraic inspection. The essence of this situation is expressed by Hougaard (in Reid, 1988, p.231), who states that "To be provocative, if the (saddlepoint) approximation is simpler than the exact distribution, we rarely need the exact distribution... if the approximation is not substantially simpler than the exact distribution then there is no point in using it". So here we are asking whether it is possible to provide a substantially less opaque quasi-equilibrium structure than (3.7.43).

This problem is not a new one, and in an inspiring paper Whittle (1957) highlights pioneering studies in nonlinear stochastic processes, from the 1920s onwards, that have been affected by it. Fields cited are as diverse as epidemics, population processes (including classical predator–prey studies), turbulence, ferromagnetism and imperfect gases. He remarks that in the case of turbulence, difficulties caused by nonlinearity have been circumvented to some extent by assuming that the velocity components have cumulants which vanish for orders greater than the third (Chandrasekhar, 1955). Although the justification for this assumption is unclear, the method of approximation is a natural one, since "a variate whose cumulants are zero for orders greater than one is deterministic, and the construction of a deterministic model is the inevitable preliminary to the construction of a stochastic model". Whittle considers the next stage in the approximation, when one neglects all cumulants of order greater than the second, i.e. he assumes that the variate is Normal. So here, we are investigating the extension of the truncation point to higher-order cumulants.

Consider first the simplest approximation approach, namely that N is distributed as a Normal $N(\mu, \sigma^2)$ random variable. This distribution has c.g.f. $K(\theta) = \mu\theta + \sigma^2\theta^2/2$, so application of the Central Limit Theorem is effectively the implementation of the truncated cumulant procedure in its simplest form, since $\kappa_i \equiv 0$ for all $i > 2$. Intuitively, one might expect non-zero values of κ_i for $i > 2$ to have much less effect on the c.g.f., and thereby the underlying probability structure, than κ_1 and κ_2, since they enter the c.g.f. through $\kappa_i/i!$ which decays rapidly to zero with increasing i for bounded κ_i. For the Poisson(λ) distribution we have $K(\theta) = \lambda(e^\theta - 1)$ and so $\kappa_i \equiv \lambda$ for all $i \geq 1$. Retaining just $\kappa_1 = \kappa_2 = \lambda$ produces the basic Normal approximation, which works well away from the tails of the distribution. Unfortunately this is not generally the case, since κ_i involves all central moments up to order i (for explicit relationships between them see Kendall and Stuart, 1977; (2.1.37) gives the first four). For example, in the case of the highly skewed Exponential(λ) distribution $\kappa_i = (i-1)!/\lambda^i$ explodes to infinity as i increases, and although it may be approximated to any desired degree of accuracy provided the cumulants κ_i are truncated at an appropriately high value of i, it remains questionable as to whether Hougaard's above 'principle of pragmatism' applies.

Let us proceed by considering the truncated c.g.f. (Renshaw, 1998)

$$K_n(\theta) \equiv \sum_{i=1}^{n} \frac{\kappa_i \theta^i}{i!};\qquad(3.7.44)$$

that is, we replace all cumulants κ_i in

$$K(\theta,t) \equiv \ln[M(\theta,t)] \equiv \sum_{i=1}^{\infty} \frac{\kappa_i(t)\theta^i}{i!}\qquad(3.7.45)$$

for $i > n$ by zero. Then the associated truncated m.g.f. is given by

$$M_n(\theta) \equiv \exp\left\{\sum_{i=1}^{n} \frac{\kappa_i \theta^i}{i!}\right\},\qquad(3.7.46)$$

and although an exact combinatorial form for the resulting approximate probability values can be constructed by comparing coefficients of $e^{N\theta}$ in the expansion of (3.7.46), it will clearly be too opaque to be of general practical use. It is just too far removed from the general Hermite representation $\exp\{\sum_{N=1}^{\infty} a_N(e^{N\theta}-1)\}$ (see Kemp and Kemp, 1965). It is precisely situations such as these that application of the saddlepoint approximation has great potential, and our purpose is to show how its use with just the first few cumulants can lead to a fairly simple form for the inversion of the truncated c.g.f.

On substituting (3.7.44) into the saddlepoint expressions (3.7.22) we obtain the (truncated) saddlepoint approximation

$$f_n(x) = \left[2\pi \sum_{i=0}^{n-2} \frac{\kappa_{i+2}\theta_0^i}{i!}\right]^{-1/2} \exp\left\{\sum_{i=1}^{n} \frac{\kappa_i \theta_0^i}{i!} - \theta_0 x\right\}\qquad(3.7.47)$$

where

$$x = \sum_{i=0}^{n-1} \frac{\kappa_{i+1}\theta_0^i}{i!}.\qquad(3.7.48)$$

Given our objective of replacing the exact distributions $\{p_N(t)\}$, $\{\pi_N\}$ or $\{\tilde{\pi}_N\}$ by relatively simple functional forms, this representation is clearly going to be of most benefit if n is 2, 3 or 4, since (3.7.48) then yields a tractable solution for $\theta_0(x)$ which can be inserted into (3.7.47). Moreover, even if the objective is to use (accurate) sample cumulants obtained from actual data, rather than working with population cumulants from a presumed model, then the procedure is still of great value. For being a completely general technique it does not necessitate the preselection of an assumed underlying distribution whose parameters are then fitted according to some statistical goodness-of-fit criteria. In this situation, for given values of $\kappa_1, \ldots, \kappa_n$ and an appropriately chosen grid of x-values, the polynomial (3.7.48) may be solved numerically (using *Matlab*, for example) to generate the $n-1$ roots for $\theta_0(x)$. Only one of these will yield suitable values for $f_n(x)$ (this follows from Theorems 6.1 and 6.2 of Daniels, 1954), which can then be plotted against x. Alternatively, a functional solution might be developed using an algebraic package such as *Mathematica*.

Let us first consider $n = 2$, with $\kappa_1 = \mu$ and $\kappa_2 = \sigma^2$. Then $K_2(\theta) = \mu\theta + \frac{1}{2}\sigma^2\theta^2$, and (3.7.48) becomes

$$x = \mu + \sigma^2\theta_0, \quad \text{i.e.} \quad \theta_0 = (x - \mu)/\sigma^2. \quad (3.7.49)$$

Thus the saddlepoint result (3.7.47) yields

$$f_2(x) = (2\pi\sigma^2)^{-1/2}\exp\{-(x-\mu)^2/2\sigma^2\}, \quad (3.7.50)$$

which is the basic Normal approximation (Example 1 of Section 3.7.2 develops the reverse approach). Suppose we now take $n = 3$, that is we incorporate the third central moment $\mu_3 = \kappa_3$. Then (3.7.48) becomes

$$x = \mu + \sigma^2\theta_0 + \kappa_3\theta_0^2/2, \quad (3.7.51)$$

so there are now two roots for θ_0, namely

$$\theta_0 = (-\sigma^2 \pm \sqrt{\psi})/\kappa_3 \quad \text{where} \quad \psi = \sigma^4 + 2\kappa_3(x-\mu). \quad (3.7.52)$$

However, since we need to recover (3.7.49) as $\kappa_3 \to 0$, only the positive one is admissible. Thus (3.7.47) takes the form

$$f_3(x) = [2\pi(\sigma^2 + \kappa_3\theta_0)]^{-1/2}\exp\{\mu\theta_0 + \sigma^2\theta^2/2 + \kappa_3\theta_0^3/6 - \theta_0 x\},$$

which, on substituting for x and θ_0 from (3.7.51) and (3.7.52), reduces to

$$f_3(x) = (4\pi^2\psi)^{-1/4}\exp\{-(1/6\kappa_3^2)[\sigma^6 - 3\sigma^2\psi + 2\psi^{3/2}]\}. \quad (3.7.53)$$

This is a completely general, and algebraically amenable, result, which provides a considerable improvement over the Normal approximation (3.7.50) since it incorporates skewness. Note that because raising n to 4 and 5 leads to mathematically tractable cubic and quartic equations for θ_0, this approach can be refined even further, though with some loss of structural transparency.

To illustrate the effect of using this truncated saddlepoint technique, let us first consider the exceedingly simple structure of the Poisson distribution, since this highlights all the salient features of the approach. Table 3.1 shows the exact Poisson probabilities $p_i = 10^i e^{-10}/i!$ $(i = 0, \ldots, 20)$ together with the corresponding values for the saddlepoint approximation (3.7.22) and the truncated c.g.f. values (3.7.47)–(3.7.48) with cut-off point $n = 2, 3, 4$ and 6. To aid comparison the table also shows the corresponding percentage deviations from the exact values. Note first that the full saddlepoint approximation provides an accuracy to within 3% for $i > 2$, and to within 1% for $i > 9$. The breakdown at $i = 0$ stems purely from the fact that the (basic) Stirling approximation yields $0! \simeq \zeta(0) = 0$ as opposed to the defined value of 1. In proportional terms the Normal approximation $(n = 2)$ is appalling; the third-order case $(n = 3)$ is clearly far better where it exists. Indeed, for $i \geq 12$ it outperforms not only the fourth- and sixth-order approximations, but it also marginally beats the full saddlepoint approximation itself. So, contrary to intuition, incorporating extra

Table 3.1: Comparison of exact Poisson(10) probabilities with full and truncated saddlepoint approximations of orders 2, 3, 4 and 6; * denotes no available value.

i	exact	saddle		2nd		3rd		4th		6th	
0	0.0^4454	*		0.0^3850	(1772%)	*		0.0^3104	(129%)	0.0^4398	(-12%)
1	0.0^3454	0.0^3492	(8.4%)	0.0^2220	(384%)	*		0.0^3506	(11%)	0.0^3358	(-21%)
2	0.0^2227	0.0^2237	(4.2%)	0.0^2514	(127%)	*		0.0^2207	(-9.0%)	0.0^2211	(-7.0%)
3	0.0^2757	0.0^2778	(2.8%)	0.0109	(44%)	*		0.0^2683	(-9.8%)	0.0^2758	(0.1%)
4	0.0189	0.0193	(2.1%)	0.0209	(10%)	*		0.0178	(-5.7%)	0.0192	(1.5%)
5	0.0378	0.0385	(1.7%)	0.0361	(-4.5%)	0.0511	(35%)	0.0370	(-2.1%)	0.0384	(1.6%)
6	0.0631	0.0639	(1.4%)	0.0567	(-10%)	0.0645	(2.3%)	0.0630	(-0.1%)	0.0639	(1.4%)
7	0.0901	0.0912	(1.2%)	0.0804	(-11%)	0.0912	(1.3%)	0.0907	(0.7%)	0.0912	(1.2%)
8	0.1126	0.1138	(1.1%)	0.1033	(-8.3%)	0.1138	(1.1%)	0.1137	(0.9%)	0.1138	(1.1%)
9	0.1251	0.1263	(0.9%)	0.1200	(-4.1%)	0.1263	(0.9%)	0.1263	(0.9%)	0.1263	(0.9%)
10	0.1251	0.1262	(0.8%)	0.1262	(0.8%)	0.1262	(0.8%)	0.1262	(0.8%)	0.1262	(0.8%)
11	0.1137	0.1146	(0.8%)	0.1200	(5.5%)	0.1146	(0.8%)	0.1146	(0.8%)	0.1146	(0.8%)
12	0.0948	0.0954	(0.7%)	0.1033	(9.0%)	0.0954	(0.7%)	0.0955	(0.7%)	0.0954	(0.7%)
13	0.0729	0.0734	(0.6%)	0.0804	(10%)	0.0734	(0.6%)	0.0735	(0.7%)	0.0734	(0.6%)
14	0.0521	0.0524	(0.6%)	0.0567	(8.9%)	0.0524	(0.6%)	0.0525	(0.8%)	0.0524	(0.6%)
15	0.0347	0.0349	(0.6%)	0.0361	(4.1%)	0.0349	(0.5%)	0.0350	(0.8%)	0.0349	(0.6%)
16	0.0217	0.0218	(0.5%)	0.0209	(-3.9%)	0.0218	(0.4%)	0.0219	(0.9%)	0.0218	(0.5%)
17	0.0128	0.0128	(0.5%)	0.0109	(-15%)	0.0128	(0.3%)	0.0129	(0.9%)	0.0128	(0.5%)
18	0.0^2709	0.0^2712	(0.5%)	0.0^2514	(-27%)	0.0^2711	(0.2%)	0.0^2715	(0.8%)	0.0^2713	(0.5%)
19	0.0^2373	0.0^2375	(0.4%)	0.0^2220	(-41%)	0.0^2374	(0.1%)	0.0^2376	(0.6%)	0.0^2375	(0.5%)
20	0.0^2187	0.0^2187	(0.4%)	0.0^3850	(-54%)	0.0^2187	(-0.04%)	0.0^2187	(0.3%)	0.0^2187	(0.4%)

cumulants does not *automatically* improve precision. The collapse of the third-order approximation in the lower tail of the distribution is due to the necessity of having $\psi \geq 0$ in (3.7.53). Since here $\sigma^2 = \kappa_3 = \lambda = 10$, (3.7.52) gives $\psi = \lambda(2x - \lambda) > 0$ if and only if $x \geq \lambda/2 = 5$. Except for $i = 1$, the sixth-order approximation consistently outperforms the fourth-order approximation, though the difference between them becomes slight for $i > 8$.

To illustrate the effect of applying the approximation procedure to the logistic process, consider $\lambda_N = N(2.2 - 0.1N)$ and $\mu_N = N(0.2 + 0.1N)$. A simulated stochastic realization of this process is shown in Renshaw (1991), together with numerical comparisons of mean, variance and skewness approximations (improved moment estimators are developed in Matis and Kiffe, 1996). Exact quasi-equilibrium cumulants are

$$\kappa_1 = 9.30, \quad \kappa_2 = 6.55, \quad \kappa_3 = -0.52, \quad \kappa_4 = -2.69, \quad \kappa_5 = 5.26 \text{ and } \kappa_6 = -44.05,$$

so the $\tilde{\pi}_N$ are slightly skew-negative (coefficient of skewness $\gamma_1 = \kappa_3/\kappa_2^{3/2} = -0.031$) and slightly flatter than Normal (coefficient of kurtosis $\gamma_2 = \kappa_4/\kappa_2^2 = -0.063$). Both the second- and third-order approximations behave similarly within the central part of the distribution (accurate to within 2% for $4 \leq i \leq 15$), and although the third-order values offer an improvement in the tails they are still highly inaccurate for $18 \leq i \leq 22$. Using the fourth-order cumulant confers no added advantage (not surprising since γ_2 is small), but has the considerable disadvantage of reducing the admissible range of i. Introducing cumulants up to sixth-order degrades the procedure still further. Thus here the third-order approximation, though not totally ideal, is best both in terms of transparency and accuracy.

To extend this standard logistic model to the cubic logistic, suppose we keep the same basic properties by retaining the linear components together with the deterministic carrying capacity ($K = 10$) and maximum population size ($N_{max} = 22$). This yields $\lambda_N = N(2.2 - N^2/220)$ and $\mu_N = N(0.2 + (0.02 - 1/220)N^2)$, and the corresponding quasi-equilibrium cumulants are now

$$\kappa_1 = 9.27, \quad \kappa_2 = 4.66, \quad \kappa_3 = 0.29, \quad \kappa_4 = -0.35, \quad \kappa_5 = -2.22 \text{ and } \kappa_6 = -0.19.$$

Thus in this respect the distribution is more Normal than before ($\kappa_3, \kappa_4, \kappa_6 \simeq 0$); the $\tilde{\pi}_N$ are slightly skew-positive ($\gamma_1 = 0.029$) and virtually meso-kurtic ($\gamma_2 = -0.016 \simeq 0$). Note the reduction in variance, and the associated reduction in the magnitude of the tail probabilities, over the standard logistic process. It transpires that the third-order approximation outperforms the second-order in the central part of the distribution ($2 \leq i \leq 16$), but behaves substantially worse for $i > 16$; in this tail the fourth-order is better than the third-order, but worse than the second-order. Thus unlike the standard logistic for which the third-order approximation is uniformly best, for the cubic logistic the second-order (i.e. Normal) approximation is optimal even in the tails, and is only marginally worse in the centre.

The difficulty of obtaining accurate tail probabilities is easily overcome for the standard logistic, for on writing $a_1 = pb_1$ and $a_2 = qb_2$, the rates $\lambda_N = Nb_1(p - N)$ and $\mu_N = Nb_2(q + N)$, whence the quasi-equilibrium distribution (3.2.11) reduces to

$$\tilde{\pi}_N \propto \frac{(b_1/b_2)^{N-1}}{N} \binom{p+q}{p-N}. \qquad (3.7.54)$$

Thus $\{\tilde{\pi}_N\}$ behaves as $1/N$ times a binomial, and so in terms of obtaining a simple functional form the truncated saddlepoint results are really irrelevant. Their purpose here is to highlight the success or failure of including higher-order cumulants in the approximation procedure. If either p or q are non-integer, then we simply replace the factorials by Stirling's approximation.

However, a simple functional form comparable to (3.7.54) is not available for the more general power-law model (3.7.41), even when $s \neq 1$ is integer. One possibility for obtaining a simple and easily interpretable approximation for the tails of the distribution $\{\tilde{\pi}_N\}$ is to exploit the tail probability structure proposed by Lugannani and Rice (1980). Jensen (1995) provides a full account of this approach (see also Reid, 1988), whilst Daniels (1983, 1987) provides numerical examples; Daniels (1987) reviews a variety of tail area approximations using the saddlepoint method. Let $\Phi(y)$ and $\phi(y)$ denote the p.d.f. and c.g.f., respectively, of the standard $N(0,1)$ distribution, and $F(x)$ the c.d.f. of our quasi-equilibrium distribution. Then the Lugananni–Rice approximation takes the form

$$F(x) \simeq [\Phi(y) - \phi(y)](z^{-1} - y^{-1}) \qquad (3.7.55)$$

where

$$y = \pm\sqrt{2}\sqrt{\{\theta x - K(\theta)\}}, \qquad z = \theta\sqrt{\{K''(\theta)\}} \quad \text{and} \quad x = K'(\theta). \qquad (3.7.56)$$

As developed, x is a mean of n observations, but the result remains valid when $n = 1$. On replacing $K(\theta)$ by $K_2(\theta)$, the approximation reduces to $F(x) \simeq \Phi[(x - \kappa_1)/\sqrt{\kappa_2}]$, and so we recover the basic Normal appoximation. The question therefore arises as to whether using $K_3(\theta)$ results in an improvement in the tail probabilities; recall that the third-order saddlepoint results may actually be worse than their Normal counterparts. The answer for our cubic logistic example is definitely no; for example, $\tilde{\pi}_1 \simeq F(1.5) - F(0.5) = 0.0^4 494$ is substantially worse than the second, third- and fourth-order saddlepoint approximations. However, one must ask whether this apparent breakdown in constructing a simple theoretical form for tail probabilities really represents a problem in practice. For the second-order (i.e. Normal) approximation works quite well for $4 \leq i \leq 18$, i.e. -2.4 to 4.1 standard deviations from the mean, whilst the third-order holds over -3.4 to 3.1 standard deviations. Thus between them they cover all but extreme excursions away from the mean quasi-equilibrium value.

These results are therefore not as disappointing as they might at first seem. Cumulants are notoriously fickle, and the fact that the third-order saddlepoint approximation works so well (at least for the examples given here), combined with the algebraic simplicity of the completely general third-order approximation (3.7.53), are grounds for considerable comfort. So the discovery that the truncated saddlepoint approach does not progress 'smoothly' with increasing n, in the sense that it offers increasingly accurate probability values across the full admissible range, is far more of a theoretical inconvenience than a practical handicap.

The order to which one takes the cumulants to be non-zero clearly depends not only on the overall precision required, but also on whether we wish to use the distribution over its full range or just over part of it. For example, if we are investigating the spread of an animal or plant population (Renshaw, 1986, 1991), or a disease such as rabies (Mollison, 1991), attention is often concentrated on the leading edge of the distribution, and in such cases the third-order approximation may suit our needs. Other situations may demand higher-order cumulants, such as the propagation of Telegraph Waves (Jakeman and Renshaw, 1987; Renshaw, 1988). The key point is that no matter whether the cumulants are evaluated theoretically, or by symbolic or numeric calculation, employing the general result (3.7.47)–(3.7.48) enables quick and easy assessment to be made both of the overall qualitative structure of the process and the precision obtained for a given cumulant-order.

3.7.5   Final comments

On the plus side, not only does the saddlepoint approximation $g(x)$ provide an extremely powerful approximation for 'smoothly changing' distributions, but we have also seen through (3.7.24) and (3.7.25) that applying first- and second-order correction factors improves the accuracy still further. However, this does raise the question as to how this accuracy increases with the number of terms used in the asymptotic expansions (3.7.13) and (3.7.38)? Moreover, whilst using the truncated saddlepoint provides good algebraic and numeric approximations to distributions based on just the first few cumulants, we have seen that this does not necessarily mean that accuracy increases with cumulant order. So is there a way of determining the optimal number of cumulants to use without resorting to trial-and-error?

A further issue relates to the logic of placing all high-order (i.e. truncated) cumulants to zero. This assertion is essentially argued as being a natural progression from the deterministic representation ($\kappa_i \equiv 0$ for $i = 2, 3, \ldots$) and the Normal approximation ($\kappa_i \equiv 0$ for $i = 3, 4, \ldots$). Yet is this totally realistic? For with the Poisson(λ) distribution $\kappa_i \equiv \lambda$ for all $i \geq 1$, whilst for the highly skewed Exponential(λ) distribution $\kappa_i = (i-1)!/\lambda^i \to \infty$ as $i \to \infty$. It would therefore be interesting to develop and compare alternative lines of cumulant truncation, and one obvious starting point is to assume that the κ_i follow a power-law defined in terms of (for example) κ_3/κ_2. Gillespie and Renshaw (2007) explore several such avenues, including: extending the inversion of the cumulant generating function to second-order; selecting an appropriate probability structure for higher-order cumulants; and, making subtle changes to the target cumulants and then optimizing via the simplex algorithm.

On the black-art side, all our expansions are developed in terms of the mean of n random variables, yet we have quite happily taken $n = 1$. So what exactly does happen to the relative magnitudes of successive terms in these expansions when $n = 1$? Of potentially greater concern is that the analyses depend on the contribution to the contour integral (3.7.5) being negligible away from the saddlepoint. Is failure in this respect the underlying reason why the approach fails to reproduce the saw-tooth nature of the even-integer Poisson distribution (3.7.26), or is something even deeper operating here? An exact analytic investigation of what happens to the contour integral away from the immediate vicinity of the saddlepoint could well prove to be highly instructive. Moreover, examination of the asymptotic expansion (3.7.38) shows that the benefit of taking the saddlepoint approximation, i.e. placing $K'(\tau) = \bar{x}$, is that we produce a series expansion in powers of n^{-1} rather than $n^{-1/2}$. Yet does this matter if we take $n = 1$? A comparison of the Edgeworth and saddlepoint representations could prove instructive.

Clearly, there are a large number of open-ended issues that need to be researched in considerable detail before this saddlepoint approximation procedure can become accepted as a general tool. For at present models need to be analyzed and assessed on a case-by-case basis. The most important development may well come from re-examining Daniel's (1960) truly ground-breaking results. For here he provides a general saddlepoint solution for multiplicative birth–death processes. Though this is awkward in that it involves integrals which switch though breaking-points, it has immense potential for yielding a general theoretical solution. It is quite conceivable that his approach could be developed into providing a single user-friendly distribution that would encapsulate a wide range of stochastic processes each of which currently requires its own specific analysis. The challenge is there for the taking.

4
The random walk

In Chapter 3 we show that although specific results involving population size probabilities, moments, times to extinction, etc. can be developed for fairly low-level (e.g. linear) stochastic processes, the construction of general time-dependent properties is far more problematic. Essentially, the difference-differential Kolomogorov forward and backward equations (3.0.2) and (3.0.3) are just too difficult to solve exactly, which is why we are forced back onto developing approximation techniques. Whilst it is indeed true that the population size probabilities $\{p_{ij}(t)\}$ of moving from state i to state j in time t can be expressed in Fourier transform format through the orthogonal representation (3.1.21), and that this representation yields powerful general results on underlying properties of the process such as existence, uniqueness and recurrence, unfortunately it tells us virtually nothing about the algebraic nature of the $p_{ij}(t)$ themselves. Success here is restricted to constructing the equilibrium and quasi-equilibrium probabilities, π_j and $\tilde{\pi}_j$, which relate to population size probabilities at time $t = \infty$ when the effect of the initial population size i has disappeared.

However, if we are interested in learning about the opening stages of a process $\{X(t)\}$ then placing $t = \infty$ is clearly not an option, and given that it is time itself that seems to be the main cause of our theoretical difficulties an obvious strategy is simply to ignore it. Thus instead of considering the probability of moving from i to j in *time* t, we consider instead the probability of moving from i to j in n *steps*. Now we have already seen through the simulation of simple birth, death and immigration processes in Chapter 2, that because $\Pr(X(t+dt) = N+1|X(t) = N) = \lambda_N dt + o(dt)$ and $\Pr(X(t+dt) = N-1|X(t) = N) = \mu_N dt + o(dt)$, the probabilities that the next event is a birth or a death are respectively given by

$$p_N = \lambda_N/(\lambda_N + \mu_N) \quad \text{and} \quad q_N = \mu_N/(\lambda_N + \mu_N). \quad (4.0.1)$$

So on denoting $p_{ij}^{(n)}$ to be the probability that the process moves from i to j in n steps, the forward and backward Kolmogorov difference-differential equations for $p_{ij}(t)$ reduce to the much simpler difference equations

$$p_{ij}^{(n+1)} = p_{i,j-1}^{(n)} p_{j-1} + p_{i,j+1}^{(n)} q_{j+1} \quad \text{and} \quad (4.0.2)$$

$$p_{ij}^{(n+1)} = q_i p_{i-1,j}^{(n)} + p_i p_{i+1,j}^{(n)}. \quad (4.0.3)$$

That these equations are intrinsically easier to work with can be seen by supposing that $X(0) = 1$, so that $p_{11}^{(0)} = 1$. For we can then construct recursive solutions for

$\{p_{1j}^{(1)}, p_{1j}^{(2)}, \ldots\}$ over $0 \leq j \leq 2$, $0 \leq j \leq 3$, etc., in marked contrast to the (generally) intractable problem of determining the time-dependent probabilities $\{p_{1j}(t)\}$.

Care must be taken not to regard the n-step probabilities $p_{ij}^{(n)}$ and the time-dependent probabilities $p_{ij}(t)$ as being interchangeable, for they (usually) contain substantially different information. For example, taking $\lambda_N = \lambda h(N)$ and $\mu_N = \mu h(N)$, for any appropriate function $h(N)$, produces exactly the same step transition probabilities p_N and q_N regardless of the form of $h(N)$. So simulated plots of N against t may bear little resemblance to those of N against n. A classic illustration of this occurs with the explosive power-law birth process discussed in Section 2.3.5, in which $\lambda_N = \lambda N^{1+\delta}$ for $\delta > 0$. For whilst the population size N can literally become infinite in a finite time t, after n steps N simply increases by the fixed finite amount n since each step has to be a single birth.

Consider, for example, the simple continuous-time birth–death process for which $\lambda_N = \lambda N$ and $\mu_N = \mu N$ over $N = 0, 1, 2, \ldots$. Suppose that the population is initially of size n_0 at time $t = 0$, and we observe the population size just after the first, second, etc., changes of state. We then have a new process which evolves at the discrete time points $n = 1, 2, 3, \ldots$. Once the population size reaches zero it stays there since no more births can take place, so state 0 (i.e. extinction) is called an *absorbing state*. It follows from (4.0.1) that for $N > 0$ each individual step may be upwards (corresponding to a birth) or downwards (corresponding to a death) with probability $p_N = p = \lambda/(\lambda + \mu)$ or $q_N = q = \mu/(\lambda + \mu)$, respectively. Since p_N and q_N do not change with $N > 0$, we refer to this process as the *simple random walk* with an absorbing barrier at zero.

Changing the boundary conditions enables the underlying process to encompass a variety of different scenarios. For example, to determine the probability that the population ever reaches size a we can introduce another absorbing barrier at position $N = a$. If we allow immigration if and only if the population is empty, then we have a *reflecting barrier* at $N = 0$ and the population will develop indefinitely since extinction is no longer possible. Whilst if the population is restricted to a maximum size a, then $\lambda_a = 0$ and we have a reflecting barrier at $N = a$. Although the interior equations (i.e. those for $0 < N < a$) remain unchanged in all these situations, we shall see that the boundary conditions (i.e. at $N = 0$ and a) can have a profound effect on the nature of the solution.

4.1 The simple unrestricted random walk

Before developing solutions to such boundary value problems let us first consider the unrestricted process. A simple way of envisaging this is to assume that an individual moves along a linear habitat, such as a river or shore line, that has been suitably zoned into neighbouring states labelled by the integers $i = \ldots, -1, 0, 1, \ldots$. Then in the context of our simple random walk, successive moves are independent of each other, with $\Pr(\text{move from site } i \text{ to } i+1) = p$ and $\Pr(\text{move from site } i \text{ to } i-1) = q$. A direct paradigm between this geographic process and a population process can be seen by considering a population of males and females in which the population 'size' X_n at the nth step denotes the net difference between the numbers of males and females. So the

nth step $Z_n = +1$ (with probability p) if there is a male birth or a female death, whilst $Z_n = -1$ (with probability $q = 1 - p$) if there is a male death or a female birth. For this process X_n can clearly take any integer value, positive or negative.

Without any loss in generality let the process start at $X_0 = 0$. Since

$$X_n = Z_1 + \cdots + Z_n \tag{4.1.1}$$

is the sum of n independent variables, X_n follows a Binomial(n, p) distribution. That is, if the process moves $n - i$ steps upwards, and hence i steps downwards, then $X_n = n - 2i$ with

$$\Pr(X_n = n - 2i) = \binom{n}{i} p^{n-i} q^i \quad (i = 0, 1, \ldots, n). \tag{4.1.2}$$

Hence on writing $j = n - 2i$ we have

$$\Pr(X_n = j) = \binom{n}{(n-j)/2} p^{(n+j)/2} q^{(n-j)/2} \quad (j = -n, -n+2, \ldots, n). \tag{4.1.3}$$

When $p = q = 1/2$, that is we have a symmetric simple random walk, then this result shows that the probability of returning to $X_0 = 0$ after $2n$ steps (this is impossible after an odd number of steps) is given by

$$\Pr(X_{2n} = 0) = \binom{2n}{n}(1/2)^{2n}. \tag{4.1.4}$$

Whence on replacing $n!$ by Stirling's approximation, namely $n! \sim \sqrt{(2\pi)} n^{n+1/2} e^{-n}$, we obtain the attractively simple result that

$$\Pr(X_{2n} = 0) \sim 1/\sqrt{(\pi n)}. \tag{4.1.5}$$

4.1.1 Normal approximation

Provided n is not too large, computer packages such as *Minitab* can be used to compute the above binomial probabilities together with their cumulative sums. Alternatively, we can use the Normal approximation. For on letting μ and σ^2 denote the mean and variance of a jump, so that here $\mu = p - q$ and $\sigma^2 = 4pq$ (since $p + q = 1$), it follows from (4.1.1) that $\mathrm{E}(X_n) = n\mu$ and $\mathrm{Var}(X_n) = n\sigma^2$. Applying the Central Limit Theorem then yields the (general) result that

$$\Pr(a \leq X_n \leq b) \simeq \frac{1}{\sqrt{(2\pi n\sigma^2)}} \int_{a-1/2}^{b+1/2} e^{-(x-n\mu)^2/2n\sigma^2} \, dx$$

$$= \Phi[(b + 1/2 - n\mu)/\sqrt{(n\sigma^2)}] - \Phi[(a - 1/2 - n\mu)/\sqrt{(n\sigma^2)}], \tag{4.1.6}$$

where $\Phi(\cdot)$ denotes the standard Normal distribution with mean 0 and variance 1. For example, the simple random walk with $p = 0.6$ has $\mu = p - q = 0.2$ and $\sigma^2 = 4pq = 0.96$. So after 40 steps

$$\Pr(X_{40} \leq 10) \simeq \Phi[(10 + 1/2 - 8)/\sqrt{(38.4)}] = \Phi(0.4034) = 0.657,$$

which compares quite well with the exact value 0.683 (3.8% too low) calculated from (4.1.3).

This Normal result is particularly useful in telling us the range in which X_n is likely to lie. Since the probability that a Normal random variable lies further than four standard deviations from its mean is less than 0.0001, we can write

$$\Pr(-4\sigma\sqrt{n} \leq X_n - \mu \leq 4\sigma\sqrt{n}) > 0.9999. \qquad (4.1.7)$$

Hence X_n has 'drift' $n\mu$, whilst the deviation from $n\mu$ increases only with $\sqrt{n}$ and not with n. For example, the symmetric random walk $p = q = 1/2$ has mean $\mu = 0$ and variance $\sigma^2 = 4pq = 1$, whence (4.1.7) becomes $\Pr(-4\sqrt{n} \leq X_n \leq 4\sqrt{n}) > 0.9999$. Thus even after a million steps X_n will almost certainly lie within a mere 4000 units of its starting point.

4.1.2 Laws of large numbers

We can immediately deduce from (4.1.7) that with high probability

$$X_n = n\mu + O(\sqrt{n}) = n\mu[1 + O(n^{-1/2})]. \qquad (4.1.8)$$

Thus if $p > q$, i.e. the probability of a jump upwards is greater than that of a jump downwards, then since $\mu > 0$ the probability that X_n is arbitrarily large becomes arbitrarily near one as n increases. Indeed, we see from (4.1.6) that for fixed j

$$\Pr(X_n > j) \simeq 1 - \Phi[(j + 1/2 - n\mu)/\sqrt{(n\sigma^2)}] \to 1 \quad \text{as} \quad n \to \infty. \qquad (4.1.9)$$

Alternatively, we can employ the *Weak Law of Large Numbers* which states that for n independent and identically distributed variables Z_i, for large n the variable $X_n = Z_1 + \cdots + Z_n$ is likely to lie near $n\mu$. More precisely, the weak law of large numbers is said to hold for the sequence $\{Z_i\}$ if, for every $\epsilon > 0$,

$$\Pr\left\{\frac{|X_n - E[X_n]|}{n} > \epsilon\right\} \to 0. \qquad (4.1.10)$$

So no matter how small $\epsilon > 0$ is, as $n \to \infty$

$$\Pr(-\epsilon < (X_n/n) - \mu < \epsilon) \to 1. \qquad (4.1.11)$$

Note that this result does not imply that $|(X_n/n) - \mu|$ remains small for all large n; it is perfectly possible for the weak law to apply yet for $|(X_n/n) - \mu|$ to fluctuate between finite, or even infinite, limits.

To make a stronger assertion we need to employ the *Strong Law of Large Numbers* which (in the simple random walk context) states that *for any $\epsilon > 0$ the probability that X_n remains in the region $n(\mu - \epsilon) < X_n < n(\mu + \epsilon)$ for all $n > m$ can be made as close to 1 as we wish by choosing m sufficiently large*. Hence for any j we have that

$$\Pr(X_n > j, X_{n+1} > j, \ldots) \to 1 \quad \text{as} \quad n \to \infty. \qquad (4.1.12)$$

So with probability one X_n drifts off to infinity. Similarly, if $p < q$ then X_n drifts off to $-\infty$ with probability one. For a full statement of this theorem, together with a brief historical account, see Feller (1968); the theorem was first proved at this level of generality by Khintchine (1929).

The above results also apply when $p = q$, so that $\mu = 0$. In particular, the Central Limit Theorem tells us that with high probability the particle will be within a distance of order $\sqrt{n}$ from its start point after n jumps. However, in Section 4.2.2 we shall show that X_n may make arbitrarily large excursions from its start point even though it is eventually certain to return to it.

4.1.3 Using the 'reflection' principle

The simple random walk has a far-reaching effect which extends well beyond the intrinsically important results that it generates in its own right. For it is often a first encounter with this process that alerts students to the unexpected notion that theoretical conclusions may appear to defy intuition and commonsense. They discover not only that commonly accepted beliefs concerning chance fluctuations are without foundation, but also that the laws of large numbers are widely misconstrued. As Feller (1968) presents these intriguing notions in some detail, here we shall just restrict ourselves to stating the key issues. These essentially originate from an inconspicuous lemma announced by Bertrand in 1887, and are developed under the general combinatorial analysis theme of 'ballot problems' (for a historical presentation see Dvoretzky and Motzkin, 1947). The basic ideas stem naturally from the following exceedingly simple result.

Suppose that, in a ballot, candidate C scores c votes and candidate D scores d votes, where $c > d$. Then the probability that throughout the counting there are always more votes for C than for D equals $(c-d)/(c+d)$.

Every arrangement $\{Z_1, Z_2, \ldots, Z_n\}$ of the integers $Z_i = \pm 1$ represents a possible voting record, and since a visual insight into the underlying mathematics can be developed far more clearly by using a geometrical approach, rather than through the generating function type argument developed earlier in this text, we shall represent such an arrangement by a polygonal line whose vertices have abscissas $0, 1, 2, \ldots, x$ and ordinates $X_0, X_1, X_2, \ldots, X_x = y$.

If c of the Z_i are positive and d are negative, then

$$x = c + d \quad \text{and} \quad y = c - d. \tag{4.1.13}$$

Note that an arbitrary point (x, y) can be joined to the origin by a path if and only if x and y satisfy (4.1.13). In this case the c places for the positive Z_i can be chosen from the $x = c + d$ available places in

$$M_{x,y} = \binom{c+d}{c} = \binom{x}{(x+y)/2} \tag{4.1.14}$$

different ways. So if we define $M_{x,y} = 0$ whenever x, y are not of the form (4.1.13), then there exist exactly $M_{x,y}$ different paths from the origin to the point (x, y).

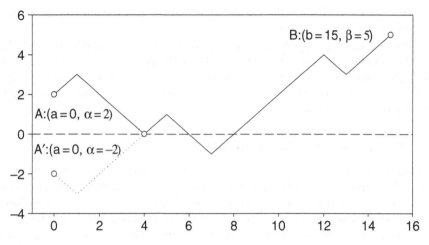

Figure 4.1 Illustrating the reflection principle.

THEOREM 4.1 *Bertrand's Ballot Theorem asserts that when $y > 0$ there exist exactly $(y/x)M_{x,y}$ paths satisfying the conditions $X_0 = 0, X_1 > 0, X_2 > 0, \ldots, X_{x-1} > 0, X_x = y$.*

To prove this result we first introduce the idea of *reflection*. Let $A = (a, \alpha)$ and $B = (b, \beta)$ be integer points with $0 \le a < b$ and $\alpha > 0$, $\beta > 0$. Then the reflection of A on the x-axis is defined to be the point $A' = (a, -\alpha)$ (see Figure 4.1).

The reflection principle: asserts that the number of paths from A to B which touch or cross the x-axis equals the number of all paths from A' to B.

Proof Consider a path $\{X_a = \alpha, \ldots, X_b = \beta\}$ from A to B having one or more vertices on the x-axis. Let $T = t$ be the abscissa of the *first* such vertex; that is, t satisfies $X_a > 0, \ldots, X_{t-1} > 0, X_t = 0$. Then $\{-X_a, \ldots, -X_{t-1}, X_t = 0\}$ is a path leading from A' to B and having $T = (t, 0)$ as its first vertex on the x-axis. Since the sections AT and $A'T$ are reflections of each other, there exists a one-to-one correspondence between all paths from A' to B and such paths from A to B that have a vertex on the x-axis (Figure 4.1 makes this obvious). □

Proof of Ballot Theorem Since $X_1 = Z_1 = \pm 1$, we can take $X_1 = 1$ for each admissible path. It therefore follows that there exist as many admissible paths as there are paths leading from the point $(1, 1)$ to (x, y) which neither touch nor cross the x-axis. Clearly, the number of admissible paths from $(1, 1)$ to (x, y) is the same as that from $(0, 0)$ to $(x - 1, y - 1)$, namely $M_{x-1, y-1}$. Whilst it follows from the reflection principle that the number of paths from $(1, 1)$ to (x, y) which touch or cross the x-axis equals the number of all paths from $(1, -1)$ to (x, y), which in turn equals the number of all paths from $(0, 0)$ to $(x - 1, y + 1)$. Hence the number of paths from $(1, 1)$ to (x, y) which neither touch nor cross the x-axis is given by $M_{x-1, y-1} - M_{x-1, y+1}$. Denote the number of positive and negative steps from $(0, 0)$ to $(x - 1, y - 1)$ and $(x - 1, y + 1)$ by c_1, d_1 and c_2, d_2, respectively. Then it follows from (4.1.13) that

$x-1 = c+d-1 = c_1+d_1$ and $y-1 = c-d-1 = c_1-d_1$, so $c_1 = c-1$ and $d_1 = d$. Similarly, $x-1 = c+d-1 = c_2+d_2$ and $y+1 = c-d+1 = c_2-d_2$, so $c_2 = c$ and $d_2 = d-1$. Whence using (4.1.14) yields

$$M_{x-1,y-1} - M_{x-1,y+1} = \binom{c+d-1}{c-1} - \binom{c+d-1}{c} = (c+d-1)!\left(\frac{c-d}{c!d!}\right)$$

$$= \binom{c+d}{c}\left(\frac{c-d}{c+d}\right) = (y/x)M_{x,y}, \qquad (4.1.15)$$

which completes the proof. □

The classical description of this result involves two gamblers: at each toss of a coin Peter either wins or loses a unit amount from Paul. The sequence $\{X_1, X_2, \ldots\}$ then represents Peter's successive gains; or, in our earlier population paradigm, the successive net differences between the number of males and females. So the Ballot Theorem refers to paths which are situated entirely above the x-axis; that is, to games in which Peter is always in profit, or populations in which the number of males always exceeds the number of females. This leads to a conclusion which runs counter to intuition, for it appears 'obvious' that in a prolonged series of coin tossings Peter should lead for roughly half the time and Paul the other half. This is entirely wrong. For we shall soon see that the lead changes at such infrequent intervals that no matter how long the series of tossings, the most probable number of changes of lead is zero; exactly one change of lead is more probable than two; whilst two changes are more probable than three, etc. So anyone not familiar with these probabilistic results who analysed the long-run case histories of individual coin-tossing games would be likely to classify the majority of coins as being maladjusted.

This observation has profound relevance to many other situations involving time-series of events. If, for example, a large portfolio comprises shares whose prices fluctuate purely at random according to a simple random walk (i.e. they are not affected by market forces or sentiment), then a surprisingly large proportion of the share trajectories will lie either above (or below) the initial price almost all the time. Only in a very few cases will they fluctuate between profit and loss in the manner that is generally expected from a well-behaved coin. As Feller (1968) comments, "... such findings should serve as a warning to those who are prone to discern secular trends and deviations from average norms".

Suppose we now turn our attention to the relatively rare event of paths which join the origin to a point $C = (2n, 0)$ on the x-axis (an odd vertex on the x-axis is clearly impossible). Denote

$$L_{2n} = \frac{1}{n+1}\binom{2n}{n}. \qquad (4.1.16)$$

Then as any such path must contain n positive and n negative steps, the total number of such paths must be $\binom{2n}{n}$. A simple argument then reveals the number of such paths which remain either (a) wholly positive, or (b) wholly non-negative, between $x = 0$ and $x = n$.

THEOREM 4.2 Among the $\binom{2n}{n}$ paths joining the origin to the point $2n$ of the x-axis there are (a) exactly L_{2n-2} paths such that

$$X_1 > 0, \quad X_2 > 0, \quad \ldots, \quad X_{2n-1} > 0, \quad X_{2n} = 0; \tag{4.1.17}$$

(b) exactly L_{2n} paths such that

$$X_1 \geq 0, \quad X_2 \geq 0, \quad \ldots, \quad X_{2n-1} \geq 0, \quad X_{2n} = 0. \tag{4.1.18}$$

That is, there are as many paths from the origin to $2n$ with all inner vertices above the x-axis as there are paths to $2n-2$ with no vertex below the x-axis.

Proof Each path satisfying condition (4.1.17) passes through the point $Y_1 = (2n-1, 1)$, whence applying Theorem 4.1 with $x = 2n-1$ and $y = 1$ shows that the number of paths to Y_1 with $X_1 > 0, \ldots, X_{2n-2} > 0$ equals

$$M_{2n-1,1} = \frac{1}{2n-1}\binom{2n-1}{n} = \frac{1}{n}\binom{2n-2}{n-1} = L_{2n-2}, \tag{4.1.19}$$

which proves (a). To prove (b) we again consider a path satisfying condition (a), but this time we omit the two end points $(0,0)$ and $(2n,0)$ and restrict attention to all inner paths joining $Y_0 = (1,1)$ to $Y_1 = (2n-1,1)$ whose vertices lie on or above the line $y = 1$. On translating the origin to Y_0, we see that these are equivalent to paths from the new origin to the point Y_1 which now takes the new coordinates $(2n-2,0)$. Moreover, none of these paths have vertices which lie below the new x-axis. We have therefore established a one-to-one correspondence between such paths and all paths satisfying (4.1.17) that join the origin to the point $2n-2$ of the x-axis. Thus the number of paths in (b) is given by the number of paths in (a) but with n increased to $n+1$, i.e. L_{2n}. □

The beauty of these results is that since they concern admissible *paths*, and not probabilities, they relate to *any* birth–death process with general rates λ_N and μ_N. For example, in terms of the simple immigration–birth–death process (Section 2.5), (4.1.17) relates to the number of paths which start from $N = 0$ and have the population being empty again *for the first time* on step $2n$. Whilst to illustrate (b), suppose we now allow the process to be *terminated* at rate β whenever it is empty, denoted by the new state $N = -1$. Then (4.1.18) relates to the number of paths which start from $N = 0$, reach $N = 0$ after $2n$ steps, and have not been involved in process termination by step $2n$. Note that in contrast to (a), the process can be empty at any of the intervening steps $2, 4, \ldots, 2n-2$.

4.1.4 First passage and return probabilities

Further insight between these two path structures can be gleaned if we remove the (usual) population process requirement that $N \geq 0$ (or $N \geq -1$ in the case of the termination process). We say that:

a return to the origin occurs at step $2n$ if $X_{2n} = 0$;
a first return to the origin occurs at step $2n$ if $X_1 \neq 0, X_2 \neq 0, \ldots, X_{2n-1} \neq 0, X_{2n} = 0$;
a first passage through $r > 0$ occurs at step n if $X_1 < r, X_2 < r, \ldots, X_{n-1} < r, X_n = r$.

A second, third, etc. return to the origin and a first passage through $r < 0$ are defined similarly.

In the case of the symmetric simple random walk (i.e. $p = q = 1/2$ on $N = \ldots, -1, 0, 1, \ldots$) all paths are equally likely to occur and so we can equate the proportion of paths exhibiting a specific characteristic to the probability of occurrence. On denoting

$$u_{2n} = \binom{2n}{n} 2^{-2n} \quad (n = 0, 1, 2, \ldots) \tag{4.1.20}$$

and

$$f_0 = 0 \quad \text{with} \quad f_{2n} = \frac{1}{2n} u_{2n-2} \quad (n = 1, 2, \ldots), \tag{4.1.21}$$

it is easily shown that

$$f_{2n} = u_{2n-2} - u_{2n} \quad (n = 1, 2, \ldots). \tag{4.1.22}$$

We can now prove three equalities between events which at first sight look to be quite different.

THEOREM 4.3 *For $n = 1, 2, \ldots$*

$$u_{2n} = \Pr(X_{2n} = 0) = \Pr(X_1 \neq 0, X_2 \neq 0, \ldots, X_{2n} \neq 0) \tag{4.1.23}$$
$$= \Pr(X_1 \geq 0, X_2 \geq 0, \ldots, X_{2n} \geq 0). \tag{4.1.24}$$

That is, the events (a) a return to the origin takes place at time $2n$, (b) no return occurs up to and including time $2n$, and (c) the path is non-negative (or equally, non-positive) between 0 and $2n$, all have the same probability u_{2n}. Moreover,

$$f_{2n} = \Pr(X_1 \neq 0, X_2 \neq 0, \ldots, X_{2n-1} \neq 0, X_{2n} = 0) \tag{4.1.25}$$
$$= \Pr(X_1 \geq 0, X_2 \geq 0, \ldots, X_{2n-2} \geq 0, X_{2n-1} < 0). \tag{4.1.26}$$

That is, the events (d) the first return to the origin takes place at time $2n$, and (e) the first passage through -1 occurs at time $2n - 1$, have the same probability f_{2n}.

Proof We see from (4.1.14) that there exist $\binom{2n}{n}$ paths joining the origin to the point $(2n, 0)$, and since the total number of possible paths is 2^{2n} the required probability is therefore given by

$$\binom{2n}{n} \div 2^{2n} = u_{2n},$$

which proves (a). Moreover, by Theorem 4.2a there are L_{2n-2} paths joining the origin to $(2n, 0)$ such that $X_1 > 0, \ldots, X_{2n-1} > 0$. Since there are twice as many paths satisfying the condition in (4.1.25) than in (4.1.17), on using (4.1.19)–(4.1.21) we see that the corresponding probability is

$$2L_{2n-2} \div 2^{2n} = \frac{2}{n}\binom{2n-2}{n-1} \div 2^{2(n-1)} 2^2 = \frac{1}{2n} u_{2n-2} = f_{2n},$$

which proves (d). To prove (e) we simply note that the condition in (4.1.26) is equivalent to $X_1 \geq 0, \ldots, X_{2n-3} \geq 0, X_{2n-2} = 0$ followed by $X_{2n-1} = -1$. Theorem 4.2b with $2n$ replaced by $2n - 2$ shows that L_{2n-2} out of $2^{2(n-1)}$ possible paths satisfy the first part, whilst $\Pr(0 \to -1) = 1/2$. So on using (4.1.19) we see that the required probability is

$$\frac{1}{n}\binom{2n-2}{n-1}2^{-2(n-1)} \times \frac{1}{2} = \frac{1}{2n}u_{2n-2} = f_{2n}.$$

With regard to (b) and (c), the probability that no zero occurs up to and including time $2n$ is clearly one minus the probability of a first return to the origin at a time $\leq 2n$. On using (4.1.25) combined with relation (4.1.22), this difference is

$$1 - f_2 - f_4 - \cdots - f_{2n} = 1 - (1 - u_2) - (u_2 - u_4) - \cdots - (u_{2n-2} - u_{2n}) = u_{2n},$$
(4.1.27)

which proves (4.1.23). Similarly, the probability (4.1.24) equals one minus the probability of a first passage through -1 before time $2n$, and on using (4.1.26) we see that this difference is again given by (4.1.27). □

Such geometrically based proofs are not only appealing in their simplicity, but they may also yield results which are much harder to derive via 'standard' analytic arguments. For example, Theorem 4.1 enumerates the paths starting at the origin for which a first passage through y occurs at time x. Since $x + y$ must be even, we may write $x = 2n - y$, whence Theorem 4.1 may be restated in the form:

THEOREM 4.4 *The probability that a first passage from the origin through $y > 0$ takes place at time $2n - y$ is given by*

$$f_{2n}^y = \frac{y}{2n-y}\binom{2n-y}{n}2^{-2n+y} \qquad (0 < y \leq n). \qquad (4.1.28)$$

The ease with which this probability result has been developed as a direct consequence of the Ballot Theorem is remarkable, and readers interested in pursuing this path-based approach further should consult the fundamental text of Feller (1968). For example, he shows that (4.1.28) also takes the following interpretation.

THEOREM 4.5 *The function f_{2n}^y defines the probability that the yth return to zero takes place at time $2n$.*

4.1.5 Probability of long leads: the First Arc Sine Law

Let us now consider the counter-intuitive properties of paths alluded to in Section 4.1.3. We say that the process spends the time from $k - 1$ to k on the positive side if the kth side of its path lies above the x-axis, that is if at least one of $X_{k-1} > 0$ and $X_k > 0$. In coin-tossing terminology this means that Peter's accumulated gain is non-negative at both the $(k - 1)$th and kth trial. Then the required path properties may be derived from the following rather neat result.

THEOREM 4.6 Let $p_{2k,2n}$ be the probability that in the time interval from 0 to $2n$ the process spends $2k$ time units on the positive side of the x-axis, and hence $2n - 2k$ time units on the negative side. Then on noting that the total time spent on the positive side is necessarily even,

$$p_{2k,2n} = u_{2k}u_{2n-2k}. \qquad (4.1.29)$$

Proof On recalling result (4.1.24) of Theorem 4.3 we see that the probability that the process keeps to the positive side during the entire time interval from 0 to $2n$ is given by u_{2n}. So $p_{2n,2n} = u_{2n}$, as asserted. For reasons of symmetry we also have $p_{0,2n} = u_{2n}$ (i.e. the process is never on the positive side), whence it remains to prove (4.1.29) for $1 \leq k \leq n - 1$. As a walk that spends $2k > 0$ time units on the positive side, and hence $2n - 2k > 0$ time units on the negative side, has to pass through zero, let $2r$ denote the time of its *first* return to zero.

First suppose that up to time $2r$ the walk keeps to the positive side, and that during the subsequent time interval from $2r$ to $2n$ it spends a further $2k - 2r \geq 0$ time units on the positive side (viz. $2k$ time units in total). Then since there are 2^{2r} possible paths in $2r$ time units, and f_{2r} is the probability of a first return to zero in $2r$ steps, it follows that there exist: (i) $2^{2r} f_{2r}$ paths of length $2r$ which return to the origin for the first time at $2r$, with half of these keeping to the positive side; and similarly, (ii) $2^{2n-2r} p_{2k-2r,2n-2r}$ paths of length $2n - 2r$ starting at $(2r, 0)$ and having exactly $2k - 2r$ sides above the x-axis. Thus the total number of such paths of length $2n$ is given by

$$(1/2) \times 2^{2r} f_{2r} \times 2^{2n-2r} p_{2k-2r,2n-2r} = 2^{2n-1} f_{2r} p_{2k-2r,2n-2r}.$$

Second, let us consider paths which keep to the negative side between 0 and $2r$, but which spend $2k$ time units on the positive side between $2r$ and $2n$. Here $2k \leq 2n - 2r$, and a repeat of the above argument shows that the number of paths in this class is

$$(1/2) \times 2^{2r} f_{2r} \times 2^{2n-2r} p_{2k,2n-2r} = 2^{2n-1} f_{2r} p_{2k,2n-2r}.$$

Thus as the total number of paths in $2n$ steps is 2^{2n}, it follows that for $1 \leq k \leq n-1$

$$p_{2k,2n} = (1/2)\sum_{r=1}^{k} f_{2r} p_{2k-2r,2n-2r} + (1/2)\sum_{r=1}^{n-k} f_{2r} p_{2k,2n-2r}, \qquad (4.1.30)$$

since $r \leq k$ for the first class and $r \leq n - k$ for the second.

We are now in a position to prove result (4.1.29) by induction. When $n = 1$, that $p_{2,2} = 1/2 = u_2 u_0 = (1/2) \times 1$ follows trivially. So suppose we assume that $p_{2k,2m} = u_{2k} u_{2m-2k}$ for all $m < n$. Then (4.1.30) gives

$$p_{2k,2n} = (1/2) u_{2n-2k} \sum_{r=1}^{k} f_{2r} u_{2k-2r} + (1/2) u_{2k} \sum_{r=1}^{n-k} f_{2r} u_{2n-2r-2k}. \qquad (4.1.31)$$

Now we see from (4.1.14) that there are $M_{2n,0} = \binom{2n}{n}$ paths connecting the origin to the point $(2n, 0)$, and $M_{2n-2r,0} = \binom{2n-2r}{n-r}$ paths connecting the point $(2r, 0)$ to $(2n, 0)$.

So as the number of paths which return to the origin for the first time at time $2r$ is $2^{2r} f_{2r}$, on summing over r we have

$$\binom{2n}{n} = \sum_{r=1}^{n} 2^{2r} f_{2r} \times \binom{2n-2r}{n-r}$$

i.e. on recalling (4.1.20),

$$u_{2n} = 2^{-2n} \binom{2n}{n} = \sum_{r=1}^{n} f_{2r} 2^{-2(n-r)} \binom{2n-2r}{n-r} = \sum_{r=1}^{n} f_{2r} u_{2n-2r}. \qquad (4.1.32)$$

Whence the first summation in (4.1.31) equals u_{2k} and the second equals u_{2n-2k}, giving

$$p_{2k,2n} = (1/2) u_{2n-2k} u_{2k} + (1/2) u_{2k} u_{2n-2k} = u_{2k} u_{2n-2k},$$

as required. □

This theorem disproves the commonly held belief that the fraction of time spent on the positive (or negative) side of the x-axis is most likely to be close to $1/2$. Indeed, the exact opposite is true. For it follows directly from (4.1.29) that possible values close to $1/2$ are least probable, and the extreme values 0 and 1 are most probable. This result is easily proved by inserting expression (4.1.20) for u_{2r} into the ratio

$$\frac{p_{2k,2n}}{p_{2k-2,2n}} = \frac{u_{2k} u_{2n-2k}}{u_{2k-2} u_{2n-2k+2}} = \frac{(2 - \frac{1}{k})}{(2 - \frac{1}{n-k+1})} \qquad (4.1.33)$$

and noting that this exceeds 1 for $2k > n+1$. The simplest way of seeing how $p_{2k,2n}$ depends on k and n is to use Stirling's approximation, namely

$$r! \sim \sqrt{2\pi} e^{-r} r^{r+1/2}, \qquad (4.1.34)$$

to obtain

$$u_{2r} \sim 1/\sqrt{\pi r}, \qquad (4.1.35)$$

and hence

$$p_{2k,2n} = u_{2k} u_{2n-2k} \sim s_{2k,2n} = \frac{1}{\pi k^{1/2} (n-k)^{1/2}}. \qquad (4.1.36)$$

Since (4.1.34) works remarkably well for $r \geq 6$, this approximation may be usefully applied across the full range of support apart from the tail regions $k, n-k = 0, \ldots, 5$. To illustrate how the probability $p_{2k,2n}$ varies with k, suppose we take $n = 50$ and $k = 25, 30, \ldots, 50$. Then on noting that $p_{2k,2n}$ is trivial to evaluate using (for example) *Maple*, we have

$2k$	50	60	70	80	90	100
$p_{2k,100}$	0.01261	0.01286	0.01373	0.01567	0.02064	0.07959
$s_{2k,100}$	0.01273	0.01300	0.01389	0.01592	0.02122	*

So $p_{100,100}$ is over 6.3 times greater than $p_{50,100}$. Moreover, $s_{2k,2n}$ clearly provides a simple and acceptably accurate approximation for $p_{2k,2n}$.

A parallel result may be developed for the associated cumulative probability density function for the fraction k/n of the time spent on the positive side of the x-axis.

THEOREM 4.7 (*The First Arc Sine Law*). *For fixed α ($0 < \alpha < 1$) and $n \to \infty$ the probability that the fraction $r = k/n$ of the time spent on the positive side is less than α tends to*

$$\frac{1}{\pi} \int_0^\alpha \frac{dx}{\sqrt{\{x(1-x)\}}} = (2/\pi) \sin^{-1}(\sqrt{\alpha}). \tag{4.1.37}$$

Proof The required probability is given by $\Pr(2k \leq 2n\alpha)$, i.e. by

$$\sum_{k=0}^{n\alpha} p_{2k,2n} \simeq \sum_{k=0}^{n\alpha} s_{2k,2n} = \frac{1}{\pi} \sum_{k=0}^{n\alpha} \frac{1}{\sqrt{\{k(n-k)\}}},$$

which may itself be approximated by the Riemann sum

$$\frac{1}{\pi} \int_0^{n\alpha} \frac{dy}{\sqrt{\{y(n-y)\}}} = \frac{1}{\pi} \int_0^\alpha \frac{dz}{\sqrt{\{z(1-z)\}}}$$

on placing $y = nz$. Putting $z = x^2$ and letting $n \to \infty$ then yields

$$\Pr(r \leq \alpha) = \frac{2}{\pi} \int_0^{\sqrt{\alpha}} \frac{dx}{\sqrt{(1-x^2)}} = \left[(2/\pi)\sin^{-1}(x)\right]_0^{\sqrt{\alpha}} = (2/\pi)\sin^{-1}(\sqrt{\alpha}),$$

as required. □

This asymptotic form is both neat and simple to apply, and in practice provides an excellent approximation even for values of n as small as 20. The integrand in (4.1.37) follows a U-shaped curve which tends to infinity at the endpoints 0 and 1. It is this that produces the striking result that the fraction of time spent on the positive (negative) side of the x-axis is much more likely to be close to zero or one than to the 'expected' value $1/2$. The practical implication is best seen by considering the probability that one of the players is in the lead for more than a proportion α of the time, namely

$$T_\alpha = 2[1 - (2/\pi)\sin^{-1}(\sqrt{\alpha})].$$

For from this we have

α	:	0.5	0.7	0.9	0.99	0.999	0.9999
T_α	:	1	0.7380	0.4097	0.1275	0.0403	0.0127

So in more than 4% (1%) of cases the more fortunate player will be in the lead for more than 99.9% (99.99%) of the time. It is such counter-intuitive behaviour that fuels the downfall of many investors in the stock-market. For a stock operating in a level market is far more likely to record continued gain or loss over the buying price than is commonly appreciated.

4.1.6 Simulated illustration

This Arc Sine Law shows that it often takes a large number of steps before the particle returns to the origin. Now it seems intuitively reasonable that if Peter and Paul toss a fair coin for a long time $2n$, then the number of ties (i.e. moments when the cumulative scores are equal) should be roughly proportional to $2n$. Yet this is not so. In reality the number of ties increases with $\sqrt{2n}$. So as the game progresses, the frequency of ties decreases rapidly with a corresponding rise in the lengths of the associated 'waves' (i.e. path segments which are wholly positive or negative). Specifically, use of the Central Limit Theorem yields the following result (see Feller, 1968 for a proof).

THEOREM 4.8 *For fixed $\alpha > 0$ the probability that the process makes fewer than $\alpha\sqrt{2n}$ returns to the origin tends as $n \to \infty$ to*

$$g(\alpha) = \sqrt{2/\pi} \int_0^\alpha e^{-s^2/2}\, ds. \qquad (4.1.38)$$

So on writing (4.1.38) in the form

$$g(\alpha) = 2\Phi(\alpha) - 1, \qquad (4.1.39)$$

we see that the median value of α is given by $2\Phi(\alpha) - 1 = 1/2$, i.e. $\alpha = \Phi^{-1}(3/4) = 0.6745$. Similarly, the lower and upper quartiles for α are given by $\alpha = \Phi^{-1}(5/8) = 0.3186$ and $\alpha = \Phi^{-1}(7/8) = 1.1504$, respectively. Thus it follows from Theorem 4.8 that the lower (Q_1), median (M) and upper (Q_3) quartiles for the total number of returns take the form

$$\Pr(\text{no. returns to origin} \leq \sqrt{n}(0.4506, 0.9539, 1.6268)) = (0.25, 0.50, 0.75). \qquad (4.1.40)$$

Simulation of the simple random walk is extremely easy, since the algorithm consists of the single step:

$$\text{if } U \leq p \text{ then } X_n = X_{n-1} + 1 \text{ else } X_n = X_{n-1} - 1 \qquad (4.1.41)$$

for $\{U\}$ a sequence of independent $U(0,1)$ random variables. Figure 4.2 shows X_n over 10^4, 10^6 and 10^8 steps for the symmetric case (i.e. $p = q = 1/2$), and it clearly illustrates the theoretical results developed above. For the realization remains fairly close to the x-axis for the first 7000 steps, making 66 returns to the origin, before moving steadily upwards. Result (4.1.40) shows that with $n = 10^4$ this value lies well within the benchmark values $Q_1 = 45$, $M = 95$ and $Q_3 = 163$. In direct contrast, we see that the process then remains wholly above the x-axis for the next 4×10^6 steps. So over the first 10^6 steps (Figure 4.2b) the number of returns is still 66 which lies well outside the inter-quartile range $(Q_1, Q_3) = (451, 1627)$. On placing $\alpha\sqrt{(2 \times 10^6)} = 66$ we obtain $\alpha = 0.0467$, whence (4.1.39) gives

$$\Pr(\text{no. returns to origin} \leq 66) = 2\Phi(0.0467) - 1 = 0.037.$$

Extending the run to $n = 10^8$ steps highlights the 'increasing wavelength phenomena': now the process makes 7116 returns to the origin which once again lies well within our guidelines of $Q_1 = 4506$, $M = 9539$ and $Q_3 = 16269$. So selection of a small part

The simple unrestricted random walk 213

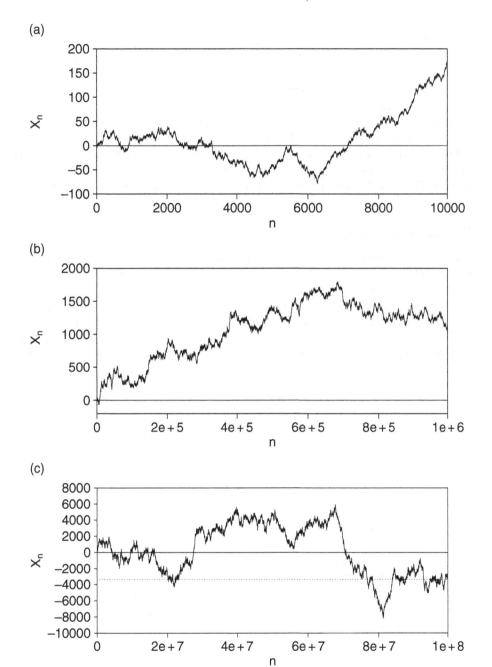

Figure 4.2 Simulated symmetric simple random walk showing: (a) 10^4 steps, (b) every 100th of 10^6 steps, (c) every 10000th of 10^8 steps.

of a run can produce behaviour that is quite different from that exhibited over the full length of the series. Indeed, how many financial analysts would claim that Figure 4.2b portrayed a bull market followed by a period of gentle decline? Yet the reality is one of pure randomness.

Care must be taken to interpret this wavelength result properly, since there is a danger here of falling into a trap. For since the steps of a simple random walk are independent of each other, the walk is just as valid when travelled backwards as forwards. So how can the expected wavelength increase as $O(\sqrt{n})$ in both directions? The fallacy of this apparent contradiction is simply due to the need to change the origin from 0 to X_n when moving backwards, and is easily seen by comparing the backwards process in Figure 4.2c with the new x-axis (dotted line) based on $X_{10^8} = -3370$.

4.2 Absorbing barriers

Whilst the above path-based approach is both powerful and mathematically appealing, the probability results developed are based on the presumption that all paths are equally likely to occur, i.e. the process is symmetric ($p = q = 1/2$). If they are not, then we have to select a different strategy. In the completely unrestricted case the problem is easily tackled since the step sizes Z_i are i.i.d. random variables. So if $E(Z_i) = \mu$ and $Var(Z_i) = \sigma^2$, since $X_n = Z_1 + \cdots + Z_n$ it follows that $E(X_n) = n\mu$ and $Var(X_n) = n\sigma^2$. Moreover, for general step size probabilities $\Pr(Z_i = j) = h_j$ ($-\infty < j < \infty$) with p.g.f.

$$H(z) \equiv \sum_{j=-\infty}^{\infty} h_j z^j \quad (|z| < 1), \tag{4.2.1}$$

the p.g.f. of X_n is given by

$$G_n(z) \equiv [H(z)]^n. \tag{4.2.2}$$

Thus the full (i.e. space-time) p.g.f. takes the simple form

$$G(z,s) \equiv \sum_{n=0}^{\infty} G_n(z) s^n = \sum_{n=0}^{\infty} [H(z) s]^n = [1 - H(z) s]^{-1} \quad (|sH(z)| < 1). \tag{4.2.3}$$

For example, suppose that for $i = 1, 2, \ldots$ we have $\Pr(Z_i = -1) = q$, $\Pr(Z_i = 0) = 1 - p - q$ and $\Pr(Z_i = 1) = p$. Then the possible positions of the process at time n are $k = 0, \pm 1, \ldots, \pm n$, and in order to reach the point k at time n the process has to make r_1 positive jumps, r_2 zero jumps and r_3 negative jumps where r_1, r_2 and r_3 are non-negative integers satisfying

$$r_1 - r_3 = k, \quad r_2 = n - r_1 - r_3. \tag{4.2.4}$$

Hence the probability that X_n equals k is given by the sum of the trinomial probabilities

$$\Pr(X_n = k) = \sum \frac{n!}{r_1! r_2! r_3!} p^{r_1} (1 - p - q)^{r_2} q^{r_3} \tag{4.2.5}$$

taken over values of r_1, r_2 and r_3 satisfying (4.2.4). Equivalently, since here $H(z) = pz + (1-p-q) + qz^{-1}$, the generating function (4.2.3), i.e.

$$G(z,s) = \frac{z}{-spz^2 + z\{1-s(1-p-q)\} - sq}, \quad (4.2.6)$$

contains all the information in the process in the sense that $\Pr(X_n = k)$ is the coefficient of $z^k s^n$ in the expansion of $G(z,s)$.

In practice the step-size probabilities will often depend, at least to some extent, on the position k. For example, suppose we are dealing with the simple birth–death process, for which $\Pr(Z_i = 1) = \lambda k/(\lambda k + \mu k) = \lambda/(\lambda+\mu)$ and $\Pr(Z_i = -1) = \mu/(\lambda+\mu)$; here $\Pr(Z_i = 0) = 0$ for $X_n = k > 0$. Now at $k=0$ these probabilities change to $\Pr(Z_i = 0) = 1$ and $\Pr(Z_i = -1) = \Pr(Z_i = 1) = 0$, since once the population size reaches zero the population becomes extinct. So the trinomial result (4.2.5) no longer holds. Moreover, it also does not hold when determining the number of steps taken for the population to reach size $a > 0$. For we evaluate the probability of absorption at a at time n by placing an absorbing barrier at a, thereby invoking the boundary condition $\Pr(X_{n+1} = a | X_n = a) = 1$.

In general, suppose we place absorbing barriers at the points a and b, and assume that $b < X_0 < a$. So motion ceases when the process reaches position a or b, and we say that absorption occurs at a and b, respectively. As $n \to \infty$ only three ultimate outcomes are possible, namely absorption at a, absorption at b, or the process wanders indefinitely between the two barriers. Though as this last outcome has probability zero, eventual absorption at one of the boundaries is certain.

To prove this last result we note that the probability that the process is still in motion at time n, i.e. that X_n occupies one of the non-absorbing states $b+1, \ldots, a-1$, cannot exceed the probability that the unrestricted process occupies one of these states at time n. For in computing the former we must exclude all possible paths which contain states lying outside $[b+1, a-1]$. Now it follows from the Strong Law of Large Numbers, i.e. result (4.1.12), that this latter probability tends to zero as $n \to \infty$. Thus the probability that absorption has not occurred by time n also tends to zero as $n \to \infty$. Moreover, it may be shown that this convergence occurs geometrically fast, i.e.

$$\Pr(b < X_n < a) = \mathrm{O}(\rho^n) \quad (4.2.7)$$

for some $0 < \rho < 1$.

4.2.1 Probability of absorption at time n

Having deduced that the probability of eventual absorption is one, we can now determine how this probability is split between the two absorbing states b and a, and also evaluate the probability distribution of the time to absorption itself. Denote

$$f_{ja}^{(n)} = \Pr(\text{process reaches } a \text{ exactly at time } n | X_0 = j). \quad (4.2.8)$$

It is useful to note that $f_{ja}^{(n)}$ also represents the probability that the unrestricted process reaches state a for the first time at time n without state b being occupied at any of the preceding times $1, 2, \ldots, n-1$, conditional on the process starting at j where $b < j < a$. That is,

$$f_{ja}^{(n)} = \Pr(b < X_1 < a, b < X_2 < a, \ldots, b < X_{n-1} < a, X_n = a | X_0 = j) \quad (n = 1, 2, \ldots). \tag{4.2.9}$$

Whilst for $n = 0$ we have the initial conditions

$$f_{aa}^{(0)} = 1 \quad \text{with} \quad f_{ja}^{(0)} = 0 \quad (j \neq a), \tag{4.2.10}$$

since if the process starts at a then absorption occurs at time 0 with probability 1; if it starts at any other point then absorption clearly cannot occur at time 0. Note that this parameterization is more general than is absolutely necessary, since without any loss of generality we could move the whole walk left or right to ensure that either the start point $j = 0$ or the lower boundary $b = 0$, or even that the upper boundary $a = 0$.

Let A_n denote the event 'absorption at a at (exactly) time n'. Then $f_{ja}^{(n)} = \Pr(A_n | X_0 = j)$. In order to develop an equation for $f_{ja}^{(n)}$ let us now parallel the construction of the Kolmogorov backward equations (4.0.3) for $p_{ij}^{(n)}$ by considering the first step of the process. If the first step is $+1$ then the process moves to $j + 1$ and, as all steps are independent, for A_n to occur we must now have A_{n-1} occurring conditional on starting at $j + 1$. Similarly, if the first step is 0 or -1 then we require A_{n-1} conditional on starting at j or $j - 1$, respectively. So as $+1$, 0 and -1 are mutually exclusive and exhaustive possibilities for the first step, we have

$$f_{ja}^{(n)} = p \Pr(A_{n-1} | \text{ start at } j+1) + (1-p-q) \Pr(A_{n-1} | \text{ start at } j)$$
$$+ q \Pr(A_{n-1} | \text{ start at } j-1)$$
$$= p f_{j+1,a}^{(n-1)} + (1-p-q) f_{ja}^{(n-1)} + q f_{j-1,a}^{(n-1)} \tag{4.2.11}$$

over $j = b+1, \ldots, a-1$ and $n = 1, 2, \ldots$. Whilst in addition to the initial conditions (4.2.10), we also have the boundary conditions

$$f_{aa}^{(n)} = f_{ba}^{(n)} = 0 \quad (n = 1, 2, \ldots). \tag{4.2.12}$$

Denote the generating function over the time variable n by

$$F_{ja}(s) \equiv \sum_{n=0}^{\infty} f_{ja}^{(n)} s^n \equiv F_j(s); \tag{4.2.13}$$

for convenience we omit the subscript a. Then multiplying equation (4.2.11) by s^n and summing over $n = 1, 2 \ldots$ yields

$$\sum_{n=1}^{\infty} f_{ja}^{(n)} s^n = p \sum_{n=1}^{\infty} f_{j+1,a}^{(n-1)} s^n + (1-p-q) \sum_{n=1}^{\infty} f_{ja}^{(n-1)} s^n + q \sum_{n=0}^{\infty} f_{j-1,a}^{(n-1)} s^n$$

i.e. $\quad F_j(s) - f_{ja}^{(0)} = psF_{j+1}(s) + (1-p-q)sF_j(s) + qsF_{j-1}(s).$

Hence as $f_{ja}^{(0)} = 0$ for $j \neq a$, we obtain the second-order linear difference equation

$$psF_{j+1}(s) + \{(1-p-q)s - 1\}F_j(s) + qsF_{j-1}(s) = 0 \qquad (4.2.14)$$

with boundary conditions given by (4.2.10) and (4.2.12), namely

$$F_a(s) = \sum_{n=0}^{\infty} f_{aa}^{(n)} s^n = f_{aa}^{(0)} = 1 \quad \text{and} \quad F_b(s) = \sum_{n=0}^{\infty} f_{ba}^{(n)} s^n = 0. \qquad (4.2.15)$$

Using the standard method of solution, we first substitute the trial function $F_j(s) = \lambda^j$ into (4.2.14) to obtain the quadratic equation

$$ps\lambda^2 - \lambda[1 - (1-p-q)s] + qs = 0, \qquad (4.2.16)$$

which yields the two roots

$$\lambda_1(s), \lambda_2(s) = (1/2ps)[1 - (1-p-q)s \pm \sqrt{\{[1-(1-p-q)s]^2 - 4pqs^2\}}]. \qquad (4.2.17)$$

Let us take s to be real and positive, and $[1-(1-p-q)s]^2 > 4pqs^2$. On taking the square root of both sides, this requires

$$0 < s < \frac{1}{(1-p-q) + 2\sqrt{(pq)}} = \frac{1}{1 - \sqrt{p} - \sqrt{q}}. \qquad (4.2.18)$$

Whence the general solution of equation (4.2.14) is given by

$$F_j(s) = A(s)\{\lambda_1(s)\}^j + B(s)\{\lambda_2(s)\}^j, \qquad (4.2.19)$$

where $A(s)$ and $B(s)$ are independent of j and are determined from the boundary conditions (4.2.15). Specifically, we have

$$F_a(s) = 1 = A\{\lambda_1(s)\}^a + B\{\lambda_2(s)\}^a$$

$$F_b(s) = 0 = A\{\lambda_1(s)\}^b + B\{\lambda_2(s)\}^b,$$

whence solving for $A(s)$ and $B(s)$ and substituting into (4.2.19) yields

$$F_{ja}(s) = F_j(s) = \frac{\{\lambda_1(s)\}^{j-b} - \{\lambda_2(s)\}^{j-b}}{\{\lambda_1(s)\}^{a-b} - \{\lambda_2(s)\}^{a-b}}. \qquad (4.2.20)$$

An identical argument based on the boundary conditions $F_{ja}(s) = 0$ and $F_{jb}(s) = 1$ yields the equivalent expression for absorption at b, namely

$$F_{jb}(s) = \frac{\{\lambda_1(s)\}^{j-a} - \{\lambda_2(s)\}^{j-a}}{\{\lambda_1(s)\}^{b-a} - \{\lambda_2(s)\}^{b-a}}, \qquad (4.2.21)$$

which simply corresponds to interchanging a and b in result (4.2.20).

Although in theory the probabilities $f_{ja}^{(n)}$ can now be obtained by expanding (4.2.20) as a power series in s, in reality the calculation is non-trivial. Let us therefore first consider the probabilities of *eventual* absorption at a and b. Since eventual absorption at a means that absorption must have taken place at some time $n = 1, 2, \ldots$, we see from (4.2.13) that

$$r_{ja} \equiv \Pr(\text{eventual absorption at } a | X_0 = j) = \sum_{n=1}^{\infty} f_{ja}^{(n)} = F_{ja}(1).$$

Now as we have taken s to be real and positive, it follows from (4.2.17) that $\lambda_1(s) \geq \lambda_2(s)$. So on placing $s = 1$ and writing $\lambda_1(1) = \lambda_1$ and $\lambda_2(1) = \lambda_2$, we have $\lambda_1 = [(p+q) + |p-q|]/2p$ and $\lambda_2 = [(p+q) - |p-q|]/2p$, i.e.

$$\begin{array}{ll} \lambda_1 = q/p > \lambda_2 = 1 & (p < q) \\ \lambda_1 = 1 > \lambda_2 = q/p & (p > q) \\ \lambda_1 = 1 = \lambda_2 & (p = q). \end{array} \qquad (4.2.22)$$

Whence substituting for λ_1 and λ_2 from (4.2.22) into (4.2.20) gives

$$r_{ja} = F_{ja}(1) = \begin{cases} [(p/q)^{a-j} - (p/q)^{a-b}]/[1 - (p/q)^{a-b}] & (p \neq q) \\ (j-b)/(a-b) & (p = q). \end{cases} \qquad (4.2.23)$$

The result for $p = q$ is obtained by writing $q/p = (1 + \epsilon)$ and then letting $\epsilon \to 0$. Note that $f_{ja}^{(n)}/r_{ja}$ is the probability of being absorbed at a at time n conditional on eventual absorption occurring at a.

To determine $r_{jb} = \Pr(\text{eventual absorption at } b | X_0 = j)$ we simply interchange a and b in expression (4.2.23) to obtain

$$r_{jb} = F_{jb}(1) = \begin{cases} [(p/q)^{b-j} - (p/q)^{b-a}]/[1 - (p/q)^{b-a}] & (p \neq q) \\ (j-a)/(b-a) & (p = q). \end{cases} \qquad (4.2.24)$$

Hence it follows that $\Pr(\text{eventual absorption} | X_0 = j) = r_{ja} + r_{jb} = 1$, as deduced earlier via the Strong Law of Large Numbers.

Returning to the more difficult problem of extracting the probabilities $f_{ja}^{(n)}$ from the p.g.f. $F_{ja}(s)$, we first show that expression (4.2.20) is a rational function of s. For algebraic simplicity let us recall that without any loss of generality we may place $j = 0$ provided $b < 0$ and $a > 0$. Denote

$$\theta = 1 - (1 - p - q)s \quad \text{and} \quad \phi = \sqrt{\{[1 - (1 - p - q)s]^2 - 4pqs^2\}}, \qquad (4.2.25)$$

whence (4.2.17) becomes $\lambda_1(s), \lambda_2(s) = (1/2ps)(\theta \pm \phi)$. Then (4.2.20) gives

$$F_{ja}(s) = \frac{(1/2ps)^{-b}[(\theta + \phi)^{-b} - (\theta - \phi)^{-b}]}{(1/2ps)^{a-b}[(\theta + \phi)^{a-b} - (\theta - \phi)^{a-b}]}$$

$$= (2ps)^a \frac{[\binom{-b}{1}\theta^{-b-1} + \binom{-b}{3}\theta^{-b-3}\phi^2 + \cdots]}{[\binom{a-b}{1}\theta^{a-b-1} + \binom{a-b}{3}\theta^{a-b-3}\phi^2 + \cdots]} \qquad (4.2.26)$$

(on cancelling ϕ top and bottom), which is s^a times a polynomial in s of degree $-b - 1$ divided by a polynomial of degree $a - b - 1$. We can therefore express $F_{0a}(s)$ in terms of the partial fraction expansion

$$F_{0a}(s) = (2ps)^a \sum_{v=1}^{a-b-1} \frac{\alpha_v}{1 - s/s_v}, \qquad (4.2.27)$$

where s_v $(v = 1, \ldots, a - b - 1)$ denote the roots of the denominator and a_v are appropriate constants.

Although the development of this approach requires some careful thought, it is useful not only in this particular case, which involves a simple random walk between two absorbing barriers, but also as a pointer towards a general method of solution for more demanding problems. To determine s_v and α_v we note from the denominator in (4.2.26) that the roots s_v must satisfy

$$(\theta + \phi)^{a-b} = (\theta - \phi)^{a-b} e^{2\pi i v},$$

since $e^{2\pi i v} = 1$ for integer v and $i = \sqrt{-1}$. Thus

$$\theta(e^{\pi i v/(a-b)} - e^{-\pi i v/(a-b)}) = \phi(e^{\pi i v/(a-b)} + e^{-\pi i v/(a-b)}).$$

That is

$$2i\theta \sin[\pi v/(a - b)] = 2\phi \cos[\pi v/(a - b)],$$

which, on substituting for θ and ϕ from (4.2.25), reduces to

$$s_v = 1/\{1 - p - q + 2\sqrt{pq} \cos[v\pi/(a - b)]\} \qquad (v = 1, 2, \ldots, a - b - 1) \qquad (4.2.28)$$

(the root $s = 0$ is also a root of the numerator in (4.2.26)). To determine $\lambda_1(s)$ and $\lambda_2(s)$ we simply parallel the above argument by noting that placing the denominator of (4.2.20) equal to zero leads to $\lambda_1(s) = k e^{\pi i v/(a-b)}$ and $\lambda_2(s) = k e^{-\pi i v/(a-b)}$ for some constant k, since $e^{\pi i v} \equiv e^{-\pi i v}$. Now we see from equation (4.2.16) that the root product $\lambda_1(s)\lambda_2(s) = (qs)/(ps) = q/p$, and so $k^2 = q/p$, i.e. $k = \sqrt{(q/p)}$, giving

$$\lambda_1(s_v), \lambda_2(s_v) = \sqrt{(q/p)} e^{\pm \pi i v/(a-b)}. \qquad (4.2.29)$$

Finally, to determine the coefficients α_v we rewrite the representation (4.2.27) in the form

$$\sum_{v=1}^{a-b-1} \frac{\alpha_v}{1 - s/s_v} = (2ps)^{-a} F_{0a}(s)$$

i.e.

$$\alpha_v + \sum_{u=1, u \neq v}^{a-b-1} \frac{\alpha_u(1 - s/s_v)}{1 - s/s_u} = (1 - s/s_v)(2ps)^{-a} F_{0a}(s).$$

Whence on taking the limit as $s \to s_v$, with $F_{0a}(s)$ given by (4.2.20) with $j = 0$, we obtain

$$\alpha_v = \lim_{s \to s_v} (1 - s/s_v)(2ps)^{-a} \frac{\lambda_1^{-b}(s) - \lambda_2^{-b}(s)}{\lambda_1^{a-b}(s) - \lambda_2^{a-b}(s)}$$

$$= (2ps_v)^{-a} (\lambda_1^{-b}(s_v) - \lambda_2^{-b}(s_v)) \lim_{s \to s_v} \frac{1 - s/s_v}{\lambda_1^{a-b}(s) - \lambda_2^{a-b}(s)}.$$

Using l'Hôpital's rule then yields

$$\alpha_v = (2ps_v)^{-a}(\lambda_1^{-b}(s_v) - \lambda_2^{-b}(s_v)) \frac{-(1/s_v)}{\frac{d}{ds}[\lambda_1^{a-b}(s) - \lambda_2^{a-b}(s)]_{s=s_v}}. \quad (4.2.30)$$

To evaluate the numerator we substitute for $\lambda_i(s_v)$ from (4.2.29), obtaining

$$\lambda_1^{-b}(s_v) - \lambda_2^{-b}(s_v) = (p/q)^{b/2}[2i\sin(\pi v b/(a-b))]. \quad (4.2.31)$$

Whilst to determine the denominator we first note that differentiation and re-use of equation (4.2.16) yields

$$\frac{d\lambda}{ds} = \frac{-\lambda}{s^2(p\lambda - q/\lambda)}. \quad (4.2.32)$$

So on using (4.2.29) we have that, at $s = s_v$,

$$\frac{d\lambda_1(s)}{ds}, \frac{d\lambda_2(s)}{ds} = \frac{\mp e^{\pm \pi i v/(a-b)}}{2ips^2 \sin(\pi v/(a-b))}. \quad (4.2.33)$$

Thus we can now write the denominator in (4.2.30) as

$$(a-b)[\lambda_1^{a-b-1}(s)\frac{d\lambda_1(s)}{ds} - \lambda_2^{a-b-1}(s)\frac{d\lambda_2(s)}{ds}]$$

evaluated at $s = s_v$, which, on substituting from (4.2.33), reduces to

$$-\frac{(a-b)(q/p)^{(a-b-1)/2}\cos(\pi v)}{ips_v^2 \sin(\pi v/(a-b))}. \quad (4.2.34)$$

Expression (4.2.30) then simplifies to give

$$\alpha_v = \frac{(-1)^{v+1}\sin(\pi b v/(a-b))\sin(\pi v/(a-b))}{(a-b)(4pq)^{(a-1)/2}s_v^{a-1}}. \quad (4.2.35)$$

Finally, we return to (4.2.27) in order to extract the coefficient of s^n in $F_{0a}(s)$, namely

$$f_{0a}^{(n)} = (2p)^a \sum_{v=1}^{a-b-1} \frac{\alpha_v}{s_v^{n-a}} \quad (n = a, a+1, \ldots)$$

$$= \frac{2\sqrt{(pq)}}{a-b}\left(\frac{p}{q}\right)^{a/2} \sum_{v=1}^{a-b-1} \frac{(-1)^{v+1}}{s_v^{n-1}} \sin\left(\frac{\pi v b}{a-b}\right) \sin\left(\frac{\pi v}{a-b}\right). \quad (4.2.36)$$

The corresponding result for $f_{0b}(n)$ follows immediately on considering the mirror-reflection of this random walk, i.e. we interchange p and q, and a and b, thereby obtaining for $n = -b, -b+1, \ldots$

$$f_{0b}^{(n)} = \frac{2\sqrt{(pq)}}{a-b}\left(\frac{q}{p}\right)^{b/2} \sum_{v=1}^{a-b-1} \frac{(-1)^{v+1}}{s_v^{n-1}} \sin\left(\frac{\pi v a}{a-b}\right) \sin\left(\frac{\pi v}{a-b}\right). \quad (4.2.37)$$

Moreover, the probability that absorption at one of the barriers takes place exactly at time n is given by

$$\Pr(N = n) = f_{0a}^{(n)} + f_{0b}^{(n)}, \qquad (4.2.38)$$

the corresponding p.g.f. being

$$E(s^N) = F_{0a}(s) + F_{0b}(s). \qquad (4.2.39)$$

Since results (4.2.36) and (4.2.37) relate to the probability of absorption at a and b, respectively, exactly at time n, to complete the description of the process we need to determine the probability $p_k^{(n)}$ that the process is in a non-absorbing state $b < k < a$ at time n. Suppose that the walk starts at $j = 0$, so that $p_0^{(0)} = 1$ and $p_k^{(0)} = 0$ for $k \neq 0$. Then on considering the last step of the process (the backward equation (4.2.11) for $f_{ja}^{(n)}$ is based on the first step), we have the forward equation

$$p_k^{(n)} = pp_{k-1}^{(n-1)} + (1 - p - q)p_k^{(n-1)} + qp_{k+1}^{(n-1)} \qquad (4.2.40)$$

over $j = -b+1, \ldots, a-1$ and $n = 1, 2, \ldots$. Define the generating function

$$P_k(s) \equiv \sum_{n=0}^{\infty} p_k^{(n)} s^n. \qquad (4.2.41)$$

Then on multiplying equation (4.2.40) by s^n and summing over $n = 1, 2, \ldots$, we obtain

$$P_k(s) - 1 = s\{pP_{k-1}(s) + (1 - p - q)P_k(s) + qP_{k+1}(s)\}, \qquad (4.2.42)$$

which is an inhomogeneous version of equation (4.2.14) for $F_j(s)$ with p and q interchanged. Although the boundary conditions

$$P_a(s) = P_b(s) = 0 \qquad (4.2.43)$$

are slightly simpler than those for $F_j(s)$, the presence of 1 in equation (4.2.42) does render the exact solution more algebraically awkward than expressions (4.2.36) and (4.2.37).

It is worth noting that even if $p + q = 1$, so that a step is now defined by a move away from the current position, then little effective algebraic simplification occurs. For although $\lambda_1(s)$ and $\lambda_2(s)$ in (4.2.17), and s_v in (4.2.28), do simplify to $\lambda_1(s), \lambda_2(s) = (1/2ps)[1 \pm \sqrt{\{1 - 4pqs^2\}}]$ and $s_v = 1/\{2\sqrt{pq}\cos[v\pi/(a-b)]\}$, respectively, since α_v (given by (4.2.35)) retains its structural form the solution (4.2.36) for $f_{0a}^{(n)}$ retains its algebraic complexity.

4.2.2 One absorbing barrier

The function $f_{0a}^{(n)}$ denotes the probability that, starting from $X_0 = 0$, the process reaches state a for the first time at time n, which clearly implies that the process never reaches state b (otherwise it would remain trapped there and so could never escape to reach state a). Let us now remove state b completely, so that the process is free to move among all states $x < a$ until it reaches state a, whereupon it remains there forever. The time to absorption (i.e. the duration of the walk), is also the first passage time from state 0 to the state a in the unrestricted random walk. Thus the probability statement (4.2.9) for $X_0 = j$ becomes

$$f_{ja}^{(n)} = \Pr(X_m < a \text{ for } 1 \leq m < n, X_n = a | X_0 = j) \quad (n = 1, 2, \ldots); \quad (4.2.44)$$

with the generating function (4.2.20) yielding

$$F_{ja}(s) = \lim_{b \to -\infty} \frac{\{\lambda_1(s)\}^{j-b} - \{\lambda_2(s)\}^{j-b}}{\{\lambda_1(s)\}^{a-b} - \{\lambda_2(s)\}^{a-b}} = \{\lambda_1(s)\}^{j-a}, \quad (4.2.45)$$

since $\lambda_1(s) > \lambda_2(s)$. Note that as $\lambda_1(s)\lambda_2(s) = q/p$, we also have

$$F_{ja}(s) = [(p/q)\lambda_2(s)]^{a-j}. \quad (4.2.46)$$

In both cases we can take $j = 0$ without any loss of generality.

If required, $f_{0a}^{(n)}$ can now be determined by a straightforward evaluation of the coefficient of s^n in the expansion of $F_{0a}(s)$. For on substituting for $\lambda_2(s)$ from (4.2.17) into (4.2.46) we see that

$$F_{0a}(s) = (1/2qs)^a [1 - (1-p-q)s - \sqrt{\{1 - s(1 - (\sqrt{p} - \sqrt{q})^2)\}}$$
$$\times \sqrt{\{1 - s(1 - (\sqrt{p} + \sqrt{q})^2)\}}]^a. \quad (4.2.47)$$

Separating out terms which either do, or do not, involve $\sqrt{\{\cdot\}}$ then enables direct expansion of $F_{0a}(s)$ as a power series in s^n. Though the resulting expression is less than pleasant, this method of solution is probably easier than letting $b \to \infty$ in solution (4.2.36).

However, there is an alternative approach which avoids the need to obtain the two-barrier solution. Suppose that $p > q$ so that there is a drift upwards. Then on starting from state $i = 0$, X_n must at some intermediate times occupy each of the intervening states $1, 2, \ldots, a-1$, since at each jump the process can move at most one unit upwards. Now we have already seen from the Strong Law of Large Numbers (result (4.1.12)) that for the unrestricted walk X_n drifts off to infinity with probability one. So the probability of ultimately reaching state 1 from state 0 must also be one. Denote this first passage time by N_1 with p.g.f. $F_1(s)$. Since each jump is independent of the time and state from which it is made, it follows that the first passage time from any given state k to the state $k+1$ also has the p.g.f. $F_1(s)$. Thus as the first passage time from 0 to a is the sum of a independent and identically distributed random variables, each with the distribution of N_1, we have

$$F_a(s) = [F_1(s)]^a. \quad (4.2.48)$$

Now on considering the first jump from state 0, the p.g.f.

$$E(s^{N_1}) = F_1(s) = pE(s^{N_1}|X_1 = 1) + (1-p-q)E(s^{N_1}|X_1 = 0) + qE(s^{N_1}|X_1 = -1)$$
$$= ps + (1-p-q)sF_1(s) + qs[F_1(s)]^2, \quad (4.2.49)$$

and solving this equation yields the two roots

$$F_1(s) = (p/q)[\lambda_1(s), \lambda_2(s)]. \tag{4.2.50}$$

Placing $s = 1$ in (4.2.17) gives $\lambda_1(1), \lambda_2(1) = (1/2p)[(p+q) \pm (p-q)]$, and since ultimate absorption at state 1 is certain we require the root satisfying $F(1) = 1$, i.e. $\lambda_2(s)$, whence

$$F_1(s) = (p/q)\lambda_2(s). \tag{4.2.51}$$

Result (4.2.46), namely

$$F_a(s) = [(p/q)\lambda_2(s)]^a, \tag{4.2.52}$$

then follows from (4.2.48).

To adapt this argument to the case $p \leq q$ we have to interpret $F_1(s)$ as the generating function conditional on the process eventually reaching state 1. Whence on denoting this latter event by A, we have

$$\Pr(A) = \sum_{n=1}^{\infty} f_{01}^{(n)} \quad \text{with} \quad F_1(s) = \sum_{n=1}^{\infty} f_{01}^{(n)} s^n = \Pr(A)\mathrm{E}(s^{N_1}|A). \tag{4.2.53}$$

Considering the first jump from state 0 then leads to the same equation (4.2.49), but now we can no longer use the condition $F_1(1) = 1$ to determine the appropriate root (4.2.50). Instead, we note that as $s \to 0$, $\lambda_1(s) \sim (1/ps) \to \infty$, and so is disqualified. Thus we again take the root (4.2.51), whence result (4.2.52) holds in all cases.

4.2.3 Number of steps to absorption

On denoting N_a to be the time to absorption at a, i.e. the first passage time from state 0 to state a, we may write

$$F_a(s) = \sum_{n=1}^{\infty} \Pr(N_a = n) s^n.$$

So for $s = 1$ we have $F_a(1) = \Pr(N_a < \infty)$, which is the probability that absorption at a ever occurs. Whence we see from (4.2.52) that

$$\Pr(N_a < \infty) = [(p/q)\lambda_2(1)]^a = \begin{cases} (p/q)^a & (p < q) \\ 1 & (p \geq q). \end{cases} \tag{4.2.54}$$

Thus when the drift is towards the barrier, or there is no drift, the duration of the walk, N_a, is finite with probability one. Whilst if the drift is away from the barrier, then N_a is finite with probability $(p/q)^a$; the remaining probability $1 - (p/q)^a$ relates to the walk continuing indefinitely, forever remaining in states below a.

The way in which the geometric probability (4.2.54) for $p < q$ decays with increasing a can be appreciated from the following example:

	p/q	0.99	0.9	0.5
	10	0.9044	0.3487	0.0010
a	100	0.3660	2.66×10^{-5}	7.89×10^{-31}
	1000	4.32×10^{-5}	1.75×10^{-46}	9.33×10^{-300}

(4.2.55)

So for p only just less than q, the walk can progress a considerable distance upwards before the drift to $-\infty$ takes hold. However, if we decrease p/q just slightly to 0.9, the walk is unlikely to reach much beyond state 10. Whilst for $q = 2p$, only 1 in 1000 such walks will even reach this state.

If $p \geq q$ then absorption at state a is certain, and since we have already shown in (4.2.52) that $F_a(s) = [F_1(s)]^a = [(p/q)\lambda_2(s)]^a$ is the p.g.f. of the number of steps, N_a, to absorption we have

$$E(N_a) = F'_a(1) = a[F_1(1)]^{a-1}F'_1(1) = aF'_1(1) = (p/q)\lambda'_2(1)$$

since $F_1(1) = 1$. Whence differentiating $\lambda_2(s)$ in (4.2.17) gives

$$E(N_a) = \begin{cases} a/(p-q) & (p > q) \\ \infty & (p = q). \end{cases} \quad (4.2.56)$$

Thus although absorption is certain when $p = q$, the distribution of the time to absorption has infinite mean (indeed, infinite moments of all orders). When $p > q$ further differentiation of $F_a(s)$ and placing $s = 1$ yields higher-order factorial moments; in particular, we obtain the variance

$$\text{Var}(N_a) = a\text{Var}(N_1) = a[F''_1(1) + F'_1(1) - \{F'_1(1)\}^2] = \frac{a[p + q - (p-q)^2]}{(p-q)^3}. \quad (4.2.57)$$

Moreover, since N_a is the sum of a independent random variables with finite variance, it follows from the Central Limit Theorem that N_a is approximately Normally distributed for appropriately large a. In terms of the mean $\mu = p - q$ and variance $\sigma^2 = p + q - (p-q)^2$ of a single jump, we may write (4.2.56) and (4.2.57) as

$$E(N_a) = a/\mu \quad \text{and} \quad \text{Var}(N_a) = a\sigma^2/\mu^3. \quad (4.2.58)$$

For example, when $p = 0.5$, $q = 0.4$ and $a = 10$ both the mean and standard deviation equal 100, indicative of a distribution with a wide spread. Increasing a to 100 changes the mean to 1000 and standard deviation to 316, whence 95% confidence limits for N_{100} are $(380, 1620)$. Reducing q to 0.2 results in a much stronger positive drift and a reduction in standard deviation; the corresponding values for the mean and standard deviation are now substantially smaller, being 33 and 19 (for $a = 10$) and 333 and 61 (for $a = 100$), respectively.

Note the singular behaviour of the process when $p = q$. For starting from state 0, X_n reaches any other given state with probability one, but the mean time to do so is infinite. Moreover, having reached this given state it will return to state 0 with probability one, again with infinite mean passage time. Thus an unrestricted walk (i.e. one with no absorbing barriers) is certain to make indefinitely large excursions from its starting point to which it is ultimately bound to return.

It is instructive to recall a general aspect of stochastic processes, that if moments alone are required then it is often possible to avoid having to construct the probability structure in order to evaluate them. Suppose we work from the outset with the expected duration of the walk (presumed finite). Denote this by d_k for given start point k. If the first step is to $k+1$ then the conditional duration of the walk from then on is d_{k+1}, so the expected (conditional) duration of the whole game is $d_{k+1} + 1$; similarly we have $d_k + 1$ or $d_{k-1} + 1$ if the first step is zero or negative. So we can write

$$d_k = p(1 + d_{k+1}) + (1 - p - q)(1 + d_k) + q(1 + d_{k-1}) \qquad (b < k < a). \qquad (4.2.59)$$

The associated (two-barrier) boundary conditions are $d_b = d_a = 0$, since if the start point is located at a boundary then absorption immediately takes place. To solve the homogeneous part of equation (4.2.59), namely

$$pd_{k+1} - (p+q)d_k + qd_{k-1} = 0,$$

we note that the auxiliary equation

$$p\lambda^2 - (p+q)\lambda + q = (p\lambda - q)(\lambda - 1) = 0$$

has roots $\lambda = 1, q/p$. So

$$d_k = A + B(q/p)^k$$

for some constants A and B. Whilst to find a particular solution try $d_k = \lambda k$. Since this satisfies (4.2.59) when $\lambda = 1/(q-p)$, we therefore have the general solution

$$d_k = k/(q-p) + A + B(q/p)^k.$$

Using the initial conditions $d_b = d_a = 0$ to determine A and B then yields

$$d_k = \left(\frac{k-a}{q-p}\right) + \left(\frac{a-b}{q-p}\right)\frac{(q/p)^a - (q/p)^k}{(q/p)^a - (q/p)^b}. \qquad (4.2.60)$$

Note that when $p > q$ and $b \to -\infty$, we have $d_0 \to a/(p-q)$ in agreement with the single barrier result (4.2.56).

Higher-order moments for the duration of the walk, D_k, from start point k can (in principle) be generated by solving the general extension of the linear equation (4.2.59), namely

$$\mathrm{E}[D_k^m] = p\mathrm{E}[(1 + D_{k+1})^m] + (1 - p - q)\mathrm{E}[(1 + D_k)^m] + q\mathrm{E}[(1 + D_{k-1})^m], \qquad (4.2.61)$$

for $\mathrm{E}[D_k^m]$ sequentially over $m = 2, 3, \ldots$.

4.2.4 Further aspects of the unrestricted random walk

The results developed in Sections 4.1.1 and 4.1.2 can be used to deduce further properties of the unrestricted random walk. For example, on placing a single barrier at a, the probability of reaching state a from state $j < a$ in the first n steps is given by

$$\Pr(\max(X_n) \geq a) = \sum_{r=1}^{n} f_{ja}^{(r)}. \qquad (4.2.62)$$

Similarly, on placing a single barrier at $b < j$ we have

$$\Pr(\min(X_n) \leq b) = \sum_{r=1}^{n} f_{jb}^{(r)}. \qquad (4.2.63)$$

Whilst imposing barriers at both a and b gives

$$\Pr(b < X_n < a) = 1 - \sum_{r=1}^{n} \left(f_{ja}^{(r)} + f_{jb}^{(r)} \right). \qquad (4.2.64)$$

Moreover, on letting $n \to \infty$ we can explore the possibility of obtaining a limiting distribution. Denote

$$U_n = \max_{0 \leq r \leq n} X_r \quad \text{and} \quad L_n = \min_{0 \leq r \leq n} X_r,$$

and consider the behaviour of U_n (say) as n increases. When the mean jump is positive $(p > q)$ the process will drift to $+\infty$ and so U_n will become indefinitely large; note that this also happens when the mean jump is zero since the process will make indefinitely large excursions (result (4.1.12)). However, when the mean jump is negative $(p < q)$ the process will drift to $-\infty$ after reaching a maximum value U in the positive direction. We call $U = \sup_n X_n$ the *supremum* of the process. Now we see from result (4.2.54) that when $p < q$ the probability that the process never reaches state a from start position $j = 0$ is $1 - (p/q)^a$. Thus

$$\Pr(U < a) = 1 - (p/q)^a \qquad (a = 1, 2, \ldots),$$

whence U follows the geometric distribution

$$\Pr(U = a) = [1 - (p/q)](p/q)^a \qquad (4.2.65)$$

with mean $E(U) = p/(q-p)$ and variance $\text{Var}(U) = pq/(q-p)^2$. The distribution of the *infimum* of the process, $L = \inf_n X_n$, follows similarly.

Another aspect of the unrestricted walk exposed by the imposition of an absorbing barrier is the phenomenon of *return to the origin*, which we have previously studied in Section 4.1.4 for the symmetric case $p = q$ by evaluating the number of suitable paths. Conditional on starting in state j, let $f_{jj}^{(n)}$ denote the probability that the first return to j occurs at time n, and $f_{jk}^{(n)}$ the probability that the process first enters state $k \neq j$ at time n. Since the steps of the random walk are independent it follows that $f_{jk}^{(n)} = f_{0,k-j}^{(n)}$, so we can place $j = 0$ without any loss of generality. Then on considering the first step we have

$$f_{00}^{(n)} = pf_{10}^{(n-1)} + qf_{-1,0}^{(n-1)} = pf_{10}^{(n-1)} + qf_{01}^{(n-1)} \qquad (n = 2, 3, \ldots). \qquad (4.2.66)$$

Note that there is no $f_{00}^{(n-1)}$ term in (4.2.66), since being in state 0 at time $n-1$ (for $n > 2$) automatically precludes being in state 0 for the first time at time n; whilst at time $n = 1$ we clearly have $f_{00}^{(1)} = 1 - p - q$. Whence taking generating functions yields

$$F_{00}(s) \equiv \sum_{n=1}^{\infty} f_{00}^{(n)} s^n = s(1 - p - q) + sp \sum_{n=1}^{\infty} f_{10}^{(n)} s^n + sq \sum_{n=1}^{\infty} f_{01}^{(n)} s^n. \quad (4.2.67)$$

Now it follows from (4.2.46) that in the second summation

$$F_{01}(s) = \sum_{n=1}^{\infty} f_{01}^{(n)} s^n = (p/q)\lambda_2(s),$$

where

$$\lambda_2(s) = (1/2ps)[1 - (1 - p - q)s - \sqrt{\{[1 - (1 - p - q)s]^2 - 4pqs^2\}}]. \quad (4.2.68)$$

Whilst the first summation involves a similar generating function taking the form $\tilde{F}_{01}(s)$ in which p and q are interchanged, namely

$$F_{10}(s) = \tilde{F}_{01}(s) = (q/p)\tilde{\lambda}_2(s), \quad (4.2.69)$$

where $\tilde{\lambda}_2(s)$ is $\lambda_2(s)$ with p and q interchanged, i.e.

$$\tilde{\lambda}_2(s) = (1/2qs)[1 - (1 - p - q)s - \sqrt{\{[1 - (1 - p - q)s]^2 - 4pqs^2\}}] = (p/q)\lambda_2(s). \quad (4.2.70)$$

Thus $F_{10}(s) = \lambda_2(s)$, whence (4.2.67) reduces to

$$F_{00}(s) = s[(1 - p - q) + 2p\lambda_2(s)]. \quad (4.2.71)$$

The probabilities $f_{00}^{(n)}$ can now be recovered by expanding (4.2.68) in powers of s.

In particular, when $p + q = 1$, so that a step involves a move away from the current position, (4.2.71) reduces to

$$F_{00}(s) = 1 - \{1 - 4pqs^2\}^{1/2} = 2pqs^2 + \sum_{n=2}^{\infty} s^{2n}(2pq)^n \frac{1.3\ldots(2n-3)}{n!}. \quad (4.2.72)$$

Thus $f_{00}^{(2)} = 2pq$ (obvious by considering the two possible paths), $f_{00}^{(2n)} = (2pq)^n 1.3\ldots(2n-3)/n!$ for $n = 2, 3, \ldots$, whilst $f_{00}^{(2n-1)} = 0$ for $n = 1, 2\ldots$ (the first return to 0 cannot occur in an odd number of steps).

Finally, we note that since $F_{00}(1) = \sum_{n=1}^{\infty} f_{00}^{(n)} = 1 - \sqrt{\{(p-q)^2\}}$ is the probability of ever returning to the origin, it follows that

$$F_{00}(1) = \begin{cases} 1 - p + q < 1 & \text{if } p > q \\ 1 & \text{if } p = q \\ 1 - q + p < 1 & \text{if } p < q. \end{cases} \quad (4.2.73)$$

So return to the origin is certain only if the walk has zero drift ($p = q$). In this case, the mean time of first return to the origin, $F'_{00}(1) = \sum_{n=1}^{\infty} n f_{00}^{(n)}$, is infinite since $\lambda'_2(1) = \infty$ in (4.2.71) when $p = q$.

4.3 Reflecting barriers

We have seen that whilst the unrestricted random walk is ideal for considering genuine movement, such as the position of an individual moving along a linear habitat zoned into discrete units $i = \ldots, -1, 0, 1, \ldots$, when i denotes population size it is far more natural to restrict attention to the non-negative integers $i = 0, 1, \ldots$. The counterexample of Section 4.1 in which X_i denotes the net difference between the number of males and females (and so can be negative as well as positive) is a trifle artificial. By placing an absorbing barrier at $i = 0$ we assert that once the population becomes empty it remains so. Whilst we can determine the probability that the process reaches size $a > 0$ by some given time by introducing an absorbing barrier at a. However, many processes do not involve the concept of absorption, but merely have lower and upper limits on population size. The simple queueing (immigration–emigration) process discussed in Section 2.6 provides an obvious example of the former, since an empty queue restarts as soon as a new customer arrives. Whilst for the latter we might simply place an upper bound a on the number of customers allowed to queue in the system. In such situations the states 0 and a correspond to *reflecting barriers*, and the process is guaranteed to continue indefinitely.

4.3.1 General equilibrium probability distribution

Before we discuss time-dependent properties of the system, let us first consider the equilibrium solution. Given that we have already shown in Section 3.2 that the continuous-time equilibrium distribution for general birth and death rates λ_N and μ_N is easily constructed, let us follow a similar approach for our discrete-time random walk. Since the effect of the initial position $X_0 = j$ disappears as the number of steps increases, let us presume that

$$\lim_{n \to \infty} \Pr(X_n = i | X_0 = j) = \pi_i \qquad (i, j = 0, 1, \ldots, a).$$

Denote the step transition probabilities of moving from state i to $i+1$, $i-1$ and i by p_i, q_i and $1 - p_i - q_i$, respectively, with $q_0 = p_a = 0$. Then on paralleling the construction of solution (3.2.7), we write down the forward equations

$$\pi_i = p_{i-1} \pi_{i-1} + (1 - p_i - q_i) \pi_i + q_{i+1} \pi_{i+1} \qquad (i = 0, 1, \ldots, a) \qquad (4.3.1)$$

and rearrange the terms to form

$$q_{i+1} \pi_{i+1} - p_i \pi_i = q_i \pi_i - p_{i-1} \pi_{i-1} = \text{constant}; \qquad (4.3.2)$$

since the two sides of this equation are identical apart from i on the left being replaced by $i - 1$ on the right. Now at $i = 0$ equation (4.3.1) gives

$$\pi_0 = \pi_0(1 - p_0) + \pi_1 q_1, \qquad \text{i.e.} \quad q_1 \pi_1 - p_0 \pi_0 = 0,$$

and so the constant equals zero. Whence (4.3.2) reduces to the balance equation

$$q_i \pi_i = p_{i-1} \pi_{i-1}. \tag{4.3.3}$$

Repeated application of this recurrence relation then yields

$$\pi_i = \frac{p_0 p_1 \cdots p_{i-1}}{q_1 q_2 \cdots q_i} \pi_0 \qquad (i = 1, \ldots, a), \tag{4.3.4}$$

where π_0 is chosen to ensure that $\sum_{i=0}^{a} \pi_i = 1$, i.e.

$$\pi_0 = \left[1 + \sum_{i=1}^{a} \frac{p_0 p_1 \cdots p_{i-1}}{q_1 q_2 \cdots q_i} \right]^{-1}. \tag{4.3.5}$$

It follows that the solution for the case of the single reflecting barrier at 0 exists provided

$$\sum_{i=1}^{\infty} \frac{p_0 p_1 \cdots p_{i-1}}{q_1 q_2 \cdots q_i} < \infty. \tag{4.3.6}$$

For the simple random walk in which $p_i = p$ and $q_i = q$ (with $q_0 = p_a = 0$), we therefore have

$$\pi_0 = 1/\sum_{i=0}^{a} (p/q)^i = \frac{1 - (p/q)}{1 - (p/q)^{a+1}} \qquad (p < q),$$

whence (4.3.4) becomes

$$\pi_i = (p/q)^i \frac{[1 - (p/q)]}{[1 - (p/q)^{a+1}]} \qquad (i = 0, \ldots, a). \tag{4.3.7}$$

When comparing the continuous-time processes developed in Chapters 2 and 3 with the discrete-time random walk, two options are open to us. The simplest is to view the random walk as the *skeleton* of the continuous-time process $\{X_N(t)\}$ by recording each change in the population size N but ignoring the time $\{t_i\}$ at which the ith change occurs. So if, for example, the population is of size N immediately following the $(i-1)$th event of a simple birth–death–immigration process, then the ith event takes the probabilities $p_i = (\alpha + \lambda N)/(\alpha + \mu N + \lambda N)$ and $q_i = \mu N/(\alpha + \lambda N + \mu N)$; here the probability of remaining in the same state i is zero ($p_i + q_i = 1$). Alternatively, we may break up the continuous time interval $t \geq 0$ into a succession of sub-intervals, e.g. $[0, h), [h, 2h), \ldots$, and presume that h is sufficiently small for us to be able to ignore the probability of more than one event occurring in any given sub-interval. This clearly works for the continuous-time logistic process with $\lambda_N = N(a_1 - b_1 N)$ and $\mu_N = N(a_2 + b_2 N)$, since the implicit condition $N \leq a_1/b_1$ guarantees that the total rate $\lambda_N + \mu_N$ remain bounded. For the simple immigration–birth–death process, however, this no longer is true, since if λ is only minutely less than μ then N may become huge thereby rendering $h \ll 1/(\alpha + \lambda N + \mu N)$ effectively zero. Thus when viewed as an approximating process, rather than as a pure event process, care is needed to ensure that the discretization makes practical sense. For example, coupling the combined simple immigration and birth rate $p_i = (\alpha + \lambda N)h$ with the simple death rate $q_i = N\mu h$ implies that the probability of remaining at i is $1 - p_i - q_i = 1 - [\alpha + (\lambda + \mu)N]h$. So to ensure that this remains positive we might impose a reflecting barrier at a, where $\Pr(X_n < a) \simeq 1$, and then choose h to ensure that $p_a + q_a = [\alpha + (\lambda + \mu)a]h \ll 1$.

As a further illustration, consider the following model of a simple queueing system. Customers may arrive at the service point only at the discrete time points $n = 1, 2, \ldots$, and they form a queue if the server is occupied. At each time point there is a probability α that a customer arrives and, independently, a probability β that the customer already being served (if there is one) completes his service. Since there is limited queueing space, the queue is restricted in size to a customers including the one being served. Thus a customer arriving to find the queue full is turned away. Let X_n denote the number of customers in the queue, including the one being served, immediately following the nth time instant, and Z_n the corresponding net change in queue size. Then for $0 \leq X_n \leq a$ ($n = 0, 1, 2, \ldots$) we have

$$X_n = X_{n-1} + Z_n; \tag{4.3.8}$$

where for $X_{n-1} = 0$

$$\Pr(Z_n = 0) = 1 - \alpha \quad \text{(no arrival)}$$
$$\Pr(Z_n = 1) = \alpha \quad \text{(arrival)}, \tag{4.3.9}$$

and for $0 < X_n < a$

$$\Pr(Z_n = -1) = (1 - \alpha)\beta \quad \text{(departure and no arrival)}$$
$$\Pr(Z_n = 0) = \alpha\beta + (1 - \alpha)(1 - \beta) \quad \text{(departure and arrival } or \tag{4.3.10}$$
$$\text{no departure and no arrival)}$$
$$\Pr(Z_n = 1) = \alpha(1 - \beta) \quad \text{(arrival and no departure)}. \tag{4.3.11}$$

Note the structural difference between the simple continuous time queue (Section 2.6) in which multiple events cannot occur in the small time interval $(t, t + dt)$ (strictly, they can occur but with negligible probability of order $o(dt)$), and the above discrete-time event changes which include the simultaneous occurrence of both an arrival and a departure. It is the admittance of such multiple events that makes discrete-time processes intrinsically more difficult to work with than their continuous-time counterparts. Moreover, we also have to take care when defining the transition probabilities at the upper barrier a. For rather than considering the probability that X_n remains at a, it is safer first to observe that for $X_n = a$ the probability of a move to $a - 1$ is

$$\Pr(Z_n = -1) = (1 - \alpha)\beta \quad \text{(departure and no arrival)}, \tag{4.3.12}$$

and then to use this to write down

$$\Pr(Z_n = 0) = 1 - (1 - \alpha)\beta \quad \text{(as the remaining probability)}. \tag{4.3.13}$$

The transition probabilities (4.3.9)–(4.3.13) give rise to the forward equilibrium equations

$$\pi_0 = (1-\alpha)\pi_0 + (1-\alpha)\beta\pi_1$$
$$\pi_1 = \alpha\pi_0 + [\alpha\beta + (1-\alpha)(1-\beta)]\pi_1 + (1-\alpha)\beta\pi_2 \qquad (4.3.14)$$
$$\pi_i = \alpha(1-\beta)\pi_{i-1} + [\alpha\beta + (1-\alpha)(1-\beta)]\pi_i + (1-\alpha)\beta\pi_{i+1} \qquad (i=2,\ldots,a-1)$$
$$\pi_a = \alpha(1-\beta)\pi_{a-1} + [1-(1-\alpha)\beta]\pi_a.$$

These clearly take the form of the general equation (4.3.1) with $p_0 = \alpha$, $p_i = \alpha(1-\beta)$ ($i = 1,\ldots,a-1$) and $q_i = (1-\alpha)\beta$ ($i = 1,\ldots,a$). Whence on writing $\rho = [\alpha(1-\beta)]/[(1-\alpha)\beta]$, solution (4.3.4)–(4.3.5) yields

$$\pi_i = \frac{\rho^i \pi_0}{1-\beta} \quad (i=1,\ldots,a) \qquad \text{where} \qquad \pi_0 = \frac{\beta-\alpha}{\beta-\alpha\rho^a}. \qquad (4.3.15)$$

Provided $\rho < 1$, i.e. the arrival rate α is less than the departure rate β, letting $a \to \infty$ in (4.3.15) shows that for a single reflecting barrier at zero

$$\pi_i = \frac{\beta-\alpha}{\beta(1-\beta)}\rho^i \quad (i=1,2,\ldots) \qquad \text{with} \qquad \pi_0 = \frac{\beta-\alpha}{\beta}. \qquad (4.3.16)$$

Similarly, on letting $a \to \infty$ in (4.3.7) we see that the equilibrium probabilities for the simple random walk with a single reflecting barrier (at zero) and $p < q$ (to ensure an overall downwards drift) are given by the geometric distribution

$$\pi_i = \left(1 - \frac{p}{q}\right)\left(\frac{p}{q}\right)^i \quad (i=0,1,\ldots). \qquad (4.3.17)$$

Since the process can only leave state 0 through the arrival of a new individual, which occurs with probability p for each time point, once the process reaches this barrier it must remain there for a time T, where T follows the geometric distribution

$$\Pr(T=r) = p(1-p)^{r-1}. \qquad (4.3.18)$$

The periods spent away from the barrier (excluding the initial one) clearly have the same distribution as the first passage time from state 1 to state 0, the p.g.f. of which is given by (4.2.69)–(4.2.70), namely

$$F_{10}(s) = \lambda_2(s) = (1/2ps)[1-(1-p-q)s - \sqrt{\{[1-(1-p-q)s]^2 - 4pqs^2\}}]. \qquad (4.3.19)$$

If $E_1, L_1, E_2, L_2, \ldots$ denote the successive times at which the process enters and leaves the reflecting barrier at 0, then $(L_1 - E_1), (L_2 - E_2), \ldots$ are independent random variables each taking the distribution (4.3.18), whilst $(E_2 - L_1), (E_3 - L_2), \ldots$ is a sequence of independent random variables each having the p.g.f. (4.3.19). Moreover, these two sequences are independent. We call the sequence of time points $E_1, L_1, E_2, L_2, \ldots$ an *alternating renewal process* (see, for example, Cox, 1962).

4.3.2 Relation between reflecting and absorbing barriers

This use of an absorbing barrier result in a reflecting barrier scenario can be taken much further, since it transpires that these two types of random walk are in essence mathematically equivalent. Specifically, if we have determined the probability of eventual absorption at one of the barriers (and hence at the other) for an arbitrary start point, then we can immediately write down the equilibrium distribution for the corresponding reflecting barrier situation. For suppose there are absorbing barriers at a and $-a$, and let $Q(x)$ denote the probability that the process is eventually absorbed at $-a$ given that the walk starts at $-a \leq x \leq a$. Let $H(x)$ denote the corresponding equilibrium distribution in the reflecting barrier case.

Following Lindley (1959) (later discussed in Cox and Miller, 1965) consider first the reflecting barrier case where the steps Z_n are continuously distributed with distribution function $F(x)$ and p.d.f. $f(x) = F'(x)$, i.e.

$$\Pr(X_n \leq X | X_{n-1} = y) = \begin{cases} 0 & x < -a \\ F(x - y) & -a \leq x < a \\ 1 & x \geq a. \end{cases}$$

Let $H_n(x)$ denote the distribution function of the position after n steps given that the walk starts at the upper barrier. Then

$$H_0(x) = \begin{cases} 1 & (x \geq a) \\ 0 & (x < a), \end{cases} \qquad (4.3.20)$$

whilst for $n \geq 1$ and $-a \leq x \leq a$

$$\Pr(X_n \leq x) = H_n(x) = \int_{-a}^{a} F(x - y) \, dH_{n-1}(y).$$

On integrating the right-hand side by parts we therefore have

$$H_n(x) = F(x - a) H_{n-1}(a) - F(x + a) H_{n-1}(-a) + \int_{-a}^{a} f(x - y) H_{n-1}(y) \, dy$$

$$= F(x - a) + \int_{-a}^{a} f(x - y) H_{n-1}(y) \, dy. \qquad (4.3.21)$$

Given that we expect an equilibrium distribution to become established as $t \to \infty$, we may write $\lim_{n \to \infty} H_n(x) = H(x)$, whence (4.3.21) yields

$$H(x) = F(x - a) + \int_{-a}^{a} f(x - y) H(y) \, dy. \qquad (4.3.22)$$

Though deriving a general solution to equation (4.3.22) in reasonably explicit terms remains a problem.

Now suppose that the barriers at $-a$ and a are absorbing, and for a walk starting at x ($-a \leq x \leq a$) let $Q_n(x)$ denote the probability that absorption occurs at the barrier $-a$ at or before the nth step. Then to construct a recurrence relation for $Q_n(x)$ we first note that absorption can occur in two mutually exclusive ways: either absorption occurs at the first step with probability $F(-x - a)$; or, the process moves to y ($-a < y < a$) at the first step and absorption occurs at one of the subsequent steps $2, 3, \ldots, n$. Thus for $n \geq 1$ and $-a < x < a$

$$Q_n(x) = F(-x-a) + \int_{-a}^{a} f(y-x) Q_{n-1}(y)\, dy \tag{4.3.23}$$

where

$$Q_0(x) = \begin{cases} 1 & (x = -a) \\ 0 & (x > -a). \end{cases}$$

Whence on rewriting (4.3.23) in terms of $-x$ we have

$$Q_n(-x) = F(x-a) + \int_{-a}^{a} f(x-y) Q_{n-1}(-y)\, dy \tag{4.3.24}$$

with

$$Q_0(-x) = \begin{cases} 1 & (x = a) \\ 0 & (x < a). \end{cases} \tag{4.3.25}$$

On comparing (4.3.20) with (4.3.22), and (4.3.22) with (4.3.24), we see that $H_n(x)$ and $Q_n(-x)$ satisfy the same initial conditions and the same recurrence relation. It therefore follows that for all $n \geq 0$

$$Q_n(-x) = H_n(x), \tag{4.3.26}$$

whence letting $n \to \infty$ yields the limiting result

$$Q(-x) = H(x). \tag{4.3.27}$$

Note that care is needed when interpreting results (4.3.26) and (4.3.27) for $x = a$ and $x = -a$, since $H_n(x)$ and $H(x)$ have discontinuities at these points.

Example The equilibrium distribution for the simple random walk with reflecting barriers at $-a$ and a is easily found by slightly changing the derivation of result (4.3.7) for barriers at 0 and a. We write $\pi_i = (p/q)^i \pi_0$ as before, but now determine π_0 from $\pi_{-a} + \cdots + \pi_a = 1$, thereby obtaining

$$\pi_i = (p/q)^{a+i} \frac{[1-(p/q)]}{[1-(p/q)^{2a+1}]} \qquad (i = -a, \ldots, a). \tag{4.3.28}$$

So the equilibrium distribution function

$$H(x) = \sum_{i=-a}^{x} \pi_i = \frac{[1-(p/q)^{a+x+1}]}{[1-(p/q)^{2a+1}]} \qquad (x = -a, \ldots, a). \tag{4.3.29}$$

Now consider the probability that eventual absorption occurs at $-a$. On replacing a and b in (4.2.24) by $a+1$ and $-a$, and j by x, we see that

$$Q(x) = \frac{[(p/q)^{-a-x} - (p/q)^{-a-(a+1)}]}{[1 - (p/q)^{-a-(a+1)}]} = \frac{[1-(p/q)^{a-x+1}]}{[1-(p/q)^{2a+1}]}. \tag{4.3.30}$$

Thus $Q(-x) = H(x)$, thereby illustrating the general result (4.3.27). Similarly, since result (4.2.37) for $f_{0b}^{(n)}$ provides the probability of absorption at barrier b exactly at time n conditional on the walk starting at 0, a simple change to b, a and 0 yields the probability $Q_n(x)$ of absorption at $-a$ at time n conditional on starting at x and having the upper absorbing barrier at $a + 1$. The distribution function of the particle's position after n steps, given that it starts at the upper reflecting barrier a, is then given by $H_n(x) = Q_n(-x)$.

4.3.3 Time-dependent probability distribution

Since this (reflecting barrier) distribution function takes the (absorbing barrier) form of a sum of geometrically weighted products of sine terms, the algebraic detail involved in deriving $H_n(x)$ directly is likely to be just as messy as that for the construction of results (4.2.36) and (4.2.37) for the absorption probabilities $f_{0a}^{(n)}$ and $f_{0b}^{(n)}$. Recall our comment following these two results which refers to similarly awkward algebra when constructing $p_k^{(n)}$, the probability of being in the non-absorbing state k at time n. Nevertheless, it is useful to indicate how a direct time-dependent solution for the reflecting barrier case might be developed.

Suppose we have reflecting barriers at 0 and a, and denote $p_i^{(n)}$ to be the probability of being in state i at time n conditional on the walk starting at the lower barrier 0. Then on considering the last move of the process, we have the forward equations

$$p_0^{(n+1)} = (1-p)p_0^{(n)} + qp_1^{(n)}$$

$$p_i^{(n+1)} = pp_{i-1}^{(n)} + (1-p-q)p_i^{(n)} + qp_{i+1}^{(n)} \quad (1 \leq i \leq a-1) \quad (4.3.31)$$

$$p_a^{(n+1)} = pp_{a-1}^{(n)} + (1-q)p_a^{(n)}.$$

Let us first develop a p.g.f. approach. On writing $G_n(z) \equiv \sum_{n=0}^{\infty} p_i^{(n)} z^n$, multiplying equations (4.3.31) by z^n, and then summing over $n = 0, 1, \ldots$, we obtain

$$(1/z)[G_0(z) - 1] = (1-p)G_0(z) + qG_1(z) \quad (4.3.32)$$

$$(1/z)G_i(z) = pG_{i-1}(z) + (1-p-q)G_i(z) + qG_{i+1}(z) \quad (1 \leq i \leq a-1) \quad (4.3.33)$$

$$(1/z)G_a(z) = pG_{a-1}(z) + (1-q)G_a(z). \quad (4.3.34)$$

Now (4.3.33) takes the auxiliary equation

$$qz\mu^2 + [z(1-p-q) - 1]\mu + ps = 0, \quad (4.3.35)$$

which yields the roots

$$\mu_1(z), \mu_2(z) = (1/2qz)[1 - (1-p-q)z \pm \sqrt{\{[1-(1-p-q)z]^2 - 4pqz^2\}}]. \quad (4.3.36)$$

Note that comparison with expression (4.2.17) for the corresponding roots of the backward equation (4.2.11) for $f_{ja}^{(n)}$ shows that $\mu_i(z) = (p/q)\lambda_i(z)$ (i.e. p and q are interchanged). The general solution of equation (4.3.33) then takes the form

$$G_i(z) = A(z)\{\mu_1(z)\}^i + B(z)\{\mu_2(z)\}^i, \quad (4.3.37)$$

where $A(z)$ and $B(z)$ are independent of i. Although these two functions are easily determined from the boundary conditions (4.3.32) and (4.3.34), the resulting solution for $G_i(z)$ is not exactly pleasant, bearing in mind that we still have to extract the coefficient of z^n in order to obtain the probability $p_i^{(n)}$. Let us therefore now consider a more direct approach based on the Method of Images.

For the purpose of illustration, suppose that $p + q = 1$ so a move from i must be to either $i - 1$ or $i + 1$. First, we have already shown (result (4.1.3)) that if the process starts at $X_0 = 0$ then for the unrestricted walk

$$\Pr(X_n = j) = q_j^{(n)} = \binom{n}{(n-j)/2} p^{(n+j)/2} q^{(n-j)/2} \qquad (j = -n, -n+2, \ldots, n). \tag{4.3.38}$$

Second, suppose that the 'interior' equation in (4.3.31), namely

$$p_i^{(n+1)} = p p_{i-1}^{(n)} + q p_{i+1}^{(n)} \qquad (i = 1, 2, \ldots, n-1), \tag{4.3.39}$$

holds for all $-\infty < i < \infty$, and that it is subject to the global boundary condition

$$p_i^{(0)} = \begin{cases} b_i & (i < 0, i > a) \\ 1 & (i = 0) \\ 0 & (0 < i \le a) \end{cases} \tag{4.3.40}$$

for some unknown constants $\{b_i\}$. Then on combining (4.3.39) and (4.3.40) we may write

$$p_i^{(n)} = \sum_{j=-\infty}^{-1} b_j q_{i-j}^{(n)} + q_i^{(n)} + \sum_{j=a+1}^{\infty} b_j q_{i-j}^{(n)}, \tag{4.3.41}$$

and our objective is to see whether we can choose $\{b_j\}$ to ensure that (4.3.41) also satisfies the boundary conditions at $i = 0$ and $i = a$. Comparing the boundary equations in (4.3.31) with the global equation (4.3.39) yields

$$p_0^{(n+1)} = q p_0^{(n)} + q p_1^{(n)} = p p_{-1}^{(n)} + q p_1^{(n)}$$

$$p_a^{(n+1)} = p p_a^{(n)} + p p_{a-1}^{(n)} = p p_{a-1}^{(n)} + q p_{a+1}^{(n)},$$

i.e. $\qquad q p_0^{(n)} = p p_{-1}^{(n)} \qquad \text{and} \qquad p p_a^{(n)} = q p_{a+1}^{(n)}. \tag{4.3.42}$

So the requirement is that the local balance equations hold at the 'connecting states' 0 and -1, and a and $a + 1$.

Now we see from (4.3.38) that

$$q_{-j}^{(n)} = (q/p)^j q_j^{(n)}, \tag{4.3.43}$$

236 *The random walk*

whence (4.3.41) gives

$$p_0^{(n)} = q_0^{(n)} + [b_{-1}q_1^{(n)} + b_{-2}q_2^{(n)} + \cdots] + [b_{a+1}q_{-a-1}^{(n)} + b_{a+2}q_{-a-2}^{(n)} + \cdots]$$
$$= q_0^{(n)} + [b_{-1}q_1^{(n)} + b_{-2}q_2^{(n)} + \cdots] + [b_{a+1}(q/p)^{a+1}q_{a+1}^{(n)} + b_{a+2}(q/p)^{a+2}q_{a+2}^{(n)} + \cdots]$$
$$p_{-1}^{(n)} = q_{-1}^{(n)} + [b_{-1}q_0^{(n)} + b_{-2}q_1^{(n)} + \cdots] + [b_{a+1}q_{-a-2}^{(n)} + b_{a+2}q_{-a-3}^{(n)} + \cdots]$$
$$= (q/p)q_1^{(n)} + [b_{-1}q_0^{(n)} + b_{-2}q_1^{(n)} + \cdots]$$
$$\quad + [b_{a+1}(q/p)^{a+2}q_{a+2}^{(n)} + b_{a+2}(q/p)^{a+3}q_{a+3}^{(n)} + \cdots].$$

Whereupon equating coefficients of $q_j^{(n)}$ for $j = 0, 1, \ldots$ in the first boundary equation (4.3.42) yields

$$\begin{aligned}
j = 0 &: \quad q = pb_{-1} \\
j = 1 &: \quad qb_{-1} = p[(q/p) + b_{-2}] \\
1 < j \leq a &: \quad qb_{-j} = pb_{-j-1} \qquad (4.3.44) \\
j = a+1 &: \quad q[b_{-a-1} + b_{a+1}(q/p)^{a+1}] = pb_{-a-2} \\
j \geq a+2 &: \quad q[b_{-j} + b_j(q/p)^j] = p[b_{-j-1} + b_{j-1}(q/p)^j].
\end{aligned}$$

Similarly, as

$$p_a^{(n)} = q_a^{(n)} + [b_{-1}q_{a+1}^{(n)} + b_{-2}q_{a+2}^{(n)} + \cdots] + [b_{a+1}q_{-1}^{(n)} + b_{a+2}q_{-2}^{(n)} + \cdots]$$
$$= q_a^{(n)} + [b_{-1}q_{a+1}^{(n)} + b_{-2}q_{a+2}^{(n)} + \cdots] + [b_{a+1}(q/p)q_1^{(n)} + b_{a+2}(q/p)^2 q_2^{(n)} + \cdots]$$
$$p_{a+1}^{(n)} = q_{a+1}^{(n)} + [b_{-1}q_{a+2}^{(n)} + b_{-2}q_{a+3}^{(n)} + \cdots] + [b_{a+1}q_0^{(n)} + b_{a+2}q_{-1}^{(n)} + \cdots]$$
$$= q_{a+1}^{(n)} + [b_{-1}q_{a+2}^{(n)} + b_{-2}q_{a+3}^{(n)} + \cdots] + [b_{a+1}q_0^{(n)} + b_{a+2}(q/p)q_1^{(n)} + \cdots],$$

equating coefficients in the second boundary equation yields

$$\begin{aligned}
j = 0 &: \quad 0 = qb_{a+1} \\
1 \leq j \leq a-1 &: \quad p[b_{a+j}(q/p)^j] = q[b_{a+j+1}(q/p)^j] \\
j = a &: \quad p[1 + b_{2a}(q/p)^a] = q[b_{2a+1}(q/p)^a] \qquad (4.3.45) \\
j = a+1 &: \quad p[b_{-1} + b_{2a+1}(q/p)^{a+1}] = q[1 + b_{2a+2}(q/p)^{a+1}] \\
j = a+r > a+1 &: \quad p[b_{-r} + b_{2a+r}(q/p)^{a+r}] = q[b_{-r+1} + b_{2a+r+1}(q/p)^{a+r}].
\end{aligned}$$

Whence solving equations (4.3.44) and (4.3.45) in pairs over $j = 0, 1, 2, \ldots$ gives

$$\begin{aligned}
j = 0 &: \quad b_{-1} = q/p & b_{a+1} = 0 \\
j = 1 &: \quad b_{-2} = (q/p)(q/p - 1) & b_{a+2} = 0 \\
j = 2 &: \quad b_{-3} = (q/p)^2(q/p - 1) & b_{a+3} = 0 \\
& \quad \vdots \quad \vdots & \vdots
\end{aligned}$$

$$(4.3.46)$$

$$j = a-1 : \quad b_{-a} = (q/p)^{a-1}(q/p-1) \qquad b_{2a} = 0$$
$$j = a : \quad b_{-a-1} = (q/p)^a(q/p-1) \qquad b_{2a+1} = (p/q)^{a+1}$$
$$j = a+1 : \quad b_{-a-2} = (q/p)^{a+1}(q/p-1) \qquad b_{2a+2} = (p/q)^{a+2}$$
$$j = a+2 : \quad b_{-a-3} = (q/p)^{a+2}(q/p-1) \qquad b_{2a+3} = (p/q-1)(p/q)^{a+3}$$
$$\vdots \qquad \vdots \qquad \qquad \vdots$$

with subsequent coefficients being given by the paired recurrence relation

$$b_{-j-1} = (q/p)b_{-j} + (q/p)^{j+1}b_j - (q/p)^j b_{j-1} \qquad (4.3.47)$$
$$b_{j+1} = (p/q)b_{a+j} + (p/q)^{j+1}b_{a-j} - (p/q)^j b_{a-j+1}.$$

Although care needs to be exercised in the evaluation of the coefficients $\{b_j\}$, this is the only real downside, and there is considerable scope for applying the approach to a whole range of random walk barrier scenarios. For example, we could relax the condition that $p + q = 1$ by replacing the binomial probabilities (4.3.38) by their trinomial counterparts, or examine the effect of incorporating more complex barrier structures, such as a lower reflecting barrier combined with an upper absorbing barrier. In such situations the key to success lies in the ratio of the unrestricted probabilities $q_j^{(n)}/q_{-j}^{(n)}$ being independent of n (as in (4.3.43)).

It is a pity that many published 'solutions' take the form of p.g.f.'s, with no guide being given as to how probabilities may be usefully extracted from them. This imaging technique removes the need for any such inversion, and to illustrate the type of solution that the recurrence relations (4.3.47) generate suppose we take $p = q$ and allow general initial conditions $p_i^{(0)} = c_i$ $(0 \le i \le a)$. Then equations (4.3.42) simplify to the mirror image conditions

$$p_0^{(n)} = p_{-1}^{(n)} \quad \text{and} \quad p_a^{(n)} = p_{a+1}^{(n)}. \qquad (4.3.48)$$

Whence substituting for

$$p_i^{(n)} = \sum_{r=-\infty}^{\infty} b_r q_{i-r}^{(n)} \qquad (4.3.49)$$

in (4.3.48) with $i = -1, 0, a, a+1$, and comparing coefficients of $\ldots, b_{-1}, b_0, b_1, \ldots$, gives $b_{-r} = b_{r+1}$ and $b_{a+r} = b_{a-r+1}$. On noting that $b_0 = c_0, \ldots, b_a = c_a$, sequential application of these two equalities then yields

$$b_{-1} = b_0 \qquad = c_0 \qquad b_{a+1} = b_a \qquad = c_a$$
$$b_{-2} = b_1 \qquad = c_1 \qquad b_{a+2} = b_{a-1} \qquad = c_{a-1}$$
$$\vdots \qquad \qquad \vdots$$

$$b_{-a-1} = b_a \quad = c_a \qquad b_{2a+1} = b_0 \quad = c_0$$
$$b_{-a-2} = b_{a+1} = c_a \qquad b_{2a+2} = b_{-1} \quad = c_0$$
$$b_{-a-3} = b_{a+2} = c_{a-1} \qquad b_{2a+3} = b_{-2} \quad = c_1$$
$$\vdots \qquad\qquad \vdots$$
$$b_{-2a-2} = b_{2a+1} = c_0 \qquad b_{3a+2} = b_{-a-1} = c_a$$
$$b_{-2a-3} = b_{2a+2} = c_0 \qquad b_{3a+3} = b_{-a-2} = c_a$$
$$\vdots \qquad\qquad \vdots$$

i.e. $\quad b_{2r(a+1)+s} = b_{2r(a+1)-1-s} = c_s \quad (-\infty < r < \infty;\ 0 \le s \le a).$ (4.3.50)

So (4.3.49) becomes

$$p_i^{(n)} = \sum_{s=0}^{a} c_s \sum_{r=-\infty}^{\infty} (q_{i-2r(a+1)-s}^{(n)} + q_{i-2r(a+1)+1+s}^{(n)}),$$ (4.3.51)

where (here) $q_j^{(n)}$ is given by the binomial coefficient (4.3.38) for $j = -n, -n+2, \ldots, n$, and zero otherwise.

4.4 The correlated random walk

Whilst in many situations the assumption of independent steps is perfectly valid, this is not universally true. For example, if the random walk relates to genuine motion, as opposed to say successive population sizes in a simple birth–death process, then a particle's momentum ensures that it is likely to travel in the same direction over several successive time steps. Interest in this problem developed following Rayleigh's (1919) solution to the problem of 'random flights'. Early discussions centred around the construction of probability distributions associated with *freely linked* random chains within a random walk setting (see Bartlett, 1949). However, it was not until Daniels' (1952) key paper that analysts started to examine the statistical behaviour of *stiff* chains, in which the orientation of a given link of the chain is influenced by those of the neighbouring links. Two early examples cited concern the elasticity of rubber (e.g. Moran, 1948), and the path of a heavy particle moving through an atmosphere of light particles where the deflection at each collision is relative to the direction of motion prior to the collision. Moyal (1950) provides comprehensive references to this multiple scattering problem. Whilst Kingman (1982) extends the process by replacing the stiffly linked chain by a continuously flexible string.

This construct has strong relevance to a surprisingly wide variety of different fields of application. For example, suppose that the walk relates to daily stock market values. Then a simplistic viewpoint would be to presume that because of market sentiment an increase (or decrease) on day n is more likely to be followed by an increase (or decrease) on day $n+1$ than a decrease (or increase). So in the simplest case we have

Pr(particle moves one unit in the same direction as the last step) $= p$
Pr(particle moves one unit in the opposite direction as the last step) $= q = 1 - p$.

Though parallel analyses can be developed to those performed previously for the independent step random walk with absorbing and reflecting barriers, here we shall just introduce basic results by examining the pure one-dimensional unrestricted process as developed in Renshaw and Henderson (1981).

The correlated random walk (c.r.w.) structure was initially introduced by Gillis (1955) for movement over a d-dimensional rectangular lattice in which each step comprises one of the $2d$ possible nearest-neighbour moves. The walk starts from the origin with all of the $2d$ possible initial directions being equally likely. So in addition to the step probabilities p and q we now also have

Pr(the walk takes any one of the directions orthogonal to the previous step) $= r$

where $p + q + 2(d-1)r = 1$. Although the extension from one- to d-dimensions does convolute the resulting algebra, important results can still be derived. For example, Henderson, Renshaw and Ford (1984) study the two-dimensional case. They first derive an expression for the characteristic function of the walk, and then use it to construct the mean and dispersion matrix of the position after n steps and derive an exact expression for the dispersion matrix of the n-step transition probabilities. A discussion of recurrence properties (introduced in Henderson, Renshaw and Ford, 1983, and examined within a general framework in Section 5.3) leads on to the development of an asymptotic expression for the expected number of distinct lattice points visited, and then to a study of the associated diffusion approximation. Gillis' aim was to show that provided $|2dp - 1| \neq 1$ the walk is recurrent if $d = 1$ or 2 and is transient if $d \geq 3$. This *Gillis conjecture* was first proved by Domb and Fisher (1958), who considered an even more general model by introducing different probabilities p_i ($-d \leq i \leq d$) of moving in different directions. Chen and Renshaw (1992) derive the n-step characteristic function of the Gillis correlated random walk, present a simpler proof of Gillis' conjecture, and generalize the Gillis–Domb–Fisher walk. Whilst Chen and Renshaw (1994) prove that the n-step characteristic function of any d-dimensional correlated random walk satisfies a recurrence formula which enables both it and the total characteristic function to be obtained.

4.4.1 Occupation probabilities: direct solution

Returning to the one-dimensional case, consider a single particle which starts at the origin, initially moves left or right each with probability $1/2$, and then performs a correlated random walk on the integers. Although the probability $p_i^{(n)}$ of being in state i after n steps can still be determined directly, we need a more detailed argument than that used in the exceedingly straightforward construction of (4.1.3) for the simple random walk. First take the number of steps to be even, and let k denote the number of changes of direction. Then

$$\Pr(\text{at } 2i \text{ after } 2n \text{ steps}) = \sum_{k=0}^{2n} \Pr(\text{at } 2i \text{ after } 2n \text{ steps} \mid \text{exactly } k \text{ turns})$$

$$\times \Pr(\text{exactly } k \text{ turns}).$$

Now each walk of $2n$ steps which involves exactly k changes of direction has probability $\frac{1}{2}q^k p^{2n-k-1}$. Let us therefore consider the number of such walks which end at position $2i$, having made, of necessity, $(n+i)$ positive and $(n-i)$ negative steps.

Suppose the first step is positive, and that the particle then moves a_1 consecutive steps (including the first) in the positive direction until the first turn, followed by a_2 consecutive steps in the negative direction until the second turn, etc. Then because we have conditioned on exactly k turns we see that the

$$\text{total number of steps} = 2n = a_1 + a_2 + \cdots + a_{k+1},$$
$$\text{total number of positive steps} = n + i = a_1 + a_3 + \cdots + a_s,$$
$$\text{total number of negative steps} = n - i = a_2 + a_4 + \cdots + a_t;$$

where $s = k+1$ and $t = k$ if k is even, else $s = k$ and $t = k+1$ if k is odd. Now the total number of ways in which $(n+i)$ can be divided into $\frac{1}{2}(s+1)$ non-empty groups is $\binom{n+i-1}{\frac{1}{2}(s+1)-1}$ (e.g. Feller, 1968). Likewise, the total number of ways of dividing $(n-i)$ into $\frac{1}{2}t$ non-empty groups is $\binom{n-i-1}{\frac{1}{2}t-1}$. Thus the total number of walks of $2n$ steps which end at $2i$ and involve exactly k changes of direction with the first step positive is

$$\binom{n+i-1}{\frac{1}{2}(s+1)-1}\binom{n-i-1}{\frac{1}{2}t-1}.$$

Similarly, if the first step is negative then there are

$$\binom{n-i-1}{\frac{1}{2}(s+1)-1}\binom{n+i-1}{\frac{1}{2}t-1}$$

such walks. Hence, on summing over all possible values of k, we have the occupation probability

$$r_{2i}(2n) = \frac{1}{2}\sum_{k=0}^{2n-2i}\left\{\binom{n+i-1}{\frac{1}{2}(s+1)-1}\binom{n-i-1}{\frac{1}{2}t-1} + \binom{n-i-1}{\frac{1}{2}(s+1)-1}\binom{n+i-1}{\frac{1}{2}t-1}\right\}q^k p^{2n-k-1}.$$
(4.4.1)

By considering the two separate cases k even and k odd, this expression may be simplified to yield the solution

$$r_{2i}(2n) = \frac{1}{2}p^{2n-1}\delta_{in} + \frac{1}{2}\sum_{j=1}^{n-i}q^{2j}p^{2n-2j-1}\left\{\binom{n+i-1}{j}\binom{n-i-1}{j-1}\right.$$
$$\left. + \binom{n-i-1}{j}\binom{n+i-1}{j-1} + 2pq^{-1}\binom{n+i-1}{j-1}\binom{n-i-1}{j-1}\right\}, \quad (4.4.2)$$

where

$$\delta_{in} = \begin{cases} 1 & (i = n) \\ 0 & (i \neq n) \end{cases} \quad (4.4.3)$$

denotes the Kronecker delta function. A similar argument produces the occupation probabilities $\{r_{2i+1}(2n+1)\}$ for an odd number of steps.

4.4.2 Occupation probabilities: p.g.f. solution

An alternative form for the occupation probabilities can be determined by using a double generating function approach, and this carries the added advantage of yielding moments of the process. We first define

$p_i(n) = \Pr(\text{particle is at } i \text{ after } n \text{ steps and leaves to the right})$
$q_i(n) = \Pr(\text{particle is at } i \text{ after } n \text{ steps and leaves to the left}).$

Then in the symmetric case, when $n = 0$ the particle is equally likely to move left or right, so $p_0(0) = q_0(0) = \frac{1}{2}$. The Kolmogorov forward equations for $\{p_i(n)\}$ and $\{q_i(n)\}$ are easily determined by noting that $p_i(n+1)$ is the probability of being at $i-1$ at time n and moving right, followed by a unit move in the same direction, plus the probability of being at $i+1$ at time n and moving left, followed by a unit move in the reverse direction, i.e.

$$p_i(n+1) = pp_{i-1}(n) + qp_{i+1}(n); \tag{4.4.4}$$

whilst a similar argument gives

$$q_i(n+1) = qp_{i-1}(n) + pq_{i+1}(n). \tag{4.4.5}$$

Now denote the p.g.f.'s

$$P(z;s) \equiv \sum_{n=0}^{\infty} \sum_{i=-\infty}^{\infty} p_i(n) z^i s^n \quad \text{and} \quad Q(z;s) \equiv \sum_{n=0}^{\infty} \sum_{i=-\infty}^{\infty} q_i(n) z^i s^n, \tag{4.4.6}$$

where $p_i(n) = q_i(n) = 0$ for $|i| > n$. Then on multiplying both sides of equations (4.4.4) and (4.4.5) by $z^i s^n$, and then summing over $s = 0, 1, 2, \ldots$ and $i = \ldots, -1, 0, 1, \ldots$, we obtain

$$P(z;s) - 1/2 = pszP(z;s) + qsz^{-1}Q(z;s) \tag{4.4.7}$$

$$Q(z;s) - 1/2 = qszP(z;s) + psz^{-1}Q(z;s). \tag{4.4.8}$$

Whence solving for $P(z;s)$ and $Q(z;s)$ yields

$$Q(z^{-1};s) = P(z;s) = \frac{1 - (p-q)sz^{-1}}{2[(1-psz)(1-psz^{-1}) - q^2s^2]}. \tag{4.4.9}$$

Since a particle at position i after n steps must leave either to the right or to the left, we may write

$$r_i(n) = p_i(n) + q_i(n). \tag{4.4.10}$$

Thus

$$R(z;s) \equiv \sum_{n=0}^{\infty} \sum_{i=-\infty}^{\infty} r_i(n) z^i s^n = P(z;s) + Q(z;s), \tag{4.4.11}$$

whence applying (4.4.9) yields

$$R(z;s) = \frac{1 - \frac{1}{2}(p-q)s(z+z^{-1})}{(1-psz)(1-psz^{-1}) - q^2s^2}. \tag{4.4.12}$$

Expanding (4.4.12) in powers of s and $(z+z^{-1})$, and collecting terms in $z^i s^n$, then gives

$$r_{-2i}(2n) = r_{2i}(2n) \tag{4.4.13}$$

$$= \sum_{j=0}^{n} p^{2j}(q-p)^{n-j}\binom{2j}{j-i}\left\{\binom{n+1}{n-j} - \frac{1}{2}(1-qp^{-1})\binom{n+j-1}{n-j}\right\}$$

and

$$r_{-2i-1}(2n+1) = r_{2i+1}(2n) \tag{4.4.14}$$

$$= \sum_{j=0}^{n} p^{2j+1}(q-p)^{n-j}\binom{2j+1}{j-i}\left\{\binom{n+j+1}{n-j} - \frac{1}{2}(1-qp^{-1})\binom{n+j}{n-j}\right\}$$

for $i \geq 0$, where we define $\binom{0}{0} = 1$. Showing that this solution matches the earlier expression (4.4.2) is non-trivial, since the proof involves the construction of the combinatorial identity

$$\sum_{j=|i|}^{n-1} \binom{n-j-1}{r-1}\binom{2j}{j-i}\binom{n+j-1}{n-j-1}(-1)^{n-j-r} \equiv \binom{n+i-1}{r-1}\binom{n-i-1}{r-1} \tag{4.4.15}$$

for odd powers of q^r, together with a slightly more complex identity for even powers (Renshaw and Henderson, 1981). This highlights the reality that applying different methods of solution to a specific problem may yield several substantially different combinatorial forms, and whilst these will be mathematically equivalent, some may be more algebraically amenable than others.

To determine the kth factorial moment of $\{r_i(n)\}$ we simply extract the coefficient of s^n in the expansion of $\partial^k R(z;s)/\partial z^k\big|_{z=1}$. In particular, we obtain

$$\mu(n) = 0 \tag{4.4.16}$$

$$\sigma^2(n) = npq^{-1} + (p-q)\{(p-q)^n - 1\}/2q^2 \tag{4.4.17}$$

for the mean and variance after n steps. That the mean, and all odd moments, should be zero is obvious from symmetry considerations; whilst for $|p-q| < 1$ we see that $\sigma^2(n) \sim npq^{-1}$ for large n. Note that when $p=0$ the particle always changes direction at each step, and so the distribution $\{r_i(n)\}$ lies entirely on $\{-1, 0, 1\}$ with $r_0(2n) = 1$ and $r_{-1}(2n+1) = r_1(2n+1) = \frac{1}{2}$. If $p = \frac{1}{2}$ then the correlated random walk is clearly indistinguishable from the simple symmetric random walk. Whilst if $p = 1$ then the particle can never change direction, so after n steps sites n and $-n$ are each occupied with probability $\frac{1}{2}$.

Similar arguments yield the probabilities $\{p_i(n)\}$ and $\{q_i(n)\}$ together with their respective (conditional) moments. For example, on expanding the p.g.f.'s (4.4.9) in powers of s and z we may show that

$$p_{2i}(2n) = q_{-2i}(2n) = \frac{1}{2}\sum_{j=0}^{n} p^{2j}(q-p)^{n-j} \qquad (4.4.18)$$

$$\times \left\{ \binom{2j}{j-i}\binom{n+j}{n-j} - (1-qp^{-1})\binom{2j-1}{j-i-1}\binom{n+j-1}{n-j} \right\};$$

the respective means and variances of the (improper) distributions being

$$\mu_p(n) = -\mu_q(n) = (p-q)\{1-(p-q)^n\}/4q \qquad (4.4.19)$$

$$\sigma_p^2(n) = \sigma_q^2(n) = \{(1+2npq) - [1-(1/2)(p-q)\{(p-q)^n - 1\}]^2\}/4q^2. \qquad (4.4.20)$$

Whilst the above combinatorial expressions for $p_i(n)$, $q_i(n)$ and $r_i(n)$ may be used to evaluate specific values for given i, n and p, the simplest numerical approach is to compute the iterative equations (4.4.4) and (4.4.5), together with (4.4.10), directly. Figure 4.3a shows $r_i(40)$ for $i = -40, -38, \ldots, 40$ (all other i-values are zero), and the change in shape as p increases, from a spike at $i=0$ when $p \simeq 0$, to a far broader distribution in the interior plus two atoms of probability at $i = \pm 40$ when $p \simeq 1$, is clearly evident. Note that the plot for $p = 0.5$ corresponds to the symmetric simple (i.e. uncorrelated) random walk and therefore has a binomial structure. When $p \simeq 0$ the walk reverses direction at almost every step, and so there is very little movement away from the origin. Conversely, when $p \simeq 1$ the walk is likely to travel in the same direction for a large number of steps before direction reversal occurs, though the probability, p^{n-1}, that states $i = \pm n$ are occupied does not become appreciable until p lies very close to one. Even with $p = 0.9$ the upturn in probability from $r_{38}(40) = 0.0057$ to $r_{40}(40) = 0.0082$ is still quite small; indeed, for $r_{40}(40) + r_{-40}(40) = p^{39} = 0.9$ (say) we require $p = 0.9^{1/39} = 0.9973 \simeq 1$. Figure 4.3b shows the corresponding plots for $p_i(40)$ ($q_i(40)$ is the mirror image since $q_i(n) = p_{-i}(n)$), and we see that their structure is essentially the same as that shown in Figure 4.3a apart from a slight shift to the right caused by the initial positive movement away from $i = 0$.

Simulations of this process are easily constructed by including the direction of travel in the simple random walk algorithm (4.1.41). For $\{U\}$ a sequence of independent uniform $U(0,1)$ random variables, with $X_0 = 0$ and $\Pr(X_1 = 1) = \Pr(X_1 = -1) = 1/2$, and the 'velocity' $v = 1$ or -1 denoting motion to the right or left, respectively, we have:

for $n = 1$ (4.4.21)

    if $U \leq 1/2$ then $X_1 = 1$ and $v = 1$, else $X_1 = -1$ and $v = -1$;

whilst for $n > 1$

    if $U \leq p$ and $v = -1$ then $X_n = X_{n-1} - 1$ and $v = -1$

    if $U > p$ and $v = -1$ then $X_n = X_{n-1} + 1$ and $v = +1$ (4.4.22)

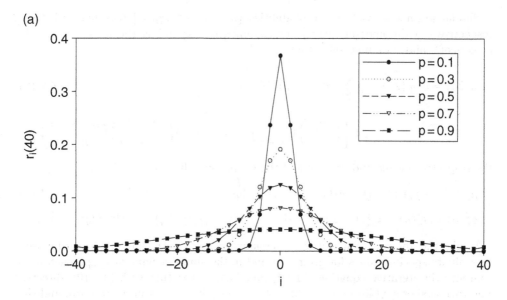

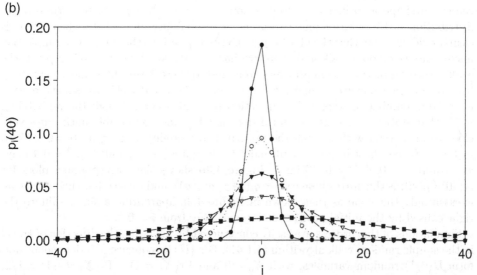

Figure 4.3 Values of (a) $r_i(40)$ and (b) $p_i(40)$ for $i = -40, -38, \ldots, 40$ and $p = 0.1, 0.3, 0.5, 0.7$ and 0.9.

if $U \leq p$ and $v = +1$ then $X_n = X_{n-1} + 1$ and $v = +1$
if $U > p$ and $v = +1$ then $X_n = X_{n-1} - 1$ and $v = -1$.

Figure 4.4 shows three realizations with (a) $p = 0.1$ (high chance of direction reversal), (b) $p = 0.5$ (simple symmetric random walk) and (c) $p = 0.9$ (low chance of direction

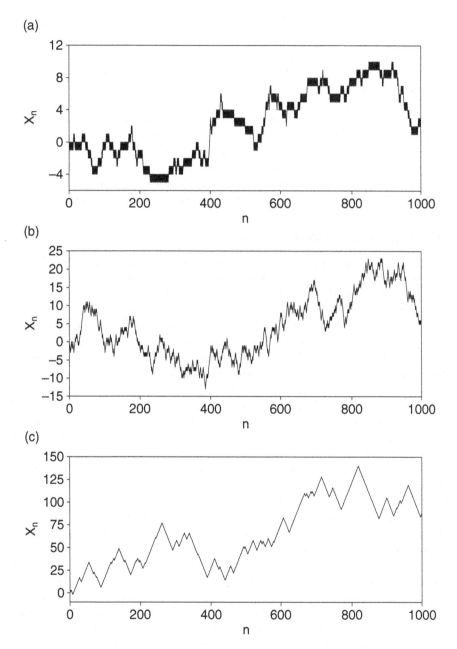

Figure 4.4 Simulated correlated random walk showing 1000 steps for (a) $p = 0.1$, (b) $p = 0.5$ and (c) $p = 0.9$.

reversal), and these highlight the different modes of behaviour of the process. In (a) we have high-frequency oscillations that hover around the start point $X_0 = 0$; (b) corresponds to Figure 4.2 for the symmetric simple random walk; whilst (c) exhibits geometrically distributed run lengths with parameter p. Note how the range of i-values covered increases with p.

In the above constructions the velocity of the process has been taken to be either $v = +1$ or -1, and an obvious generalization is to replace v by the random velocity V_n at time n. Renshaw (1987), for example, considers $V_n = V_{n-1} + \epsilon_n$ where $\{\epsilon_n\}$ are i.i.d. Bernoulli random variables; though many other forms are clearly possible and would be well worth investigating. Furthermore, modelling velocity, rather than displacement, gives rise to the Ornstein–Uhlenbeck process which plays a key role in the analysis of Markov processes in continuous space and time (see Section 6.3).

4.4.3   Application to share trading

Early studies of share price movement were based on the simple random walk, and hence presumed that price changes are uncorrelated over time. The Bachelier–Osbourne model (Bachelier, 1900; Osbourne, 1959), for example, uses this construct with Normally distributed price changes, whilst in the Black–Scholes (1973) model the value of the underlying stock of a financial option behaves as Brownian motion. However, such processes do not take into account the notion that share prices move with inertia driven by short-term market sentiment, and by employing the correlated random walk we can take such momentum into consideration. For Figure 4.4c clearly bears some degree of similarity to the time-series, $\{X_n\}$, of daily share prices. So given that these often show local upward or downward trends over periods of a few days, the use of this albeit very simple correlated random walk process in the development of buy/sell strategies offers considerable scope for application in the financial markets. Purely as an instructional tool, let us therefore see whether individual investors might be able to profit by using the c.r.w. in order to exploit short to medium range oscillations in share price.

Using this approach to model share prices in financial markets is an area of research that is receiving increasing interest (see, for example, Allaart, 2004a&b; Allaart and Monticino, 2001). For it allows us to investigate the effects of applying trading rules based on different assumptions relating to initial conditions, transaction charges, discount rates, buying and selling rules, etc. The main hurdle to investing money profitably lies in the cost of making a buy or sell transaction, which comprises broker costs, stamp duty and bid-sell spread prices. The two main investment strategies used by many private investors exemplify the way in which such charges can have a dramatic effect on overall profit or loss. 'Buy-and-hold' (BH) is the simplest strategy since shares are purchased at time zero, held in the interim, and then sold at some later time n. In contrast, employing a 'trailing stop-loss' (TSL) allows for multiple trades during a given time period, with specific rules dictating when shares are to be bought and sold. In a roughly level market which displays fairly low volatility, BH should outperform TSL since the potential gain expected from short-term profit taking is likely to be outweighed by transaction costs. In contrast, TSL not only prevents the chance of potential financial ruin during a market crash (e.g. the 'financial crunch' caused by a run on the banks in 2009), but it also allows profit to be made during the subsequent market recovery.

Now any successful strategy requires a reasonable rate of return just to avoid losing money. This cannot happen under a simple random walk model since even without costs the expected profit is zero. So does the situation change under a correlated random walk scenario? Intuitively, turning a profit should be possible since if $p \simeq 1$ then the share price is likely to exhibit long (geometrically distributed) run lengths before changing direction. Hence buying or selling after each upturn or downturn should prove to be profitable, provided we can determine where these change points occur. So key questions include: how large does p need to be in order to overcome transaction charges; what is the expected rate of return for a given value of p and amount invested; how do we estimate p when market sentiment is constantly changing; and, under what conditions does TSL produce a higher expected profit than BH? These and other important issues, such as making the model more applicable to daily share price movement by incorporating multiple velocities $v = 0, \pm1, \pm2, \ldots$ (here we simple take $v = \pm1$) open up a huge range of problems for future study (see Hutton, 2007).

Since bull (rising) and bear (falling) markets may possess different underlying characteristics, let us consider the asymmetric version of the c.r.w. defined through (4.4.4)–(4.4.5). That is, there are four possible daily price changes, namely

$$\Pr(X_n = i+1 | X_{n-1} = i, X_{n-2} = i-1) = p_1$$
$$\Pr(X_n = i-1 | X_{n-1} = i, X_{n-2} = i-1) = q_1 = 1 - p_1$$
$$\Pr(X_n = i-1 | X_{n-1} = i, X_{n-2} = i+1) = p_2 \quad (4.4.23)$$
$$\Pr(X_n = i+1 | X_{n-1} = i, X_{n-2} = i+1) = q_2 = 1 - p_2,$$

i.e. single unit changes corresponding to up-up, up-down, down-down and down-up. Though more realistic general velocity models can be constructed (e.g. allowing $v = 0$ enables the share price to remain unchanged, whilst taking $|v| = 0, 1, 2, \ldots$ allows for a general daily price change distribution), system (4.4.23) is ideal for illustrating the basic procedure. At time $n = 0$ suppose that the investor has an amount $I(0)$ to invest, with buy and sell transactions both costing C units. Then if at time n he has cash $I(n)$ and wishes to buy, then the number of shares purchased at price $X(n)$ is

$$S(n) = (I(n) - C)/X(n); \quad (4.4.24)$$

whilst if he wishes to sell $S(n)$ shares then the cash raised is

$$I(n) = S(n)X(n) - C. \quad (4.4.25)$$

Whence combining (4.4.23)–(4.4.25) in order to form a c.r.w. trading algorithm enables us to simulate the system and thereby determine how choice of parameters affects overall profit/loss. Readers who are seriously interested in the technical aspects of buy/sell strategies are advised to consult a handbook of trading tools, such as Achelis (2001), before following this investment route, since informed decision making usually

Table 4.1: Final share or cash value under c.r.w. strategies (1)/(2) and end price at time $n = 500$.

p_1	0.5	0.6	0.7	0.8	0.9
0.5	169/744 (116)	365/915 (160)	834/1406 (240)	1542/1750 (300)	3114/3555 (440)
0.6	16/424 (40)	81/338 (74)	588/820 (190)	951/1127 (248)	2445/2871 (416)
p_2 0.7	274/502 (*)	115/147 (*)	577/617 (118)	981/848 (190)	2116/2029 (364)
0.8	649/772 (*)	357/402 (*)	693/232 (*)	1203/877 (152)	2294/2182 (308)
0.9	690/912 (*)	611/804 (*)	695/634 (*)	947/996 (*)	1035/1037 (*)

involves far more than just charting share prices. However, as many punters base their buy/sell decisions on very basic information, consider two strategies based purely on local share price change:

(1) buy on down-up-up and sell on up-down-down;
(2) buy on down-up-up-up and sell on up-down-down-down.

Option (1) corresponds to instinctive short-term trading, whilst option (2) is more conservative since it relates to breaking through trading levels. Suppose that $I(0) = 1000$ units (e.g. pounds, dollars, euros, etc.) and that each transaction costs $C = 10$ units. Then since we wish to consider both bear and bull markets let p_1 and p_2 vary over the values $0.5, 0.6, \ldots, 0.9$ under both options. Since this exercise is purely illustrative we shall consider a single realization (using the same initial seed in each case) over 500 trading days following the first buy, for each of these possibilities. Table 4.1 shows the ultimate value (in cash or share value) together with the final price; we stop the process, denoted by (*), if $I(n)$ drops to $C = 10$. Four main points are immediately apparent from this single run.

(a) Unless p_1 substantially exceeds 0.5 it would be better not to trade at all. For the cost of constantly trading lies not just in the transaction charges, but also in that a downturn immediately following a buy causes a loss of 2 and 3 share price points under options (1) and (2), respectively. So short-term oscillations can severely affect short-term profit, which explains why small investors are advised to buy-and-hold shares for the medium-term (i.e. weeks or months) rather than for days, or even hours.
(b) When the share price collapses substantially less money is lost by using TSL option (1) or (2) rather than by 'hanging-on' under BH. Though in practice use of a pure stop-loss 90% or 85% 'get-out' strategy (SL), in which shares are sold as soon as they drop by 10% or 15% of their maximum (or purchase) value, and the money obtained invested elsewhere, would come into play.
(c) If the underlying structure is generated by a c.r.w., then avoid trading in a bear market.
(d) Real-life values of p_1 and p_2 are far more likely to lie nearer to 0.5 than 1, and it is in this regime that option (2) appears to work better.

Expanding this exercise into a proper simulation experiment with say 100 replications over the grid of parameter values $p_1, p_2 = 0, 0.05, 0.1, \ldots, 1.0$ enables a full empirical study to be made of the investment surface as a function of p_1 and p_2. In particular, the way in which the median, quartile and range change can prove to be especially informative. Moreover, it indicates how small p_1 and p_2 can be for the long-term gain of a strategy to be positive. Indeed, simulation provides the obvious way of studying and comparing a large set of potential investment strategies, especially when these are more realistic, and hence more complex, than the above simple example. For switching between different versions usually involves little more than changing a few lines of code. However, it is still useful to derive analytic results wherever possible, since these can shed additional light on overall behaviour.

Consider, for example, the extreme case where $p_1, p_2 \simeq 1$. Although the expected price change over all possible paths is zero, any given path will comprise large geometrically distributed increasing/decreasing runs. So following a down-up-up movement, the expected price rise before the up-run terminates is large. Thus the strategy of buying (say) on down-up-up and selling on up-down-down leads to a large expected long-term profit. Though note that if a buy on down-up-up is immediately followed by a price reversal due to the next two steps being down-down, then there is a loss of 1 unit. Now under strategy (1) only an oscillating down-up-down movement avoids a buy or a sell, and leaves the overall gain unchanged, so a formal analysis based on sample paths is perfectly feasible. This approach is still possible under strategy (2), though the analysis is more involved. So let us investigate using a TSL strategy under the assumption that at each time step $t = 1, 2, \ldots$ the share price $X(t)$ changes value by ± 1 units. We buy if capital is available and $X(t) = X(min) + \alpha$ and sell if shares are held and $X(t) = X(max) - \beta$, where $X(min)$ and $X(max)$ denote the local minimum and maximum share price (i.e. since the last trade). Let $S(t)$ and $I(t)$ denote the number of shares and amount of capital held at time t, and suppose there is a fixed transaction charge C for each buy or sell trade. If the time at which shares are first purchased is t_1 with share price $X(t_1)$, i.e. $X(t) = X(min) + \alpha$ for the first time at $t = t_1$, then the number of shares bought is

$$S(t_1) = (I(0) - C)/X(t_1). \tag{4.4.26}$$

So the associated value is $I(0) - C$ since the immediate effect of the trade is to reduce the value of the asset by C. These shares are then held until we reach the threshold β below X_{max}, at which point we sell. So if this occurs at time $t = t_2$ then the capital raised by the sale is

$$I(t_2) = S(t_1)X(t_2) - C = (I(0) - C)[X(t_2)/X(t_1)] - C, \tag{4.4.27}$$

which clearly shows the impact that C makes on potential profits. Now bearing in mind the complexity of the p.d.f. $\Pr(X_n = x_n)$ given in solution (4.4.2), and the iterative nature of the expressions for $I(t_3), I(t_4), \ldots$, determination of the full probability structure for $I(t_n)$ is clearly an unrealistic proposition even for this simple model. However, useful progress can be made if we restrict our attention to expected values.

For under our basic c.r.w. model alternating upward and downward runs are geometrically distributed with expected lengths p_1/q_1 and p_2/q_2, respectively. Let $\{x\}$ denote the smallest integer $\geq x$. Then since each share price cycle comprises an upward run, a downward run, and two intervening steps where it changes its direction of movement, the expected time taken to complete one cycle is $T = \{p_1/q_1\} + \{p_2/q_2\} + 2$. Whilst the expected change in share price is $\Delta X = \{p_1/q_1\} - \{p_2/q_2\}$. Whence combining these two results with the total buy-and-sell transaction charge, $2C$, yields the expected profit/loss over a trading cycle as a function of p_1, p_2 and C. We can examine volatility in a similar fashion, since $\text{Var}(T) = \text{Var}(\Delta X) = p_1/q_1^2 + p_2/q_2^2$. For the purpose of illustration take $\alpha = \beta = 1$ and assume that at time $t = 0$ no shares are held and that the first step is an increase to $X(1) = X(0) + 1$, so $X(min) = X(0)$. Thus since $\alpha = 1$ we purchase

$$S(1) = (I(0) - C)/X(1) \tag{4.4.28}$$

shares. In terms of expected values, the price then rises for $\{p_1/q_1\}$ time units to $X(max) = X(\{p_1/q_1\} + 1)$ before there is a decrease of size $\beta = 1$. At this point the shares are sold at time $\{p_1/q_1\} + 2$, with the capital raised being

$$I(\{p_1/q_1\} + 2) = S(1)X(\{p_1/q_1\} + 2) - C. \tag{4.4.29}$$

The price then continues to fall for $\{p_2/q_2\}$ time units, during which period no shares are purchased, and at the end of which it reaches a new local minimum. The price then rises by 1 unit, at which point a new cycle starts at time $T + 1$ with the purchase of a fresh round of shares priced at $X(\{p_1/q_1\} + \{p_2/q_2\} + 3)$. Whence continuing in this vein yields the recurrence relation for the expected value of investment capital held at the end of the kth cycle, namely

$$I(kT+1) = [I((k-1)T+1) - C]\left(1 + \frac{\{p_1/q_1\} - 1}{(k-1)\Delta X + 1} - C\right). \tag{4.4.30}$$

This easily extends to cover general integers α and β, whence an induction argument yields the solution (Hutton, 2007)

$$I(kT + t_0) = (I(0) - C)\prod_{i=1}^{k}\left(1 + \frac{\{p_1/q_1\} + 1 - (\alpha+\beta)}{(k-i)\Delta X + X_0}\right)$$

$$- 2C\sum_{m=1}^{k-1}\prod_{i=1}^{k}\left(1 + \frac{\{p_1/q_1\} + 1 - (\alpha+\beta)}{(k-i)\Delta X + X_0}\right) - C \tag{4.4.31}$$

where

$$(t_0, X_0) = \begin{cases} (\alpha, X(0) + \alpha) & \text{if first step is upwards} \\ (1 + \{p_2/q_2\} + \alpha, X(0) - 1 - \{p_2/q_2\} + \alpha) & \text{if first step is downwards.} \end{cases} \tag{4.4.32}$$

Thus, as would be expected, the larger the length of the upward price runs (i.e. the larger the value of $\{p_1/q_1\}$), the higher the expected return is for fixed k. Moreover, the numerator $\{p_1/q_1\} + 1 - (\alpha + \beta)$ in (4.4.31) is maximized by minimizing $(\alpha + \beta)$, which occurs at $\alpha = \beta = 1$. Whilst the denominator is minimized when X_0 takes its minimum value, again corresponding to $\alpha = 1$. Thus the most profitable TSL strategy, with or without transaction costs, occurs under the rule $\alpha = \beta = 1$. Moreover, placing $I(kT + t_0) \geq I(0)$ yields a critical threshold for the charge to investment ratio, namely $C/I(0)$, which must not be exceeded if the investor is expected to show a profitable return. These, and other results by Hutton (2007), provide a good indication of when using TSL or BH, or just investing through a 'safe' savings institution, are likely to offer the best option for given p_1 and p_2 estimated from historical data on share price development. Though probability, rather than expected value, questions relating to final profit/loss, worst and best case scenarios, etc. are best tackled through subsequent simulation analyses. This combination of a theoretical analysis, followed by a simulation study, offers a powerful way forward for private investors, many of whom will not be expert traders, to develop optimal strategies for buying and selling in the stock market. Also, using the resulting preferred algorithms on historical time-series for specific market sectors provides a measure of confidence on the c.r.w. approach before any money is committed. This is especially relevant in sectors where the daily price change, v, can show a large degree of variation. For then we have to allow the random variable v to follow a p.d.f. over $v = 0, \pm 1, \pm 2, \ldots$, and the ensuing multiple velocity correlated random walk structure is substantially more complex to investigate. This fusion of stochastic modelling and financial trading is fast developing into a potentially huge growth industry, and the study of c.r.w. related algorithms which enable the inclusion of market sentiment clearly has a valuable role to play.

5
Markov chains

In the simple random walk not only does the position X_t after t steps depend solely on the previous position X_{t-1}, but the size increment Z_t is restricted to the three values -1, 0 and 1. Whilst this is fine for describing successive events in say a simple birth–death–immigration process, occasions will often arise where this basic structure breaks down. Obvious examples include the birth of twins, mass immigration, annihilation of the whole population, or describing population size over a coarse grid of event times in which the probability of more than one event in the time interval $(t, t+h)$ cannot be ignored. It then becomes necessary to construct a general transition structure over the integers in which a transition from state i to state j occurs with probability p_{ij}. Such a process is called a *Markov chain* if for any discrete set of values $a, b, c, \ldots, i, j$ we have

$$\Pr(X_{t+1} = j | X_0 = a, X_1 = b, \ldots, X_t = i) = \Pr(X_{t+1} = j | X_t = i). \tag{5.0.1}$$

That is, the conditional distribution of X_{t+1} given $X_0, X_1, \ldots, X_t$ depends only on the value of X_t and not on the earlier values $X_0, X_1, \ldots, X_{t-1}$.

The simple random walk is clearly equivalent to a Markov chain, since the transition probabilities

$$p_{ij} = \begin{cases} q & j = i-1 \\ r = 1-p-q & j = i \\ p & j = i+1 \\ 0 & j < i-1 \text{ and } j > i+1 \end{cases} \tag{5.0.2}$$

over the doubly infinite range $-\infty < i, j < \infty$ correspond to the transition matrix

$$\mathbf{P} \equiv \begin{pmatrix} \vdots & \vdots & \vdots & \vdots & \vdots \\ \cdots & q & r & p & 0 & 0 & \cdots \\ \cdots & 0 & q & r & p & 0 & \cdots \\ \cdots & 0 & 0 & q & r & p & \cdots \\ \vdots & \vdots & \vdots & \vdots & \vdots \end{pmatrix}. \tag{5.0.3}$$

Whilst for our random walk examples over the finite state space $i = 0, 1, \ldots, a$ we have

$$\mathbf{P} = \begin{pmatrix} p_{00} & p_{01} & 0 & 0 & \cdots & 0 & 0 & 0 \\ q & 1-p-q & p & 0 & \cdots & 0 & 0 & 0 \\ 0 & q & 1-p-q & p & \cdots & 0 & 0 & 0 \\ \vdots & \vdots & \vdots & \vdots & \cdots & \vdots & \vdots & \vdots \\ 0 & 0 & 0 & 0 & \cdots & q & 1-p-q & p \\ 0 & 0 & 0 & 0 & \cdots & 0 & p_{a,a-1} & p_{aa} \end{pmatrix}, \qquad (5.0.4)$$

where: (i) $p_{00} = p_{aa} = 1$ and $p_{01} = p_{a,a-1} = 0$ for absorbing barriers at 0 and a (Section 4.2); and, (ii) $p_{00} = 1-p$, $p_{01} = p$, $p_{a,a-1} = q$ and $p_{aa} = 1-q$ for reflecting barriers at 0 and a (Section 4.3).

The tri-diagonal structure of these two examples means that off-diagonal elements are zero. Now a common industrial situation for which the matrix is full, in the sense that most of its elements are positive, concerns *system reliability*. For this may be increased by including redundant components which can take over from a failed operating component. Suppose, for example, that an aircraft guidance system comprises two identical units, each of which operates independently of the other. In each time unit each working component can fail with probability q and each failed component can be mended with probability p. Then if i and j denote the number of working components at the beginning and end of each time period, we have $p_{00} = (1-p)^2$, $p_{12} = p(1-q)$, etc., whence

$$\mathbf{P} = \begin{pmatrix} (1-p)^2 & 2p(1-p) & p^2 \\ (1-p)q & pq + (1-p)(1-q) & p(1-q) \\ q^2 & 2q(1-q) & (1-q)^2 \end{pmatrix}. \qquad (5.0.5)$$

Whilst engineers monitoring the amount of water each week in a small reservoir that supplies a hydro-electric scheme might first choose the states

less than 10%	critically low	(CL)
10% – 24%	poor	(P)
25% – 49%	fair	(F)
50% – 74%	good	(G)
75% – 94%	excellent	(E)
95% – overflowing	critically full	(CF)

and then observe from historical data that the associated transition probabilities between these six states (CL,P,F,G,E,CF) are given by

$$\mathbf{P} = \begin{pmatrix} 0.45 & 0.34 & 0.13 & 0.05 & 0.02 & 0.01 \\ 0.28 & 0.39 & 0.19 & 0.08 & 0.04 & 0.02 \\ 0.13 & 0.23 & 0.35 & 0.18 & 0.07 & 0.04 \\ 0.08 & 0.16 & 0.21 & 0.32 & 0.15 & 0.08 \\ 0.05 & 0.09 & 0.12 & 0.22 & 0.39 & 0.13 \\ 0.00 & 0.02 & 0.90 & 0.17 & 0.46 & 0.26 \end{pmatrix}. \qquad (5.0.6)$$

Important operational questions then relate to how often the water level will be in each of these six states, with CL and CF being especially significant in terms of potential reduced power output and downstream flooding, respectively.

Network systems bridge the gap between sparse population processes such as (5.0.3) and full matrix systems like (5.0.6), since they often contain a mixture of positive and zero elements. In this scenario the state space comprises a collection of nodes connected by a set of paths. For example, arrivals at a hospital can present either at Accident and Emergency or at Outpatients, and whilst the transition rates and connected paths between say reception, triage, consultation, immediate treatment, X-ray, scan, operating theatre, admittance to an orthopaedic or medical ward, and exit, will clearly differ between the two modes of entry, these two subsystems are clearly interconnected. Construction of an accurate model of the full system then enables detailed studies to be made of how the hospital layout can be changed so that the health service there can maximize its efficiency.

The definition (5.0.1) of a Markov chain is surprisingly flexible. For example, in the case of the correlated random walk one might think that condition (5.0.1) is violated, since the transition probabilities take the form

$$\Pr(X_{t+1} = j | X_t = j-1, X_{t-1} = j-2) = p$$
$$\Pr(X_{t+1} = j | X_t = j+1, X_{t-1} = j+2) = p \qquad (5.0.7)$$
$$\Pr(X_{t+1} = j | X_t = j-1, X_{t-1} = j) = q$$
$$\Pr(X_{t+1} = j | X_t = j+1, X_{t-1} = j) = q,$$

and these involve the previous two events and not just the last one. However, if we consider pairs of states $\{(i,j)\}$, then we can transform the resulting two-dimensional state space into a one-dimensional space $\{[s]\}$ by making the correspondence

$$\begin{aligned}
&[0] = (0,0); \\
&[1] = (0,1); \quad [-1] = (0,-1); \\
&[2] = (1,2); \quad [-2] = (-1,-2); \quad [3] = (1,0); \quad [-3] = (-1,0); \\
&[4] = (2,3); \quad [-4] = (-2,-3); \quad [5] = (2,1); \quad [-5] = (-2,-1); \quad \text{etc.}
\end{aligned} \qquad (5.0.8)$$

That is, for $i > 1$ we have $[2i] = (i, i+1)$, $[2i+1] = (i, i-1)$, $[-2i] = (-i, -i-1)$ and $[-2i-1] = (-i, -i+1)$. Thus the **P**-matrix for $i > 4$ has entries

$$\Pr\{(i, i+1) \to (i+1, i+2)\} = \Pr\{[2i] \to [2i+2]\} = p$$
$$\Pr\{(i, i+1) \to (i+1, i)\} = \Pr\{[2i] \to [2i+1]\} = q$$
$$\Pr\{(i, i-1) \to (i-1, i-2)\} = \Pr\{[2i+1] \to [2i-1]\} = p$$
$$\Pr\{(i, i-1) \to (i-1, i)\} = \Pr\{[2i+1] \to [2i-2]\} = q,$$

with, for example, values for the central part of **P** over $i = -5, \ldots, 5$; $j = -7, \ldots, 7$ taking the form

$$\begin{pmatrix}
\cdot & \cdot & \cdot & \cdot & p & q & \cdot & \cdot & \cdot & \cdot & \cdot & \cdot \\
q & p & \cdot & \cdot & \cdot & \cdot & \cdot & \cdot & \cdot & \cdot & \cdot & \cdot \\
\cdot & \cdot & \cdot & \cdot & \cdot & q & \cdot & p & \cdot & \cdot & \cdot & \cdot \\
\cdot & \cdot & q & p & \cdot & \cdot & \cdot & \cdot & \cdot & \cdot & \cdot & \cdot \\
\cdot & \cdot & \cdot & \cdot & q & p & \cdot & \cdot & \cdot & \cdot & \cdot & \cdot \\
\cdot & \cdot & \cdot & \cdot & \tfrac{1}{2} & \cdot & \tfrac{1}{2} & \cdot & \cdot & \cdot & \cdot & \cdot \\
\cdot & \cdot & \cdot & \cdot & \cdot & \cdot & p & q & \cdot & \cdot & \cdot & \cdot \\
\cdot & \cdot & \cdot & \cdot & \cdot & \cdot & \cdot & \cdot & p & q & \cdot & \cdot \\
\cdot & \cdot & \cdot & \cdot & \cdot & p & \cdot & q & \cdot & \cdot & \cdot & \cdot \\
\cdot & \cdot & \cdot & \cdot & \cdot & \cdot & \cdot & \cdot & \cdot & \cdot & p & q \\
\cdot & \cdot & \cdot & \cdot & \cdot & \cdot & \cdot & \cdot & q & p & \cdot & \cdot
\end{pmatrix} \quad (5.0.9)$$

In general, the structure of **P** can be as complicated as we like, but since the fundamental ideas of working with this matrix construct can be developed by recourse to a simple 2×2 matrix, we shall first fix our attention on the basic binary process with states 0 and 1. However, before we proceed with this analysis it is worth commenting on the striking difference between Markov processes in continuous and discrete time. Consider, for example, the simple continuous-time birth–death process (Section 2.4). Here the transition probabilities relate to the infinitesimal time period $(t, t+dt)$, and take the simple form

$$p_{ij} = \begin{cases} \lambda i\, dt & j = i+1 \\ 1 - (\lambda + \mu)i\, dt & j = i \\ \mu i\, dt & j = i-1 \\ o(dt) & \text{otherwise.} \end{cases} \quad (5.0.10)$$

Thus to order $o(dt)$ only three types of event are possible, namely a single birth, a single death and no change. In contrast, suppose we take the corresponding discrete-time process in which population change occurs at the integer points $t = 1, 2, \ldots$, and each individual has *probability* λ of giving birth, and μ of dying, at each time point. Then whereas for the continuous-time process the *rates* $\lambda \geq 0$ and $\mu \geq 0$ are unbounded, we now have the restriction that Pr(an individual neither gives birth nor dies at a given time point t) $= 1 - \lambda - \mu \geq 0$. Moreover, instead of just three possible events listed in (5.0.10) we now have many more, since for $r, s = 0, 1, 2, \ldots$

$$\Pr(r \text{ births and } s \text{ deaths at time } t+1 \mid X_t = n) = \frac{n!}{r!s!(n-r-s)!}\lambda^r \mu^s (1-\lambda-\mu)^{n-r-s}$$

$$\equiv q(r, s; n) \quad (\text{say}), \quad (5.0.11)$$

where $0 \leq r+s \leq n$ and $q(r, s; n) = 0$ otherwise. So the corresponding transition probabilities take the substantially more messy form

$$p_{n_t, n_{t+1}} = \Pr(X_{t+1} = n_{t+1} \mid X_t = n_t) = \sum_{r=0}^{n_t} q(r, r - n_{t+1} + n_t; n_t). \quad (5.0.12)$$

That said, whereas the continuous-time probability structure necessitates solving the Kolmogorov forward or backward equations in terms of either the probability generating function or Laplace transforms, and then inverting these to obtain the probabilities themselves, in the discrete-time case we can write down the full probability solution directly. For given the transition probabilities (5.0.12), which are defined in terms of the trinomial probabilities $\{q(r,s;n)\}$ in (5.0.11), we have the full joint probabilities

$$\Pr(X_1 = n_1, X_2 = n_2, \ldots, X_t = n_t | X_0 = n_0) = \prod_{m=0}^{t-1} p_{n_m, n_{m+1}} \qquad (5.0.13)$$

together with the occupation probabilities

$$\Pr(X_t = n_t | X_0 = n_0) = \sum_{n_1, \ldots, n_{t-1}=0}^{\infty} \prod_{m=0}^{t-1} p_{n_m, n_{m+1}}. \qquad (5.0.14)$$

Although this last expression is rather opaque, it does generate an immediate solution without the need for solving any intermediate equations. Moreover, it is relatively straightforward to extend these results to cover the general transition matrix $\mathbf{P}$, and thereby derive maximum likelihood estimates of the transition rates. Readers interested in estimation and inference techniques for Markov chains will find a concise introduction in Bhat and Miller (2002).

5.1   Two-state Markov chain

Suppose we describe a population process not in terms of its size, but in terms of whether it is 'empty' or 'not empty' after the nth event. Then the simplest way of regarding this purely binary skeleton is to envisage it as a sequence of independent Bernoulli trials in which the probability of 'success' or 'failure' at each trial depends solely on the outcome of the previous trial. We let success (state 1) correspond to one of the outcomes, say not empty, whence failure (state 0) corresponds to the other, namely empty.

Denote the probability of success at the $(n+1)$th trial given that the nth trial resulted in failure by α, and the probability of failure at the $(n+1)$th trial given that the nth trial resulted in success by β. Then the corresponding probabilities of failure and success are $1-\alpha$ and $1-\beta$, respectively, whence the associated transition probability matrix

$$\mathbf{P} = (p_{ij}) = \begin{pmatrix} 1-\alpha & \alpha \\ \beta & 1-\beta \end{pmatrix} \qquad (i,j=0,1). \qquad (5.1.1)$$

Note that we are making the assumption that α and β are independent of time. Also, we exclude the two trivial cases: (i) $\alpha = \beta = 0$, since the system remains forever in its initial state; and, (ii) $\alpha = \beta = 1$, when the system alternates deterministically between 0 and 1.

5.1.1 Occupation probabilities

Let the row vector $\mathbf{p}^{(n)} = (p_0^{(n)}, p_1^{(n)})$ denote the *n-step occupation probabilities* of being in state 0 or 1 at time n; so at time $n = 0$ the initial probabilities of the two states are given by $\mathbf{p}^{(0)} = (p_0^{(0)}, p_1^{(0)})$. Then on constructing the forward equations we have

$$\Pr(X_n = 0) = \Pr(X_{n-1} = 0)p_{00} + \Pr(X_{n-1} = 1)p_{10}$$
$$\Pr(X_n = 1) = \Pr(X_{n-1} = 0)p_{01} + \Pr(X_{n-1} = 1)p_{11},$$

i.e.
$$p_0^{(n)} = p_0^{(n-1)}(1-\alpha) + p_1^{(n-1)}\beta$$
$$p_1^{(n)} = p_0^{(n-1)}\alpha + p_1^{(n-1)}(1-\beta). \tag{5.1.2}$$

These take the compact matrix form

$$\mathbf{p}^{(n)} = \mathbf{p}^{(n-1)}\mathbf{P}, \tag{5.1.3}$$

which iterates to yield the solution

$$\mathbf{p}^{(n)} = \mathbf{p}^{(n-2)}\mathbf{P}^2 = \cdots = \mathbf{p}^{(0)}\mathbf{P}^n. \tag{5.1.4}$$

This is a totally general result, in that for any transition matrix $\mathbf{P}$ we can write down the n-step transition probabilities $p_i^{(n)}$ in terms of the relation (5.1.4). Any problems in constructing the solution therefore centre wholly around finding the matrix power $\mathbf{P}^n$. For once the (j,k)th elements of $\mathbf{P}^n$, namely $p_{jk}^{(n)} = \Pr(\text{in state } k \text{ at time } n | \text{ in state } j \text{ at time } 0)$, are determined, they yield the solution to (5.1.4) as

$$p_0^{(n)} = p_0^{(0)} p_{00}^{(n)} + p_1^{(0)} p_{10}^{(n)}$$
$$p_1^{(n)} = p_0^{(0)} p_{01}^{(n)} + p_1^{(0)} p_{11}^{(n)}. \tag{5.1.5}$$

The $\{p_{jk}^{(n)}\}$ are called the *n-step transition probabilities*.

For example, on taking $\alpha = 0.4$ and $\beta = 0.2$, we have (to three decimal places)

$$\mathbf{P}^2 = \begin{pmatrix} 0.44 & 0.56 \\ 0.28 & 0.72 \end{pmatrix} \quad \text{and} \quad \mathbf{P}^5 = \begin{pmatrix} 0.340 & 0.660 \\ 0.330 & 0.670 \end{pmatrix}.$$

Whilst on computing $\mathbf{P}^{10}$, since $p_0^{(0)} + p_1^{(1)} = 1$ we have (to five decimal places)

$$\mathbf{p}^{(10)} = (p_0^{(0)}, p_1^{(0)}) \begin{pmatrix} 0.33340 & 0.66660 \\ 0.33330 & 0.66670 \end{pmatrix} \simeq (1/3, 2/3).$$

So we see that for most practical purposes the initial condition ceases to have any effect after $n = 10$ steps.

Suppose, therefore, we assume that the system settles down to a condition of statistical equilibrium as the number of steps n increases, i.e. $\mathbf{p}^{(n)} \to \boldsymbol{\pi} = (\pi_0, \pi_1)$ as $n \to \infty$. Then it follows from result (5.1.4) that $\boldsymbol{\pi}$ satisfies

$$\boldsymbol{\pi} = \boldsymbol{\pi}\mathbf{P}. \tag{5.1.6}$$

Thus

$$\pi_0 = \pi_0(1-\alpha) + \beta\pi_1, \qquad \text{i.e.} \quad \pi_0\alpha = \pi_1\beta$$
$$\pi_1 = \pi_0\alpha + \pi_1(1-\beta), \qquad \text{i.e.} \quad \pi_0\beta = \pi_0\alpha, \qquad (5.1.7)$$

which, on using the condition $\pi_0 + \pi_1 = 1$, yields the equilibrium solution

$$\pi_0 = \frac{\beta}{\alpha+\beta}, \quad \pi_1 = \frac{\alpha}{\alpha+\beta}. \qquad (5.1.8)$$

So in the above example with $\alpha = 0.4$ and $\beta = 0.2$ we have $\pi_0 = 1/3$ and $\pi_0 = 2/3$, in agreement with the numerical iteration.

If the initial probability distribution $\mathbf{p}^{(0)} = \boldsymbol{\pi}$, then repeated use of (5.1.3) yields

$$\mathbf{p}^{(1)} = \boldsymbol{\pi}\mathbf{P} = \boldsymbol{\pi}, \quad \mathbf{p}^{(2)} = \mathbf{p}^{(1)}\mathbf{P} = \boldsymbol{\pi}\mathbf{P} = \boldsymbol{\pi}, \quad \text{etc.,}$$

and so

$$\mathbf{p}^{(n)} = \boldsymbol{\pi} \qquad (n = 1, 2, \ldots). \qquad (5.1.9)$$

Thus $\mathbf{p}^{(n)}$ is a *stationary distribution* if $\mathbf{p}^{(0)} = \boldsymbol{\pi}$, i.e. it does not change with time. Care needs to be exercised if $\alpha = \beta = 1$, since (5.1.8) yields $\pi_0 = \pi_1 = 1/2$ and so $\boldsymbol{\pi} = (1/2, 1/2)$ is a stationary distribution. Yet it is not an equilibrium distribution, since if (say) $p_0^{(0)} = 1$ then the process oscillates between the two states 0 and 1 with $p_0^{(2n)} = p_1^{(2n+1)} = 1$ for $n = 1, 2, \ldots$. A parallel situation occurs when $\alpha = \beta = 0$, for in this case the system remains forever in the initial state.

5.1.2 Matrix solution 1

In order to determine the time-dependent probabilities (5.1.4), namely $\mathbf{p}^{(n)} = \mathbf{p}^{(0)}\mathbf{P}^n$, we have to be able to evaluate $\mathbf{P}^n$. For small matrices numerical values can be found quite conveniently simply by computing the powers of $\mathbf{P}$ directly. However, for large matrices, or where algebraic analysis is required, we need to employ the spectral representation of $\mathbf{P}$. Suppose that $\mathbf{P}$ is an $m \times m$ matrix with distinct eigenvalues $\omega_1, \ldots, \omega_m$. Then it is a standard result of matrix theory that we can find an $m \times m$ matrix $\mathbf{Q}$ such that

$$\mathbf{P} = \mathbf{Q} \begin{pmatrix} \omega_1 & 0 & \cdots & 0 \\ 0 & \omega_2 & \cdots & 0 \\ \vdots & & \cdots & \vdots \\ 0 & 0 & \cdots & \omega_m \end{pmatrix} \mathbf{Q}^{-1}. \qquad (5.1.10)$$

The columns of $\mathbf{Q}$ are the column eigenvectors $\mathbf{q}_1, \ldots, \mathbf{q}_m$ of $\mathbf{P}$, i.e. the solutions of the equations

$$\mathbf{P}\mathbf{q}_i = \omega_i \mathbf{q}_i \qquad (i = 1, \ldots, m), \qquad (5.1.11)$$

and the eigenvalues ω_i are solutions of the determinantal equation

$$|\mathbf{P} - \omega\mathbf{I}| = 0 \qquad (5.1.12)$$

where **I** denotes the identity matrix. Thus
$$\mathbf{P}^2 = \mathbf{Q}\text{diag}\{w_i\}\mathbf{Q}^{-1} \times \mathbf{Q}\text{diag}\{w_i\}\mathbf{Q}^{-1} = \mathbf{Q}\text{diag}\{w_i^2\}\mathbf{Q}^{-1},$$
and in general
$$\mathbf{P}^n = \mathbf{Q}\begin{pmatrix} w_1^n & 0 & \cdots & 0 \\ 0 & w_2^n & \cdots & 0 \\ \cdot & \cdot & \cdots & \cdot \\ 0 & 0 & \cdots & w_m^n \end{pmatrix}\mathbf{Q}^{-1}. \tag{5.1.13}$$

So for our 2×2 matrix (5.1.1), equation (5.1.12) gives
$$(1 - \alpha - w)(1 - \beta - w) - \alpha\beta = 0,$$
which yields the two roots $w_1 = 1$ and $w_2 = 1 - \alpha - \beta$. Thus $w_1 \neq w_2$ provided $\alpha + \beta \neq 0$, i.e. we avoid the trivial case $\alpha = \beta = 0$ where the process never leaves its initial state. Equation (5.1.11) for $\mathbf{q}_i = (u, v)^T$ then becomes
$$\begin{pmatrix} 1 - \alpha & \alpha \\ \beta & 1 - \beta \end{pmatrix}\begin{pmatrix} u \\ v \end{pmatrix} = w_i \begin{pmatrix} u \\ v \end{pmatrix} \quad (i = 1, \ldots, m),$$
which reduces to $u = v$ when $w_1 = 1$, and $\beta u + \alpha v = 0$ when $w_2 = 1 - \alpha - \beta$. Since u and v can only be determined to within a multiplicative constant, a user-friendly choice for $\mathbf{Q}$ is to take
$$\mathbf{Q} = \begin{pmatrix} 1 & \alpha \\ 1 & -\beta \end{pmatrix} \quad \text{for which} \quad \mathbf{Q}^{-1} = \frac{1}{\alpha + \beta}\begin{pmatrix} \beta & \alpha \\ 1 & -1 \end{pmatrix}.$$
Whence (5.1.13) yields
$$\mathbf{P}^n = \frac{1}{\alpha + \beta}\begin{pmatrix} 1 & \alpha \\ 1 & -\beta \end{pmatrix}\begin{pmatrix} 1 & 0 \\ 0 & (1 - \alpha - \beta)^n \end{pmatrix}\begin{pmatrix} \beta & \alpha \\ 1 & -1 \end{pmatrix}$$
$$= \frac{1}{\alpha + \beta}\begin{pmatrix} \beta & \alpha \\ \beta & \alpha \end{pmatrix} + \frac{(1 - \alpha - \beta)^n}{\alpha + \beta}\begin{pmatrix} \alpha & -\alpha \\ -\beta & \beta \end{pmatrix}, \tag{5.1.14}$$
and for any initial vector $\mathbf{p}^{(0)}$ we can now use (5.1.4) to find $\mathbf{p}^{(n)}$.

On recalling (5.1.8) we see from (5.1.14) that
$$p_0^{(n)} = \pi_0 + (1 - \alpha - \beta)^n(\pi_1 p_0^{(0)} - \pi_0 p_1^{(0)})$$
$$p_1^{(n)} = \pi_1 + (1 - \alpha - \beta)^n(\pi_0 p_1^{(0)} - \pi_1 p_0^{(0)}). \tag{5.1.15}$$

So provided $|1 - \alpha - \beta| < 1$, the time-dependent probabilities $\mathbf{p}^{(n)}$ approach the equilibrium probabilities $\boldsymbol{\pi}$ geometrically fast. This provides a formal proof that the equilibrium distribution does indeed exist. We shall see shortly that a parallel conclusion can be drawn for the general $m \times m$ matrix. For our numerical example with $\alpha = 0.4$ and $\beta = 0.2$, for which $\boldsymbol{\pi} = (1/3, 2/3)$, we have the convergence factor $(1 - \alpha - \beta) = 0.4$. Since $0.4^{10} = 0.000105 \simeq 0$, this verifies our previous conclusion that the initial condition ceases to have any real effect after as few as $n = 10$ steps.

5.1.3 Matrix solution 2

Although the matrix $\mathbf{Q}^{-1}$ can be written down straightaway in the 2×2 case, this will not be true in more general $m \times m$ situations, and so it is useful to develop an equivalent procedure which does not involve inverting a matrix. Paralleling (5.1.11), consider a new matrix $\mathbf{R}$ whose rows are the row eigenvectors $\mathbf{r}_1, \ldots, \mathbf{r}_m$ of $\mathbf{P}$, i.e. the solutions of the equations

$$\mathbf{r}_i \mathbf{P} = \omega_i \mathbf{r}_i \quad (i = 1, \ldots, m). \tag{5.1.16}$$

Then not only are the row and column eigenvectors orthogonal, in that

$$\mathbf{r}_i \mathbf{q}_j = 0 \quad (i, j = 1, \ldots, m; i \neq j), \tag{5.1.17}$$

but we can also multiply the $\mathbf{r}_i$ by non-zero constants to ensure that

$$\mathbf{r}_i \mathbf{q}_i = 1 \quad (i = 1, \ldots, m). \tag{5.1.18}$$

So the matrices

$$\mathbf{A}_i = \mathbf{q}_i \mathbf{r}_i \quad (i = 1, \ldots, m) \tag{5.1.19}$$

satisfy the relations

$$\mathbf{A}_i \mathbf{A}_j = \begin{cases} \mathbf{q}_i \mathbf{r}_i \mathbf{q}_j \mathbf{r}_j = \mathbf{q}_i \times 0 \times \mathbf{r}_j = 0 & \text{for } i \neq j \\ \mathbf{q}_i \mathbf{r}_i \mathbf{q}_i \mathbf{r}_i = \mathbf{q}_i \times 1 \times \mathbf{r}_i = \mathbf{A}_i & \text{for } i = j. \end{cases} \tag{5.1.20}$$

Whence on using the standard matrix result

$$\mathbf{P} = \sum_{i=1}^{m} \omega_i \mathbf{A}_i, \tag{5.1.21}$$

it follows that

$$\mathbf{P}^2 = \sum_{i,j=0}^{m} \omega_i \omega_j \mathbf{A}_i \mathbf{A}_j = \sum_{i=0}^{m} \omega_i^2 \mathbf{A}_i,$$

and in general

$$\mathbf{P}^n = \sum_{i=0}^{m} \omega_i^n \mathbf{A}_i. \tag{5.1.22}$$

Since determining $\mathbf{r}_i$ is no harder than finding $\mathbf{q}_i$, this representation provides a useful alternative to (5.1.13) as it avoids the extra, potentially awkward, step of determining $\mathbf{Q}^{-1}$.

Continuing our 2×2 example, the row eigenvector equation (5.1.16) for $\mathbf{r}_i = (u, v)$ becomes

$$(u, v) \begin{pmatrix} 1 - \alpha & \alpha \\ \beta & 1 - \beta \end{pmatrix} = \omega_i (u, v) \quad (i = 1, 2),$$

which reduces to $\alpha u = \beta v$ when $w_1 = 1$, and $u + v = 0$ when $w_2 = 1 - \alpha - \beta$. Thus

$$\mathbf{R} = \begin{pmatrix} c_1 \beta & c_1 \alpha \\ c_2 & -c_2 \end{pmatrix}$$

where the constants c_1 and c_2 are chosen to ensure that $\mathbf{r}_i \mathbf{q}_i = 1$ ($i = 1, 2$). Now

$$\mathbf{r}_1 \mathbf{q}_1 = c_1(\beta, \alpha) \begin{pmatrix} 1 \\ 1 \end{pmatrix} = c_1(\alpha + \beta) \quad \text{and} \quad \mathbf{r}_2 \mathbf{q}_2 = c_2(1, -1) \begin{pmatrix} \alpha \\ -\beta \end{pmatrix} = c_2(\alpha + \beta),$$

so $c_1 = c_2 = 1/(\alpha + \beta)$. Whence (5.1.19) becomes

$$\mathbf{A}_1 = \mathbf{q}_1 \mathbf{r}_1 = \frac{1}{\alpha + \beta} \begin{pmatrix} 1 \\ 1 \end{pmatrix} (\beta, \alpha) = \frac{1}{\alpha + \beta} \begin{pmatrix} \beta & \alpha \\ \beta & \alpha \end{pmatrix}$$

$$\mathbf{A}_2 = \mathbf{q}_2 \mathbf{r}_2 = \frac{1}{\alpha + \beta} \begin{pmatrix} \alpha \\ -\beta \end{pmatrix} (1, -1) = \frac{1}{\alpha + \beta} \begin{pmatrix} \alpha & -\alpha \\ -\beta & \beta \end{pmatrix},$$

and applying (5.1.22) yields

$$\mathbf{P}^n = \mathbf{A}_1 + (1 - \alpha - \beta)^n \mathbf{A}_2 = \frac{1}{\alpha + \beta} \begin{pmatrix} \beta & \alpha \\ \beta & \alpha \end{pmatrix} + \frac{1}{\alpha + \beta} \begin{pmatrix} \alpha & -\alpha \\ -\beta & \beta \end{pmatrix}, \quad (5.1.23)$$

in exact agreement with (5.1.14).

5.1.4 The Discrete Telegraph Wave

We essentially met the 2×2 transition matrix $\mathbf{P}$ in the construction of the correlated random walk (Section 4.4), albeit implicitly. For the matrix (5.1.1) corresponds to the step transition matrix, with states 0 and 1 corresponding to movement to the left and right, respectively. So in this context α (β) denote the probabilities of moving to the right (left) given that the previous step was to the left (right). The trajectory $\{X_n\}$ of the process generated by $\mathbf{P}$ is therefore of considerable interest, since its integral relates to the position taken by the associated random walk. We call $\{X_n\}$ the *Discrete Telegraph Wave*.

Suppose we are initially in state 0, and denote the random variable T_0 to be the time of first return to state 0. Thus $T = n$ if at times $1, 2, \ldots, n-1$ state 1 is occupied and at time n the process returns to state 0. We call T_0 the *recurrence time* of state 0, and denote its probability distribution by $\{f_{00}^{(n)}; n = 1, 2, \ldots\}$. Clearly, in this case, $T_0 - 1$ is the length of time the process spends continuously in state 1. Thus

$$f_{00}^{(n)} = \Pr(T_0 = n) = \alpha(1 - \beta)^{n-2} \beta \quad (n = 2, 3, \ldots) \quad (5.1.24)$$

with

$$f_{00}^{(1)} = 1 - \alpha. \quad (5.1.25)$$

This is a geometric-type distribution, and it is easily seen that T_0 has mean

$$E(T_0) = \frac{\alpha + \beta}{\beta} = \frac{1}{\pi_0}. \quad (5.1.26)$$

Similarly, the recurrence time of state 1, T_1, has probability distribution

$$f_{11}^{(n)} = \Pr(T_1 = n) = \beta(1-\alpha)^{n-2}\alpha \quad (n=2,3,\ldots) \tag{5.1.27}$$

with

$$f_{00}^{(1)} = 1 - \beta, \tag{5.1.28}$$

and mean

$$E(T_1) = \frac{\alpha + \beta}{\alpha} = \frac{1}{\pi_1}. \tag{5.1.29}$$

So in our example with $\alpha = 0.4$ and $\beta = 0.2$, for which $\pi_0 = 1/3$ and $\pi_1 = 2/3$, we have $E(T_0) = 3$ and $E(T_1) = 1.5$.

A similar analysis yields the distributions of empty and not-empty spells, where an empty spell of length E is defined as E empty trials followed by a not-empty trial. For given that state 1 is occupied at time 1, E is the length of time up to (but not including) the next occurrence of state 0. Hence

$$\Pr(E = n) = \Pr(X_1 = \cdots = X_n = 1, X_{n+1} = 0 | X_1 = 1)$$
$$= (1-\beta)^{n-1}\beta = 0.8^{n-1}0.2, \tag{5.1.30}$$

which is a pure geometric distribution (unlike $f_{00}^{(n)}$ and $f_{11}^{(n)}$) with mean

$$E(E) = \sum_{n=1}^{\infty} n(1-\beta)^{n-1}\beta = 1/\beta = 5. \tag{5.1.31}$$

Similarly, for the length, F, of a non-empty spell, we have

$$\Pr(F = n) = (1-\alpha)^{n-1}\alpha = 0.6^{n-1}0.4, \tag{5.1.32}$$

with

$$E(F) = 1/\alpha = 2.5. \tag{5.1.33}$$

Whence on defining a cycle, C, to be an empty spell followed by a non-empty spell, the length distribution of C is the convolution of two independent geometric distributions with mean $E(C) = E(E) + E(F) = 7.5$.

The moments of the Telegraph Wave provide a useful insight into the structure of the associated Markov process. For example, expressions (5.1.31) and (5.1.33) for $E(F)$ and $E(E)$ lead directly to estimates for α and β. Whilst if state 1 represents an amalgam of all states $N = 1, 2, \ldots$, for example the busy states in a discrete-time $M/M/1$ queue, then the variance and skewness of T_0, T_1, E and F may tell us a great deal about the slack and busy periods of the process.

5.1.5 Relation to the continuous-time process

Before we leave this 2×2 representation it is worth commenting on the relation between discrete-time and continuous-time processes. Denote θ and ϕ to be the continuous-time rates of moving from state 0 to 1, and 1 to 0, respectively, with

α and β the probabilities of success at time $(n+1)h$ given failure at time nh, and failure at time $(n+1)h$ given success at time nh, respectively. Then provided h is sufficiently small to render the probability of two or more events occurring in a time interval $(t, t+h)$ negligible, on placing $\alpha = \theta h$ and $\beta = \phi h$ we see that for all practical purposes the two processes are equivalent. Consider, for example, a single telephone line which is either free (state 0) or busy (state 1), and assume that any calls which arise when the line is busy are lost. Then the forward equations for the probabilities $\{p_0(t), p_1(t)\}$ are

$$dp_0(t)/dt = -\theta p_0(t) + \phi p_1(t)$$
$$dp_1(t)/dt = -\phi p_1(t) + \theta p_0(t), \quad (5.1.34)$$

with the solution

$$p_0(t) = p_0(0)e^{-(\theta+\phi)t} + [\phi/(\theta+\phi)]\{1 - e^{-(\theta+\phi)t}\}$$
$$p_1(t) = p_1(0)e^{-(\theta+\phi)t} + [\theta/(\theta+\phi)]\{1 - e^{-(\theta+\phi)t}\}. \quad (5.1.35)$$

So when $t \to \infty$ we recover the discrete-time equilibrium distribution, namely

$$p_0(\infty) = \phi/(\theta+\phi) = \beta/(\alpha+\beta) = \pi_0$$
$$p_1(\infty) = \theta/(\theta+\phi) = \alpha/(\alpha+\beta) = \pi_1. \quad (5.1.36)$$

Moreover, on substituting for α and β in the discrete solution (5.1.15), it follows that on writing $t = nh$ and then letting $h \to 0$ and $n \to \infty$, we obtain

$$p_0^{(n)} = \left(\frac{\phi}{\theta+\phi}\right) + \{1 - (\theta+\phi)h\}^{t/h}[\left(\frac{\theta}{\theta+\phi}\right)p_0^{(0)} - \left(\frac{\phi}{\theta+\phi}\right)(1 - p_0^{(0)})] \quad (5.1.37)$$

$$\to \left(\frac{\phi}{\theta+\phi}\right) + e^{-(\theta+\phi)t}[p_0^{(0)} - \left(\frac{\phi}{\theta+\phi}\right)] \quad (5.1.38)$$

with

$$p_1^{(n)} \to \left(\frac{\theta}{\theta+\phi}\right) + e^{-(\theta+\phi)t}[p_1^{(0)} - \left(\frac{\theta}{\theta+\phi}\right)], \quad (5.1.39)$$

in agreement with the continuous-time solution (5.1.35).

Let us now exploit this relationship between the discrete- and continuous-time Telegraph Waves to see whether collapsing the states of a process to a $(0, 1)$-structure might be a reasonable procedure in general. Consider, for example, the simple immigration–death process (Section 2.5) starting in state 0. Then we see from the coefficient of z^0 in expression (2.5.7) for the p.g.f. $G(z;t)$ that

$$p_0(t) = \exp\{-(\alpha/\mu)(1 - e^{-\mu t})\}. \quad (5.1.40)$$

Whence letting $t \to \infty$ yields the equilibrium probability

$$p_0(\infty) = e^{-\alpha/\mu}. \quad (5.1.41)$$

Comparing (5.1.41) with the Telegraph result (5.1.36) then shows that we require

$$\phi/(\theta + \phi) = e^{-\alpha/\mu}. \tag{5.1.42}$$

If μ is large relative to α, then the probability of being in states $2, 3, \ldots$ is fairly small, and the immigration–death process should be well-approximated by the Telegraph Wave. If it is not, then the success of the approximation clearly depends on the purpose it is being used for, since the p.d.f. of the time spent away from state 0 cannot be exponentially distributed as required for the two-state Markov process. In this situation we could adopt a renewal process approach (see Section 2.7.3) and use the mean recurrence time of state 0 as a second relation in order to express θ and ϕ in terms of α and μ.

As a second example, suppose we are interested in the probability that the simple immigration–birth–death process starting from state 0 is empty at time $t > 0$. Placing $n_0 = 0$ in (2.5.14) gives the exact probability

$$p_0(t) = [(\lambda e^{(\lambda-\mu)t} - \mu)/(\lambda - \mu)]^{-\alpha/\lambda}. \tag{5.1.43}$$

So if we simply wish to know whether $N = 0$ or $N > 0$ then the question arises as to whether the two-state process defined through the forward equations (5.1.34) can offer a reasonable approximation to (5.1.43)? On expanding this expression we see that for large t and $\lambda < \mu$

$$p_0(t) \sim [\mu/(\mu - \lambda)]^{-\alpha/\lambda}[1 + (\alpha/\mu)e^{(\lambda-\mu)t}]. \tag{5.1.44}$$

Whence on placing $p_0(0) = 1$ in (5.1.35) it follows that the two $p_0(t)$ probabilities match at $t = \infty$ if

$$\phi/(\phi + \theta) = [(\mu - \lambda)/\mu]^{\alpha/\lambda}. \tag{5.1.45}$$

One way of determining a second equation for (θ, ϕ) would be to equate the two exponents, that is to place $\theta + \phi = \mu - \lambda$, thereby obtaining

$$\theta = (\mu - \lambda)[1 - (1 - \lambda/\mu)^{\alpha/\lambda}] \quad \text{and} \quad \phi = (\mu - \lambda)(1 - \lambda/\mu)^{\alpha/\lambda}. \tag{5.1.46}$$

The success of this continuous-time Telegraph Wave representation could then be judged by comparing coefficients of the second-order term, namely the degree to which $\theta/(\theta + \phi)$ matches $[\mu/(\mu - \lambda)]^{-\alpha/\lambda}(\alpha/\mu)$, i.e. how close $(1 - \lambda/\mu)^{\alpha/\lambda}(1 + \alpha/\mu)$ lies to 1. This criterion produces a set of (λ, μ, α)-values over which the approximation may be deemed to be 'reasonable', though any other appropriate measure, e.g. the maximum distance between the two $p_0(t)$ functions over $0 < t < \infty$, could also be used. This *clipping procedure* is clearly of general applicability, and provides a powerful weapon for developing a greater understanding of single-state probabilities in complex stochastic systems.

This technique is closely related to the concept of *lumpability* in which the states of an elaborate model are lumped together when the identification of some of the states is deemed to be unnecessary. Care needs to be taken when making such a reduction since the underlying dependence structure means that it may result in a totally different process to the one envisaged. Markov chains in which states can be lumped together without losing the basic Markov property are called *lumpable Markov chains*. A necessary and sufficient condition for a Markov chain to be lumpable with respect to a partition $\tilde{S} = \{S_1, S_2, \ldots, S_m\}$ of the state space S is that for every pair of sets S_i and S_j the associated probabilities

$$\tilde{p}_{ij} = \sum_{r \in S_j} p_{kr} \qquad (i, j = 1, 2, \ldots, m) \tag{5.1.47}$$

for all $k \in S_i$ (Bhat and Miller, 2002). Unfortunately, this condition is quite restrictive, so lumpability cannot be used to simplify exact analyses in many situations. Though 'almost lumpable' situations clearly lend themselves to approximation possibilities. For a discussion of inherent problems and an appropriate statistical test for lumpability see Thomas and Barr (1977).

5.2 Examples of m-state Markov chains

In principle the extension of these ideas from two states to m states is straightforward, since we simply replace the 2×2 **P**-matrix (5.1.1) by its $m \times m$ counterpart. For the solution for the state occupation probabilities is given by (5.1.4), namely $\mathbf{p}^{(n)} = \mathbf{p}^{(0)} \mathbf{P}^n$. However, in order to develop $\mathbf{P}^n$ we need to evaluate the eigenvalues ω_i and the column eigenvectors $\mathbf{q}_i$; result (5.1.13) then yields the required solution provided we can determine the matrix inverse $\mathbf{Q}^{-1}$. An alternative approach is to calculate the row eigenvectors $\mathbf{r}_i$ and hence the $\mathbf{A}_i$-matrices (5.1.19), whence $\mathbf{P}^n$ now follows through (5.1.22). So because the general polynomial equation generated by (5.1.12), namely $|\mathbf{P} - \omega \mathbf{I}| = 0$, can be solved in closed algebraic form only when $m \leq 4$, both methods are problematic for $m \geq 5$ unless $\mathbf{P}$ is highly structured. Nevertheless, even if the ω_i cannot be solved explicitly the approach is still of considerable use. For define $\omega_1, \ldots, \omega_m$ to be the eigenvalues arranged in decreasing order of magnitude. Then since in many real-life population processes we have $\omega_1 = 1$, it follows from (5.1.22) that

$$\mathbf{p}^{(n)} = \mathbf{p}^{(0)} [\mathbf{A}_1 + \omega_2^n \mathbf{A}_2 + \mathrm{O}(\omega_3^n)]. \tag{5.2.1}$$

Thus we not only know the equilibrium probabilities, but also the manner and speed at which they are attained.

5.2.1 The Ehrenfest model

To illustrate the way in which this approach can be implemented, consider the classic Ehrenfest model of diffusion for the stochastic motion of molecules moving between two connected containers C and D (following Kac, 1947). Suppose there are $2a$ molecules distributed between C and D, and that the nth event involves a molecule being chosen at random from among the $2a$ molecules present and transferred to the other container. Denote the state of the system by the number of molecules, $a + i$, in container C ($i = -a, -a+1, \ldots, a$). Then as the probability that a C-molecule is chosen at the next event (in which case the system moves from state i to $i - 1$) is $(a+i)/2a$, and similarly the probability that a D-molecule is chosen is $(a-i)/2a$ (so the system moves from i to $i+1$), we have the transition probabilities $p_{i,i-1} = (a+i)/2a$ and $p_{i,i+1} = (a-i)/2a$. Note that if the system reaches a barrier then it is certain to move to the adjacent interior position at the next step. So for $i,j = -a,\ldots,a$, the transition matrix takes the form

$$\mathbf{P} = \begin{pmatrix} 0 & 1 & 0 & 0 & \cdots & 0 & 0 & 0 \\ \frac{1}{2a} & 0 & \frac{2a-1}{2a} & 0 & \cdots & 0 & 0 & 0 \\ 0 & \frac{2}{2a} & 0 & \frac{2a-2}{2a} & \cdots & 0 & 0 & 0 \\ \vdots & \vdots & \vdots & \vdots & \vdots & \vdots & \vdots & \vdots \\ 0 & 0 & 0 & 0 & \cdots & \frac{2a-1}{2a} & 0 & \frac{1}{2a} \\ 0 & 0 & 0 & 0 & \cdots & 0 & 1 & 0 \end{pmatrix}. \quad (5.2.2)$$

Since the chain will always move away from its current position i at the next step, an equilibrium situation clearly cannot exist, so here the solution to the vector-matrix equation $\boldsymbol{\pi} = \boldsymbol{\pi}\mathbf{P}$ corresponds to the stationary distribution (see the comment following (5.1.9)). The individual components of this stationary equation are given by

$$\pi_{-a} = (1/2a)\pi_{-a+1} \quad (5.2.3)$$
$$\pi_i = [(a-i+1)/2a]\pi_{i-1} + [(a+i+1)/2a]\pi_{i+1} \quad (i=-a+1,\ldots,a-1) \quad (5.2.4)$$
$$\pi_a = (1/2a)\pi_{a-1}. \quad (5.2.5)$$

Equations (5.2.3) and (5.2.4) yield

$$\pi_{-a+1} = (2a)\pi_{-a} = \binom{2a}{1}\pi_{-a} \quad \text{and} \quad \pi_{-a+2} = (2a/2)[(2a)-1]\pi_{-a} = \binom{2a}{2}\pi_{-a}, \quad (5.2.6)$$

whence writing $i = -a+j$ and substituting the trial solution

$$\pi_i = \binom{2a}{j}\pi_{-a} \quad (5.2.7)$$

into equation (5.2.4) shows (by induction) that (5.2.7) holds for all $i = -a,\ldots,a$. Note that equation (5.2.5) is satisfied automatically. To determine π_{-a} we simply use the fact that

$$1 = \sum_{i=-a}^{a} \pi_i = \sum_{j=0}^{2a} \binom{2a}{j}\pi_{-a} = (1+1)^{2a}\pi_{-a}, \quad \text{i.e.} \quad \pi_{-a} = 2^{-2a}.$$

Thus (5.2.7) corresponds to the symmetric binomial distribution

$$\pi_i = \binom{2a}{a+i} 2^{-2a} \quad (i=-a,\ldots,a). \quad (5.2.8)$$

An even simpler derivation follows from the (stationary) balance equations $\pi_i p_{i,i+1} = \pi_{i+1} p_{i+1,i}$, i.e.

$$\pi_{-a+i}[(2a-2)/2a] = \pi_{-a+i+1}[(i+1)/2a], \tag{5.2.9}$$

through sequential application of

$$\pi_{-a+i+1} = [(2a-i)/(i+1)]\pi_{-a+i} \qquad (i = 0, \ldots, 2a-1). \tag{5.2.10}$$

Suppose, for example, that $2a = 10^6$ molecules (a very small physical number), and consider the probability that the number of molecules in container C, say X, lies more than 1% away from the mean value of 500,000. Since the standard deviation of X is $\sqrt{(2a/4)} = \sqrt{(250,000)} = 500$, this is equivalent to lying further than (a massive) 10 standard deviations away from the mean value. So even departures as small as 1% from a 50:50 split between the two containers are extremely unlikely to occur. Indeed, direct use of the Central Limit Theorem yields

$$\Pr(X < 495,000 \text{ or } X > 505,000) = 2[1 - \Phi(10)] = 0.152 \times 10^{-22}. \tag{5.2.11}$$

Though as this theorem holds only in the centre of the distribution and not in the tails, (5.2.11) should be viewed in a purely qualitative context. *Maple* (for instance) provides a good way of assessing the tail accuracy of the Normal approximation, since it enables efficient computation of the binomial probabilities.

A formal proof of the symmetric binomial result (5.2.7) can be neatly tied in with the general time-dependent probability solution

$$\mathbf{p}^{(n)} = \mathbf{p}^{(0)} \mathbf{P}^n; \tag{5.2.12}$$

where from (5.1.13)

$$\mathbf{P}^n = \mathbf{Q}\,\mathrm{diag}\{\omega_0^n, \ldots, \omega_{2a}^n\}\mathbf{Q}^{-1}, \tag{5.2.13}$$

and $\mathbf{Q}$ denotes the matrix of column eigenvectors. For on denoting $\mathbf{R}$ to be the matrix of row eigenvectors $\mathbf{r}_i$ where

$$\mathbf{RP} = \mathbf{\Lambda R}, \qquad \text{i.e.} \quad \mathbf{P} = \mathbf{R}^{-1}\mathbf{\Lambda R}, \tag{5.2.14}$$

and $\mathbf{\Lambda} = \mathrm{diag}\{\omega_0, \ldots, \omega_{2a}\}$, we also have the parallel result that

$$\mathbf{P}^n = \mathbf{R}^{-1}\mathrm{diag}\{\omega_0^n, \ldots, \omega_{2a}^n\}\mathbf{R}. \tag{5.2.15}$$

Now on writing $\mathbf{r}_i = (x_0, \ldots, x_{2a})$ we see that the equations for $\mathbf{r}_i$, namely $\mathbf{r}_i\mathbf{P} = \omega_i \mathbf{r}_i$ ($i = 0, \ldots, 2a$), are given by

$$x_1 = 2a\omega x_0$$

$$2ax_0 + 2x_2 = 2a\omega x_1$$

$$(2a-1)x_1 + 3x_3 = 2a\omega x_2$$
$$\vdots \qquad \vdots \qquad (5.2.16)$$
$$2x_{2a-2} + 2ax_{2a} = 2a\omega x_{2a-1}$$
$$x_{2a-1} = 2a\omega x_{2a}.$$

Whence on denoting the generating function $G(z) \equiv \sum_{j=0}^{2a} x_j z^j$, system (5.2.16) can be written in the form

$$\sum_{j=0}^{2a} j x_j z^{j-1} + \sum_{j=0}^{2a}(2a-j)x_j z^{j+1} = 2a\omega \sum_{j=0}^{2a} x_j z^j,$$

i.e.
$$G'(z) + 2azG(z) - z^2 G'(z) = 2a\omega G(z). \qquad (5.2.17)$$

So
$$(1-z^2)\frac{dG(z)}{dz} = 2a(\omega - z)G(z), \qquad (5.2.18)$$

which integrates to give

$$\ln[G(z)] = a\int \left\{\frac{(\omega-1)}{1-z} + \frac{(\omega+1)}{1+z}\right\} dz = \text{constant} + \ln\{(1+z)^{a(1+\omega)}(1-z)^{a(1-\omega)}\}.$$

Hence the general solution to the differential equation (5.2.17) is

$$G(z) = C(1+z)^{a(1+\omega)}(1-z)^{a(1-\omega)} \qquad (5.2.19)$$

for some constant C.

Given the nature of the p.g.f. $G(z)$, we are clearly interested only in solutions that are polynomials of degree $2a$ in z. Now on writing $\omega = k/a$, we see from (5.2.19) that

$$G(z) = C(1+z)^{a+k}(1-z)^{a-k}, \qquad (5.2.20)$$

and this form yields $2a+1$ such polynomials provided we take $k = -a, -a+1, \ldots, a$. Hence the (descending) ω-values

$$\omega_0 = a/a = 1, \; \omega_1 = (a-1)/a, \; \ldots, \; \omega_{2a} = -a/a = -1 \qquad (5.2.21)$$

are the (distinct) eigenvalues of $\mathbf{P}$. The corresponding elements, x_j, of the row eigenvector $\mathbf{r}_i = (x_0, \ldots, x_{2a})$ are now given by the coefficients of z^j in (5.2.20).

Placing the eigenvalues (5.2.21) into result (5.2.15) yields (for large n) the limiting solution

$$\mathbf{p}^{(n)} \sim \mathbf{R}^{-1}\text{diag}\{1, 0, \ldots, 0, (-1)^n\}\mathbf{R}, \qquad (5.2.22)$$

which highlights the oscillatory nature of the process. Moreover, since the row eigenvector equation $\mathbf{r}_i\mathbf{P} = \omega_i \mathbf{r}_i$ transforms into the stationary equation $\boldsymbol{\pi} = \boldsymbol{\pi}\mathbf{P}$ at $\omega = \omega_0 = 1$ (remember that the persistent oscillations prevent the development of an equilibrium distribution), provided we make the correspondence $\pi_{-a+i} = x_i$ expression (5.2.19) with $\omega = 1$ yields the stationary p.g.f.

$$H(z) = z^{-a}G(z) = Cz^{-a}(1+z)^{2a}. \tag{5.2.23}$$

To determine C we simply use $H(1) = 1 = C2^{2a}$, i.e. $C = 2^{-2a}$, whence extracting the coefficient of z^i in (5.2.23) recovers the symmetric binomial probabilities (5.2.8).

Having determined $\mathbf{R}$, the solution for $\mathbf{p}^{(n)}$ is given by (5.2.15) provided we can evaluate $\mathbf{R}^{-1}$. Denote the matrices $(r_{jk}) = \mathbf{R}$ and $(r^{jk}) = \mathbf{R}^{-1}$ $(j, k = -a, \ldots, a)$. Then

$$\sum_{i=-a}^{a} r^{ji} r_{ik} = \delta_{jk},$$

where δ_{jk} denotes the Kronecker delta function, so

$$z^j = \sum_{k=-a}^{a} \delta_{jk} z^k = \sum_{i=-a}^{a} r^{ji} \sum_{k=-a}^{a} r_{ik} z^k. \tag{5.2.24}$$

Noting that the p.g.f. $H(z)$ in (5.2.23) corresponds to taking $\omega = \omega_0 = 1$ in (5.2.19), let

$$H_i(z) \equiv 2^{-2a} z^{-a} (1+z)^{a+i} (1-z)^{a-i} = \sum_{k=-a}^{a} r_{ik} z^k \tag{5.2.25}$$

denote the corresponding p.g.f.'s over the full set of eigenvalues $\omega_0 = 1, \ldots, \omega_{2a} = -1$. Then (5.2.24) takes the form

$$z^j = \left(\frac{1-z^2}{4z} \right)^a \sum_{i=-a}^{a} r^{ji} \left(\frac{1+z}{1-z} \right)^i. \tag{5.2.26}$$

Let $\theta = (1+z)/(1-z)$, i.e. $z = -(1-\theta)/(1+\theta)$. Then on putting $i = -j$ in (5.2.25), expression (5.2.26) gives

$$\sum_{i=-a}^{a} r^{ji} \theta^i = (-1)^{a+j} (1-\theta)^{a+j} (1+\theta)^{a-j} \theta^{-a} = (-1)^{a+j} 2^{2a} H_{-j}(\theta),$$

from which it follows that

$$r^{ji} = (-1)^{a+j} 2^{2a} r_{-j,i}. \tag{5.2.27}$$

Thus we have now found all the elements in the three matrices $\mathbf{R}^{-1}$, $\mathbf{\Lambda}$ and $\mathbf{R}$, whence direct matrix multiplication in (5.2.15) yields the solution for the time-dependent probabilities $\mathbf{p}^{(n)}$.

Whilst it might seem likely that a similar approach would yield the column eigenvectors $\mathbf{q}_i$, this turns out not to be the case. In essence, whilst the row eigenvector equations $\mathbf{r}_i \mathbf{P} = \omega_i \mathbf{r}_i$ mimic the stationary probability equations $\pi \mathbf{P} = \pi$ at $\omega_0 = 1$, the equivalent column eigenvector equation $\mathbf{P} \mathbf{c}_0 = \omega_0 \mathbf{c}_0 = \mathbf{c}_0$ has no such clear interpretation. For (in stationary probability terms) this would correspond to $\mathbf{P} \pi^T = \pi^T$, which runs counter to $\pi = \pi \mathbf{P}$ unless $\mathbf{P}$ is symmetric.

Algebraically, on writing $\mathbf{q}_j = (y_0, \ldots, y_{2a})$, the equation for $\mathbf{q}_j$ is given by

$$2ay_1 = 2a\omega y_0$$
$$y_0 + (2a-1)y_2 = 2a\omega y_1$$
$$2y_1 + (2a-2)y_3 = 2a\omega y_2$$
$$\vdots \qquad \vdots$$
$$iy_{i-1} + (2a-i)y_{i+1} = 2a\omega y_i \qquad (5.2.28)$$
$$\vdots \qquad \vdots$$
$$(2a-1)y_{2a-2} + y_{2a} = 2a\omega y_{2a-1}$$
$$2ay_{2a-1} = 2a\omega y_{2a}. \qquad (5.2.29)$$

Whence on denoting the generating function $J(z) \equiv \sum_{i=0}^{2a} y_i z^i$, multiplying system (5.2.28) by z^i and summing over $i = 0, 1, \ldots, 2a$ results in an equation that contains not only $J(z)$ and $J'(z)$, but also y_0 and y_{2a}. One method of solution would be to replace $J(z)$ by the full range p.g.f. $\tilde{J}(z) \equiv \sum_{i=-\infty}^{\infty} y_i z^i$, and then employ the Method of Images to determine appropriate values of y_i for $i < 0$ and $i > 2a$. However, this route for determining $\mathbf{Q}$, and thereby $\mathbf{Q}^{-1}$, would be substantially more tricky than determining the components (5.2.27) of $\mathbf{R}^{-1}$. Thus using (5.2.15) to form the row eigenvector solution

$$\mathbf{p}^{(n)} = \mathbf{p}^{(0)} \mathbf{R}^{-1} \mathrm{diag}\{\omega_0^n, \ldots, \omega_{2a}^n\} \mathbf{R} \qquad (5.2.30)$$

is considerably easier than using either the parallel column eigenvector solution (5.2.13), namely

$$\mathbf{p}^{(n)} = \mathbf{p}^{(0)} \mathbf{Q} \mathrm{diag}\{\omega_0^n, \ldots, \omega_{2a}^n\} \mathbf{Q}^{-1}, \qquad (5.2.31)$$

or the mixed row and column solution (5.1.22), i.e.

$$\mathbf{p}^{(n)} = \mathbf{p}^{(0)} \sum_{i=0}^{2a} \omega_i^n \mathbf{q}_i \mathbf{r}_i \qquad \text{(where } \mathbf{r}_i \mathbf{q}_i = 1; \ i = 0, \ldots, 2a\text{)}. \qquad (5.2.32)$$

This illustrates that in any given situation some thought must be given as to which is the 'best' approach to employ. Moreover, in spite of this Ehrenfest process having an inherently simple mathematical structure, the algebra involved in the determination of $\mathbf{p}^{(n)}$ is nontrivial. So for more general processes, it is clearly important to ensure that the most amenable method of solution is chosen at the outset.

Moments of the Ehrenfest model can be determined fairly easily. For as $\mathbf{p}^{(n+1)} = \mathbf{p}^{(n)} \mathbf{P}$ we have

$$2ap_0^{(n+1)} = p_1^{(n)}$$
$$2ap_1^{(n+1)} = 2ap_0^{(n)} + 2p_2^{(n)}$$

$$2ap_2^{(n+1)} = (2a-1)p_1^{(n)} + 3p_3^{(n)}$$

$$\vdots \qquad \vdots \qquad (5.2.33)$$

$$2ap_{2a-1}^{(n+1)} = 2p_{2a-2}^{(n)} + 2ap_{2a}^{(n)}$$

$$2ap_{2a}^{(n+1)} = p_{2a-1}^{(n)},$$

which are similar in structure to equations (5.2.16) for the components $(x_0, \ldots, x_{2a})$ of the row eigenvector $\mathbf{r}_0$ since $\omega_0 = 1$. Whence on paralleling the development of equation (5.2.17), but this time using the time-dependent p.g.f. $H_n(z) \equiv \sum_{i=0}^{2a} p_i^{(n)} z^n$, we may write equations (5.2.33) as

$$2aH_{n+1}(z) = (1-z^2)H_n'(z) + 2azH_n(z). \qquad (5.2.34)$$

Differentiating w.r.t. z gives

$$2aH_{n+1}'(z) = (1-z^2)H_n''(z) + 2(a-1)zH_n'(z) + 2aH_n(z), \qquad (5.2.35)$$

so the mean position, $\mu_n = H_n'(1)$, satisfies the recurrence relation

$$\mu_{n+1} = (1-1/a)\mu_n + 1. \qquad (5.2.36)$$

Whence substituting the trial solution

$$\mu_n = A(1-1/a)^n + B$$

for constants A and B, and using the initial condition $\mu_0 = i_0$, yields the solution

$$\mu_n = a + (i_0 - a)(1-1/a)^n. \qquad (5.2.37)$$

To determine the associated variance, $\sigma_n^2 = H_n''(1) + \mu_n - \mu_n^2$, we first differentiate (5.2.35) and place $z = 1$ to form

$$2aH_{n+1}''(1) = (2a-4)H_n''(1) + (4a-2)H_n'(1). \qquad (5.2.38)$$

Replacing μ_{n+1} by (5.2.36) then leads to the relation

$$\sigma_{n+1}^2 = (1-2/a)\sigma_n^2 - (\mu_n/a - 1)^2 + 1. \qquad (5.2.39)$$

Substituting the trial expression

$$\sigma_n^2 = A(1-2/a)^n + B(1-1/a)^{2n} + C$$

into (5.2.39) for (new) constants A, B and C, and equating coefficients of $(1-1/a)^{2n}$ and also terms independent of n, gives

$$\sigma_n^2 = A(1-2/a)^n - (i_0 - a)^2(1-1/a)^{2n} + a/2.$$

Finally, to determine A we simply use the initial condition $\sigma_0^2 = 0$, obtaining

$$\sigma_n^2 = (i_0 - a)^2[(1-2/a)^n - (1-1/a)^{2n}] + (a/2)[1 - (1-2/a)^n]. \qquad (5.2.40)$$

So as $n \to \infty$ we have $\mu_n \to a$ and $\sigma_n^2 \to a/2$, corresponding to the symmetric stationary binomial distribution (5.2.8) taken over $i = 0, 1, \ldots, 2a$.

5.2.2 The Perron–Frobenius Theorem

That the Ehrenfest process always moves away from its current state is encapsulated in the eigenvalues (5.2.21). For although $\omega_0 = 1$ and $|\omega_i| < 1$ for $i = 1, \ldots, 2a-1$, we see that $\omega_{2a} = -1$. Whence expression (5.2.22) reveals the oscillatory nature of the probabilities. However, in general there will often be a genuine approach to equilibrium, in which case we must have $\omega_0 = 1$ and $|\omega_i| < 1$ for all other i in order for $\mathbf{P}^n$ to converge. Since $\mathbf{P}$ is a transition probability matrix, it follows that all its elements $p_{ij} \geq 0$, and so $\mathbf{P} \geq 0$. Moreover, if any state can be reached from any other, i.e. *all states connect*, then $\mathbf{P}$ is deemed to be *irreducible*. An illustration of a *reducible* process is provided by the simple random walk over the states $i = 0, 1, \ldots$ with an absorbing barrier at $i = 0$ (see Section 4.2.2). In this situation, if $p < q$ the process ends up at $i = 0$ with probability 1. The trivial nature of this equilibrium distribution, namely $\boldsymbol{\pi} = (1, 0, 0, \ldots)$, suggests that (as far as equilibrium is concerned) we will miss little by focussing attention on irreducible processes. Note that it follows from results (4.2.23) and (4.2.24) that for two absorbing barriers (at $i = a$ and $i = b < a$) the absorption probabilities $p_a(\infty)$ and $p_b(\infty)$ are dependent on the initial position $j = b+1, \ldots, a-1$. So in this case the limiting distribution $\boldsymbol{\pi} = (r_{jb}, 0, \ldots, 0, r_{ja})$ does not correspond to an equilibrium distribution. In general, the basic result is as follows.

THE PERRON–FROBENIUS THEOREM *If $\mathbf{P}$ is an irreducible and non-negative square matrix, then*

(a) *$\mathbf{P}$ has a real positive eigenvalue ω_0 with the following properties:*

(b) *corresponding to ω_0 there is an column eigenvector $\mathbf{q}$ all of whose elements may be taken as positive, i.e. there exists a vector $\mathbf{q} > 0$ such that $\mathbf{Pq} = \omega_0 \mathbf{q}$;*

(c) *any other eigenvalue ω of $\mathbf{P}$ satisfies $|\omega| \leq \omega_0$;*

(d) *ω_0 increases when any element of $\mathbf{P}$ increases;*

(e) *ω_0 is a simple root of the determinantal equation $|\omega \mathbf{I} - \mathbf{P}| = 0$, where $\mathbf{I}$ denotes the identity matrix;*

(f) *ω_0 is bounded above by*

$$\omega_0 \leq \max_j \left(\sum_k p_{jk}\right) \quad \text{and} \quad \omega_0 \leq \max_k \left(\sum_j p_{jk}\right); \qquad (5.2.41)$$

(g) *if $\mathbf{P}$ has exactly t eigenvalues equal in modulus to ω_0, then these values are all different and are the roots of the equation $\omega^t - \omega_0^t = 0$.*

If $\mathbf{P}$ is irreducible, then its transpose $\mathbf{P}^T$ must be irreducible, and conversely. It therefore follows that $\mathbf{P}$ also has a positive row eigenvector corresponding to ω_0.

In general, the importance of result (5.2.41) is that the largest row or column sum automatically defines an upper bound to ω_0. However, since for (honest) probability transition matrices the row sums all equal one, this result just implies that $\omega_0 \leq 1$. Moreover, increasing one element of $\mathbf{P}$ must be counterbalanced by a simultaneous decrease in another element in the same row.

As an illustration of how the matrix decomposition works in practice, consider the discrete logistic paradigm of the continuous logistic model discussed earlier. Whereas the continuous process has the birth and death rates (3.2.35), namely $\lambda_N = N(a_1 - b_1 N)$ and $\mu_N = N(a_2 + b_2 N)$, in the discrete state space analogue we have the probabilities

$$p_{i,i+1} = i(a_1 - b_1 i) \quad (i = 1, 2, \ldots, i_{max})$$
$$p_{i,i-1} = i(a_2 + b_2 i) \quad \text{with} \quad p_{ii} = 1 - p_{i,i-1} - p_{i,i+1} \quad (5.2.42)$$

where i_{max} is the integer part of a_1/b_1. So we not only require $p_{10} = 0$ in order to have a genuine equilibrium distribution (otherwise we have the trivial solution $\pi_0 = 1$ with $\pi_i = 0$ for $i > 0$), but we also need the parameters a_1, b_1, a_2 and b_2 to be sufficiently small to ensure that $0 \leq p_{ij} \leq 1$ over $i, j = 1, 2, \ldots, i_{max}$.

Consider, for example, the birth-death transition probabilities $\lambda_i = 0.3i(1 - 0.2i)$ and $\mu_i = 0.1i(1 + 0.1i)$. Then the associated state space is $i = 0, \ldots, 5$, whence on replacing $\mu_1 = 0.11$ by 0 in order to enable the existence of a proper equilibrium distribution over the reduced state space $i = 1, \ldots, 5$, we have the transition probability matrix

$$\mathbf{P} = \begin{pmatrix} 0.76 & 0.24 & 0 & 0 & 0 \\ 0.24 & 0.40 & 0.36 & 0 & 0 \\ 0 & 0.39 & 0.25 & 0.36 & 0 \\ 0 & 0 & 0.56 & 0.20 & 0.24 \\ 0 & 0 & 0 & 0.75 & 0.25 \end{pmatrix}. \quad (5.2.43)$$

Clearly $\mathbf{P}$ is irreducible, and as $p_{ii} > 0$ for all $i = 1, \ldots, 5$ the process is also aperiodic (i.e. has period 1) since the process can remain where it is at each step. Hence there is just one eigenvalue of modulus 1, namely $\omega_1 = 1$, with $|\omega_i| < 1$ for $i = 2, \ldots, 5$. Using *Maple* (for example) to extract these remaining eigenvalues gives, in decreasing order of magnitude, $\omega_2 = 0.8053$, $\omega_3 = 0.4613$, $\omega_4 = -0.4461$ and $\omega_5 = 0.0400$. Since $\omega_2 > 0$, it follows that, as n increases, $\mathbf{p}^{(n)}$ moves in a non-oscillatory manner (unlike the Ehrenfest process) towards π at rate 0.8053^n. To determine the occupation probabilities

$$\mathbf{p}^{(n)} = \mathbf{p}^{(0)} \mathbf{Q} \text{diag}\{1, \omega_2^n, \ldots, \omega_5^n\} \mathbf{Q}^{-1}, \quad (5.2.44)$$

we simply form the corresponding matrix of column eigenvectors $\mathbf{Q} = (\mathbf{q}_1, \ldots, \mathbf{q}_5)$ and then evaluate (5.2.44) for given initial condition $\mathbf{p}^{(0)}$.

5.3 First return and passage probabilities

Suppose that the chain is initially in state j, and denote $f_{jj}^{(n)}$ to be the probability that the next occurrence of j is at time n. Thus $f_{jj}^{(1)} = p_{jj}$, and $f_{jj}^{(n)} = \Pr(X_r \neq j, r = 1, \ldots, n-1; X_n = j | X_0 = j)$ for $n = 2, 3, \ldots$. We call $f_{jj}^{(n)}$ the *first return probability* to state j at time n, and in a similar way define the *first passage probability* from j to k at time n as being the probability $f_{jk}^{(n)}$ that state k is avoided at times $1, \ldots, n-1$ and entered at time n, given that state j was initially occupied at time 0. Thus $f_{jk}^{(1)} = p_{jk}$ and $f_{jk}^{(n)} = \Pr(X_r \neq k, r = 1, \ldots, n-1; X_n = k | X_0 = j)$ for $n = 2, 3, \ldots$. We have already met these concepts in the context of the simple random walk (Section 4.1.4), and here we are simply extending them to cover transitions from a given state j to any other state k, i.e. not just to the nearest neighbours $j - 1$ and $j + 1$.

5.3.1 Classification of states

The states of a Markov chain fall into distinct types according to their limiting behaviour. If ultimate return to the initial state j is certain then state j is called *recurrent*, and in this case the time of the first return to j is a random variable called the *recurrence time*. The state is then called *positive recurrent* or *null recurrent* depending on whether the mean recurrence time is finite or infinite. Conversely, if ultimate return to j has probability less than one then state j is called *transient*. We normally exclude from discussion the trivial case where $p_{ij} = 0$ for every $i \neq j$. For once a chain leaves such an *ephemeral* state j it can never return there. In matrix terms, the column of the transition matrix $\mathbf{P}$ corresponding to an ephemeral state consists entirely of zeros.

Given that the chain starts in state j, the sum

$$f_j = \sum_{j=1}^{\infty} f_{jj}^{(n)} \qquad (5.3.1)$$

is the probability that state j is eventually re-entered. So if $f_j = 1$ state j is recurrent, whilst if $f_j < 1$ it is transient. Thus, conditional on starting in a transient state j there is a positive probability $1 - f_j$ that state j will never be re-entered, whereas for a recurrent state eventual return is certain. So for a recurrent state $\{f_{jj}^{(n)}\}$ $(n = 1, 2, \ldots)$ is a proper probability distribution; its mean,

$$\mu_j = \sum_{n=1}^{\infty} n f_{jj}^{(n)}, \qquad (5.3.2)$$

is called the *mean recurrence time*. If μ_j is infinite then state j is null recurrent. Similarly, the sum

$$f_{jk} = \sum_{n=1}^{\infty} f_{jk}^{(n)} \qquad (5.3.3)$$

is the (first passage) probability of ever reaching state k from j. If $f_{jk} = 1$, so that eventual passage from j to k is certain, then

$$\mu_{jk} = \sum_{n=1}^{\infty} n f_{jk}^{(n)} \tag{5.3.4}$$

denotes the *mean first passage time* from state j to k.

Returning to the simple 2×2 matrix example (5.1.1), namely

$$\mathbf{P} = (p_{ij}) = \begin{pmatrix} 1-\alpha & \alpha \\ \beta & 1-\beta \end{pmatrix} \qquad (i,j = 0,1), \tag{5.3.5}$$

we see that provided $\alpha, \beta > 0$ movement from state 0 to 1 and state 1 to 0 is always possible, i.e. states 0 and 1 *communicate* with each other. So the process is clearly recurrent. In contrast, if (say) $\alpha = 0$ then once state 0 is occupied the process can never leave it, so state 1 must be transient. Now we have already determined the mean recurrence times $\mu_0 = 1/\pi_0$ and $\mu_1 = 1/\pi_1$ in the context of the Telegraph Wave (expressions (5.1.26) and (5.1.29)). Using a parallel approach we see that $f_{01}^{(n)}$ is the probability that, on starting in state 0 at time $n = 0$, the process remains there for the next $n-1$ steps when it then moves to state 1. Thus

$$f_{01}^{(n)} = (1-\alpha)^{n-1} \alpha, \tag{5.3.6}$$

and so the mean passage time

$$\mu_{01} = \alpha \sum_{n=1}^{\infty} n(1-\alpha)^{n-1} = 1/\alpha. \tag{5.3.7}$$

Likewise,

$$f_{10}^{(n)} = (1-\beta)^{n-1} \beta \qquad \text{and} \qquad \mu_{10} = 1/\beta. \tag{5.3.8}$$

To consider movement over an infinite state space, let us return to the unrestricted simple random walk discussed in Chapter 4. Now we have already seen from result (4.2.73) that

$$f_0 = \sum_{n=1}^{\infty} f_{00}^{(n)} = \begin{cases} 1-p+q < 1 & \text{if } p > q \\ 1 & \text{if } p = q \\ 1-q+p < 1 & \text{if } p < q. \end{cases} \tag{5.3.9}$$

So return to the origin is certain if the walk has zero drift (i.e. $p = q$). Moreover, in this case the mean recurrence time μ_0 is infinite. In Markov chain terminology this means that for a symmetric random walk ($p = q$) every state is null recurrent. For an asymmetric random walk ($p \neq q$) return to the start state is uncertain and so every state is transient.

Suppose $p + q = 1$, so that each step is either $+1$ or -1. If the walk starts at the origin then after one step it is either in state 1 or -1. It can, however, return to the origin at the second step, and since return to the origin can occur only at the even times $n = 2, 4, 6, \ldots$, state 0, and hence all other states, are *periodic* with period 2.

276 Markov chains

In general, if a chain starts in state j and subsequent occupations of j can occur only at times $t, 2t, 3t, \ldots$, for integer $t > 1$, then state j is called periodic with period t. A state which is not periodic is called *aperiodic*, and essentially has period 1. Finally, we call a positive recurrent aperiodic state *ergodic*. So states 0 and 1 in the above 2×2 example (5.3.5) are ergodic provided $\alpha, \beta > 0$.

5.3.2 Relating first return and passage probabilities

Whilst probability expressions may be easily determined for these two-state chain and unrestricted simple random walk examples, in general this is not the case. For determining closed expressions for the eigenvalues $\omega_1, \omega_2, \ldots$ of $\mathbf{P}$ in order to evaluate the occupation probabilities $\mathbf{p}^{(n)} = \mathbf{p}^{(n-1)}\mathbf{P}^n$ is only possible if $n \leq 4$. However, we have already seen in our discussion of the random walk (Chapter 4), that p.g.f. solutions are more user-friendly. So let us denote

$$F_{jj}(s) \equiv \sum_{n=1}^{\infty} f_{jj}^{(n)} s^n \quad \text{and} \quad P_{jj}(s) \equiv \sum_{n=1}^{\infty} p_{jj}^{(n)} s^n \quad (5.3.10)$$

to be the generating functions of the first return probabilities $f_{jj}^{(n)}$ and n-step transition probabilities $p_{jj}^{(n)}$, respectively. Then by deriving a relation between them we can obtain an important classification criterion for the state j based purely on the $p_{jj}^{(n)}$. This means that there is no need to evaluate the $f_{jj}^{(n)}$ in order to classify the states of the process.

Now the first return to state j must occur at one of the times $1, 2, \ldots, n$, and these are mutually exclusive possibilities. So on defining $p_{jj}^{(0)} = 1$, it follows that the probability that the chain is at j at time n, given that it started there, is given by

$$p_{jj}^{(n)} = f_{jj}^{(1)} p_{jj}^{(n-1)} + f_{jj}^{(2)} p_{jj}^{(n-2)} + \cdots + f_{jj}^{(n-1)} p_{jj}^{(1)} + f_{jj}^{(n)} \quad (n = 1, 2, \ldots). \quad (5.3.11)$$

Whence multiplying both sides by s^n and summing over $n = 1, 2, \ldots$ yields

$$P_{jj}(s) = F_{jj}(s) + F_{jj}(s) P_{jj}(s). \quad (5.3.12)$$

On rearranging terms we therefore have the key relations

$$F_{jj}(s) = \frac{P_{jj}(s)}{1 + P_{jj}(s)} \quad \text{and} \quad P_{jj}(s) = \frac{F_{jj}(s)}{1 - F_{jj}(s)}. \quad (5.3.13)$$

The argument for the first passage probabilities proceeds in a similar fashion. We define

$$F_{jk}(s) \equiv \sum_{n=1}^{\infty} f_{jk}^{(n)} s^n \quad \text{and} \quad P_{jk}(s) \equiv \sum_{n=1}^{\infty} p_{jk}^{(n)} s^n, \quad (5.3.14)$$

and note that moving from state j to k in n steps necessarily involves reaching k from j for the first time at some time $1, 2, \ldots, n$. So

$$p_{jk}^{(n)} = f_{jk}^{(1)} p_{kk}^{(n-1)} + f_{jk}^{(2)} p_{kk}^{(n-2)} + \cdots + f_{jk}^{(n-1)} p_{kk}^{(1)} + f_{jk}^{(n)} \quad (n = 1, 2, \ldots); \quad (5.3.15)$$

which leads to
$$P_{jk}(s) = F_{jk}(s) + F_{jk}(s)P_{kk}(s), \quad (5.3.16)$$
and hence
$$F_{jk}(s) = \frac{P_{jk}(s)}{1 + P_{kk}(s)}. \quad (5.3.17)$$

We can now relate the classification of the states of a Markov chain as being recurrent, transient, etc. in terms of the limiting properties of these p.g.f.'s. For on combining (5.3.1) and (5.3.10) we see that $f_j = F_{jj}(1)$ is the probability that state j, once entered, is ever re-entered. Whilst it follows from (5.3.2) that $\mu_j = F'_{jj}(1)$, whether μ_j is finite or infinite. Hence state j is transient if $F_{jj}(1) < 1$ and recurrent if $F_{jj}(1) = 1$; in the latter case state j is positive recurrent if $F'_{jj}(1) < \infty$ and null recurrent if $F'_{jj}(1) = \infty$.

To relate these results to the n-step transition probabilities, we first note that for any states j and k we have $F_{kk}(1) \leq 1$ and $F_{jk} \leq 1$. Now we see from (5.3.13) that $F_{kk}(1) < 1$ if and only if $P_{kk}(1) < \infty$. Thus a necessary and sufficient condition for k to be a transient state is that the series $\sum_n p_{kk}^{(n)}$ is divergent. In this case it follows from (5.3.17) that, for each j, $\sum_n p_{jk}^{(n)}$ is convergent.

We also see from (5.3.13) (with k replacing j) that $F_{kk}(1) = 1$ if and only $P_{kk}(1) = \infty$. So a necessary and sufficient condition for k to be a recurrent state is that $\sum_n p_{kk}^{(n)}$ diverges. If j *communicates* with k then $F_{jk}(1) > 0$, whence it follows from (5.3.17) that $P_{kk}(1) = \infty$ implies that $P_{jk}(1) = \infty$. Hence if k is recurrent then $\sum_n p_{jk}^{(n)}$ diverges for each state j which communicates with k.

Suppose k is a recurrent, aperiodic state (i.e. it is ergodic). Then a key result for Markov chains, known as the *Ergodic Theorem*, is that in the limit as $n \to \infty$, the n-step transition probability $p_{kk}^{(n)}$ is the inverse of the mean recurrence time of state k, i.e.
$$\lim_{n \to \infty} p_{kk}^{(n)} = 1/\mu_k. \quad (5.3.18)$$

The proof of this important relation (see, for example, Cox and Miller, 1965) involves applying a power-series theorem of Erdös, Feller and Pollard (1949), a proof of which is contained in Feller (1968), to (5.3.13) written in the form
$$1 + P_{kk}(s) = \frac{1}{1 - F_{kk}(s)}.$$

For any other state j, we have from (5.3.13) and (5.3.16) that
$$P_{jk}(s) = \frac{F_{jk}(s)}{1 - F_{kk}(s)}, \quad (5.3.19)$$
whence a similar argument leads to
$$\lim_{n \to \infty} p_{jk}^{(n)} = F_{jk}(1)/\mu_k. \quad (5.3.20)$$

Moreover, if k is a recurrent periodic state with period t, and j communicates with k, then a parallel development leads to

$$\lim_{n\to\infty} p_{kk}^{(nt)} = t/\mu_k.$$

Given that it may well be possible to determine an exact expression for the p.g.f. of the n-step transition probabilities $p_{jk}^{(n)}$, but not the probabilities themselves, these limiting results are clearly important in helping us to understand the fundamental behaviour of a process when the $p_{jk}^{(n)}$ are intractable to direct evaluation. It is therefore useful to collate them into the following theorem.

MARKOV CHAIN LIMIT THEOREM *For an arbitrary fixed state k:*

(i) *k is transient if and only if $\sum_n p_{kk}^{(n)}$ converges (i.e. $P_{kk}(1) < \infty$) and in this case $\sum_n p_{jk}^{(n)}$ converges for each j;*

(ii) *k is recurrent if and only if $\sum_n p_{kk}^{(n)}$ diverges (i.e. $P_{kk}(1) = \infty$) and in this case $\sum_n p_{jk}^{(n)}$ diverges for every j which communicates with k;*

(iii) *if k is recurrent then it has mean recurrence time*

$$\mu_k = \sum_n n f_{kk}^{(n)} = F'_{kk}(1), \tag{5.3.21}$$

where we define $1/\mu_k = 0$ if $\mu_k = \infty$;

(iv) *if k is ergodic then*

$$\lim_{n\to\infty} p_{kk}^{(n)} = 1/\mu_k \quad \text{and} \quad \lim_{n\to\infty} p_{jk}^{(n)} = F_{jk}(1)/\mu_k \tag{5.3.22}$$

for each state j which communicates with k;

(v) *whilst if k is recurrent and periodic with period t then*

$$\lim_{n\to\infty} p_{kk}^{(nt)} = t/\mu_k. \tag{5.3.23}$$

To demonstrate the power of these results consider a simple random walk starting at $j = 0$. Now we have already shown in (4.2.71) that the first passage p.g.f.

$$F_{00}(s) = s[(1 - p - q) + 2p\lambda_2(s)]$$

where $\lambda_2(s)$ is given by (4.2.68). Hence it follows from (5.3.13) that the n-step transition p.g.f. takes the form

$$P_{00}(s) = \frac{F_{00}(s)}{1 - F_{00}(s)} = \frac{s[(1 - p - q) + 2p\lambda_2(s)]}{1 - s[(1 - p - q) + 2p\lambda_2(s)]}. \tag{5.3.24}$$

Moreover, as seen earlier in (4.2.73), $F_{00}(1) < 1$ unless $p = q$, in which case $F_{00}(1) = 1$ and $F'_{00}(1) = \infty$. So the process is transient if $p \neq q$ and null recurrent if $p = q$. Alternatively, on substituting for $F_{00}(1)$ in (5.3.24), we see that $P_{00}(1) < \infty$ if $p \neq q$ and $P_{00}(1) = \infty$ if $p = q$; so we could also use part (i) of the Markov Chain Limit Theorem to prove transience for $p \neq q$ and part (ii) to prove recurrence for $p = q$.

First return and passage probabilities 279

Now in practice there is often a single goal of deriving either the n-step transition probabilities or the first passage probabilities. So bearing in mind the simplicity of relations (5.3.13) and (5.3.17) that link them together, it is worthwhile working with the more mathematically tractable of the two and then, if necessary, transforming to the other. To illustrate this linkage let us return to the two-state Markov chain, where in Section 5.1.1 we determined the occupation probabilities $p_{jk}^{(n)}$ ($j, k = 1, 2$) using a matrix approach. Now the first return to say state 0 at time n necessitates transferring from 0 to 1 at the first step, remaining in state 1 for the next $n-2$ time units, and then finally returning to state 0 at the nth step. So the first return probabilities

$$f_{00}^{(1)} = (1-\alpha) \quad \text{and} \quad f_{00}^{(n)} = \alpha(1-\beta)^{n-2}\beta \quad (n = 2, 3, \ldots), \quad (5.3.25)$$

giving

$$F_{00}(s) = (1-\alpha)s + \alpha\beta \sum_{n=0}^{\infty} (1-\beta)^n s^{n+2} = (1-\alpha)s + \frac{\alpha\beta s^2}{1-(1-\beta)s}. \quad (5.3.26)$$

Similarly,

$$f_{10}^{(n)} = (1-\beta)^{n-1}\beta \quad (n = 1, 2, \ldots), \quad (5.3.27)$$

giving

$$F_{10}(s) = \beta \sum_{n=1}^{\infty} (1-\beta)^n s^{n+1} = \frac{\beta s}{1-(1-\beta)s}. \quad (5.3.28)$$

Whence on using (5.3.13) it follows that

$$P_{00}(s) = \frac{F_{00}(s)}{1-F_{00}(s)} = \frac{(1-\alpha)s - (1-\alpha-\beta)s^2}{(1-s)[1-(1-\alpha-\beta)s]}$$

which, on splitting into partial fractions and expanding in powers of s^n, yields

$$p_{00}^{(n)} = [\beta/(\alpha+\beta)] + [\alpha/(\alpha+\beta)](1-\alpha-\beta)^n. \quad (5.3.29)$$

On recalling that the equilibrium probabilities $(\pi_0, \pi_1) = (\beta, \alpha)/(\alpha+\beta)$, and that here $p_0^{(0)} = 1$ and $p_1^{(0)} = 0$, we see that (5.3.29) is in exact agreement with solution (5.1.15). Though the latter derivation, being based on the eigenvalue and eigenvector decomposition of the transition matrix, is less amenable to work with. Whilst on using (5.3.19), we can use the first return p.g.f.'s $F_{10}(s)$ and $F_{00}(s)$ to evaluate the first passage p.g.f.

$$P_{10}(s) = \frac{F_{10}(s)}{1-F_{00}(s)} = \frac{\beta s}{(1-s)[1-(1-\alpha-\beta)s]}. \quad (5.3.30)$$

This expands to give

$$p_{10}^{(n)} = [\beta/(\alpha+\beta)][1-(1-\alpha-\beta)^n], \quad (5.3.31)$$

which again agrees with (5.1.15), this time with $p_1^{(0)} = 1$ and $p_0^{(0)} = 0$.

At first sight it might appear as though classifying states would be an arduous task, since the number of possible steps to be examined increases with n^2. Fortunately, there is a 'class solidarity' among the different types of state in the sense that a state of a given type can only intercommunicate with other states of the same type (Chung, 1960).

5.3.3 Closed sets of states

Let j and k be two intercommunicating states, and r and t be two integers such that $p_{jk}^{(r)} > 0$ and $p_{kj}^{(t)} > 0$. Then since the number of sample paths leading from state j back to j in $n = r + s + t$ steps must be at least as great as the number of paths leading first from j to k in r steps, remaining there for s steps, and then moving back from k to j in t steps, it follows that

$$p_{jj}^{(n)} \geq p_{jk}^{(r)} p_{kk}^{(s)} p_{kj}^{(t)}. \tag{5.3.32}$$

Similarly,

$$p_{kk}^{(n)} \geq p_{kj}^{(t)} p_{jj}^{(s)} p_{jk}^{(r)}. \tag{5.3.33}$$

Hence there are constants $A > 0$ and $B > 0$, and an integer $v = r + t$, such that

$$p_{jj}^{(n)} \geq A p_{kk}^{(n-v)} \quad \text{and} \quad p_{kk}^{(n)} \geq B p_{jj}^{(n-v)} \quad (n > v). \tag{5.3.34}$$

Thus the transition probabilities $p_{jj}^{(n)}$ and $p_{kk}^{(n)}$ have the same asymptotic behaviour. Moreover, if j is periodic, then so is k with the same period, and conversely.

Suppose that j is a recurrent state and that k is another state with which j communicates. Then there is an integer r such that $p_{jk}^{(r)} > 0$. Since j is recurrent, k must communicate with j since otherwise there would be a positive probability of never returning to j. Hence j and k must intercommunicate, whence it follows from the above that they must be of the same type and period.

Finally, we say that a set S of states is *closed* if each state in S communicates only with other states in S; once entered it is never vacated. So if j belongs to a closed set S then $p_{jk} = 0$ for all k outside S. In particular, a single state forming a closed set is an *absorbing state*. In summary, we have the:

MARKOV DECOMPOSITION THEOREM *The states of a Markov chain may be divided into two sets (one of which may be empty), in which one set comprises all the recurrent states, and the other all the transient states. The recurrent states may be decomposed uniquely into closed sets. Within each closed set, all states intercommunicate, and they are all of the same type and period. No communication is possible between any two closed sets.*

To illustrate these results consider a simple random walk on the states $0, 1, 2, \ldots$ with a reflecting barrier at 0 (Section 4.3), and to this add an additional state at -1 which, once left, is never revisited. Then the associated probability transition matrix over the states $-1, 0, 1, \ldots$ takes the form

$$\mathbf{P} = \begin{pmatrix} 1-p & p & 0 & 0 & 0 & \cdots \\ 0 & 1-p & p & 0 & 0 & \cdots \\ 0 & q & 1-p-q & p & 0 & \cdots \\ 0 & 0 & q & 1-p-q & p & \cdots \\ \vdots & \vdots & \vdots & \vdots & \vdots & \end{pmatrix}. \quad (5.3.35)$$

The states $0, 1, 2, \ldots$ clearly form a closed set, since any one of these states can be reached from any other, whilst state -1 is transient (provided $p > 0$). As the simple random walk is transient if $p > q$, and null recurrent if $p = q$, it follows that the same holds true for the process (5.3.35). Moreover, if $p < q$ then at each step there is a downwards drift towards state 0, and so all states (except -1) are recurrent; they are also ergodic if $p + q < 1$ (i.e. the process is not periodic). In constrast, for the simple random walk with absorbing barriers at b and a (Section 4.2) the interior states $b+1, \ldots, a-1$ are all transient. They cannot be recurrent since each communicates with the absorbing states a and b, each of which is a closed set consisting of a single state.

This theorem has two important implications. First, a finite chain cannot consist only of transient states, since if it did then all the transition probabilities $p_{jk}^{(n)}$ would tend to zero as $n \to \infty$. Yet this is impossible since the finite sum $\sum_k p_{jk}^{(n)} = 1$. Second, a finite chain cannot have any null recurrent states, since the one-step transition probabilities within a closed set of null recurrent states would form a finite stochastic matrix $\mathbf{P}$ such that $\mathbf{P}^n \to 0$ as $n \to \infty$, which is again impossible.

5.3.4 Irreducible chains

An extremely important class of Markov chains is the class of *irreducible chains* in which all states intercommunicate. Moreover, since an irreducible chain forms a single closed set and all its states are of the same type, it is perfectly in order to talk of the *chain* or *process* as being recurrent, transient, periodic, etc. So whilst the simple unrestricted random walk is an irreducible chain, that of example (5.3.35) is not since this process can never return to state -1 once it has left it. It follows from the Markov Decomposition Theorem that, in general, a Markov chain is composed of transient states (if any) together with a number of irreducible sub-chains. Once the system enters such a sub-chain it never leaves it.

Now we have already remarked that an irreducible finite chain is necessarily positive recurrent. If it is also aperiodic then it is ergodic, and there is a unique row vector $\boldsymbol{\pi}$ of limiting occupation probabilities called the *equilibrium distribution*. In such cases we can not only employ the results of the Perron–Frobenius Theorem (Section 5.2.2), but we can also recall that these probabilities are given by the inverses of the mean recurrence times. So as $\lim_{n \to \infty} \mathbf{P}^n = (\boldsymbol{\pi}, \ldots, \boldsymbol{\pi})^T = \mathbf{1}\boldsymbol{\pi}$, where $\mathbf{1}$ denotes the column vector of 1's, it follows that for any initial probability distribution $\mathbf{p}^{(0)}$

$$\lim_{n \to \infty} \mathbf{p}^{(n)} = \lim_{n \to \infty} \mathbf{p}^{(0)} \mathbf{P}^n = \mathbf{p}^{(0)} \mathbf{1} \boldsymbol{\pi} = \boldsymbol{\pi}.$$

282 Markov chains

Thus a finite ergodic system will eventually settle down to a condition of statistical equilibrium independent of the initial conditions. In practice the question of determining when equilibrium has (for all intents and purposes) been achieved, i.e. the length of the burn-in period, is a non-trivial one. Though if we know the first three eigenvalues in decreasing order of magnitude, namely $\omega_1 = 1$, ω_2 and ω_3, on rewriting result (5.2.1) as

$$\mathbf{p}^{(n)} = \boldsymbol{\pi} + \mathbf{p}^{(0)}\omega_2^n \mathbf{A}_2 + \mathrm{O}(\omega_3^n), \qquad (5.3.36)$$

we see that to achieve a desired degree of accuracy, say $\mathbf{p}^{(n)} = \boldsymbol{\pi} + \mathrm{O}(10^{-r})$, we may take $n \geq n_{burn\,in} \simeq -r/\log_{10}\omega_2$.

For infinite irreducible processes the limiting behaviour is even richer, since such systems may be null recurrent as well as transient or positive recurrent. Ergodic systems retain the property of having unique equilibrium distributions and these are also unique stationary distributions. In this case too, the equilibrium probabilities are the inverses of the mean recurrence times. Indeed, we have seen that they also possess that property that in a long realization they are, with probability one, the proportions of times spent in the corresponding states. This is an extremely important result when we are investigating a population process through a simulation experiment. For it enables us to examine the properties of a chain from a *single realization* rather than from a large number of independent realizations.

5.4 Branching processes

Although we have considered the theoretical implications of moving in a single step from state j to any other general state k with probability p_{jk}, our examples have so far centred around random walk type processes in which only movement to the neighbouring states $j-1$, j and $j+1$ is possible. In essence, this is because the problem of determining the matrix power $\mathbf{P}^n$ is difficult enough for tri-diagonal matrices; for richer matrix structures it is likely to be mathematically intractable. In population process scenarios this is no great hardship, since for the continuous-time models considered so far the population size N can change only to either $N+1$ at the next step (e.g. birth or immigration) or to $N-1$ (e.g. death or emigration). So if we are examining the *skeleton* of a Markov process, that is we record the series of successive event types but disregard the times at which they occur, then the resulting Markov chain will often possess a tri-diagonal transition probability matrix $\mathbf{P}$.

Obvious exceptions include the mass annihilation of individuals due to catastrophic events, and batch immigration in queueing situations; here the tri-diagonal structure of $\mathbf{P}$ also has non-zero elements in the first row and column. However, there is a far more general situation for which all elements p_{jk} may be non-zero. For instead of considering a Markov chain as being the discrete-time skeleton of a continuous-time Markov process, suppose that the population genuinely develops over discrete generations. This can be achieved either by recording the specific generation an individual belongs to, or by considering species that reproduce on say a yearly basis. Then whereas in the continuous-time birth–death process the *population size* N simply changes to

$N-1$, N or $N+1$ in the infinitesimal time interval $(t, t+dt)$ with probabilities $\mu N dt$, $1-(\mu+\lambda)Ndt$ and $\lambda N dt$, the corresponding discrete-time transition mechanism is substantially more complex. For at every generation *each individual* has probability μ, λ and $1-\mu-\lambda \geq 0$ of either dying, giving birth or remaining inactive. So the resulting one-step transition probabilities of moving from population size j to k take the trinomial form

$$p_{jk} = \sum_{r,s} \frac{j!}{r!s!(j-r-s)!} \lambda^r \mu^s (1-\lambda-\mu)^{j-r-s}, \qquad (5.4.1)$$

where the summation is over $r, s \geq 0$ with $k = j + r - s$. Not only is this p.d.f. cumbersome, but to obtain the n-step transition probabilities $p_{jk}^{(n)}$ we still have to insert (5.4.1) into a general summation over all possible intervening states $(i_1, \ldots, i_{n-1})$, namely

$$p_{jk}^{(n)} = \sum_{i_1,\ldots,i_{n-1}} p_{ji_1} p_{i_1 i_2} \cdots p_{i_{n-1} k}. \qquad (5.4.2)$$

The contrast with the simple continuous-time probability, $p_{jk}(t)$, is stark; for not only is the corresponding result (2.4.18) remarkably simple in comparison, but the geometric solution (2.4.14)–(2.4.15) for $j = n_0 = 1$ is positively trivial. This surely dispels any preconceived notion that discrete-time Markov chains might be easier to analyse than continuous-time Markov processes because they involve only size and not time.

Consider a population of individuals, each of which develops independently and has probability $g_0, g_1, g_2, \ldots$ of producing $i = 0, 1, 2, \ldots$ offspring at the next generation. We can either assume that the parent dies immediately after giving birth; or else we can interpret $i = 0$ as corresponding to individual parent death, with $i = 1$ corresponding to no change, $i = 2$ to one progeny birth, $i = 3$ to twins, etc. Initial individuals are regarded as belonging to the zeroth generation, their offspring as belonging to the first generation, the offspring of the first generation as comprising the second generation, and so on. If all the individuals of a generation fail to reproduce then the population becomes *extinct*. Two particularly striking examples are the survival of family surnames, where the population comprises the number of male descendants, and nuclear fission. The former was considered by Lotka (1931) for males in the USA, whilst the latter formed the basis for the (theoretical) development of the atomic bomb in the 1940s at Los Alamos. This basic model is called the *simple branching process*, and in subsequent years tremendous development has taken place. So here all we can do is to introduce the fundamental ideas, and then encourage the reader to probe deeper via classic books such as Asmussen and Hering (1983), Athreya and Jagers (1996), Athreya and Ney (1972) and Harris (1963).

5.4.1 Population size moments

Let X_n denote the number of individuals in the nth generation. Then if $X_n = j$, it follows that X_{n+1} is the sum of j independent random variables $Z_1, \ldots, Z_j$ each having the distribution $\{g_i\}$. On denoting

$$G(z) \equiv \sum_{i=0}^{\infty} g_i z^i \tag{5.4.3}$$

to be the p.g.f. of the distribution of the number of offspring per individual, we therefore have

$$\Pr(X_{n+1} = k | X_n = j) = \Pr(Z_1 + \cdots + Z_j = k)$$
$$= \text{coefficient of } z^k \text{ in } \{G(z)\}^j. \tag{5.4.4}$$

Clearly, if $X_n = 0$ then $X_{n+1} = 0$ since no further offspring can be produced. Hence state 0 is an absorbing state, with the probability of absorption being the probability that the denumerably infinite Markov chain $\{X_n\}$ becomes extinct. Note that since individuals develop independently, we do not lose any generality by taking $j = 1$. For if $X_0 = j > 1$, we simply regard k as being the total population size at time n of j unconnected branching processes which develop in parallel with each other.

So let us take the initial condition $X_0 = 1$, and denote $p_k^{(n)} = \Pr(X_n = k)$. Then for given n, $\{p_k^{(n)}\}$ is the distribution of the number of nth generation descendants of a single individual. Since

$$p_k^{(n)} = \sum_{j=0}^{\infty} p_j^{(n-1)} p_{jk}, \tag{5.4.5}$$

it follows from (5.4.4) that

$$p_k^{(n)} = \text{coefficient of } z^k \text{ in } \sum_{j=0}^{\infty} p_j^{(n-1)} \{G(z)\}^j. \tag{5.4.6}$$

Whence on defining the p.g.f. of the size of the nth generation to be

$$F_n(z) \equiv \sum_{k=0}^{\infty} p_k^{(n)} z^k, \tag{5.4.7}$$

we have

$$F_n(z) = \sum_{j=0}^{\infty} p_j^{(n-1)} \{G(z)\}^j = F_{n-1}(G(z)). \tag{5.4.8}$$

The recurrent nature of this solution makes it abundantly clear that apart from a few trivial cases it will impossible to obtain an explicit closed form expression for $F_n(z)$. Hence although relation (5.4.8) is simple to write down, the associated probabilities are anything but, since they involve an extremely nasty extension from the trinomial structure (5.4.1)–(5.4.2).

Such difficulties may be appreciated by considering two examples of the offspring distribution $\{g_i\}$ which do yield amenable expressions for the p.g.f. $G(z)$. Suppose first that g_i takes the geometric form $(1-p)p^i$ ($i = 0, 1, \ldots$). Then $G(z) = (1-p)/(1-pz)$, whence for a single initial parent at time $n = 0$ we have $F_0(z) = z$, which gives rise to the continued fraction representation

$$F_1(z) = G(z) = \frac{1-p}{1-pz}$$

$$F_2(z) = G(G(z)) = \frac{1-p}{1-p\left(\frac{1-p}{1-pz}\right)} \quad (5.4.9)$$

$$F_3(z) = G(G(G(z))) = \frac{1-p}{1-p\left(\frac{1-p}{1-p\left(\frac{1-p}{1-pz}\right)}\right)}, \quad \text{etc..}$$

Whilst letting g_i take the Poisson form $\lambda_i e^{-\lambda}/i!$ $(i = 0, 1, \ldots)$, i.e. $G(z) = \exp\{-\lambda(1-z)\}$, yields the continued exponential representation

$$F_1(z) = \exp\{-\lambda(1-z)\}$$

$$F_2(z) = \exp\{-\lambda(1 - \exp\{-\lambda(1-z)\})\}, \ldots \quad (5.4.10)$$

Since neither (5.4.9) nor (5.4.10) yield a tractable closed form for the coefficient of z^n, the idea of determining neat expressions for population size probability structure is ruled out. However, this still leaves us with the development of moments and the probability of ultimate extinction.

To construct moments we first need to rewrite the recurrence relation (5.4.8) in a slightly different form. On taking $n = 1, 2, \ldots$ we see (as in (5.4.9)) that since $F_0(z) = z$ we have $F_1(z) = G(z)$, $F_2(z) = G(G(z))$, ... So in general

$$F_n(z) = G(G(\ldots G(G(z))\ldots)) = G(F_{n-1}(z)). \quad (5.4.11)$$

Write $z = e^\theta$, and define the one-step and n-step cumulant generating functions

$$K(\theta) \equiv \ln[G(e^\theta)] \quad \text{and} \quad K_n(\theta) \equiv \ln[F_n(e^\theta)]. \quad (5.4.12)$$

Then (5.4.11) becomes

$$\ln[F_n(e^\theta)] = \ln[G(F_{n-1}(e^\theta))],$$

i.e.
$$K_n(\theta) = K(K_{n-1}(\theta)). \quad (5.4.13)$$

Denote the mean and variance of the number of offspring per individual by $m = K'(0)$ and $\sigma^2 = K''(0)$, and the mean and variance of the size of the nth generation by m_n and σ_n^2, respectively. Then on differentiating (5.4.13) with respect to θ, and noting that $K_{n-1}(0) = 0$, we obtain

$$m_n = K'_n(0) = K'_{n-1}(0)K'[K_{n-1}(0)] = mm_{n-1} \quad (5.4.14)$$

and

$$\sigma_n^2 = K''_n(0) = K''_{n-1}(0)K'[K_{n-1}(0)] + [K'_{n-1}(0)]^2 K''[K_{n-1}(0)]$$

$$= m\sigma_{n-1}^2 + \sigma^2 m^{2n-2}. \quad (5.4.15)$$

Equation (5.4.14) immediately gives

$$m_n = m^n; \quad (5.4.16)$$

whilst successive use of relation (5.4.15) yields

$$\sigma_n^2 = m^2\sigma_{n-2}^2 + \sigma^2(m^{2n-2} + m^{2n-3})$$
$$= m^3\sigma_{n-3}^2 + \sigma^2(m^{2n-2} + m^{2n-3} + m^{2n-4}) = \cdots$$
$$= m^{n-1}\sigma_1^2 + \sigma^2(m^{2n-2} + \cdots + m^n) = \sigma^2 m^{n-1}(1-m^n)/(1-m), \quad (5.4.17)$$

since $\sigma_1^2 = \sigma^2$. So for our geometric example,

$$K(\theta) = \ln[G(e^\theta)] = \ln(1-p) - \ln(1-pe^\theta),$$

which differentiates to yield $m = K'(0) = p/q$ and $\sigma^2 = K''(0) = p/q^2$. Whence (5.4.16) and (5.4.17) give

$$m_n = (p/q)^n \quad \text{and} \quad \sigma_n^2 = (p/q)^n[1 - (p/q)^n]/(q-p). \quad (5.4.18)$$

It follows that in the *sub-critical* case ($m < 1$) both m_n and σ_n^2 approach zero with m^n; in the *critical* case ($m = 1$) we have $m_n = 1$ for all n, and σ_n^2 increases linearly with nm; whilst in the *super-critical* case ($m > 1$) both m_n and σ_n^2 increase geometrically fast with m^n and m^{2n}, respectively. Moreover, we see from (2.4.19) that when $n_0 = 1$ the simple birth–death mean $m(t) = e^{(\lambda-\mu)t}$. Whence equating m to the mean number of offspring produced by time $t = 1$ yields the equivalence

$$m \iff m(1) = e^{(\lambda-\mu)}. \quad (5.4.19)$$

So at the discrete times $t = n = 0, 1, 2, \ldots$

$$m_n \iff m(n) = [m(1)]^n \iff m^n, \quad (5.4.20)$$

in agreement with (5.4.16). Similarly, by equating σ^2 to the variance of the number of offspring produced by time $t = 1$, on using (2.4.20) we have the second equivalence

$$\sigma^2 \iff V(1) = \left(\frac{\lambda+\mu}{\lambda-\mu}\right) e^{(\lambda-\mu)}(e^{(\lambda-\mu)} - 1), \quad (5.4.21)$$

whence

$$\sigma_n^2 \iff V(n) = \left(\frac{\lambda+\mu}{\lambda-\mu}\right) [e^{(\lambda-\mu)}]^n \{[e^{(\lambda-\mu)}]^n - 1\} \iff \sigma^2 m^{n-1}[(m^n - 1)/(m-1)], \quad (5.4.22)$$

in agreement with (5.4.17). Thus we have a direct equivalence between the mean and variance of the simple birth–death and branching processes, which begs the question as to whether there is a similar relationship between the two probabilities of extinction?

5.4.2 Probability of extinction

Although $p_0(t)$ for the linear birth-death process with $n_0 = 1$ takes the simple form (2.4.23), namely

$$p_0(t) = [\mu - \mu e^{-(\lambda-\mu)t}]/[\lambda - \mu e^{-(\lambda-\mu)t}], \quad (5.4.23)$$

it transpires that the corresponding form for the branching process is not so appealing. From (5.4.7) we see that the probability that there are no individuals in the nth generation, i.e. that extinction occurs at or before the nth generation, is given by $p_0^{(n)} = \Pr(X_n = 0 | X_0 = 1) = F_n(0)$. Whence on noting from (5.3.20) that

$$\lim_{n \to \infty} p_0^{(n)} = \lim_{n \to \infty} p_{10}^{(n)} = F_{10}(1)/\mu_0, \qquad (5.4.24)$$

we see that $p_0^{(n)}$ tends to some limit ξ as $n \to \infty$, namely the *probability of ultimate extinction*. It then follows that since letting $n \to \infty$ in (5.4.11) at $z = 0$ gives

$$F_n(0) = G(F_{n-1}(0)), \qquad (5.4.25)$$

ξ must satisfy the equation

$$\xi = G(\xi). \qquad (5.4.26)$$

To determine ξ we must therefore examine the roots of the equation

$$x = G(x), \qquad (5.4.27)$$

bearing in mind that since ξ is a probability we must seek roots satisfying $0 \leq x \leq 1$. Now the solutions of (5.4.27) are the values of x at which the curve $y = G(x)$ and $y = x$ intersect, and to examine these intersection points we note that:
(i) $G(0) = g_0 > 0$, since if $g_0 = 0$ then death, and hence extinction, cannot occur;
(ii) $G'(x) = \sum_{i=0}^{\infty} i g_i x^{i-1} > 0$; and, (iii) $G''(x) = \sum_{i=0}^{\infty} i(i-1) g_i x^{i-2} > 0$. Thus the curve $y = G(x)$ is a positive convex increasing function, and so can intersect the line $y = x$ in at most two places. Moreover, since $G(1) = \sum_{i=1}^{\infty} g_i = 1$, it is clear that $x = 1$ is a point of intersection. Let $\xi_1 \leq \xi_2$ denote the two positive roots of (5.4.27). Then since $m = G'(1)$, the curve $y = G(x)$ crosses the line $y = x$ at $x = 1$ from above if $m < 1$, tangentially if $m = 1$, and from below if $m > 1$. Hence

$$\begin{array}{ll} \xi_1 = 1 < \xi_2 & \text{if } m < 1, \\ \xi_1 = \xi_2 = 1 & \text{if } m = 1, \\ \xi_1 < \xi_2 = 1 & \text{if } m > 1. \end{array} \qquad (5.4.28)$$

Since $G(x)$ is increasing, it follows that for any positive root ξ_i of (5.4.27),

$$F_1(0) = G(0) < G(\xi_i) = \xi_i$$
$$F_2(0) = G(G(0)) < G(\xi_i) = \xi_i, \quad \text{etc.},$$

so in general,

$$F_n(0) < \xi \quad (n = 1, 2, \ldots; i = 1, 2). \qquad (5.4.29)$$

As $F_n(0)$ is the probability of extinction at or before the nth generation, this means that the probability of ultimate extinction $\xi = \lim_{n \to \infty} F_n(0)$ must be the smallest positive root of (5.4.27). Whence on using (5.4.28) we have the following result.

BRANCHING EXTINCTION THEOREM   *The probability ξ of ultimate extinction in a simple discrete branching process is given by the smallest positive root of the equation $x = G(x)$. When the mean number of offspring per* *individual $m \leq 1$, then $\xi = 1$ and extinction is certain; whilst if $m > 1$, then $\xi < 1$ and the probability that the population grows indefinitely is $1 - \xi$.*

Given that we have already proved this last statement concerning the possibility of indefinite population growth when extinction is uncertain for the simple continuous-time birth–death process, it seems eminently sensible that the result will also apply to the general discrete-time branching scenario. However, since the mean $m_n = m^n \to \infty$ as $n \to \infty$, for completeness we need to provide a formal proof. Because the zero state is absorbing, it follows from the Markov Decomposition Theorem (Section 5.3.3) that all other states $k = 1, 2, \ldots$ are also transient since they communicate with the zero state. Thus

$$\lim_{n \to \infty} p_k^{(n)} = 0 \quad (k = 1, 2, \ldots),$$

whilst we have already shown that

$$\lim_{n \to \infty} p_0^{(n)} = \xi.$$

Now for any $m = 1, 2, \ldots$

$$\Pr(X_n > m) + \sum_{k=1}^{m} p_k^{(n)} + p_0^{(n)} = 1.$$

So

$$\lim_{n \to \infty} \Pr(X_n > m) + \xi = 1.$$

Hence for any m, no matter how large, $\Pr(X_n > m) \to 1 - \xi$ as $n \to \infty$, which proves that when $\xi < 1$ there is a positive probability $1 - \xi$ of indefinite growth. □

For illustration, consider the simple example $G(x) = g_0 + (1 - g_0)x^2$. That is, in each generation an individual produces either no offspring or twins. Then on taking $p_0^{(0)} = 1$ we have $p_1^{(0)} = G(0) = g_0$, whence successive application of the recurrence relation (5.4.25) leads to

$$p_0^{(2)} = G(p_0^{(1)}) = g_0 + (1 - g_0)g_0^2 \tag{5.4.30}$$

$$p_0^{(3)} = G(p_0^{(2)}) = g_0 + (1 - g_0)g_0^2[1 + (1 - g_0)g_0]^2, \quad \text{etc.,}$$

which swiftly becomes unmanageable as n increases. In contrast, the probability of ultimate extinction, ξ, is simply given by the smallest positive root of (5.4.27), namely $\xi = 1$ or $g_0/(1 - g_0)$. Thus $\xi = 1$ if $g_0 \geq 1/2$, whilst $\xi = g_0/(1 - g_0)$ if $g_0 < 1/2$. Note that this is in direct accord with the Branching Extinction Theorem since $m = 2(1 - g_0) < 1$ if $g_0 < 1/2$.

Figure 5.1 shows the intersection of $y = x$ with $y = g_0 + (1 - g_0)x^2$, and we see that when

$$g_0 = 0.4, \quad \xi = \min(2/3, 1) = 2/3$$
$$g_0 = 0.5, \quad \xi = \min(1, 1) = 1$$
$$g_0 = 0.6, \quad \xi = \min(1, 3/2) = 1.$$

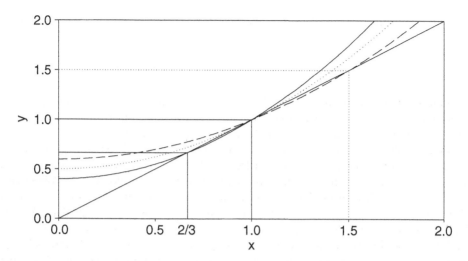

Figure 5.1 Intersection of $y = x$ and $y = g_0 + (1-g_0)x^2$ for $g_0 = 0.4$ (—), 0.5 (···) and 0.6 (- - -).

To examine how fast $p_0^{(n)} \to \xi$ as n increases, successive application of relation (5.4.30) gives

		g_0	
n	0.4	0.5	0.6
0	0.4000	0.5000	0.6000
1	0.4960	0.6250	0.7440
2	0.5476	0.6953	0.8214
5	0.6173	0.8004	0.9259
10	0.6523	0.8708	0.9785
25	0.6662	0.9359	0.9993
50	0.6667	0.9649	1.0000
∞	0.6666	1	1

$p_0^{(n)}$:

So whilst the processes with $g_0 = 0.4$ and 0.6 converge to $2/3$ and 1 quite quickly, convergence with $g_0 = 0.5$, i.e. $m = 1$, proceeds more slowly.

In order to investigate whether a comparison can be made with the continuous-time result (2.4.24)–(2.4.25), namely $p_0(\infty) = 1$ if $\lambda \leq \mu$ and $p_0(\infty) = \mu/\lambda$ if $\lambda > \mu$, we first need to express λ and μ in terms of m and σ^2. Now solving equations (5.4.19) and (5.4.21) for λ and μ gives

$$\lambda = \frac{\ln(m)}{2}\left[\frac{\sigma^2}{m(m-1)} + 1\right] \quad \text{and} \quad \mu = \frac{\ln(m)}{2}\left[\frac{\sigma^2}{m(m-1)} - 1\right]. \quad (5.4.31)$$

Since in our example $m = 2(1 - g_0)$ and $\sigma^2 = 4g_0(1 - g_0) = m(2 - m)$, we therefore have

$$\lambda = [1/2(m-1)]\ln(m) \quad \text{and} \quad \mu = [(3-2m)/2(m-1)]\ln(m). \quad (5.4.32)$$

Thus although $\lambda > 0$ for all $0 < g_0 < 1$, i.e. $0 < m < 2$, we see that $\mu < 0$ for $g_0 < 0.25$, i.e. $m > 1.5$. So even in this extremely simple case the continuous result (5.4.23) breaks down as a basic approximation to the branching extinction probability.

Moreover, since in general not all values of m and σ^2 will lead to non-negative values for the continuous-time rates λ and μ, it follows that the extinction result (5.4.23) cannot even be used as a general crude approximation. For example, if $X \sim \text{Poisson}(m)$, so that $\sigma^2 = m$, then although (5.4.31) reduces to $\lambda = [m \ln(m)]/[2(m-1)] > 0$ for all $m > 0$, we see that $\mu = [(2-m)\ln(m)]/[2(m-1)] < 0$ when $m > 2$. Though given that the simple birth–death process relates to single births, whilst an unbounded number can be produced with the Poisson offspring distribution, this failure is hardly surprising. Thus the general opaqueness of applied branching process results is just something we have to learn to live with, since approximations developed through the much simpler continuous process paradigm are (generally speaking) not good enough.

5.5 A brief note on martingales

Within our applied probability setting, which mainly concentrates on developing results through the construction and solution of Kolmogorov forward and backward equations, limit theorems play a purely supporting role. However, towards the purer end of the probability spectrum they feature much more strongly, and in this context extensive use is made of martingales. Since this field is huge in its own right, all we shall do here is to provide a brief introduction to the underlying principles and to illustrate their usefulness by application to a few specific examples.

Possibly the most well-known martingale scenario is that of a gambler who wages say £1 on an evens bet. If he loses then he wagers £2, £2^2, £2^3, etc. on each successive play until he wins. Thus his inevitable ultimate win will cover his lost stakes and profit him by £1. The inherent danger is, of course, that since both he and the casino have limited resource then one or the other may well go bust before this occurs. If the gambler starts with initial capital S_0, and S_n denotes his capital after n plays, then the sequence $\{S_n\}$ is called a *martingale*. Although the gambler knows the values of $S_0, \ldots, S_n$ before his $(n+1)$th wager, he can only guess at the value of S_{n+1}. If the game is fair then, conditional on the results of the past plays, he will expect no change in his present capital on average. That is,

$$\mathrm{E}(S_{n+1}|S_0, S_1, \ldots, S_n) = S_n. \quad (5.5.1)$$

DEFINITION *If the sequence of plays is $X_1, X_2, \ldots$, then the sequence $\{S_n\}$ is a martingale with respect to $\{X_n\}$ if, for all $n \geq 1$,*

$$E(S_{n+1}|X_1,\ldots,X_n) = S_n \quad \text{and} \quad E(S_n < \infty). \qquad (5.5.2)$$

So (5.5.1) shows that the time-series of the gambler's capital is a martingale with respect to itself. The extra level of generality induced by $\{X_n\}$ in (5.5.2) is particularly useful in that it enables us to deal with a sequence $\{X_n\}$ of random variables, such as a Markov chain, that may itself not be a martingale. For it is often possible to find some function ψ such that $\{S_n = \psi(X_n)\}$ is a martingale. The property (5.5.2) then becomes the assertion that, given the values of $X_1,\ldots,X_n$, the expected value of $S_{n+1} = \psi(X_{n+1})$ is $S_n = \psi(X_n)$. The simple unrestricted random walk (4.1.1) discussed in Section 4.1 provides an excellent illustration of this generalization, since it is clear that the position $S_n = X_1 + \cdots + X_n$ of the individual after n steps satisfies $E(|S_n|) \le n$ and

$$E(S_{n+1}|X_1,\ldots,X_n) = S_n + (p-q). \qquad (5.5.3)$$

Whence it is easily seen that $Y_n = S_n - n(p-q)$ defines a martingale with respect to $\{X_n\}$.

Example 1a (simple branching process) We have shown through the Branching Extinction Theorem in Section 5.4.2 that the probability ξ of ultimate extinction is given by the smallest positive root of the equation $x = G(x)$ where $G(z)$ denotes the p.g.f. of the number of offspring per individual. There are two martingales associated with this process. First, conditional on the size of the nth generation $X_n = x_n$, X_{n+1} is the sum of x_n independent variables, and hence for mean family size $m = G'(1)$

$$E(X_{n+1}|X_n = x_n) = mx_n, \qquad (5.5.4)$$

in agreement with (5.4.14) obtained by differentiating the iterated c.g.f. Thus by the Markov property,

$$E(X_{n+1}|X_1,\ldots,X_n) = mX_n. \qquad (5.5.5)$$

Now define

$$W_n = X_n/E(X_n). \qquad (5.5.6)$$

Then on recalling from (5.4.16) that $E(X_n) = m^n$, it follows that

$$E(W_{n+1}|X_1,\ldots,X_n) = W_n, \qquad (5.5.7)$$

and so $\{W_n\}$ is a martingale (with respect to $\{X_n\}$).

Example 1b (simple branching process) To construct the second martingale let X_i now denote the number of members of the $(n+1)$st generation which stem from the ith member of the nth generation. Then

$$Z_{n+1} = X_1 + X_2 + \cdots + X_{Z_n} \qquad (5.5.8)$$

is the sum of a random number of variables which are each independent and identically distributed with the same distribution as the number Z_1 of first generation offspring of the original individual. Write $U_n = \xi^{Z_n}$ where ξ is the probability of ultimate extinction for a process starting from a single individual. Thus

$$E(U_{n+1}|Z_1,\ldots,Z_n) = E(\xi^{(X_1+\cdots+X_{Z_n})}|Z_1,\ldots,Z_n)$$

$$= \prod_{i=1}^{Z_n} E(\xi^{X_i}|Z_1,\ldots,Z_n) \qquad \text{(by independence)}$$

$$= \prod_{i=1}^{Z_n} E(\xi^{X_i}) = \prod_{i=1}^{n} G(\xi) = \xi^{Z_n} \qquad (\text{since } \xi = G(\xi))$$

$$= U_n.$$

So $\{U_n\}$ is also a martingale.

Example 2a (sequence of partial sums) Let $X_1, X_2, \ldots$ be independent variables with zero means. Then the sequence of partial sums $S_n = X_1 + X_2 + \cdots + X_n$ is a martingale (with respect to $\{X_n\}$). For

$$E(S_{n+1}|X_1,\ldots,X_n) = E(S_n + X_{n+1}|X_1,\ldots,X_n)$$

$$= E(S_n|X_1,\ldots,X_n) + E(X_{n+1}|X_1,\ldots,X_n)$$

$$= S_n + 0 \qquad \text{(by independence)}.$$

Example 2b (sequence of partial sums) Suppose that we now work in terms of $T_n = S_n^2$. Then provided the X_i have finite variances we have

$$E(T_{n+1}|X_1,\ldots,X_n) = E[(S_n + X_{n+1})^2|X_1,\ldots,X_n]$$

$$= E(S_n^2 + 2S_n X_{n+1} + X_{n+1}^2|X_1,\ldots,X_n)$$

$$= T_n + 2E(X_{n+1})E(S_n|X_1,\ldots,X_n) + E(X_{n+1}^2)$$

$$= T_n + 0 + E(X_{n+1}^2) \geq T_n.$$

Hence $\{T_n\}$ is not a martingale, since the equality "$= S_n$" in (5.5.2) is replaced by the inequality "$\geq S_n$". It is, however, a *submartingale* and has properties similar to those of a martingale.

These examples show that martingales are imbedded in applied probability studies, even if we are not always aware of their underlying presence. They are extremely useful since they not only lead to general theorems which inject mathematical rigour into many applied probability analyses, but they also always converge subject to an appropriate condition being placed on their moments. With regard to the latter the basic result is the:

MARTINGALE CONVERGENCE THEOREM *If $\{S_n\}$ is a martingale and satisfies the condition that $E(S_n^2) < M < \infty$ for some M and all n, then there* exists a random variable S such that S_n converges to S and in mean square.

Whence an investigation into the properties of S is likely to reveal limit characteristics of the process under study. Whilst an example of the former is the:

DOOB–KOLMOGOROV INEQUALITY *If $\{S_n\}$ is a martingale with respect to $\{X_n\}$, then for all $\epsilon > 0$*

$$\Pr(\max_{1 \le i \le n} |S_i| \ge \epsilon) \le \epsilon^{-2} \mathrm{E}(S_n^2). \tag{5.5.9}$$

To illustrate these two theorems let us recall that example 1b concerns the martingale $W_n = X_n/\mathrm{E}(X_n)$. Now we know via result (5.4.18) that with $m = p/q \ne 1$ and $\sigma^2 = p/q^2$, i.e. $p = m^2/\sigma^2$ and $q = m/\sigma^2$,

$$\mathrm{Var}(X_n) = \mathrm{E}(X_n^2) - m^{2n} = m^{2n}\sigma^2(1 - m^{-n})/[m(m-1)]. \tag{5.5.10}$$

So

$$\mathrm{E}(W_n^2) = 1 + \sigma^2(1 - m^{-n})/[m(m-1)] \qquad (m \ne 1). \tag{5.5.11}$$

Thus there exists a random variable W such that $W_n \to W$ almost surely as $n \to \infty$, whence $W_n \to W$ in distribution also. Applying the Martingale Convergence Theorem then shows that the generating functions associated with W_n tend to those of W, and hence makes the following 'hand-wavy' limiting argument rigorous. Suppose $m > 1$, so that the probability of extinction $\xi < 1$. Then there is a non-zero probability, $1 - \xi$, that the population never becomes extinct. First, note that since $\mathrm{E}(W_n) = 1$ it automatically follows that $\mathrm{E}(W) = 1$, whilst expression (5.5.11) shows that the variance

$$\mathrm{Var}(W_n) = \sigma^2(1 - m^{-n})/[m(m-1)] \to \sigma^2/[m(m-1)] \qquad (\text{for } m > 1). \tag{5.5.12}$$

Now define the m.g.f.

$$g_n(z) = \mathrm{E}(z^{W_n}). \tag{5.5.13}$$

Then on recalling result (5.4.11) we see that

$$g_n(z) = \mathrm{E}(z^{X_n m^{-n}}) = F_n(z^{m^{-n}}), \tag{5.5.14}$$

and so g_n satisfies the functional recurrence relation

$$g_n(z) = G(g_{n-1}(z^{1/n})). \tag{5.5.15}$$

Thus as $n \to \infty$, $W_n \to W$ and $g_n(z) \to g(z) = \mathrm{E}(z^w)$, thereby yielding the limiting functional relationship

$$g(z) = G(g(z^{1/m})). \tag{5.5.16}$$

Whilst using the Doob–Kolmogorov Inequality (5.5.9) in conjunction with (5.5.11) shows that

$$\Pr(\max_{1\leq i\leq n} |W_i| \geq \epsilon) \leq \epsilon^{-2}\{1 + \sigma^2(1 - m^{-n})/[m(m-1)]\}$$

$$\to \epsilon^{-2}\{1 + \sigma^2/[m(m-1)]\} \qquad \text{as } n \to \infty \quad \text{for } m > 1. \quad (5.5.17)$$

The above examples clearly expose just a tiny fragment of the highly interesting and powerful field of martingales, and readers wishing to gain a deeper and richer insight are advised to delve into some of the excellent introductory treatises on offer. Two good starting points are Grimmett and Stirzaker (1992; especially chapter 12), and Williams (1991) who takes a more measure-theoretic standpoint yet still provides useful hints and suggestions for the 'working probabilist'.

6
Markov processes in continuous time and space

So far we have just considered populations that develop over the discrete set of states $N = 0, 1, 2, \ldots$, with Chapters 2 and 3 involving continuous time over $t \geq 0$ and Chapters 4 and 5 discrete time points $t = 0, 1, 2, \ldots$. A comparison of the results obtained under these two time regimes shows them to be quite different in character. In brief, continuous-time results are more transparent for fairly basic models, such as the simple birth–death process, since the corresponding forward and backward equations are amenable to direct solution. Whilst discrete-time solutions are better suited to more complex nonlinear transition rates, since these fit more easily within a general branching process structure for which theoretical results are known and so can be applied to any given situation. This suggests that we should pay specific regard to a third such possibility, namely population processes that develop both in continuous time and space. Though there is a fourth option, namely continuous state space processes developing in discrete time, this is less relevant to practical application and so shall not be considered here.

Situations in which this third option makes considerable sense often involve populations that have the potential for containing huge numbers of individuals. For example, suppose marine organisms have an average density of one individual per mm^3. Then a cubic kilometre of ocean will contain $\simeq 10^{18}$ such individuals, whence the idea of tracking *individual population size* changes over say $N = 0, 1, 2, \ldots, 10^{19}$ becomes ludicrous. What makes far more sense is to study change in *population density*, say $X = N/10^{19}$, since we may then regard X as being a continuous random variable operating over a far smaller range, in this case $0 \leq X \leq 1$. So as well as visualizing continuous time (as before) as the limit as $dt \to 0$ of the discrete time space $0, dt, 2dt, \ldots$, we shall now also envisage continuous population size or density as being the limit as $dx \to 0$ of the discrete state space $0, dx, 2dx, \ldots$.

This space–time incremental approach was first developed in order to describe the motion of particles suspended in a fluid, and moving under the rapid, successive random impacts of neighbouring particles. A graph plot showing the resulting track of a particle would show an erratic, but continuous, path of short straight line segments. This physical phenomenon is known as Brownian motion following its discovery by the botanist Robert Brown in 1827. Though its theoretical development had to wait until 1905 when Einstein first advanced an explanation, later made rigorous by Wiener in 1923. Because particles are observed to *diffuse* through their containing medium, we call the resulting Markov motion a *diffusion process*. On denoting $p(x;t)dx = \Pr(X = x$ at time $t)$ over $x = \ldots, -dx, 0, dx, 2dx, \ldots$ and $t = 0, dt, 2dt, \ldots$, the corresponding Kolmogorov equation takes the form

$$p(x; t+dt) = \int_{-\infty}^{\infty} p(y;t)\gamma(y,x;t)\,dy, \qquad (6.0.1)$$

where the transition probability function

$$\gamma(y,x;t)dx = \Pr(\text{particle is at } x \text{ at time } t+dt|\text{ at } y \text{ at time } t).$$

Since a full discussion of this equation is way beyond the scope of this book, we shall just consider some useful special cases that are amenable to direct solution. In effect we shall be paralleling the approach already taken for discrete state space Markov population processes, though whereas these give rise to differential-diffusion equations in time and space, now we shall be dealing with purely partial differential equations.

The idea of using a continuous population variable is not new to us, since in Section 3.6 we regard the population size N as being a continuous variable over $N \geq 0$ in order to exploit the fact that analysis of models with continuous N is intrinsically less demanding than that required for discrete N. There we showed that taking a crude second-order Taylor series expansion in the forward equation produces a second-order partial differential equation for the probabilities $p_N(t)$ which yields the particularly neat form (3.6.11) for the equilibrium probabilities $\{\pi_N\}$. We saw that although this technique works quite well in the centre of the distribution, it performs poorly in the tails, so we illustrated how perturbation methods can circumvent this problem. Moreover, if we can construct the associated cumulative generating function then the resulting saddlepoint approximation, given by (3.7.14) with $n = 1$, is 'optimal'; whilst if we know only the first few cumulants we can still employ the truncated saddlepoint representation (3.7.47). In this chapter we return to the diffusion approach, and examine its close links with the related field of stochastic differential equations (s.d.e.'s).

6.1 The basic Wiener process

In Section 3.6 we worked in terms of the second-order Taylor series expansion of $p_{N-1}(t)$, $p_{N+1}(t)$, λ_{N-1} and μ_{N+1}, and here we follow the same route. Though rather than regarding the approach as a crude approximating device operating over $N = 0, 1, 2, \ldots$, we shall now use a limiting argument involving the variable $X = \ldots, -dx, 0, dx, 2dx, \ldots$ which may be taken to be continuous over $-\infty < x < \infty$ in the limit as $dx \to 0$. To set the scene let us consider the most basic type of diffusion process, called the *Wiener process*, which is simply the limit of the simple random walk as the step size, dx, and time increment, dt, each tend to zero in some particular way. For recall from result (4.1.3) that with transition probabilities $\Pr(Z_i = 1) = p$ and $\Pr(Z_i = -1) = q = 1 - p$ the occupation probabilities take the binomial form

$$\Pr(X_n = k) = \binom{n}{(n-k)/2} p^{(n+k)/2} q^{(n-k)/2} \qquad (k = -n, -n+2, \ldots, n). \qquad (6.1.1)$$

Whence it appears intuitively reasonable to presume that in the diffusion limit these probabilities approach the corresponding Normal approximation.

Now in time t there are $n = t/dt$ steps, each of size $\pm dx$. Moreover, each step has mean $p(+dx) + q(-dx) = (p-q)dx$ and variance $p(+dx)^2 + q(-dx)^2 - [(p-q)dx]^2 = 4pq(dx)^2$, since $p + q = 1$. So using the Brownian motion paradigm, on assuming that $x = 0$ at time $t = 0$ the mean and variance of the particle's position at time $t \geq 0$ are given by

$$m(t) = (p-q)t(dx/dt) \quad \text{and} \quad \sigma^2(t) = 4pqt(dx^2/dt). \quad (6.1.2)$$

If we now allow $dx \to 0$ and $dt \to 0$, then both $m(t)$ and $\sigma^2(t)$ must remain finite if the process is to make practical sense. In particular, suppose we require the limiting process to have mean μ and variance σ^2 in unit time. Then dx and dt must tend to zero in such a way that

$$m(1) = (p-q)(dx/dt) \to \mu$$
$$\sigma^2(1) = 4pq(dx^2/dt) \to \sigma^2, \quad (6.1.3)$$

and these conditions are satisfied by taking

$$dx = \sigma\sqrt{dt} \quad (6.1.4)$$

along with

$$p = (1/2)\{1 + (\mu\sqrt{dt}/\sigma)\} \quad \text{and} \quad q = (1/2)\{1 - (\mu\sqrt{dt}/\sigma)\}. \quad (6.1.5)$$

On applying the Central Limit Theorem, the process $\{X(t)\}$ then follows the Normal distribution with mean μt and variance $\sigma^2 t$, namely

$$p(x;t)dx = \Pr(x \leq X(t) < x + dx) = (2\pi\sigma^2 t)^{-1/2}\exp\{-(x-\mu t)^2/2\sigma^2 t\}dx. \quad (6.1.6)$$

This limiting process is known as the *Wiener process* with drift μ and variance parameter σ^2; if $\mu = 0$ so that $m(t) = 0$ then we have the *Brownian motion process*. Since the step size $\pm dx$ at time t is independent of all previous locations $X(s)$ for all $s \leq t$, the process is clearly Markov.

Conditions (6.1.4) and (6.1.5) imply that p and q must lie close to $1/2$, and that dx must be of order $\sqrt{dt}$. Thus the particle's velocity, being

$$dx/dt = \sigma\sqrt{dt}/dt = \sigma/\sqrt{dt}, \quad (6.1.7)$$

becomes infinite in the limit as $dt \to 0$. In spite of this lacking practical credence, the Brownian motion process has proved to be an excellent representation of random movement in continuous media, and it forms an important basis for understanding the dynamics of population growth and spread. Readers interested in pursuing an early discourse in this wide-ranging field of diffusion processes should consult the entertaining text by Okubo (1980).

As a stochastic process, the Wiener process is perfectly proper. Moreover, it is easy to visualize the appearance of a realization for fixed $dx > 0$, no matter how small dx is, since $X(t)$ is simply a rescaled simple random walk. For example, Figure 6.1 shows three realizations starting from $x(0) = 0$ for $\mu = 1$ and $\sigma = 2$ with $dt = 1$, 0.01 and

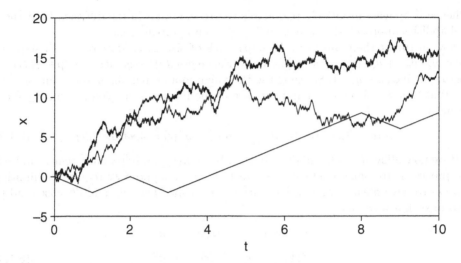

Figure 6.1 Simulated Wiener process with $\mu = 1$ and $\sigma = 2$ over $0 \leq t \leq 10$ with $dt = 1.0$ (——) lower, 0.01 (- - -) middle and 0.0001 ($\cdots$) upper.

0.0001. It follows from (6.1.4) that the corresponding step size increments are $dx = 2$, 0.2 and 0.02; whilst the step probabilities (6.1.5) are given by $(p, q) = (0.75, 0.25)$, $(0.525, 0.475)$ and $(0.5025, 0.4975)$. So in random walk terms there is a large visual difference between the first two realizations, but not between the last two. However, in diffusion terms this discrepancy is really illusory, since as $dx \propto \sqrt{dt}$ the process is essentially fractal with respect to this scaling. Indeed, as the same start seed was used in each case, zooming in on the first 1% of realizations 2 and 3 produces realizations 1 and 2, respectively.

Where visualization problems do occur is in the limit $dx = dt = 0$, since then the Wiener process, although continuous, has an uncountably infinite number of small spikes in any finite time interval and is therefore nowhere differentiable. In practice, however, quantum theory considerations ensure that dx (and hence dt) is strictly positive, and so such mathematical niggles are of no real practical concern. The impossible task of simulating a true Wiener process therefore never arises.

6.1.1 Diffusion equations for the Wiener process

As well as using a Central Limit argument to derive the Normal solution (6.1.6), based on the sum of $n = t/dt$ independent Bernoulli variables each having the distribution (6.1.5), we can also apply the above limiting procedure to the forward Kolmogorov equation

$$p_{jk}^{(n+1)} = p_{j,k-1}^{(n)} p + p_{j,k+1}^{(n)} q \tag{6.1.8}$$

for the simple random walk. Denote $x_0 = jdx$, $x = kdx$ and $t = ndt$, and let $p(x_0, x; t)dx$ be the conditional probability that the particle is at x at time t given that it starts at x_0 at time $t = 0$. Then on rescaling the integer state space from

$(n = 0, 1, \ldots; x = \ldots, -1, 0, 1, \ldots)$ to $(t = 0, dt, \ldots; x = \ldots, -dx, 0, dx, \ldots)$, we see that equation (6.1.8) becomes

$$p(x_0, x; t + dt) = pp(x_0, x - dx; t) + qp(x_0, x + dx; t) \tag{6.1.9}$$

since the factor dx cancels throughout. Whence taking the Taylor series expansion of each term gives

$$p + dt \frac{\partial p}{\partial t} = p[p - dx \frac{\partial p}{\partial x} + \frac{dx^2}{2} \frac{\partial^2 p}{\partial x^2} - \cdots] + q[p + dx \frac{\partial p}{\partial x} + \frac{dx^2}{2} \frac{\partial^2 p}{\partial x^2} + \cdots], \tag{6.1.10}$$

i.e., as $p + q = 1$,

$$dt \frac{\partial p}{\partial t} = (q - p) dx \frac{\partial p}{\partial x} + \frac{dx^2}{2} \frac{\partial^2 p}{\partial x^2} + O(dx^3).$$

Substituting for dt, p and q from (6.1.4) and (6.1.5), namely $dt = dx^2/\sigma^2$ and $q - p = -\mu\sqrt{dt}/\sigma = -\mu dx/\sigma^2$, and letting $dx \to 0$, then yields

$$\frac{\partial p(x_0, x; t)}{\partial t} = -\mu \frac{\partial p(x_0, x; t)}{\partial x} + \frac{\sigma^2}{2} \frac{\partial^2 p(x_0, x; t)}{\partial x^2}. \tag{6.1.11}$$

In physics, this partial differential equation, which is of first order in t and second order in x, has two familiar interpretations. First, it describes the one-dimensional equation for diffusion under an external field of force (e.g. gravity), in which $p(x_0, x; t)$ denotes the density of particles at position x at time t. Second, with $\mu = 0$ it corresponds to the equation for heat conduction, in which $p(x_0, x; t)$ represents temperature and $\sigma^2/2$ temperature conductivity. This highlights the (often forgotten) strong connection between stochastic processes and applied mathematics, physics and, indeed, electrical engineering; placing a difficult problem in one of these fields in the context of one of the others can sometimes lead to surprisingly swift progress.

The backward version of the forward equation (6.1.8) is

$$p_{jk}^{(n+1)} = pp_{j+1,k}^{(n)} + qp_{j-1,k}^{(n)}, \tag{6.1.12}$$

whence similar limiting operations to the above yield the dual backward diffusion equation

$$\frac{\partial p(x_0, x; t)}{\partial t} = \mu \frac{\partial p(x_0, x; t)}{\partial x_0} + \frac{\sigma^2}{2} \frac{\partial^2 p(x_0, x; t)}{\partial x_0^2}. \tag{6.1.13}$$

Note that equations (6.1.11) and (6.1.13) are identical apart from the sign of μ. This is to be expected, since the probability of moving from x_0 to x in time t under drift μ is clearly the same as that of moving the same distance in the reverse direction (i.e. from x to x_0) under drift $-\mu$.

It is a simple matter to verify by direct substitution that the Normal solution (6.1.6) with start point x_0, namely

$$p(x_0, x; t) = \frac{1}{\sigma\sqrt{2\pi t}} \exp\left\{ \frac{-(x - x_0 - \mu t)^2}{2\sigma^2 t} \right\}, \tag{6.1.14}$$

satisfies both equations (6.1.11) and (6.1.13). However, in order to provide a guide to handling more complicated problems, it is instructive to derive result (6.1.14) by solving one of these equations directly. Various methods of attack, such as using the separation of variables, are open to us, but given the simple nature of the Normal solution the easiest approach is to work directly with the associated moment generating function. Let

$$M(\theta; t) \equiv \int_{-\infty}^{\infty} p(x_0, x; t) e^{\theta x} \, dx \qquad (6.1.15)$$

be the m.g.f. of $p(x_0, x; t)$. Now

$$\int_{-\infty}^{\infty} \frac{\partial p}{\partial x} \, dx = [pe^{\theta x}]_{-\infty}^{\infty} - \int_{-\infty}^{\infty} p\theta e^{\theta x} \, dx = -\theta M,$$

and (similarly)

$$\int_{-\infty}^{\infty} \frac{\partial^2 p}{\partial x^2} e^{\theta x} \, dx = \theta^2 M.$$

Whence multiplying the forward equation (6.1.11) by $e^{\theta x}$ and integrating with respect to x yields

$$\frac{\partial M}{\partial t} = (\mu\theta + \sigma^2\theta/2)M.$$

Since $M(\theta; 0) = e^{x_0 \theta}$ at $t = 0$, we therefore have

$$M(\theta; t) = \exp\{(x_0 + \mu t)\theta + \sigma^2 t\theta^2\}, \qquad (6.1.16)$$

which is the m.g.f. of a Normal distribution with mean $x_0 + \mu t$ and variance $\sigma^2 t$, in agreement with (6.1.14).

The increment $X(t) - X(0) \sim N(\mu t, \sigma^2 t)$ is clearly independent of the value of $X(0)$. Thus, in addition to the Markov property, the process $X(t)$ also has the stronger property that for any non-overlapping time intervals (t_1, t_2) and (t_3, t_4) the random variables $X(t_2) - X(t_1)$ and $X(t_4) - X(t_3)$ are independent. Moreover, $X(t_2) - X(t_1)$ is Normally distributed with mean $\mu(t_2 - t_1)$ and variance $\sigma^2(t_2 - t_1)$.

6.1.2 Wiener process with reflecting barriers

In Section 4.3 we showed that the equilibrium probability distribution for the simple random walk operating between two reflecting barriers at $i = 0$ and a is given by expression (4.3.7), namely

$$\pi_i = (p/q)^i \frac{[1 - (p/q)]}{[1 - (p/q)^{a+1}]} \qquad (i = 0, \ldots, a). \qquad (6.1.17)$$

Now on substituting for p, q and $\sqrt{dt}$ from (6.1.4) and (6.1.5), with $i \to x/dx$ and $a \to a/dx$, we have as $dx \to 0$

$$\left(\frac{p}{q}\right)^i = \left(\frac{1 + (\mu/\sigma^2)dx}{1 - (\mu/\sigma^2)dx}\right)^{x/dx} \to e^{(2\mu/\sigma^2)x}. \qquad (6.1.18)$$

Similarly,
$$(p/q)^a \to e^{(2\mu/\sigma^2)a}. \tag{6.1.19}$$

Whence as $(q-p)/p \simeq -(2\mu/\sigma^2)dx$, combining (6.1.17)–(6.1.19) yields the truncated exponential p.d.f.

$$\pi(x) = \frac{(2\mu/\sigma^2)e^{(2\mu/\sigma^2)x}}{e^{(2\mu/\sigma^2)a} - 1} \qquad (0 \le x \le a). \tag{6.1.20}$$

So in the case of a single reflecting barrier at $x = 0$, with $\mu < 0$ to ensure that an equilibrium distribution exists, on letting $a \to \infty$ we have the pure exponential p.d.f.

$$\pi(x) = -(2\mu/\sigma^2)e^{(2\mu/\sigma^2)x} \qquad (x \ge 0, \mu < 0). \tag{6.1.21}$$

However, whilst changing the reflecting barrier condition at $i = 0$ has a direct effect on π_i, making a comparable change at $x = 0$ has no effect whatsoever. For although over the discrete space $i = 0, 1, 2, \ldots$ the proportion of time the process spends in state $i = 0$ is $\pi_0 > 0$, the proportion of time the continuous space process spends at $x = 0$ is $\pi(0)dx \to 0$ as $dx \to 0$. This exposes a considerable difference between discrete and continuous state space models, which clearly needs to be handled with care. For suppose we genuinely want to spend a non-zero proportion of time at $x = 0$. On imposing the transition probabilities $p_{01} = p'$ and $p_{00} = 1 - p'$, at $i = 0$ we have the balance equation $\pi_0 p' = \pi_1 q$, so $p' = \pi_1 q/\pi_0$. Thus to incorporate an atom of probability $\pi(0) = \pi_0$ at $x = 0$, we need to reduce (6.1.21) to

$$\pi(x) = (1 - \pi_0)(-2\mu/\sigma^2)e^{(2\mu/\sigma^2)x} \qquad (x > 0; \mu < 0). \tag{6.1.22}$$

Whence on replacing π_1 by $\pi(dx)$, it follows that

$$p' = \pi(dx)q/\pi_0$$
$$= [(1-\pi_0)/\pi_0][(-2\mu/\sigma^2)e^{(2\mu/\sigma^2)dx}][(1/2)\{1 - (\mu dx/\sigma^2)\}]$$
$$\simeq [(1-\pi_0)/\pi_0](-\mu dx/\sigma^2). \tag{6.1.23}$$

This means that we have to take $p' = O(dx)$ rather than $p = 1/2 + O(dx)$. For example, to retain the probability that $i = 0$ for a simple random walk with a single reflecting barrier at 0, we substitute for $\pi(0) = \pi_0 = 1 - (p/q)$ (obtained by letting $a \to \infty$ in (6.1.17)). Whilst for the M/M/1 queue with arrival rate $p = \alpha$, service rate $q = \beta$ and traffic intensity $\rho = \alpha/\beta$ (Section 2.6.1) we place $\pi_0 = \rho$.

Even though we may change the boundary condition at 0 to ensure that $\pi(0) = \pi_0$, the associated trajectories are radically different in the discrete- and continuous-time scenarios. For we have already seen that the velocity of the process for $x > 0$ becomes infinite as $dx = \sigma\sqrt{dt} \to 0$. So if the process departs $x = 0$ by moving to $x = dx$ it can immediately return to $x = 0$ again where it spends another non-zero length of time. Thus, in contrast to the discrete-time case, further action is needed if the particle is to be able to leave the vicinity of the reflecting barrier without undue difficulty. One possible solution would be to make its one-step move from 0 to say 1, rather than to dx, and we shall progress this idea further in Section 6.2.

302 *Markov processes in continuous time and space*

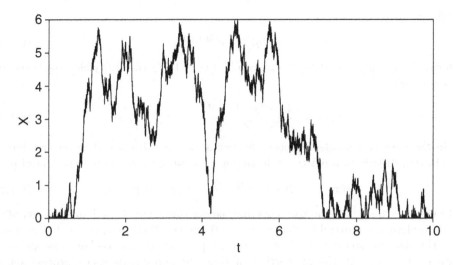

Figure 6.2 Simulated Wiener process $x(t)$ ($0 \leq t \leq 10$) starting from $x(0) = 0$ with a modified reflecting barrier at $x = 0$ corresponding to $\pi(0) = 0.2$; $\mu = -1$ and $\sigma = 2$ with $dt = 0.0001$.

Figure 6.2 shows a realization starting from $x(0) = 0$ for $\mu = -1$ and $\sigma = 2$ with $dt = 0.0001$. Here we have chosen $\pi(0) = 0.2$, so as $dx = \sigma\sqrt{dt} = 0.02$ it follows from (6.1.23) that $p' \simeq 0.02$. Hence as successive moves are independent, the number of steps spent in state 0 before leaving for state 0.02 is geometrically distributed with mean 50, which corresponds to the short time period $50dt = 0.005$. However, the process is easily sucked back into state 0, and the figure shows that, once state 0 is reached, there is a good chance that the process will remain trapped near there until such time as it manages to 'break out'. Here $x(t)$ remains near 0 until time $t \simeq 0.7$, at which point it behaves as an unrestricted process until $t \simeq 7.1$ where it once again comes under the magnetic influence of the reflecting barrier. Note that over $0 \leq t \leq 1000$ the realization spends 20.17% of its time at $x = 0$, which is very close to the target value $\pi_0 = 0.2$.

In order to determine the time-dependent properties of the Wiener process starting from $X(0) = x_0 > 0$, we need to solve the forward diffusion equation (6.1.11) subject to the initial condition

$$p(x;0) = \delta(x - x_0). \qquad (6.1.24)$$

Whilst to determine the boundary equation at $x = 0$, we first note that for an infinite number of trajectories all starting at x_0, the probability $p(x;t)dx$ is the proportion of trajectories which lie in the interval $(x, x + dx)$ during the time interval $(t, t + dt)$. Since the existence of a reflecting barrier at 0 means that no trajectory can pass below 0, we therefore have

$$\int_0^\infty p(x;t)\,dt = 1. \qquad (6.1.25)$$

Consequently,

$$\frac{\partial}{\partial t} \int_0^\infty p(x;t)\, dx = \int_0^\infty \frac{\partial p(x;t)}{\partial t}\, dx = 0. \qquad (6.1.26)$$

Whence on integrating (6.1.11) with respect to x it follows that

$$0 = \int_0^\infty \frac{\partial p(x;t)}{\partial t}\, dx = \int_0^\infty \frac{\partial}{\partial x}\left[-\mu p(x;t) + \frac{\sigma^2}{2}\frac{\partial p(x;t)}{\partial x}\right] dx,$$

which gives rise to the boundary condition

$$\left[\frac{\sigma^2}{2}\frac{\partial p(x;t)}{\partial x} - \mu p(x;t)\right]_{x=0} = 0. \qquad (6.1.27)$$

Equation (6.1.11) may be solved by using the Method of Images (e.g. Sommerfield, 1949) in which the process is allowed to operate over $-\infty < x < \infty$ by reflecting it at $x = 0$. To satisfy condition (6.1.27) we place a point image at the negative initial point $x = x_0$ together with a continuous system of images in the range $x < -x_0$. So suppose

$$p(x;t) = \frac{1}{\sigma\sqrt{2\pi t}}\left[\exp\left\{\frac{-(x-x_0-\mu t)^2}{2\sigma^2 t}\right\} + c\exp\left\{\frac{-(x+x_0-\mu t)^2}{2\sigma^2 t}\right\}\right.$$
$$\left. + \int_{-\infty}^{-x_0} \exp\left\{\frac{-(x-\xi-\mu t)^2}{2\sigma^2 t}\right\} h(\xi)\, d\xi\right]. \qquad (6.1.28)$$

Then as the Normal solution (6.1.14) satisfies the basic equation (6.1.11) irrespective of whether $x_0 > 0$, $x_0 < 0$ or $x_0 = \xi < 0$, and whether it is scaled by 1, c or $h(\xi)d\xi$, it follows that the linear combination (6.1.28) must also satisfy (6.1.11). Thus all that remains is to find values of c and $h(\xi)$ that enable (6.1.28) to satisfy the boundary condition (6.1.27), since the initial condition (6.1.24) is automatically satisfied.

Now differentiating (6.1.28) and placing $x = 0$ gives

$$0 = (\sigma\sqrt{2\pi t})\left[\frac{\sigma^2}{2}\frac{\partial p(x;t)}{\partial x} - \mu p(x;t)\right]_{x=0}$$
$$= \left(\frac{x_0 - \mu t}{2t}\right)\exp\left\{\frac{-(x_0+\mu t)^2}{2\sigma^2 t}\right\} - c\left(\frac{x_0+\mu t}{2t}\right)\exp\left\{\frac{-(x_0-\mu t)^2}{2\sigma^2 t}\right\}$$
$$+ \int_{-\infty}^{-x_0}\left(\frac{(\xi+\mu t)}{2\sigma^2 t} - \mu\right)\exp\left\{\frac{-(\xi+\mu t)^2}{2\sigma^2 t}\right\} h(\xi)\, d\xi.$$

Whence on integrating the first integral by parts, we have

$$0 = \exp\left\{\frac{-(x_0^2+\mu^2 t^2)}{2\sigma^2 t}\right\}\left[\left(\frac{x_0-\mu t}{2t}\right)e^{-x_0\mu/\sigma^2} - c\left(\frac{x_0+\mu t}{2t}\right)e^{x_0\mu/\sigma^2}\right.$$
$$\left. - (\sigma^2/2)h(-x_0)e^{x_0\mu/\sigma^2}\right] + \int_{-\infty}^{-x_0}\left[\frac{\sigma^2}{2}\frac{dh(\xi)}{d\xi} - \mu h(\xi)\right] d\xi.$$

Separating out the two components of this expression then yields the equations

$$\frac{\sigma^2}{2}\frac{dh(\xi)}{d\xi} = \mu h(\xi) \qquad (6.1.29)$$

and

$$(x_0 - \mu t)e^{-x_0\mu/\sigma^2} - c(x_0 + \mu t)e^{x_0\mu/\sigma^2} - (\sigma^2 t)h(-x_0)e^{x_0\mu/\sigma^2} = 0. \qquad (6.1.30)$$

On integrating (6.1.29) to form

$$h(\xi) = (2\mu/\sigma^2)e^{2\mu\xi/\sigma^2}, \qquad (6.1.31)$$

and taking

$$c = e^{-2x_0\mu/\sigma^2}, \qquad (6.1.32)$$

we see that (6.1.30) is automatically satisfied. Whilst on substituting for $h(\xi)$ from (6.1.31), the integral part of (6.1.28) becomes

$$(2\mu/\sigma^2)e^{2x\mu t/\sigma^2}\Phi\left(\frac{-x_0 - \mu t - x}{\sigma\sqrt{t}}\right),$$

where $\Phi(\cdot)$ denotes the standard Normal c.d.f. Thus the solution (6.1.28) takes the final form

$$p(x;t) = \frac{1}{\sigma\sqrt{2\pi t}}\left[\exp\left\{\frac{-(x - x_0 - \mu t)^2}{2\sigma^2 t}\right\} + \exp\left\{\frac{-4x_0\mu t - (x + x_0 - \mu t)^2}{2\sigma^2 t}\right\}\right]$$

$$+ \frac{2\mu}{\sigma^2}\exp\left\{\frac{2\mu x}{\sigma^2}\right\}\left[1 - \Phi\left(\frac{x + x_0 + \mu t}{\sigma\sqrt{t}}\right)\right]. \qquad (6.1.33)$$

Note that if $\mu < 0$, so that the process drifts left towards the reflecting barrier at $x = 0$, then letting $t \to \infty$ and replacing μ by $|\mu|$ recovers the equilibrium exponential distribution (6.1.21), namely

$$p(x;\infty) = (2|\mu|/\sigma^2)\exp\{-2|\mu|x/\sigma^2\} \qquad (x > 0, \mu < 0). \qquad (6.1.34)$$

In contrast, if $\mu > 0$ then the process drifts towards $+\infty$, and $p(x;t) \to 0$ as $t \to \infty$ for all $x > 0$.

At this point it is worth recalling the relatively tortuous recurrence relation-based solution developed in Section 4.3.3 for the simple random walk with two reflecting barriers. For the structure of (6.1.33) is far more transparent in nature, being based on two Normal exponents and an exponentially weighted Normal integral. This highlights one of the main differences between the discrete and continuous time approaches; whereas the former usually depends on the construction of potentially messy combinatorial relationships, the latter is far more likely to yield solutions involving standard functions with known properties.

6.1.3 Wiener process with absorbing barriers

Suppose we now parallel the simple random walk scenario with absorbing barriers, developed in Section 4.2. Consider first the case of a single absorbing barrier at

$x = a > 0$, and take $X(0) = x_0 = 0$. We retain the forward equation (6.1.11) and the initial condition (6.1.24), i.e. $p(x; 0) = \delta(x)$, but now we have to impose the boundary condition

$$p(a; t) = 0 \quad (t > 0). \tag{6.1.35}$$

To see where this comes from, consider the boundary equation at $x = a > 0$, namely

$$p(a; t + dt) = p(a; t) + pp(a - dx; t). \tag{6.1.36}$$

On expanding (6.1.36) as a Taylor series, with $p = (1/2)\{1 + (\mu\sqrt{dt}/\sigma)\}$ and $dt = dx^2/\sigma^2$, it follows that at $x = a$

$$p + \frac{dx^2}{\sigma^2}\frac{\partial p}{\partial t} = p + (1/2)(1 + \frac{\mu dx}{\sigma^2})[p - dx\frac{\partial p}{\partial x} + \frac{dx^2}{2}\frac{\partial^2 p}{\partial x^2} - \cdots]. \tag{6.1.37}$$

Thus unlike expansion (6.1.10), which gives rise to the p.d.e. (6.1.11), here we can only extract the much simpler expression

$$p(a; t) + O(dx) = 0,$$

which leads to (6.1.35) as $dx \to 0$.

As in the unrestricted case we base our solution on the Normal form (6.1.14), but this time employ the Method of Images by first imagining the barrier at $x = a$ to be a mirror and then placing an image source at $x = 2a$, since this point corresponds to the perceived location of $x = 0$. Then the solution

$$p(x; t) = \frac{1}{\sigma\sqrt{2\pi t}}\left[\exp\left\{\frac{-(x - \mu t)^2}{2\sigma^2 t}\right\} + A\exp\left\{\frac{-(x - 2a - \mu t)^2}{2\sigma^2 t}\right\}\right] \tag{6.1.38}$$

not only automatically satisfies the unrestricted equation (6.1.11) for $x < a$, but the boundary equation (6.1.35) is also satisfied provided the constant A is chosen to ensure that the probability flux flowing upwards from the first Normal component is exactly counterbalanced by the flux flowing downwards from the second. To achieve this we simply place $x = a$ in (6.1.38) to obtain

$$A = -\exp(2a\mu/\sigma^2). \tag{6.1.39}$$

So we can regard this solution for $p(x; t)$ as a superposition of a source of unit strength at the origin and a source of strength $-\exp(2\mu a/\sigma^2)$ (i.e. a 'sink') at $x = 2a$.

To parallel the first passage time results of Section 4.2 for the simple random walk, where we consider the number of steps n to reach state $x = a$ for the first time starting from $x = 0$, write

$$\Pr(X(s) < a \text{ for all } 0 < s < t, X(t) \leq x | X(0) = 0) = \int_{-\infty}^{x} p(y; t)\, dy \equiv P(x; t). \tag{6.1.40}$$

Then $P(a; t)$ is the probability that absorption has not yet occurred at a by time $T = t$, i.e. $P(a; t) = \Pr(T > t)$. Whence the corresponding p.d.f. of the absorption time T is given by

$$g(t) = \frac{d}{dt}[1 - P(a;t)] = -\frac{d}{dt}P(a;t). \quad (6.1.41)$$

If we now assume that $\mu \geq 0$, so that the probability of eventual absorption at a is one, then it follows from solution (6.1.38)–(6.1.39) that equation (6.1.41) takes the form

$$g(t) = -\frac{d}{dt}\left\{\Phi\left(\frac{a - \mu t}{\sigma\sqrt{t}}\right) - \exp\left(\frac{2\mu a}{\sigma^2}\right)\Phi\left(\frac{-a - \mu t}{\sigma\sqrt{t}}\right)\right\}, \quad (6.1.42)$$

where $\phi(\cdot)$ and $\Phi(\cdot)$ denote the standard Normal p.d.f. and c.d.f., respectively. This yields

$$g(t) = \frac{a}{\sigma\sqrt{2\pi t^3}}\exp\left\{-\frac{(a - \mu t)^2}{2\sigma^2 t}\right\} \quad (6.1.43)$$

as the p.d.f. of T for given $a > x_0 = 0$.

Note that for a process with zero drift, i.e. $\mu = 0$, $g(t)$ is of order $O(t^{-3/2})$ as $t \to \infty$, so in this case T has no finite moments. For $\mu > 0$, rather than determining moments by direct integration of (6.1.43) let us adopt an easier approach by first developing the associated m.g.f.

$$g^*(s|x_0, a) = \int_0^\infty e^{-st} g(t|x_0, a)\, dt. \quad (6.1.44)$$

Here we have taken care to highlight the dependence of g^* on x_0. For this can be exploited by taking Laplace transforms on both sides of the backward equation (6.1.13) to obtain

$$sg^*(s|x_0, a) - p(x_0, x; 0) = \mu \frac{dg^*(s|x_0, a)}{dx_0} + \frac{\sigma^2}{2}\frac{d^2 g^*(s|x_0, a)}{dx_0^2}, \quad (6.1.45)$$

bearing in mind the relation between $p(x_0, x; t)$ and $g(t)$ through (6.1.40)–(6.1.41). Since $p(x_0, x; 0) = \delta(x - x_0)$, for $x \neq x_0$ this is a second-order differential equation with constant coefficients, the general solution of which is given by

$$g^*(s|x_0, a) = Ae^{\lambda_1(s)x_0} + Be^{\lambda_2(s)x_0} \quad (6.1.46)$$

for some constants A and B. The values $\lambda_1(s)$ and $\lambda_2(s)$ are the roots of the auxiliary equation

$$(\sigma^2/2)\lambda^2 + \mu\lambda - s = 0, \quad \text{i.e.}$$

$$\lambda_1(s), \lambda_2(s) = (1/\sigma^2)[-\mu \pm \sqrt{\mu^2 + 2s\sigma^2}]. \quad (6.1.47)$$

Thus for $s > 0$ we have $\lambda_2(s) < 0 < \lambda_1(s)$. Now we see from (6.1.44) that

$$g^*(s|x_0, a) = \int_0^\infty e^{-st} g(t|x_0, a)\, dt \leq \int_0^\infty g(t|x_0, a)\, dt \leq 1,$$

whence it follows that $B = 0$ since otherwise $g^*(s)$ becomes unbounded as $x_0 \to -\infty$. Moreover, when $x_0 = a$ absorption occurs immediately, and so

$$g^*(s|a, a) = 1 - Ae^{\lambda_1(s)a},$$

which yields the solution

$$g^*(s|x_0, a) = e^{(x_0-a)\lambda_1(s)} \qquad (x_0 \leq a) \qquad (6.1.48)$$

for the Laplace transform of the first passage time distribution from x_0 to a.

Before we invert this expression it is worth noting that placing $s = 0$ in (6.1.44) gives

$$g^*(0|x_0, a) = \int_0^\infty g(t|x_0, a)\, dt = \pi_0(x_0, a),$$

i.e. the probability $\pi(x_0, a)$ of ever reaching a when starting from $x_0 < a$. Whence we see from (6.1.47) and (6.1.48) that

$$\pi(x_0, a) = \exp\{(x_0 - a)(-\mu + |\mu|)/\sigma^2\} = \begin{cases} 1 & : \mu \geq 0 \\ e^{-2\mu(x_0-a)/\sigma^2} & : \mu < 0. \end{cases} \qquad (6.1.49)$$

So this situation is directly analogous to that of the simple random walk (Section 4.2.2), in that when there is either a drift towards the barrier ($\mu > 0$) or no drift ($\mu = 0$) then the probability of ultimate absorption at a is one; whilst when there is a drift away from the barrier ($\mu < 0$) then there is a positive probability of never reaching the barrier.

When $x_0 = 0$ result (6.1.49) gives the distribution of the maximum possible displacement, X^{max}, namely

$$\Pr(X^{max} \leq a) = 1 - \pi(0, a) = \begin{cases} 0 & : \mu \geq 0 \\ 1 - e^{-2a|\mu|/\sigma^2} & : \mu < 0. \end{cases} \qquad (6.1.50)$$

So for $\mu < 0$, X^{max} follows an exponential distribution with mean $\sigma^2/(2|\mu|)$. Moreover, because the Wiener process is a limiting simple random walk, this result also follows directly from the corresponding expression for the simple random walk, namely on using (6.1.19)

$$\Pr(X^{max} \leq a) = 1 - F_{0a}(1) = 1 - (p/q)^a \to 1 - e^{2a\mu/\sigma^2} \qquad (\mu < 0)$$

as $dx = \sigma\sqrt{dt} \to 0$. Here we have used result (4.2.45) to obtain $F_{0a}(1) = 1 - \{\lambda_1(1)\}^{-a}$, where in the terminology of (4.2.22) $\lambda_1(1) = \lambda_1 = q/p$ for $p < q$ (since $\mu < 0$).

Similar methods can be applied to the two absorbing barrier case, where in addition to the barrier at $a > 0$ we now also have a second one at $-b < 0$ (note that for convenience we use $-b$, rather than b as in Section 4.2). The solution must satisfy the diffusion equation (6.1.45) for $-b < x < a$; whilst in addition to the initial condition $p(x; 0) = \delta(x)$ and boundary condition (6.1.35) we must also have a parallel boundary condition for $-b$, namely

$$p(-b, t) = 0 \qquad (t > 0). \qquad (6.1.51)$$

308 *Markov processes in continuous time and space*

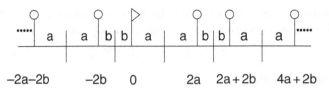

Figure 6.3 Extension of the Wiener process on $-b \leq x(t) \leq a$ starting at $x(0) = 0$, with mirrors placed at the absorbing barriers $-b$ and a extending the image range to $(-\infty, \infty)$; point sources are placed at $2n(a+b)$ and $2a + 2n(a+b)$ for $n = 0, \pm1, \pm2, \ldots$.

One method of solution is to extend the earlier Method of Images based on (6.1.38) to a doubly infinite series of images. Here we place mirrors at both a and $-b$, observe the doubly infinite number of reflections of the point probability source at the origin, and then place a new point source at each one. Figure 6.3 shows that these point sources are located at $\{2a, 2a + 2b, 4a + 2b, 4a + 4b, 6a + 4b, \ldots\}$ and $\{-2b, -2a - 2b, -2a - 4b, -4a - 4b, -4a - 6b, \ldots\}$, i.e. at $2n(a+b)$ and $2a + 2n(a+b)$ over $n = 0, \pm1, \pm2, \ldots$. The reflections of the two mirrors themselves occur at $a + n(a+b)$. Sources with strengths $\exp(\theta y_n)$ are placed at the points $y_n = 2n(a+b)$ (this yields the required unit source at $x = 0$), and strengths $d \exp(\theta z_n)$ at the points $z_n = 2a + 2n(a+b)$ for some constant d. Then on using the principle of the Reflection of Images, the required solution is the linear superposition of solutions for each such point source weighted by the corresponding strength, i.e. on paralleling (6.1.38),

$$p(x;t) = \frac{1}{\sigma\sqrt{2\pi t}} \sum_{n=-\infty}^{\infty} \left[\exp\left\{\theta y_n - \frac{(x - x_n - \mu t)^2}{2\sigma^2 t}\right\} + d\exp\left\{\theta z_n - \frac{(x - z_n - \mu t)^2}{2\sigma^2 t}\right\}\right].$$
(6.1.52)

This series expression is absolutely convergent and differentiable term by term, and since θ, y_n and z_n are independent of both x and t, each of the Normal component terms in (6.1.52) satisfies the forward equation (6.1.11) with $x_0 = 0$. Thus so does the trial solution (6.1.52) for $p(x;t)$. Moreover, on taking $d = -1$ and $\theta = \mu/\sigma^2$ we see that the source at y_n annuls that at z_{-n}. Since $z_{-n} = 2a - 2n(a+b) = 2a - y_n$, the second term

$$d\exp\left\{\theta z_{-n} - \frac{(a - z_{-n} - \mu t)^2}{2\sigma^2 t}\right\}$$

may be written as

$$-\exp\left\{\frac{2\mu t(2a - y_n) - [(a - y_n - \mu t) + 2\mu t]^2}{2\sigma^2 t}\right\},$$

which, after a little rearrangement, cancels with the first term. A similar argument shows that at $x = -b$ the source at z_{-n-1} annuls that at y_n. Thus the boundary conditions (6.1.35) and (6.1.51) are satisfied.

Although this approach works extremely well, it is interesting to demonstrate yet another method of attack, this time based on the Method of Separation of Variables.

Suppose we consider a Fourier series expansion for $p(x;t)$ in which each term changes exponentially in both t and x. Specifically, let us assume that

$$p(x;t) = e^{kx} \sum_{n=1}^{\infty} e^{-\lambda_n t} a_n \sin\left\{\frac{n\pi(x+b)}{a+b}\right\}. \qquad (6.1.53)$$

This pure sine expansion is chosen since it vanishes at $x = -b$ and $x = a$, thus guaranteeing that the boundary conditions (6.1.35) and (6.1.51) are automatically satisfied. Direct substitution shows that (6.1.53) satisfies the forward equation (6.1.11) provided

$$k = \frac{\mu}{\sigma^2} \quad \text{and} \quad \lambda_n = \frac{\mu^2}{2\sigma^2} + \frac{1}{2}\left(\frac{n\pi\sigma}{a+b}\right)^2. \qquad (6.1.54)$$

Finally, the coefficients a_n are determined from the boundary condition $p(x;0) = \delta(x_0)$ (if $x_0 \neq 0$ we simply move the barriers to $a - x_0$ and $-b - x_0$). The obvious solution is

$$a_n = \frac{2}{a+b} \sin\left(\frac{n\pi b}{a+b}\right), \qquad (6.1.55)$$

since

$$\sum_{n=1}^{\infty} \sin\left(\frac{n\pi b}{a+b}\right) \sin\left(\frac{n\pi(x+b)}{a+b}\right) = \begin{cases} 0 & x \neq 0 \\ \infty & x = 0, \end{cases}$$

whilst the coefficient $2/(a+b)$ ensures that $\int_{-b}^{a} p(x;t)\,dx = 1$. So although the series solutions (6.1.52) and (6.1.53) are radically different in appearance, they are, nevertheless, alternative representations of the same function.

6.2 The Fokker–Planck diffusion equation

Whilst this basic development is ideal for considering the motion of a particle where $\Pr(x \to x+dx)$ and $\Pr(x \to x-dx)$ are independent of x, or (in population process terms) an immigration–emigration process for a large number of molecules in a container, in practice these two-step probabilities will often depend on x. So to develop the diffusion equivalent to say a logistic birth–death process we need to take a more general approach.

Let us therefore consider a homogeneous Markov chain in which we can move from state k to states $k+1$ and $k-1$ with probabilities θ_k and ϕ_k, respectively, whilst with probability $1 - \theta_k - \phi_k$ we remain at k. Denote the corresponding n-step transition probabilities by $p_{jk}^{(n)}$. Then the Kolmogorov forward equation is given by

$$p_{jk}^{(n+1)} = p_{j,k-1}^{(n)} \theta_{k-1} + p_{jk}^{(n)}(1 - \theta_k - \phi_k) + p_{j,k+1}^{(n)} \phi_{k+1}. \qquad (6.2.1)$$

Now consider the process as a particle taking small steps of amount $-dx$, 0 or dx in small time intervals of length dt, where (for the time being) dx and dt are assumed fixed. So if the particle is at x at time t then the probabilities that it will be at $x + dx$, x and $x - dx$ at time $t + dt$ are respectively $\theta(x)$, $1 - \theta(x) - \phi(x)$ and $\phi(x)$.

Let $p(x_0, x; t)dx$ be the conditional probability that the particle is at x at time t given that it starts at x_0 at time $t = 0$. Then the close connection between diffusion and Markov population processes may be seen through the following useful (albeit non-rigorous) approach which parallels the analysis of (6.1.8).

In terms of the size increment dx, equation (6.2.1) takes the form

$$p(x_0, x; t+dt)dx = p(x_0, x - dx; t)dx\{\theta(x - dx)\} + p(x_0, x; t)dx\{1 - \theta(x) - \phi(x)\}$$
$$+ p(x_0, x + dx; t)dx\{\phi(x + dx)\}. \qquad (6.2.2)$$

Note that dx cancels throughout. To ensure that the process is proper, we clearly need to place some restrictions on $\theta(x)$ and $\phi(x)$ as dx and dt approach zero. Write the instantaneous mean and variance per unit time of the change in $X(t) = x$, as $\beta(x)$ and $\alpha(x)$, respectively. That is,

$$\beta(x) = \lim_{dt \to 0} \frac{E[X(t + dt) - X(t) | X(t) = x]}{dt} \qquad (6.2.3)$$

$$\alpha(x) = \lim_{dt \to 0} \frac{\text{Var}[X(t + dt) - X(t) | X(t) = x]}{dt}. \qquad (6.2.4)$$

Then in the small time interval $(t, t + dt)$ the mean change in position is

$$\{\theta(x) - \phi(x)\}dx, \qquad (6.2.5)$$

whilst the variance of the change in position is

$$[\theta(x) + \phi(x) - \{\theta(x) - \phi(x)\}^2](dx)^2. \qquad (6.2.6)$$

The relations (6.2.3) and (6.2.4) thus become

$$\beta(x) = \lim_{dx, dt \to 0} \{\theta(x) - \phi(x)\}\frac{dx}{dt} \qquad (6.2.7)$$

$$\alpha(x) = \lim_{dx, dt \to 0} [\theta(x) + \phi(x) - \{\theta(x) - \phi(x)\}^2]\frac{(dx)^2}{dt}. \qquad (6.2.8)$$

To determine the allowable forms for $\theta(x)$ and $\phi(x)$, together with admissible ways for dx and dt to approach zero, suppose that the instantaneous variance $\alpha(x)$ is a bounded function satisfying

$$\alpha(x) \leq A \qquad (6.2.9)$$

for all x and some constant $A > 0$. If we now take $(dx)^2 = Adt$, then from (6.2.7)

$$\theta(x) - \phi(x) = \beta(x)dx/A, \qquad (6.2.10)$$

whence from (6.2.8)

$$\theta(x) + \phi(x) = \alpha(x)/A + \{\beta(x)dx/A\}^2 \approx \alpha(x)/A. \qquad (6.2.11)$$

So
$$\theta(x) = (1/2A)\{\alpha(x) + \beta(x)dx\} \tag{6.2.12}$$
$$\phi(x) = (1/2A)\{\alpha(x) - \beta(x)dx\}, \tag{6.2.13}$$

whilst condition (6.2.9) ensures that

$$\Pr(\text{move in } (t, t+dt)|X(t) = x) = \theta(x) + \phi(x) \le 1. \tag{6.2.14}$$

Now on forming the Taylor series expansion of the forward equation (6.2.2), to order $o(dx^2, dt)$

$$p + dt\frac{\partial p}{\partial t} = [p - dx\frac{\partial p}{\partial x} + (1/2)(dx)^2\frac{\partial^2 p}{\partial x^2}][\theta - dx\theta' + (1/2)dx^2\theta''] + p(1 - \theta - \phi)$$
$$+ [p + dx\frac{\partial p}{\partial x} + (1/2)(dx)^2\frac{\partial^2 p}{\partial x^2}][\phi + dx\phi' + (1/2)(dx)^2\phi'']$$

(where ' denotes $\partial/\partial x$). Using relations (6.2.9)–(6.2.13), and letting $dx, dt \to 0$, then yields the forward diffusion equation

$$\frac{\partial p}{\partial t} = [\alpha''(x)/2 - \beta'(x)]p + [\alpha'(x) - \beta(x)]\frac{\partial p}{\partial x} + [\alpha(x)/2]\frac{\partial^2 p}{\partial x^2}. \tag{6.2.15}$$

A neater, equivalent, representation is

$$\frac{\partial p}{\partial t} = \frac{1}{2}\frac{\partial^2}{\partial x^2}\{\alpha(x)p\} - \frac{\partial}{\partial x}\{\beta(x)p\}, \tag{6.2.16}$$

known as the Focker–Planck equation.

The solution $p(x_0, x; t)$ is defined solely by the instantaneous mean $\beta(x)$ and variance $\alpha(x)$, which warns us that care is needed when using the diffusion process as an approximation to a population process. For an uncountably infinite number of population processes possess the same mean and variance, and hence the same diffusion 'equivalent', yet each has its own unique behaviour defined through the population transition rates $\{\lambda_N, \mu_N\}$.

When making a formal comparison between this approach and that developed earlier for the Wiener process (Section 6.1), we need to ensure that we are comparing like with like. For unlike the current (Fokker–Planck) approach, in the Wiener process the particle moves at *every* step, which implies that $\theta(x) + \phi(x) \equiv 1$. Now we see from (6.2.12) and (6.2.13) that this requires $\alpha(x) \equiv A$, and on placing $\beta(x) = \mu$ and $\alpha(x) = \sigma^2$ the forward equation (6.2.16) reduces to (6.1.11). Thus $A = \sigma^2$ and $\beta(x) \equiv \mu$, whence the associated transition probabilities are

$$\theta(x) = (1/2)\{1 + \beta(x)dx/A\} = (1/2)\{1 + \mu\sqrt{dt}/\sigma\} \tag{6.2.17}$$
$$\phi(x) = (1/2)\{1 - \beta(x)dx/A\} = (1/2)\{1 - \mu\sqrt{dt}/\sigma\}, \tag{6.2.18}$$

which are in exact agreement with (6.1.5).

The corresponding backward equation follows similarly, and is given by expanding

$$p(x_0, x; t+dt)dx_0 = \theta(x_0)p(x_0+dx_0, x; t)dx_0 + (1-\theta(x_0)-\phi(x_0))p(x_0, x; t)dx_0$$
$$+ \phi(x_0)p(x_0-dx_0, x; t)dx_0 \qquad (6.2.19)$$

as a Taylor series in x_0; here it is the increment dx_0 that cancels throughout. Writing $(dx_0)^2 = A dt$ then yields

$$\frac{\partial p}{\partial t} = \frac{1}{2}\alpha(x_0)\frac{\partial^2 p}{\partial x_0^2} + \beta(x_0)\frac{\partial p}{\partial x_0}. \qquad (6.2.20)$$

So in the forward equation (6.2.16) the backward variable x_0 is essentially constant and enters only through the initial conditions; whilst in the backward equation (6.2.20) it is the forward variable x that is held in check.

The forward and the simpler backward representations also hold if we allow the transition mechanism to depend not only on the state variable x but also on the time t. In this case we are led to define $\beta(x,t)$ and $\alpha(x,t)$, both depending on x and t, as the instantaneous mean and variance per unit time of the change in $X(t) = x$, but we shall not pursue this level of generality here.

From a practical point of view both equations are useful in particular circumstances, as we have seen earlier when considering discrete state space processes. If we are interested in the probability distribution for a given initial value of $X(t_0) = x_0$, then clearly the forward equation is the appropriate one to use. Conversely, if we wish to determine say the first passage time distribution to a fixed state a as a function of the initial position x_0, then we should use the backward equation. It can be shown that the backward equation is the more general of the two in that it is also satisfied by the probability distribution function

$$P(x_0, t_0; x, t) = \Pr(X(t) \leq x | X(t_0) = x_0) = \int_{-\infty}^{x} p(x_0, t_0; y, t)\, dy. \qquad (6.2.21)$$

Moreover, unlike the forward equation its form remains unaffected by the type of boundary condition that may be imposed on the process. For example, suppose we have a diffusion process which is developing inside a finite interval and that when the diffusing particle reaches a boundary it is held there for an exponential distributed time s. It then returns instantaneously to an interior point x having p.d.f. $h(x)$, whereupon it immediately starts diffusing again. Such a process is still Markov, but when we try to formulate the forward equation we see that during the small time interval $(t, t+dt)$ the point x can be reached not only from neighbouring points but also from a point on the boundary. So the transitions are no longer local in character, and this feature destroys the simple character of the forward equation; though it does not change the form of the backward equation. Thus the forward and backward equations are no longer dual processes.

6.2.1 Simulation of the simple immigration–death diffusion process

To illustrate the Focker–Planck formulation consider the simple immigration–death process for which the general population birth and death rates take the form $\lambda_N = \alpha$ and $\mu_N = \mu N$, respectively. On replacing $N = 0, 1, 2, \ldots$ by $x \geq 0$ we have

$$E[\{X(t+dt) - X(t)\}|X(t) = x] = (\lambda_x - \mu_x)dt = (\alpha - \mu x)dt \quad (6.2.22)$$
$$\text{Var}[\{X(t+dt) - X(t)\}|X(t) = x] = (\lambda_x + \mu_x)dt - [(\lambda_x - \mu_x)dt]^2 = (\alpha + \mu x)dt + o(dt).$$

Whence using (6.2.3) and (6.2.4) yields

$$\beta(x) = \alpha - \mu x \quad \text{and} \quad a(x) = \alpha + \mu x. \quad (6.2.23)$$

It then follows from (6.2.12) and (6.2.13) that the associated transition probabilities are

$$\theta(x) = (1/2A)\{(\alpha + \mu x) + (\alpha - \mu x)dx\} \quad (6.2.24)$$
$$\phi(x) = (1/2A)\{(\alpha + \mu x) - (\alpha - \mu x)dx\}. \quad (6.2.25)$$

Note the considerable structural difference between the population and diffusion transition rates, forced by the need to ensure that $X(t)$ neither explodes nor remains constant in a finite time period when the population step size changes from ± 1 to $\pm dx$. For whilst the population birth and death rates are quite different in form, being $\lambda_N = \alpha$ and $\mu_N = \mu N$, to first order the corresponding diffusion rates, $\theta(x)$ and $\phi(x)$, both equal $(1/2A)(\alpha + \mu x)$. Even their second-order components have the same magnitude, namely $(1/2A)(\alpha - \mu x)dx$; it is only their sign that changes.

Inspection of (6.2.23) highlights the restrictive nature of condition (6.2.9), for here the instantaneous variance $a(x) = \alpha + \mu x$ is clearly unbounded. Note that this concern is not shared by the population process $\{X_N\}$, since $\lambda_N = \alpha$ and $\mu_N = \mu N$ are probability *rates* and so may take any non-negative values. So to convert a continuous-time population process on $N = 0, 1, 2, \ldots$ to a diffusion process on $0 \leq x < \infty$, with an unbounded $a(x)$, we need to define a 'pragmatic maximum', X^{max}, for $X(t)$. This is easily defined for positive recurrent processes simply by taking (say)

$$X^{max} = \max\{X(0), \kappa_1(\infty) + 6\sqrt{\kappa_2(\infty)}\}, \quad (6.2.26)$$

since it is unlikely that $X(t)$ will lie more than six standard deviations above the equilibrium mean within any reasonably long time interval $[0, T]$. However, for null recurrent and transient processes we need to use time-dependent moments; whilst even more care is needed when working with explosive, i.e. dishonest, processes (see Section 2.3.5).

Suppose we wish to simulate the simple immigration–death diffusion process with $\alpha = 1$ and $\mu = 0.1$, starting from $X(0) = 1$. Then as the equilibrium probabilities $\pi_i \sim \text{Poisson}(\alpha/\mu)$ (see (2.5.9)), result (6.2.26) suggests taking $X^{max} = 10 + 6\sqrt{10} = 28.97$. Whence it follows from (6.2.9) and (6.2.23) that $A \geq \max_{0 \leq x \leq X^{max}}(\alpha + \mu x) = 3.90 \simeq 4$. Thus even choosing a fairly large displacement increment $dx = 0.01$, and taking $A = 4$, gives rise to a very small time increment $dt = dx^2/A = 0.000025$. In its simplest form the simulation algorithm is an easy extension of (4.1.41) for the simple random walk. Namely, for $X(t) = x$, $\{U\}$ a sequence of i.i.d. $U(0,1)$ random variables, and $t = 0, dt, 2dt, \ldots, T$,

314 *Markov processes in continuous time and space*

$$\begin{aligned}
&\text{if } U \leq \theta(x) &&\text{then } X(t+dt) = x + dx \\
&\text{else if } U \leq \theta(x) + \phi(x) &&\text{then } X(t+dt) = x - dx \\
&\text{else } X(t+dt) = x.
\end{aligned} \qquad (6.2.27)$$

Figure 6.4a shows both a diffusion and a population realization starting at $X(0) = 1$, and at the macro level their broad structure appears to be remarkably similar, in

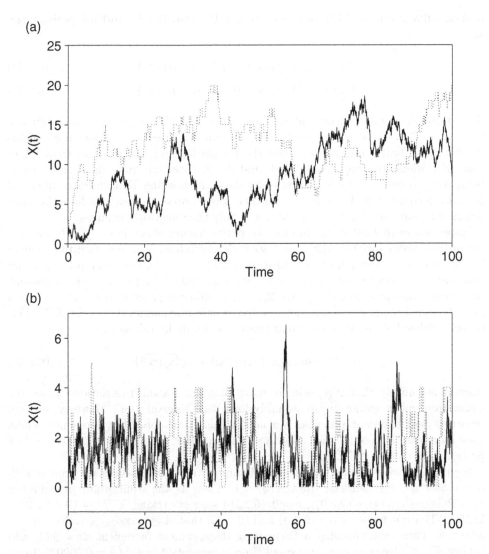

Figure 6.4 Immigration–death diffusion (——) and population ($\cdots$) processes starting from $X(0) = 1$, with step increments $dx = 0.01$ and 1, respectively, for $\alpha = 1$ and: (a) $\mu = 0.1$ and $A = 4$; (b) $\mu = 1.0$ and $A = 8$ with $X(t) = \max[0, X(t)]$.

spite of the rapid local diffusion fluctuations being very different from the much longer Exponential($\lambda_N + \mu_N$) distributed horizontal steps of the population process. However, although this diffusion plot over $t = 0(0.01)100$ with $dx = 0.01$ and $dt = 0.000025$ shows $X(t) > 0$, in another realization $X(t)$ dipped below zero in the opening stages. So there is clearly a behavioural discrepancy between the population process and its diffusion approximation. For whilst the population death rate $\mu_0 = 0$ automatically precludes negative population sizes, the associated diffusion rate $\phi(x) = 0$ at $\alpha(x) = \beta(x)dx$. Specifically, using (6.2.23), we see that the lower diffusion bound occurs at $(\alpha + \mu x) = (\alpha - \mu x)dx$, i.e. at

$$X^{min} = -(\alpha/\mu)[(1 - dx)/(1 + dx)] \simeq -(\alpha/\mu) < 0, \qquad (6.2.28)$$

irrespective of the size of the small increment dx.

Though Figure 6.4a suggests that $\Pr(X(\infty) < 0) \simeq 0$, this boundary probability becomes much more important when $\alpha/\mu \simeq 1$. For then $\kappa_1(\infty) \simeq 1$ lies much closer to zero, and the population probability of being there, namely $\pi_0 = e^{-1} = 0.368$ if $\alpha/\mu = 1$, is large. So what happens to the associated diffusion process? Since $\theta(0) = (\alpha/2A)(1 + dx)$ is only marginally larger than $\phi(0) = (\alpha/2A)(1 - dx)$, for $X(t) \simeq 0$ the process behaves as an (almost) symmetric simple random walk, which (with high probability) results in excursions below zero (see Section 4.1). Several options are open to us. First, we could allow $X(t)$ to become negative, and regard $X(t) > -\alpha/\mu = -1$ as not being too unrealistic. Second, we can simply replace the death rate $\phi(0)$ by zero. Figure 6.4b illustrates this latter scenario, and visual comparison between the diffusion and population simulations is encouraging. Third, we could parallel results (6.1.22) and (6.1.23) for the Wiener process with a reflecting barrier at zero. Or, if we wish to regard restarting the process from empty and immigration as being two different mechanisms, and hold an empty process for an Exponential(γ) time before restarting it in an interior position, δ, away from zero, we could replace lines 1 and 2 of algorithm (6.2.27) (for $x = 0$) by

$$\text{if } U \leq \gamma dt \qquad \text{then } X(t + dt) = \delta > 0. \qquad (6.2.29)$$

Placing $\gamma = \alpha$ recovers the Exponential(α) distribution corresponding to a new immigrant restarting the population process from empty, though we could equally well select any other value. Moreover, whilst taking $\delta = 1$ mimics the population process in that an arriving immigrant immediately raises $X(t)$ from 0 to 1, taking $\delta > 1$ allows for mass immigration (see Section 2.7). Figure 6.5 illustrates this last situation, showing a realization of a diffusion process for $\alpha = 1$, $\mu = 5$, $dx = 0.01$ and $A = 8$, with $\gamma = 0.5$ and $\delta = 10$. The problem of how we might best simulate diffusion processes under severe (population) boundary constraints clearly merits further investigation.

If computing efficiency is an issue, then one way of making the algorithm faster is to parallel the population approach taken for the continuous-time general birth–death process. For there the time to the next event follows the Exponential($\lambda_N + \mu_N$) distribution, whilst for a diffusion simulation the number of steps M to the next event follows a geometric distribution with $\Pr(M = m) = pq^{m-1}$ ($m = 1, 2, \ldots$). Here $p = \theta(x) + \phi(x)$ and $q = 1 - p$ respectively denote the probability of change and no

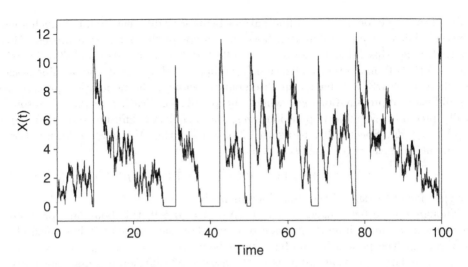

Figure 6.5 Immigration–death diffusion process starting from $X(0) = 1$ with $\alpha = 1$, $\mu = 0.5$, $dx = 0.01$ and mass immigration at rate $\gamma = 0.5$ into state $\delta = 10$ when empty.

change. For the simulation shown in Figure 6.4a, for which $\alpha = 1$, $\mu = 0.1$, $dx = 0.01$ and $A = 4$, we see from (6.2.24) and (6.2.25) that

$$p = (1/A)(\alpha + \mu x) \simeq \begin{cases} 1 & \text{for } x \simeq X^{max} = 30 \\ 0.5 & \text{for } x \simeq \kappa_1(\infty) = \alpha/\mu = 10 \\ 0.25 & \text{for } x \simeq 0 \\ 0 & \text{for } x \simeq X^{min} = -\alpha/\mu = -10. \end{cases}$$

Though since zero lies $\sqrt{10} > 3$ standard deviations below $\kappa_1(\infty)$ it is highly unlikely that $X(\infty) < 0$, so the lower bound (6.2.28) is really irrelevant. However, it can prove useful if we wish to implement a severe boundary condition at $x = 0$. For example, in the case of Figure 6.5 we have $p = \gamma dt = \gamma dx^2/A = 1/160000$, so on average 160000 successive steps will result in no change, which is a huge computational waste. Now

$$\Pr(M \leq m) = p(1 + q + \cdots + q^{m-1}) = p(1 - q^m)/(1 - q) = 1 - q^m.$$

So for a sequence $\{U\}$ of i.i.d. $U(0,1)$ pseudo-random variables, we choose

$$M = 1 \quad \text{if } U \leq 1 - q,$$
$$\text{else} \quad M = 2 \quad \text{if } U \leq 1 - q^2, \ldots, \quad \text{etc.}$$

Thus $M = m$ if

$$1 - q^{m-1} < U \leq 1 - q^m,$$

which, on replacing U by $1 - U$, reduces to

$$m - 1 < \ln(U)/\ln(q) \leq m,$$

i.e. $\qquad m = 1 + \text{integer part}\{\ln(U)/\ln(q)\}.$ \hfill (6.2.30)

Hence if $X(t) = 0$ with $X(t - dt) > 0$, then $X(t)$ moves away from zero at time $t + m\,dt$.

6.2.2 Equilibrium probability solution

To develop the associated probability structure, let us return to the forward equation (6.2.16) and assume that a non-trivial equilibrium (or quasi-equilibrium) solution exists. Then placing $\partial p/\partial t = 0$ gives

$$\frac{1}{2}\frac{\partial^2}{\partial x^2}\{\alpha(x)p(x;\infty)\} - \frac{\partial}{\partial x}\{\beta(x)p(x;\infty)\} = 0, \qquad (6.2.31)$$

which has already been developed through a less formal argument in Section 3.6.1 (with $f(N)$ and $g(N)$ replacing $\beta(x)$ and $\alpha(x)$, respectively). Let $X^{min} \leq x \leq X^{max}$, where X^{min} is the solution of $\phi(x) = 0$ and X^{max} is chosen to ensure that $\alpha(x) \leq A$ (condition (6.2.9)). Then equation (6.2.31) integrates to yield

$$\frac{\partial}{\partial x}\{\alpha(x)p(x;\infty)\} - 2\beta(x)p(x;\infty) = c \qquad (6.2.32)$$

for some constant c. On placing $p(x;\infty) = 0$ at $x = X^{min}$ we see that $c = 0$. Whence equation (6.2.32) yields the general equilibrium diffusion solution

$$p(x;\infty) = \frac{k}{\alpha(x)}\exp\left\{2\int\frac{\beta(x)}{\alpha(x)}\,dx\right\} \qquad (X^{min} \leq x \leq X^{max}), \qquad (6.2.33)$$

where the normalizing constant k is chosen to ensure that

$$\int_{X^{min}}^{X^{max}} p(x;\infty)\,dx = 1. \qquad (6.2.34)$$

Note that we previously developed this result in terms of the informal diffusion approximation (3.6.5) for π_N.

Equation (6.2.33) can also be derived directly through the *diffusion balance equation*

$$p(x;\infty)\theta(x) = p(x + dx;\infty)\phi(x + dx). \qquad (6.2.35)$$

For on substituting for $\theta(x)$ and $\phi(x)$ from (6.2.12) and (6.2.13), and removing the common factor $(1/2A)$, equation (6.2.35) reduces to

$$[\alpha(x) + \beta(x)dx]p(x;\infty) = [\alpha(x + dx) - \beta(x + dx)dx]p(x + dx;\infty).$$

Whence on expanding to order $O(dx)$ we obtain (with $'$ denoting d/dx)

$$[\alpha(x) + \beta(x)dx]p(x;\infty) = [\alpha(x) + \alpha'(x)dx - \beta(x)dx][p(x;\infty) + p'(x;\infty)dx],$$

i.e.
$$\frac{d}{dx}[\alpha(x)p(x;\infty)] = 2\beta(x)p(x;\infty), \qquad (6.2.36)$$

which is identical to (6.2.32) with $c = 0$.

To illustrate this result, let us return to our simple immigration–death example for which (6.2.23) gives $\beta(x) = \alpha - \mu x$ and $\alpha(x) = \alpha + \mu x$. On noting from (6.2.28) that $X^{min} = -\alpha/\mu$, solution (6.2.33) yields the gamma p.d.f.

318 *Markov processes in continuous time and space*

$$p(x;\infty) = ke^{-2x}(\alpha + \mu x)^{(4\alpha/\mu)-1} \qquad (-\alpha/\mu \leq x \leq X^{max}). \qquad (6.2.37)$$

So on relaxing the condition $x \leq X^{max}$, it follows that $Y = X + \alpha/\mu$ is a Gamma(a, θ) random variable with p.d.f.

$$p_Y(y) = \frac{y^{a-1}e^{-y/\theta}}{\Gamma(a)\theta^a} \qquad (y \geq 0; a > 0, \theta > 0), \qquad (6.2.38)$$

where $a = 4\alpha/\mu$ and $\theta = 1/2$. Thus Y has mean $a\theta = 2\alpha/\mu$, variance $a\theta^2 = \alpha/\mu$ and skewness $2a\theta^3 = \alpha/\mu$. Now we have already seen in Section 2.5.1 that the corresponding equilibrium population random variable N follows the Poisson(α/μ) p.d.f. (2.5.9), i.e.

$$p_N(\infty) = \frac{(\alpha/\mu)^N e^{-\alpha/\mu}}{N!} \qquad (N = 0, 1, 2, \ldots). \qquad (6.2.39)$$

Hence the effect of taking the diffusion limit is to replace the Poisson distribution by the shifted Gamma$(4\alpha/\mu, 1/2)$ distribution. Given that the mean, variance and skewness of both N and $X = Y - \alpha/\mu$ all equal α/μ, there is clearly a high degree of similarity between these two p.d.f.'s. Figure 6.6 shows a comparison between them for $\alpha/\mu = 10$, and the only real difference is that the diffusion modal value is slightly lower.

Although the general population process equilibrium solution (3.2.5)–(3.2.6), namely

$$\pi_N = \frac{(\lambda_0\lambda_1 \ldots \lambda_{N-1})/(\mu_1\mu_2 \ldots \mu_N)}{\sum_{i=0}^{N}(\lambda_0\lambda_1 \ldots \lambda_{i-1})/(\mu_1\mu_2 \ldots \mu_i)}, \qquad (6.2.40)$$

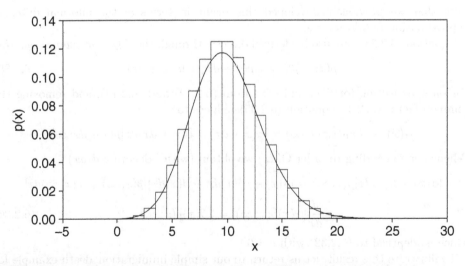

Figure 6.6 Comparison of the immigration–death diffusion p.d.f. (6.2.37) for $\alpha = 1$ and $\mu = 0.1$ (—) and the Poisson(10) p.d.f.

is exact, it is combinatorially opaque in all but the simplest of processes. In contrast, the integral $\int [\beta(x)/\alpha(x)]\, dx$ in the diffusion solution (6.2.33) will often yield an algebraically amenable result. Consider, for example, the case of the logistic process (Section 3.2.1) with $\lambda_N = N(a_1 - b_1 N)$ and $\mu_N = N(a_2 + b_2 N)$, for which we have already obtained (see result (3.2.37)) the exact quasi-equilibrium solution

$$\tilde{\pi}_N = \tilde{\omega}_N / \sum_{i=1}^{N^{max}} \tilde{\omega}_i \quad \text{where} \quad \tilde{\omega}_1 = 1 \quad \text{and} \quad \tilde{\omega}_N = \frac{[a_1 - b_1]\ldots[a_1 - (N-1)b_1]}{[a_2 + 2b_2]\ldots[a_2 + Nb_2]N}$$

(6.2.41)

over $N = 1, \ldots, N^{max} =$ integer part$[a_1/b_1]$. To obtain the corresponding diffusion limit we simply replace λ_N and μ_N by $\lambda_x = x(a_1 - b_1 x)$ and $\mu_x = x(a_2 + b_2 x)$, and use (6.2.3) and (6.2.4) with (6.2.22) to construct $\beta(x) = x(a_1 - b_1 x) - x(a_2 + b_2 x)$ and $\alpha(x) = x(a_1 - b_1 x) + x(a_2 + b_2 x)$. On writing $a = a_1 - a_2$, $b = b_1 + b_2$, $c = a_1 + a_2$ and $d = b_1 - b_2$, substituting $\alpha(x)$ and $\beta(x)$ into the general solution (6.2.33) then gives

$$p(x; \infty) = (k/x)e^{2bx/d}(c - dx)^{[2(bc-ad)/d^2]-1}.$$

(6.2.42)

This is a neater representation than (6.2.41) because of the much simpler normalizing constant.

In spite of the considerable difference in structure between the birth and death rates for the simple immigration–death and logistic processes, the two diffusion solutions (6.2.37) and (6.2.42) are remarkably similar. For the logistic result takes the form of $(1/x)$ times a Gamma p.d.f., and so is approximately Gamma for x lying near to the deterministic carrying capacity $K = (a_1 - a_2)/(b_1 + b_2) = a/b$, i.e. near $\beta(x) = 0$.

Although the range of support for the logistic population process, namely $1 \leq N \leq a_1/b_1$, is well-defined, the lower quasi-equilibrium limit has been guaranteed by imposing the condition that $\mu_1 = 0$. However, this is not possible for the diffusion process, which is defined solely through the two functions $\alpha(x)$ and $\beta(x)$, and hence does not allow any adjustment to be made to say μ_1. So some care is required in this case. Indeed, if we place no lower restriction on x, then since (6.2.42) shows that $p(x; \infty) \sim k/x$ for $x \sim 0$, the distribution $\{p(x; \infty)\}$ is clearly not proper over $x \geq 0$. The reason for this is that if $x \sim 0$, then $\alpha(x) \sim 0$ and $\beta(x) \sim 0$, and so the process becomes trapped at zero. The simplest option is to take $x \geq 1$, thereby ensuring a direct parallel with the population process. In terms of an upper limit, for $dx \simeq 0$ we have $\theta(x) = \phi(x) = 0$ at $\alpha(x) = 0$, so if $b_1 - b_2 > 0$ then there is also an absorbing state at $x = c/d$. Whilst if $b_1 < b_2$ then $\alpha(x) \to \infty$ as $x \to \infty$ and so (in theory) we should impose an upper bound $x \leq X^{max}$ to ensure that the variance $\alpha(x)$ remains bounded. In practice, however, this is not needed, since the process suffers an increasingly strong negative pull as x rises above K.

To illustrate the extent to which realizations of the population and diffusion logistic processes differ, let us return to the $s = 1$ example of Figure 3.1 (Section 3.2.2) for which $\lambda_N = N(2.2 - 0.1N)$ and $\mu_N = N(0.2 + 0.1N)$. Here $a_1 = 2.2$, $a_2 = 0.2$ and

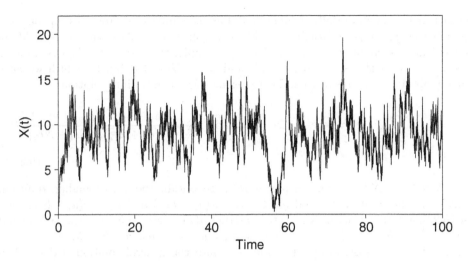

Figure 6.7 Realization of the logistic diffusion process at times $t = 0, 0.01, \ldots, 100$ for $\lambda_x = x(2.2 - 0.1x)$, $\mu_x = x(0.2 + 0.1x)$, $A = 100$ and $dx = 0.01$, starting from $X(0) = 1$.

$b_1 = b_2 = 0.1$, so $a = 2$, $b = 0.2$, $c = 2.4$ and $d = 0$. Since $d = 0$ the integrand $\beta(x)/\alpha(x)$ is now linear, and expression (6.2.33) yields

$$p(x; \infty) = (k'/x)\exp\{-(b/c)(x - a/b)^2\} = (k'/x)\exp\{-(x-10)^2/12\} \quad (6.2.43)$$

for constant k'. Thus for this special case the Gamma component of (6.2.42) changes to a Normal($a/b, c/2b$), i.e. $N(10, 6)$, form. On disregarding the factor $1/x$ (which induces slight negative skewness), we see that the lower bound 0 and upper (population process) bound $a_1/b_1 = 22$ are both more than four standard deviations from the deterministic equilibrium value $K = a/b = 10$, so over reasonable time lengths neither is likely to be breached. Moreover, since $\alpha(a_1/b_1) = 52.8$, taking $A = 100$ places a more than adequate upper bound on $\alpha(x)$. Figure 6.7 shows a realization of this diffusion logistic process with $dx = 0.01$, and hence $dt = dx^2/A = 10^{-6}$, at times $t = 0, 0.01, \ldots, 100$, i.e. at every 10^4th step. The process moves swiftly away from its start position at $X(0) = 1$ towards $K = 10$, whereupon it mostly oscillates within the range $(5, 15)$. Note, however, the presence of diffusion excursions towards both the lower and upper population limits, namely $X(56.35) = 0.33$ and $X(74.13) = 19.59$. Whilst in terms of basic structure the $(s = 1)$ population realization shown in Figure 3.1 is broadly compatible with its diffusion soulmate (Figure 6.7), the latter exhibits much more high-frequency structure. This is not too surprising, for whereas Figure 3.1 involves a mere 367 events, the diffusion simulation comprises 10^8 events and thereby enables a far wider scale of pattern.

Finally, since $\theta(x) = 0.0121x - 0.00001x^2$ and $\phi(x) = 0.0119x + 0.00001x^2$, for $x \simeq 0$ we have $p = \Pr(x \to x + dx | \text{event occurs}) = \theta(x)/[\theta(x) + \phi(x)] \simeq 0.5042$ and $q = 1 - p = 0.4958$. So near the lower barrier the process behaves as an almost symmetric random walk, whence it follows that a move to zero is quite likely. Thus a simulation is likely to become absorbed at zero unless $X(0)$ lies several dx's away from it. Indeed, because $\theta(0) = 0$, the process can never leave zero once it arrives there; note the contrast with the immigration–death process for which $\theta(0) = (\alpha/2A)(1 - dx) \simeq \alpha/2A$, and even $\theta(-\alpha/\mu) = \alpha dx/A > 0$ provided $dx > 0$. Moreover, we can use the simple random walk result (4.2.54) to determine an approximate value for the probability that this process is absorbed at 0 before it moves towards K. Consider the skeleton diffusion process, i.e. the subset $\{X(t)\}$ where $X(t) \neq X(t - dt)$, and rewrite (4.2.54) in the form

$$\rho[X(0)] = \Pr(\text{absorption at 0 before 'burn-in'}) \simeq (q/p)^{X(0)/dx}. \qquad (6.2.44)$$

Then $\rho(0.01) = 0.9833$, $\rho(0.05) = 0.9191$, $\rho(0.1) = 0.8454$, $\rho(0.4) = 0.5107$ and $\rho(1.0) = 0.1864$. So unless $X(0) > 0.05$, most realizations will become absorbed at 0 before the process can become established. Even with $X(0) = 0.4$ only half will fully develop; whilst at the lower quasi-equilibrium bound $X(0) = 1$, roughly 20% will fail to do so.

6.2.3 Boundary conditions

We have already seen that solutions for the simple Wiener process do not emerge too readily when we impose boundary conditions, so analysis of the general Focker–Planck equation with boundary conditions is likely to prove even more awkward. Hence all we can do here is to point out possible lines of enquiry. For example, in Section 6.1.2 we took an appropriate limit of the equilibrium distribution (6.1.17) for the simple random walk between two reflecting barriers at 0 and a to form the exponential p.d.f. (6.1.21). Though imposing a different boundary condition at say $x = 0$ exposed two fundamental problems. First, we had to change the probability of a move from 0 to dx from $1/2 + O(dx)$ to $O(dx)$ in order to hold the process at 0 for an exponentially distributed (i.e. non-zero) length of time before moving away. Then we had to accept that to avoid becoming trapped at 0 as $dx, dt \to 0$, the one-step move from 0 would have to be to say δ, rather than dx. Thus the equilibrium distribution over $x \geq 0$ comprises an exponential distribution at $x = 0$, coupled with a continuous p.d.f. over $x > 0$ which involves the *transient* development of the process from δ until such time as state 0 is once again reached. Exactly the same considerations apply to the general Focker–Planck process, though the resulting algebra is even more opaque.

If we simply wish to have a straightforward reflecting barrier at a, then the required boundary condition is

$$\left[\frac{1}{2}\frac{\partial}{\partial x}\{\alpha(x)p(x;t)\} - \beta(x)p(x;t)\right]_{x=a} = 0. \qquad (6.2.45)$$

For consider the forward equation (6.2.16) for $p(x;t)$, namely

$$\frac{\partial p(x;t)}{\partial t} = \frac{1}{2}\frac{\partial^2}{\partial x^2}\{\alpha(x)p(x;t)\} - \frac{\partial}{\partial x}\{\beta(x)p(x;t)\}. \qquad (6.2.46)$$

Now for a large number of individuals starting from x_0, all independently undergoing the same diffusion process, $p(x;t)dx$ is the proportion of individuals in the interval

$(x, x + dx)$ at time t. As the existence of a reflecting barrier at $a > x_0$ means that no individual can pass beyond a, we must have

$$\int_{-\infty}^{a} p(x; t)\, dx = 1. \tag{6.2.47}$$

Consequently,

$$\frac{\partial}{\partial t} \int_{-\infty}^{a} p(x; t)\, dx = \int_{-\infty}^{a} \frac{\partial p(x; t)}{\partial t}\, dx = 0.$$

So it follows from (6.2.46) that

$$0 = \frac{\partial}{\partial t} \int_{-\infty}^{a} p(x; t)\, dx = \int_{-\infty}^{a} \frac{\partial}{\partial x}\left[\frac{1}{2}\frac{\partial}{\partial x}\{\alpha(x)p(x; t)\} - \beta(x)p(x; t)\right] dx$$

$$= \left[\frac{1}{2}\frac{\partial}{\partial x}\{\alpha(x)p(x; t)\} - \beta(x)p(x; t)\right]_{x=a},$$

whence we obtain the condition (6.2.45).

For an absorbing barrier at $a > x_0$, provided the infinitesimal variance $\alpha(x)$ does not vanish at $x = a$ the 'intuitive' boundary condition is to ensure that $p(x; t)$ vanishes at $x = a$ for all $t > 0$. That is,

$$p(a; t) \equiv 0 \quad \text{(for all } t > 0\text{)}, \tag{6.2.48}$$

since then neither $p(a + dx; t + dt)$ nor $p(a - dx; t + dt)$ can be reached from $p(a; t)$. To prove this formally, suppose to the contrary that in the interval $(a, a - dx)$ we have $p(x; t) \geq \eta > 0$ within some interval $t_1 \leq t \leq t_2$. Then the probability $h(t)dt$ that the individual is absorbed is the probability that the individual is near a at time t and that diffusion carries the individual beyond a. Now

$$h(t)dt \geq \eta dx \Pr[X(t + dt) - X(t) > dx | a - dx < X(t) < a]. \tag{6.2.49}$$

Since $\alpha(x) > 0$ near $x = a$, we may therefore conclude that

$$\alpha(x)dt = \text{Var}[dX(t)|X(t) = x] \geq kdt \tag{6.2.50}$$

for some $k > 0$, and hence that $dX(t)$ will exceed $\epsilon\sqrt{dt}$ with non-zero probability q (say), for some $\epsilon > 0$. Whence taking $dx = \epsilon\sqrt{dt}$ in (6.2.49) gives

$$h(t)dt \geq \eta \epsilon q \sqrt{dt} \quad (t_1 < t < t_2). \tag{6.2.51}$$

This implies that $h(t)$ is infinite for $t_1 < t < t_2$, which directly contradicts the fact that $h(t)$ is the probability density of the first passage time to a. Thus $p(a; t)$ must be zero.

6.2.4 Time-dependent probability solutions

The success of the general equilibrium result (6.2.33) is due the loss of the time variable t, for this enables us to integrate the resulting forward differential component (6.2.31) to form the much more straightforward first-order differential equation (6.2.36). However, if we are to study time-dependence then such luxury is no longer available to us and we have to analyse the full forward and backward equations (6.2.16) and (6.2.20) as they stand. Fortunately, theoretical solutions can be obtained by first separating out the two variables involved, namely x and t (forward), or x_0 and t (backward), and then developing a series-based solution in x or x_0 (see, for example, Piaggio, 1962). However, since the resulting expressions are likely to be algebraically opaque this approach has dubious merit for analysing population processes; though it is highly relevant in mathematical physics.

One compromise would be to include a time-dependent term in the equilibrium solution (6.2.33), thereby forming the new distribution $\tilde{p}(x;t)$. Ideally, $\tilde{p}(x;t)$ should be a proper distribution with $\tilde{p}(x;0) = p(x;0)$ and $\tilde{p}(x;\infty) = p(x;\infty)$. To illustrate this approximation procedure let us once again consider the simple immigration–death process. On substituting for $y = x + \alpha/\mu$ in equation (6.2.16) with $\alpha(x) = \alpha + \mu x$ and $\beta(x) = \alpha - \mu x$ from (6.2.23), we see that the p.d.e. for $p_Y(y;t)$ takes the form

$$\frac{\partial p_Y}{\partial t} = \frac{\mu}{2}\frac{\partial^2}{\partial y^2}\{y p_Y\} - \frac{\partial}{\partial y}\{(2\alpha - \mu y)p_Y\}, \tag{6.2.52}$$

which is not too user-friendly. However, expanding expression (2.5.7) for the immigration–death p.g.f. $G(z;t)$ shows that for a population starting from empty, i.e. $n_0 = 0$, $p_i(t) = \gamma^i e^{-\gamma}/i!$ where $\gamma = (\alpha/\mu)(1 - e^{-\mu t})$. So an obvious approximation to the diffusion solution is to replace the parameter $a = 4\alpha/\mu$ in the Gamma p.d.f. (6.2.38) by the time-dependent version $a = (4\alpha/\mu)(1 - e^{-\mu t})$. In general, decisions as to whether to work with the population process itself, or its diffusion 'equivalent', will depend on the relative ease with which: (i) the exact solution can be obtained for the population p.g.f. $G(z;t)$, and the coefficient z^i subsequently extracted from it; and, (ii) the corresponding diffusion p.d.e. can be solved.

6.2.5 The associated stochastic differential equation

An equivalent, and directly parallel, way of visualizing a diffusion process may be obtained by defining the corresponding stochastic differential equation (s.d.e.). Since this research field has spawned a huge literature in its own right, all we can do here is simply to note the correspondence between the population, diffusion and s.d.e. approaches. The pity is that whilst both population processes and s.d.e.'s have attracted a great deal of interest over the past half century, virtually no attempt has been made to link them and thereby open up the potential for further developing both research areas through a two-way transfer of properties.

Consider a purely random process $Z(t)$ with $\mathrm{E}\{Z(t)\} = 0$, $\mathrm{Var}\{Z(t)\} = 1$ and $\mathrm{E}\{|Z(t)|^3\} = o(1/\sqrt{dt})$. Now write

$$dX(t) = \beta(x,t)dt + Z(t)\sqrt{\{\alpha(x,t)dt\}}, \tag{6.2.53}$$

which holds conditionally on $X(t) = x$. Thus given that $X(t) = x$, the increment $dX(t)$ in a small time interval $(t, t+dt)$ has mean $\beta(x,t)dt$ and variance $\alpha(x,t)dt$ and is independent of all previous increments. So this equation may be written in terms of the moment generating function

$$\mathrm{E}[e^{-\theta dX(t)}|X(t)=x] = 1 - \theta\{\beta(x,t)dt\} + \frac{\theta^2}{2}\{\alpha(x,t)dt\} + o(dt), \qquad (6.2.54)$$

where we assume that the term $o(dt)$ is uniform in x. Denote

$$\psi(\theta;t) = \mathrm{E}[e^{-\theta X(t)}] = \int_{-\infty}^{\infty} e^{-\theta x} p(x;t)\, dt, \qquad (6.2.55)$$

where $p(x;t)$ is the p.d.f. of $X(t)$ subject to some given initial conditions. Then on differentiating (6.2.55) we obtain

$$\frac{\partial \psi}{\partial t} = \int_{-\infty}^{\infty} e^{-\theta x} \frac{\partial p(x;t)}{\partial t}\, dx. \qquad (6.2.56)$$

We shall now consider an alternative derivation for $\partial \psi/\partial t$, and thereby show the equivalence of the diffusion and s.d.e. approaches by matching up the two resulting expressions. First note that

$$\frac{\partial \psi}{\partial t} = \lim_{dt\to 0} \frac{1}{dt}\{\psi(\theta;t+dt) - \psi(\theta;t)\} = \lim_{dt\to 0} \frac{1}{dt} \mathrm{E}[e^{-\theta\{X(t)+dX(t)\}} - e^{-\theta X(t)}]. \qquad (6.2.57)$$

The expected value on the right can be written as

$$\mathrm{E}[\{e^{-\theta dX(t)} - 1\}e^{-\theta X(t)}] = \int_{-\infty}^{\infty} \mathrm{E}[\{e^{-\theta dX(t)} - 1\}e^{-\theta X(t)}|X(t)=x] p(x;t)\, dx$$

$$= \int_{-\infty}^{\infty} \mathrm{E}[\{e^{-\theta dX(t)} - 1\}|X(t)=x] e^{-\theta x} p(x;t)\, dx.$$

Whence using (6.2.54) and letting $dt \to 0$ gives

$$\frac{\partial \psi}{\partial t} = \int_{-\infty}^{\infty} \{-\theta\beta(x,t) + \frac{1}{2}\theta^2\alpha(x,t)\}e^{-\theta x} p(x;t)\, dx.$$

Integrating the first component by parts once, and the second twice, then yields

$$\frac{\partial \psi}{\partial t} = \int_{-\infty}^{\infty} e^{-\theta x}[\frac{1}{2}\frac{\partial^2}{\partial x^2}\{\alpha(x,t)p(x;t)\} - \frac{\partial}{\partial x}\{\beta(x,t)p(x;t)\}]\, dx \qquad (6.2.58)$$

$$+ \left[e^{-\theta x}\{\beta(x,t)p(x;t) - (\theta/2)(\alpha(x,t)p(x;t)) - (1/2)\frac{\partial}{\partial x}\{\alpha(x,t)p(x;t)\}\}\right]_{-\infty}^{\infty}.$$

Thus provided the last three terms all equal zero at $x=-\infty$ and $x=\infty$, direct comparison of the integrands of (6.2.56) and (6.2.58) recovers the Fokker–Planck equation (6.2.16).

It is important to note that although both approaches give rise to equation (6.2.16), they are fundamentally different. For we see from (6.2.2) that each step of the diffusion process is restricted to the three values dx, 0 and $-dx$, with probabilities $\theta(x)$, $1 - \theta(x) - \phi(x)$ and $\phi(x)$, whilst the step size in the s.d.e. (6.2.53) can take any real value (to order $o(\sqrt{dt})$). This considerable difference in incremental structure should not be forgotten when we examine and compare the stochastic behaviour of diffusion and s.d.e. processes which possess the same Focker–Planck representation (discussed further in Section 9.3). Indeed, as the forward and backward equations are defined solely in terms of the instantaneous mean and variance, there is no reason why the effect of higher-order incremental moment structure should not be substantial. Nevertheless, given that the s.d.e. and diffusion approaches are equivalent to second-order, and that the latter may provide a useful description of a stochastic population process, the application of s.d.e. theory to population processes is both timely and ripe for development.

6.3 The Ornstein–Uhlenbeck process

We have previously remarked that in spite of the velocity $dx/dt = \sigma/\sqrt{dt}$ becoming infinite as dt tends to zero, the Wiener process proves to be an excellent representation of random movement in discontinuous media. However, if velocity plays an important role in the underlying dynamics of the process under study, whether it be say the rate of increase or decrease in the number of malaria-infected individuals in a given region, or the rate of change in a share price, then this defect clearly needs to be corrected.

The most sensible way forward is to model the velocity $U(t)$ of the process itself, rather than the displacement (as above), and we have already seen this construct used in the development of the correlated random walk (Section 4.4). For there we presumed that a particle's momentum ensures that it keeps travelling in the same direction with velocity ± 1, until it suffers an impact at some random time which causes it to reverse its direction and hence have velocity ∓ 1. The classic approach developed by Uhlenbeck and Ornstein (1930) effectively generalizes the correlated random walk procedure by allowing $U(t)$ to take any real value and not just $+1$ or -1. In a small time interval there are two factors which affect the change in momentum, and thereby velocity. First, the frictional resistance, β, of the surrounding medium induces a change in $U(t)$ proportional to $U(t)$. Second, the particle suffers random impacts from neighbouring particles whose effect in successive small time intervals can be represented by independent random variables with zero mean; i.e. increments of a Wiener process without drift. Thus we have the formal stochastic differential equation representation

$$dU(t) = -\beta U(t)dt + dY(t), \qquad (6.3.1)$$

where $Y(t)$ is an unrestricted Wiener process with variance parameter σ^2. We denote $X(t)$ to be the displacement at time t, and $dX(t) = X(t+dt) - X(t)$ its incremental change between times t and $t+dt$; whilst $dU(t) = U(t+dt) - U(t)$ denotes the corresponding change in velocity.

Since we have already seen through (6.1.6) that $Y(t) \sim N(\mu t, \sigma^2 t)$, the general s.d.e. representation (6.2.53) simplifies to

$$dY(t) = \mu dt + \sigma Z(t)\sqrt{dt} \qquad (6.3.2)$$

where $Z(t)$ is a random Gaussian (i.e. Normal) process with zero mean and unit variance, often called pure or white noise. The s.d.e. (6.3.2) is interpreted as stating that the change in $X(t)$ during the small time interval $(t, t+dt)$ is a Normal variate with mean μdt and variance $\sigma^2 dt$, and is independent of $X(t)$ and of the change during any other time interval. Thus $\mathrm{E}[dY(t)] = \mu dt$, $\mathrm{Var}[dY(t)] = \sigma^2 dt$ and $\mathrm{Cov}[Y(t), Y(s)] = 0$ ($t \neq s$).

Now conditional on $U(t) = u$, and remembering that here $\mu = 0$, we may write (6.3.1)–(6.3.2) as

$$dU(t) = -\beta u dt + \sigma Z(t) \sqrt{dt}. \tag{6.3.3}$$

Whence comparison with the general s.d.e. (6.2.53) shows that $U(t)$ is a diffusion process with infinitesimal mean $\beta(u, t) = -\beta u$ and variance $\alpha(u, t) = \sigma^2$. Thus the (forward) Focker–Planck equation (6.2.16) becomes

$$\frac{\partial p}{\partial t} = \frac{\sigma^2}{2} \frac{\partial^2 p}{\partial u^2} + \beta \frac{\partial}{\partial u} \{up\}, \tag{6.3.4}$$

where $p(u; t)$ is the p.d.f. of the velocity $U(t)$ at time t. Denote the m.g.f. of $U(t)$ by

$$M(\theta; t) = \int_{-\infty}^{\infty} e^{u\theta} p(u; t) \, du, \tag{6.3.5}$$

and note that

$$\frac{\partial M}{\partial \theta} = -\int_{-\infty}^{\infty} e^{u\theta} (up) \, du.$$

Then on applying the transform (6.3.5) to each component of equation (6.3.4) in turn, and integrating by parts, we see that $M(\theta; t)$ satisfies the first-order Lagrange linear partial differential equation

$$\frac{\partial M}{\partial t} = \frac{1}{2} \sigma^2 \theta^2 M - \beta \theta \frac{\partial M}{\partial \theta}. \tag{6.3.6}$$

In terms of the c.g.f. $K(\theta; t) \equiv \ln[M(\theta; t)]$, this equation simplifies to

$$\frac{\partial K}{\partial t} + \beta \theta \frac{\partial K}{\partial \theta} = \frac{1}{2} \sigma^2 \theta^2, \tag{6.3.7}$$

which, for an initial velocity $U(0) = u_0$, yields the solution

$$K(\theta; t) = [u_0 e^{-\beta t}] \theta + [(\sigma^2/2\beta)(1 - e^{-2\beta t})](\theta^2/2). \tag{6.3.8}$$

This quadratic form is the signature of a Normal distribution for $U(t)$ with

$$\mathrm{E}[U(t)] = u_0 e^{-\beta t} \quad \text{and} \quad \mathrm{Var}[U(t)] = (\sigma^2/2\beta)(1 - e^{-2\beta t}). \tag{6.3.9}$$

When $t \to \infty$ we therefore obtain a Normal$(0, \sigma^2/2\beta)$ equilibrium distribution for the velocity $U(\infty)$, which is in accordance with Maxwell's law for the velocity of particles in equilibrium. The process $U(t)$ is called the *Ornstein–Uhlenbeck* process, and it is both Gaussian and Markov. Unlike the Wiener process it does not have independent increments.

To obtain the displacement distribution for $X(t)$, we observe that in a small time interval $(t, t + dt)$

$$dX(t) = U(t)dt, \qquad (6.3.10)$$

which integrates to

$$X(t) - X(0) = \int_0^t U(s)\, ds. \qquad (6.3.11)$$

Although $X(t)$ is not a Markov process, the two-dimensional process $\{X(t), U(t)\}$ is Markov, with joint increments defined by (6.3.3) and (6.3.10). This stochastic integral may be interpreted in two ways. First, we can regard a realization $\{U(s)\}$ $(0 \le s \le t)$ as being drawn from the ensemble of all possible realizations. Then as each member of this ensemble is a continuous function, the integral (6.3.11) exists and defines a random variable. Second, we can envisage the integral as an appropriate limit as $dt \to 0$ of the approximating linear combination of dependent variables, viz:

$$X(t) - X(0) = \sum_{i=1}^{t/dt} [U(idt)|U((i-1)dt)]dt. \qquad (6.3.12)$$

Hence the fact that $U(t)$ is Normal implies that $X(t) - X(0)$ is also Normal. Moreover,

$$\mathrm{E}[X(t) - X(0)] = \int_0^t \mathrm{E}[U(s)]\, ds \qquad (6.3.13)$$

and

$$\mathrm{Var}[X(t) - X(0)] = \int_0^t \int_s^t \mathrm{Cov}[U(s)U(w)]\, dw\, ds. \qquad (6.3.14)$$

Suppose that the process $\{U(t)\}$ is in statistical equilibrium. Then since $U(t)$ has zero mean and variance $\sigma^2/2\beta$, it follows from (6.3.13) that $\mathrm{E}[X(t) - X(0)] = 0$; whilst

$$\mathrm{Cov}[U(s), U(w)] = \mathrm{E}[U(s), U(w)] = \mathrm{E}[U(s)\mathrm{E}\{U(w)|U(s)\}] \qquad (w > s).$$

Now (6.3.9) gives

$$\mathrm{E}[U(w)|U(s)] = U(s)e^{-\beta(w-s)},$$

so

$$\mathrm{Cov}[U(s)U(w)] = \mathrm{E}[\{U(s)\}^2]e^{-\beta(w-s)} = \mathrm{Var}[U(s)]e^{-\beta(w-s)} = (\sigma^2/2\beta)e^{-\beta(w-s)}. \qquad (6.3.15)$$

Thus in equilibrium, the correlation coefficient between $U(s)$ and $U(w)$ $(w > s)$ is $e^{-\beta(w-s)}$.

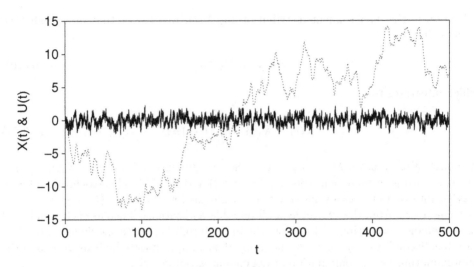

Figure 6.8 Realization of the Ornstein–Uhlenbeck process at times $t = 0, 0.01, \ldots, 500$ for $\beta = 1$ and $\sigma = 1$, with $dx = dt = 0.01$, showing $X(t)$ ($\cdots$) and $U(t)$ (——) starting from $X(0) = U(0) = 0$.

Finally, substituting (6.3.15) into (6.3.14) yields

$$\text{Var}[X(t) - X(0)] = (\sigma^2/2\beta) \int_0^t e^{\beta s} \int_s^t e^{-\beta w} \, dw \, ds = (\sigma^2/2\beta)[\beta t - 1 + e^{-\beta t}].$$

(6.3.16)

Thus whilst $\text{Var}[X(t) - X(0)] \simeq (\sigma^2/2)t$ for large t, i.e. it increases with t as happens for the Wiener process, its local behaviour is far different. For it follows from (6.3.16) that when t is small, $\text{Var}[X(t) - X(0)] \simeq (\sigma^2 \beta/4)t^2$ increases with t^2. This reflects the fact that whilst for the Wiener process the instantaneous velocity $dx/dt = \sigma/\sqrt{dt} \to \infty$ as $dt \to 0$, that for the Ornstein–Uhlenbeck (OU) process, namely $U(t)$, remains (effectively) bounded.

Figure 6.8 shows a realization of the OU process at times $t = 0, 0.01, \ldots, 500$ for $\beta = \sigma = 1$ and $dx = dt = 0.01$, and the marked difference in behaviour between the constrained autoregressive velocity process $U(t)$ and its integral, namely the displacement process $X(t)$, is clearly evident. Though the time-dependent nature of the variance (6.3.16) automatically prevents the OU process from providing an approximation to an equilibrium population process, in contrast to the Focker–Planck formulation (e.g. Figures 6.4 and 6.5 for the immigration–death processes, and Figure 6.7 for the diffusion process), it does provide an extremely useful framework for many other processes whose variance increases with t. The movement of share prices is one likely area of useful application, especially since upwards or downwards drift can easily be related to velocity.

6.3.1 The OU process as a time-transformed Wiener process

Although the Wiener and OU processes have different moment structures, given that both are Gaussian it seems reasonable to surmise that we can transform one into the other. Let $Y(t)$ be a Wiener process with zero drift and unit variance parameter (i.e. $\mu = 0$ and $\sigma^2 = 1$ in (6.1.6)), so $Y(t) \sim N(0, \sigma^2 t)$. Then for continuous functions $g(t)$ and $h(t)$ the process $X(t) = g(t)Y\{h(t)\}$ is also a Gaussian diffusion process; for we have simply transformed the scale of time measurement. Moreover,

$$dX(t) = g'(t)Y\{h(t)\}dt + g(t)dY\{h(t)\}$$
$$= [g'(t)/g(t)]X(t)dt + g(t)Z\{h(t)\}\sqrt{h'(t)dt}, \qquad (6.3.17)$$

where $Z(\cdot)$ is a pure white noise process with zero mean and unit variance. Whence direct comparison with (6.2.53) shows that $X(t)$ has infinitesimal mean $\beta(x,t) = [g'(t)/g(t)]x$ and variance $\alpha(x,t) = [g(t)]^2 h'(t)$.

For example, consider the trial process

$$X(t) = e^{-\beta t}Y[(\sigma^2/2\beta)e^{2\beta t}] \qquad (\beta > 0; -\infty < t < \infty), \qquad (6.3.18)$$

for which $g(t) = e^{-\beta t}$ and $h(t) = (\sigma^2/2\beta)e^{2\beta t}$. Then the above result shows that the infinitesimal mean and variance of $X(t)$ are given by $-\beta x$ and σ^2, respectively. Thus $X(t)$ is an OU process, and results from transforming the Wiener process $Y(t)$ via the simple change of scale (6.3.18). Conversely, on inverting (6.3.18) to form

$$Y(t) = \sqrt{(\sigma^2/2\beta t)}X[(1/2\beta)\ln(2\beta t/\sigma^2)], \qquad (6.3.19)$$

we see that if $X(t)$ is an OU process then $Y(t)$ is a Wiener process. This neat relationship between Wiener and OU processes is one that merits further exploitation in practical modelling situations.

6.3.2 Rapid oscillations of the Wiener and OU processes

Both Figure 6.1 (middle) for the Wiener process and Figure 6.8 for the OU process exhibit rapid oscillations even though dt takes the relatively high value 0.01, so the question arises as to whether an infinite number of oscillations will occur in any short time period as $dt \to 0$? Specifically, we can show that both processes possess the interesting property that when a passage of $Y(t)$ (or $X(t)$) through a value x occurs at time t_0, then $Y(t)$ (or $X(t)$) takes the value x infinitely often in the interval ($t_0 < t < t_0 + \epsilon$) no matter how small ϵ may be.

To deduce this result we first observe that (6.1.49) with $\mu = 0$ implies that for the Wiener process in $0 \leq t < \infty$ every x is visited infinitely often with probability one, since a first passage from any x_0 to any other x occurs with probability one. Now consider the two processes

$$W_1(t) = t^{-1/2}Y(t) \quad \text{and} \quad W_2(t) = t^{1/2}Y(1/t),$$

and note that their probability structures are trivially identical (just replace t by $1/t$). Since $Y(t)$ crosses the origin infinitely often during $0 \leq t < \infty$, so must $W_1(t)$, and hence also $W_2(t)$. Thus $Y(1/t)$ crosses the origin infinitely often as $t \to \infty$, thereby implying that $Y(t)$ must do likewise as $t \to 0$, which proves the result.

This phenomenon explains why we found earlier for the Wiener (and hence OU) process that if we impose a condition involving a visit to a barrier, with the visit having finite duration with say an exponential distribution, then we cannot have continuous movement (i.e. diffusion) away from the barrier. For that would imply an infinite number of returns to the barrier in a small time interval, and since at each return we have a visit of finite duration, movement away from the barrier would never occur. As demonstrated in Figure 6.5, this can only be achieved by superimposing a jump process there.

7
Modelling bivariate processes

So far we have just considered single-species population dynamics. However, individuals often do not exist in isolation but they coexist with individuals from many other species, and thereby exhibit between-species interaction. In biology, for example, species may compete with each other for common resources that are in short supply, such as food or space, or it may be that individuals from different species attack each other directly. Whilst in epidemiology, interaction involves an infected individual infecting a susceptible, whereupon the infected and susceptible population sizes simultaneously increase and decrease by one, respectively.

Since we are able to make substantial progress with the analysis of single-species populations, one might think that the extension to two-species (or even multi-species) populations would just involve replacing scalar variables by vectors, and accepting some increase in algebraic complexity. Unfortunately this is not the case, since the resulting bivariate nonlinear stochastic equations are, in general, currently intractable to direct solution, and are likely to remain as such for many decades to come. Developing a greater understanding of such processes therefore involves far greater reliance on simulation and approximation techniques than we have seen so far. However, before we proceed down this path, let us first develop a simple bivariate process for which the preceding methods of solution do carry across, in order to illustrate the basic approaches involved.

7.1 Simple immigration–death–switch process

In Section 2.5.1 we construct the Lagrange equation for the p.g.f. $G(z;t)$ of the simple immigration–death process, and show that it leads fairly easily to a product (2.5.7) of binomial and Poisson p.g.f.'s. In particular, if the population is initially empty, i.e. $X(0) = 0$, then the population size p.d.f. takes the basic Poisson form

$$p_N(t) = [(\alpha/\mu)(1 - e^{-\mu t})]^N / N!. \tag{7.1.1}$$

Suppose we now extend this process by considering two species with immigration and death rates α_i and μ_i ($i = 1, 2$), and allow type-i individuals to change into type-j individuals at rate ν_i. Then the forward Kolmogorov equation corresponding to (2.5.3) for the population size probabilities $\Pr(X_1(t) = M, X_2(t) = N)$ is given by

$$dp_{MN}(t)/dt = \alpha_1 p_{M-1,N}(t) + \alpha_2 p_{M,N-1}(t) + \mu_1(M+1)p_{M+1,N}(t) + \mu_2(N+1)p_{M,N+1}(t)$$
$$+ \nu_1(M+1)p_{M+1,N-1}(t) + \nu_2(N+1)p_{M-1,N+1}$$
$$- [\alpha_1 + \alpha_2 + M(\mu_1 + \nu_1) + N(\mu_2 + \nu_2)]p_{MN}(t). \quad (7.1.2)$$

So does this equation lead to a similarly simple Poisson p.d.f.?

On rewriting (7.1.2) in terms of the bivariate p.g.f.

$$G(z_1, z_2; t) \equiv \sum_{M=0}^{\infty} \sum_{N=0}^{\infty} p_{MN}(t) z_1^M z_2^N, \quad (7.1.3)$$

we obtain the Lagrange form

$$\frac{\partial G}{\partial t} = \sum_{i=1, j \neq i}^{2} \{[\mu_i(1 - z_i) + \nu_i(z_j - z_i)]\frac{\partial G}{\partial z_i} + \alpha_i(z_i - 1)G\}. \quad (7.1.4)$$

The solution of this type of p.d.e. follows the natural extension of the one-species system (2.5.4), so we first write down the associated auxiliary equations

$$\frac{dt}{1} = \frac{dz_1}{\mu_1(z_1 - 1) + \nu_1(z_1 - z_2)} = \frac{dz_2}{\mu_2(z_2 - 1) + \nu_2(z_2 - z_1)}$$
$$= \frac{dG}{G[\alpha_1(z_1 - 1) + \alpha_2(z_2 - 1)]}. \quad (7.1.5)$$

Since these equations are linear in z_1 and z_2, integration is fairly straightforward. Denote $\sigma_i = \mu_i + \nu_i$ and write $y_i = 1 - z_i$ $(i = 1, 2)$. Then combining the first two equations in (7.1.5) gives, for any constant b,

$$dt = \frac{d(y_1 - by_2)}{y_1(\sigma_1 + \nu_2 b) - y_2(\nu_1 + \sigma_2 b)}. \quad (7.1.6)$$

Thus to make the numerator the differential of the denominator we need to take

$$b = (\nu_1 + \sigma_2 b)/(\sigma_1 + \nu_2 b),$$

which has roots

$$b_1, b_2 = (1/2\nu_2)[(\sigma_2 - \sigma_1) \pm \sqrt{\{(\sigma_1 - \sigma_2)^2 + 4\nu_1 \nu_2\}}]. \quad (7.1.7)$$

In which case (7.1.6) integrates to give

$$y_1 - b_i y_2 = B_i e^{r_i t} \quad (i = 1, 2), \quad (7.1.8)$$

where

$$r_1, r_2 = (1/2)[(\sigma_1 + \sigma_2) \pm \sqrt{\{(\sigma_1 - \sigma_2)^2 + 4\nu_1 \nu_2\}}] \quad (7.1.9)$$

and B_1, B_2 are constants of integration. Whence solving (7.1.8) for y_1 and y_2 yields

$$y_1 = (b_2 - b_1)^{-1}(b_2 B_1 e^{r_1 t} - b_1 B_2 e^{r_2 t})$$
$$y_2 = (b_2 - b_1)^{-1}(B_1 e^{r_1 t} - B_2 e^{r_2 t}). \quad (7.1.10)$$

Now on returning to the first and fourth components of the auxiliary equations (7.1.5), we see that

$$dG/G = -(\alpha_1 y_1 + \alpha_2 y_2)dt. \tag{7.1.11}$$

Whence substituting for y_1 and y_2 from (7.1.10) gives the general solution

$$G(z_1, z_2; t) = f(B_1, B_2) \exp\{(b_1 - b_2)^{-1}[r_1^{-1}(\alpha_1 b_2 + \alpha_2)B_1 e^{r_1 t} - r_2^{-1}(\alpha_1 b_1 + \alpha_2)B_2 e^{r_2 t}]\} \tag{7.1.12}$$

for some arbitrary function f. To determine f we use the initial condition $(X_1(0), X_2(0)) = (m_0, n_0)$, together with (7.1.10) for $t = 0$, to form

$$f(B_1, B_2) = [1 - (b_2 - b_1)^{-1}(b_2 B_1 - b_1 B_2)]^{m_0}[1 - (b_2 - b_1)^{-1}(B_1 - B_2)]^{n_0}$$
$$\times \exp\{-(b_1 - b_2)^{-1}[r_1^{-1}(\alpha_1 b_2 + \alpha_2)B_1 - r_2^{-1}(\alpha_1 b_1 + \alpha_2)B_2]\}. \tag{7.1.13}$$

Finally, we replace the arbitrary constants B_1 and B_2 by expressions (7.1.8), namely

$$B_i = [(1 - z_1) - b_i(1 - z_2)]e^{-r_i t} \quad (i = 1, 2),$$

thereby obtaining the solution (Renshaw, 1973b)

$$G(z_1, z_2; t) = [1 - (b_2 - b_1)^{-1}\{(1 - z_1)(b_2 e^{-r_1 t} - b_1 e^{-r_2 t}) - (1 - z_2)b_1 b_2(e^{-r_1 t} - e^{-r_2 t})\}]^{m_0}$$
$$\times [1 - (b_2 - b_1)^{-1}\{(1 - z_1)(e^{-r_1 t} - e^{-r_2 t}) - (1 - z_2)(b_1 e^{-r_1 t} - b_2 e^{-r_2 t})\}]^{n_0}$$
$$\times \exp\{(b_2 - b_1)^{-1}[r_1^{-1}(\alpha_1 b_2 + \alpha_2)\{(1 - z_1) - b_1(1 - z_2)\}(e^{-r_1 t} - 1)$$
$$- r_2^{-1}(\alpha_1 b_1 + \alpha_2)\{(1 - z_1) - b_2(1 - z_2)\}(e^{-r_2 t} - 1)]\}. \tag{7.1.14}$$

Although expression (7.1.14) looks slightly messy, it is a direct parallel of the corresponding single-species result (2.5.7), namely

$$G(z; t) = [1 + (z - 1)e^{-\mu t}]^{n_0} \exp\{(\alpha/\mu)(z - 1)(1 - e^{-\mu t})\}. \tag{7.1.15}$$

For we see that this convolution of a binomial and a Poisson p.d.f. becomes the convolution of two trinomials and two independent Poisson p.d.f.'s. The trinomial forms result from the development of initial populations of sizes m_0 and n_0, whilst the Poisson forms result from the development of individuals who arrive after time $t = 0$. Moreover, after a little simplification, it follows that in equilibrium the two species behave as two independent immigration–death processes, with immigration rates $(\alpha_2 \nu_2 + \alpha_1 \sigma_2)$ and $(\alpha_1 \nu_1 + \alpha_2 \sigma_1)$ and common death rate $(\sigma_1 \sigma_2 - \nu_1 \nu_2)$. So the equilibrium probability structure does not distinguish between (a) two independent immigration–death processes and (b) two immigration–death processes with type-change. Thus the precise process under study is reflected only in the initial transient phase of population development.

7.1.1 Generating moments

Whilst extracting the coefficient of $z_1^M z_2^N$ in $G(z_1, z_2; t)$ yields the probabilities $\{p_{MN}(t)\}$, differentiating this p.g.f. yields the bivariate factorial moments

$$\mu_{(rs)} = \left. \frac{\partial^{r+s} G(z_1, z_2; t)}{\partial z_1^r z_2^s} \right|_{z_1=z_2=1}. \tag{7.1.16}$$

Moreover, in a direct extension of the general univariate results developed in Section 2.1.3, differentiating the corresponding m.g.f. $M(\theta_1, \theta_2; t) = G(e^{\theta_1}, e^{\theta_2}; t)$ and placing $\theta_1 = \theta_2 = 0$ yields the bivariate raw moments μ'_{rs}, whilst differentiating the c.g.f. $K(\theta_1, \theta_2; t) = \ln[M(\theta_1, \theta_2; t)]$ yields the bivariate cumulants κ_{rs}. This last approach is often the most useful, since it provides the simplest way of obtaining the central moments μ_{rs}.

For illustration, suppose we wish to determine the mean number of type-i individuals at time t, i.e.

$$m_i(t) = \left. \frac{\partial G(z_1, z_2; t)}{\partial z_i} \right|_{z_1=z_2=1} \qquad (i = 1, 2).$$

Then on differentiating both sides of expression (7.1.14) with respect to z_i, and denoting

$$\omega_1, \omega_2 = (1/2)[-(\sigma_1 + \sigma_2) \pm \sqrt{\{(\sigma_1 - \sigma_2)^2 + 4\nu_1\nu_2\}}], \tag{7.1.17}$$

we obtain

$$m_1(t) = (\omega_1 - \omega_2)^{-1} \{ e^{\omega_1 t}[m_0(\omega_1 + \sigma_2) + \alpha_1 + \nu_2 n_0 + \omega_1^{-1}(\alpha_2 \nu_2 + \alpha_1 \sigma_2)]$$
$$- e^{\omega_2 t}[m_0(\omega_2 + \sigma_2) + \alpha_1 + \nu_2 n_0 + \omega_2^{-1}(\alpha_2 \nu_2 + \alpha_1 \sigma_2)] \}$$
$$+ (\omega_1 \omega_2)^{-1}(\alpha_2 \nu_2 + \alpha_1 \sigma_2), \tag{7.1.18}$$

together with a similar expression for $m_2(t)$.

However, this presumes that we have already determined $G(z_1, z_2; t)$, and this may be difficult to achieve in more complex situations. Fortunately, by differentiating the Lagrange p.d.e. we can obtain equations for the moments directly which may be much easier to solve. For example, differentiating (7.1.4) with respect to z_i and placing $z_1 = z_2 = 1$ yields

$$dm_i(t)/dt = \alpha_i - (\mu_i + \nu_i)m_i(t) + \nu_j m_j(t) \qquad (i, j = 1, 2; j \neq i). \tag{7.1.19}$$

Although this is identical to the deterministic equation, in general this will not be the case (we shall return to this point later on). Two methods of solution are available to us. Either we eliminate say $m_2(t)$ to form a standard second-order ordinary differential equation in $m_1(t)$, or else we apply the Laplace transform

$$\mathcal{L}[m_i(t)] \equiv m_i^*(s) = \int_0^\infty e^{-st} m_i(t)\, dt \tag{7.1.20}$$

to equations (7.1.19), thereby reducing them to the simultaneous form
$$sm_i^*(s) - m_i(0) = \alpha_i s^{-1} - (\mu_i + \nu_i)m_i^*(s) + \nu_j m_j^*(s). \tag{7.1.21}$$

For with $m_1(0) = m_0$ and $m_2(0) = n_0$ these two equations easily solve to yield
$$m_1^*(s) = [m_0 s + (\alpha_1 + m_0\sigma_2 + \nu_2 n_0) + s^{-1}(\alpha_2 \nu_2 + \alpha_1 \sigma_2)]/(s - w_1)(s - w_2). \tag{7.1.22}$$

Whence expanding the right-hand side of this expression into partial fractions, and then applying the inversion result
$$(s - a)^{-1} = \mathcal{L}(e^{at}), \tag{7.1.23}$$
recovers (7.1.18).

In general nonlinear situations, the algebra arising from using the above techniques is likely to be daunting, and use of an algebraic solver, such as *Mathematica*, might well prove appealing, especially when higher-order moments are being determined. However, if the differential moment equations are linear we may employ a standard matrix approach. For example, on evaluating $\partial^2 K(\theta_1, \theta_2; t)/\partial\theta_1\partial\theta_2$ at $\theta_1 = \theta_2 = 0$, we obtain the equations for the variance-covariances $\mathbf{V}(t) = (V_{11}(t), V_{12}(t), V_{22}(t))^T$ which may be written in the vector-matrix form
$$d\mathbf{V}(t)/dt = \mathbf{R}\mathbf{V}(t) + \mathbf{f}(t) \tag{7.1.24}$$

where
$$\mathbf{R} = \begin{pmatrix} -2\sigma_1 & 2\nu_2 & 0 \\ \nu_1 & -(\sigma_1 + \sigma_2) & \nu_2 \\ 0 & 2\nu_1 & -2\sigma_2 \end{pmatrix} \quad \text{and} \quad \mathbf{f}(t) = \begin{pmatrix} \sigma_1 m_1(t) + \nu_2 m_2(t) + \alpha_1 \\ -\nu_1 m_1(t) - \nu_2 m_2(t) \\ \sigma_2 m_2(t) + \nu_1 m_1(t) + \alpha_2 \end{pmatrix}. \tag{7.1.25}$$

Now equation (7.1.24) integrates (see Theorem 1.4 of Bellman, 1953) to give
$$\mathbf{V}(t) = \mathbf{y}(t) + \int_0^t \mathbf{Y}(t - s)\mathbf{f}(s)\,ds, \tag{7.1.26}$$
where $\mathbf{y}(t)$ and $\mathbf{Y}(t)$ are the solutions of the equations
$$d\mathbf{y}(t)/dt = \mathbf{R}\mathbf{y}(t) \quad \text{and} \quad d\mathbf{Y}(t)/dt = \mathbf{R}\mathbf{Y}(t), \tag{7.1.27}$$
and $\mathbf{Y}(0) = \mathbf{I}$ is the identity matrix. On placing $\mathbf{y}(0) = \mathbf{V}(0) = \mathbf{0}$, it follows from (7.1.27) that $\mathbf{y}(t) \equiv \mathbf{0}$ for all $t \geq 0$. To solve the second of equations (7.1.27), we first note from (7.1.25) that the eigenvalues of $\mathbf{R}$ are given by $\zeta_1 = 2w_1$, $\zeta_2 = w_1 + w_2$ and $\zeta_3 = 2w_2$ where w_1 and w_2 are defined in (7.1.17). By reference to standard results in matrix theory (e.g. Bellman, 1960) it then follows that provided the ζ_i are distinct (i.e. $w_1 \neq w_2$) this equation integrates to
$$\mathbf{Y}(t) = \sum_{k=1}^{3} \mathbf{A}_k e^{\zeta_k t}. \tag{7.1.28}$$

Here the matrices $\mathbf{A}_k$ are given by

$$\mathbf{A}_k = \mathbf{q}_k \mathbf{r}_k \quad (k=1,2,3), \tag{7.1.29}$$

where $\mathbf{q}_k$ and $\mathbf{r}_k$ are the column and row eigenvectors of $\mathbf{r}$ normalized so that

$$\mathbf{r}_k \mathbf{q}_k = 1 \quad (k=1,2,3). \tag{7.1.30}$$

On writing $D = \sqrt{\{(\sigma_1 - \sigma_2)^2 + 4\nu_1\nu_2\}}$, it is straightforward to show that $\mathbf{q}_1 = (2\nu_2/(\sigma_1 - \sigma_2 + D), 1, 2\nu_1/(\sigma_2 - \sigma_1 + D))^T$, $\mathbf{r}_1 = (\nu_1/(\sigma_1 - \sigma_2 + D), 1, \nu_2/(\sigma_2 - \sigma_1 + D))$, etc. Whence on applying the normalization (7.1.30), we see that (7.1.29) yields

$$\mathbf{A}_k = \begin{pmatrix} 2\nu_1\nu_2(\sigma_1 - \sigma_2 + lD)^{-2} & 2\nu_2(\sigma_1 - \sigma_2 + lD)^{-1} & 2\nu_2^2[(lD)^2 - (\sigma_1 - \sigma_2)^2]^{-1} \\ \nu_1(\sigma_1 - \sigma_2 + lD)^{-1} & 1 & \nu_2(\sigma_2 - \sigma_1 + lD)^{-1} \\ 2\nu_1^2[(lD)^2 - (\sigma_1 - \sigma_2)^2]^{-1} & 2\nu_1(\sigma_2 - \sigma_1 + lD)^{-1} & 2\nu_1\nu_2(\sigma_2 - \sigma_1 + lD)^{-2} \end{pmatrix} \phi_k \tag{7.1.31}$$

where $l = 2 - i$ and $\phi_1 = \phi_3 = 2\nu_1\nu_2 D^{-2}$ with $\phi_2 = (\sigma_1 - \sigma_2)^2 D^{-2}$.

On combining (7.1.26), (7.1.28) and (7.1.31) it then follows that

$$\mathbf{V}(t) = \sum_{k=1}^{3} \mathbf{A}_k \int_0^t e^{\zeta_k(t-s)} \mathbf{f}(s)\, ds. \tag{7.1.32}$$

Whence writing $\mathbf{f}(s)$ in expression (7.1.25) as

$$\mathbf{f}(s) = \mathbf{f}_0 + \mathbf{f}_1 e^{\omega_1 s} + \mathbf{f}_2 e^{\omega_2 s}, \tag{7.1.33}$$

where $\mathbf{f}_0$, $\mathbf{f}_1$ and $\mathbf{f}_2$ are constant column vectors, integrating expression (7.1.32) yields the solution (for $\omega_1 \neq \omega_2$)

$$\mathbf{V}(t) = \sum_{k=1}^{3} \mathbf{A}_k [\mathbf{f}_0(-\zeta_k)^{-1}(1 - e^{\zeta_k t}) + \mathbf{f}_1(\omega_1 - \zeta_k)^{-1}(e^{\omega_1 t} - e^{\zeta_k t})$$
$$+ \mathbf{f}_2(\omega_2 - \zeta_k)^{-1}(e^{\omega_2 t} - e^{\zeta_k t})]. \tag{7.1.34}$$

Since the total population size $X_1(t) + X_2(t)$ is bounded above by a simple immigration–death process with rates $\max(\alpha_i)$ and $\min(\mu_i)$, it follows that an equilibrium solution must exist. Whence on letting $t \to \infty$ in (7.1.18) we see that $m_1(\infty) = (\omega_1\omega_2)^{-1}(\alpha_2\nu_2 + \alpha_1\sigma_2)$ with $m_2(\infty) = (\omega_2\omega_1)^{-1}(\alpha_1\nu_1 + \alpha_2\sigma_1)$. Showing that $\omega_1, \omega_2 < 0$ in (7.1.17) is trivial, since this reduces to the condition $\nu_1\nu_2 < (\mu_1 + \nu_1)(\mu_2 + \nu_2)$ which is bound to occur unless both death rates are zero. Similarly, after a little algebraic simplification, (7.1.34) yields $V_{12}(\infty) = 0$ together with $V_{ii}(\infty) = m_i(\infty)$, which is in accord with the two species developing as two independent processes with immigration rates $\alpha'_i = (\alpha_j\nu_j + \alpha_i\sigma_j)$ $(j \neq i)$ and common death rate $\mu' = (\sigma_1\sigma_2 - \nu_1\nu_2)$, as previously noted following (7.1.15).

7.2 Count-dependent growth

The above immigration–death–switch process is mathematically tractable and gives rise to reasonably transparent expressions for probabilities and moments. This is essentially because the population is both closed, i.e. none of the individuals give birth, and linear. So before we proceed to a general discussion of nonlinear open processes it is worthwhile briefly describing a simple example which highlights the fundamental problems that can be encountered in systems that are only slightly more complicated in structure.

Aphids are small, sap-sucking insects which create a serious global pest problem for agricultural crops and forest trees. They secrete honeydew to attract other insects, but this is a double-edged sword as it forms a weak cover on each infested leaf, thereby preventing aphids from further sucking and moving which causes them to starve to death. Prajneshu (1998) justifies the assumptions of a linear birth rate and of a death rate which is proportional to the product of the current population size $N(t)$ and the cumulative size $F(t) = \int_0^t N(s)\,ds$. Whence the deterministic equation of population growth is given by

$$dN(t)/dt = \lambda N(t) - \mu N(t) \int_0^t N(s)\,ds. \qquad (7.2.1)$$

This differential-integral equation can be reduced to a purely differential form by differentiating (7.2.1) to obtain (with N' denoting $dN(t)/dt$)

$$N'' = -\mu N^2 + (N')^2/N. \qquad (7.2.2)$$

On writing $u = N'/N$, this further reduces to

$$u(du/dN) = -\mu,$$

which integrates to give

$$u(N) = \sqrt{\{k - 2\mu N\}}. \qquad (7.2.3)$$

To determine the constant of integration k, we see from (7.2.1) that at $t = 0$, $N'(0) = \lambda N(0) = \lambda N_0$ (say), and so $u(0) = N'(0)/N(0) = \lambda$. Whence it follows from (7.2.3) that $k = \lambda^2 + 2\mu N_0$. Thus (7.2.3) becomes

$$N' = N\sqrt{(k - 2\mu N)}, \qquad (7.2.4)$$

which, on using the transformation $z = \sqrt{(k - 2\mu N)}$, integrates to yield the solution

$$N(t) = ae^{-bt}(1 + de^{-bt})^{-2} \qquad (7.2.5)$$

where $b = \sqrt{k} = \sqrt{(\lambda^2 + 2\mu N_0)}$, $d = (b + \lambda)/(b - \lambda)$ and $a = 2b^2 d/\mu$. Note that at $t = 0$ the function (7.2.5) reduces to $N(0) = N_0$, as required.

7.2.1 Stochastic representation

Although the nonlinear deterministic equation (7.2.1) yields the simple solution (7.2.5), the fact that several steps are needed to obtain it suggests that the corresponding stochastic equations might be difficult to solve. To preserve the underlying Markov structure of the process we need to consider change in both $N(t)$ and $F(t)$, rather than $N(t)$ alone as in (7.2.1). On writing (7.2.1) in the bivariate form

$$dN(t)/dt = \lambda N(t) - \mu N(t)F(t)$$
$$dF(t)/dt = N(t), \qquad (7.2.6)$$

we see that for $N(t) = n$ and $F(t) = f$ the required transition rates take the form

$$\Pr[N(t+dt) = n+1, F(t+dt) = f+1] = \lambda dt \qquad (7.2.7)$$
$$\Pr[N(t+dt) = n-1] = \mu n f dt. \qquad (7.2.8)$$

Thus the forward equation is given by

$$dp_{nf}(t)/dt = \lambda(n-1)p_{n-1,f-1}(t) - (\lambda n + \mu n f)p_{nf}(t) + \mu(n+1)f p_{n+1,f}(t). \qquad (7.2.9)$$

Multiplying both sides by $z^n s^f$ and summing over $n, f = 0, 1, 2, \ldots$ then yields the second-order p.d.e.

$$\frac{\partial G(z,s;t)}{\partial t} = \lambda z(zs-1)\frac{\partial G(z,s;t)}{\partial z} - \mu s(z-1)\frac{\partial^2 G(z,s;t)}{\partial z \partial s} \qquad (7.2.10)$$

for the p.g.f.

$$G(z,s;t) \equiv \sum_{n=0}^{\infty}\sum_{f=0}^{\infty} p_{nf}(t) z^n s^f.$$

Now the only difference between this aphid process and the linear birth–death process developed in Section 2.4 is that the death rate μn has been augmented to $\mu n f$. Yet the effect on the probability structure is considerable, for the basic first-order p.d.e. (2.4.7) for the birth–death p.g.f. has become much more complex. Yet in practical terms, the inclusion of the product-pair nf in the forward equations is relatively simple. Some progress can be made by considering moments, but even this is fraught with difficulty. For on differentiating (7.2.10) with respect to z and s, in order to construct the factorial moments (7.1.16), we obtain the equations

$$d\mu_{(1.)}(t)/dt = \lambda \mu_{(1.)}(t) - \mu \mu_{(11)}(t)$$
$$d\mu_{(.1)}(t)/dt = \lambda \mu_{(1.)}(t). \qquad (7.2.11)$$

Two features are immediately apparent. First, unlike the immigration–death–switch process, the stochastic update equations (7.2.11) for the mean population sizes $\mu_{(1.)}(t)$ and $\mu_{(.1)}(t)$ now differ from their deterministic counterparts (7.2.6). For the death component $\mu N(t)F(t)$ is replaced not by its paradigm $\mu \mu_{(1.)}(t)\mu_{(.1)}(t)$ but by the second-order term $\mu \mu_{(11)}(t)$. So the expected and deterministic solutions are fundamentally different. This feature is common in nonlinear systems and highlights the danger in believing (wrongly) that, provided population sizes are large, a deterministic solution corresponds to the mean value of a stochastic process. Whilst the difference between them can, indeed, sometimes be negligible, at other times it may be considerable, and it is generally necessary to examine both the deterministic and mean solutions on a case-by-case basis in order to discover what the difference actually is. Second, differentiating equation (7.2.10) once more to generate the second-order moments brings in third-order terms, etc. So in direct contrast to the construction of the closed variance-covariance equations (7.1.24) it is not possible to develop a closed set of equations for our aphid moments. This gives rise to the classic *moment closure problem* which we shall now consider in some detail.

7.3 Bivariate saddlepoint approximation

Although our analysis of the simple immigration–death–switch process in Section 7.1 yields exact solutions for the population size probabilities, means and variance-covariances, Section 7.2 shows that adding even a small amount of complexity to a system is likely to lead to depressingly difficult algebra. Moreover, even if we could derive exact bivariate stochastic solutions, their resulting structure is likely to be sufficiently opaque to prevent them from revealing any meaningful interpretation. Indeed, for nonlinear systems, exact solutions are likely to prove intractable in all but the simplest scenarios. So there is clearly a strong need to generate 'good' approximate stochastic solutions, and given our previous success in using the univariate saddlepoint approximation (Section 3.7) the obvious way forward is to extend this approach to the multivariate case. This route offers several key advantages: no distributional assumptions are required; it works regardless of the moment-order deemed appropriate; and, we obtain an algebraic form for the associated p.d.f. irrespective of whether or not we have complete knowledge of the cumulants. This latter feature is especially important, since unlike the univariate case no multivariate families of distributions currently exist which embrace all cumulants up to any given order. We shall see that, in general, the technique allows swift convergence to the required p.d.f., though analysis of a severe test case illustrates its operational limit.

In terms of straight practical application, extending the truncated saddlepoint approach from single- to multi-variable scenarios simply involves replacing (3.7.22) by the m-variable form

$$f(x_1,\ldots,x_m) \simeq \frac{\exp\{K(\theta_1,\ldots,\theta_m) - \theta_1 x_1 - \cdots - \theta_m x_m\}}{(2\pi)^{m/2}\sqrt{|K''(\theta_1,\ldots,\theta_m)|}} \quad (7.3.1)$$

where

$$\partial K(\theta_1,\ldots,\theta_m)/\partial \theta_i = x_i \quad (i=1,\ldots,m) \quad (7.3.2)$$

and $|K''(\cdot)|$ is the determinant of second derivatives. So here the univariate c.g.f. $K(\theta) = \ln[M(\theta)]$ is replaced by the multivariate equivalent

$$K(\theta_1,\ldots,\theta_m) = \ln[\int_{-\infty}^{\infty}\ldots\int_{-\infty}^{\infty} e^{(\theta_1 x_1 + \cdots + \theta_m x_m)} f(x_1,\ldots,x_m)\,dx_1\ldots dx_m]. \quad (7.3.3)$$

We shall remain with two-variable processes for reasons of algebraic simplicity; no additional theoretical difficulties occur when $m > 2$.

7.3.1 Simple illustrations

Example 1 Let us first parallel Example 1 of Section 3.7.2 by supposing that the variables (X, Y) follow a bivariate Normal distribution with zero means, variances σ_1^2 and σ_2^2, and correlation ρ. Then the associated c.g.f.

$$K(\theta_1, \theta_2) = (1/2)(\theta_1^2 \sigma_1^2 + 2\rho \sigma_1 \sigma_2 \theta_1 \theta_2 + \theta_2^2 \sigma_2^2). \tag{7.3.4}$$

Whence equations (7.3.2) become

$$\theta_1 \sigma_1^2 + \rho \sigma_1 \sigma_2 \theta_2 = x$$
$$\theta_2 \sigma_2^2 + \rho \sigma_1 \sigma_2 \theta_1 = y, \tag{7.3.5}$$

and for given (x, y) these solve to give

$$\theta_1 = \frac{(x\sigma_2 - y\rho\sigma_1)}{\sigma_1^2 \sigma_2 (1 - \rho^2)} \quad \text{and} \quad \theta_2 = \frac{(y\sigma_1 - x\rho\sigma_2)}{\sigma_1 \sigma_2^2 (1 - \rho^2)}. \tag{7.3.6}$$

Inserting (7.3.6) into (7.3.1) with $m = 2$ and $(x_1, x_2) = (x, y)$ then yields the exact p.d.f. of the bivariate Normal distribution, namely

$$f(x, y) = \frac{1}{2\pi \sigma_1 \sigma_2 \sqrt{(1 - \rho^2)}} \exp \left\{ \frac{-1}{2(1 - \rho^2)} \left(\frac{x^2}{\sigma_1^2} - \frac{2\rho xy}{\sigma_1 \sigma_2} + \frac{y^2}{\sigma_2^2} \right) \right\}. \tag{7.3.7}$$

Example 2 Now suppose that X and Y are i.i.d. Poisson(1) variables, and let $U = X$ and $V = X + Y$. Then

$$\Pr(U = u, V = v) = \frac{e^{-2}}{u!(v - u)!} \tag{7.3.8}$$

with c.g.f.

$$K(\theta_1, \theta_2) = e^{\theta_1 + \theta_2} + e^{\theta_2} - 2. \tag{7.3.9}$$

Solving (7.3.2) with (7.3.9) yields $e^{\theta_1} = u/(v - u)$ and $e^{\theta_2} = v - u$, whence (7.3.1) becomes

$$f(u, v) \simeq \frac{e^{v-2}}{2\pi u^{u+1/2}(v - u)^{v-u+1/2}}. \tag{7.3.10}$$

On replacing $n!$ in the exact solution (7.3.8) by $\sqrt{(2\pi)}e^{-n}n^{n+1/2}$ we immediately recover (7.3.10). So here the success of the saddlepoint approximation is limited only by the accuracy of Stirling's approximation.

Example 3 We shall now show how the saddlepoint approach can be applied to a stochastic process, as opposed to a given p.d.f. such as (7.3.8). Let pairs of type-1 and type-2 individuals arrive as a Poisson process at rate α, with individuals dying independently from each other according to two simple death processes with rates μ_1 and μ_2, respectively. Then on denoting $p_{ij}(t) = \Pr(\text{population is of size } (i, j) \text{ at time } t)$, the forward equations are given by

$$dp_{ij}(t)/dt = \alpha p_{i-1,j-1}(t) + \mu_1(i+1)p_{i+1,j}(t) + \mu_2(j+1)p_{i,j+1}(t) - (\alpha + i\mu_1 + j\mu_2)p_{ij}(t),$$
(7.3.11)

whence the associated equation for the p.g.f. $G(z_1, z_2; t) \equiv \sum_{i,j=0}^{\infty} p_{ij}(t) z_1^i z_2^j$ takes the form

$$\frac{\partial G}{\partial t} - \mu_1(1-z_1)\frac{\partial G}{\partial z_1} - \mu_2(1-z_2)\frac{\partial G}{\partial z_2} = -\alpha(1 - z_1 z_2)G. \quad (7.3.12)$$

This equation solves in the standard way, via the auxiliary equations

$$\frac{dt}{1} = \frac{-dz_1}{\mu_1(1-z_1)} = \frac{-dz_2}{\mu_2(1-z_2)} = \frac{-dG}{\alpha G(1-z_1 z_2)}, \quad (7.3.13)$$

to yield the solution

$$G(z_1, z_2; t) = \exp\{-(\alpha/\mu_1)(1-z_1)(1-e^{-\mu_1 t}) - (\alpha/\mu_2)(1-z_2)(1-e^{-\mu_2 t})$$
$$+ [\alpha/(\mu_1 + \mu_2)](1-z_1)(1-z_2)(1-e^{-(\mu_1+\mu_2)t})\}; \quad (7.3.14)$$

here we have assumed that the population is initially empty, i.e. $G(z_1, z_2; 0) \equiv 1$. On considering each of the three component parts of (7.3.14), and denoting $\phi_1 = (\alpha/\mu_1)(1-e^{-\mu_1 t})$, $\phi_2 = (\alpha/\mu_2)(1-e^{-\mu_2 t})$ and $\phi_{12} = [\alpha/(\mu_1 + \mu_2)](1-e^{-(\mu_1+\mu_2)t})$, we see that (X, Y) is distributed as the three-way convolution

$$\text{Poisson}(X, \phi_1 - \phi_{12}) \star \text{Poisson}(Y, \phi_2 - \phi_{12}) \star \text{Poisson}(XY, \phi_{12}). \quad (7.3.15)$$

So the exact population size probabilities are given by

$$p_{ij}(t) = \sum_{s=0}^{\min(i,j)} \frac{(\phi_1 - \phi_{12})^{i-s}(\phi_2 - \phi_{12})^{j-s}\phi_{12}^s \exp\{-\phi_1 - \phi_2 + \phi_{12}\}}{(i-s)!(j-s)!s!}. \quad (7.3.16)$$

To evaluate the associated saddlepoint values we first note that as the c.g.f. is given by (7.3.14) with $K(\theta_1, \theta_2; t) = \ln[G(e^{\theta_1}, e^{\theta_2}; t)]$, the saddlepoint equations (7.3.2) become

$$(\phi_1 - \phi_{12})e^{\theta_1} + \phi_{12}e^{\theta_1+\theta_2} = x$$
$$(\phi_2 - \phi_{12})e^{\theta_2} + \phi_{12}e^{\theta_1+\theta_2} = y. \quad (7.3.17)$$

For algebraic convenience denote these equations as

$$aw + bwz = x \quad \text{and} \quad cz + bwz = y, \quad (7.3.18)$$

where $w = e^{\theta_1}$, $z = e^{\theta_2}$, $a = \phi_1 - \phi_{12}$, $b = \phi_{12}$ and $c = \phi_2 - \phi_{12}$. Then

$$w = (1/2ab)\{-[ac - (x-y)b] + \sqrt{[ac - (x-y)b]^2 + 4abcx}\}$$
$$z = (1/2bc)\{-[ac - (y-x)b] + \sqrt{[ac - (y-x)b]^2 + 4abcy}\}. \quad (7.3.19)$$

Now the determinant

$$|K''(\theta_1, \theta_2; t)| = xy - (bwz)^2. \quad (7.3.20)$$

342 *Modelling bivariate processes*

So the full saddlepoint form (7.3.1) is given by

$$f(x,y) = \frac{\exp\{-a(1-w) - c(1-z) - b(1-wz)\}w^{-x}z^{-y}}{2\pi\sqrt{xy - (bwz)^2}}, \qquad (7.3.21)$$

which is clearly far more transparent than the exact summation (7.3.16).

To illustrate this result consider the equilibrium process (i.e. at $t = \infty$) with $\alpha = 6$, $\mu_1 = 1$ and $\mu_2 = 2$. Then from (7.3.15) we see that the process behaves as the threefold Poisson convolution

$$\text{Poisson}(X, 4) \star \text{Poisson}(Y, 1) \star \text{Poisson}(XY, 2), \qquad (7.3.22)$$

and although the associated exact probabilities (7.3.16) take the fairly amenable form

$$p_{ij}(\infty) = e^{-7} \sum_{r=0}^{\min(i,j)} \frac{4^{i-r}2^r}{(i-r)!(j-r)!r!}, \qquad (7.3.23)$$

the behaviour of this multi-termed expression is still much more difficult to visualize than that of the single-termed saddlepoint approximation (7.3.21), namely

$$f(x,y) = \frac{w^{-x}z^{-y}e^{4w+y-7}}{2\pi\sqrt{xy - 4(wz)^2}}. \qquad (7.3.24)$$

To assess the numerical accuracy of (7.3.24), on letting x and y take the respective marginal (Poisson) mean values $\alpha/\mu_1 = 6$ and $\alpha/\mu_2 = 3$ we have $p_{6,3}(\infty) = 0.041068$ (to six decimal places), so the saddlepoint value $f(6,3) = 0.042536$ (3.57% too high) compares well. Whilst to examine a worst case scenario let us take $x = y = 1$; for since Stirling's approximation to 1! is 2.32, we might expect (7.3.24) to perform poorly. In fact, $f(1,1) = 0.006579$ compares quite favourably with $p_{1,1}(\infty) = 0.005471$.

7.3.2 Cumulant truncation

Although the saddlepoint approximation is totally defined by the c.g.f. $K(\theta_1, \theta_2)$, in many situations this function is either very difficult, or impossible, to determine, as illustrated by the count dependent growth process of Section 7.2. However, our previous success in applying the univariate truncation scheme to the logistic process (Section 3.7.4) suggests that a good way of circumventing this problem would be to extend our partial c.g.f. technique to multivariate scenarios. Since we have already shown through Table 3.1 that for the univariate Poisson(10) process placing $\kappa_i = 0$ for $i > 3$ does not work for $x < 5$, to keep our (bivariate) options open we shall consider $\kappa_{ij} \equiv 0$ for all $i + j > 4$. Thus the full c.g.f. $K(\theta_1, \theta_2) \equiv \sum_{i,j=1}^{\infty} \kappa_{ij}\theta_1^i\theta_2^j/(i!j!)$ is replaced by

$$K(\theta_1, \theta_2) = \kappa_{10}\theta_1 + \kappa_{01}\theta_2 + \kappa_{20}\theta_1^2/2 + \kappa_{11}\theta_1\theta_2 + \kappa_{02}\theta_2^2/2$$
$$+ \kappa_{30}\theta_1^3/6 + \kappa_{21}\theta_1^2\theta_2/2 + \kappa_{12}\theta_1\theta_2^2/2 + \kappa_{03}\theta_2^3/6 \qquad (7.3.25)$$
$$+ \kappa_{40}\theta_1^4/24 + \kappa_{31}\theta_1^3\theta_2/6 + \kappa_{22}\theta_1^2\theta_2^2/4 + \kappa_{13}\theta_1\theta_2^3/6 + \kappa_{04}\theta_2^4/24.$$

Whence the saddlepoint equation (7.3.2) for (θ_1, θ_2) becomes

$$g(\theta_1, \theta_2) \equiv \kappa_{10} + \kappa_{20}\theta_1 + \kappa_{11}\theta_2 + \kappa_{30}\theta_1^2/2 + \kappa_{21}\theta_1\theta_2 + \kappa_{12}\theta_2^2/2$$
$$+ \kappa_{40}\theta_1^3/6 + \kappa_{31}\theta_1^2\theta_2/2 + \kappa_{22}\theta_1\theta_2^2/2 + \kappa_{13}\theta_2^3/6 - x = 0$$
$$h(\theta_1, \theta_2) \equiv \kappa_{01} + \kappa_{11}\theta_1 + \kappa_{02}\theta_2 + \kappa_{21}\theta_1^2/2 + \kappa_{12}\theta_1\theta_2 + \kappa_{03}\theta_2^2/2 \quad (7.3.26)$$
$$+ \kappa_{31}\theta_1^3/6 + \kappa_{22}\theta_1^2\theta_2/2 + \kappa_{13}\theta_1\theta_2^2/2 + \kappa_{04}\theta_2^3/6 - y = 0,$$

whilst $|K''(\theta_1, \theta_2)|$ becomes $|g_1 h_2 - g_2 h_1|$ where

$$g_1 = \partial g/\partial \theta_1 = \kappa_{20} + \kappa_{30}\theta_1 + \kappa_{21}\theta_2 + \kappa_{40}\theta_1^2/2 + \kappa_{31}\theta_1\theta_2 + \kappa_{03}\theta_2^2/2$$
$$g_2 = \partial g/\partial \theta_2 = \partial h/\partial \theta_1 = h_1 = \kappa_{11} + \kappa_{21}\theta_1 + \kappa_{12}\theta_2 + \kappa_{31}\theta_1^2/2 + \kappa_{22}\theta_1\theta_2 + \kappa_{13}\theta_2^2/2$$
$$h_2 = \partial h/\partial \theta_2 = \kappa_{02} + \kappa_{12}\theta_1 + \kappa_{03}\theta_2 + \kappa_{22}\theta_1^2/2 + \kappa_{13}\theta_1\theta_2 + \kappa_{04}\theta_2^2/2. \quad (7.3.27)$$

Numerical solutions to equations (7.3.27) may be computed efficiently by using, for example, bivariate Newton–Raphson with the iterates

$$(\theta_1^{(n+1)}, \theta_2^{(n+1)}) = (\theta_1^{(n)} + \alpha_n, \theta_2^{(n)} + \beta_n), \quad (7.3.28)$$

where we use (7.3.27) at $\theta_1 = \theta_1^{(n)}$ and $\theta_2 = \theta_2^{(n)}$ to form

$$\alpha_n = \frac{hg_2 - gh_2}{g_1 h_2 - g_2 h_1} \quad \text{and} \quad \beta_n = \frac{hg_1 - gh_1}{g_1 h_2 - g_2 h_1}. \quad (7.3.29)$$

Once the required level of convergence has been achieved, the corresponding saddlepoint approximation is then given by

$$\tilde{f}(x, y) = \lim_{n \to \infty} \frac{\exp\{K(\theta_1^{(n)}, \theta_2^{(n)}) - \theta_1^{(n)} x - \theta_2^{(n)} y\}}{2\pi\sqrt{(g_1 h_2 - g_2 h_1)}}. \quad (7.3.30)$$

For this saddlepoint procedure to yield full support to the underlying process, the quadratic surface generated by (7.3.26) must involve real (θ_1, θ_2) for all appropriate (x, y)-values. One (rather extreme) way of achieving this is to place $\kappa_{31} = \kappa_{22} = \kappa_{13} = 0$, for we see from (7.3.26) that for large $|\theta_i|$, $g \sim \kappa_{40}\theta_1^3/6 - x$ and $h \sim \kappa_{04}\theta_2^3/6 - y$, which independently sweep out all possible x and y values. So a saddlepoint solution will always exist. However, given that there is no unique way of relaxing the zero condition on κ_{31}, κ_{22} and κ_{13}, a key question arises as to how $\max_{x,y} |[\tilde{f}(x, y) - f(x, y)]/f(x, y)|$ or $\max_{x,y} |\tilde{f}(x, y) - f(x, y)|$ changes with $\kappa_{40}, \ldots, \kappa_{04}$. At present this is an open problem (see Gillespie and Renshaw, 2005).

To illustrate how to apply this truncation procedure let us follow the example in Renshaw (2000) by constructing a p.d.f. which possesses the third-order cumulant structure

$$\begin{aligned} &\kappa_{10} = 10, \ \kappa_{01} = 10, \\ &\kappa_{20} = 9, \ \kappa_{11} = 4, \ \kappa_{02} = 16, \\ &\kappa_{30} = 15, \ \kappa_{21} = 10, \ \kappa_{12} = 15, \ \kappa_{03} = 25 \end{aligned} \quad (7.3.31)$$

344 *Modelling bivariate processes*

over $0 \leq x, y \leq 100$. Now given that saddlepoint p.d.f.'s operate over $-\infty < x, y < \infty$, we have to accept that non-negligible probability mass may accrue outside our chosen bounded region. Thus not only will we have to scale the resulting p.d.f. in order to ensure that probabilities sum to one, but the ensuing saddlepoint cumulants, $\tilde{\kappa}_{ij}$, may well differ from the target cumulants, κ_{ij}. The development of a universally optimal procedure for achieving this goal presents another open problem, but the following approach works well. The five free-ranging fourth-order cumulants are used to achieve the best least squares fit; here we select the initial values $\tilde{\kappa}_{40} = \tilde{\kappa}_{04} = 100$ (which are high enough to ensure that the iterated p.d.f.'s have full support over $0 \leq x, y \leq 100$) together with $\tilde{\kappa}_{31} = \tilde{\kappa}_{22} = \tilde{\kappa}_{13} = 0$. In general, deriving a p.d.f. to fit nth-order cumulants would use the $n+2$ $(n+1)$-th order cumulants. For $\{Z\}$ a set of independent uniformly distributed psuedo-random numbers on $(-0.5, 0.5)$, we therefore have the following simulation procedure.

Truncated saddlepoint algorithm (A7.1)

(i) set initial saddlepoint cumulants to the target values, for example (7.3.31), i.e. $\tilde{\kappa}_{10} = \kappa_{10}, \ldots, \tilde{\kappa}_{03} = \kappa_{03}$, and choose appropriate values for $\tilde{\kappa}_{40}, \ldots, \tilde{\kappa}_{04}$

(ii) increment a randomly chosen cumulant $\tilde{\kappa}_{ij}$ by δZ, for appropriate δ

(iii) (a) evaluate the resulting saddlepoint probabilities $\bar{f}(x, y)$ via the iterative procedure (7.3.26)–(7.3.30), and then (b) rescale them to form $\hat{f}(x, y)$ to ensure that (here) $\sum_{i,j=0}^{100} \hat{f}(x, y) = 1$

(iv) evaluate the cumulants $\hat{\kappa}_{ij}$ corresponding to $\hat{f}(x, y)$

(v) determine $S = \sum_{i,j=0}^{3}(\hat{\kappa}_{ij} - \kappa_{ij})^2$, and update $\tilde{\kappa}_{ij}$ if S is reduced and $g_1 h_2 - g_2 h_1 > 0$ (to ensure (7.3.30) is real)

(vi) print $\hat{f}(x, y)$ when S reaches say 10^{-6}, then stop

(vii) return to (ii)

To enable $\tilde{\kappa}_{40}, \ldots, \tilde{\kappa}_{04}$ to 'bed-in', the initial choice of δ was kept relatively high at $\delta = 10$, but even such coarse-tuning swiftly led to $\max|\hat{\kappa}_{ij} - \kappa_{ij}| < 0.03$. Switching to a fine-tuning regime with $\delta = 0.1$ then quickly produced $\max|\hat{\kappa}_{ij} - \kappa_{ij}| < 0.0006$. If required, further accuracy could be achieved with micro-tuning using say $\delta = 0.001$; here this gives $\max|\hat{\kappa}_{ij} - \kappa_{ij}| \sim 0.00001$. The algorithm is clearly precise, and it is also quick to converge. The resulting 'working' $\tilde{\kappa}_{ij}$-values which give rise to the best-fit values $\hat{\kappa}_{ij} = \kappa_{ij}$ are given by

$\tilde{\kappa}_{10} = 9.704, \quad \tilde{\kappa}_{01} = 9.810,$

$\tilde{\kappa}_{20} = 8.651, \quad \tilde{\kappa}_{11} = 4.640, \quad \tilde{\kappa}_{02} = 16.579,$

$\tilde{\kappa}_{30} = 20.157, \quad \tilde{\kappa}_{21} = 10.157, \quad \tilde{\kappa}_{12} = 14.867, \quad \tilde{\kappa}_{03} = 20.944,$

$\tilde{\kappa}_{40} = 86.058, \quad \tilde{\kappa}_{31} = 4.988, \quad \tilde{\kappa}_{22} = 3.542, \quad \tilde{\kappa}_{31} = 3.526, \quad \tilde{\kappa}_{04} = 118.078,$

and their distance from κ_{ij} highlights the need for rescaling. The associated fourth-order cumulant estimates $\hat{\kappa}_{40}, \ldots, \hat{\kappa}_{04}$ are given by

$\hat{\kappa}_{40} = 71.715, \hat{\kappa}_{31} = 13.427, \hat{\kappa}_{22} = 11.694, \hat{\kappa}_{13} = 16.984, \hat{\kappa}_{04} = 118.078,$

and so in this example κ_{40} and κ_{04} play a dominant role.

Although κ_1 and κ_2 are respectively 3.33 and 2.5 standard deviations above zero, the presence of skewness leads to substantial probability mass outside $0 \leq x, y \leq 100$ (in step (iiia) $\sum_{i,j=0}^{100} \tilde{f}(x,y) = 0.9151$), so rescaling is essential. The required p.d.f. $\hat{f}(x,y)$ (step (iiib)) is therefore $\tilde{f}(x,y)/0.9151$ over $0 \leq x, y \leq 100$ (Figure 7.1a), where $\tilde{f}(x,y)$ is given by (7.3.30) and g_i, h_i by (7.3.27), with κ_{ij} being replaced by the above values of $\tilde{\kappa}_{ij}$. Note that of the first nine $\tilde{\kappa}_{ij}$-values, seven are in fairly close agreement with κ_{ij}; only $\tilde{\kappa}_{30}$ and $\tilde{\kappa}_{03}$ are out of line. Also, $\tilde{\kappa}_{40}$ and $\tilde{\kappa}_{04}$ are substantially larger than $\tilde{\kappa}_{31}$, $\tilde{\kappa}_{22}$ and $\tilde{\kappa}_{13}$. As we shall soon see, under certain conditions this last observation can prove to be of crucial importance if we are to achieve support over the full range of x, y.

To obtain the corresponding rescaled Normal p.d.f. we can employ the same algorithm, but this time with $\tilde{\kappa}_{30}, \ldots, \tilde{\kappa}_{03}, \tilde{\kappa}_{40}, \ldots, \tilde{\kappa}_{04}$ fixed at zero (the second-order saddlepoint approximation is guaranteed to exist). Initial coarse-tuning is unnecessary, since third- and fourth-order cumulants no longer feature, and fine-tuning swiftly produces excellent accuracy (maximum absolute error is < 0.00004): subsequent micro-tuning yields an accuracy of 0.000001. In the absence of skewness, very little probability leakage now occurs (in step (iiia) $\sum_{i,j=0}^{100} \tilde{f}(x,y) = 0.9946$). Nevertheless, the tiny amount of rescaling employed in step (iiib) does have an appreciable effect on $\hat{\kappa}_{30}, \ldots, \hat{\kappa}_{04}$. The p.d.f. $\hat{f}(x,y)$ takes the scaled bivariate Normal form $(1/0.9946)N(9.9822, 9.9378; 9.0665, 4.1723, 16.6494)$ over $0 \leq x, y \leq 100$ (Figure 7.1b), and comparison with Figure 7.1a shows that this p.d.f. is less peaked, having a modal value of 0.0138 compared to 0.0173 under fourth-order truncation.

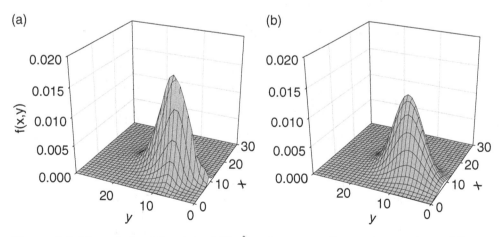

Figure 7.1 Truncated saddlepoint p.d.f.'s $\hat{f}(x,y)$ corresponding to the cumulants of Example (7.3.31): (a) least squares fit to $\kappa_{10}, \ldots, \kappa_{03}$ with free-ranging $\kappa_{40}, \ldots, \kappa_{04}$; (b) least squares (Normal) fit to $\kappa_{10}, \ldots, \kappa_{02}$ with $\tilde{\kappa}_{30} = \cdots = \tilde{\kappa}_{04} = 0$.

7.3.3 A cautionary tale!

This highly successful example bodes well for many practical situations, though caution must be exercised if the proposed cumulant structure exhibits awkward characteristics. As an extreme example, if the variances and covariance are zero then the p.d.f. must comprise a single spike of mass one at the mean, so all central moments are zero. Thus in this case any attempt to construct a p.d.f. with say non-negative third-order cumulants is doomed to failure. A less obvious situation is where one (or both) of the means lies close to an axis. For the resulting constriction on shape may mean that not all sets of $\{\kappa_{10}, \ldots, \kappa_{03}\}$ can lend themselves to an associated p.d.f. To illustrate this scenario Renshaw (2000) uses the Matis, Zheng and Kiffe (1995) cumulants, given in column 2 of Table 7.1, for a model with swarming, multiple births and nonexponential birth intervals. For here the ratios $\kappa_{10}/\sqrt{\kappa_{20}} = 1.21$ and (especially) $\kappa_{01}/\sqrt{\kappa_{02}} = 0.83$ are sufficiently low to force substantial skewness into the p.d.f. Moreover, when $\kappa_{40} = \cdots = \kappa_{04} = 0$, $\partial K/\partial\theta_1$ and $\partial K/\partial\theta_2$ are minimized at $\theta_1 = \theta_2 = 0$, with values $\kappa_{10} = 7.64$ and $\kappa_{01} = 3.29$, respectively. So the third-order truncated saddlepoint approximation exists only for $x \geq 8, y \geq 4$, and we therefore have to select large fourth-order cumulants in order to ensure existence over all $x, y \geq 0$.

Table 7.1 shows that although the best fit estimates lie in the same ball-park as the target cumulants, the fitting procedure fails to produce anything like the same degree of accuracy as in the previous example. For strong leakage out of the positive quadrant has resulted in an extremely high level of cumulant sensitivity; i.e. change

Table 7.1: Comparison of target cumulants $\{\kappa_{ij}\}$ and best fit least squares cumulants $\{\hat{\kappa}_{ij}\}$ (over $0 \leq x \leq 200, 0 \leq y \leq 120$), together with the working cumulants $\{\tilde{\kappa}_{ij}\}$, for the cumulant structure of Matis et al. (1995).

ij	κ_{ij}(target)	$\hat{\kappa}_{ij}$(best fit)	$\tilde{\kappa}_{ij}$(working)
10	7.64	8.11	5.20
01	3.29	3.97	1.75
20	40.17	39.03	50.94
11	15.32	18.07	31.55
02	15.91	18.46	20.11
30	388.9	388.86	357.69
21	148.3	150.87	166.93
12	107.2	103.89	131.09
03	138.9	138.08	133.30
40		4409.40	7715.35
31		1199.63	12.96
22		665.13	−352.59
13		629.29	35.27
04		1214.95	2993.33

to one cumulant can severely affect the others. Moreover, this also badly affects the rate of convergence of the algorithm. So not only does this example provide a tough test for the truncation procedure, but it is by no means certain that an appropriate p.d.f. can be generated which possesses sufficiently precise first-, second- and third-order cumulants. The problem is that we have implicitly taken fifth- and higher-order cumulants to be zero, even though the true (but unknown) values may be extremely large. So it does not automatically follow that manipulation of the first- to fourth-order cumulants will enable this potentially huge discrepancy to be counterbalanced. This dilemma is central to all moment closure problems.

Attempts to fit a scaled Normal p.d.f. (i.e. with $\tilde{\kappa}_{30} = \cdots = \tilde{\kappa}_{04} = 0$) over $x, y \geq 0$ failed completely due to the low value of $\kappa_{01}/\sqrt{\kappa_{02}} = 0.83$. For this forces a substantial part of the resulting p.d.f. to lie outwith the quadrant $x, y \geq 0$, whence too little structure remains to accommodate all five first- and second-order moments. Even allowing $\tilde{\kappa}_{30}, \ldots, \tilde{\kappa}_{04}$ to be free-ranging results in convergence difficulties, with the means becoming trapped away from their target values. Numerical investigation shows that the general problem of applying the truncated saddlepoint approach when either $\kappa_{10}/\sqrt{\kappa_{10}}$ or $\kappa_{01}/\sqrt{\kappa_{02}}$ is small is non-trivial, and alternative search strategies, probably based on efficient hill-climbing procedures, need to be developed. Moreover, some degree of computational sophistication may well be required, since expressions (7.3.25)–(7.3.30) impose strong functional relations between the cumulants.

7.3.4 A spatial example

We have seen through example (7.3.31) that provided the means are not too close to the axes, the saddlepoint approach provides a powerful way of constructing p.d.f.'s which possess a desired cumulant structure. What we shall now show is that it also provides an excellent way of deriving (approximate) probability solutions to otherwise intractable stochastic processes.

One of the earliest such models to receive extensive investigation (see Renshaw, 1986) was the two-colony birth–death–immigration-migration process (discussed in some detail later on in Section 9.2). Although the non-spatial birth–death–immigration process is relatively simple to solve (Section 2.5), this two-colony version is analytically intractable to direct solution and so provides a good test-bed for the saddlepoint approach. Whilst fairly crude approximate probability expressions can be derived (see Renshaw, 1973a), exact closed form theoretical solutions cannot be obtained. So if the saddlepoint approach yields a reasonably concise algebraic form for the p.d.f., then we will have made a substantial advance. Individuals in colony $i = 1, 2$ are assumed to give birth at rate λ_i, die at rate μ_i and migrate to colony $j \neq i$ at rate ν_i, and new individuals arrive at i at rate α_i. So this process is equivalent to the switch process of Section 7.1, except that we now allow births and also interpret the change of an individual from 'type' i to j as being a physical migration from 'colony' i to j for $i \neq j = 1, 2$. This construct generates a spatial process, which, being a wide-ranging and important field of research in its own right, will be analysed later in Chapters 9

348 Modelling bivariate processes

and 10. For the moment, however, we shall restrict our attention to the simple two-colony process.

Including the birth rates λ_i in the forward equation (7.1.4) generates the corresponding equation

$$\frac{\partial K}{\partial t} = [\lambda_1(e^{\theta_1} - 1) + \mu_1(e^{-\theta_1} - 1) + \nu_1(e^{\theta_2 - \theta_1} - 1)]\frac{\partial K}{\partial \theta_1} + \alpha_1(e^{\theta_1} - 1)$$

$$+ [\lambda_2(e^{\theta_2} - 1) + \mu_2(e^{-\theta_2} - 1) + \nu_2(e^{\theta_1 - \theta_2} - 1)]\frac{\partial K}{\partial \theta_2} + \alpha_2(e^{\theta_2} - 1) \quad (7.3.32)$$

for the c.g.f. $K(\theta_1, \theta_2; t) \equiv G(e^{\theta_1}, e^{\theta_2}; t)$. However, whereas equation (7.1.4) may be solved relatively easily, equation (7.3.32) is intractable to direct solution even though it represents one of the simplest types of bivariate stochastic scenarios. This suggests that multiplicative multivariate stochastic population processes (see Renshaw, 1986, for examples) will not generally have solutions capable of being expressed in closed form. This, of course, poses monumental problems in ecology and biology, since most processes of practical relevance will involve birth in one form or another, and so are likely to be intractable to a full stochastic analysis. This difficulty explains the need for developing a suite of approximation techniques.

Even if numerical, rather than algebraic, solutions are required, all may not be plain sailing. In principle, numerical solutions for the population size probabilities $\{p_{ij}(t)\}$ $(i, j = 0, 1, \ldots; t \geq 0)$ can be derived by writing the Kolmogorov forward difference-differential equations in the discrete form

$$[p_{ij}(t + dt) - p_{ij}(t)]/dt = (\alpha_1 + (i - 1)\lambda_1)p_{i-1,j}(t) + (\alpha_2 + (j - 1)\lambda_2)p_{i,j-1}(t)$$
$$+ (i + 1)\mu_1 p_{i+1,j}(t) + (j + 1)\mu_2 p_{i,j+1}(t)$$
$$+ (i + 1)\nu_1 p_{i+1,j-1}(t) + (j + 1)\nu_2 p_{i-1,j+1}(t)$$
$$- [\alpha_1 + \alpha_2 + i(\lambda_1 + \mu_1 + \nu_1) + j(\lambda_2 + \mu_2 + \nu_2)]p_{ij}(t) \quad (7.3.33)$$

for $t = dt, 2dt, \ldots$ and appropriately small dt. However, as i, j cover all non-negative integers we have to bound the solution over a finite range $i = 0, \ldots, M; j = 0, \ldots, N$, and M and N may need to be so large to obtain the required level of precision that they result in massive compute time and storage demands. So computational feasibility is by no means guaranteed. Moreover, to retain a set of closed equations we need to change the process by modifying it at the boundaries, here by prohibiting birth, migration and immigration into $i = M$ or $j = N$. If all we require are the equilibrium probabilities $\pi_{ij} = p_{ij}(\infty)$ then we may speed up convergence by taking dt to be as large as possible subject to the $p_{ij}(t)$ remaining positive. Whilst to obtain values for $p_{ij}(t)$ we must choose dt to balance precision against compute time.

In both cases moments can be obtained directly from the calculated probabilities. Though since here the process is linear (unlike (7.2.9)) these can be calculated directly. For differentiating (7.3.32) to form $\partial^{r+s} K/\partial \theta_1^r \partial \theta_2^s$, and then placing $\theta_1 = \theta_2 = 0$, yields a set of first-order linear differential equations for the cumulants κ_{rs} whose solution does not necessitate moment closure. For example, on writing $\xi_i = \lambda_i - \mu_i - \nu_i$, the equilibrium cumulants up to third-order (which are also the central moments) take the form

$$\kappa_{10} = (\alpha_2\nu_2 - \alpha_1\xi_2)/(\xi_1\xi_2 - \nu_1\nu_2)$$
$$\kappa_{11} = [\alpha_1\nu_1(\lambda_2\nu_2\xi_1 - \lambda_1\xi_2^2) + \alpha_2\nu_2(\lambda_1\nu_1\xi_2 - \lambda_2\xi_1^2)]/(\xi_1 + \xi_2)(\xi_1\xi_2 - \nu_1\nu_2)^2$$
$$\kappa_{20} = -(\nu_2/\xi_1)\kappa_{11} + (\xi_1 - \lambda_1)(\alpha_2\nu_2 - \alpha_1\xi_2)/\xi_1(\xi_1\xi_2 - \nu_1\nu_2), \qquad (7.3.34)$$

together with the equations

$$3\xi_1\kappa_{30} + 3\nu_2\kappa_{21} = -3(\lambda_1 + \mu_1 + \nu_1)\kappa_{20} - 3\nu_2\kappa_{11} - \xi_1\kappa_{10} - \nu_2\kappa_{01} - \alpha_1$$
$$\nu_1\kappa_{30} + (2\xi_1 + \xi_2)\kappa_{21} + 2\nu_2\kappa_{12} = 2\nu_1\kappa_{20} - (\lambda_1 + \mu_1 + \nu_1 - 2\nu_2)\kappa_{11} \qquad (7.3.35)$$
$$- \nu_2\kappa_{02} - \nu_1\kappa_{10} + \nu_2\kappa_{01};$$

expressions for κ_{01}, κ_{02}, κ_{12} and κ_{03} are symmetric versions of the above. Solving for $\kappa_{30},\ldots,\kappa_{03}$ and placing these expressions, together with those from (7.3.34), into the third-order truncated equilibrium c.g.f. $K_3(\theta_1, \theta_2; \infty) = \sum_{r,s=1}^{3} \kappa_{rs}\theta_1^r\theta_2^s/(r!s!)$ then yields the required probability surface. However, care must be taken not to use this procedure on automatic pilot, since a numerical study (Renshaw, 2000) shows that third-order truncation does not yield saddlepoint values over the full range of i, $j = 0, 1, \ldots$. To achieve that we need fourth-order truncation, whence the corresponding saddlepoint approximation, $\tilde{f}(x, y; \infty)$, is probably best obtained through the use of a symbolic algebra package such as *Mathematica*.

7.4 Counting processes

Although the presence of birth (i.e. multiplicative events) greatly complicates theoretical analysis in all but the simplest scenarios, in many situations it is still possible to gain a considerable understanding of an underlying process by excluding it. In Section 7.1, for example, we were able to develop analytic solutions for the probability and moment structures of a change-type process simply by allowing immigration and death. Such models are of great importance because they facilitate the study of interesting generic processes without causing any loss of mathematical tractability. One area of application that has been particularly successful is quantum optics, where population models have been routinely used to describe the quantum nature of electromagnetic radiation (Srinivasan, 1988). An especially interesting problem concerns the stochastic evolution of populations of photons within optical cavities. This field has been studied since the inception of the laser (Shimoda, Takahasi and Townes, 1957), and understanding the counting statistics of photo-electron pulses registered by detectors such as photomultiplier tubes has been essential for the interpretation of experimental measurements (Saleh, 1978). In particular, Jakeman and Shepherd (1984) and Shepherd (1984) monitor the cavity population of interest via the counting statistics of emigration, modelled as a simple death process. Properties of the number of emigrants leaving the population in a fixed time interval correspond precisely to the experimentally measured photon-counting statistics, and so provide an indirect measure of the evolution of the cavity population. This approach not only provides additional insight into the quantum formulation of the problem, but it also enables

350 Modelling bivariate processes

the interchange of models and techniques with those in the field of classical population statistics. In this context, Jakeman, Phayre and Renshaw (1995) draw attention to the unusual properties of an exactly solvable population model which is generic to an area of quantum optics involving 'non-classical light'.

For many years following the invention of the laser, it was generally accepted that an adequate representation of photodetection was provided by the doubly stochastic Poisson process. In this classical situation the probability of registering c pulses during a time interval of fixed length t (see Mandel, 1959; Cox and Lewis, 1966) is given by

$$p_c(t) = \int_0^\infty \frac{I^c e^{-I}}{c!} f(I)\, dI, \qquad (7.4.1)$$

where I is the instantaneous light intensity integrated over the interval $(0,t)$ and $f(I)$ is its probability density function. It follows from (7.4.1) that the variance to mean ratio, ρ, of c (called the Fano factor) must always be greater than or equal to that of the Poisson distribution, for which $\rho = 1$. Light with counting statistics which can be represented in this way, such as coherent (laser) light, which is Poisson distributed, or thermal light, which satisfies a geometric distribution, is termed *classical* light. However, it is easy to construct discrete models for the incident photon flux which cannot be derived through the representation (7.4.1). Whilst sub-Poissonian models ($\rho < 1$) clearly cannot be so represented, neither can a wide range of super-Poissonian models ($\rho > 1$). Such light is now termed *non-classical*.

Intense activity in the development of experimental methods for generating non-classical light has produced a range of techniques which provide overwhelming evidence for its existence. Since most applications require consideration of both wave and particle properties, theoretical treatments have generally avoided a classical population statistics approach. One of the few exceptions is the Jakeman, Phayre and Renshaw (1995) investigation, which involves the simultaneous emission of pairs of photons at rate α_2 and the subsequent death of each individual single photon at rate μ. This was itself stimulated by the burgeoning area of non-classical light in the late 1980s (Louden and Knight, 1987). Although the equilibrium distribution is super-Poissonian, 'odd-even' effects (explained below) ensure the breakdown of (7.4.1), so that the resulting light is non-classical. Not only does the simplicity of this model mean that it is analytically amenable, but it reflects one of the earliest mechanisms used to produce non-classical light through parametric down-conversion in a nonlinear crystal (Burnham and Weinberg, 1970), and so is realizable experimentally.

7.4.1 Paired-immigration–death process

Such odd-even effects can be easily developed by constructing, and then solving, the Kolmogorov forward equation for the probability that the population is of size $X = n$ at time t. For on replacing single-immigration by paired-immigration in equation (2.5.3) (with $\lambda = 0$), we see that

$$dp_n(t)/dt = \mu(n+1)p_{n+1}(t) - (\alpha_2 + \mu n)p_n(t) + \alpha_2 p_{n-2}(t) \qquad (n = 0, 1, 2, \ldots)$$
$$(7.4.2)$$

for $p_{-1}(t) = p_{-2}(t) = 0$. This equation yields a tractable expression for the associated p.g.f. $G(z;t) \equiv \sum_{n=0}^{\infty} z^n p_n(t)$, namely

$$\frac{\partial G(z;t)}{\partial t} + \mu(z-1)\frac{\partial G(z;t)}{\partial t} = \alpha_2(z^2-1)G(z;t), \quad (7.4.3)$$

from which the probabilities, $p_n(t)$, and their corresponding rth-order moments, may be extracted. For (7.4.3) is almost identical to (2.5.4) for the simple immigration–death process; the only difference being that $\alpha(z-1)G(z;t)$ is replaced by $\alpha_2(z^2-1)G(z;t)$. Whence on assuming a zero initial population size, a parallel analysis via the corresponding auxiliary equations yields

$$G(z;t) = \exp\{(\alpha_2/2\mu)[(z^2+2z) - (e^{-\mu t}(z-1)+1)(e^{-\mu t}(z-1)+3)]\}. \quad (7.4.4)$$

The structure of this solution is revealed by denoting

$$Q_k(z;\lambda) \equiv \sum_{n=0}^{\infty} z^{kn} \frac{\lambda^n e^{-\lambda}}{n!} = \exp\{\lambda(z^k-1)\}. \quad (7.4.5)$$

For we then observe from (7.4.4) that

$$G(z;t) = Q_2[z; \alpha_2(1-e^{-2\mu t})/2\mu]Q_1[z; \alpha_2(1-e^{-\mu t})^2/\mu]. \quad (7.4.6)$$

Thus the paired immigration–death process is a product of two Poisson p.g.f.'s, one being over the even integers $n = 0, 2, 4, \ldots$ and the other over the non-negative integers $n = 0, 1, 2, \ldots$. It is the relative strength of the former over the latter that determines the intensity of the odd-even effect.

This phenomenon is probably best appreciated through the equilibrium p.g.f.

$$G(z;\infty) = \exp\{(\alpha_2/2\mu)(z^2+2z-3)\}. \quad (7.4.7)$$

First recall that the generating function for the Hermite polynomials $H_n(x)$ is given by

$$\sum_{n=0}^{\infty} H_n(x) z^n/n! \equiv \exp(2xz - z^2) \quad (7.4.8)$$

(Abramowitz and Stegun 1970, §22.9.17), and that Hermite polynomials are related to the Laguerre polynomials

$$L_n^\phi(x) \equiv \sum_{m=0}^{n} (-1)^m \binom{n+\phi}{n-m} \frac{x^m}{m!} \quad (\phi > -1) \quad (7.4.9)$$

(see, for example, Erdélyi, 1953) through

$$H_{2n}(x) \equiv (-1)^n 2^{2n} n! L_n^{-1/2}(x^2) \quad \text{and} \quad H_{2n+1}(x) \equiv (-1)^n 2^{2n+1} n! x L_n^{1/2}(x^2) \quad (7.4.10)$$

(Abramowitz and Stegun 1970, §22.5.40/41). Then on denoting the mean population size by $\theta = 2\alpha_2/\mu$, direct comparison of (7.4.7)–(7.4.10) shows that the rth-order factorial moments

$$\mu_{(r)} \equiv \mathrm{E}[N(N-1)\ldots(N-r+1)] = \left.\frac{\partial^r G(z;\infty)}{\partial z^r}\right|_{z=1} \quad (7.4.11)$$

take the form

$$\mu_{(2r)}(\infty) = r!\theta^r L_r^{-1/2}(-\theta)$$
$$\mu_{(2r+1)}(\infty) = r!\theta^{r+1} L_r^{1/2}(-\theta). \quad (7.4.12)$$

Since these expressions are virtually identical in r except for the switch from $L_r^{-1/2}(\cdot)$ to $L_r^{1/2}(\cdot)$, they result in a saw-tooth (i.e. up-and-down) effect as r increases. This phenomenom is also manifest in the equilibrium probabilities

$$p_n(\infty) = \left.\frac{\partial^r G(z;\infty)}{\partial z^r}\right|_{z=0}, \quad (7.4.13)$$

with

$$p_{2n}(\infty) = \theta^n \exp(-3\theta/4) L_n^{-1/2}(-\theta/4) n!/(2n!)$$
$$p_{2n+1}(\infty) = \theta^{n+1} \exp(-3\theta/4) L_n^{1/2}(-\theta/4) n!/[2(2n+1)!]. \quad (7.4.14)$$

It is this odd-even effect which causes the generation of non-classical light.

This approach not only readily yields the time-dependent moments $\mu_{(r)}(t)$ and probabilities $p_n(t)$, but it also easily extends to cover both single- and paired-immigration with rates α_1 and α_2 (Gillespie and Renshaw, 2008). For example, extracting the coefficients of z^{2n} and z^{2n+1} in $G(z;t)$ yields for $k = 0, 1$

$$p_{2n+k}(t) = \frac{n! 2^{2n}}{(2n+k)!} \left(\frac{\alpha_2(1-e^{-2\mu t})}{2\mu}\right)^n \exp\left(\frac{(2\alpha_1 + \alpha_2(3-e^{-\mu t}))(e^{-\mu t}-1)}{2\mu}\right)$$
$$\times L_n^{k-1/2}\left(\frac{(\alpha_1 + \alpha_2(1-e^{-\mu t}))^2(1-e^{-\mu t})^2}{2\alpha_2\mu(e^{-2\mu t}-1)}\right) f_1(k), \quad (7.4.15)$$

where $f_1(0) = 1$ and $f_1(1) = (\alpha_1 + \alpha_2(1-e^{-\mu t}))(1-e^{-\mu t})/\mu$. Whence comparison with (7.4.14) not only shows that the odd-even effect has been preserved in spite of the addition of single immigrants, but also that it exists for all time $t \geq 0$ and not just in equilibrium. Factorial moments also show a simple portrayal of the odd-even effect. For forming the higher derivatives $G^{(2r)}(1;t)$ and $G^{(2r+1)}(1;t)$ gives, for $k = 0, 1$,

$$\mu_{(2r+k)}(t) = r!\left(\frac{2\alpha_2(1-e^{-\mu t})}{\mu}\right)^r L_r^{k-1/2}\left(\frac{\mu\mu_{(1)}^2(t)}{2\alpha_2(e^{-2\mu t}-1)}\right) f_2(k), \quad (7.4.16)$$

where $f_2(0) = 1$ and $f_2(1) = \mu_1(t)$. So once again, moving between states $2r$ and $2r+1$ essentially just involves a switch from $L_r^{-1/2}(\cdot)$ to $L_r^{1/2}(\cdot)$. As regards second-order moments, the variance

$$\mathrm{Var}[N(t)] = \{2\alpha_2(1-e^{-2\mu t}) + [\alpha_1 + \alpha_2(1-e^{-\mu t})](1-e^{-\mu t})\}/\mu \quad (7.4.17)$$

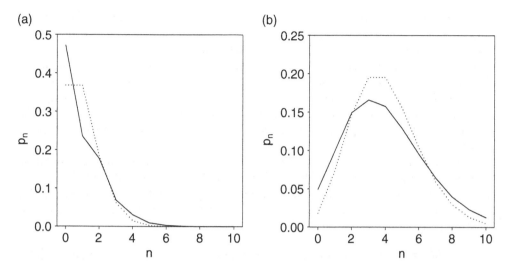

Figure 7.2 Equilibrium probabilities $p_n(\infty)$ for the paired-immigration–death process (—) and a Poisson process with the same mean (· · ·): (a) $\alpha_2 = 1$, $\mu = 2$ (mean 1); (b) $\alpha_2 = 4$, $\mu = 2$ (mean 4).

is linear in α_1 and α_2. Whilst the autocorrelation (e.g. Chatfield, 1991) between the population sizes $N(t)$ at time t and $N(t+s)$ at the later time $t+s$ is given by

$$\rho(t,s) = \frac{e^{-\mu s}(1 - e^{-\mu t})[\alpha_1 + \alpha_2(3 + e^{-\mu t})]/\mu}{\sqrt{\text{Var}[N(t)]\text{Var}[N(t+s)]}}. \qquad (7.4.18)$$

Note that in equilibrium $\rho(\infty, s) = e^{-\mu s}$, which (perhaps surprisingly) does not depend on either α_1 or α_2.

Figure 7.2 compares the equilibrium probabilities (7.4.14) for the paired-immigration–death process (i.e. $\alpha_1 = 0$) with $\mu = 2$ and $\alpha_2 = 1$ and 4, to Poisson probabilities having the same mean (1 and 4). Though the odd-even effect is clearly noticeable when the mean $\theta = 2\alpha_2/\mu \simeq 1$, it is not visually obvious for values of θ much greater than unity. However, the variance of the equilibrium solution is strongly super-Poissonian for all values of θ. For we see from (7.4.17) that the variance-mean ratio

$$\text{Var}[N(\infty)]/\theta = (3\alpha_2/\mu)/(2\alpha_2/\mu) = 3/2 > 1,$$

irrespective of the values of α_2 and μ. Thus when viewed in an optical physics context this process is clearly non-classical.

7.4.2 Single-paired-immigration–death counting process

Now suppose that the population is monitored purely *externally* by counting the number of individuals emigrating during a fixed time interval $(0, t)$, with emigration being modelled as a pure death process with rate η. Note that here there is a fine distinction between death and emigration. For as well as μ denoting individual death rate and η the individual 'escape' rate from the system (with all escapees being subsequently counted), we could also have an individual death rate of $\mu+\eta$ and 'inefficient counting', with each individual who dies having probability $\eta/(\mu+\eta)$ of having their death recorded. The raison d'être which underlies this approach is of fundamental significance, since it provides a means of answering a key general question. Namely, "If a stochastic process is developing within a hidden system, with the only information being provided by the event times of escaping individuals, can the properties of the hidden process be inferred purely from knowledge of these counting statistics?"

Let $p_{nc}(t)$ denote the joint probability that there are n individuals in the population at time t and that c individuals have been counted in the time interval $(0,t)$. Also, to inject a little more generality, we continue to assume that immigrants can arrive either singly at rate α_1 or in pairs at rate α_2. Then the univariate Kolmogorov forward equation (7.4.2) takes the bivariate form

$$dp_{nc}(t)/dt = \mu(n+1)p_{n+1,c}(t) + \eta(n+1)p_{n+1,c-1}(t) + \alpha_1 p_{n-1,c}(t) + \alpha_2 p_{n-2,c}(t)$$
$$- (\mu n + \eta n + \alpha_1 + \alpha_2)p_{nc}(t), \quad (7.4.19)$$

and this leads to an eminently tractable solution for the associated p.g.f. We can then use this to derive joint, conditional and marginal probability distributions together with the corresponding moment and correlation properties. The result is that a surprising amount of information on the development of the population size $n(t)$ can be gleaned from knowledge of the counting variable $c(t)$. This is of immense benefit across a wide range of fields. Important examples include: quantum optics, where externally based radiation information enables the elucidation of the quantum nature of the underlying electromagnetic radiation process; and, the study of epidemic disease, where we might only be able to record those individuals who show symptoms of the disease, are isolated from the general population, or who die.

The bivariate equations (7.4.19) can be solved by employing the same standard procedures as before, namely either by taking the Laplace transform with respect to t and then solving the resulting differential equation in s and inverting the solution, or else by solving the Lagrange equations. Taking the latter approach, on applying the generating function

$$R(z,s;t) \equiv \sum_{n,c=0}^{\infty} z^n s^c p_{nc}(t) \quad (7.4.20)$$

to equations (7.4.19) we have

$$\frac{\partial R}{\partial t} = [\mu(1-z) + \eta(s-z)]\frac{\partial R}{\partial z} - \alpha_1(1-z)R - \alpha_2(1-z^2)R, \quad (7.4.21)$$

which gives rise to the associated auxiliary equations

$$\frac{dt}{-1} = \frac{dz}{\mu(1-z) + \eta(s-z)} = \frac{dR}{[\alpha_1(1-z) + \alpha_2(1-z^2)]R}. \quad (7.4.22)$$

Now this population counting process is fundamentally different from the pure population processes we have considered previously. For since counts can never decrease, the marginal counting probabilities $p_{.c}(t)$ cannot tend to an equilibrium distribution as $t \to \infty$. Two options are therefore open to us. First, we can take $n(0) = 0$ (say), and hence $R(z, s; 0) = 1$, and thereby obtain a transient solution in both n and c. Note, though, that as t increases, the marginal population size probabilities $p_{n.}(t)$ still approach the equilibrium values $p_n(\infty)$ given by (7.4.14). Second, we can assume that the population size is already in equilibrium at $t = 0$, whence $c(t)$ now tells us how counts are distributed with time for a population process that has *already reached* equilibrium (Jakeman, Phayre and Renshaw, 1995). So in the case of $\alpha_1 = 0$, the initial condition $R(z, s; 0)$ takes the form (7.4.7).

On taking the first approach (i.e. $n = c = 0$ at $t = 0$), and denoting $\delta = \mu + \eta$ and $\xi = \delta z - \eta s - \mu$, we obtain the solution

$$R(z,s;t) = \exp\left\{\frac{\alpha_1}{\delta^2}(\xi(1-e^{-\delta t}) + \delta\eta t(s-1))\right\} \times \exp\left\{\frac{\alpha_2}{2\delta^3}(2\delta t(\eta s + \mu)^2 - 2\delta^3 t)\right\}$$

$$\times \exp\left\{\frac{\alpha_2}{2\delta^3}(\xi^2(1-e^{-2\delta t}) + 4\xi(1-e^{-\delta t})(\eta s + \mu))\right\}. \qquad (7.4.23)$$

Although it may look rather opaque, since the exponents of (7.4.23) are linear in α_1 and α_2 this expression for $R(z, s; t)$ is simply the product of a pure singleton and a pure pairs p.g.f.

The counting probabilities, $p_{.c}(t)$, can now be determined by setting $z = 1$ in (7.4.23) and extracting the coefficient of s^c. This leads for $k = 0, 1$ to

$$p_{.2c+k}(t) = \frac{c!}{(2c+k)!}\left[\frac{2\alpha_2\eta^2}{\delta^3}[2\delta t - (1-e^{-\delta t})(3-e^{-\delta t})]\right]^c \exp\left\{\frac{\alpha_1\eta}{\delta^2}[1 - \delta t - e^{-\delta t}]\right\}$$

$$\times L_c^{k-1/2}\left(-\frac{\{\alpha_1\delta[\delta t + e^{-\delta t} - 1] + \alpha_2[2(\eta-\mu)(1-e^{-\delta t}) - \eta(1-e^{-2\delta t}) + 2\delta\mu t]\}^2}{2\delta^3\alpha_2[(1-e^{-2\delta t}) - 4(1-e^{-\delta t}) + 2\delta t]}\right)$$

$$\times \exp\left\{\frac{\alpha_2}{2\delta^3}[\eta(4\mu(1-e^{-\delta t}) + \eta(1-e^{-2\delta t})) + 2\delta t(\mu^2 - \delta^2)]\right\} f_3(k) \qquad (7.4.24)$$

where $f_3(0) = 1$ and

$$f_3(1) = \frac{\alpha_1\eta}{\delta^2}(\delta t + e^{-\delta t} - 1) + \frac{\alpha_2\eta}{\delta^3}[2(\eta-\mu)(1-e^{-\delta t}) - \eta(1-e^{-2\delta t}) + 2\delta\mu t].$$

So in exactly the same way as we have seen for $p_{n.}(t)$, the counting probabilities $p_{.2c}(t)$ and $p_{.2c+1}(t)$ essentially differ only by the powers of the Laguerre polynomial, namely $L_c^{-1/2}(\cdot)$ and $L_c^{1/2}(\cdot)$.

Furthermore, the factorial moments $\{\mu_{(r)}^c(t)\}$ of the counting distribution may also be obtained directly, this time by successively differentiating $R(z, s; t)$ to form $\partial R^{2r+k}(1, s; t)/\partial s^{2r+k}$ at $s = 1$ for $k = 0, 1$. This procedure yields

$$\mu_{(2r+k)}^c(t) = r!2^{2r}\left[\frac{\alpha_2\eta^2}{2\delta^3}[2\delta t - (1-e^{-\delta t})(3-e^{-\delta t})]\right]^r$$

$$\times L_r^{k-1/2}\left(\frac{(\alpha_1 + 2\alpha_2)^2(\delta t + e^{-\delta t} - 1)^2}{2\alpha_2\delta[(1-e^{-\delta t})(3-e^{-\delta t}) - 2\delta t]}\right) f_4(k) \qquad (7.4.25)$$

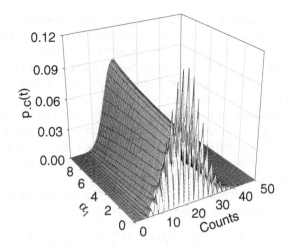

Figure 7.3 Plot of the counting probabilities $p_{.c}(t)$ with $\alpha_1 + 2\alpha_2 = 10$ for varying α_1, with $\mu = 0$, large η and fixed time $t = 2$.

where $f_4(0) = 1$ and $f_4(1) = (\eta/\delta^2)(\alpha_1 + 2\alpha_2)(\delta t + e^{-\delta t} - 1)$. So here the counting moments $\mu^c_{(2r)}(t)$ and $\mu^c_{(2r+1)}(t)$ essentially differ only in the powers of $L_r^{-1/2}(\cdot)$ and $L_r^{1/2}(\cdot)$.

To appreciate the nature of the odd-even effect it is useful to take $\alpha_1 = 0$ and $\alpha_2 \gg \delta = \mu + \nu$. For then immigrants arrive in pairs and so $p_{2n+1.}(t) \equiv 0$ and $p_{2n.}(t) = (\alpha_2 t)^n e^{-\alpha_2 t}/n!$ until either the first death or emigration occurs. However, as soon this happens the 'even-Poisson' structure degrades. Similar degradation occurs following the first single immigrant arrival if we allow $\alpha_1 > 0$. For example, suppose that $\mu = 0$ and ν is extremely large. Then as soon as immigrants arrive they will be counted virtually instantaneously. To illustrate this degradation suppose we take $t = 2$ with a total immigration rate of 10, i.e. $\alpha_1 + 2\alpha_2 = 10$. Figure 7.3 shows how changing the balance between α_1 and α_2 affects the counting probabilities $p_{.c}(2)(t)$, and it is clear that the extreme saw-tooth shape of the distribution is only (visually) present when $\alpha_1 \simeq 0$. This can be explained by placing $s = 0$, $z = 1$, $\alpha_2 = 0$ and $\mu = 0$ in (7.4.23) in order to obtain

$$\tilde{p}_0(t) = \Pr(\text{no counts from single immigrants by time } t)$$
$$= \exp\{(\alpha_1/\eta)(1 - e^{-\eta t} - \eta t)\}. \quad (7.4.26)$$

For when $\alpha_1/\eta \simeq 0$, (7.4.26) simplifies to $\tilde{p}_0(t) \simeq e^{-\alpha_1 t}$, and since this swiftly decays to zero with increasing $\alpha_1 t$ the intensity of the odd-even effect must decay with it. Indeed, the time required to reduce $\tilde{p}_0(t)$ to (say) 0.5, namely $\ln(2)/\alpha_1$, reduces inversely with α_1.

Figure 7.4 illustrates the parallel temporal degradation, showing how the counting probabilities $p_{.c}(t)$ change through time, where now $\alpha_1 = 0.075$ is taken to be both fixed and substantially smaller than $\alpha_2 = 1.2$. Whilst a strong odd-even effect is clearly

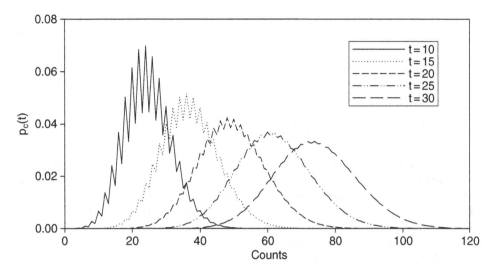

Figure 7.4 Plot of the counting probabilities $p._c(t)$ with $\alpha_1 = 0.075$, $\alpha_2 = 1.2$ and varying times $t = 10, 15, \ldots, 30$, with $\mu = 0$ and large η.

present at times $t = 10$ and $t = 15$, as t increases further the saw-tooth shape becomes much less pronounced. Quantitatively, we see from (7.4.26) that the probability of zero single immigrant counts reduces from $\tilde{p}_0(10) = 0.472$ to $\tilde{p}_0(30) = 0.105$, resulting in a much reduced odd-even effect.

7.4.3 Batch-immigration–death counting process

We have observed that when single and paired immigrants enter the population then not only is there an odd-even effect exhibited by both the population and counting probabilities, but also that this phenomenon becomes increasingly dominated by single immigrants as α_1 increases relative to α_2. So what happens in the totally general case when k-tuple immigration occurs at rate α_k ($k = 1, 2, \ldots; \sum_k k\alpha_k < \infty$)? Though this scenario leads to interesting mathematics in its own right, here we are primarily interested in exhibiting the type of behaviour that we can expect to observe in the general k-tuple case.

Ignoring counts for the moment, paralleling the single-paired immigration equations (7.4.19) leads to the Kolmogorov forward equation for the population size probabilities $p_{n.}(t)$ (now denoted by $p_n(t)$ for convenience), namely

$$dp_n(t)/dt = \mu(n+1)p_{n+1}(t) + \alpha \sum_{k=1}^{\infty} q_k p_{n-k}(t) - (\mu n + \alpha)p_n(t), \qquad (7.4.27)$$

where $p_{n-k}(t) \equiv 0$ for all $k > n$. Whence on multiplying equations (7.4.27) by z^n and summing over n in the usual manner, we see that the equation for $G(z;t)$ now takes the form

$$\frac{\partial G}{\partial t} = \mu(1-z)\frac{\partial G}{\partial z} + \alpha G \sum_{k=1}^{\infty} q_k(z^k - 1). \qquad (7.4.28)$$

This can be solved either by using Lagrange's technique, or by taking Laplace transforms, and yields

$$G(z;t) = \exp\left\{\frac{\alpha}{\mu}\sum_{k=1}^{\infty} q_k \sum_{j=1}^{k}\left[\frac{z^j}{j} - \frac{(1+(z-1)e^{-\mu t})^j}{j}\right]\right\} \qquad (7.4.29)$$

for an initial population of size zero at time $t = 0$. Not only may moments and probabilities be derived directly from this solution, but Matthews, Hopcraft and Jakeman (2003) show that by making a judicious choice for the mass immigration p.g.f. it is possible to tailor the process to take specific distributional forms.

For example, when $\alpha q_3 = \alpha_3$ and $q_k = 0$ otherwise, we have a pure triple-immigration–death process, and extracting the coefficient of z^n in (7.4.29) leads (after a little algebra) to the solution

$$p_n(t) = e^{\xi_0} \sum_{r=0}^{[n/3]} \frac{\xi_3^r(-\xi_2)^{(n-r)/2}}{r!(n-r)!} H_{n-r}(\xi_1/2\sqrt{-\xi_2}). \qquad (7.4.30)$$

Here the Hermite polynomials $H_n(x)$ are defined through the generating function (7.4.8), whilst $\xi_0 = -\xi_3 - \xi_2 - \xi_1$ where

$$\xi_3 = \frac{\alpha_3}{3\mu}(1 - e^{-3\mu t}), \quad \xi_2 = \frac{\alpha_3}{2\mu}(1 - e^{-\mu t})^2(1 + 2e^{-\mu t}) \quad \text{and} \quad \xi_1 = \frac{\alpha_3}{\mu}(1 - e^{-\mu t})^3. \qquad (7.4.31)$$

It is the presence of $[n/3]$ (i.e. the integer part of $n/3$) in the upper limit of the summation in (7.4.30) that induces the triple effect. Moreover, on forming the c.g.f. $\ln[G(e^\theta;t)]$ and extracting the coefficient of $\theta^i/i!$, it follows from (7.4.29) that the ith cumulant is given by

$$\kappa_i(t) = \xi_1 + 2^i\xi_2 + 3^i\xi_3. \qquad (7.4.32)$$

Note that this cumulant structure exhibits the characteristic that $\kappa_{i+1}(t)/\kappa_i(t) \to 3$ as $i \to \infty$, in contrast to the value 1 for the Poisson process generated by the simple immigration–death process. In general, for batches of k immigrants this ratio tends to k, so we can select a required level of "super-Poissonness" by choosing an appropriate value for k. A corresponding analysis based on the marginal counting p.g.f. yields parallel results for the counting probabilities.

Although we have so far just discussed single-, paired- and triple-immigration scenarios, in general k-tuple behaviour can exhibit similar manifestations in both the associated probability and moment structures. However, since we have already shown in Section 2.1.6 that an infinite number of probability distributions can give rise to the same set of moments, there is no reason to presuppose that such similarity will hold universally. Let us therefore now consider a radically different example in which we exploit the classic phenomenon constructed by Schoenberg (1983).

Schoenberg-Poisson immigration

First recall from (2.1.57) that not only does the Poisson distribution $p_n(t) = e^{-2}2^m/m!$ defined over $n = 2^m$ for $m = 0, 1, 2, \ldots$ generate the rth raw moment $\mu'_r = \exp\{2(2^r - 1)\}$ (see expression (2.1.58)), but so does the infinitely large family of distributions $p^\epsilon_{2^m} = e^{-2}2^m(1 + \epsilon a_m)/m!$ (see expression (2.1.65)) for $-1 \leq \epsilon \leq 1$. Here a_m is defined through (2.1.63), so with $q = 2$

$$a_m = m![(1-2)(1-2^2)\ldots(1-2^m)]^{-1}. \tag{7.4.33}$$

Suppose we now insert this into our stochastic population model by taking the immigration rate $q_i = (e^{-2}2^k/k!)(1 + \epsilon a_k)$ for $i = 2^k$ over $k = 0, 1, 2, \ldots$, and $q_i = 0$ otherwise. Then the forward Kolmogorov equation (7.4.27) takes the form

$$dp_n(t)/dt = \mu(n+1)p_{n+1}(t) + \alpha \sum_{k=0}^{\infty} \frac{e^{-2}2^k}{k!}(1 + \epsilon a_k)p_{n-2^k}(t) - (\mu n + \alpha)p_n(t), \tag{7.4.34}$$

which leads (using standard procedures) to the generating function solution

$$G(z;t) = \exp\{\frac{\alpha}{\mu}\sum_{k=0}^{\infty}\frac{e^{-2}2^k}{k!}(1+\epsilon a_k)\sum_{i=1}^{2^k}\frac{z^i - (1+(z-1)e^{-\mu t})^i}{i}\}. \tag{7.4.35}$$

Expanding (7.4.35) in powers of z^n then yields the population size probabilities $\{p_n(t)\}$; readers interested in the details of this algebraic manipulation should examine the derivation of the solution for general $\{q_k\}$ presented in Gillespie and Renshaw (2005).

To show that the associated moments are independent of ϵ, we first substitute for $z = 1 + z'$ in (7.4.35) to form

$$G(1+z';t) = \exp\{\frac{\alpha}{\mu}\sum_{k=0}^{\infty}\frac{e^{-2}2^k}{k!}(1+\epsilon a_k)\sum_{i=1}^{2^k}\frac{1}{i}\sum_{j=0}^{i}\binom{i}{j}(1-e^{-j\mu t})(z')^j\}. \tag{7.4.36}$$

Whence collecting terms in $(z')^m$ within the exponent gives

$$G(1+z';t) = \exp\left(\sum_{m=1}^{\infty}\varphi_m(z')^m\right), \tag{7.4.37}$$

where

$$\varphi_m = \frac{\alpha}{\mu}\sum_{k=0}^{\infty}\frac{e^{-2}2^k}{k!}(1+\epsilon a_k)\sum_{i=1}^{2^k}\frac{1}{i}\binom{i}{m}(1-e^{-m\mu t})$$

$$= \frac{\alpha(1-e^{-m\mu t})}{m\mu}\sum_{k=0}^{\infty}\frac{e^{-2}2^k}{k!}(1+\epsilon a_k)\binom{2^k}{m}. \tag{7.4.38}$$

360 *Modelling bivariate processes*

Now it may be shown (after a little algebra) that for $r = 0, 1, 2, \ldots$

$$\sum_{k=0}^{\infty} \frac{e^{-2} 2^k}{k!} a_k 2^{rk} \equiv 0. \tag{7.4.39}$$

So as $\binom{2^k}{m}$ is a polynomial in 2^k it follows that

$$\varphi_m = \frac{\alpha(1 - e^{-m\mu t})}{m\mu} \sum_{k=0}^{\infty} \frac{e^{-2} 2^k}{k!} \binom{2^k}{m}, \tag{7.4.40}$$

which, being independent of ϵ, proves that the factorial moments of the process (7.4.34) are independent of ϵ. Thus our system of Schoenberg–Poisson immigration–death processes generates an infinity of different probability distributions all of which possess the same moment structure. In particular, they share a common mean and variance, namely

$$\kappa_1(t) = (\alpha e^2/\mu)(1 - e^{-\mu t}) \quad \text{and} \quad \kappa_2(t) = (\alpha e^2/2\mu)[(1 - e^{-\mu t})^2 + e^4(1 - e^{-2\mu t})]. \tag{7.4.41}$$

Figure 7.5a depicts three equilibrium probability distributions, $\{\pi_n^\epsilon\}$, for the two extreme values $\epsilon = -1$ and 1, together with the centre value $\epsilon = 0$, and shows how π_n^{-1} and π_n^1 both oscillate around, and bound, π_n^0. Whilst a simulation (Figure 7.5b) of the process for $\epsilon = 0$ highlights the strong population surges induced by a sudden influx of fresh immigrants (here of size > 60 at times $t \simeq 10$ and 65) which are immediately followed by their swift decline. It is the existence of these surges that causes the variance, $\kappa_2(\infty) = e^2(1 + e^4)/2$, to be large in comparison to the overall mean, $\kappa_1(\infty) = e^2$.

Derivation of the associated counting probability structures follows a similar route. For suppose that individuals emigrate from the system at rate η, and denote $\delta = \mu + \eta$. Then in exactly the same way that we previously constructed the forward single-paired population-count equation (7.4.19), we can write down the corresponding bivariate Schoenberg equation

$$\frac{dp_{nc}(t)}{dt} = \eta(n+1)p_{n+1,c-1}(t) + \mu(n+1)p_{n+1,c}(t)$$

$$+ \alpha \sum_{k=0}^{\infty} \frac{e^{-2} 2^k}{k!}(1 + \epsilon a_k)p_{n-2^k,c}(t) - (\delta n + \alpha)p_{nc}(t), \tag{7.4.42}$$

which yields the solution

$$R(z, s; t) = \exp\{\alpha t e^{-2} \sum_{k=0}^{\infty} \frac{2^k}{k!} \frac{(\eta s + \mu)^{2^k}}{\delta^{2^k}}(1 + \epsilon a_k) - \alpha t\} \tag{7.4.43}$$

$$\times \exp\{\sum_{k=0}^{\infty} \frac{\alpha e^{-2} 2^k}{k! \delta^{2^k + 1}} \sum_{r=1}^{2^k} \frac{1}{r}\binom{2^k}{2^k - r}(\eta s + \mu)^{2^k - r}(\delta z - \eta s - \mu)^r (1 - e^{-r\delta t})\}$$

(a)

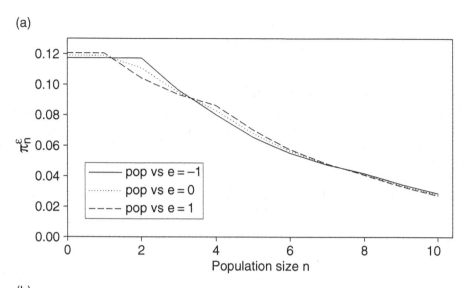

(b)

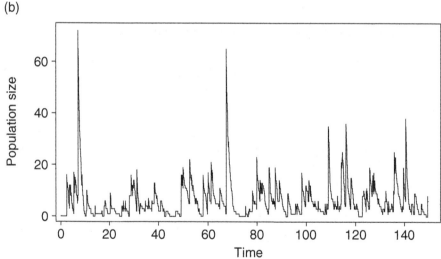

Figure 7.5 (a) Equilibrium probability distributions of the Schoenberg–Poisson immigration–death process for $\alpha = 1$, $\mu = 1$ and $\epsilon = -1$ (—), 0 ($\cdots$) and 1 (– –) (the corresponding moments of these distributions are identical); (b) a stochastic simulation for $\epsilon = 0$.

for the associated p.g.f. Whence on placing $z = 1$ we obtain the marginal counting p.g.f., and on further substituting for $s = 1 + s'$ the factorial p.g.f. Employing a similar approach to that used on the population generating function (7.4.36) then shows that the counting moments are also independent of ϵ. In particular, the mean and variance are given by

$$\kappa_1^c(t) = (\alpha\eta e^2/\delta^2)(\delta t + e^{-\delta t} - 1), \tag{7.4.44}$$

$$\kappa_2^c(t) = (\alpha\eta e^2/\delta^3)[\delta^2 t - (1 - e^{-\delta t})(\eta e^4 + \mu) + \eta(e^4 - 1)(\delta t - (1 - e^{-\delta t})^2/2)]. \tag{7.4.45}$$

General batch immigration (with $\eta = 0$)

Given that we have already developed the general equilibrium population size p.g.f. in Section 2.7.2, let us take this opportunity to develop the (non-counting) theory a little further. A less opaque representation of result (7.4.29) can be constructed by defining

$$\zeta_i(j) = \frac{1}{i} - \left(\frac{e^{-\mu t}}{1 - e^{-\mu t}}\right)^i \sum_{k=1}^{j} \binom{k}{i} \frac{(1 - e^{-\mu t})^k}{k} \qquad (j > 0) \tag{7.4.46}$$

with

$$\zeta_0(j) = -\sum_{i=1}^{j} \zeta_i(j). \tag{7.4.47}$$

For we can then write the generating function (7.4.29) in the form

$$G(z;t) = \exp\left(\sum_{i=0}^{\infty} \xi_i z^i\right) \tag{7.4.48}$$

where

$$\xi_0 = -\sum_{j=1}^{\infty} \xi_j \quad \text{and} \quad \xi_i = \frac{\alpha}{\mu} \sum_{j=i}^{\infty} q_j \zeta_i(j) \quad (i > 0). \tag{7.4.49}$$

On denoting

$$Q_i(\lambda) \equiv \sum_{n=0}^{\infty} \frac{\lambda^n e^{-\lambda}}{n!} z^{in} = \exp\{\lambda(z^i - 1)\}, \tag{7.4.50}$$

it then follows that $G(z;t)$ has the pure product representation

$$G(z;t) = Q_1(\xi_1) Q_2(\xi_2) \ldots Q_i(\xi_i) \ldots. \tag{7.4.51}$$

Taking logarithms and placing $z = e^\theta$ in the p.g.f. (7.4.51) leads to the associated c.g.f.

$$K(\theta;t) \equiv \sum_{i=1}^{\infty} \frac{\kappa_i \theta^i}{i!} = \sum_{i=1}^{\infty} \frac{\theta^i}{i!} \sum_{k=1}^{\infty} k^i \xi_k. \tag{7.4.52}$$

This is a particularly simple and transparent form which yields the ith cumulant

$$\kappa_i(t) = \xi_1 + 2^i \xi_2 + \cdots + k^i \xi_k + \cdots. \tag{7.4.53}$$

In particular, the mean and variance are

$$\kappa_1(t) = \frac{\alpha}{\mu}(1 - e^{-\mu t}) \sum_{j=1}^{\infty} j q_j \qquad (7.4.54)$$

$$\kappa_2(t) = \frac{\alpha}{2\mu} \sum_{j=1}^{\infty} j[j(1 - e^{-2\mu t}) + (1 - e^{-\mu t})^2] q_j. \qquad (7.4.55)$$

Moreover, letting $t \to \infty$ in (7.4.54) and (7.4.55) gives

$$\kappa_1(\infty) = \frac{\alpha}{\mu} \sum_{j=1}^{\infty} j q_j \quad \text{and} \quad \kappa_2(\infty) = \frac{\alpha}{\mu} \sum_{j=1}^{\infty} j(j+1) q_j. \qquad (7.4.56)$$

So the equilibrium mean and variance exist provided $\sum_{j=1}^{\infty} j q_j < \infty$ and $\sum_{j=1}^{\infty} j(j+1) q_j < \infty$, respectively. In general, the existence of the rth cumulant $\kappa_r(\infty)$ requires $\sum_{j=1}^{\infty} j^r q_j < \infty$. This implies that by making a judicious choice for the $\{q_j\}$ we can generate processes for which only the first r limiting cumulants exist. For example, letting $\{q_j\}$ involve the Riemann zeta function through

$$q_j = 1/[j^{r+2} \sum_{k=1}^{\infty} \frac{1}{k^{r+2}}] \qquad (7.4.57)$$

results in the first r cumulants being finite and the remainder infinite. For a full and general discussion of results for mass immigration (and annihilation) processes relating to existence, uniqueness, p.g.f., resolvent and equilibrium issues see Chen and Renshaw (1990, 1993a&b, 1995, 1997, 2000, 2004) and Renshaw and Chen (1997).

To find the population size probabilities $p_n(t)$ we extract the coefficient of z^n in the generating function (7.4.29), and this calculation is facilitated by noting that

$$\exp\left(\sum_{i=0}^{\infty} \xi_i z^i\right) = e^{\xi_0} \times \sum_{i_1=0}^{\infty} \xi_1 \frac{z^{i_1}}{i_1!} \times \sum_{i_2=0}^{\infty} \xi_2 \frac{z^{2i_2}}{i_2!} \times \cdots \times \sum_{i_k=0}^{\infty} \xi_k \frac{z^{k i_k}}{i_k!} \times \cdots.$$

For we see that $p_0 = e^{\xi_0}$, whilst for $n > 0$

$$p_n(t) = \sum_{j_n=0}^{[m_n/n]} \frac{\xi_n^{j_n}}{j_n!} \sum_{j_{n-1}=0}^{[m_{n-1}/(n-1)]} \frac{\xi_{n-1}^{j_{n-1}}}{j_{n-1}!} \cdots \sum_{j_3=0}^{[m_3/3]} \frac{\xi_3^{j_3}}{j_3!} \sum_{j_2=0}^{[m_2/2]} \frac{\xi_2^{j_2}}{j_2!} \frac{\xi_1^{m_1}}{m_1!} e^{\xi_0}, \qquad (7.4.58)$$

where $[x]$ denotes the integer part of x, $m_i = 0$ for $i > n$, $m_n = n$ and $m_i = m_{i+1} - (i+1)j_{i+1}$ for $i = 1, 2, \ldots, n-1$. Although we shall not reproduce the results here in order to avoid combinatorial overload, parallel expressions can also be derived for the associated factorial moments and counting probabilities (for details see Gillespie and Renshaw, 2005).

The representation (7.4.58) is especially useful for showing that the odd-even effect previously demonstrated for pair-immigration through the probabilities (7.4.15) for $p_{2n}(t)$ and $p_{2n+1}(t)$ extends naturally to the k-batch immigration–death process (i.e. $q_i = 0$ for $i \neq k$). For considering the basic $k = 2$ case first, we see that solution (7.4.58) reduces to

$$p_n(t) = \sum_{j_2=0}^{[m_2/2]} \frac{\xi_2^{j_2} \xi_1^{m_1}}{j_2! \, m_1!} e^{\xi_0}. \tag{7.4.59}$$

Hence as the only terms in (7.4.59) generating a non-zero contribution are those with $j_3 = \cdots = j_n = 0$, and since $m_i = m_{i+1} - (i+1)j_{i+1}$, it follows that $m_2 = m_3 = \cdots = m_n = n$. Moreover, as $m_1 = m_2 - 2j_2 = n - 2j_2$, we have, on replacing n by $2n$ and $2n+1$, respectively, that

$$p_{2n}(t) = \sum_{j_2=0}^{n} \frac{\xi_2^{j_2} \xi_1^{2n-2j_2}}{j_2! \, (2n-2j_2)!} e^{\xi_0} \tag{7.4.60}$$

and

$$p_{2n+1}(t) = \sum_{j_2=0}^{n} \frac{\xi_2^{j_2} \xi_1^{2n-2j_2}}{j_2! \, (2n-2j_2)!} e^{\xi_0} \times \left(\frac{\xi_1}{2n+1-2j_2}\right), \tag{7.4.61}$$

from which the structure of the odd-even effect is clearly apparent. Similarly, on taking $k=3$ and $n = 3s + r$ over $s = 0, 1, 2, \ldots$ and $r = 0, 1, 2$, we have

$$p_{3s+r}(t) = \sum_{j_3=0}^{s} \frac{\xi_3^{j_3}}{j_3!} \sum_{j_2=0}^{[(3(s-j_3)+r)/2]} \frac{\xi_2^{j_2} \xi_1^{m_1}}{j_2! \, m_1!} e^{\xi_0} \tag{7.4.62}$$

where $m_1 = 3(s - j_3) + r - 2j_2$. So for given s the dominant first summation remains unchanged over $r = 0, 1, \ldots, k-1$; only the second summation is affected. Likewise, for general k and $n = ks + r$ over $r = 0, 1, \ldots, k-1$, for given s the dominant summation $\sum_{j_k=0}^{s} \xi_k^{j_k}/j_k!$ remains unchanged as r sweeps through $0, 1, \ldots, k-1$, thereby giving rise to k-tuple saw-tooth behaviour centred around this driving term (Gillespie and Renshaw, 2005). Parallel results can be developed for both the corresponding moments and the counting measures.

7.4.4 Summary and further developments

The above analyses are based on the fact that although complex systems often have the potential for exhibiting extremely rich dynamic behaviour, gaining a *direct* understanding of such behavioural structure may not be possible if the system remains hidden. As the multiple immigration–death process is a basic stochastic process, in the sense that no new individuals are produced within the system through multiplicative events (i.e. birth), it is sufficiently mathematically tractable to enable us to show that a surprisingly deep level of information on the hidden population process can be inferred *indirectly* from the associated counting process. The question therefore arises as to whether the counting process can still provide a high level of information content on the development of the unobservable population process when more complex models are considered?

Now a simple and transparent way of modelling multiplicative events is to employ the linear birth–death (BD) process of Section 2.4, since its inherent mathematical tractability suggests that the associated counting probabilities may not be too opaque to prevent the extraction of process structure. Moreover, on including immigration (Section 2.5), the resulting birth–immigration–death (BDI) process not only provides an immediate extension to the above immigration–death counting model, but it is also directly associated with a kind of behaviour commonly encountered in optics (Jakeman, Hopcraft and Matthews, 2003). This paradigm between stochastic population models and quantum optics is an intriguing one, and highlights the potentially large benefits to be gained by linking together what appear to be totally disparate processes. For when the birth and immigration rates are equal, thermal or Bose–Einstein statistics are predicted through the geometric form of the p.d.f.

$$p_n(t) = (\mu - \lambda)(\mu - \lambda e^{(\lambda-\mu)t})^2[\lambda(1 - e^{-(\lambda-\mu)t})]^n \qquad (7.4.63)$$

obtained by placing $\alpha = \lambda$ in solution (2.5.15). This structural feature is found to characterize the photon statistics of a laser operating below its threshold (Shimoda, Takahasi and Townes, 1957) and also Gaussian speckle noise generated when laser light is scattered by particles or rough surfaces (Bertolotti, 1974). In the case of the laser model, spontaneous and stimulated emissions within a population of photons in a cavity are analogous to immigration and birth, respectively, and absorption is the analogue of death. Now we have already shown in Section 2.7.2 that in equilibrium the mass immigration–death (MID) process (2.7.16) results in a negative binomial distribution which is identical to that of the BDI process. So in population process terms, the only way of differentiating between the two models is to compare properties of individual realizations. It was the appreciation of this key identifiability problem, initially highlighted through study of population processes, that led Jakeman, Hopcraft and Matthews (2003) to develop the physics paradigm by devising an optimal strategy for distinguishing between the BDI and MID scenarios when measurement is restricted to the external monitoring scene.

Here we shall just concentrate on the simpler BD process, since the analysis for the BDI model involves more tedious (though straightforward) algebraic development, and so adds little to our overall understanding. On replacing paired-immigration at rate α_2 by birth at rate λ, we see that for $n, c = 0, 1, 2, \ldots$ the forward equation (7.4.19) takes the form

$$dp_{nc}(t)/dt = \lambda(n-1)p_{n-1,c}(t) + \mu(n+1)p_{n+1,c}(t) + \eta(n+1)p_{n+1,c-1}(t)$$
$$- n(\lambda + \mu + \eta)p_{nc}(t), \qquad (7.4.64)$$

where $p_{n,-1}(t) \equiv 0$. Whence on paralleling the construction of solution (7.4.23) for the paired-immigration–death process, it follows that the joint p.g.f. (7.4.20) now takes the form

$$R(z, s; t) = \left(\frac{\theta_2 z e^{\lambda(\theta_1-\theta_2)t} - \theta_1 \theta_2 e^{\lambda(\theta_1-\theta_2)t} - \theta_1 z + \theta_1 \theta_2}{z e^{\lambda(\theta_1-\theta_2)t} - \theta_1 e^{\lambda(\theta_1-\theta_2)t} - z + \theta_2} \right)^{n_0}, \qquad (7.4.65)$$

where

$$\theta_1, \theta_2 = [\lambda + \eta + \mu \mp \sqrt{(\lambda + \eta + \mu)^2 - 4\lambda(\mu + \eta s)}]/(2\lambda). \qquad (7.4.66)$$

Note that setting $s = 1$ in (7.4.65) recovers the generating function (2.4.11) for the simple birth-death process.

Since we have already studied the population features of the BD process in Section 2.4, here we shall restrict our analysis to the determination of counting measures. Now inspection of $R(z, 1; t)$ shows that although the population dummy variable z appears in a simple product of binomial and negative binomial p.g.f.'s, the marginal counting p.g.f. $R(1, s; t)$ involves a much more complicated structure. However, although extraction of the counting probabilities $p_{.c}(t)$ is algebraically messy, and involves rather awkward functions of modified Bessel Functions, forming $[\partial R(1, s; t)/\partial s]_{s=1}$ easily leads to the counting mean

$$\kappa_1^c(t) = \begin{cases} \eta n_0 (e^{(\lambda-\mu-\eta)t} - 1)/(\lambda - \mu - \eta) & \text{for } \lambda \neq \mu + \eta \\ \eta n_0 t & \text{for } \lambda = \mu + \eta. \end{cases} \quad (7.4.67)$$

Whence allowing $t \to \infty$ shows that

$$\kappa_1^c(\infty) = \begin{cases} n_0 \eta/(\mu + \eta - \lambda) & \text{for } \lambda < \mu + \eta \\ \infty & \text{for } \lambda \geq \mu + \eta. \end{cases} \quad (7.4.68)$$

For when the population explodes, i.e. $\lambda \geq \mu + \eta$, the counts explode. Conversely, when the population is certain to become extinct, i.e. $\lambda < \mu + \eta$, the counting mean approaches the limit $n_0 \eta/(\mu + \eta - \lambda)$.

Similarly, forming $[\partial^2 Q(1, s; t)/\partial s^2]_{s=1}$ leads to the counting variance

$$\kappa_2^c(t) = n_0 \eta \{ e^{(\lambda-\eta-\mu)t} [(\lambda - \mu)^2 - \eta^2 - 4\lambda \eta t (\lambda - \eta - \mu)]$$
$$+ e^{2(\lambda-\eta-\mu)t} \eta(\lambda + \eta + \mu) - (\lambda - \mu)^2$$
$$- \eta(\lambda + \mu)\}/(\lambda - \eta - \mu)^3 \quad \text{for } \lambda \neq \eta + \mu$$
$$\kappa_2^c(t) = n_0 \eta t [3(1 - t\eta) + 2t^2 \lambda \eta]/3 \quad \text{for } \lambda = \eta + \mu. \quad (7.4.69)$$

Whence for large t it follows that

$$\kappa_2^c(t) \simeq n_0 \eta^2 e^{2(\lambda-\eta-\mu)t} (\lambda + \eta + \mu)/(\lambda - \eta - \mu)^3 \quad \text{for } \lambda > \eta + \mu$$
$$\kappa_2^c(t) \simeq 2n_0 t^3 \lambda \eta^2 / 3 \quad \text{for } \lambda = \eta + \mu$$
$$\kappa_2^c(t) = n_0 \eta [(\lambda - \mu)^2 + \eta(\lambda + \mu)]/(-\lambda + \eta + \mu)^3 \quad \text{for } \lambda < \eta + \mu. \quad (7.4.70)$$

Thus unlike the population variance which tends to zero as $t \to \infty$ for $\lambda < \mu + \eta$, the counting variance approaches a finite limit. For ultimate extinction is certain to occur (see Section 2.4.2), at which point there are no longer any individuals available for counting. Indeed, placing $z = s = 1$ in (7.4.65) shows that for $n_0 = 1$ the probability of extinction

$$R(1, 1; \infty) = \sum_{c=0}^{\infty} p_{.c}(\infty) = \begin{cases} 1 & \text{for } \lambda \leq \mu + \eta \\ (\eta + \mu)/\lambda & \text{for } \lambda > \mu + \eta. \end{cases} \quad (7.4.71)$$

Finally, for $\lambda < \mu + \eta$, expression (7.4.65) simplifies dramatically when we form the marginal equilibrium counting p.g.f., namely $R(1, s; \infty) = \theta_1^{n_0}$, from which it is easy to obtain the equilibrium counting probabilities (for $n_0 = 1$)

$$p_0(\infty) = \frac{\lambda + \eta - \mu - \sqrt{(\lambda + \eta + \mu)^2 - 4\lambda\mu}}{\lambda - \eta - \mu - \sqrt{(\lambda + \eta + \mu)^2 - 4\lambda\mu}}$$

$$p_c(\infty) = \left(\frac{-\sqrt{(\lambda + \eta + \mu)^2 - 4\lambda\mu}}{2\lambda}\right)\binom{1/2}{c}\left(\frac{-4\lambda\eta}{(\lambda + \eta + \mu)^2 - 4\lambda\mu}\right)^c \quad (c = 1, 2, \ldots).$$

(7.4.72)

7.5 Applying MCMC to hidden event times

Although we have been able to make substantial analytic progress in Section 7.4 in terms of constructing moment and probability structures for counting distributions, the intrinsically simple batch-immigration–death and birth–death processes considered still give rise to algebraically awkward expressions. So if we wish to use these counting measures in order to make statistical inferences on the underlying (unobserved) population process then we have real problems. For computation of the associated counting likelihood is plagued by the necessity to integrate the joint likelihood with respect to the missing data. Moreover, relatively little work has been undertaken on developing least squares moment procedures even though the approach appears to be potentially tractable (see Särkkä and Renshaw, 2006). Theoretical advances may be made for the simple immigration–death scenario, but even slight extensions to this structure yield likelihood integrals that are too complicated to be analytically tractable. For example, we have seen in Section 7.4.2 that merely replacing single immigrants by paired immigration yields non-trivial counting probabilities expressed in terms of Laguerre polynomials. Construction of the associated likelihood therefore requires a far more tangential approach (detailed in Gibson and Renshaw, 2001b) based on labelling each death as being the first, or the second, of an immigrant pair to die. However, introducing multiplicative events effectively renders even this strategy intractable to analytic solution. For construction of the likelihood not only becomes markedly worse when we include birth (witness the relative complexity of the birth–death–counting p.g.f. (7.4.65), but it becomes acutely more so when we extend our analyses to the logistic and power-law logistic population processes developed in Chapter 3. So we need to abandon a purely algebraic approach, and turn, instead, towards constructing a computational procedure which has the capability of operating across a wide spectrum of different processes and degrees of complexity.

7.5.1 Introducing Markov chain Monte Carlo

In a comprehensive review of contributions to statistical theory and general methodology from 1901 to 2000 Davison (2001) emphasizes the vital role played by simulation during the last century. It has been used to guide theoretical work ever since "Student's" (1908) derivation of the t-statistic, and following the appearance of the celebrated algorithm of Metropolis, Rosenbluth, Teller and Teller (1953) (later generalized by Hastings, 1970) simulation became a standard statistical procedure in fields such as statistical mechanics and physical chemistry. Though we had to wait until the 1970s and 1980s for its importance to be fully appreciated by statisticians tackling problems in spatial statistics and image analysis. The possibilities that the Metropolis–Hastings algorithm offers for investigating posterior distributions and likelihood functions were initially realized around 1990. The latter has led to a revolution in the advancement of modern day statistics through the development and application of Markov chain Monte Carlo (MCMC) methods. Smith and Roberts (1993) and Tierney (1994) provide an excellent starting point for readers who are interested in investigating this huge and rapidly expanding field.

MCMC techniques not only play a central role in Bayesian statistics, but they offer a powerful means of fitting models and estimating parameters in our missing data scenarios. In essence, the approach operates as follows. Suppose we wish to fit a stochastic population model, parameterized by a set A, to an incomplete set of observations X; for example, we might have known count times but unknown immigration and death times. Let Z denote the event space for the missing data components, and for $\mathbf{x} \in X$ and $\mathbf{z} \in Z$ let $f(\mathbf{x}, \mathbf{z}|\mathbf{a})$ denote the probability density of complete outcomes from the stochastic model for parameters $\mathbf{a} \in A$. Then given the observation $\mathbf{x} \in X$, and using $\pi(\cdot)$ to denote posterior densities, we can express the posterior parameter density through the relation

$$\pi(\mathbf{a}) = \int_{A \times Z} \pi(\mathbf{a}|\mathbf{z}')\pi(\mathbf{a}', \mathbf{z}')\, d\mathbf{a}'\, d\mathbf{z}'. \qquad (7.5.1)$$

Now although the absolute value of the density $\pi(\mathbf{a}, \mathbf{z}|\mathbf{x})$ is unknown, it is proportional to $\pi(\mathbf{a})f(\mathbf{x}|\mathbf{a})$, which enables a Markov chain to be constructed with equilibrium distribution $\pi(\mathbf{x}, \mathbf{z}|\mathbf{a})$. This means that the integral (7.5.1) can be estimated by generating a sequence of samples $\{(\mathbf{a}_i, \mathbf{z}_i)\}$ from this Markov chain and then forming (for example) a histogram.

The technique is of general applicability, and is especially important when considering fundamentally nonlinear processes involving infection, predation, competition, etc. which we shall soon be meeting in Chapter 8. For in such situations we often have information only on individuals who have died or show signs of infection, and so all other information relating to times of immigration, birth, infection, etc. have to be regarded as missing. Gibson (1997a,b), for example, uses it to fit space–time models for the spread of a virus in a citrus orchard, thereby eliciting evidence which suggests the simultaneous presence of both primary and secondary infection processes. Here individuals are of two types, namely either susceptible to infection or infected, and Gibson and Renshaw (1998) extend this process to include exposed and removed states; only birth and death times are recorded, with the times of all other transitions being regarded as missing. They achieve this by using the powerful idea of reversible jump MCMC (Green, 1995), which is a major methodological advance specifically tailored to enable the analysis of processes in which the chain explores a state space of components of varying dimension. This construction is a vital concept when considering population processes, since the dimension of the system increases and decreases in line with the changing population size. The ability to use MCMC methods to infer properties of stochastic population processes from partial information on numbers of emigrants is particularly striking. Indeed, it is surprising how much knowledge can be gleaned from processes for which relatively little information is recorded. O'Neill and Roberts (1997), for example, use MCMC to estimate parameters in an epidemic model for smallpox from data in which only the times of removals from the process are known; no data are available on infection times. These examples suggest that the methods used might well provide a general technique for inferring process structure for other stochastic processes where observations are necessarily incomplete. Renshaw and Gibson (1998) therefore investigate this question for a range of stochastic processes of varying complexity: for it cannot be assumed that the success achieved in the above analyses automatically extends to models of arbitrary complexity, since the mixing properties of the chains will become poorer as the complexity of the state space increases. Their results show that the approach can be successfully applied to stochastic models with practical relevance well beyond that for simple epidemic models. Here we shall demonstrate how the technique is developed and used by applying it to our single/paired-immigration–death process. Algorithms to analyse more complex processes can be developed by natural extension.

7.5.2 Fitting the simple immigration–death process to incomplete observations

To illustrate the basic procedure, let us revert back to the simple immigration–death model developed in Section 2.5.1. Thus for the time being we shall disregard batch immigration. Suppose we start observing the process at time $w_0 = 0$, and record m deaths at times $w_1, \ldots, w_m$ together with n_i immigrants between w_{i-1} and w_i at times $t_{i,1}, \ldots, t_{i,n_i}$. Then the set of values

$$\mathbf{y} = \{w_i; t_{i,1}, \ldots, t_{i,n_i} | i = 1, \ldots, m\} \tag{7.5.2}$$

fully describes the observed realization. It is straightforward to derive the likelihood density, $\phi(\mathbf{y}|\alpha, \mu)$, for the observations $\mathbf{y}$ given the parameters α and μ. For $\mathbf{t} = (t_1, \ldots, t_n)$, let $L(r, n; t_0, \mathbf{t}, w)$ denote the likelihood density that, starting with r individuals at t_0, n immigrants arrive at $t_1, \ldots, t_n$ followed by a death at w. On appealing to the Markov property we may multiply the probabilities of successive events and inter-event times to obtain

$$L(r, n; t_0, \mathbf{t}, w) = \alpha e^{-(\alpha + r\mu)(t_1 - t_0)} \times \cdots \times \alpha e^{-(\alpha + (r+n-1)\mu)(t_n - t_{n-1})}$$
$$\times (r + n)\mu e^{-(\alpha + (r+n)\mu)(w - t_n)}$$
$$= (r + n)\mu \alpha^n e^{-\alpha(w - t_0)} \exp\{\mu[rt_0 + (t_1 + \cdots + t_n) - (r+n)w]\}. \tag{7.5.3}$$

Multiplying terms of this form corresponding to each of the m deaths then yields

$$\phi(\mathbf{y}|\alpha,\mu) \equiv \prod_{i=1}^{m} L(n_1 + \cdots + n_{i-1} - (i-1), n_i; w_{i-1}, t_{i,1}, \ldots, t_{i,n_i}, w_i)$$

$$= \{\prod_{i=1}^{m}(n_1 + \cdots + n_i - (i-1))\}\mu^m \alpha^{\sum_{i=1}^{m} n_i} e^{-\alpha w_m} \quad (7.5.4)$$

$$\times \exp\{\mu \sum_{i=1}^{m}(t_{i,1} + \cdots + t_{i,n_i})\} \exp\{-\mu[\sum_{i=1}^{m} w_i + w_m \sum_{i=1}^{m}(n_i - 1)]\},$$

which, in spite of looking rather messy, is easy to compute.

Following the Renshaw and Gibson (1998) example, suppose we take $\alpha = 1.0$ and $\mu = 0.05$, and assume that the population is empty at $t = 0$. Then result (2.5.7) shows that when $n_0 = 0$ the population size follows a Poisson$[(\alpha/\mu)(1 - e^{-\mu t})]$ distribution with an equilibrium mean value of $\alpha/\mu = 20$. Figure 7.6 shows a realization of this process for the first 100 deaths, and depicts the immigration and death times together with the associated change in population size. Here the transient component has clearly dissipated by $t \simeq 50$; thereafter the population size is in equilibrium, and with high probability remains within three standard deviations, i.e. $(\alpha/\mu) \pm 3\sqrt{(\alpha/\mu)} \simeq 20 \pm 13$, of the equilibrium mean. Figure 7.7 shows the likelihood $\phi(\mathbf{y}|\alpha,\mu)$ for this realization over the parameter space $A = \{(\alpha,\mu)|0.5 \leq \alpha \leq 1.5, 0.02 \leq \mu \leq 0.07\}$, where ϕ has been normalized so that its maximum value over A is unity. Now it is straightforward to deduce from (7.5.4) that $\phi(\mathbf{y}|\alpha,\mu)$ is maximized by

$$\check{\alpha} = (\sum_{i=1}^{m} n_i)/w_m = 0.9892 \quad (7.5.5)$$

$$\check{\mu} = m/[-\sum_{i=1}^{m}(t_{i,1} + \cdots + t_{i,n_i}) + \sum_{i=1}^{m} w_i + w_m \sum_{i=1}^{m}(n_i - 1)] = 0.0445;$$

whilst contour and mesh plots (Figures 7.7a&b) show that this joint likelihood is contained within a steeply rising narrow bell which is virtually elliptical in cross-section. Hence the associated marginal likelihoods (Figures 7.7c&d) take the classic symmetric bell-shaped form. This structure provides a useful touchstone against which we can compare the parameter likelihood obtained using the MCMC procedure when observations are incomplete.

Suppose that instead of knowing the full event-times (7.5.2), all we observe are the death-times $w_1, \ldots, w_m$. Then in the notation of Section 7.5.1, $\mathbf{x}$ represents the observed $\{w_1, \ldots, w_m\}$, whilst $\mathbf{z}$ denotes the missing immigration times $\{t_{i,1}, \ldots, t_{i,n_i} | i = 1, \ldots, n\}$. To repeat the above likelihood approach, we therefore need to integrate the likelihood (7.5.4) over the set of unknown observations $\mathbf{z}$. That is, we compute the integral

$$\phi(\mathbf{x}|\alpha,\mu) = \int_{\mathbf{z} \in Z} \phi(\mathbf{x}, \mathbf{z}|\alpha,\mu)\, d\mathbf{z}. \quad (7.5.6)$$

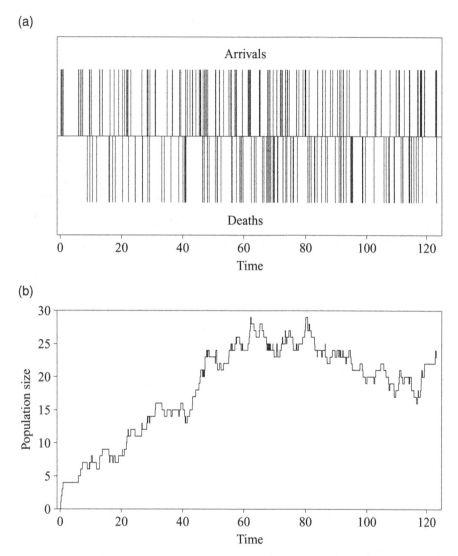

Figure 7.6 Simulated realization over 100 deaths of a simple immigration–death process with $\alpha = 1.0$ and $\mu = 0.05$ showing: (a) event-times, with spikes of $+1$ and -1 corresponding to immigration and death times, respectively; and, (b) population size against time.

However, before we develop this computational technique it is worth noting that the immigration–death process is unusual in that it is sufficiently simple to admit some analytic progress. For let $q_{nr}(w) = \Pr(n$ arrivals and no deaths in $(0, w)|N(0)=r)$, where $N(t)$ denotes the population size at time t. Then on proceeding via the Kolmogorov forward equation (7.4.19) with $\alpha_2 = \eta = 0$ and $p_{r0}(0) = 1$, it can be shown that

372 *Modelling bivariate processes*

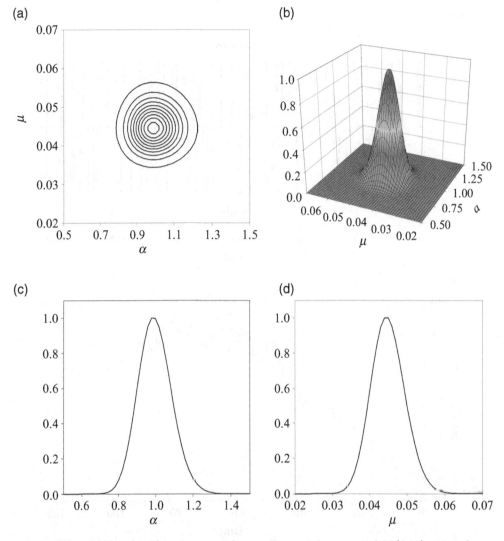

Figure 7.7 Likelihood surface corresponding to Figure 7.6 over $\alpha = 0.50(0.02)1.50$ and $\mu = 0.020(0.001)0.070$ showing: (a) contour levels $0.05(0.15)0.95$ and (b) mesh plot; together with marginal likelihoods for (c) α and (d) μ.

$$q_{nr}(w) = p_{n+r,0}(w) = (\alpha/\mu)^n e^{-(\alpha+r\mu)w}(1 - e^{-\mu w})^n/n!. \tag{7.5.7}$$

Hence the likelihood density that the first death occurs in $(w, w + dw)$ after n arrivals in $(0, w)$ is

$$p_{nr}(w) = q_{nr}(w)(n+r)\mu. \tag{7.5.8}$$

Summing over $n = 0, 1, 2, \ldots$ then gives the density of the first death occurring in $(w, w + dw)$ for $N(0) = r$, namely

$$\sum_{n=0}^{\infty} p_{nr}(w) = \mu e^{-(\alpha + r\mu)w} [(\alpha/\mu)(1 - e^{-\mu w}) + r] \exp\{(\alpha/\mu)(1 - e^{-\mu w})\}. \qquad (7.5.9)$$

However, although extending this analysis to the m-death situation is mathematically tractable, the resulting summations are extremely convoluted, and so any intuitive understanding this theoretical approach might bring is lost. We therefore need to employ an alternative procedure, and for this the MCMC approach is ideally suited.

Suppose we take the prior density of the parameters (α, μ) to be uniform over the set A. Then under this assumption it follows that $\phi(\mathbf{x}|\alpha, \mu)$ is proportional to the posterior density $\pi(\alpha, \mu|\mathbf{x})$, whence we can use the MCMC approach to explore the relative values of $\phi(\mathbf{x}|\alpha, \mu)$ over A. Suppose that after i iterations of the algorithm we have a set of arrival times $\mathbf{z}_i$ and parameter values $\mathbf{a}_i = (\alpha_i, \mu_i)$. Then the $(i+1)$th step comprises the following two parts.

MCMC simulation algorithm (A7.2)

(1) Propose a modification, $\mathbf{z}'$, to $\mathbf{z}_i$, as follows:

    (i) with probability b a new immigration event is proposed with arrival time drawn uniformly from $(0, w_m)$;

    (ii) with probability d a randomly selected immigrant is deleted;

    (iii) with probability $s = 1 - b - d$ a randomly selected arrival time is moved to a new time selected uniformly from $(0, w_m)$. The proposed $\mathbf{z}'$ is then accepted (in which case $\mathbf{z}_{i+1} = \mathbf{z}'$) with probability

$$\min\left\{1, \frac{\phi(\mathbf{x}, \mathbf{z}'|\mathbf{a}_i) \times d \times w_m}{\phi(\mathbf{x}, \mathbf{z}_i|\mathbf{a}_i) \times b \times (N_i + 1)}\right\}, \quad \text{for case (i);} \qquad (7.5.10)$$

$$\min\left\{1, \frac{\phi(\mathbf{x}, \mathbf{z}'|\mathbf{a}_i) \times b \times N_i}{\phi(\mathbf{x}, \mathbf{z}_i|\mathbf{a}_i) \times d \times w_m}\right\}, \quad \text{for case (ii);} \qquad (7.5.11)$$

$$\min\left\{1, \frac{\phi(\mathbf{x}, \mathbf{z}'|\mathbf{a}_i)}{\phi(\mathbf{x}, \mathbf{z}_i|\mathbf{a}_i)}\right\}, \quad \text{for case (iii).} \qquad (7.5.12)$$

Here N_i denotes the number of arrivals in $\mathbf{z}_i$. If the change is rejected then $\mathbf{z}_{i+1} = \mathbf{z}_i$.

(2) A new set of parameters $\mathbf{a}'$ is proposed from a suitable proposal distribution $g(\mathbf{a}', \mathbf{a}_i)$. The new parameter vector is accepted (whence $\mathbf{a}_{i+1} = \mathbf{a}'$) with probability

$$\min\left\{1, \frac{\phi(\mathbf{x}, \mathbf{z}_{i+1}|\mathbf{a}')g(\mathbf{a}_i, \mathbf{a}')}{\phi(\mathbf{x}, \mathbf{z}_{i+1}|\mathbf{a}_i)g(\mathbf{a}', \mathbf{a}_i)}\right\}. \qquad (7.5.13)$$

If the proposed $\mathbf{a}'$ is rejected then $\mathbf{a}_{i+1} = \mathbf{a}_i$.

To start the procedure we need to select both the initial parameter values $\mathbf{a}_0$ and a set of arrival times $\mathbf{z}_0$. We first ensure that the parameter space A is chosen to include 'all likely values' of α and μ, and then select $\mathbf{a}_0$ uniformly from A. Whilst $\mathbf{z}_0$ can be obtained by simulating an arrival time from $U(w_{i-1}, w_i)$ for each $i = 1, \ldots, m$ (where $w_0 = 0$). For this ensures that $\phi(\mathbf{x}, \mathbf{z}_0|\mathbf{a}_0) > 0$, since each death is preceded by an immigration. To select new resampled parameters $(\tilde{\alpha}, \tilde{\mu})$ based on $\mathbf{a}_i = (\alpha_i, \mu_i)$, we may take the proposal distribution $g(\mathbf{a}'|\mathbf{a}_i)$ to be the narrow moving/fixed uniform window

$$\begin{array}{ll} U(\tilde{\alpha} - \delta, \tilde{\alpha} + \delta) \text{ if } \tilde{\alpha} \geq \delta & \text{and } U(0, 2\delta) \text{ otherwise} \\ U(\tilde{\mu} - \epsilon, \tilde{\mu} + \epsilon) \text{ if } \tilde{\mu} \geq \epsilon & \text{and } U(0, 2\epsilon) \text{ otherwise,} \end{array} \quad (7.5.14)$$

for appropriately small δ and ϵ. So if $\tilde{\alpha} < \delta$ or $\tilde{\mu} < \epsilon$ then the new proposed values are selected from the fixed uniform windows $U(0, 2\delta)$ or $U(0, 2\epsilon)$, respectively. Thus with probability 0.5 the next resampled value is taken from the moving uniform window: such overlap is required to prevent the resampled values becoming trapped near zero. The advantage of using this resampling scheme, over one based on a completely static window, is that we do not have to prejudge what the admissible parameter range might be. The uniform distribution is used since we are presuming no prior information on α and μ.

Figure 7.8 shows the change in the resampled values α_i and μ_i, together with the log-likelihood $\ln[\phi(\mathbf{w}, \mathbf{z}_i|\alpha_i, \mu_i)]$ computed via (7.5.4), for $\alpha = 1.0$ and $\mu = 0.05$, based on the simulated realization of 100 deaths shown in Figure 7.6. The MCMC realization was run for 10^6 accepted events (this required 1.30×10^6 proposed events), and every 10th one was recorded. Choosing $b = d = 0.3$ and $s = 0.4$ allows for a roughly equal balance between 'add', 'delete' and 'move' transitions. Whilst taking $\delta = 0.05$ and $\epsilon = 0.0025$ enables good movement across the range of parameter values without incurring a high percentage of rejected proposed values. Two features are immediately apparent. First, all three plots swiftly burn in, i.e. the effect of the initial parameter values $\mathbf{a}_0$ soon disappears. Second, because the proposal distribution (7.5.14) does not have bounded support the process is subject to sudden upwards excursions. Visual comparison of the three series shows that high peaks in $\ln[\phi(\cdot)]$ are coincident with sudden upsurges in α_i and μ_i. For as only the death times are known, a process with considerably different parameter values and arrival times from the given ones may well have a much higher likelihood. The beauty of the reversible jump MCMC approach is that it always pulls such aberrant processes back into the realms of reality. Strictly speaking, however, implementation of the above algorithm with (7.5.14) allowing unrestricted α and μ is not valid, since if an improper uniform prior distribution is used (as here), then the resulting posterior distribution is not proper for this model. To overcome this difficulty we therefore need to impose upper bounds on $\tilde{\alpha}$ and $\tilde{\mu}$ to ensure that the equilibrium density of the Markov chain is indeed proper.

Direct inspection of Figure 7.8 suggests taking the values 1.5 and 0.2 as appropriate upper bounds for $\tilde{\alpha}$ and $\tilde{\mu}$, respectively. To effect this let the window (7.5.14) involve the additional component

$$\begin{array}{ll} \tilde{\alpha} \sim U(1.5 - 2\delta, 1.5) & \text{if } \tilde{\alpha} \geq 1.5 - \delta \\ \tilde{\mu} \sim U(0.2 - 2\epsilon, 0.2) & \text{if } \tilde{\mu} \geq 0.2 - \epsilon. \end{array} \quad (7.5.15)$$

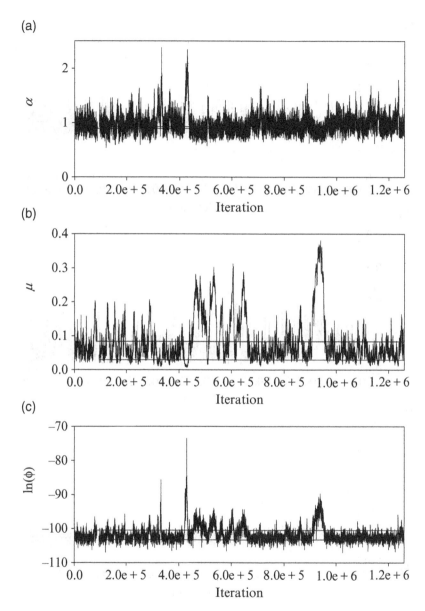

Figure 7.8 Resampled values of (a) $\tilde{\alpha}_i$, (b) $\tilde{\mu}_i$ and (c) $\ln[\phi(\mathbf{w}, \mathbf{z}_i|\alpha_i, \mu_i)]$ against recorded event number for the MCMC procedure based on Figure 7.6 with unbounded parameters.

The resulting stabilization of $\ln[\phi(\cdot)]$ (Figure 7.9c) is clearly apparent, as is the need to retain 10^6 accepted events (1.25×10^6 proposed events) to enable the parameter space to be fully covered (Figures 7.9a&b). This tail truncation has no effect on the modal values of the marginal histograms (Figures 7.10c&d), namely 0.93 for α and

376 *Modelling bivariate processes*

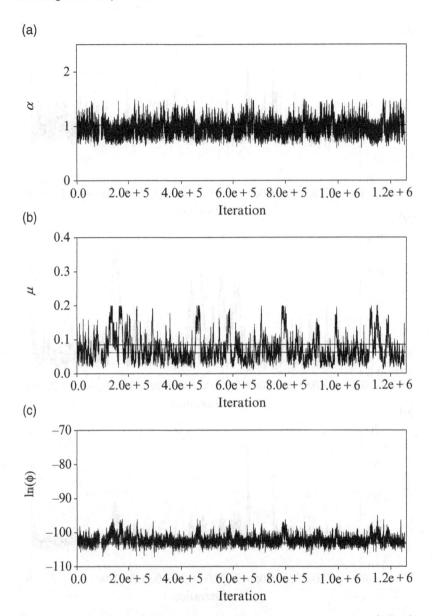

Figure 7.9 As Figure 7.8, but using the bounded parameter range (7.5.15).

0.044 for μ, both of which lie close to the maximum likelihood parameter estimates of 0.989 and 0.0445 for the case in which both immigration and death times are observed. Figures 10a&b show the relative values of the joint likelihood ϕ, and there is clearly a marked contrast between this case, which embodies substantial loss of information due to immigration times being unknown, and the case of complete information shown

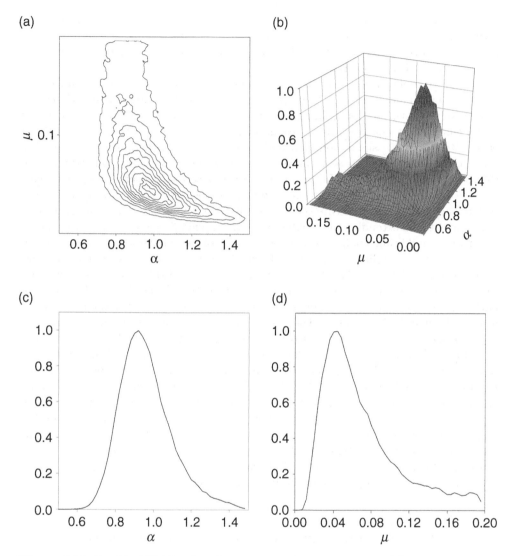

Figure 7.10 Empirical likelihood surface corresponding to Figure 7.9: (a) contour plot with levels 0.05(0.1)0.95, and (b) mesh plot over $\alpha = 0.5(0.02)1.5$, $\mu = 0(0.004)0.2$; together with marginal likelihoods for (c) α and (d) μ.

in Figures 7.7a&b. Moreover, the joint likelihood surface is now 'banana'-shaped, and this feature highlights the inverse relationship between α and μ. The fact that ϕ does not vanish in the limit as μ becomes large is reflected in the long tails of the posterior marginal distribution for μ, which indicates a potential problem regarding the use of marginal densities for estimating parameters in this regime. For because of this tail, selecting a very high upper bound for μ will lead to a marginal distribution for α

which is biased downwards. This exemplifies a general problem which may arise when fitting stochastic models to incomplete observations, namely that likelihood surfaces may exhibit severe departures from the convenient bell-shaped likelihoods which arise when observing the complete process.

For our example with 100 deaths the (bounded) marginal interquartile ranges are $(0.89, 1.07)$ for $\tilde{\alpha}$ and $(0.041, 0.100)$ for $\tilde{\mu}$, compared with $(0.92, 1.04)$ and $(0.041, 0.047)$, respectively, for the complete marginal likelihoods based on both immigration and death times. This analysis raises various issues. First note the bias resulting from basing it on a single simulation run; and, second, that considerably more than 100 death times would need to be recorded if we are to effect a substantial reduction in the width of these intervals. This latter point raises a serious concern since, as we have seen earlier, much of the information content in a stochastic process which converges towards an equilibrium structure is contained in the opening (i.e. transient) part of its development, and extending the number of deaths essentially involves information gain on only the equilibrium component. Since here this follows a Poisson(α/μ) distribution, it follows that in equilibrium α/μ is sufficient for the distribution of population size values. It is only by using the additional information provided by the inter-event times that we can disentangle α and μ. Third, the geometry of the likelihood surface, in particular the long ridges extending out from the maximum, presents a further problem in that the marginal parameter distributions are heavily influenced by the choice of upper bounds for $\tilde{\alpha}$ and $\tilde{\mu}$. A detailed theoretical description of the likelihood surface, including a deeper analysis of the long-ridge structures, is contained in Gibson and Renshaw (2001a&b). Fourth, there is clearly more than one way of implementing the parameter resampling procedure. For although we have sampled and then tested the $\tilde{\alpha}$- and $\tilde{\mu}$-values separately, an alternative approach would be to resample and test them together, though this would reduce the density of accepted changes. Another possibility might be to reparameterize in terms of α and α/μ, though this procedure develops more slowly. Whilst variations on the arrive-delete-move strategy for exploring the space of hidden immigration times include employing birth–death, and split–combine, strategies (see, for example, Cappé, Robert and Rydén, 2003).

7.5.3 Extension to the single/paired-immigration–death process

The problem of inferring properties of a population process purely from data on removals alone is generic to many areas of science, classic examples being epidemiology, ecology and quantum optics. So it is natural to ask whether the above success in handling the simple immigration–death process will continue when the number of parameters is increased? Fortunately the answer is yes, with the MCMC approach performing surprisingly well even when applied to the logistic process developed in Section 3.2 (Renshaw and Gibson, 1998). Given that this scenario requires four parameters to be estimated, it clearly presents a tough challenge. In reality, this degree of complexity probably represents the limit of what can meaningfully be achieved. For example, the extension to the power-law logistic process raises problems due to the estimates of the power-law coefficient exhibiting fairly strong bias. Nevertheless, even here the (qualitative) performance of the MCMC procedure is still surprisingly good, bearing in mind the huge amount of process information that has been discarded.

Given that there is clearly a limit as to how far the MCMC approach can be pushed, let us illustrate the potential difficulties that can arise when we analyse a multiparameter system. Suppose we extend the above simple immigration–death model by allowing both single and paired immigrants to arrive at rates α_1 and α_2, respectively. For although this structure is simpler in nature than that of the power-law logistic process, it is fundamentally different in that the 'birth' component, i.e. the overall arrival rate $\alpha_1 + 2\alpha_2$, is not affected by the population size N. This raises the question as to the extent to which the MCMC procedure can differentiate between the single and paired immigration rates? For with the birth–death–immigration process, for example, the overall population birth rate $\alpha + \lambda N \simeq \alpha$ and $\simeq \lambda N$ when N is small and large, respectively. Similarly, the individual components comprising the power-law logistic birth and death rates $N(a_1 N - b_1 N^s)$ and $N(a_2 + b_2 N^s)$ can be distinguished as N changes. In direct contrast, no such differentiation is possible here since $\alpha_1 + 2\alpha_2$ is independent of N.

This single/paired-immigration–death process is also intrinsically interesting for a totally different reason. For if μ is very small relative to α_1 and α_2, then there will generally be a large number of intervening deaths between the arrival of an individual and its subsequent death, and so the linkage between a given death and its associated immigration event will be minimal. All we can really determine is the overall rate $\alpha_1 + 2\alpha_2$. Conversely, as $\mu \to \infty$ single- and paired-immigrants will immediately be followed by their single- and paired-deaths, so the information on death times *de facto* provides full and separate information on α_1 and α_2. As μ increases from 0 through to ∞, we move between these two extremes.

Continuing with our example, suppose we take $\alpha_1 = 1.0$ and $\alpha_2 = 0.5$ (so that the expected numbers of each immigrant type balance) and $\mu = 0.05$; together with $\tilde{\alpha}_1 \leq 3.0$, $\tilde{\alpha}_2 \leq 1.5$ and $\tilde{\mu} \leq 0.25$, in conjunction with uniform sliding windows of size 0.1, 0.05 and 0.005, respectively. As in the previous analysis, the initial simulated arrival configuration comprises a single immigrant in each inter-death interval, so no immigrant pairs are initially included (i.e. $\tilde{\alpha}_2$ is presumed to be zero). Figure 7.11a–c shows the resulting resampled marginal histograms for $\tilde{\alpha}_1$, $\tilde{\alpha}_2$ and $\tilde{\mu}$ over 2×10^6 accepted events; every 50th of the 6547500 events generated is recorded. Rather perversely, we see that the MCMC analysis implies that $\tilde{\alpha}_1 \simeq 0$ and $\tilde{\alpha}_2 \simeq 2\alpha_2$, i.e. that a quite different stochastic process (i.e. pure paired-immigration–death) is operating to the one which generated the initial simulated process (i.e. balanced single/paired-immigration–death). Moreover, the log-likelihood values produced greatly exceed the true log-likelihood value, so the estimated model has a far higher likelihood than the model used to generate the data in the first place. Clearly, the doubts expressed in Section 7.5.2 have been realized.

Moreover, this large difference between true and inferred process is even worse than is exhibited in Figure 7.11. Since not only does the marginal density of $\tilde{\mu}$ (Figure 7.11c) rise linearly towards the imposed upper bound 0.25, but removing this restriction has virtually no effect on $\tilde{\alpha}_1$ and $\tilde{\alpha}_2$; $\tilde{\mu}$ now takes values up to 0.75, with the log-

380 *Modelling bivariate processes*

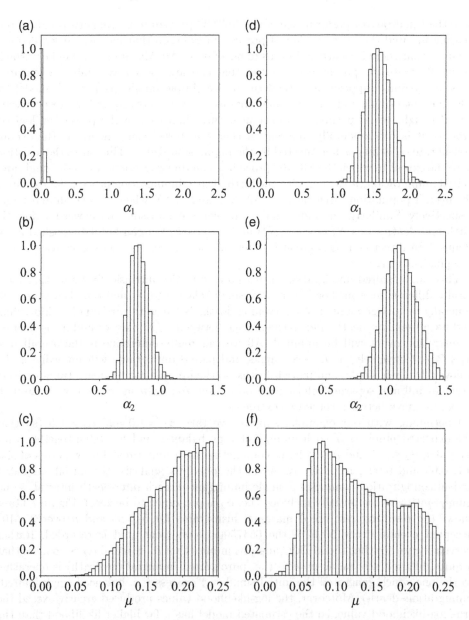

Figure 7.11 Single/paired-immigration–death process with $\alpha_1 = 1.0$, $\alpha_2 = 0.5$ and $\mu = 0.05$ ($\tilde{\mu} \leq 0.25$) showing marginal MCMC distributions of: (a) α_1, (b) α_2 and (c) μ for a randomized initial arrival configuration: (d)–(f) as (a)–(c) but with the true initial arrival configuration.

likelihood reaching correspondingly high values. This strongly suggests that there exist virtually unconnected regions of the Markov chain state space that can effectively only be reached from particular types of initial configuration. Figure 7.11d–f shows the result of replacing the initial random configuration by the true one. Not only do the single immigrants now feature strongly, but $\tilde{\alpha}_1$ actually exceeds α_1. Note that this fairly strong upwards bias in $\tilde{\alpha}_1$ and $\tilde{\alpha}_2$ is counterbalanced by upwards bias in $\tilde{\mu}$, implying that the process is more active than it really is. Comparing the two log-likelihood distributions shows that there is no overlap between them, so these two processes inferred from the same data set are effectively disconnected. This behaviour is an example of the problems which can arise due to poor mixing of a Markov chain. For with the Metropolis sampler constructed here, the conditional distribution of the missing events is heavily dependent on the model parameters and vice versa. Thus although the chain is theoretically irreducible over the full state-space, in practice it is disconnected for simulation runs of computationally reasonable length. This means that inferences are strongly dependent on initial conditions. So the most secure way of constructing credible inferences from this MCMC approach in multi-parameter situations is to conduct a suite of simulation experiments over the full range of possible initial parameter and event configurations. For then the most appropriate model choice can be made based on available prior knowledge of the system characteristics.

7.5.4 A comparison of Metropolis Q- and direct P-matrix strategies

The huge upsurge in the development of techniques based on the Metropolis–Hastings algorithm has centred almost exclusively around the construction of MCMC procedures. Practical implementation (e.g. Gilks, Richardson and Spiegelhalter, 1996) has expanded into a wide variety of different types of statistical problems, often concerned with parameter estimation. Whilst theoretical developments have made considerable advances in areas such as the construction of convergence diagnostic techniques (e.g. Cowles and Carlin, 1996; Brooks and Roberts, 1998), computational schemes to speed up the MCMC approach, and issues concerning model selection. Studies relating to stochastic processes have generally involved parameter estimation based on incomplete observations, as illustrated above. However, the rationale which underlies the original development of the algorithm, namely the construction of stochastic processes which possess a given equilibrium distribution, has been left virtually ignored.

In principle, provided we know an associated transition matrix $\mathbf{P}$ satisfying $\boldsymbol{\pi} = \boldsymbol{\pi}\mathbf{P}$, it is straightforward to construct a simulated realization from a Markov chain which has a given equilibrium distribution $\boldsymbol{\pi} = (\pi_0, \pi_1, \ldots, \pi_N)$.

Basic Markov chain algorithm (A7.3)
  (i) note current state i
  (ii) generate a pseudo-uniform random variable U on $(0, 1)$
  (iii) select a potential next step j via
      if $U \leq P_{i0} = p_{i0}$ then $j = 0$, else
      if $U \leq P_{i1} = p_{i0} + p_{i1}$ then $j = 1$, else, $\ldots$,

if $U \leq P_{i,N-1} = p_{i0} + \cdots + p_{i,N-1}$ then $j = N-1$, else $j = N$
(iv) return to (ii)

Selecting values of $\{p_{ij}\}$ for a Markov population chain involving single birth or death is clearly trivial, since we can make direct use of the balance equations $\mu_{i+1} = \lambda_i \pi_i / \pi_{i+1}$ and then scale λ_i to ensure that $0 \leq \lambda_i, \mu_i \leq 1$ for all i. So if we are content to remain within this basic population scenario then algorithm A7.3 works well. However, for a more general transition network direct computation of $\mathbf{P}$ can be complicated. Even assuming reversibility, which means that for given p_{ij} ($i < j$) all the p_{ij} for $i > j$ are defined through $p_{ji} = \pi_i p_{ij}/\pi_j$, does not offer much in the way of computational simplification. It was the need to simulate complex systems in statistical physics that inspired Metropolis, Rosenbluth, Teller and Teller (1953) to introduce the pioneering algorithm A7.2 which completely dispenses with the need to determine $\mathbf{P}$. For by selecting a 'user-friendly' symmetric transition matrix $\mathbf{Q}$, recovery of $\mathbf{P}$ occurs automatically through the addition of a simple extra step in A7.3, namely:

Metropolis algorithm (A7.4)
  (i) note current state i
  (ii) generate a pseudo-uniform random variable U on $(0,1)$
  (iii) *'propose'* the next step j via
    if $U \leq Q_{i0} = q_{i0}$ then $j = 0$, else
    if $U \leq Q_{i1} = q_{i0} + q_{i1}$ then $j = 1$, else, ... ,
    if $U \leq Q_{i,N-1} = q_{i0} + \cdots + q_{i,N-1}$ then $j = N-1$, else $j = N$
  (iv) generate a new U
  (v) *'accept'* the proposed move via
    if $U \leq \min(1, \pi_j/\pi_i)$ then $i \to j$, else remain at i
  (vi) return to (ii)

The inverse problem involves determining whether there is an optimal method of selecting a proposal $\mathbf{Q}$-matrix corresponding to a given set of equilibrium probabilities, both in terms of trajectory characteristics and the coverage of the range of support. It must be stressed that the construction of stochastic trajectories which possess target distributional properties is far removed from the usual issues concerning the implementation of MCMC, so problems involving diagnostic convergence techniques, etc. do not arise. Instead, questions relate far more to the nature of the transitions between different states, and how well the trajectory space is covered. Choice of the matrices $\{p_{ij}\}$ and $\{q_{ij}\}$ is clearly governed by the extent to which we have knowledge of the underlying system. For example, if we wish to take an extreme view and disregard system structure entirely, then simulation of the Markov chain $\{X_n\}$ is effectively trivial since we may take $p_{ij} = \pi_j$ which is independent of i. Thus regardless of the value of X_n, $X_{n+1} = \min_j\{U \leq \pi_0 + \cdots + \pi_j\}$. In direct contrast, suppose we know that the transition matrix is tridiagonal, that is we have a basic birth–death chain with rates λ_i and μ_i. Then $X_{n+1} = X_n + 1$ if $U \leq \lambda_i$, $X_{n+1} = X_n - 1$ if $\lambda_i < U \leq \lambda_i + \mu_i$, else X_n remains unchanged. Clearly neither situation requires the use of the Metropolis–Hastings algorithm. More complex systems, however, are likely to be algebraically less forgiving, and employing a $\mathbf{Q}$-matrix approach may be far more efficient.

The extreme positive skewness of the discrete Cauchy distribution $\pi_i = \phi/(1+i^2)$ over $i = 0, 1, 2, \ldots$ (recall that no moments exist) makes it an excellent test case for comparing the degrees of 'success' generated by specific proposal matrices $\mathbf{Q}$ covering the required range of support. For the range of i-values over which a typical simulation might run may be huge. The normalization constant $\phi = [(j/2)(\Psi(-j) - \Psi(j))]^{-1}$ ($j = \sqrt{(-1)}$) involves the digamma function $\Psi(\cdot)$ (Abramowitz and Stegun, 1970, §6.3), and is easily evaluated to any desired degree of numerical accuracy via *Maple* (to 6 decimal places $\phi = 0.481539$). Renshaw (2004) uses this to examine a variety of trial schemes which are summarized below in increasing order of complexity.

(a) Metropolis simple random walk (MSRW) $\mathbf{Q}$ takes the tri-diagonal form of a symmetric simple random walk with a reflecting barrier at 0 having rates

$$q_{00} = 0.6, q_{01} = 0.4 \quad \text{and} \quad q_{i,i-1} = 0.4, q_{ii} = 0.2, q_{i,i+1} = 0.4 \quad (i > 0). \quad (7.5.16)$$

Here the process switches between hovering near zero and undertaking mild upward excursions. A simulation over 10,000 steps shows that although there are occasional large excursions which then collapse back towards zero, the state space is not covered sufficiently well to allow the exact and empirical p.d.f.'s to be in good agreement for values of $i > 60$.

(b) Geometric Metropolis (GM) The simple random walk component is now replaced by the geometrically distributed proposed step size

$$q_{i,i+r} = \beta(1-\theta)\theta^{|r|-1} \quad (r \neq 0) \quad \text{for} \quad r \geq -i. \quad (7.5.17)$$

This distribution is easy to simulate, and the range of i-values covered is much greater than that under (a). For example, a 10,000 step run starting from $i = 0$ with $\beta = 0.5$ and $\theta = 0.9$ exhibited full support up to $i = 319$ and partial support up to $i = 604$. So if the underlying population dynamics rationale is equally strong for both the MSRW and GM schemes then the second is clearly preferable. Note that since the p.d.f. $(1-\theta)\theta^{r-1}$ ($r = 1, 2, \ldots$) has mean $1/\theta$, one might think that the range of support would increase indefinitely as θ rises towards 1. However, although raising θ from say 0.9 to 0.99 does indeed result in a rough doubling of $\max(i)$, it does so at the expense of a corresponding decrease in the number of excursions experienced in the simulation. For the larger the value of $r > 0$, the smaller the acceptance probability $(1+i^2)/(1+(i+r)^2)$.

(c) Metropolis Small World (MSW) If we can accept that the range of i is, for all practical purposes, bounded (e.g. $0 \leq i < M$), then we may combine the (a) local and (b) global (when $\theta \sim 1$) step transitions by taking

$$q_{i,i-1} = q_{i,i+1} = 1/2 - \delta \quad \text{with} \quad q_j = \epsilon \quad \text{otherwise.} \quad (7.5.18)$$

The symmetry and simplicity of the $\mathbf{Q}$-matrix is preserved by replacing the linear state space by a circular one in which states 0 and $M-1$ are neighbours. We require $\delta = (M-2)\epsilon/2$ to ensure row sums equal one, and $\epsilon \leq 1/(M-2)$ to maintain non-negative entries. Thus at each step we move to either of the two neighbouring states with equal probability $[1 - (M-2)\epsilon]/2$, and to each non-neighbouring state with probability ϵ. This matrix is of considerable importance in its own right, for it relates directly to the Small World scenario observed by Watts and Strogatz (1998) in which many complex randomized networks can exhibit two apparently contradictory characteristics Highham (2007, 2008). For not only do such networks exhibit 'nearest-neighbour clustering', but they are also 'small worlds' in that any two nodes can typically be connected by a relatively short path. The classic example of this phenomenon is the 'Six Degrees of Separation' effect in which any two people can almost always be connected through at most six acquaintances. The small world property may be measured by expressing the average mean hitting time in terms of the average number of shortcuts per random walk, and results obtained agree closely with those arising from the mean-field network theory of Newman, Moore and Watts (2000). This fascinating phenomenon appears to be ubiquitous in real-life, and further examples cited by Higham include: authors connected by joint papers; hyperlinked web pages; and, nerve systems (which are based on neuron connections). The innate simplicity of this **Q**-matrix scheme means that the proposal step (iii) in algorithm A7.4 can be constructed efficiently. Though a careful balance needs to be struck between M and ϵ. For increasing M, and thereby the potential range of i, automatically decreases ϵ, and thereby the probability of moving from one local region on the circle $\{0, 1, \ldots, M-1\}$ to another at the next step. Thus the greater the number of sites M considered, the more patchy the overall coverage becomes.

Whilst the Metropolis algorithm is clearly excellent for simulating realizations directly from an equilibrium p.d.f. for moderate values of M, in general large **Q**-matrices may well involve massive compute times. For up to $M-1$ 'if'-checks (viz. $U \leq q_{i0}, U \leq q_{i0} + q_{i1}$, etc.) may be needed to determine the next step j; though storing the cumulative transition matrix and ratio checks $\{\min(1, \pi_j/\pi_i)\}$ at the outset helps to reduce this. However, if we are prepared to replace the general **Q**-matrix regime by a more restricted scenario, we may dispense with the Metropolis acceptance step (v) in algorithm 7.4. For we can then use a procedure based purely on the transition matrix **P**, which means that the algorithm is far faster to run. To illustrate this approach let us return to our tri-diagonal form with $q_{i,i+1} = \lambda_i$, $q_{ii} = 1 - \lambda_i - \mu_i$ and $q_{i,i-1} = \mu_i = c$ for some constant $c > 0$. The balance equations now yield $\lambda_i = c\pi_{i+1}/\pi_i$ ($i = 0, 1, \ldots$) provided we ensure that $\lambda_0 = c\pi_1/\pi_0 < 1$ and $\lambda_i + \mu_i = c(1 + \pi_{i+1}/\pi_i) < 1$ for $i > 0$, i.e.

$$c < \min_{i>0} \left\{ \frac{\pi_0}{\pi_1}, \frac{\pi_i}{\pi_i + \pi_{i+1}} \right\}. \tag{7.5.19}$$

For many equilibrium p.d.f.'s there will exist an integer $i' > 0$ such that $\pi_{i+1} < \pi_i$ for all $i > i'$, whence $\lambda_i < c = \mu_i$ ($i > i'$), thereby guaranteeing a general downwards drift in the upper tail of the random walk. Note, however, that care must be exercised with p.d.f.'s which exhibit sawtooth equilibrium behaviour (as in Figure 7.4), since this simple feature no longer applies.

(d) Mass Annihilation/Immigration (MAI) Although this construction has simple mathematical features, it can take a prohibitively long time to cover a reasonable range of support. So simulated realizations may well remain compromised by the choice of initial start value. This drawback can be circumvented by recalling that in our discussion of continuous-time Markov processes in Section 2.7 we examined the concepts of 'catastrophy', i.e. 'annihilation' of the entire population of size i at rate β_i, and 'mass immigration' into an empty population of j immigrants at rate α_j. This suggests that for discrete-time Markov chains we should expand the tri-diagonal form into the transition probability matrix

$$\mathbf{P} = \begin{pmatrix} 1 - \lambda_0 - \sum_{n=1}^{\infty} \alpha_n & \lambda_0 + \alpha_1 & \alpha_2 & \alpha_3 & \alpha_4 & \cdots \\ \mu_1 + \beta_1 & q_{11} - \beta_1 & \lambda_1 & 0 & 0 & \cdots \\ \beta_2 & \mu_2 & q_{22} - \beta_2 & \lambda_2 & 0 & \cdots \\ \beta_3 & 0 & \mu_3 & q_{33} - \beta_3 & \lambda_3 & \cdots \\ \vdots & \vdots & \vdots & \vdots & \vdots & \end{pmatrix}. \quad (7.5.20)$$

So now the process performs a series of local random walks and global jumps; at each step it can move from state i to 0 with probability β_i and then move onwards to state j in a different region of state space with probability α_j. Changing the magnitude of the β_i changes the length of time spent performing each local random walk. This process offers considerable mathematical opportunities, since allowing annihilation to be followed swiftly by mass immigration enables a process to switch into a totally different region of the spate space virtually instantaneously without losing the underlying Markov structure. The equilibrium equations corresponding to (7.5.20) are given by

$$\pi_0 \left(\sum_{j=1}^{\infty} \alpha_j + \lambda_0 \right) = \pi_1 \mu_1 + \sum_{i=1}^{\infty} \pi_i \beta_i$$

$$\pi_1(\lambda_1 + \mu_1 + \beta_1) = \pi_0(\lambda_0 + \alpha_1) + \pi_2 \mu_2 \quad (7.5.21)$$

$$\pi_i(\lambda_i + \mu_i + \beta_i) = \pi_0 \alpha_i + \pi_{i-1}\lambda_{i-1} + \pi_{i+1}\mu_{i+1} \quad (i = 2, 3, \ldots),$$

and for any given $\boldsymbol{\pi}$ the choice of admissible $\{\lambda_i, \mu_i, \alpha_i, \beta_i\}$ is clearly huge. The only restriction is that all elements of the transition matrix (7.5.20) must be non-negative. So suppose we split the detailed balance equations into the local and global component parts

$$\pi_i \lambda_i = \pi_{i+1} \mu_{i+1} \ (i \geq 0) \quad \text{and} \quad \alpha_i \pi_0 = \beta_i \pi_i \ (i \geq 1); \quad (7.5.22)$$

for if these hold true then the full balance equations (7.5.21) must also hold true. Retaining $\mu_i = c$ automatically ensures that $\lambda_i = c\pi_{i+1}/\pi_i$, with

$$\alpha_i = \frac{\beta_i \pi_i}{\pi_0} = \beta_i \times \frac{\lambda_{i-1} \lambda_{i-2} \ldots \lambda_0}{\mu_i \mu_{i-1} \ldots \mu_1} \quad (i \geq 1). \quad (7.5.23)$$

Hence if catastrophe out of state i occurs at constant rate $\beta_i = \beta$ then

$$\alpha_i = \beta \pi_i / \pi_0 \ (i > 0), \quad (7.5.24)$$

and the transition probability from 0 to i is directly proportional to π_i. An alternative strategy (for a bounded matrix) would be to ensure that all regimes of the state space have an equal chance of being reached from $i = 0$ by taking $\alpha_i = \alpha$ and $\beta_i = \alpha\pi_0/\pi_i$ ($i > 0$); for an unbounded matrix we could let $\alpha_i = \alpha\psi^i$ and $\beta_i = \alpha\psi^i\pi_0/\pi_i$ ($i > 0$) for some ψ just less than 1. Note that different **P**-matrix strategies may be compared by extending the ideas developed in Sections 4.2.3 and 5.3.2 in order to determine the expected number of steps, and the associated first passage probabilities, required to move from a small i-value, e.g. 0 or 1, to a much larger one, e.g. $j = 10^n$ for $n = 3, \ldots, 6$.

To compare MAI to the GM and MSW schemes suppose we take $\beta_i = \beta$ so that $\alpha_i = \beta\pi_i/\pi_0 = \beta/(1+i^2)$. Then on combining these global jump rates with the local random walk rates (7.5.16), the transition probability matrix (7.5.20) becomes

$$\mathbf{P} = \begin{pmatrix} 1 - 0.4h_0 - \sum_{i=1}^{\infty} \frac{\beta}{1+i^2} & 0.4h_0 + \frac{\beta}{1+1^2} & \frac{\beta}{1+2^2} & \frac{\beta}{1+3^2} & \frac{\beta}{1+4^2} & \cdots \\ 0.4 + \beta & k_1 & 0.4h_1 & 0 & 0 & \cdots \\ \beta & 0.4 & k_2 & 0.4h_2 & 0 & \cdots \\ \beta & 0 & 0.4 & k_3 & 0.4h_3 & \cdots \\ \vdots & \vdots & \vdots & \vdots & \vdots & \end{pmatrix}$$

(7.5.25)

where $k_i = 0.6 - 0.4h_i - \beta$ and $h_i = \pi_{i+1}/\pi_i = (1+i^2)/[1+(i+1)^2]$. So for $i \geq 2$ step (iii) in the Markov chain algorithm A7.3 becomes

$$j = 0 \text{ if } U \leq \beta, \qquad \text{else } j = i - 1 \text{ if } U \leq 0.4 + \beta,$$
$$\text{else } j = i \text{ if } U \leq 1 - 0.4h_i, \qquad \text{else } j = i + 1. \tag{7.5.26}$$

Whilst for $i = 1$ we have

$$j = 0 \text{ if } U \leq 0.4 + \beta, \quad \text{else } j = 1 \text{ if } U \leq 1 - 0.4h_1, \quad \text{else } j = 2. \tag{7.5.27}$$

Although both of these steps are quick to compute, using the first row in the matrix (7.5.25) to determine the step-size j away from $i = 0$, via $j = \min_j\{U \leq p_{00} + p_{01} + \cdots + p_{0j}\}$, does involve a potentially large compute time penalty. For not only has $\sum_{i=n}^{\infty} 1/(1+i^2)$ no closed form solution, but the Cauchy distribution has infinite mean. Nevertheless, the probability of being in state 0 and moving to $i \geq 2$ through mass immigration, namely $\pi_0\beta\sum_{i=2}^{\infty}[1/(1+i^2)] = \pi_0\beta(\phi^{-1} - 1.5)$, is a mere 0.0277 when $\beta = 0.1$. So provided we can keep β fairly small this 'slow' part of the algorithm may well operate fairly infrequently.

Figure 7.12 shows the empirical log p.d.f.'s $\{\log(\hat{\pi}_i)\}$ for the GM, MSW and MAI schemes over 10^6 iterations, together with the exact p.d.f. $\pi_i = \phi/(1+i^2)$. Here $\hat{\pi}_i$ is simply the proportion of time state i is occupied following burn-in. For the GM scheme (7.5.17) with $\theta = 0.9$ and $\beta = 0.5$ the simulated and exact log-probabilities (Figure 7.12a) show good agreement up to $i \sim 200$ and reasonable agreement up to $i \sim 319$. Thereafter some of the i-states remain empty, with none being occupied beyond $i = 604$. The MSW scheme (7.5.18) is shown for $\epsilon = 0.0008$, $\delta = 0.3992$ and $N = 1000$

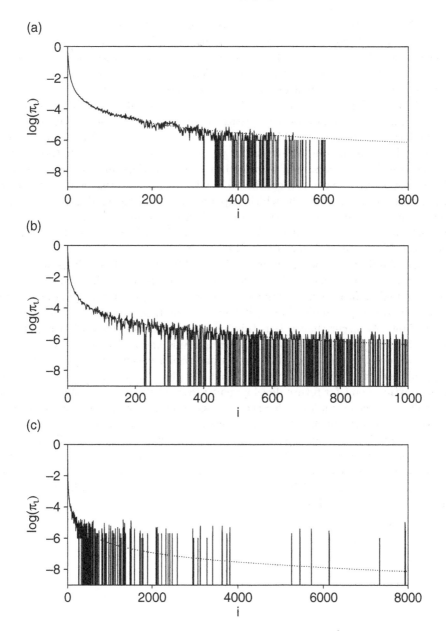

Figure 7.12 Empirical equilibrium p.d.f.'s $\hat{\pi}_i$ evaluated over 10^6 steps (—), and exact Cauchy p.d.f. $\pi_i = \phi/(1+i^2)$ ($\cdots$), generated via the (a) geometric Metropolis and (b) Metropolis Small World **Q**-matrix and (c) mass annihilation/immigration **P**-matrix approaches.

(Figure 7.12b). So the nearest-neighbour rates $q_{i,i-1} = q_{i,i+1} = 0.1008$ and the small world rates $q_{ij} = 0.0008$ ($j \neq i-1, i+1$). Although the fixed range $0, 1, \ldots, 1000$ is better covered than under GM, this is achieved at the cost of less precise probability estimates over values of $i \leq 200$. Nevertheless, use of an appropriate smoothing function would enable estimates to be obtained up to $i = 1000$. The corresponding MAI scheme (7.5.20), given by (7.5.25) with $\beta = 0.1$, illustrates how introducing mass annihilation/immigration can greatly extend the range of i-values covered by the simulation (Figure 7.12c). Moreover, the ability to alter the balance between the local and global jump components of the process by changing the values 0.4, 0.6 and β in the transition matrix, and thereby the precision of estimates corresponding to different regions of i-space, is a considerable advantage. Note that with 10^6 iterations the smallest $\log(\hat{\pi}_i) > 0$ is -6, which explains why the expected gaps between the 'population spikes' steadily increase. For the histogram bin-width increases with i as the exact Cauchy log-probabilities $\{\log(\pi_i)\} \downarrow -\infty$. As before, these spikes could easily be smoothed by using an appropriate filter.

For this simulation experiment $\max\{X_n\} \sim 8000$ under MAI, which is a 100-fold improvement in the range of X_n-values covered compared with MSRW, and more than a 10-fold improvement compared to GM. Although similar coverage could in principle be developed under MSW, it has very small acceptance probabilities when j is substantially larger than i. It is this increase in the range of trajectories that are likely to be generated, coupled with the considerable knowledge already developed for the (continuous-time) mass annihilation/immigration process, that makes MAI such an appealing choice. We are not claiming that it is globally optimal, but it does provide a good starting point for future development provided that moves from i to j through the origin are deemed plausible for the process under study. Because the transit time at zero can be made vanishingly small by letting $\sum \alpha_i \to \infty$, even this issue can easily be circumvented. In specific applications it is likely that information will be available on the process in addition to the equilibrium probabilities, such as estimates of first return and passage probabilities. So we can use this to help us choose the parameters λ_i and α_i in the matrix (7.5.20), with μ_i and β_i then being determined through the split balance equations (7.5.22).

8
Two-species interaction processes

In nature individuals seldom exist in single-species communities, but they live alongside individuals from other species. Whilst a large number of these species will be unaffected by the presence or absence of one another, in some cases two or more will interact. So to analyse such scenarios we need to extend our study of population growth from single-species to multi-species processes. In theory the number of different ways that such interaction can occur is immense, but here we shall restrict our discussion to three truly interactive two-species processes of fundamental importance, namely *competition, predation* and *infection*. Though a fourth type of process, *symbiosis*, occurs in farming, it is rarely seen in natural populations.

8.1 Competition processes

Let us first consider competition, since this provides a natural bivariate extension to the univariate Verhulst–Pearl logistic process developed in Section 3.2. If individuals compete for common resources that are in short supply, such as food or space, then as the population size of one species increases, it will have an adverse effect on the net growth rate of the other. Denote $N_1(t)$ and $N_2(t)$ to be the number of individuals of species 1 and 2, respectively, present at time $t \geq 0$. Then the obvious way of extending the univariate deterministic logistic equation (3.2.29), namely

$$dN(t)/dt = N(t)[r - sN(t)], \tag{8.1.1}$$

to the bivariate case is to write

$$dN_1(t)/dt = N_1(t)[r_1 - s_{11}N_1(t) - s_{12}N_2(t)]$$
$$dN_2(t)/dt = N_2(t)[r_2 - s_{21}N_1(t) - s_{22}N_2(t)] \tag{8.1.2}$$

for rates $r_i > 0$ and $s_{ij} > 0$ $(i,j = 1,2)$. Since the presence of $N_2(t) > 0$ decreases $dN_1(t)/dt$, and similarly $N_1(t) > 0$ decreases $dN_2(t)/dt$, it follows that the population is bounded above by two independent logistic processes with rates r_i and s_{ii}. Thus overall deterministic stability is assured. Nevertheless, the question remains as to whether both species can coexist in harmony, or whether one will destroy the other, with the surviving species then developing as a simple stochastic logistic process.

In general, the whole issue of multi-species stability is a difficult one to tackle. For at the practical level there is considerable evidence to suggest that population stability is typically greater in communities with many interacting species than in those that effectively comprise single species. However, species integration in a complex community is a highly nonlinear affair, and quite remarkable instabilities can result from the introduction or removal of a single species (May, 1971a,b). So to avoid the fascinating world of chaotic dynamics, which for continuous-time processes generally involves three or more interacting species, we shall concentrate on purely two-species interactions.

Before we proceed to develop the bivariate competition structure associated with (8.1.2), it is worthwhile repeating Park's (1954) warning that the functional existence of inter-species competition may be inferred from a body of data even when no such inter-species dependence exists. The danger is that if two species are correlated in such a way that the population increase of one tends to be associated with the decrease of the other, then competition between the species may be suspected as a causal agent. But other possibilities must also be considered. For example, if each species eats entirely different food and the availability of each food type alternates temporally, then it might be expected that the two population sizes would respond with similar alternations. Observed correlations can therefore be environmental as well as competitive in nature, and this must be borne in mind when working with field observations.

8.1.1 Deterministic analysis

Although the single-species deterministic logistic equation (8.1.1) is easily solved, yielding for $s = r/K$ the neat representations (3.2.30)–(3.2.33), in general the same does not hold true either for the bivariate simultaneous equations (8.1.2) or their natural m-species extension

$$dN_i/dt = N_i(r_i - s_{i1}N_1 - \cdots - s_{im}N_m) \qquad (i = 1,\ldots,m). \qquad (8.1.3)$$

Here each population grows logistically if it is alone, with parameters r_i and s_{ii}; whilst s_{ij} ($j \neq i$) measures the extent to which the presence of species j affects the growth of species i.

However, some analytic progress can be made if the inhibitory effects of species 1 and 2 are the same for both populations, that is, we can write

$$dN_1/dt = N_1[r_1 - s_1(N_1 + pN_2)]$$
$$dN_2/dt = N_2[r_2 - s_2(N_1 + pN_2)]. \qquad (8.1.4)$$

Thus an individual of either species behaves as though it were competing with a population of size $N = N_1 + pN_2$, where the parameter p allows each species to differ in its inhibitory effect. If individuals of species 2 make smaller inroads on available resources then $p < 1$, otherwise $p > 1$. Now on comparing the general representation (8.1.2) with this special case (8.1.4), we see that $s_{11} = s_1$, $s_{12} = ps_1$, $s_{21} = s_2$ and $s_{22} = ps_2$. Hence as

$$p = s_{12}/s_{11} = s_{22}/s_{21}, \qquad \text{i.e.} \quad s_{11}s_{22} = s_{12}s_{21}, \qquad (8.1.5)$$

the product of the within-species inhibitory growth rates exactly balances the product of the between-species inhibitory growth rates. Equations (8.1.4) do not yield a neat closed form solution, but they do enable some simplification. For eliminating $N_1 + pN_2$ and integrating gives

$$s_2[\ln\{N_1(t) - N_1(0)\}] - s_1[\ln\{N_2(t) - N_2(0)\}] = t(s_2 r_1 - s_1 r_2).$$

Whence taking the exponential of both sides yields

$$\frac{[N_1(t)]^{s_2}}{[N_2(t)]^{s_1}} = \frac{[N_1(0)]^{s_2}}{[N_2(0)]^{s_1}} \exp\{t(s_2 r_1 - s_1 r_2)\}. \tag{8.1.6}$$

Whilst this is not a complete solution, in that it does not detail the separate development of $N_1(t)$ and $N_2(t)$, it does provide valuable information on their relative magnitudes. For on noting from (8.1.4) that dN_1/dt and dN_2/dt are both negative if $N = N_1 + pN_2$ exceeds r_1/s_1 and r_2/s_2, we see that $N_1(t)$ and $N_2(t)$ have maximum permissible sizes which are not greater than their individual carrying capacities $K_1 = r_1/s_1$ and $K_2 = r_2/ps_2$. Thus (8.1.6) shows that

(i) if $s_2 r_1 > s_1 r_2$ then $N_2(t) \to 0$,
(ii) if $s_2 r_1 < s_1 r_2$ then $N_1(t) \to 0$,
(iii) if $s_2 r_1 = s_1 r_2$ then neither $\to 0$.

So unless the exceptional case (iii) occurs, only one of the species will persist and the other will eventually die out. If $s_2 r_1 > s_1 r_2$ then species 1 will ultimately win, and conversely if $s_2 r_1 < s_1 r_2$; the speed at which these events happen is governed by $\exp\{(t(s_2 r_1 - s_1 r_2)\}$. Once the losing species has become extinct the winner then grows in accordance with the corresponding single-species logistic process. Note, however, that initial examination of the growth of these two populations can lead to the wrong conclusion. For if $N_1(t)$ and $N_2(t)$ are both small compared to their carrying capacities K_1 and K_2, then in the opening stages of the process it follows from (8.1.4) that $N_1(t) \simeq N_1(0)e^{r_1 t}$ and $N_2(t) \simeq N_2(0)e^{r_2 t}$. Thus if $r_2 > r_1 > (s_1/s_2) r_2$ (i.e. condition (i) holds true), then although species 2 has the initial competitive edge it eventually succumbs to a greater competitive pressure. This confirms the well-known ecological fact that having a higher reproductive ability is no guarantee of eventual success.

In general, assumption (8.1.5), on which this analysis is based, will not hold true, and so even the incomplete solution (8.1.6) will not be available to us. Though for given situations with known parameter values we can always obtain numerical solutions to equations (8.1.2) (or (8.1.3) for m-species situations). These can be developed either by using standard numerical integration procedures over $t \geq 0$, or by constructing the natural extension to single-species logistic growth over the discrete time points $t = 0, 1, 2, \ldots$. For we have already shown that equation (8.1.1) may be represented in the form (3.2.33), i.e.

$$N(t+1) = e^r N(t)/[1 + N(t)(e^r - 1)(s/r)], \tag{8.1.7}$$

and it is easily confirmed that the equivalent two-species representation is given by the pair of difference equations

$$N_i(t+1) = a_i N_i(t)/[1 + b_i N_i(t) + c_i N_j(t)] \qquad (i, j = 1, 2; j \neq i) \tag{8.1.8}$$

where $a_i = e^{r_i}$, $b_i = s_{ii}(a_i - 1)/r_i$ and $c_i = s_{ij}(a_i - 1)/r_i$ (Pielou, 1977). This discrete representation is especially useful when sketching the trajectories of $N_1(t)$ and $N_2(t)$ both against t and each other.

To appreciate the types of behaviour that may be observed in practice it is worth recalling three classic biological experiments to investigate between-species competition: (i) Gause (1932) on two competing species of yeast, *Saccharomyces cerevisiae* and *Schizosaccharomyces kefir*; (ii) Gause (1934) on *Paramecium aurelia* and *Paramecium caudatum*; and, (iii) Birch (1953) on mixed populations of two types of grain beetle developing within maize or wheat, namely *Calandra oryzae* (both large and small strains) and *Rhizopertha dominica* (for details see Renshaw, 1991). In experiment (i) Gause obtained the parameter estimates

$$r_1 = 0.2183 \quad s_{11} = 0.0168 \quad s_{12} = 0.0529 \quad (Sacch.)$$
$$r_2 = 0.0607 \quad s_{22} = 0.0105 \quad s_{21} = 0.0046 \quad (Schiz.).$$

So whilst *Saccharomyces*, $N_1(t)$, initially increases much faster than *Schizosaccharomyces*, $N_2(t)$, (since $r_1/r_2 = 3.57$), *Schiz.* has a far stronger relative competitive effect on *Sacch.* ($s_{12}/s_{11} = 3.15$) than *Sacch.* has on *Schiz.* ($s_{21}/s_{22} = 0.439$). Indeed, once the total competitive effect on $N_1(t)$, namely $s_{11}[N_1(t) + 3.15N_2(t)]$, exceeds r_1, then $dN_1(t)/dt$ becomes negative and $N_1(t)$ subsequently decays to zero. Note that $N_2(t)$ has 3.15 times the effect on $N_1(t)$ than $N_1(t)$ exerts on itself. Moreover, solving equations (8.1.2) numerically shows that after reaching the maximum value $N_1(41) = 9.1$, $N_1(t)$ changes from a state of rapid growth to moderate decline towards zero ($N_1(280) = 0.001$). In contrast, although the initial growth of $N_2(t)$ is severely restricted by the rapid rise of $N_1(t)$, even though $s_{21}/s_{22} = 0.439$ is substantially less than one, once $N_1(t)$ starts to decay then the competitive effect of $N_1(t)$ on $N_2(t)$ decays with it and so $N_2(t)$ is able to continue increasing towards its own carrying capacity; the two curves cross at $t = 180$.

In the follow-up experiment (ii) over much longer time-periods, although each species of *Paramecium* grew successfully when grown separately, when they were raised together *P. aurelia* always drove out *P. caudatum*. This competitive situation is not as simple as the yeast experiment (i), for not only do the values of s_{12}/s_{11} and s_{21}/s_{22} vary with the age of the culture, but also the bacteria that were the food source produced a toxin that harmed *P. caudatum* more than *P. aurelia*. Nevertheless, the simple representation (8.1.2) clearly provides a realistic approximation to reality.

Care must be taken, however, not to presume that it is a forgone conclusion as to which species eventually wins. For the later experiments (iii) showed that a relatively slight change in environment could alter which of the two competing species became extinct. In replicated studies of mixed populations of grain beetles developing within either wheat or maize, whilst *C. oryzae* (large) always drove *C. oryzae* (small) to extinction in maize at 29.1°C and 13% humidity, the converse happened in wheat at 29.1°C and 14% humidity. Similarly, *C. oryzae* (small) always drove *R. dominica* to extinction in wheat at 29.1°C and 14% humidity, with the converse happening when the temperature was increased to 32.3°C. In each of the 60 cases examined, the species which held the initial advantage always won; though as shown by the *Saccharomyces* experiment (i), this does not hold in general. Moreover, whilst the above analyses revolve around purely deterministic considerations, in practice stochastic fluctuations are likely to play an important role. So what is really meant by population stability? Is one species always destined to become extinct, or is there a range of parameter values over which two competing species can successfully coexist, with eventual extinction of either species being due solely to natural variation?

8.1.2 Stability

A deterministic answer to this question may be easily constructed by returning to equations (8.1.2). For suppose there exists a positive equilibrium solution, i.e. $dN_1/dt = dN_2/dt = 0$ at $(N_1, N_2) = (N_1^*, N_2^*) > 0$. Then

$$0 = r_1 - s_{11}N_1^* - s_{12}N_2^*$$
$$0 = r_2 - s_{21}N_1^* - s_{22}N_2^*, \qquad (8.1.9)$$

whence

$$N_1^* = \frac{r_1 s_{22} - r_2 s_{12}}{s_{11} s_{22} - s_{12} s_{21}} \quad \text{and} \quad N_2^* = \frac{r_2 s_{11} - r_1 s_{21}}{s_{11} s_{22} - s_{12} s_{21}}. \qquad (8.1.10)$$

It now remains to determine whether this equilibrium point (N_1^*, N_2^*) is *stable* or *unstable*. That is, if the population values N_1 and N_2 are slightly displaced from this point, will (N_1, N_2) return to (N_1^*, N_2^*) or will it move away?

Let us consider the geometry of equations (8.1.9). From the first, $N_1^* = 0$ when $N_2^* = r_1/s_{12}$ and $N_2^* = 0$ when $N_1^* = r_1/s_{11}$; whilst from the second, $N_1^* = 0$ when $N_2^* = r_2/s_{22}$ and $N_2^* = 0$ when $N_1^* = r_2/s_{21}$. So four different types of behaviour are possible depending on whether

$$r_1/s_{12} < \text{ or } > r_2/s_{22} \quad \text{and} \quad r_1/s_{11} < \text{ or } > r_2/s_{21}. \qquad (8.1.11)$$

Denote by (a) and (b) the two lines generated by equations (8.1.9), and note from (8.1.2) that dN_1/dt is greater than, equal to or less than zero depending on whether (N_1, N_2) lies above, on or below line (a); and similarly for dN_2/dt with respect to line (b). Figure 8.1 shows a pictorial representation of this stability structure, and it is clear that:

(i) if $r_1/r_2 > s_{12}/s_{22}$ and s_{11}/s_{21} then species 1 wins,
(ii) if $r_1/r_2 < s_{12}/s_{22}$ and s_{11}/s_{21} then species 2 wins,

and in neither case does a positive equilibrium solution (N_1^*, N_2^*) exist; whilst

(iii) if $s_{12}/s_{22} < r_1/r_2 < s_{11}/s_{21}$ then (N_1^*, N_2^*) is a stable equilibrium point, and
(iv) if $s_{11}/s_{21} < r_1/r_2 < s_{12}/s_{22}$ then (N_1^*, N_2^*) is an unstable equilibrium point.

That is, if N_1 and N_2 are displaced from it by an arbitrarily small amount then they continue to move away and one of the species will ultimately become extinct; which one depends on the start point $(N_1(0), N_2(0))$.

These simple graphical arguments show that a stable deterministic equilibrium can exist if and only if $r_1 s_{22} > r_2 s_{12}$ and $r_2 s_{11} > r_1 s_{21}$, and a way of interpreting this condition is to denote $t_{ij} = s_{ij}/r_i$ $(i,j = 1,2)$ as the competitive effect of species j on species i relative to the natural growth rate of species i. For then (8.1.11) is equivalent

394 *Two-species interaction processes*

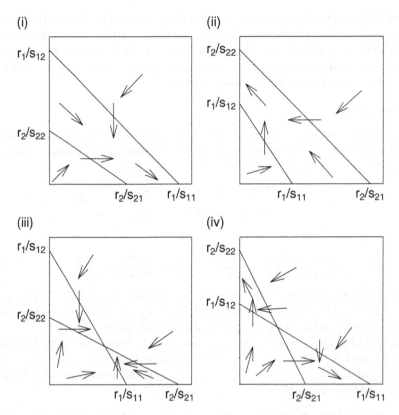

Figure 8.1 Stability diagrams for two-species competition: (i) N_1 always wins; (ii) N_2 always wins; (iii) stable equilibrium; and, (iv) unstable equilibrium with the winning species depending on the start position.

to $t_{11} > t_{21}$ and $t_{22} > t_{12}$, i.e. the relative competitive effect of a species on itself is greater than its effect on the other species. Comparison of Figures 8.1iii&iv shows that condition (8.1.11) is also equivalent to the requirement that the equilibrium point (N_1^*, N_2^*) lies above the line joining $K_1 = r_1/s_{11}$ on the N_1-axis and $K_2 = r_2/s_{22}$ on the N_2-axis, where K_i denotes the carrying capacity of species i in the absence of species j.

Local stability

In order to develop a more mathematically rigorous approach, suppose we restrict our attention to *local* stability, in the sense that any *small* deterministic deviation from (N_1^*, N_2^*) decays to zero. Write

$$N_i(t) = N_i^*[1 + n_i(t)] \qquad (i = 1, 2), \tag{8.1.12}$$

where the $n_i(t)$ are assumed to be appropriately small. Then substituting expressions (8.1.12) into equations (8.1.2), ignoring the second-order terms $n_1^2(t)$, $n_2^2(t)$ and $n_1(t)n_2(t)$, and using the equilibrium relations (8.1.9), gives

$$dn_1(t)/dt = -(s_{11}N_1^*)n_1(t) - (s_{12}N_2^*)n_2(t)$$
$$dn_2(t)/dt = -(s_{21}N_1^*)n_1(t) - (s_{22}N_2^*)n_2(t). \tag{8.1.13}$$

Unlike the quadratic form (8.1.2), these linearized differential equations are easy to solve and yield the solution

$$n_1(t) = Ae^{w_1 t} + Be^{w_2 t} \tag{8.1.14}$$
$$n_2(t) = -[A(w_1 + s_{11}N_1^*)e^{w_1 t} + B(w_2 + s_{11}N_1^*)e^{w_2 t}]/(s_{12}N_2^*),$$

where

$$w_1, w_2 = (1/2)[-(s_{11}N_1^* + s_{22}N_2^*) \pm \sqrt{\{(s_{11}N_1^* - s_{22}N_2^*)^2 + 4s_{12}s_{21}N_1^*N_2^*\}}] \tag{8.1.15}$$

and the constants of integration A and B are determined from the initial values $n_i(0) = (N_i(0)/N_i^*) - 1$.

If an equilibrium solution $(N_1^*, N_2^*) > 0$ exists, then the square root in expression (8.1.15) contains two positive terms. So w_1 and w_2 are both real, which means that cyclic convergence or divergence, to or from equilibrium, cannot arise. Moreover, since local stability requires both $n_1(t)$ and $n_2(t)$ to decay to zero as t increases, this requires $w_1, w_2 < 0$. Now on writing (8.1.15) as

$$w_1, w_2 = (1/2)(s_{11}N_1^* + s_{22}N_2^*)[-1 \pm \sqrt{\{1 - 4(s_{11}s_{22} - s_{12}s_{21})N_1^*N_2^*/(s_{11}N_1^* + s_{22}N_2^*)\}}],$$

it is apparent that this can occur only if $s_{11}s_{22} - s_{12}s_{21} > 0$. Whence it follows from (8.1.10) that if this condition holds then $N_1^*, N_2^* > 0$ if $r_1 s_{22} - r_2 s_{12} > 0$ and $r_2 s_{11} - r_1 s_{12} > 0$, which proves the local stability condition (iii) above. Conversely, if $s_{11}s_{22} - s_{12}s_{21} < 0$, so that $w_1 > 0$, i.e. $n_1(t)$ and $n_2(t)$ grow exponentially, then $N_1^*, N_2^* > 0$ if $r_1 s_{22} - r_2 s_{12} < 0$ and $r_2 s_{11} - r_1 s_{12} < 0$. This proves the local instability condition (iv).

Global stability

Of the four possible cases shown in Figure 8.1, only the unstable equilibrium case (iv) allows for either N_1 or N_2 to win. The winning outcome depends on the start position, and whilst the linearized solution (8.1.14) shows that N_1 wins if $A > 0$ and N_2 wins if $A < 0$, the only real way of determining the exact set of initial winning states $\Omega_i = \{(N_1(0), N_2(0))\}$ for species i, that is those that lead to $N_i(\infty) = K_i$ and $N_j(\infty) = 0$ $(j \neq i)$, is to solve equations (8.1.2) numerically for given parameter values. However, a good feel for how the process develops in each case can be gained by considering the ratio of equations (8.1.2), namely

$$\frac{dN_1}{dN_2} = \frac{N_1(r_1 - s_{11}N_1 - s_{12}N_2)}{N_2(r_2 - s_{21}N_1 - s_{22}N_2)}. \tag{8.1.16}$$

For constructing the vector (dN_1, dN_2), i.e. one proportional to $(N_1[r_1 - s_{11}N_1 - s_{12}N_2], N_2[r_2 - s_{21}N_1 - s_{22}N_2])$, at each point (N_1, N_2) shows the strength and direction of deterministic motion away from these points. Figure 8.2 shows the four resulting individual-based phase plots corresponding to the equilibrium values $N_1^* = N_2^* = 4$

396 *Two-species interaction processes*

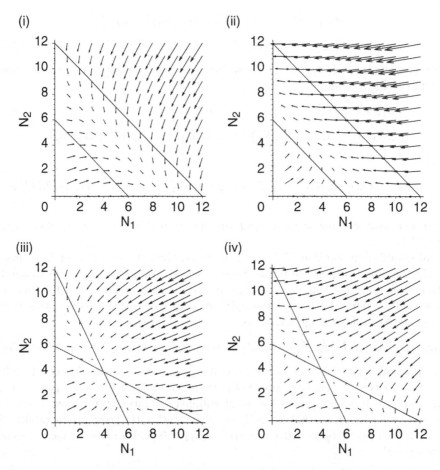

Figure 8.2 *Maple* phase plots corresponding to the unstable regimes (i), (ii), (iv) and stable regime (iii) of Figure 8.1; arrow length is proportional to the individual growth rates. Also shown are the lines of zero flux from $K_1 = r_1/s_{11}$ to $t_{12}^{-1} = r_1/s_{12}$ for $dN_1/dt = 0$ (——), and from $K_2 = r_2/s_{22}$ to $t_{21}^{-1} = r_2/s_{21}$ for $dN_2/dt = 0$ (- - -).

and carrying capacities K_1 and K_2 equal to 6 or 12; here the length of the arrows relates to the *individual* growth rate vector $(r_1 - s_{11}N_1 - s_{12}N_2, r_2 - s_{21}N_1 - s_{22}N_2)$ in order to keep the range of arrow lengths within sensible visual limits. On taking the intrinsic growth rate $r_1 = 0.24$ for N_1 to be twice that of $r_2 = 0.12$ for N_2, it follows that the within- and between-species interaction parameters are given by

case	s_{11}	s_{12}	s_{21}	s_{22}
(i)	0.02	0.02	0.02	0.02
(ii)	0.04	0.04	0.01	0.01
(iii)	0.04	0.02	0.01	0.02
(iv)	0.02	0.04	0.02	0.01

(8.1.17)

These arrow plots not only highlight the paths taken by the deterministic trajectories as the process develops, but they also indicate the strength of the force-field at specific points in the (N_1, N_2)-plane.

General stability

An obvious generalization of the simple process (8.1.2), that allows for natural growth and competition both within and between species, takes the form

$$dN_i/dt = f_i(N_1, N_2) \quad (i = 1, 2) \tag{8.1.18}$$

where f_1 and f_2 are given functions of N_1 and N_2. If $(N_1^*, N_2^*) > 0$ is an equilibrium solution of these equations we then have

$$f_i(N_1^*, N_2^*) = 0 \quad (i = 1, 2). \tag{8.1.19}$$

Whilst if the population size of species i becoming zero implies extinction of that species (e.g. there is no immigration component) then

$$f_1(0, N_2) = 0 = f_2(N_1, 0). \tag{8.1.20}$$

In Section 3.2.3, for example, we saw that the multi-equilibrium single-species process

$$dN/dt = -(N - 20)(N - 50)(N - 100) \tag{8.1.21}$$

gives rise to the stable deterministic equilibrium points $N^* = 20$ and 100, and the unstable equilibrium point $N^* = 50$. This suggests that an appropriate form for the equivalent two-species case can be constructed by considering cubic or higher-order nonlinear forms for $f_i(N_1, N_2)$. For example, the process

$$dN_1/dt = N_1(12 - N_1 - N_2)$$
$$dN_2/dt = N_2(32 - N_1 N_2) \tag{8.1.22}$$

has two equilibrium points, namely $(8, 4)$ and $(4, 8)$.

To determine whether such equilibrium points are stable or unstable, let us parallel our earlier stability analysis by considering small deviations about them. On substituting $N_i(t) = N_i^*(1 + n_i(t))$ into the general equations (8.1.18), and then taking Taylor series expansions, we recover the linear approximation (8.1.13), where now

$$s_{ij} = -\left(\frac{1}{N_i^* N_j^*}\right) \frac{\partial f_i}{\partial n_j}\bigg|_{n_1 = n_2 = 0} = -\left(\frac{1}{N_i^*}\right) \frac{\partial f_i}{\partial N_j}\bigg|_{N_1 = N_1^*, N_2 = N_2^*}. \tag{8.1.23}$$

Thus the previous stability analysis, with its four regimes (i)–(iv), extends to the general process, and we simply examine each equilibrium point in turn. So provided $N_i^* > 0$, the above condition for local stability, namely $s_{11} s_{22} - s_{12} s_{21} > 0$, becomes

$$\theta(N_1, N_2) \equiv \left[\frac{\partial f_1}{\partial N_1} \frac{\partial f_2}{\partial N_2} - \frac{\partial f_1}{\partial N_2} \frac{\partial f_2}{\partial N_1}\right]_{N_1 = N_1^*, N_2 = N_2^*} > 0. \tag{8.1.24}$$

Whence applying this result to (8.1.22) shows that since $\theta(8,4) = 128 > 0$ and $\theta(4,8) = -128 < 0$ the points $(8,4)$ and $(4,8)$ correspond to stable and unstable equilibrium values, respectively.

8.1.3 Stochastic behaviour

In the earlier chapters we paid relatively little attention to deterministic analyses, primarily because in single-variable situations it is often possible to make a considerable amount of progress through stochastic analyses. With two (or more) variables this becomes more difficult, and so deterministic studies play a stronger role in determining overall behaviour, especially when inferences are made in conjunction with simulation-based stochastic analyses. Since biological experimentation was possible long before computer simulation was feasible, many of the early inferences were based on series of experiments, and although the experimenters were often very well regarded the highly specific nature of their studies led, almost inevitably, to incorrect general conclusions. As these are described in some detail in Renshaw (1991), here we shall just provide a brief summary in order to highlight the main issues involved. The fundamental point is that random effects can cause a process to produce radically different outcomes from identical start and environmental conditions. So although deterministic behaviour may well provide valuable qualitative information on process development, it should always carry a 'health warning'.

For readers interested in the biological aspects of species interaction, Krebs (1985) provides several laboratory examples illustrating various competitive outcomes. For example, Park's (1954) studies on the effect of competition between two types of flour beetle under various conditions of temperature and humidity showed that in some regimes the outcome was uncertain, which reinforces the need to undertake stochastic analyses. Moreover, in one of the more extreme regimes, the stronger survivor was not the competitor which appeared to have the advantage when judged purely on single-species experiments; this parallels the deterministic prediction in Section 8.1.1 based on Cause's (1932) study of competition between two species of yeast. So judgements as to the likely winner must be based on time scales long enough to ensure extinction of one of the species, and not merely on short-term observations. This, of course, presupposes that we are working within regimes (i), (ii) or (iv) of Figure 8.2; the existence of the stable regime (iii) shows that in some situations it is perfectly possible for two species to live together in harmony, and this may happen even if they differ only slightly in their requirements. Although this seems to contradict Gause's (1934) competitive exclusion principle, namely that "as a result of competition two similar species scarcely ever occupy similar niches, but displace each other in a manner that each takes possession of certain peculiar kinds of food and modes of life in which it has an advantage over its competitor", for some species such differences in habitat or requirements need only be very slight for their niches to be effectively different. Mathematically speaking, this means that no matter how small the angle between the two isoclines is in Figure 8.2iii, provided they intersect in the positive quadrant then stability, and hence coexistence, is guaranteed.

Now when we constructed the stochastic transition rates for the two-parameter single-species logistic deterministic model (8.1.1), we generated the four-parameter structure (3.2.35) for the birth and death rates λ_N and μ_N. Although this is perfectly manageable, a parallel extension for our competition model would involve up to 12 parameters (three for each of the four rates), and so for illustrative simplicity let us just work with the positive and negative components of the deterministic competition equations (8.1.2) and form the six-parameter system

$$B_i(N_1, N_2) = r_i N_i \quad \text{and} \quad D_i(N_1, N_2) = N_i(s_{ii} N_i + s_{ij} N_j) \quad (i, j = 1, 2; j \neq i). \tag{8.1.25}$$

So in the unstable equilibrium case exhibited in Figure 8.2iv, at the equilibrium point $(4, 4)$ we have $B_1 = D_1 = 0.96$ and $B_2 = D_2 = 0.48$ (the individual species rates must exhibit zero net flow at the stable equilibrium point). Thus

$$\Pr[(4, 4) \to (5, 4), (3, 4), (4, 5), (4, 3)] = (1/6)(2, 2, 1, 1),$$

and so the first step is equally likely to take the process into the deterministic regime in which N_1 wins or N_2 wins. Whilst on starting from $(1, 1)$ in the stable equilibrium regime (iii), since $B_1 = 0.24$, $D_1 = 0.06$, $B_2 = 0.12$ and $D_2 = 0.03$ we have

$$\Pr[(1, 1) \to (2, 1), (0, 1), (1, 2), (1, 0)] = (1/15)(8, 2, 4, 1).$$

Hence the process involves the immediate extinction of species 1 with probability $2/15$ and species 2 with probability $1/15$; in which case the process then converges towards the single-species stable equilibrium points $(6, 0)$ and $(0, 6)$. In both examples, stochastic variation is clearly of paramount importance.

Stochastic simulation

The easiest way to demonstrate stochastic behaviour is through simulation, especially since the algorithm just involves a straightforward extension of that already constructed for the general single-species birth–death process (Section 3.2.2). For suppose we write

$$R(N_1, N_2) = B_1(N_1, N_2) + D_1(N_1, N_2) + B_2(N_1, N_2) + D_2(N_1, N_2). \tag{8.1.26}$$

Then for $\{U\}$ a sequence of independent $U(0, 1)$ random variable, the inter-event times τ are exponentially distributed with parameter $R(N_1, N_2)$, i.e.

$$\tau = \ln(U)/R(N_1, N_2). \tag{8.1.27}$$

Whilst the probability that the next event is a birth or death of a member of species i is B_i/R or D_i/R, respectively. This gives rise to the following two-species procedure.

General two-species simulation algorithm (A8.1)
    (i) read in parameters
    (ii) set $t = 0$, $N_1 = N_1(0)$ and $N_2 = N_2(0)$
    (iii) print t, N_1 and N_2
    (iv) generate a new $U \sim U(0, 1)$ random number

(v) evaluate B_1, D_1, B_2, D_2 and R
(vi) if $U \leq B_1/R$ then $N_1 \to N_1 + 1$
else if $U \leq (B_1 + D_1)/R$ then $N_1 \to N_1 - 1$
else if $U \leq (B_1 + D_1 + B_2)/R$ then $N_2 \to N_2 + 1$
else $N_2 \to N_2 - 1$
(vii) generate a new $U \sim U(0,1)$ random number
(viii) update t to $t - [\ln(U)]/R$
(ix) return to (iii)

Steps (v) and (vi) are easily extended to enable the simulation of general m-species interaction processes.

In order to compare stochastic trajectories with their corresponding deterministic phase plots (shown in Figure 8.2), let us consider a variety of start points and trajectory regimes. We retain the parameters used for Figure 8.2, except that r_1 and r_2 are doubled to 0.48 and 0.24, respectively, in order to avoid early movement onto an axis.

Non-equilibrium case (i) Figures 8.3a&b show the trajectory and time-series plot for $B_1 = 0.48N_1$, $D_1 = N_1(0.02N_1 + 0.02N_2)$, $B_2 = 0.24N_2$ and $D_2 = N_2(0.02N_1 + 0.02N_2)$, and start point $N_1(0) = N_2(0) = 30$. Since here $D_1 = D_2$ for $N_1 = N_2$, we see that $N_1(t)$ and $N_2(t)$ swiftly decline together, until the birth rates roughly balance the death rates. At this point $N_2(t)$ gradually declines to zero (in line with the deterministic Figure 8.2i), and $N_1(t)$ rises back up to wander around its own carrying capacity value of $r_1/s_{11} = 24$. The *non-equilibrium case (ii)* follows similarly.

Stable equilibrium case (iii) In contrast, using the same start point with $B_1 = 0.48N_1$, $D_1 = N_1(0.04N_1 + 0.02N_2)$, $B_2 = 0.24N_2$ and $D_2 = N_2(0.01N_1 + 0.02N_2)$ swiftly leads to N_2 winning (Figure 8.3c) due to its dominant effect on N_1 at large values. Note that this runs contrary to the deterministic trajectory shown in Figure 8.2iii. Only by substantially reducing $N_2(0)$, here to 5, can this be prevented, with the process then progressing towards a stochastic meander around the deterministic equilibrium value $(8,8)$ (Figures 8.3c&d).

Non-stable equilibrium case (iv) Finally, Figure 8.3e shows two stochastic realizations starting from the neighbouring points $(30, 15)$ and $(30, 14)$, based on $B_1 = 0.48N_1$, $D_1 = N_1(0.02N_1 + 0.04N_2)$, $B_2 = 0.24N_2$ and $D_2 = N_2(0.02N_1 + 0.01N_2)$. Although the former leads towards N_2 winning (see Figure 8.3f) and the latter to N_1 winning, as suggested by the deterministic Figure 8.2iv, since the two trajectories initially cross this deterministic prediction is misleading. Also shown is a trajectory starting from $(2, 2)$ (Figure 8.3e), which roughly parallels the deterministic prediction (Figure 8.2iv) of initial growth of both N_1 and N_2 followed soon after by collapse onto the N_1-axis.

These examples show that stochastic behaviour is likely to be far more variable and less intuitively predictable than its deterministic counterpart, which highlights the danger of placing too much strategic importance on the deterministic approach.

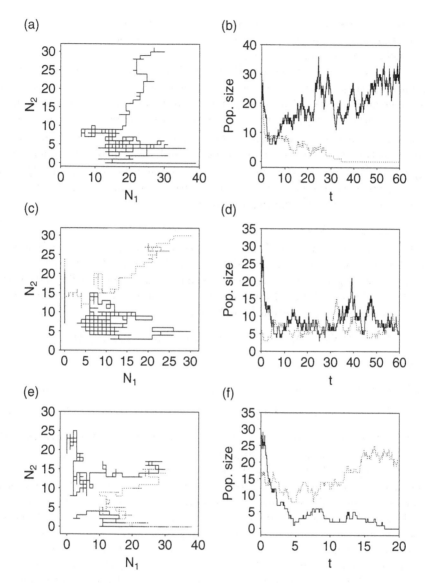

Figure 8.3 Stochastic trajectory and time-series plots paralleling the deterministic phase plots of Figure 8.2: (a) N_1 wins in non-equilibrium from start point $(30, 30)$, and (b) $N_1(t)$ (—) and $N_2(t)$ (⋯) against t; (c) N_2 wins and N_1, N_2 coexist in stable equilibrium from start points $(30, 30)$ and $(30, 5)$, and (d) $N_1(t)$ and $N_2(t)$ against t for the coexisting case; (e) N_2 and N_1 win from neighbouring start points $(30, 15)$ and $(30, 14)$, plus a trajectory from $(2, 2)$, and (f) $N_1(t)$ and $N_2(t)$ against t for N_2 winning.

Indeed, given the ease with which stochastic simulations can be generated, one might well argue that a deterministic population-based analysis should always be reinforced by a parallel stochastic simulation exercise.

As well as demonstrating the type and range of behavioural characteristics likely to be exhibited by trajectories under a given model structure, this simulation approach also provides an ideal way of obtaining numerical estimates $\hat{p}_{ij}(t)$ of the occupation probabilities $p_{ij}(t)$. For suppose we simulate the process n times using the same start position but different seeds, and record the number of times, $r_{ij}^{(n)}(t)$, that state (i,j) is occupied at time t. Then since realizations are independent of each other it follows that $\hat{p}_{ij}(t) = r_{ij}^{(n)}(t)/n$, with (binomial) estimated standard error (e.s.e.) $\sqrt{\{\hat{p}_{ij}(t)(1-\hat{p}_{ij}(t))/n\}}$. Here n is chosen to ensure that (within the limits of available compute time) the e.s.e.'s lie within some required bound.

Population size probabilities

To develop an analytic approach we first need to construct the two-species Kolmogorov equation for the population size probabilities $p_{ij}(t) = \Pr(X_1(t) = i, X_2(t) = j)$. This directly parallels that for the single-species equations developed earlier. For example, by considering the probability of no or one event in the small time interval $(t, t + dt)$, the single-species forward equation (3.3.1), namely

$$dp_i(t)/dt = \lambda_{i-1}p_{i-1}(t) - (\lambda_i + \mu_i)p_i(t) + \mu_{i+1}p_{i+1}(t), \tag{8.1.28}$$

becomes

$$dp_{ij}(t)/dt = B_1(i-1,j)p_{i-1,j}(t) + D_1(i+1,j)p_{i+1,j}(t) + B_2(i,j-1)p_{i,j-1}(t)$$
$$+ D_2(i,j+1)p_{i,j+1}(t) - R(i,j)p_{ij}(t). \tag{8.1.29}$$

The backward equation follows in exactly the same way, as does construction of the equation for the general multi-species process.

However, except in the most trivial cases equation (8.1.29) is intractable to direct solution, which means that we are forced back into obtaining either numerical or approximate theoretical solutions. Unfortunately, both of these approaches carry complications. Taking the numerical route first, we see from (8.1.29) that the range of support is infinite if $p_{ij}(t) > 0$ for all $0 \leq i,j < \infty$ and $t > 0$. The rates (8.1.25) clearly give rise to this situation. So in order to solve (8.1.29) numerically we need to impose the condition that for suitably chosen I, J we take $B_1(I,j) \equiv 0 \equiv B_2(i,J)$ for all $0 \leq j \leq J$ and $0 \leq i \leq I$. If I and J are too large, then $\{p_{ij}(t)\}$ will be effectively zero well within this rectangular region and so substantial computing time will be spent unnecessarily; whilst if I and J are too small, then some of the $p_{ij}(t)$ for $i > I$ or $j > J$ will be non-negligible, thereby giving rise to substantial numerical error during the calculation of the bounded probabilities. Unless highly accurate numerical solutions are required, which is often not the case, then it is usually sufficient to evaluate equations (8.1.29) over the successive time-points $t = dt, 2dt, \ldots$ for suitably small dt (see Renshaw, 1991), rather than employing a more sophisticated numerical integration procedure.

Extinction probabilities

Given that the downwards pressure on the two population sizes increases with (large) i and j under all four regimes shown in Figure 8.2, then provided $D_1(1,j) > 0$ and $D_2(i,1) > 0$ (for some i and j) one of the species is certain to become extinct, to be followed by eventual extinction of the other species. Whilst in principle the probability that species 1 (say) becomes extinct first may be evaluated by solving equation (8.1.29) numerically in order to form $[\sum_{j=1}^{J} p_{0j}(t)]/[1 - p_{00}(t)]$ for large t, in practice this computational procedure is likely to be highly inefficient.

A more expedient approach is to determine the extinction probabilities directly. Denote

$$q_{ij} = \Pr\{\text{species 1 becomes extinct first} | X_1(0) = i, X_2(0) = j\}, \quad (8.1.30)$$

and replace the rates $B_1(i,j), \ldots$ by the associated next-event probabilities $\tilde{B}_1(i,j) = B_1(i,j)/R(i,j)$, etc. Then on using the backward equation approach to consider the first move away from (i,j), in contrast to the forward equation (8.1.29) which considers the last move towards (i,j), we see that

$$q_{ij} = \tilde{B}_1(i,j)q_{i+1,j} + \tilde{D}_1(i,j)q_{i-1,j} + \tilde{B}_2(i,j)q_{i,j+1} + \tilde{D}_2(i,j)q_{i,j-1}. \quad (8.1.31)$$

To solve equations (8.1.31) numerically within the bounded region ($0 \leq i \leq I, 0 \leq j \leq J$) we may employ the following procedure.

Bounded population algorithm (A8.2)
(i) select appropriate values for I and J
(ii) put $q_{0j}^{(0)} = 1$ (species 1 loses) for $j = 1, \ldots, J$, and $q_{i0}^{(0)} = 0$ (species 1 wins) for $i = 1, \ldots, I$
(iii) place $q_{ij}^{(0)} = 0.5$ (for example), for $1 \leq i \leq I, 1 \leq j \leq J$
(iv) solve equations (8.1.31) over $i = 1, \ldots, I; j = 1, \ldots, J$ in the iterative form

$$q_{ij}^{(n+1)} = \tilde{B}_1(i,j)q_{i+1,j}^{(n)} + \tilde{D}_1(i,j)q_{i-1,j}^{(n)} + \tilde{B}_2(i,j)q_{i,j+1}^{(n)} + \tilde{D}_2(i,j)q_{i,j-1}^{(n)}, \quad (8.1.32)$$

for $n = 1, 2, \ldots$ until the $\{q_{ij}^{(n)}\}$-values converge; the boundary values may be determined via the linear interpolation approximation

$$q_{I+1,j}^{(n+1)} = 2q_{Ij}^{(n)} - q_{I-1,j}^{(n)} \quad (j = 1, \ldots, J)$$
$$q_{i,J+1}^{(n+1)} = 2q_{iJ}^{(n)} - q_{i,J-1}^{(n)} \quad (i = 1, \ldots, I). \quad (8.1.33)$$

To illustrate this procedure let us return to the parameter values (8.1.17) used for the deterministic phase plots of Figure 8.2. Suppose we choose boundary values $I = J = 20$, and deem the algorithm to have converged when $\sum_{i=0}^{I}\sum_{j=0}^{J}|q_{ij}^{(n+1)} - q_{ij}^{(n)}| < 10^{-6}$. Figure 8.4 shows the resulting $\{q_{ij}\}$-values, i.e. the probability that species 2 wins for start point (i,j), corresponding to these four scenarios. Specifically, the birth rates $B_1 = 0.24i$ and $B_2 = 0.12j$, whilst the death rates are given by

(i) $D_1 = i(0.02i + 0.02j)$ $D_2 = j(0.02i + 0.02j)$
(ii) $D_1 = i(0.04i + 0.04j)$ $D_2 = j(0.01i + 0.01j)$
(iii) $D_1 = i(0.04i + 0.02j)$ $D_2 = j(0.01i + 0.02j)$ (8.1.34)
(iv) $D_1 = i(0.02i + 0.04j)$ $D_2 = j(0.02i + 0.01j)$.

Under the two non-equilibrium regimes (i) and (ii) we see that the deterministic and stochastic predictions are in close agreement since $q_{ij} \simeq 0$ and 1, respectively, provided that in (i) i and in (ii) j do not lie close to zero. However, if they do, then there is a substantial level of disagreement, since $\tilde{D}_1(i,j) = i(i+j)/[12i + 6j + (i+j)^2]$ and $\tilde{D}_2(i,j) = j(i+j)/[24i + 12j + (4i+j)(i+j)]$ are both large enough to result in the early extinction of (i) species 1 and (ii) species 2, respectively, contrary to the behaviour predicted in Figure 8.2. For the stable equilibrium case (iii), the deterministic plot clearly provides little information on the q_{ij}, except to imply early extinction of species 1 if $i \simeq 1$ and $j \gg 1$, and conversely for species 2. However, with probability one, one of the species must become extinct first, and Figure 8.4c highlights that for most values of j this is likely to be species 1 unless $i \simeq 1$. Whilst for the unstable equilibrium case (iv), although the deterministic plots are quite definitive in their prediction unless (i,j) lies very close to the equilibrium point $(4,4)$, the outcome of the stochastic analysis is far less certain with q_{ij} remaining away from both 0 and 1 over a fairly wide range of (i,j)-values.

Number of steps to extinction

Only a slight change is required to convert this procedure into one that determines the expected number of steps to extinction. For extending the approach developed in Section 4.2.3 for the number of steps to absorption, let t_{ij} be the expected number of steps to extinction given $X_1(0) = i$, $X_2(0) = j$. Suppose, for example, that the first event is a move from (i,j) to $(i+1,j)$. Then the expected number of events from then on is $t_{i+1,j}$, whence the total expected duration of the competition process is $1 + t_{i+1,j}$. Thus by considering all four possible moves, and noting that $\tilde{B}_1 + \tilde{D}_1 + \tilde{B}_2 + \tilde{D}_2 = 1$, we obtain the general equation

$$t_{ij} = \tilde{B}_1(i,j)t_{i+1,j} + \tilde{D}_1(i,j)t_{i-1,j} + \tilde{B}_2(i,j)t_{i,j+1} + \tilde{D}_2(i,j)t_{i,j-1} + 1. \quad (8.1.35)$$

Now apart from the extra '1' on the right-hand side, equation (8.1.35) is identical to (8.1.31). Only the boundary conditions are substantially different, taking the four forms:

(a) $t_{0j} = 0$ $(j = 1, 2, \ldots)$ species 1 becomes extinct
(b) $t_{i0} = 0$ $(i = 1, 2, \ldots)$ species 2 becomes extinct
(c) $t_{i0} = t_{0j} = 0$ $(i,j = 1, 2, \ldots)$ either species becomes extinct
(d) $t_{00} = 0$ both species become extinct.

Whence modifying operations (ii) to (iv) in algorithm A8.2 leads to:

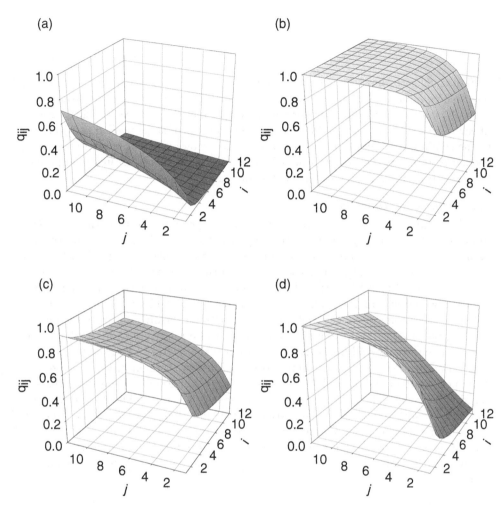

Figure 8.4 Probability, $\{q_{ij}\}$, of species 1 becoming extinct first for start positions $0 \leq i, j \leq 12$, corresponding to the four deterministic scenarios of Figure 8.2.

Bounded population algorithm (A8.3)
 (i) select appropriate values for I and J
 (ii) put $t_{i0}^{(0)} = t_{0j}^{(0)} = 0$ for $i = 1, \ldots, I$ and $j = 1, \ldots, J$
 (iii) set start values $t_{ij} = 1$ (for example), for $1 \leq i \leq I, 1 \leq j \leq J$
 (iv) solve equations (8.1.35) over $i = 1, \ldots, I; j = 1, \ldots, J$ in the iterative form

$$t_{ij}^{(n+1)} = \tilde{B}_1(i,j)t_{i+1,j}^{(n)} + \tilde{D}_1(i,j)t_{i-1,j}^{(n)} + \tilde{B}_2(i,j)t_{i,j+1}^{(n)} + \tilde{D}_2(i,j)t_{i,j-1}^{(n)} + 1 \tag{8.1.36}$$

for $n = 1, 2, \ldots$ until the $\{t_{ij}^{(n)}\}$-values converge; the boundary values may be determined via the linear interpolation approximation

406 *Two-species interaction processes*

$$t_{I+1,j}^{(n+1)} = 2t_{Ij}^{(n)} - t_{I-1,j}^{(n)} \qquad (j = 1, \ldots, J)$$

$$t_{i,J+1}^{(n+1)} = 2t_{iJ}^{(n)} - t_{i,J-1}^{(n)} \qquad (i = 1, \ldots, I). \tag{8.1.37}$$

Figures 8.5a&b show the expected number of steps to extinction for the stable and unstable equilibrium examples corresponding to Figures 8.4c&d. Apart from a steep drop-off near the axes, reflecting early extinction, Figure 8.5a shows a gradual, roughly linear, increase with i and j which reflects the number of steps needed to revert back to the equilibrium region. In contrast, Figure 8.5b shows that not only are fewer steps needed to reach extinction (as would be expected since this is an unstable process), but away from the axes the surface is less planar.

Expected time to extinction

An almost identical procedure gives rise to the expected time to extinction, $t(i,j)$, from start position (i,j). For as the exponentially distributed time to the next event from (i,j) has mean $1/R(i,j)$, we simply replace the $+1$ term in (8.1.36) by $+1/R(i,j)$. Figures 8.5c&d show these expected times to extinction corresponding to Figures 8.5a&b. Since $E[t(i,j)] = 1/R(i,j)$ decreases swiftly with increasing i and j, Figure 8.5c shows a marked difference to the step-plot of Figure 8.5a. In particular, note how the surface no longer continually rises with increasing i and j. In the unstable equilibrium case, as inter-event times are likely to be relatively short for large i and j, start points lying well above the equilibrium value give rise to a large number of short inter-event times until extinction occurs. Thus the time-plot, Figure 8.5d, is far more peaked than the step-plot shown in Figure 8.5b. Finally, comparing the step-based Figures 8.5a&b with the time-based Figures 8.5c&d highlights the danger of placing too much temporal interpretation on the underlying Markov chain, since this can exhibit substantially different structure from the full Markov process. Note that since successive events are independent, a similar procedure can be used to obtain the variance of the time to extinction, together with higher-order moments, should these be required.

Quasi-equilibrium probabilities

To develop the corresponding set of equations for the quasi-equilibrium probabilities, we first have to revisit our earlier univariate discussion in Section 3.2 concerning the two associated definitions, namely the exact version $\pi_i^{(Q)} = \lim_{t\to\infty} p_i(t)/[1-p_0(t)]$ from (3.2.13), and the pragmatic version $\tilde{\pi}_i = \tilde{\omega}_i/\sum_{k=1}^{\infty}\tilde{\omega}_k$ from (3.2.10) and (3.2.13). For our general bivariate situation the natural extension to the former is

$$\pi_{ij}^{(Q)} = \lim_{t\to\infty}\{p_{ij}(t)/[1-\sum_{i=0}^{\infty}p_{i0}(t)-\sum_{j=1}^{\infty}p_{0j}(t)]\} \qquad (i,j=1,2,\ldots). \tag{8.1.38}$$

Whilst the latter is given by the solution to the equations

$$\tilde{\pi}_{ij} = \tilde{\pi}_{i-1,j}\tilde{B}_1(i-1,j) + \tilde{\pi}_{i+1,j}\tilde{D}_1(i+1,j) + \tilde{\pi}_{i,j-1}\tilde{B}_2(i,j-1) + \tilde{\pi}_{i,j+1}\tilde{D}_2(i,j+1) \tag{8.1.39}$$

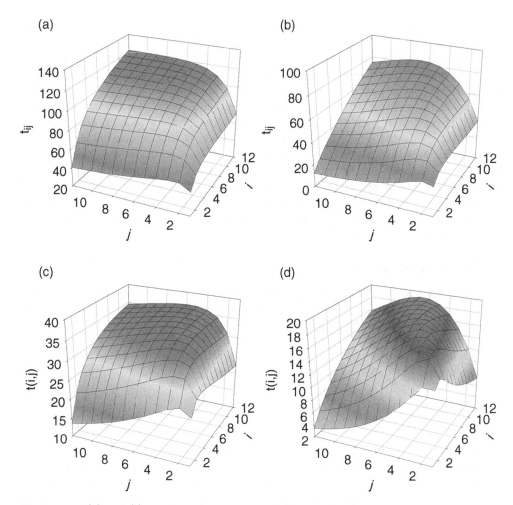

Figure 8.5 (a) and (b) number of steps, t_{ij}, and (c) and (d) times to extinction, $t(i,j)$, of either species 1 or 2 for start positions $0 \leq i, j \leq 12$, corresponding to the stable and unstable equilibrium scenarios, respectively, of Figure 8.4.

under the presumption that $D_1(1,j) = D_2(i,1) = 0$ for all $i, j = 1, 2, \ldots$. Now to generate $\Pr(X_1(t) = i, X_2(t)) = j)$ for large t, given that neither species has become extinct, we have to consider the last move made by the process, as distinct from (8.1.32) for $\{q_{ij}\}$, and (8.1.35) for $\{t_{ij}\}$ plus its counterpart $\{t(i,j)\}$, which are based on the first move. For in quasi-equilibrium the effect of the start position has disappeared, and so the first move cannot be considered. As (8.1.39) is the dual (forward) equation to the backward equation (8.1.31), it can be solved numerically using a parallel iterative procedure with $\{\tilde{\pi}_{ij}^{(n)}\}$ replacing $\{q_{ij}^{(n)}\}$ over $i = 1, \ldots, I, j = 1, \ldots, J$. The only real difference lies in the initial and boundary conditions, since we now have to take $\tilde{D}_1(1,j) = 0 = \tilde{D}_2(i,1)$ for $i, j > 0$ to ensure that neither species becomes extinct, together with (for example) the start probabilities $\tilde{\pi}_{ij}^{(0)} = 1/IJ$ over $i = 1, \ldots, I$ and $j = 1, \ldots, J$.

At this point it is worthwhile stressing that the increased dimensionality of the process renders direct theoretical determination of the $\{\tilde{\pi}_{ij}\}$ impossible. For whereas in the single-species process elements in the one-dimensional equilibrium probability vector can be constructed through successive application of the balance equations (3.2.4), i.e. $\mu_i \pi_i = \lambda_{i-1} \pi_{i-1}$, the two-species balance equations (8.1.39) do not reduce to a directly solvable recurrence relation. The reason is that in one dimension there is only one way that a process can reach $i+1$ from i, whilst in two (and higher) dimensions there are an infinity of different paths connecting two neighbouring states. This problem lies at the heart of analysing multi-species processes, and was raised by physicists as early as the 1920s in connection with the famous Ising problem, and rapidly gained the interest of statisticians and probabilists from the mid-1960s onwards (see, for example, Bartlett, 1974).

8.1.4 Moment equations

An alternative procedure, which does not require the imposition of zero death rates at $i = 1$ and $j = 1$, is to solve the full time-dependent forward equations (8.1.29) numerically; the simple device of replacing $dp_{ij}(t)/dt$ by $[p_{ij}(t + dt) - p_{ij}(t)]/dt$, for suitably small dt, is usually sufficient for most practical purposes. For the quasi-equilibrium distribution (8.1.38) can then be extracted by rescaling the resulting probabilities $p_{ij}(t)$ to form (for appropriately large t)

$$\pi_{ij}^{(Q)} \simeq \pi_{ij}^{(Q)}(t) = p_{ij}(t) / \sum_{r=1}^{I} \sum_{s=1}^{J} p_{rs}(t) \quad (0 < i \leq I, 0 < j \leq J; t \gg 1). \quad (8.1.40)$$

Moreover, knowledge of $p_{ij}(t)$ confers the added benefit of highlighting how the four extinction-related probability structures, namely extinction of neither species, species 1 only, species 2 only, and both species, change relative to each other with increasing t. Note that if all we require are the $\pi_{ij}^{(Q)}$, then we can take the time-increment dt to be (surprisingly) large without losing convergence of the algorithm.

For example, suppose we retain the death rates (8.1.34) but increase the individual birth rates to $r_1 = 1.2$ and $r_2 = 0.6$, thereby preventing early extinction since the new (deterministic) stable equilibrium point $(20, 20)$ now lies reasonably distant from both axes. Figure 8.6 shows the probabilities $p_{ij}(t)$ evaluated from (8.1.40) at time $t = 100$, with $dt = 0.001$, using the (equilibrium) start point $i = j = 20$. Here we placed $B_1(i, 50) = B_2(50, j) = 0$ over $0 \leq i, j \leq 50$ to ensure that probability did not leak from the system at the upper boundaries. Note that this distribution comprises three separate features, namely: a central unimodal distribution, which provides a reasonable indication of the shape of the quasi-equilibrium distribution (for $i, j > 0$); and two unimodal single-species logistic distributions centred around the *individual* deterministic equilibrium points $i^* = 30$ and $j^* = 30$. The latter result from extinction of species 2 ($i > 0, j = 0$) and 1 ($i = 0, j > 0$), respectively, before $t = 100$. Note that

Competition processes 409

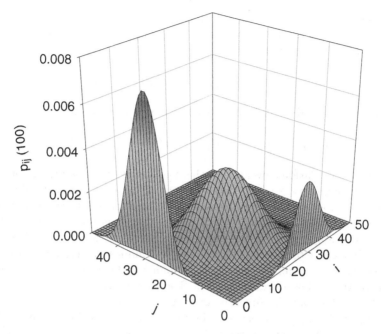

Figure 8.6 Probabilities $p_{ij}(100)$ over $i, j = 0, 1, \ldots, 50$ for $B_1 = 1.2i$, $D_1 = i(0.04i + 0.02j)$, $B_2 = 0.6j$ and $D_2 = j(0.01i + 0.02j)$.

these two univariate distributions are sited far enough away from zero to ensure that early extinction of *both* species, i.e. $p_{00}(100)$, is negligible.

Derivation of the associated moment structure has limited appeal in such situations, since it relates to a combined measure across three distributions (four if $p_{00}(t)$ is non-negligible). A far more useful exercise would be to determine the conditional moment structures based on (i) $i, j > 0$ (through quasi-equilibrium), (ii) $i > 0, j = 0$ and (iii) $i = 0, j > 0$; the last two have already been considered via the single-species logistic process in Section 3.2. Nevertheless, it is instructive to illustrate how moments can be derived, both directly from the probability equations, and indirectly via construction of the associated generating functions. Inserting the birth and death rates (8.1.25) into the forward equations (8.1.29) yields

$$dp_{ij}(t)/dt = r_1(i-1)p_{i-1,j}(t) + (i+1)[s_{11}(i+1) + s_{12}j]p_{i+1,j}(t)$$
$$+ r_2(j-1)p_{i,j-1}(t) + (j+1)[s_{21}i + s_{22}(j+1)]p_{i,j+1}(t) \quad (8.1.41)$$
$$- [r_1i + r_2j + i(s_{11}i + s_{12}j) + j(s_{21}i + s_{22}j)]p_{ij}(t).$$

Whence on extending the univariate definition (2.1.28) for the rth raw moment to the bivariate (r, s)th raw moment

$$\mu'_{rs}(t) \equiv E[(X_1(t))^r (X_2(t))^j] = \sum_{i,j=0}^{\infty} i^r j^s p_{ij}(t), \quad (8.1.42)$$

multiplying equation (8.1.41) by $i^r j^s$ and then summing over $i, j = 0, 1, \ldots$ leads to a series of recurrence equations for $\{\mu'_{rs}(t)\}$. For example, taking $r = 1$ and $s = 0$ yields the first-order stochastic update equation

$$d\mu'_{10}(t)/dt = r_1\mu'_{10}(t) - s_{11}\mu'_{20}(t) - s_{12}\mu'_{11}(t). \tag{8.1.43}$$

Whilst taking $r = 2$ and $s = 0$ leads to the second-order equation

$$d\mu'_{20}/dt = r_1(2\mu'_{20} + \mu'_{10}) + s_{11}(\mu'_{20} - 2\mu'_{30}) + s_{12}(\mu'_{11} - 2\mu'_{21}). \tag{8.1.44}$$

Inspection of these relations reveals that there are always terms on the right-hand side of higher-order (namely $r + s + 1$) than exist on the left ($d\mu'_{rs}/dt$ is of order $r + s$). So no matter what order we choose to truncate this series of equations, there will always be more variables than equations, which means that it is not possible to develop a direct solution through this recurrence system. This once again illustrates the general problem of moment closure when dealing with nonlinear transition rates; the only way of proceeding is to replace moments of higher order than some $r + s$ by either zero, or a function of lower-order moments. The construction of optimal strategies for achieving this latter route is currently an open problem.

Equation (8.1.43) exposes another major issue. For on denoting the means by $m_1 = \mu'_{10}$ and $m_2 = \mu'_{01}$, and the variances and covariance by $\sigma_1^2 = \mu'_{20} - (\mu'_{10})^2$, $\sigma_2^2 = \mu'_{02} - (\mu'_{01})^2$ and $\sigma_{11} = \mu'_{11} - \mu'_{10}\mu'_{01}$, it takes the form

$$dm_1/dt = m_1[r_1 - (s_{11}m_1 + s_{12}m_2)] - s_{11}\sigma_1^2 - s_{12}\sigma_{11}, \tag{8.1.45}$$

with a similar expression for dm_2/dt. Comparison with the corresponding deterministic equations (8.1.2) now exposes a wide chasm between the deterministic and stochastic analyses. For the presence of second-order moments in (8.1.45) shows that the resulting solutions may be radically different. Only when the second-order central moments are zero will the two scenarios produce the same results. Nevertheless, one must be careful not to be too critical of deterministic analyses. For example, using the first-order equations (8.1.43) to determine $m_1(t) = \mu'_{10}(t)$ and $m_2(t) = \mu'_{01}(t)$ necessitates complete knowledge of the second-order raw moments $\mu'_{20}(t)$, etc. Yet because of the moment closure problem these can only be determined approximately via (8.1.44), and determining the level of the resulting induced inaccuracy is non-trivial. Moreover, if $m_1(t)$ and $m_2(t)$ remain large throughout the time period $(0, T)$ of interest, then it may transpire that $\sigma_1^2(t)$, $\sigma_{11}(t)$ and $\sigma_2^2(t)$ are consistently small relative to them. In such situations the effect of using the deterministic equations may be no worse than using the truncated second-order stochastic ones.

An alternative way of generating moments, which avoids constructing a separate equation for each $\mu'_{rs}(t)$, is to proceed via the p.g.f. $G(z_1, z_2; t) \equiv \sum_{i,j=0}^{\infty} p_{ij}(t)z_1^i z_2^j$. First multiply equations (8.1.41) by $z_1^i z_2^j$ and sum over $i, j = 0, 1, 2, \ldots$. For on noting that $\partial G/\partial z_1 = \sum_{i,j=0}^{\infty} i p_{ij}(t) z_1^{i-1} z_2^j$, $\partial^2 G/\partial z_1^2 = \sum_{i,j=0}^{\infty} i(i-1) p_{ij}(t) z_1^{i-2} z_2^j$, etc., this leads to the partial differential equation

$$\frac{\partial G}{\partial t} = (r_1 z_1 - s_{11})(z_1 - 1)\frac{\partial G}{\partial z_1} + (r_2 z_2 - s_{22})(z_2 - 1)\frac{\partial G}{\partial z_2} \qquad (8.1.46)$$

$$+ s_{11} z_1 (1-z_1)\frac{\partial^2 G}{\partial z_1^2} + [s_{12} z_2(1-z_1) + s_{21} z_1(1-z_2)]\frac{\partial^2 G}{\partial z_1 \partial z_2} + s_{22} z_2(1-z_2)\frac{\partial^2 G}{\partial z_2^2}.$$

Even if a closed-form solution could be obtained from this equation, it would be too complex and opaque to be useful. For although (in principle) we could differentiate this solution (r,s) times and place $z_1 = z_2 = 1$ in order to form the bivariate factorial moments

$$\mu_{(rs)}(t) \equiv \mathrm{E}[X_1(t)(X_1(t)-1)\ldots(X_1(t)-r+1)X_2(t)(X_2(t)-1)\ldots(X_2(t)-s+1)]$$

$$= \sum_{i=r}^{\infty}\sum_{j=s}^{\infty} i(i-1)\ldots(i-r+1)j(j-1)\ldots(j-s+1)p_{ij}(t)$$

$$= \left.\frac{\partial G^{r+s}(z_1, z_2; t)}{\partial z_1^r \partial z_2^s}\right|_{z_1=z_2=1}, \qquad (8.1.47)$$

in practice this would not be feasible. We can, however, still perform the same operation on the differential equation (8.1.46). For example, differentiating this once with respect to z_1 yields

$$d\mu_{(10)}/dt = (r_1 - s_{11})\mu_{(10)} - s_{11}\mu_{(20)} - s_{12}\mu_{(11)}. \qquad (8.1.48)$$

Whence on noting that $\mu_{(10)} = m_1$ and $\mu_{(01)} = m_2$ with $\mu_{(20)} = \sigma_1^2 + m_1^2 - m_1$, $\mu_{(11)} = \sigma_{11} + m_1 m_2$ and $\mu_{(02)} = \sigma_2^2 + m_2^2 - m_2$, we recover the central moment equation (8.1.45). Whether one prefers to follow the direct moment equation route, or else proceed by differentiating the p.g.f. equation, is clearly a matter of personal taste.

Whilst some situations (as found, for example, in quantum optics) lend themselves naturally to the construction of factorial moments, other scenarios (often found in statistics) are best interpreted in terms of raw or central moments. Fortunately, only slight changes need to be made to the underlying p.g.f. equation in order to effect this; we simply employ a bivariate extension to the univariate case previously discussed in Section 2.1.3. Considering the raw moments (8.1.42) first, place $z_1 = e^{\theta_1}$ and $z_2 = e^{\theta_2}$. Then the m.g.f.

$$M(\theta_1, \theta_2; t) \equiv G(e^{\theta_1}, e^{\theta_2}; t) = \sum_{i,j=0}^{\infty} e^{i\theta_1 + j\theta_2} p_{ij}(t) \qquad (8.1.49)$$

$$= \sum_{r,s=0}^{\infty} \frac{\theta_1^r \theta_2^s}{r!s!} \sum_{i,j=0}^{\infty} i^r j^s p_{ij}(t) = \sum_{r,s=0}^{\infty} \frac{\mu'_{rs}(t)\theta_1^r \theta_2^s}{r!s!}.$$

Thus μ'_{rs} is the coefficient of $\theta_1^r \theta_2^s/r!s!$ in the expansion of $M(\theta_1, \theta_2; t)$. Equivalently, we can differentiate $M(\theta_1, \theta_2; t)$ to obtain

$$\mu'_{rs} \equiv \left.\frac{\partial^{r+s} M(\theta_1, \theta_2; t)}{\partial \theta_1^r \partial \theta_2^s}\right|_{\theta_1=\theta_2=0} \qquad (8.1.50)$$

as a direct extension of the univariate situation (2.1.30). To construct the parallel p.g.f. equation to (8.1.46), note that for $i = 1, 2$

$$\frac{\partial G}{\partial t} = \frac{\partial M}{\partial t}, \quad \frac{\partial G}{\partial z_i} = e^{-\theta_i}\frac{\partial M}{\partial \theta_i}, \quad \frac{\partial^2 G}{\partial z_i^2} = e^{-2\theta_i}\left[\frac{\partial^2 M}{\partial \theta_i^2} - \frac{\partial M}{\partial \theta_i}\right] \quad \text{and}$$

$$\frac{\partial^2 G}{\partial z_1 \partial z_2} = e^{-(\theta_1+\theta_2)}\frac{\partial^2 M}{\partial \theta_1 \partial \theta_2}.$$

So (8.1.46) becomes

$$\frac{\partial M}{\partial t} = r_1(e^{\theta_1} - 1)\frac{\partial M}{\partial \theta_1} + r_2(e^{\theta_2} - 1)\frac{\partial M}{\partial \theta_2} - s_{11}(1 - e^{-\theta_1})\frac{\partial^2 M}{\partial \theta_1^2}$$

$$- s_{22}(1 - e^{-\theta_2})\frac{\partial^2 M}{\partial \theta_2^2} - [s_{12}(1 - e^{-\theta_1}) + s_{21}(1 - e^{-\theta_2})]\frac{\partial^2 M}{\partial \theta_1 \partial \theta_2}, \quad (8.1.51)$$

which is marginally simpler. Differentiating (8.1.51) to form $\partial^{r+s}M/\partial\theta_1^r\partial\theta_2^s$ at $\theta_1 = \theta_2 = 0$ then yields a set of equations for the raw moments $\{\mu'_{rs}\}$. For example, differentiating once with respect to θ_1 gives

$$d\mu'_{10}/dt = r_1\mu'_{10} - s_{11}\mu'_{20} - s_{12}\mu'_{11}, \quad (8.1.52)$$

and on substituting for $\mu'_{20} = \sigma_1^2 + m_1^2$, $\mu'_{11} = \sigma_{11} + m_1 m_2$ and $\mu'_{02} = \sigma_2^2 + m_2^2$ we recover the central moment equation (8.1.45).

A more straightforward way of developing central moments is to extend the univariate c.g.f. (2.1.35) to the bivariate form

$$K(\theta_1, \theta_2; t) \equiv \ln[M(\theta_1, \theta_2; t)] = \sum_{r,s=0}^{\infty} \frac{\kappa_{rs}\theta_1^r\theta_2^s}{r!s!}. \quad (8.1.53)$$

Thus the (r, s)th cumulant, κ_{rs}, is the coefficient of $\theta_1^r\theta_2^s/r!s!$ in the expansion of $K(\theta_1, \theta_2; t)$. Or equivalently, we can differentiate $K(\theta_1, \theta_2; t)$ to obtain

$$\kappa_{rs} \equiv \left.\frac{\partial^{r+s}K(\theta_1, \theta_2; t)}{\partial\theta_1^r\partial\theta_2^s}\right|_{\theta_1=\theta_2=0}. \quad (8.1.54)$$

Taking logarithms on both sides of (8.1.49) shows that

$$\ln[M(\theta_1,\theta_2;t)] = [(\mu'_{10}\theta_1 + \mu'_{01}\theta_2) + (\mu'_{20}\theta_1^2/2 + \mu'_{11}\theta_1\theta_2 + \mu'_{02}\theta_2^2/2) + \cdots]$$
$$- (1/2)[(\mu'_{10}\theta_1 + \mu'_{01}\theta_2) + \cdots]^2 + \cdots,$$

and comparing this relation with (8.1.53) recovers the central moments $\kappa_{10} = m_1$, $\kappa_{01} = m_2$, $\kappa_{20} = \sigma_1^2$, $\kappa_{11} = \sigma_{11}$ and $\kappa_{02} = \sigma_2^2$. Indeed, there is even equality to third-order, with $\mu_{30} = \kappa_{30}$ and $\mu_{21} = \kappa_{21}$; though differences do emerge at the fourth- and higher-orders (see Kendall and Stuart, 1977). For example,

$$\mu_{40} = \kappa_{40} + 3\kappa_{20}^2, \quad \mu_{31} = \kappa_{31} + 3\kappa_{20}\kappa_{11} \quad \text{and} \quad \mu_{22} = \kappa_{22} + \kappa_{20}\kappa_{02} + 2\kappa_{11}^2. \tag{8.1.55}$$

Moreover, on noting that $M = e^K$, equation (8.1.51) takes the form

$$\frac{\partial K}{\partial t} = r_1(e^{\theta_1} - 1)\frac{\partial K}{\partial \theta_1} + r_2(e^{\theta_2} - 1)\frac{\partial K}{\partial \theta_2}$$

$$- s_{11}(1 - e^{-\theta_1})\left[\frac{\partial^2 K}{\partial \theta_1^2} + \left(\frac{\partial K}{\partial \theta_1}\right)^2\right] - s_{22}(1 - e^{-\theta_2})\left[\frac{\partial^2 K}{\partial \theta_2^2} + \left(\frac{\partial K}{\partial \theta_2}\right)^2\right]$$

$$- [s_{12}(1 - e^{-\theta_1}) + s_{21}(1 - e^{-\theta_2})]\left[\frac{\partial^2 K}{\partial \theta_1 \partial \theta_2} + \left(\frac{\partial K}{\partial \theta_1}\right)\left(\frac{\partial K}{\partial \theta_2}\right)\right]. \tag{8.1.56}$$

So although the quadratic differential terms do increase the degree of complexity of this equation over the m.g.f. version (8.1.51), the resulting cumulants immediately yield exact central moments to low order (see above), and lie 'closer' to them than raw or factorial moments for higher orders. Thus less algebraic manipulation is required to construct the central moments. Differentiating (8.1.56) and then using (8.1.54) generates the set of cumulant equations. So with $r = 1$ and $s = 0$, for example, we recover the first central moment equation (8.1.45); working through raw moment equations, such as (8.1.43), is slightly more involved.

In principle we could use this procedure to generate the first few orders of cumulants, possibly through a symbolic programming language such as *Mathematica*, and then insert these expressions into the saddlepoint approximation developed in Section 7.3. However, this poses several problems. First, the underlying c.g.f. equation (8.1.56) is sufficiently complex to ensure that, as r and s increase, cumulant expressions will rapidly become too opaque to be of real practical use. Second, the quadmodal nature of the p.d.f. $p_{ij}(t)$ (see Figure 8.6) means that any unimodal saddlepoint p.d.f. is likely to be suspect. Third, one would have to investigate whether achieving moment closure by replacing all higher-order moments by zero is optimal in this situation. The first two points suggest that theoretical p.d.f. results will be of limited use, and so studies should be restricted to specific investigations based on numerical solution of the Kolmogorov equations (8.1.29), and either the resulting (true) quasi-extinction probabilities (8.1.38) or the pragmatic version developed through equations (8.1.39). This third point is well worth further study, and suggested ways forward include taking higher-order moments to be functions of the lower-order moments (Gillespie and Renshaw, 2007).

8.2 Predator–prey processes

Although our study of two interacting populations has so far been restricted to direct competition within and between species, the number of possible variations on this theme is considerable. For even with the basic quadratic structure (8.1.2), each of the parameters r_i and s_{ij} ($i, j = 1, 2$) may be positive or negative, or even zero. Yet this structure is itself only a simple case from a huge raft of possible functional relationships. If analytic progress is to be made we must therefore reduce the set of target models to worthwhile ones which relate directly to real-life problems. For example, suppose we denote the structure of our competition model though the configuration

$$\left(\begin{array}{c|cc} r_1 & s_{11} & s_{12} \\ r_2 & s_{21} & s_{22} \end{array}\right) = \left(\begin{array}{c|cc} + & - & - \\ + & - & - \end{array}\right), \quad (8.2.1)$$

where $+$ denotes enhancement, $-$ impedance, and 0 no effect. Then the predator–prey process, in which one species uses the other as a food supply and is of equal biological importance, takes the form

$$\left(\begin{array}{c|cc} + & 0 & - \\ - & + & 0 \end{array}\right). \quad (8.2.2)$$

Whilst if the second population not only lives on the first but also cultivates it in a symbiotic relationship (e.g. humans farming animals), we might take

$$\left(\begin{array}{c|cc} - & 0 & + \\ - & + & 0 \end{array}\right). \quad (8.2.3)$$

Moreover, if species 1 now corresponds to individuals who are susceptible to a disease and species 2 to those who are infected, then the associated epidemic process may be modelled through the basic representation

$$\left(\begin{array}{c|cc} 0 & 0 & - \\ - & + & 0 \end{array}\right). \quad (8.2.4)$$

Note that this is identical to the predator–prey form (8.2.2), apart from r_1 switching from $+$ to 0. The fundamental difference between these two structures lies in their interpretation, but before we investigate this further in Section 8.3 let us first develop the main ideas which underpin predator–prey processes. For early accounts of model development see Keyfitz (1977) and Hallam (1986).

8.2.1 The Lotka–Volterra process

Four types of predation may be distinguished, each of which is equivalent when expressed in its simplest mathematical form (Renshaw, 1991). Herbivores prey on plants or their fruits or seeds which may be damaged in the process. Carnivores prey on herbivores or other carnivores, and it is this behaviour that is most commonly understood as predation. Parasitism is a variant on predation, and involves the parasite laying eggs either on or in the host, which is then subsequently eaten. Cannibalism is a particular form of predation involving just one species; the predator and prey are usually the adults and young, respectively.

The earliest representations of predator–prey behaviour were constructed independently by Lotka (1925) and Volterra (1926). Although their points of departure were similar, Volterra's analysis was considerably deeper. Several possible cases were considered, with the simplest taking the configuration (8.2.2). One species, on finding sufficient food in its environment, multiplies indefinitely when left to itself, whilst the other perishes for lack of nourishment if left alone; but the second feeds on the first, and so the two species can coexist together. This simple process provides not only an excellent introductory illustration to predator–prey behaviour, but also a useful touchstone against which more complex modelling structures can be compared.

Let $N_1(t)$ denote the number of prey (or hosts), and $N_2(t)$ the number of predators (or parasites), at time t. Suppose that in the absence of predators, prey increase at rate r_1, whilst in the absence of prey, predators die at rate r_2. Then paralleling the deterministic competition scheme (8.1.2), the simplest representation is

$$dN_1/dt = N_1(r_1 - b_1 N_2)$$
$$dN_2/dt = N_2(-r_2 + b_2 N_1). \qquad (8.2.5)$$

Here b_1 denotes the death rate of a given prey being eaten by a given predator, b_2 the birth rate of a given predator due to killing and eating a given prey, and $N_1 N_2$ is the number of contact pairs, i.e. the number of different ways that the N_1 prey can interact with the N_2 predators. Thus in terms of the representation (8.1.2) we see that $r_1 > 0$, $s_{11} = 0$ and $s_{12} = b_1$, with $r_2 < 0$, $s_{21} = -b_2$ and $s_{22} = 0$. So if we ignore within-species interaction, then the competition and predator–prey models differ only in the second equation, which involves a sign change. Note that we have assumed that prey and predators mix homogeneously; the introduction of inhomogeneous mixing induces substantial complexity, and we shall (briefly) return to this type of 'spatial' process later.

Before we proceed to study equations (8.2.5) it is worthwhile injecting a note of caution concerning terminology. For whilst in population dynamics terms these are universally called the classic Lotka–Volterra equations, within the field of stochastic differential equations this name refers to the system

$$dN_1(t)/dt = N_1(t)[r_1 - s_{11} N_1(t) + s_{12} N_2(t)]$$
$$dN_2(t)/dt = N_2(t)[r_2 - s_{22} N_2(t) + s_{21} N_1(t)] \qquad (8.2.6)$$

for positive parameters r_i and s_{ij} ($i, j = 1, 2$). Thus here in the absence of interaction both species undergo separate logistic processes, whilst in its presence each species enhances the other. So we now have the configuration

$$\begin{pmatrix} r_1 & s_{11} & s_{12} \\ r_2 & s_{21} & s_{22} \end{pmatrix} = \begin{pmatrix} + & - & + \\ + & + & - \end{pmatrix}, \qquad (8.2.7)$$

which is substantially different from the ones outlined above. Since the system (8.2.6) is given considerable attention in the s.d.e. literature we shall not discuss it further here, but simply note that access is easily obtained through the discussion of asymptotic behaviour in Mao, Sabanis and Renshaw (2003) and numerical convergence issues in Marion, Mao and Renshaw (2002).

Equations (8.2.5) combine to form the single equation

$$\frac{dN_1}{dN_2} = \frac{N_1(r_1 - b_1 N_2)}{N_2(-r_2 + b_2 N_1)}, \tag{8.2.8}$$

which, on writing in the form

$$[-(r_2/N_1) + b_2]dN_1 = [(r_1/N_2) - b_1]dN_2,$$

integrates directly to give

$$r_2 \ln(N_1) - b_2 N_1 + r_1 \ln(N_2) - b_1 N_2 = \text{constant}. \tag{8.2.9}$$

This expression represents a family of closed curves in which each member corresponds to a different value of the constant, this being determined by the initial position $(N_1(0), N_2(0))$.

The fact that this deterministic solution is neutrally stable, i.e. it neither converges nor diverges, means that stochastic variation may well cause trajectories to switch across different family members. This implies that (8.2.9) may yield little insight into stochastic behaviour. Nevertheless, although we shall soon see that this is indeed true in terms of the shape of individual stochastic trajectories, the deterministic solution is still able to expose useful properties of the predator–prey system. Three such curves are illustrated in Figure 8.7 for the process

$$dN_1/dt = N_1(1.5 - 0.1N_2)$$
$$dN_2/dt = N_2(-0.25 + 0.01N_1) \tag{8.2.10}$$

with start points $N_1 = 1, 10, 20$ and $N_2 = 15$. Each trajectory follows a closed path in an anticlockwise direction indefinitely; there is neither damping towards the

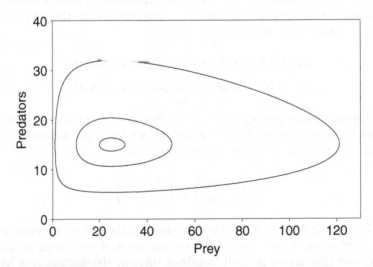

Figure 8.7 Family of closed curves for the deterministic Lotka–Volterra system (8.2.10) with $N_1(0) = 1, 10, 20$ and $N_2(0) = 15$.

deterministic equilibrium point $N_1^* = r_2/b_2 = 25$, $N_2^* = r_1/b_1 = 15$, nor an outwards drift towards the axes. The trajectory for $N_1(0) = 20$ is almost elliptical in shape, that for $N_1(0) = 10$ is more egg-shaped, whilst the trajectory for $N_1(0) = 1$ is squashed by the N_2-axis since it is not allowed to cross it.

Suppose we concentrate our attention on the inner curve, which corresponds to small departures from the equilibrium point (N_1^*, N_2^*), by writing either

$$N_i(t) = N_i^*[1 + n_i(t)] \quad \text{(multiplicative effect) or} \quad (8.2.11)$$

$$N_i(t) = N_i^* + n_i(t) \quad \text{(additive effect).} \quad (8.2.12)$$

Since $N_i(t)$ is integer, the former corresponds to $n_i(t) = 0, \pm 1/N_i^*, \pm 2/N_i^*, \ldots$ and the latter to $n_i(t) = \pm 1, \pm 2, \ldots$. So (8.2.11) better suits the concept of a 'small deviation' from equilibrium. Indeed, although both transformations yield the same first-order approximation, this does not apply for higher-order approximations, so retaining (8.2.11) for general use, rather than (8.2.12), is probably safer. Substituting this form into equations (8.2.10) and ignoring terms in $n_1(t)n_2(t)$ gives the first-order linear equations

$$dn_1/dt = -r_1 n_2 (1 + n_1) \simeq -r_2 n_2$$

$$dn_2/dt = r_2 n_1 (1 + n_2) \simeq r_1 n_1, \quad (8.2.13)$$

which easily integrate to yield the linearized solution

$$N_1(t) = (r_2/b_2)[1 + \alpha \cos\{t\sqrt{(r_1 r_2)} + \beta\}]$$

$$N_2(t) = (r_1/b_1)[1 + \alpha \sqrt{(r_2/r_1)} \sin\{t\sqrt{(r_1 r_2)} + \beta\}]. \quad (8.2.14)$$

The constants α and β are determined from the known initial values $N_1(0) = (r_2/b_2)[1 + \alpha \cos(\beta)]$ and $N_2(0) = (r_1/b_1)[1 + \alpha \sqrt{(r_2/r_1)} \sin(\beta)]$.

As $\cos^2(\beta) + \sin^2(\beta) = 1$, the ellipsoidal shape of the local trajectories, namely

$$n_1^2 r_2 + n_2^2 r_1 = \alpha^2 r_2, \quad (8.2.15)$$

follows immediately. Alternatively, we can substitute (8.2.11) directly into the closed curve solution (8.2.9) and expand the log-term, thereby obtaining

$$r_2[n_1^2/2 - n_1^3/3 + \cdots] + r_1[n_2^2/2 - n_2^3/3 + \cdots] = \text{constant}. \quad (8.2.16)$$

For on taking terms to $O(n_i^2)$ we recover the first-order curves (8.2.15), whilst to obtain the corresponding rth-order curves we simply take terms to $O(n_i^{r+1})$.

Various deterministic properties can be extracted directly from the local solutions (8.2.14):

(i) both populations vary sinusoidally with period $T = 2\pi/\sqrt{r_1 r_2}$ (so in example (8.2.10) $T = 10.26$);
(ii) the populations are always $\pi/2$ (i.e. 90 degrees) out of phase, prey leading;
(iii) the prey and predator amplitudes are $\alpha r_2/b_2$ and $(\alpha/b_1)\sqrt{r_1 r_2}$, respectively
– note that both their ratio and T are independent of the start point $(N_1(0), N_2(0))$;

(iv) because of the symmetry of the ellipse, the mean population sizes taken around a cycle are equal to the equilibrium values.

However, since these properties are based not only on a deterministic argument, but also on a linearized one, the extent to which they relate to stochastic behaviour is clearly questionable.

Stochastic simulation

The stochastic analyses developed for the competition process in Section 8.1.3 also apply here, since they relate to totally general birth and death rates $B_i(N_1, N_2)$ and $D_i(N_1, N_2)$. Moreover, the most appropriate way of extracting the four stochastic transition rates from the two deterministic equations (8.2.5) is to equate positive terms to birth and negative ones to death, thereby yielding

$$B_1 = r_1 N_1, \quad D_1 = b_1 N_1 N_2, \quad B_2 = b_2 N_1 N_2 \quad \text{and} \quad D_2 = r_2 N_2. \qquad (8.2.17)$$

Note that whilst these prey rates are the same as those for species 1 in the competition decomposition (8.1.25), B_2 and D_2 are interchanged when switching between species 2 and predators. Simulation proceeds exactly as for the competition algorithm A8.1 in Section 8.1.3, and Figure 8.8 shows two stochastic trajectories corresponding to the deterministic model of Figure 8.7, i.e. $r_1 = 1.5$, $r_2 = 0.25$, $b_1 = 0.1$ and $b_2 = 0.01$. The first run starts at the deterministic equilibrium point $(25, 15)$, and since near this point B_i and D_i are in approximate balance the trajectory initially wanders about it before a random excursion takes it deeper into the influence of the squashed elliptical paths (8.2.9), when subsequent divergence onto the N_2-axis results in prey extinction. Now the further a trajectory moves from its equilibrium point the more 'deterministic force' it appears to experience. The second run demonstrates this effect

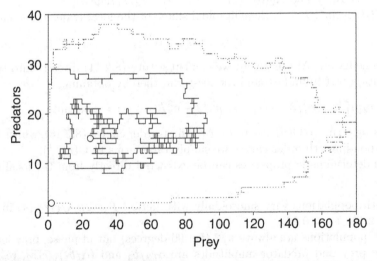

Figure 8.8 Two stochastic simulations corresponding to the deterministic Lotka–Volterra process (8.2.10) shown in Figure 8.7, with start points $(15, 10)$ (—) and $(2, 2)$ ($\cdots$).

clearly, showing decisive behaviour from the start value $(2,2)$ which lies well away from the equilibrium point. The trajectory moves in a fairly smooth manner in accord with the deterministic prediction almost until its end, when the prey again become extinct.

Partial insight into why the stochastic trajectories appear to spiral outwards, in contrast to the closed curve deterministic behaviour, can be gleaned by considering the expected update equations. Suppose that after k events a trajectory has expected position (m_k, n_k) at expected time t_k. Then as the time to the next event is exponentially distributed with parameter $R(m_k, n_k)$ given by (8.1.26), we see that

$$\tau_{k+1} = 1/R(m_k, n_k) \qquad (8.2.18)$$

together with

$$m_{k+1} = m_k + (B_1 - D_1)/R \quad \text{and} \quad n_{k+1} = n_k + (B_2 - D_2)/R. \qquad (8.2.19)$$

Figure 8.9 shows these expected update trajectories corresponding to Figures 8.7 and 8.8, and demonstrates an initial gradual unwinding until the process moves far away from (N_1^*, N_2^*) when the unwinding accelerates. Ultimately, the discretization leads to m_k or n_k becoming negative; though in practice once m_k or n_k approach 1 natural variation is likely to carry the trajectory onto an axis (as demonstrated in Figure 8.8).

This behavioural difference between deterministic closed cycles and stochastic divergent spirals may be explained by considering the respective gradients at (m_k, n_k). For although they are equal at this point, since

$$\frac{dN_2}{dN_1} = \frac{B_2 - D_2}{B_1 - D_1} = \frac{n_{k+1} - n_k}{m_{k+1} - m_k}$$

the expected update trajectory moves away from (m_k, n_k) in a linear piecewise fashion, whilst the deterministic trajectory follows a smooth convex path which must therefore lie inside it. So if the latter follows a closed loop, then the former must spiral outwards.

8.2.2 The Volterra process

Some modification to the stochastic Lotka–Volterra model is clearly required if it is to mimic the sustained stochastic cycles often observed in practice. Moreover, the assumption that in the absence of predators the prey population grows exponentially as a pure birth process with rate r_1 is also unrealistic. Fortunately, both issues are easily resolved by the trivial device of introducing the logistic term cN_1^2 into the prey equation. So the deterministic equations (8.2.5) now become

$$dN_1/dt = N_1(r_1 - cN_1 - b_1 N_2)$$
$$dN_2/dt = N_2(-r_2 + b_2 N_1), \qquad (8.2.20)$$

a form which was developed in considerable generality by Volterra (1931). The presence of c prevents exponential prey growth when $N_2 = 0$, and with it the associated boom-and-bust dynamics which leads to extinction of the predator species. Furthermore, by causing the deterministic trajectories to follow converging spirals, it allows the expected update trajectories to lie within the Lotka–Volterra (i.e. $c = 0$) deterministic closed loop trajectories.

To demonstrate the change in behaviour induced by this small parameter change, consider the (symmetric) Lotka–Volterra model with $r_1 = r_2 = 1$ and $b_1 = b_2 = 0.01$ (with $c = 0$). For since the associated deterministic equilibrium point $(N_1^*, N_2^*) = (100, 100)$ now lies much further away from the axes than the $(25, 15)$-value associated with Figures 8.7 and 8.8, we might anticipate a higher chance of maintaining at least a few full cycles before extinction occurs. Figure 8.10a confirms that this does indeed happen. After an initial movement away from (N_1^*, N_2^*), nine cycles are generated before a high prey population excursion increases the amplitude of the cycles which then leads to predator extinction four cycles later. Note the structural similarity between these stochastic trajectories and the expected update trajectories of Figure 8.9. Moreover, 12 cycles occur between the first and last prey peaks at $t = 4.45$ and 86.85, and the resulting average cycle length of 6.87 is in fair accord with that predicted earlier by linear deterministic theory, namely $T = 2\pi/\sqrt{r_1 r_2} = 2\pi \simeq 6.28$.

Comparing Figures 8.10a&b highlights the stabilizing effect induced by the small logistic parameter $c = 0.001$. For although the resulting prey carrying capacity of $r_1/c = 1000$ is much higher than the maximum prey value of 500 observed in Figure 8.10a, it is still sufficient to prevent a prey boom. Moreover, in spite of the Volterra simulations being far less smooth than the Lotka–Volterra ones, a spectral or autocorrelation analysis confirms the presence of a strong cyclic behaviour. The prey equilibrium point remains unchanged at $N_1^* = r_2/b_2$, but the presence of c reduces the predator equilibrium point to $N_2^* = (1/b_1)(r_1 - cr_2/b_2)$. This implies an effective upper bound on c, since $N_2^* > 0$ only if $c < r_1 b_2/r_2$ (here 0.01). If we wish to maintain the value of N_2^* as c rises we could increase the prey birth rate to $r_1' = r_1 + cr_2/b_2$ (here $r_1' = 1 + 100c$).

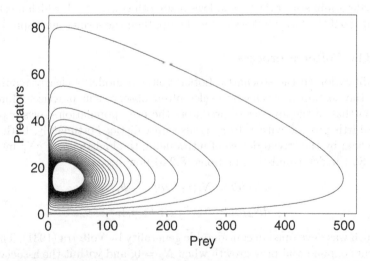

Figure 8.9 Expected update trajectories corresponding to the stochastic simulations of Figure 8.8 with a start point of $(15, 10)$.

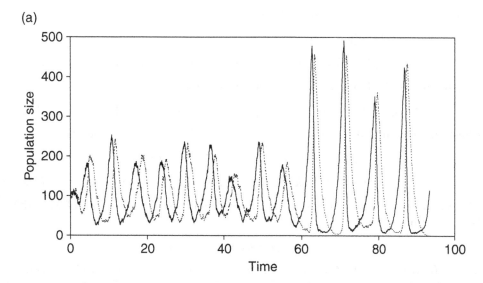
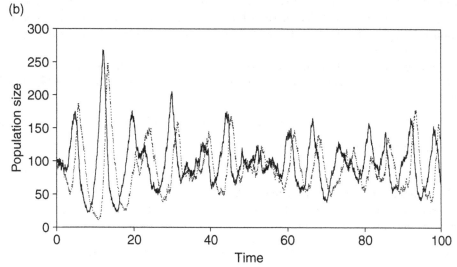

Figure 8.10 Stochastic simulations of the deterministic Volterra process (8.2.20) with $r_1 = r_2 = 1$ and $b_1 = b_2 = 0.01$ for (a) $c = 0$ and (b) $c = 0.001$ showing prey (—) and predators ($\cdots$) starting from $(100, 100)$.

Determining the properties of the linearized Volterra equations is trivial. For the general linear equations

$$dn_1/dt = k_{11}n_1 + k_{12}n_2$$
$$dn_2/dt = k_{21}n_1 + k_{22}n_2 \qquad (8.2.21)$$

with constant k_{ij} take the standard solution

$$n_1(t) = A\exp(\lambda_1 t) + B\exp(\lambda_2 t)$$
$$n_2(t) = (1/k_{12})[A(\lambda_1 - k_{11})\exp(\lambda_1 t) + B(\lambda_2 - k_{11})\exp(\lambda_2 t)] \qquad (8.2.22)$$

(e.g. Keyfitz, 1977; Renshaw, 1991). The eigenvalues

$$\lambda_1, \lambda_2 = (1/2)[(k_{11} + k_{22}) \pm \sqrt{\{(k_{11} - k_{22})^2 + 4k_{12}k_{21}\}}], \qquad (8.2.23)$$

whilst the constants A and B are easily determined via the initial conditions

$$n_1(0) = A + B$$
$$n_2(0) = (1/k_{12})[A(\lambda_1 - k_{11}) + B(\lambda_2 - k_{11})]. \qquad (8.2.24)$$

The associated behavioural characteristics are dominated by the values of λ_1 and λ_2. For on defining

$$\Delta = (k_{11} - k_{22})^2 + 4k_{12}k_{21}, \qquad (8.2.25)$$

we see from (8.2.23) that if $\Delta < 0$ then λ_1 and λ_2 are complex, which results in damped cycles if $k_{11} + k_{22} < 0$ (i.e. $\mathcal{R}(\lambda_i) < 0$) and divergent cycles if $k_{11} + k_{22} > 0$ (i.e. $\mathcal{R}(\lambda_i) > 0$). Whilst if $\Delta > 0$ then both λ_1 and λ_2 are real, resulting in straight exponential decay if both $\lambda_1 < 0$ and $\lambda_2 < 0$, and straight exponential growth if either $\lambda_1 > 0$ or $\lambda_2 > 0$.

To apply these results to our Volterra process we simply insert $N_i(t) = N_i^*(t)[1 + n_i(t)]$ ($i = 1, 2$) into equations (8.2.20) and retain the linear components, thereby obtaining

$$k_{11} = -cN_1^*, \quad k_{12} = -b_1 N_2^*, \quad k_{21} = b_2 N_1^* \quad \text{and} \quad k_{22} = 0. \qquad (8.2.26)$$

Placing these values in (8.2.23) gives

$$\lambda_1, \lambda_2 = (1/2)[-cN_1^* \pm \sqrt{\{(cN_1^*)^2 - 4b_1 b_2 N_1^* N_2^*\}}]. \qquad (8.2.27)$$

So for values of c small enough to ensure $\Delta < 0$, since $N_1^* = r_2/b_2$ the spiral trajectories converge inwards with $\exp\{\lambda_1 t\} - \exp\{-ct(r_2/2b_2)\}$. Note from (8.2.25) that as c increases towards its upper limit of $c_{max} = r_1 b_2/r_2$ (at higher values $N_2^* < 0$),

$$\Delta = (cN_1^*)^2 - 4b_1 b_2 N_1^* N_2^* = (cr_2/b_2)^2 - 4r_2(r_1 - cr_2/b_2)$$

switches from being complex to real when $c > c' = 2b_2[-1 + \sqrt{(1 + r_1/r_2)}]$.

For example, taking $r_1 = r_2 = 1$, $b_1 = b_2 = 0.01$ and $c = 0.001$ (as in Figure 8.10b) gives $N_1^* = r_2/b_2 = 100$ and $N_2^* = (1/b_1)(r_1 - cr_2/b_2) = 100(1 - 100c)$. So $c' = 0.02(\sqrt{2} - 1) \simeq 0.0083 < c_{max} = 0.01$ and the eigenvalues $\lambda_1, \lambda_2 = -50c \pm \sqrt{(2500c^2 + 100c - 1)}$. Whence it follows from (8.2.22) that for $k = 1, 2$ and constants A_k, B_k,

$$n_k(t) = e^{-50ct}[A_k \cos\{t\sqrt{(1 - 100c - 2500c^2)}\} + B_k \sin\{t\sqrt{(1 - 100c - 2500c^2)}\}]. \qquad (8.2.28)$$

Hence the associated cycle period

$$T = 2\pi/\sqrt{(1 - 100c - 2500c^2)}. \qquad (8.2.29)$$

Predator–prey processes 423

So including the small Volterra term $c = 0.001$ has little effect on the cycle period, merely increasing T from the Lotka–Volterra value of 6.2832 to 6.6323. However, over the time period $(0, T)$ it does cause the amplitude of the $n_k(t)$ to reduce by a factor of 0.737. It is this inwards (deterministic) spiralling force that enables stochastic realizations to avoid both axes, and hence prey or predator extinction. Note that $T \to \infty$ as c approaches c'.

Linearized second-order moments

Suppose we wish to choose c to be as tiny as possible consistent with the probability of extinction of either species being small over the time period of interest. A simple way of determining the relationship between c and the probability of extinction is to convert the deterministic equations (8.2.20) into stochastic equations by allowing for chance fluctuations in population size. That is, we add noise components $dZ_1(t)$ and $dZ_2(t)$ to produce

$$X_1(t+dt) - X_1(t) = X_1(t)[r_1 - cX_1(t) - b_1 X_2(t)]dt + dZ_1(t)$$
$$X_2(t+dt) - X_2(t) = X_2(t)[-r_2 + b_2 X_1(t)]dt + dZ_2(t). \quad (8.2.30)$$

Now let the stochastic process $\{X_1(t), X_2(t)\}$ remain close to the deterministic equilibrium point (N_1^*, N_2^*). Then in the small time interval $(t, t+dt)$, on extracting the birth and death components from equations (8.2.20) we may take

$$\Pr(dZ_1 = +1) = N_1^*(r_1 - cN_1^*)dt \quad \text{(prey birth)}$$
$$\Pr(dZ_1 = -1) = N_1^* b_1 N_2^* dt \quad \text{(prey death)}$$
$$\Pr(dZ_2 = +1) = N_2^* b_2 N_1^* dt \quad \text{(predator birth)} \quad (8.2.31)$$
$$\Pr(dZ_2 = -1) = N_2^* r_2 dt \quad \text{(predator death)}.$$

Thus under this local assumption $dZ_1(t)$ and $dZ_2(t)$ are independent of t; they are also presumed to be independent both of each other and also of $dZ_1(s)$ and $dZ_2(s)$ for all times $s \neq t$. Note that $\mathrm{E}[dZ_1] = \mathrm{E}[dZ_2] = 0$, together with

$$\mathrm{Var}[dZ_1] = 2b_1 N_1^* N_2^* dt \quad \text{and} \quad \mathrm{Var}[dZ_2] = 2b_2 N_1^* N_2^* dt. \quad (8.2.32)$$

Let us now parallel the single-species argument developed in Section 3.5.2 by writing

$$X_i(t) = N_i^*[1 + n_i(t)] \quad (i = 1, 2) \quad (8.2.33)$$

for small variables $n_i(t)$. Then on substituting (8.2.33) into (8.2.30), and assuming that the quadratic terms n_1^2, $n_1 n_2$ and n_2^2 may be considered negligible in comparison with n_1 and n_2, we obtain the linearized equations

$$n_1(t+dt) = n_1(t) - [cN_1^* n_1(t) + b_1 N_2^* n_2(t)]dt + dZ_1(t)/N_1^*$$
$$n_2(t+dt) = n_2(t) + [b_2 N_1^* n_1(t)]dt + dZ_2(t)/N_2^*. \quad (8.2.34)$$

Next we square and cross-multiply equations (8.2.34), take expectations on both sides of the resulting equations, and then ignore terms in dt^2. Whence on presuming that the process is in a state of quasi-equilibrium, for which $\text{Var}[n_1(t+dt)] = \text{Var}[n_1(t)]$, etc., writing $\sigma_i^2 = \text{Var}(n_i)$ and $\sigma_{12} = \text{Cov}(n_1, n_2)$ leads to

$$\sigma_1^2 = \sigma_1^2 - 2(cN_1^*\sigma_1^2 + b_1 N_2^* \sigma_{12})dt + \text{Var}(dZ_1/N_1^*)$$
$$\sigma_2^2 = \sigma_2^2 + 2b_2 N_1^* \sigma_{12} dt + \text{Var}(dZ_2/N_2^*) \qquad (8.2.35)$$
$$\sigma_{12} = \sigma_{12} + (b_2 N_1^* \sigma_1^2 - cN_1^* \sigma_{12} - b_1 N_2^* \sigma_2^2)dt.$$

Using the variance results (8.2.32) then yields the second-order (quasi-equilibrium) moments

$$\sigma_1^2 = [1 + (N_2^*/N_1^*)](b_1/cN_1^*)$$
$$\sigma_2^2 = [r_2 \sigma_1^2 + c(N_1^*/N_2^*)]/(b_1 N_2^*) \qquad (8.2.36)$$
$$\sigma_{12} = -1/N_2^*,$$

with $\text{Var}(X_i) = (N_i^*)^2 \sigma_i^2$ and $\text{Cov}(X_1, X_2) = N_1^* N_2^* \sigma_{12} = -N_2^*$.

Since $X_i = N_i^*(1+n_i)$, the smaller the standard deviation of n_i the less is the chance of extinction (i.e. $X_i < 0$) within a given time period. Suppose we place $\sigma_i = 1/k_i$ for $i = 1, 2$. Then in the case of our example with $r_1 = r_2 = 1$ and $b_1 = b_2 = 0.01$ we have

$$\sigma_1^2 = \frac{2 - 100c}{10000c} = 1/k_1^2 \quad \text{and} \quad \sigma_2^2 = \frac{(2 - 100c)/10000c + c/(1 - 100c)}{1 - 100c} = 1/k_2^2. \qquad (8.2.37)$$

So whilst the very low value $c = 0.000198$ implies early prey extinction, since it corresponds to $k_1 = 1$, only a slight rise in c to 0.000392 gives $k_1 = 2$ (medium-term prey extinction); whilst taking $c = 0.000583$ and 0.000769 results in $k_1 = 3$ and 4, which suggests long-term and effectively no prey extinction, respectively. The corresponding c-values for predators are only marginally higher, namely 0.000203, 0.000409, 0.000621 and 0.000840 for $k_2 = 1, 2, 3$ and 4, so predators are as likely to become extinct as prey. For our (even higher) earlier value $c = 0.001$ we have $k_1 = 5.26$ and $k_2 = 4.71$, which explains both the stability of the simulation shown in Figure 8.10b and the strong deterministic cyclic damping factor of 0.737.

Because any process possessing a quasi-equilibrium structure can be linearized, the approach is totally general in its applicability. Though the approximate nature of this linearization technique means that results should be viewed more from a qualitative than a quantitative perspective. Nevertheless, they are still very useful in providing substantial insight into the nature of the likely behavioural characteristics of a process, especially when they are combined with parallel simulation exercises. In the above example, a simulation run corresponding to $k_1 = 1$ (i.e. $c = 0.000198$) resulted in predator extinction at time 232.1, though the predator and prey populations had dropped to one at times 168.1 and 164.6, respectively. Increasing k_1 to 2 (i.e. $c = 0.000392$) extended the time until prey extinction to 415.4. Whilst taking $k_1 = 3$ and 4 (i.e. $c = 0.000583$ and 0.000769) only raised the times to prey extinction to 462.2 and 512.4. For once the amplitude of the process becomes large, the local linearization

Mean time to extinction

An approximate theoretical result for the mean time to extinction is easily obtained provided we assume (purely for extinction purposes) that each species can be treated independently. For on returning to the single-species result (3.3.40), we see that the mean time to extinction for a population of size one is given by

$$T_1 = \sum_{i=1}^{\infty} q_i \quad \text{where} \quad q_1 = \frac{1}{\mu_1} \quad \text{and} \quad q_i = \frac{\lambda_1 \lambda_2 \ldots \lambda_{i-1}}{\mu_1 \mu_2 \ldots \mu_i} \quad (i > 1). \tag{8.2.38}$$

Moreover, comparison with (3.2.10) shows that $\tilde{\omega}_i = \mu_1 q_i$. Whence on ignoring the distinction between pragmatic and exact quasi-equilibrium probabilities, it follows from (3.2.20) that since $\pi_1^{(Q)} = q_1 / \sum_{i=1}^{\infty} q_i$ result (8.2.38) also takes the form

$$T_1 = q_1 / \pi_1^{(Q)}. \tag{8.2.39}$$

So continuing our example, suppose we consider prey extinction by taking $X_2 \simeq N_2^* = 100(1 - 100c)$ (for predator extinction we take $X_1 \simeq N_1^*$). Then on placing

$$\lambda_i = \begin{cases} i(1 - ci) : i = 1, \ldots, [1/c] \\ 0 \quad : \text{otherwise} \end{cases} \quad \text{and} \quad \mu_i = 0.01 i X_2 = i(1 - 100c), \tag{8.2.40}$$

applying (8.2.38) gives

$$T_1 = \frac{1}{1 - 100c} + \sum_{i=2}^{[1/c]} \frac{(1-c)(1-2c)\ldots[1-(i-1)c]}{i(1-100c)^i}. \tag{8.2.41}$$

Whence the values of T_1 corresponding to the above c-values are

c:	0.000198	0.000392	0.000583	0.000769	0.001
T_1:	8.97	15.20	29.76	64.80	194.17

Although comparison with our previous simulated values suggests that this approach downweights T_1, it does nevertheless provide a useful feel for how the Volterra c-parameter affects the time to extinction in spite of the procedure involving a high degree of approximation. For extinction is essentially associated with boom-and-bust dynamics, in which a high surge in predator numbers causes a corresponding crash in the prey population, and this behaviour is far removed from taking either $X_1 = N_1^*$ or $X_2 = N_2^*$. Note that if we wish to apply (8.2.39) rather than (8.2.38) then we could determine $\pi_1^{(Q)}$ by using the Normal approximation

$$\pi_1^{(Q)} \simeq \Pr(0.5 \leq X_1 \leq 1.5) \simeq \Phi[(1.5 - N_i^*)/(N_i^* \sigma_i)] - \Phi[(0.5 - N_i^*)/(N_i^* \sigma_i)]. \tag{8.2.42}$$

8.2.3 A model for prey cover

One obvious way of obtaining sustained cyclic behaviour without losing the simplicity of the Lotka–Volterra model is to ensure that not all of the prey can become extinct. For then the stochastic divergent behaviour shown in Figure 8.8 automatically avoids the prey axis and thereby greatly enhances the chance of yielding sustained cycles. So suppose a constant number, k, of prey can take some form of cover which makes them inaccessible to predators. Then the N_2 predators can attack only $N_1 - k$ of the prey and the basic representation (8.2.5) changes to

$$dN_1/dt = r_1 N_1 - b_1(N_1 - k)N_2$$
$$dN_2/dt = -r_2 N_2 + b_2(N_1 - k)N_2 \qquad (8.2.43)$$

(Maynard Smith, 1974). As

$$dN_1/dt = 0 \quad \text{at} \quad r_1 N_1/b_1 = N_2(N_1 - k)$$
$$dN_2/dt = 0 \quad \text{at} \quad r_2 N_2/b_2 = N_2(N_1 - k), \qquad (8.2.44)$$

the equilibrium values are $N_1^* = (r_2/b_2) + k$ and $N_2^* = (r_1/b_1) + k(r_1 b_2/r_2 b_1)$, which increase linearly with the cover parameter k.

To determine the stability of (N_1^*, N_2^*) we note from (8.2.44) that the two isoclines (Section 8.1.2) $N_1 = k + r_2/b_2$ and $N_2 = (r_1/b_1)N_1/(N_1 - k)$ yield a phase-plane diagram (Figure 6.12 of Renshaw, 1991) with a similar local structure to Figure 8.1iii. So (N_1^*, N_2^*) corresponds to a stable equilibrium point. More formally, on writing equations (8.2.43) in the general form (8.1.18) with

$$dN_1(t)/dt = f_1(N_1, N_2) = r_1 N_1 - b_1(N_1 - k)N_2$$
$$dN_2(t)/dt = f_2(N_1, N_2) = -r_2 N_2 + b_2(N_1 - k)N_2, \qquad (8.2.45)$$

we see that at $N_i = N_i^*$

$$\partial f_1/\partial N_1 = -k r_1 b_2/r_2 \qquad \text{and} \qquad \partial f_1/\partial N_2 = -b_1 r_2/b_2$$
$$\partial f_2/\partial N_1 = (b_2/b_1)[r_1 + k r_1 b_2/r_2] \qquad \text{and} \qquad \partial f_2/\partial N_2 = 0.$$

Whence on applying result (8.1.24), it follows that because $\theta(N_1^*, N_2^*) = r_1(r_2 + kb_2) > 0$, (N_1^*, N_2^*) is a stable equilibrium point. Thus the cover parameter k clearly has a stabilizing effect, since it transforms the neutrally stable (i.e. closed) deterministic Lotka–Volterra cycles into a converging spiral.

To examine the nature of this convergence, let us follow the usual local linearization route by writing $N_i(t) = N_i^*[1 + n_i(t)]$. Then for small $n_i(t)$, equations (8.2.43) become

$$dn_1/dt = -k(r_1 b_2/r_2)n_1 - r_1 n_2$$
$$dn_2/dt = b_2 N_1^* n_1. \qquad (8.2.46)$$

Since these equations are identical to the general linear equations (8.2.21) with

$$k_{11} = -kr_1b_2/r_2, \quad k_{12} = -r_1$$
$$k_{21} = b_2 N_1^*, \quad k_{22} = 0, \qquad (8.2.47)$$

expression (8.2.25) becomes

$$\Delta = (kr_1b_2/r_2)^2 - 4b_2 r_1 N_1^*. \qquad (8.2.48)$$

Whence substituting these values into (8.2.23) yields the eigenvalues

$$\lambda_1, \lambda_2 = (1/2)[-(kr_1b_2/r_2) \pm \sqrt{\Delta}], \qquad (8.2.49)$$

and the general solution (8.2.22)–(8.2.24) may now be written down directly. So provided k is appropriately small (to ensure $\Delta < 0$), we have damped cycles around the equilibrium point. Moreover, the rate of convergence increases swiftly with k, since the $n_i(t)$ converge to zero at a speed determined by the damping factor $\exp(-kr_1b_2t/2r_2)$. Note that the presence of even a tiny amount of prey cover transforms neutrally stable deterministic cycles into convergent ones.

The period of oscillation, $T = 4\pi/\sqrt{(-\Delta)}$, decreases as $-\Delta$ increases. Now on writing (8.2.48) as

$$\Delta = (r_1b_2/r_2)^2(k - 2r_2^2/r_1b_2)^2 - 4r_2(r_1 + r_2), \qquad (8.2.50)$$

we see that Δ takes the minimum value $\Delta_{\min} = -4r_2(r_1 + r_2)$ when $k = 2r_2^2/r_1b_2$, which corresponds to a minimum period of oscillation of $T_{\min} = 2\pi/\sqrt{\{r_2(r_1 + r_2)\}}$. Whilst at $k = 0$ (i.e. no prey cover) $\Delta = -4r_1r_2$, so $T = T_0 = 2\pi/\sqrt{(r_1r_2)}$. Thus although increasing k first reduces T to $T_{\min}$, thereafter T increases indefinitely until $\Delta = 0$, at which point oscillations cease and straight exponential damping takes over.

Stochastic behaviour

Having already shown for the Volterra process how convergent deterministic cycles turn into persistent ones when stochastic variability is introduced (Figure 8.10), we might well anticipate that a similar situation happens for this process. Indeed, whereas the Volterra model raises a critical question concerning how large the logistic parameter c needs to be in order to ensure a low probability of extinction within a given time interval, here we can pose the same question in terms of the cover parameter k. To parallel our earlier Lotka–Volterra and Volterra analyses, we simply extract the stochastic birth and death rates from the deterministic equations (8.2.43), namely

$$B_1 = r_1 N_1, \quad D_1 = b_1(N_1 - k)N_2, \quad B_2 = b_2(N_1 - k)N_2 \text{ and } D_2 = r_2 N_2, \qquad (8.2.51)$$

and then proceed as before. Retaining the parameters $r_1 = r_2 = 1$ and $b_1 = b_2 = 0.01$, Figure 11a shows a simulation run for $k = 1$ starting from $(5, 5)$, and this demonstrates that protecting even a single prey can induce a strong stabilizing effect on the process. Moreover, whereas in this simulation predator extinction occurred at time $t = 243$, increasing k to 5 led to extinction at time 2552, whilst with $k = 10$ the process was still going strong at time 10,000. Comparison with the corresponding deterministic trajectories (Figure 11b) highlights the substantial difference between the stochastic

428 *Two-species interaction processes*

(a)

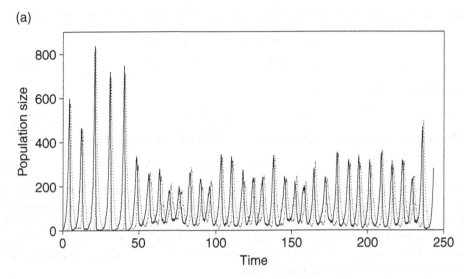

(b)

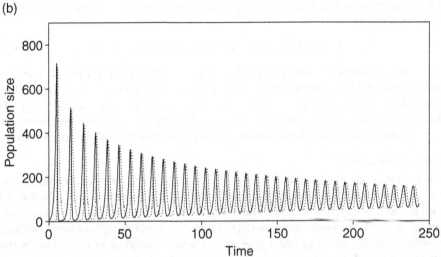

Figure 8.11 (*a*) Simulations of the Lotka–Volterra process with prey cover for $r_1 = r_2 = 1$, $b_1 = b_2 = 0.01$ and $k = 1$, showing prey (—) and predators ($\cdots$) starting from $(5,5)$; (*b*) shows the corresponding deterministic trajectories.

and deterministic behaviour, with the latter showing gradual convergence towards the equilibrium point $(N_1^*, N_2^*) = (101, 101)$.

8.2.4 Sustained deterministic and stochastic limit cycles

The strong differences exhibited between deterministic and stochastic behaviour for all three of the above models highlight a fundamental and serious issue. For a deterministic analyst, on being presented with a cyclic predator-prey data set, may well favour the simple Lotka–Volterra model on the grounds that it yields sustained deterministic cycles in contrast to the convergent cycles generated by the Volterra and prey-cover models. Whilst a stochastic analyst is likely to take a diametrically opposite view, since only the last two models can produce sustained stochastic cycles; the Lotka–Volterra model exhibits divergent oscillations onto an axis. Ideally both approaches should always be considered, with any gross disparity between their outcomes being given critical attention. In general, the best way forward is to adopt a pragmatic modelling approach, with a balance being struck between using key information about the process under consideration and the overall transparency of the resulting model.

Suppose, for example, that we wish to generate sustained stable limit cycles under both the deterministic and stochastic regimes. Then based on practical experience we know that there is an upper limit to how many prey a single predator may kill per unit time no matter how abundant they are. So following Leslie and Gower (1960), let us make the predator equation more realistic: first, assume that predators have an individual growth rate r_2 (i.e. they do not rely exclusively on a single species of prey for food); second, take the individual predator death rate to be $b_2(N_2/N_1)$ in order to let it depend on the *relative* sizes of the two populations. For the larger N_2/N_1 becomes, the smaller are the number of prey available to each predator, and consequently the resource available for predator growth declines. Thus the deterministic predator equation changes to

$$dN_2/dt = N_2[r_2 - b_2(N_2/N_1)]. \tag{8.2.52}$$

Extending this approach to the prey population, let us retain the underlying Volterra logistic process in order to keep the prey population under control in the absence of predators, and take the predator kill rate of an individual prey to be $f(N_1, N_2) = wN_2/(D + N_1)$ for some threshold value D. If the number of prey is small (i.e. $N_1 \ll D$) then we effectively return to the earlier situation with $f(N_1, N_2) \simeq (w/D)N_2$ (i.e. $b_1 = w/D$). Whilst conversely, if $N_1 \gg D$ then $f(N_1, N_2) \simeq wN_2/N_1$. Holling (1965) shows that this type of functional response is characteristic of invertebrate predators; that for vertebrate predators differs because they can learn to search for a prey species that has become more abundant. Note that the size of D relates to the prey's ability to evade attack: the more elusive the prey, the greater D becomes. Thus the corresponding prey equation is

$$dN_1/dt = N_1[r_1 - cN_1 - \{wN_2/(D + N_1)\}]. \tag{8.2.53}$$

This linear fractional form for predator–prey interaction is an extremely simple structure that (as we shall soon see) delivers the required sustained cyclic behaviour and is easy to interpret. If required, more complex expressions could easily be developed and tested by using polynomial fractional forms or an exponential structure such as $f(N_1, N_2) = w(1 - \exp\{-aN_1/w\})$.

Deterministic behaviour

It follows from (8.2.52) and (8.2.53) that the equilibrium point (N_1^*, N_2^*) is the solution to the two isocline equations

$$(r_1 - cN_1)(D + N_1) = wN_2 \quad \text{(when } dN_1/dt = 0) \tag{8.2.54}$$
$$r_2 N_1 = b_2 N_2 \quad \text{(when } dN_2/dt = 0). \tag{8.2.55}$$

That is,

$$N_1^* = (1/2c)[r_1 - cD - (wr_2/b_2) + \sqrt{\{[r_1 - cD - (wr_2/b_2)]^2 + 4cr_1 D\}}]$$
$$N_2^* = (r_2/b_2)N_1^*; \tag{8.2.56}$$

since $4cr_1 D > 0$ we discount the negative root as this yields a negative equilibrium point. Although the prey isocline (8.2.54) is a parabola, and the predator isocline (8.2.55) is a straight line through the origin, the analysis of the phase-plane diagram (Figure 8.12) directly parallels that of the simple competition process exhibited in Figure 8.1 where we had to consider four separate cases. For Figure 8.12 suggests three potentially different behavioural regimes depending on whether the point of intersection lies some distance to the left or right of the parabola peak or lies near to it.

Readers who are interested in the original stability analysis should refer to May (1974) and Tanner (1975), though the essence of the argument can be seen through direct inspection of equations (8.2.52) and (8.2.53). First note that since $dN_i/dt \to 0$ as $N_i \to 0$, the trajectories can never reach the axes. Moreover, for $N_1 \simeq 0$

$$dN_2/dt \simeq N_2(r_2 - b_2 N_2/N_1) < 0 \quad \text{for} \quad N_2 > r_2 N_1/b_2$$
$$dN_1/dt \simeq N_1(r_1 - wN_2/D) > 0 \quad \text{as soon as} \quad N_2 < r_1 D/w,$$

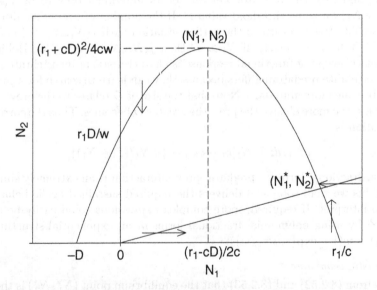

Figure 8.12 Stability diagram for the Holling–Tanner predator–prey model.

which ensures that N_2 decays until N_1 moves away from the axis. Similarly, for $N_2 \simeq 0$

$$dN_1/dt \simeq N_1(r_1 - cN_1)$$
$$dN_2/dt \simeq N_2 r_2 > 0,$$

so N_1 behaves as a single-species logistic process until N_2 has moved away from its axis. Finally, for large N_1 and N_2 we see that $dN_1/dt < 0$, with $dN_2/dt < 0$ as soon as $N_1 < b_2 N_2/r_2$, and so (N_1, N_2) must remain bounded for any given (positive) start condition. Since we have shown that neither extinction nor population explosion is possible, the existence of limit cycles is therefore guaranteed provided that the equilibrium point (N_1^*, N_2^*) is unstable.

On placing $N_i(t) = N_i^*[1 + n_i(t)]$, for small $n_i(t)$ equations (8.2.52) and (8.2.53) become

$$dn_1/dt = n_1[-cN_1^* + N_1^*(r_1 - cN_1^*)/(D + N_1^*)] - n_2(r_1 - cN_1^*)$$
$$dn_2/dt = n_1 r_2 - n_2 r_2, \qquad (8.2.57)$$

and whilst the coefficients $k_{21} = (r_1 - cN_1^*)$, $k_{21} = r_2$ and $k_{22} = -r_2$ take a simple form, we see that $k_{11} = -cN_1^* + N_1^*(r_1 - cN_1^*)/(D + N_1^*)$ is more involved. It is this feature that makes the local stability analysis more awkward than that for the parallel Volterra and prey-cover situations considered earlier. For now (8.2.25) becomes

$$\Delta = [-cN_1^* + N_1^*(r_1 - cN_1^*)/(D + N_1^*) + r_2]^2 - 4(r_1 - cN_1^*)r_2, \qquad (8.2.58)$$

and for local instability (implying stable limit cycles) we require $\Delta < 0$. Although it is not immediately apparent from expression (8.2.58) what this condition might imply, we can construct a pictorial representation of the stability structure around the intersection of the two isoclines similar to that shown in Figure 8.1 for the competition process (see Figure 6.9 of Renshaw, 1991). This shows that if (N_1^*, N_2^*) lies to the far left or right of the parabola maximum $(N_1', N_2') = ((r_1 - cD)/2c, (r_1 + cD)^2/(4cw))$ then the process is stable, and only when the slope of the second isocline (r_2/b_2) takes more middling values can limit cycles occur.

As a simple illustration, suppose we fix r_1, c, D and w (which determines the parabola in Figure 8.12). Then when $r_2 \simeq 0$ (i.e. the second isocline is almost horizontal) we have $N_1^* \simeq r_1/c$ and $N_2^* \simeq 0$, so $\Delta \simeq r_1^2 > 0$ which guarantees stability. Whilst on allowing r_2 to increase, for appropriate parameter values Δ will first dip below zero before tending to infinity as $r_2 \to \infty$ (when the second isocline is now vertical). For example, on taking $r_1 = 1$, $c = 0.001$, $D = 15$ and $w = 2$ we have $(N_1', N_2') = (492.5, 128.8)$. Whence at:

$r_2 = 0$	$N_1^* = 1000 \gg N_1'$	and $N_2^* = 0 \ll N_2'$	with $\Delta = 1 > 0$;
$r_2 = 0.078713$	$N_1^* = 691.8 > N_1'$	and $N_2^* = 108.9 < N_2'$	with $\Delta = 0$;
$r_2 = 0.13073$	$N_1^* = 492.5 = N_1'$	and $N_2^* = 128.8 = N_2'$	with $\Delta = -0.2483 < 0$;
$r_2 = 2.01$	$N_1^* = 2.11 \ll N_1'$	and $N_2^* = 8.54 \ll N_2'$	with $\Delta = \Delta(\min) = -3.4786$;
$r_2 = 3.8679$	$N_1^* = 1.04 \ll N_1'$	and $N_2^* = 8.01 \ll N_2'$	with $\Delta = 0$;
$r_2 = 5$	$N_1^* = 0.79 \ll N_1'$	and $N_2^* = 7.89 \ll N_2'$	with $\Delta = 5.510 > 0$.

So limit cycles occur when $0.13073 < r_2 < 3.8679$, with convergence to deterministic equilibrium otherwise. Note that Δ takes its minimum value at $r_2 = 2.01$ where (N_1^*, N_2^*) lies close to zero. Also, for all practical purposes, a large r_2 giving rise to $\Delta > 0$ can be discounted since the resulting values for N_i^* are impracticably low. However, care must be taken not to read too much into this *local* analysis, since limit cycle behaviour is essentially a *global* phenomenon and the isocline geometry away from the equilibrium point can be markedly different to that near to it.

Figure 8.13a shows a deterministic trajectory for parameter values $r_1 = 1$, $r_2 = 0.5$, $w = 1.8$, $D = 15$, $b_2 = 0.5$ and $c = 0.001$, starting from $(N_1(0), N_2(0)) = (15, 15)$. In the absence of predators the prey have carrying capacity $r_1/c = 1000$, whilst (8.2.56) yields the equilibrium values $N_1^* = N_2^* = 18.01$. Since (8.2.58) gives $\Delta = -0.928$, the local analysis suggests fairly strong limit cycle behaviour. The trajectory initially unwinds quite slowly about the equilibrium value, with the maximum value of $N_1(t)$ rising from 24.29 on the first cycle, to a mere 33.56 on the sixth. Only then does the pace of unwinding rapidly increase, with maxima of 91.84 on the tenth cycle, 377.59 on the twelve and 481.41 on the fourteenth, by which time the trajectory lies close to the limit cycle itself. In contrast, trajectories starting from high $N_i(0)$-values spiral in rapidly towards the limit cycle from the outset.

Stochastic behaviour

This particular example has been chosen to demonstrate not only the considerable range in population size, with $2.8 \leq N_1(t) \leq 499$ and $5.1 \leq N_2(t) \leq 246$, but also that when $N_1(t)$ and $N_2(t)$ are both small all cycles lie close to each other. Hence any stochastic perturbation in this region may well result in a severe change in behaviour, with large changes in amplitude and the corresponding possibility of extinction. The problem is exacerbated by having to define exactly what we mean by 'the corresponding stochastic process' to a given deterministic model, for there are an infinite number of them. In the case of the Volterra process (Figure 8.10), for example, we took the birth and death rates to be $B_1 = N_1(r_1 - cN_1)$, $D_1 = b_1 N_1 N_2$, $B_2 = b_2 N_1 N_2$ and $D_2 = r_2 N_2$. However, we could just as easily have selected $B_1 = r_1 N_1$ and $D_1 = N_1(cN_1 + b_1 N_2)$, which causes the N_1-population to switch from being bounded above by r_1/c to being unbounded. The stochastic properties of these two scenarios are clearly different, although the underlying deterministic structures are identical. Indeed, we could place

$$B_1 = N_1(r_1 - cN_1) + f_1(N_1, N_2), \qquad D_1 = b_1 N_1 N_2 + f_1(N_1, N_2),$$
$$B_2 = b_2 N_1 N_2 + f_2(N_1, N_2), \qquad D_2 = r_2 N_2 + f_2(N_1, N_2) \qquad (8.2.59)$$

for any positive functions $f_i(N_1, N_2)$. The best way of avoiding this trap when constructing the deterministic process is first to define the birth and death rates, and only then go on to construct the deterministic equation. Otherwise, we face the problem of how to decompose it in order to form the stochastic rates.

Exactly the same considerations apply to the Holling–Tanner process, so let us specifically choose the birth and death rates to be $B_1 = N_1(r_1 - cN_1)$, $D_1 = wN_1N_2/(D+N_1)$, $B_2 = r_2N_2$ and $D_2 = b_2 N_2^2/N_1$. Figure 8.13b shows a stochastic

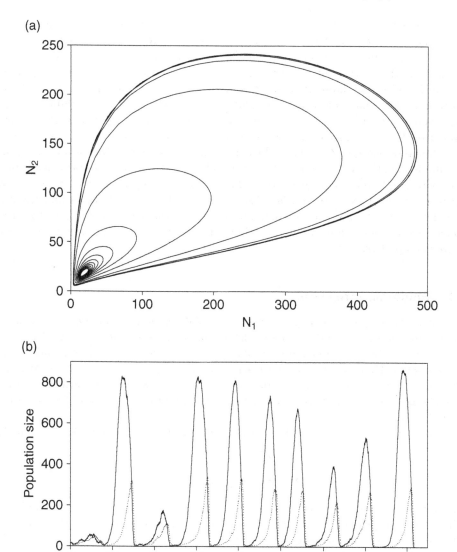

Figure 8.13 The Holling–Tanner process with parameter values $r_1 = 1$, $r_2 = 0.5$, $w = 1.8$, $D = 15$, $b_2 = 0.5$ and $c = 0.001$: (a) deterministic trajectory starting from $N_1(0) = N_2(0) = 15$; (b) corresponding stochastic simulation for prey (—) and predators ($\cdots$).

simulation, and although the amplitude of the first cycle is small, that of the second is large, in direct contrast to the gradual unwinding experienced by the deterministic trajectories shown in Figure 8.13a. The N_1-population drops several times to $N_1 = 1$, from where prey extinction is quite likely, since $B_1 = 0.999$ and $D_1 = 1.8 N_2/16 > B_1$

if $N_2 > 9$, with $D_2 = 0.1125$ at $N_2 = 1$. Note that since $B_2 = 0.5N_2$ and $D_2 = 0.5N_2(N_2/N_1) \ll B_2$ for $N_2 \ll N_1$, the chance of predators becoming extinct before prey is small. Our parameters have, of course, been deliberately selected in order to highlight this critical behaviour, and only a minor tweak is needed to generate sustained cycles over large time periods with high probability. For our simulation run nine cycles occur before prey become extinct, and the process exhibits smooth cyclic behaviour with the predator maxima being roughly coincident with the prey population half-way through its decay period. Thus the prey lead the predators by approximately one quarter-cycle, which is in line with the (substantially different) deterministic Lotka–Volterra model. There is also a wide variation in prey peaks from 58 to 866. The latter is not far short of the maximum carrying capacity $r_1/c = 1000$, which occurs when $N_2 = 0$, and is substantially greater than the deterministic limit cycle maximum. So whilst in general qualitative terms the deterministic and stochastic models exhibit similar limit cycle behaviour, specific realizations do possess markedly different properties.

8.3 Epidemic processes

Although example (8.2.59) exposes the issue that an infinite number of stochastic models are associated with a single deterministic representation, virtually all of these will give birth to monsters of purely mathematical imagination and so can be discounted. Even so, it is worth restressing the importance of first constructing the stochastic rates, since not only do these define the associated deterministic rates uniquely, but their construction concentrates the mind on how the different species really do interact.

Consider, for example, the deterministic Lotka–Volterra equations (8.2.5), namely

$$dN_1/dt = N_1(r_1 - b_1 N_2)$$
$$dN_2/dt = N_2(-r_2 + b_2 N_1), \qquad (8.3.1)$$

and suppose we make the correspondence: $N_1(t)$ (number of prey) $\to x(t)$ (number of individuals susceptible to a disease); $N_2(t)$ (number of predators) $\to y(t)$ (number of individuals infected by the disease). Then on taking $r_1 = 0$, $b_1 = b_2 = \beta$, and $r_2 = \gamma$, equations (8.3.1) may be written in the epidemic form

$$dx/dt = -\beta xy$$
$$dy/dt = \beta xy - \gamma y. \qquad (8.3.2)$$

Here β denotes the infection rate, γ the death or removal rate of infectives, and xy the number of possible contact-pairs between susceptibles and infectives. The fundamental difference between these two sets of equations lies in their interpretation. For whilst in (8.3.2) the death of a susceptible automatically gives rise to the birth of an infective, since it involves a transfer of state for the same individual, in (8.3.1) prey do not become predators but merely act as food for them. So the birth rate $B_2\{(N_1, N_2) \to (N_1, N_2 + 1)\} = b_2 N_1 N_2$ and death rate $D_1\{(N_1, N_2) \to$

$(N_1-1, N_2)\} = b_1 N_1 N_2$ amalgamate to form the single infection rate $\{(x,y) \to (x-1, y+1)\} = \beta xy$. The fact that these two radically different rate structures give rise to the same deterministic equations is clearly a cause for concern for the unwary. Note that if susceptibles give birth, then we place $r_1 = \lambda > 0$, which gives rise to the more general birth–epidemic process.

8.3.1 Simple epidemic

In the simplest type of epidemic model infection spreads by contact between members of a community, and infected individuals are not removed from circulation by recovery, isolation or death; so $\gamma = 0$ in equation (8.3.2). Ultimately, all individuals susceptible to the disease must therefore become infected. Although these assumptions are (from an epidemiological point of view) extremely simple, they are nevertheless applicable to some milder types of infection.

Suppose we have a homogeneously mixing group of $n+1$ individuals, and that at time $t = 0$ the epidemic starts with just one infected individual. The remaining n individuals are all assumed to be susceptible to infection. Then in a direct parallel to the competition and predator–prey processes, there are $x(t)y(t)$ possible contact-pairs between susceptibles and infectives at time t, each of which can turn a susceptible into an infective at rate β. It therefore follows that the deterministic rate of decline of susceptibles is given by $dx(t)/dt$ in equations (8.3.2) where $x(t) + y(t) = n + 1$. On eliminating $x(t)$, we see that the corresponding rate of increase of infectives is given by

$$dy/dt = \beta y(n+1-y), \tag{8.3.3}$$

which is identical to the logistic equation (3.2.29) with intrinsic rate of growth $r = \beta(n+1)$ and carrying capacity $K = n+1$. Whence use of the logistic solution (3.2.30) yields

$$y(t) = (n+1)/[1 + n\exp\{-(n+1)\beta t\}], \tag{8.3.4}$$

which implies that the number of remaining individuals is

$$x(t) = n + 1 - y(t) = n(n+1)/[n + \exp\{(n+1)\beta t\}]. \tag{8.3.5}$$

A major feature of practical interest is the rate, $u(t)$, at which new infectives occur, namely

$$u(t) = \frac{dy(t)}{dt} = \beta x(t)y(t) = \frac{\beta n(n+1)^2 \exp\{-(n+1)\beta t\}}{[1 + n\exp\{-(n+1)\beta t\}]^2}. \tag{8.3.6}$$

This function is called the *epidemic curve* and reaches its maximum value when $du/dt = 0$, i.e. when $\beta t = [\ln(n)]/(n+1)$. At this point $u_{max} = \beta(n+1)^2/4$, and as $x = y = (n+1)/2$ the epidemic is half over.

Stochastic behaviour

From a stochastic viewpoint the apparently simple nature of this process proves to be surprisingly deceptive. Since the number of infectives, Y, can never decrease, the death rate $\mu_Y = 0$, whilst the birth rate takes the form $\lambda_Y = \beta Y(n+1-Y)$. Unfortunately, this quadratic birth process still leaves a less than friendly system of equations. On denoting $p_r(t) = \Pr(Y(t) = r)$ the forward equations

$$dp_1/dt = -n\beta p_1 \tag{8.3.7}$$

$$dp_r/dt = (r-1)(n-r+2)\beta p_{r-1} - r(n-r+1)\beta p_r \qquad (r = 2, \ldots, n+1)$$

initially appear to be amenable to a sequential solution. For with $p_1(0) = 1$ we have

$$p_1(t) = e^{-n\beta t} \tag{8.3.8}$$

which, when substituted into the next equation (i.e. $r = 2$), yields

$$p_2(t) = [(n+1)/(n-2)][e^{-n\beta t} - e^{-(2n-2)\beta t}]. \tag{8.3.9}$$

However, the simplistic nature of this solution soon evaporates when we attempt to solve for higher values of r. To see this, since equations (8.3.7) give rise to linear sums of exponentials whose exponents are non-positive integer multiples of βt, let us write

$$p_r(t) \equiv \sum_{s=0}^{\infty} a_{rs} e^{-s\beta t}. \tag{8.3.10}$$

Then substituting (8.3.10) into (8.3.7) yields the recurrence relation

$$a_{rs} = \left\{ \frac{(r-1)(n-r+2)}{r(n-r+1) - s} \right\} a_{r-1,s} \qquad (r = 2, 3, \ldots), \tag{8.3.11}$$

where we see from (8.3.8) that at $r = 1$ we have $a_{1,-n} = 1$ and $a_{1s} = 0$ otherwise. Unfortunately, since s is integer some of the denominators in (8.3.11) are zero, which complicates this series expansion approach.

This combinatorial issue is better understood by working in terms of the Laplace transform

$$p_r^*(\theta) = \int_0^{\infty} e^{-\theta t} p_r(t)\, dt. \tag{8.3.12}$$

For on applying (8.3.12) to equations (8.3.7) we obtain

$$\theta p_1^* - 1 = -n\beta p_1^* \tag{8.3.13}$$

$$\theta p_r^* = (r-1)(n-r+2)\beta p_{r-1}^* - r(n-r+1)\beta p_r^* \qquad (r = 2, \ldots, n+1),$$

which yields the solution

$$p_r^*(\theta) = \frac{(r-1)!n!\beta^{r-1}}{(n-r+1)!} \prod_{j=1}^{r} [\theta + j(n-j+1)\beta]^{-1} \qquad (r = 1, \ldots, n+1). \tag{8.3.14}$$

In theory, we need only invert (8.3.14) in order to derive $p_r(t)$. For the right-hand side can be expressed as a sum of partial fractions containing terms like $[\theta + j(n-j+1)\beta]^{-1}$ and $[\theta + j(n-j+1)\beta]^{-2}$; the latter occurs when there are repeated factors, which happens when $j > (n+1)/2$. These two terms give rise to the inverses $e^{-j(n-j+1)\beta t}$ and $te^{-j(n-j+1)\beta t}$, respectively. Unfortunately, the work involved in evaluating the partial fraction expansions and determining the various coefficients

Epidemic processes

is quite laborious, and results in very inelegant expressions (see Bailey, 1957). What makes the calculation worse, is that different formulae are required according to whether n is odd or even.

The main obstacle to an easy manipulation of the algebra is the presence of the repeated factors in (8.3.14) when $r > (n+1)/2$, and this complication can be prevented by pretending that n is no longer integer. For in doing so, we avoid the repeated factor problem. So let us slightly modify the epidemic model by changing the chance of a new infection in $(t, t+dt)$ from $r(n-r+1)\beta dt$ to $r(n+\epsilon-r+1)\beta dt$ where $0 < \epsilon \ll 1$. Then equations (8.3.7) and (8.3.13) remain unchanged, as does their solution (8.3.14), provided we interpret the factorials in terms of their gamma function equivalents. Now one of the reasons for presenting this analysis is to demonstrate that this simple (and effectively univariate) process leads to closed expressions for the probabilities $p_r(t)$ that are almost too awkward to be able to advance our understanding of the underlying process, and thereby to convince readers that the development of parallel results for genuine bivariate processes has little practical worth. As such, we shall merely cite the solutions elegantly presented in Bailey (1975), and refer readers to that excellent text for specific analytic details.

Since Bailey (1975) works in terms of the number of susceptibles $x(t)$, on denoting

$$q_r(t) = \Pr[x(t) = r] = \Pr[y(t) = n+1-r] = p_{n+1-r}(t)$$

we see from (8.3.14) that the Laplace transform

$$q_r^*(\theta) = \frac{(n-r)!n!\beta^{n-r}}{r!} \prod_{k=r}^{n} [\theta + k(n-k+1)\beta]^{-1} \qquad (r = 0, \ldots, n). \qquad (8.3.15)$$

The safest way of making the change of infection rate from $r(n+1-r)\beta$ to $r(m+1-r)\beta$, where $m = n + \epsilon$, is to replace the forward equations (8.3.7) for the number of infectives by the associated susceptible equations

$$dq_r/dt = (r+1)(m-r)\beta q_{r+1} - r(m-r+1)\beta q_r \qquad (r=0, \ldots, n-1)$$

$$dq_n/dt = -n(m-n+1)\beta q_n. \qquad (8.3.16)$$

For this clearly shows where n changes to m. The corresponding solution to (8.3.15) is therefore

$$q_r^*(\theta) = \frac{n!(m-r)\ldots(m-n+1)\beta^{n-r}}{r!} \prod_{k=r}^{n} [\theta + k(m-k+1)\beta]^{-1} \qquad (r=0, \ldots, n). \qquad (8.3.17)$$

Since m is not integer the product term in (8.3.17) no longer contains any repeated factors, so we can write $q_r^*(\theta)$ in the form

$$q_r^*(\theta) = \sum_{k=r}^{n} \frac{c_{rk}}{\theta + k(m-k+1)} \qquad (r=0, \ldots, n), \qquad (8.3.18)$$

where

$$c_{rj} = \lim_{\theta \to -k(m-k+1)} \{\theta + k(m-k+1)\} q_r^*(\theta)$$

$$= \frac{(-1)^{k-r}(m-2k+1)n!(m-r)!}{r!(k-r)!(n-k)!(m-n)!(m-k-r+1)\ldots(m-k-n+1)}. \quad (8.3.19)$$

For convenience we have placed $\beta = 1$, since this just implies a change in time-scale to $\tau = \beta t$. As the individual terms in (8.3.18) are simply the Laplace transforms of exponentials, the probabilities $q_r(\tau)$ are given by

$$q_r(\tau) = \sum_{k=r}^{n} c_{rk} e^{-k(m-k+1)\tau} \quad (r = 0, \ldots, n). \quad (8.3.20)$$

Whence letting $\epsilon \to 0$ yields the required inverse transform of (8.3.15). Although minor algebraic complications arise because one of the factors in the product $(m-k-r+1)\ldots(m-k-n+1)$ may equal ϵ, so that $c_{rk} \to \infty$ as $\epsilon \to 0$, a finite limit is obtained by pairing off terms involving c_{rk} and $c_{r,n-k+1}$.

The degree of complexity involved in this solution can be seen by considering specific results for n even (see Bailey, 1975). First, if $r > n/2$ we have

$$q_r(\tau) = \sum_{k=1}^{n-r+1} c_{rk} e^{-k(n-k+1)\tau} \quad (r > n/2), \quad (8.3.21)$$

where

$$c_{rk} = \frac{(-1)^{k-1}(n-2k+1)n!(n-r)!(r-k-1)!}{r!(k-1)!(n-k)!(n-r-k+1)!}. \quad (8.3.22)$$

However, if $r \leq n/2$ then there are repeated factors and the algebra becomes more awkward. Denote $\delta(0) = 1$ and $\delta(r) = 0$ $(r > 0)$. Then we must now use

$$q_r(\tau) = \delta(r) + \sum_{k=1}^{n/2} c_{rk} e^{-k(n-k+1)\tau} + \tau \sum_{k=r}^{n/2} d_{rk} e^{-k(n-k+1)\tau} \quad (0 \leq r \leq n/2) \quad (8.3.23)$$

where, if $j < r$, we retain the value of c_{rk} given by (8.3.22), but if $k \geq r$ we replace this with

$$c_{rk} = \frac{(-1)^r(n-2k+1)n!(n-r)!}{r!(k-1)!(n-k)!(n-r-k+1)!(k-r)!}$$

$$\times \left\{ \sum_{w=k}^{n-k} w^{-1} + \sum_{w=k-r+1}^{n-k-r+1} w^{-1} - \frac{2}{n-2k+1} \right\} \quad (j \geq r), \quad (8.3.24)$$

subject to the proviso that when $r = 1$ the first summation in (8.3.24) does not arise and the second is absent when $r = n/2$. Whilst

$$d_{rk} = \frac{(-1)^{r+1}(n-2k+1)^2 n!(n-r)!}{r!(k-1)!(n-k)!(n-k-r+1)!(k-r)!}. \quad (8.3.25)$$

The corresponding formulae for n odd involve some modification to the above results.

This discussion should be sufficient to demonstrate the degree of complexity involved in the analysis of even the simplest type of epidemic process. The problem arises from the quadratic nature of the contact function, and since predator–prey and competition processes involve more complex bivariate contact structures one can appreciate that trying to pursue exact closed expressions for population size probabilities in such scenarios is a road to nowhere. For even if one did generate such solutions, what could usefully be done with them?

At this point it is worth re-stressing that although the deterministic infective equation is identical to the logistic equation (3.2.29) (with $r = \beta(n+1)$ and $K = n+1$), the associated stochastic infective process is a particularly specialized case since the death rate is always zero. So unlike the stochastic logistic process, which possesses a mathematically amenable quasi-equilibrium distribution, the infection population moves ever-upwards towards the carrying capacity K. Thus apart from determining $q_r(t)$ $(= p_{n-r+1}(t))$, the key probabilistic feature of interest concerns the time to extinction of the susceptible population.

Moments and epidemic curve

If all we require are closed expressions for the jth raw moment, $\mu'_j(\tau)$, of the number of susceptibles at time τ, then employing the series solution (8.3.20) yields

$$\mu'_j(\tau) \equiv \sum_{r=0}^{n} r^j q_r(\tau) = \sum_{r=0}^{n} r^j \sum_{k=r}^{n} c_{rk} e^{-k(m-k+1)\tau}. \qquad (8.3.26)$$

However, a slightly more elegant result can be developed by considering the p.g.f. of the ϵ-modified process. For on denoting

$$G(z;\tau) \equiv \sum_{r=0}^{n} z^r q_r(\tau) \qquad (8.3.27)$$

and using (8.3.18), its transform

$$G^*(z,\theta) \equiv \sum_{r=0}^{n} z^r q_r^*(\tau) = \sum_{r=0}^{n}\sum_{k=r}^{n} \frac{c_{rk} z^r}{\theta + k(m-k+1)}$$

$$= \sum_{k=0}^{n}\sum_{r=0}^{k} \frac{c_{rk} z^r}{\theta + k(m-k+1)} = \sum_{k=0}^{n} \frac{1}{\theta + k(m-k+1)} \sum_{r=0}^{k} c_{rk} z^r. \qquad (8.3.28)$$

On taking the inverse transform we therefore have

$$G(z;\tau) = \sum_{k=0}^{n} e^{-k(m-k+1)\tau} G_k(z) \qquad \text{where} \qquad G_k(z) \equiv \sum_{r=0}^{k} c_{rk} z^r. \qquad (8.3.29)$$

So if we can relate $G_k(z)$ to some well-known function then we have a compact expression for $G(z,\tau)$ with known properties. Now on using (8.3.19), the ratio of the $(r+1)$th and rth coefficients in $G(z;\tau)$ is

440 *Two-species interaction processes*

$$\frac{c_{r+1,k}}{c_{rk}} = \frac{(-k+r)(k-m-1+r)}{(-m+r)}. \tag{8.3.30}$$

Whence it follows that

$$G_k(z) = c_{0k} F(-k, k-m-1; -m, z), \tag{8.3.31}$$

where the terminating hypergeometric series

$$F(-k, k-m-1; -m, z) \equiv \sum_{r=0}^{k} \frac{(-k)\ldots(-k+r-1) \times (k-m-1)\ldots(k-m+r-2)}{(-m)\ldots(-m+r-1)} z^r. \tag{8.3.32}$$

Since putting $r = 0$ in (8.3.19) yields

$$c_{0k} = \frac{(-1)^k (m-2k+1) n! m!}{k!(n-k)!(m-n)!(m-k+1)\ldots(m-k-n+1)}, \tag{8.3.33}$$

the p.g.f. solution (8.3.29) takes the neat form

$$G(z; \tau) = \sum_{k=0}^{n} c_{0k} e^{-k(m-k+1)\tau} F(-k, k-m-1; -m, z). \tag{8.3.34}$$

Note that $c_{00} = 1$, as expected; whilst $G(z; \infty) = 1$ since all the susceptibles must eventually become infected.

The special case $m = n$ is again obtained by letting $\epsilon \to 0$, with the problem of c_{0k} becoming infinite being overcome by combining the terms for k and $n - k + 1$. Although this procedure does not yield a particularly elegant expression for the limiting form of $G(z; \tau)$, it does give rise to a reasonably transparent result for the stochastic mean $\mu'_1(\tau)$. As Bailey remarks, the greatest tractability is achieved through the device of retaining the non-integer m until the latest possible stage, and only then letting $m \to n$. Whence differentiating (8.3.34) to form the stochastic mean

$$\mu'_1(t) \equiv \left.\frac{\partial G(z;t)}{\partial z}\right|_{z=1} = \sum_{k=0}^{n} c_{0k} e^{-k(m-k+1)\tau} \frac{k(k-m-1)}{m} F(-k+1, k-m; -m+1, 1), \tag{8.3.35}$$

employing the standard result that

$$F(-a+1, a-b; -b+1, 1) = \Gamma(b-a+1)\Gamma(a)/\Gamma(b), \tag{8.3.36}$$

and then substituting for c_{0k} from (8.3.33), leads to the fairly simple expression

$$\mu'_1(\tau) = \sum_{k=1}^{n} \frac{(-1)^{k+1}(m-2k+1) n!}{(n-k)!(m-n)\ldots(m-n-k+1)} e^{-k(n-k+1)\tau}. \tag{8.3.37}$$

It at this point that we let $\epsilon \to 0$. Although the factor $(m-n)$ in the denominator of (8.3.37) yields a term involving ϵ^{-1}, combining the terms for k and $n-k+1$ leads to a finite limit, and hence Haskey's (1954) result that

$$\mu'_1(\tau) = \sum_{k=1}^{L} \frac{n!}{(n-k)!(k-1)!} \{(n-2k+1)^2\tau + 2 - (n-2k+1)\sum_{l=k}^{n-k} l^{-1}\} e^{-k(n-k+1)\tau}.$$
(8.3.38)

For n even the upper limit $L = n/2$; whilst for n odd $L = (n+1)/2$ together with the introduction of a factor of $1/2$ into the term corresponding to $k = (n+1)/2$. Haskey's original derivation took a different, and more arduous, route than the above limiting procedure. He started with the transformed probabilities (8.3.15) for $q_r^*(\theta)$, and then pursued an extremely laborious, albeit highly skillful, series of partial fraction expansions.

To illustrate this result, taking $n = 10$ yields

$$\mu'_1(\tau) = e^{-10\tau}(810\tau - 234\frac{17}{28}) + e^{-18\tau}(4410\tau - 902\frac{1}{4}) + e^{-24\tau}(9000\tau - 1247\frac{1}{7})$$
$$+ e^{-28\tau}(7560\tau + 126) + e^{-30\tau}(1260\tau + 2268),$$
(8.3.39)

and as τ increases, the first exponential term (i.e. when $k = 1$) soon becomes dominant. Comparison with the corresponding deterministic expression highlights the strong difference between the stochastic mean and deterministic number. For on placing $n = 10$ and $\tau = \beta t$ in (8.3.4), we see that

$$x(\tau) = 1210e^{-11\tau}/[1 + 10e^{-11\tau}]$$
(8.3.40)

takes a very different form to (8.3.39). This difference becomes even more apparent when we consider the associated epidemic curves, namely the rates of change $w(\tau) = -d\mu'_1(\tau)/d\tau$ (stochastic) and $u(\tau) = dy(\tau)/d\tau = -dx(\tau)/d\tau$ (deterministic) for the number of infectives. Figure 8.14 shows these two forms for $n = 10$; whilst $u(\tau)$ is symmetric about the maximum ordinate, $w(\tau)$ has a lower maximum value, is skewed right and falls away more slowly. Nevertheless, the two maxima do occur at virtually the same time, which is comforting from a public health perspective.

Further insight between the deterministic and expected analyses can be gleaned by switching our attention back onto the forward (infective) equations (8.3.7). For as

$$dp_r/d\tau = (r-1)(n-r+2)p_{r-1} - r(n-r+1)p_r \quad (0 \le r \le n),$$
(8.3.41)

on multiplying both sides by r we obtain

$$rdp_r/d\tau = [(r-1)^2(n+1) - (r-1)^3 + (r-1)(n+1) - (r-1)^2]p_{r-1}$$
$$- r^2[(n+1) - r]p_r,$$
(8.3.42)

whence summing over $r = 1, \ldots, n+1$ gives

$$d\nu'_1/d\tau = (n+1)\nu'_1 - \nu'_2 = \nu'_1(n+1-\nu'_1) - \nu_2.$$
(8.3.43)

Here $\nu'_i(\tau)$ and $\nu_i(\tau)$ denote the ith raw and central moments, respectively, of the number of infectives at the scaled time $\tau = \beta t$. Comparison with the corresponding deterministic equation shows that (8.3.3) is (8.3.43) with the variance $\nu_2(\tau)$ replaced by zero. This reinforces our earlier point that deterministic equations are equivalent to the first-order truncation of (stochastic) moment equations. So here any similarity between the two would reflect not that population sizes are small, but that the variance is

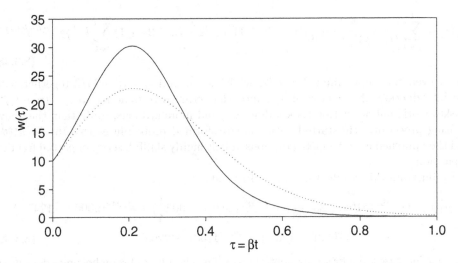

Figure 8.14 Deterministic, $u(\tau) = dy(\tau)/d\tau$ $(\cdots)$, and stochastic, $w(\tau) = -d\mu'_1(\tau)/d\tau$ (—), epidemic curves for $n = 10$.

negligible in comparison to the mean. The fact that $\nu_2(\tau) > 0$ ensures that $d\nu_1(\tau)/d\tau < dy(\tau)/d\tau$, which explains why the stochastic curve in Figure 8.14 is flatter than its deterministic counterpart.

Higher-order raw moment equations may be developed in a similar manner by forming

$$d\nu'_i/d\tau = \sum_{r=1}^{n+1} r^i dp_r/d\tau$$

$$= \sum_{r=1}^{n+1} \{[1+(r-1)]^i(r-1)[(n+1)-(r-1)]p_{r-1} - r^{i+1}[(n+1)-r]p_r\}$$

$$= (n+1)\nu'_1 + \sum_{j=2}^{i}\left[(n+1)\binom{i}{j-1} - \binom{i}{j-2}\right]\nu'_j - i\nu'_{i+1}. \quad (8.3.44)$$

So when $i = 1$ we recover (8.3.43), when $i = 2$

$$d\nu'_2/dt = (n+1)\nu'_1 + (2n+1)\nu'_2 - 2\nu'_3, \quad (8.3.45)$$

whilst when $i = 3$

$$d\nu'_3/dt = (n+1)\nu'_1 + (3n+2)\nu'_2 + 3n\nu'_3 - 3\nu'_4, \quad \text{etc.} \quad (8.3.46)$$

Thus in principle, having determined an exact solution for $\nu'_1(\tau) = n+1-\mu'_1(\tau)$ via (8.3.38), we can now solve equations (8.3.44) recursively for the raw moments $\nu'_i(\tau)$ over $i = 2, 3, \ldots$. However, the work involved in using this approach is far outweighed not only by the resulting opaqueness of the generated solutions, but also by the lack

Epidemic processes

of a general closed form expression for them. Extending the analysis to power-law logistic processes would be even more daunting.

Although exact analytic progress can be made by developing the associated generating functions in terms of hypergeometric functions and Jacobi polynomials (see Bailey, 1975), and also by employing transformation and perturbation techniques, one has to question whether such routes, at least for our purposes, are worthwhile. For not only is the simple epidemic a one-dimensional process, rather than a genuinely two-dimensional construct like the predator–prey and competition processes, but the only reason that even this analytic development is successful is that unlike its parent logistic process the number of susceptibles (infectives) can only decrease (increase). Nevertheless, it is useful to develop the associated generating function equations since these can then be used to derive approximate moments through moment closure, and thereby approximated p.d.f.'s though the saddlepoint approximation.

In terms of the general univariate Kolmogorov equation (3.4.1), matching the power-law and infective transition rates $\lambda_Y = Y(a_1 - b_1 Y^s) = Y(n+1-Y)$ and $\mu_Y = Y(a_2 + b_2 Y^s) = 0$ gives $a_1 = n+1$, $b_1 = -1$ and $s = 1$ together with $a_2 = b_2 = 0$. So equation (3.4.6) yields the infective m.g.f.

$$\frac{\partial M(\theta;\tau)}{\partial \tau} = (e^\theta - 1)[(n+1)\frac{\partial M(\theta;\tau)}{\partial \theta} - \frac{\partial^2 M(\theta;\tau)}{\partial \theta^2}] \quad \text{(infectives)}. \quad (8.3.47)$$

Whence on forming $\partial^2 M(\theta;\tau)/\partial\theta^2$ and $\partial^3 M(\theta;\tau)/\partial\theta^3$ at $\theta = 0$ we easily recover the first two raw moment equations (8.3.45) and (8.3.46). Whilst to obtain the general ith-order equation (8.3.44) we employ Liebnitz' differentiation rule: if $f(x) = g(x)h(x)$ then the nth derivative

$$f^{(n)}(x) = \sum_{k=0}^{n} \binom{n}{k} g^{(n-k)}(x) h^{(k)}(x). \quad (8.3.48)$$

Similarly, to construct the parallel susceptible equation we now match $\lambda_X = X(a_1 - b_1 X^s) = 0$ and $\mu_X = X(a_2 + b_2 X^s) = X(n+1-X)$ at $s = 1$ to obtain $a_1 = b_1 = 0$ with $a_2 = n+1$ and $b_2 = -1$. This yields

$$\frac{\partial M(\theta;\tau)}{\partial \tau} = (e^{-\theta} - 1)[(n+1)\frac{\partial M(\theta;\tau)}{\partial \theta} - \frac{\partial^2 M(\theta;\tau)}{\partial \theta^2}] \quad \text{(susceptibles)}. \quad (8.3.49)$$

Note that equations (8.3.47) and (8.3.49) are identical apart from e^θ being replaced by $e^{-\theta}$; for the addition of an infective is the same as the removal of a susceptible. Moreover, since the simple epidemic process is a special case of the general logistic process, the techniques developed in Chapter 3 relating to moment closure, cumulant construction, together with saddlepoint and perturbation approximations, can be applied directly.

Duration time

Results that are both exact and algebraically amenable can, however, be developed by exploiting the similarity between the simple (i.e. linear) birth process and the nonlinear (quadratic) infection process. For given the number of infectives $Y = y$, the time, Z_y, to the next infection is exponentially distributed with parameter $\beta y(n+1-y)$. So as the sojourns in successive states are independent, the time taken for the epidemic to reach size m is $T_m = Z_1 + Z_2 + \cdots + Z_{m-1}$, with mean

$$E(T_m) = \sum_{i=1}^{m-1} E(Z_i) = \frac{1}{\beta} \sum_{i=1}^{m-1} \frac{1}{i(n+1-i)} \qquad (8.3.50)$$

and variance

$$\text{Var}(T_m) = \sum_{i=1}^{m-1} \text{Var}(Z_i) = \frac{1}{\beta^2} \sum_{i=1}^{m-1} \frac{1}{[i(n+1-i)]^2}. \qquad (8.3.51)$$

In particular, if we are interested in the total duration time $T = T_{n+1}$ of the epidemic, i.e. the time taken for all the susceptibles to become infected, then splitting (8.3.50) and (8.3.51) into partial fractions yields

$$E(T) = \frac{2}{(n+1)\beta} \sum_{i=1}^{n} i^{-1} \quad \text{and} \quad \text{Var}(T) = \frac{2}{(n+1)^2\beta^2} \left\{ \frac{2}{(n+1)} \sum_{i=1}^{n} i^{-1} + \sum_{i=1}^{n} i^{-2} \right\}. \qquad (8.3.52)$$

So if n is large, on using results (2.2.14) and (2.2.17), namely

$$\sum_{i=1}^{n} i^{-1} \sim \gamma + \ln(n) \quad \text{and} \quad \sum_{i=1}^{n} i^{-2} \sim \pi^2/6 \qquad (8.3.53)$$

where $\gamma = 0.577216\ldots$ denotes Euler's constant, we see that

$$E(T) \sim \frac{2[\gamma + \ln(n)]}{(n+1)\beta} \sim \frac{2\ln(n)}{n\beta} \qquad (8.3.54)$$

and

$$\text{Var}(T) \sim \frac{2}{(n+1)^2\beta^2} \left\{ \frac{2[\gamma + \ln(n)]}{n+1} + \frac{\pi^2}{6} \right\} \sim \frac{\pi^2}{3n^2\beta^2}. \qquad (8.3.55)$$

Thus the larger the initial number of susceptibles, the more quickly the epidemic is completed, with the growth in the maximum infection rate, namely $dy/dt = \beta(n+1)^2/4$ at $y = (n+1)/2$, far outstripping the number of susceptibles, n, to be infected. Moreover, combining (8.3.54) and (8.3.55) shows that the coefficient of variation

$$\text{C.Var}(T) = \sqrt{\text{Var}(T)}/E(T) \sim \pi/[2\sqrt{3}\ln(n)] \qquad (8.3.56)$$

declines quite slowly as n increases. This implies that substantial differences in epidemiological behaviour can occur between separate, but otherwise identical, groups of individuals. So care is required when attributing fast infective growth to a virulent disease since the results obtained might simply be due to chance fluctuations.

The SIS model
The simple epidemic process lacks a (non-trivial) equilibrium solution because every susceptible must eventually become infected. However, for mild diseases, such as the

common cold, it is reasonable to suppose that an infected individual will eventually cease to be infected and thereby return to the susceptible state at rate δ. Whence the deterministic equations now take the form

$$dx/dt = -\beta xy + \delta y$$
$$dy/dt = \beta xy - \delta y, \qquad (8.3.57)$$

with $x(t) + y(t) = n + 1$ as before. These equations integrate to yield

$$y(t) = \frac{(n+1) - \delta/\beta}{1 + (n - \delta/\beta)e^{-[(n+1)\beta - \delta]t}}, \qquad (8.3.58)$$

which gives rise to two main outcomes. (i) If $\delta > (n+1)\beta$ then $y(\infty) = 0$ and the infection dies out; since for all $0 < y < n+1$ infectives are returned to the susceptible state faster than they can re-infect susceptibles. (ii) Conversely, if $\delta < (n+1)\beta$ then $y(t)$ approaches the asymptotic value $y(\infty) = (n+1) - \delta/\beta$; so the number of susceptibles converges to δ/β, and not zero has happens under the SI model. So which of these two deterministic outcomes occurs depends on the *critical population size* $n_{crit} = \delta/\beta - 1$. There are also two special cases in which $y(t) = 1$ for all $t > 0$, namely: (iii) $\delta = (n+1)\beta$ (i.e. $n = n_{crit}$) and (iv) $\delta = n\beta$.

Taking a stochastic perspective, if just one single infective is present and this dies before it is able to infect a susceptible, then the infective population becomes extinct thereafter. However, if $\delta \ll n\beta$ then the chance of this occurring within a reasonably long time period is small and so a quasi-equilibrium distribution exists. Whence on denoting $\tilde{\pi}_i$ to be the quasi-equilibrium probability that the population comprises i infectives, we have the balance equation

$$(n + 1 - i)i\beta\tilde{\pi}_i = (i+1)\delta\tilde{\pi}_{i+1} \qquad (i = 1, \ldots, n), \qquad (8.3.59)$$

which has the solution

$$\tilde{\pi}_i = \frac{n!}{i(n-i+1)!}(\beta/\delta)^{i-1}\tilde{\pi}_1 \qquad (i = 1, \ldots, n+1) \qquad (8.3.60)$$

where

$$\tilde{\pi}_1 = \left[\sum_{i=1}^{n+1} \frac{n!}{i(n-i+1)!}(\beta/\delta)^{i-1}\right]^{-1}. \qquad (8.3.61)$$

So the modal value occurs at $i = m$ where $\tilde{\pi}_i \simeq \tilde{\pi}_{i+1}$, i.e.

$$m \simeq (1/2)\{[(n+1) - \delta/\beta] + \sqrt{[(n+1) - \delta/\beta]^2 - \delta/\beta}\}. \qquad (8.3.62)$$

Moments of the process may be constructed by using the techniques developed in Section 3.5.

8.3.2 General epidemic

In order to have a genuinely bivariate epidemic process we need to remove the direct linkage between the number of susceptibles and infectives, and this can be done in a variety of different ways. The simplest approach is to allow infectives to be removed from circulation through either isolation or death, which gives rise to the so-called *general epidemic* process; extensions which allow both infectives and susceptibles to be subject to immigration–birth–death processes involve a higher degree of complexity. Given that the general public perceives epidemics as being more pertinent than competition or predator–prey scenarios, since the latter relate more to plants and animals than to humans, it is not surprising that epidemic theory has received the lions' share of theoretical development. Here we can only scratch the surface of what has been achieved, but the techniques described should at least wet the appetite of researchers wishing to investigate this broad field in greater depth. However, it should not be forgotten that since deterministic development is far easier than stochastic analysis, the former dominates the applied literature, even though it has the potential for yielding results poles apart from reality. The direct impact that epidemiology has on everyday life clearly highlights the inherent danger in not conducting a holistic approach involving both deterministic and stochastic, as well as non-spatial and spatial, considerations.

To determine deterministic behaviour let $z(t)$ denote the number of infected individuals who have been removed from a population of total size n (not $n+1$ as per the simple epidemic). Then equations (8.3.2) expand to

$$dx/dt = -\beta xy$$
$$dy/dt = \beta xy - \gamma y \qquad (8.3.63)$$
$$dz/dt = \gamma y.$$

Note that since $x(t) + y(t) + z(t) = n$ for all $t \geq 0$, this is a two-species system. Moreover, for convenience we often work in terms of the *relative removal rate* $\rho = \gamma/\beta$. Suppose that the epidemic starts at $t = 0$ with $x(0) = x_0$, $y(0) = y_0$ and $z(0) = 0$. Then we see from (8.3.63) that at time $t = 0$

$$dy/dt = \beta y_0(x_0 - \rho), \qquad (8.3.64)$$

and so an epidemic can only develop (i.e. have $dy/dt > 0$) if $x_0 > \rho$. Thus $x_0 = \rho$ defines a *deterministic threshold density* of susceptibles below which an epidemic cannot occur, since infectives are removed at a faster rate than new infectives can be produced.

The epidemic curve

Some analytic progress can be made by first eliminating y to form

$$dx/dz = -x/\rho. \qquad (8.3.65)$$

For this equation integrates to yield

$$x = x_0 e^{-z/\rho}, \qquad (8.3.66)$$

whence the last of equations (8.3.63) can now be written purely as a function of z, namely

$$dz/dt = \gamma[n - z - x_0 e^{-z/\rho}]. \qquad (8.3.67)$$

Not surprisingly (bearing in mind our previous experiences with the competition and predator–prey paradigms) this equation cannot be solved directly, though an approximate solution can be obtained by assuming that z/ρ remains small. For then $e^{-z/\rho} \simeq 1 - (z/\rho) + (z/\rho)^2/2$, whence (8.3.67) reduces to the first-order quadratic o.d.e.

$$dz/dt \simeq \gamma\{n - x_0 + [(x_0/\rho) - 1]z - (x_0/2\rho^2)z^2]\} \tag{8.3.68}$$

which takes the standard solution

$$z(t) = (\rho^2/x_0)\{(x_0/\rho) - 1 + \alpha\tanh[(\alpha\gamma t/2) - \phi]\}. \tag{8.3.69}$$

Here the amplitude $\alpha = \sqrt{\{[(x_0/\rho) - 1]^2 + 2x_0 y_0/\rho^2\}}$ and the phase $\phi = \tanh^{-1}\{[(x_0/\rho) - 1]/\alpha\}$, where $\tanh(\theta) = (e^\theta - e^{-\theta})/(e^\theta + e^{-\theta})$.

So based on the approximation (8.3.68), the epidemic curve, which we now define in terms of the removals $z(t)$, and not the infectives $y(t)$ as for the simple epidemic, can be obtained by differentiating (8.3.69) to form

$$dz/dt \simeq (\gamma\alpha^2\rho^2/2x_0)\text{sech}^2[(\alpha\gamma t/2) - \phi] \tag{8.3.70}$$

where $\text{sech}(\theta) = 2/(e^\theta + e^{-\theta})$. Expression (8.3.70) describes a symmetric bell-shaped curve centred around the maximum value of $\gamma\alpha^2\rho^2/2x_0$ at time $t = 2\phi/\alpha\gamma$, and successfully mimics the way in which in many epidemics the daily number of new reported cases first climbs to a peak and then decays.

The Deterministic Threshold Theorem

To obtain the total size of the epidemic, namely the total number of removals when there are no longer any infectives left in the population, we simply let $t \to \infty$ in (8.3.69) to obtain

$$z(\infty) \simeq (\rho^2/x_0)[(x_0/\rho) - 1 + \alpha]. \tag{8.3.71}$$

Now if in addition to assuming that z/ρ remains small, we can also assume that $2x_0 y_0/\rho^2 \ll [(x_0/\rho) - 1]^2$, then $\alpha \simeq (x_0/\rho) - 1$ (for $x_0 > \rho$). Thus

$$z(\infty) \simeq 2\rho[1 - (\rho/x_0)]. \tag{8.3.72}$$

So if $x_0 > \rho$, on writing $x_0 = \rho + v$ it follows from (8.3.72) that for $0 < v \ll \rho$

$$z(\infty) \simeq 2\rho v/(\rho + v) \simeq 2v. \tag{8.3.73}$$

Hence the initial number of susceptibles, $\rho + v$, is ultimately reduced to $\rho - v$, i.e. to a value as far below the threshold ρ as it was initially above. This is the celebrated KERMACK AND MCKENDRICK (1927) THRESHOLD THEOREM, and in spite of being based on two deterministic approximations it provides considerable insight into the mechanics governing the outbreak of epidemic disease. Unfortunately this remarkably prescient result attracted little attention for three decades before its worth was fully appreciated.

The reliability of this theorem, and hence its practical effectiveness, clearly depends on the accuracy of the fairly strong approximations used to obtain it. Rather than studying this through a direct analytic robustness approach involving fairly turgid algebraic manipulation, Kendall (1956) develops an elegant inverse argument. For he regards expression (8.3.69) as being the exact number of removals for a *new* epidemic process in which the infection rate β is now assumed to depend on z, namely

$$\beta(z) = 2\beta/[(1 - z/\rho) + (1 - z/\rho)^{-1}]. \qquad (8.3.74)$$

So although $\beta(0) = \beta$, we have $\beta(z) < \beta$ for $0 < z < \rho$. Thus the result of assuming the approximations in Kermack and McKendrick's approach is to underestimate the infection rate β, and hence the total size of the epidemic. If, for example, $z/\rho = 0.1$ or 0.2 then $\beta(z) = 0.9945\beta$ or 0.9756β and so the approximations have negligible effect; though once $z/\rho > 0.5$ then $\beta(z) < 0.8\beta$ and the underestimation becomes far more severe. Indeed, should at any time $z > \rho$, then the infection rate becomes negative and so the model becomes totally unrealistic. To prove result (8.3.74) we first note that (8.3.66) has to be replaced by

$$x = x_0 \exp\{-(1/\gamma) \int_0^z \beta(w)\, dw\}, \qquad (8.3.75)$$

whence (8.3.67) becomes

$$dz/dt = \gamma[n - z - x_0 \exp\{-(1/\gamma) \int_0^z \beta(w)\, dw\}]. \qquad (8.3.76)$$

Inserting (8.3.74) into equation (8.3.76) then recovers the approximating equation (8.3.68).

This approach exposes another fundamental difference between deterministic and stochastic mathematics. For whilst the stochastic epidemic starts with (at least) one infective and concludes as soon as the last susceptible becomes infected, the deterministic epidemic not only concludes at $t = +\infty$ but can also be envisaged as starting (i.e. $y_0 = 0$) at $t = -\infty$. On returning to the exact equation (8.3.67), we see that $dz/dt = 0$ at

$$n - z - x_0 e^{-z/\rho} = 0. \qquad (8.3.77)$$

Let the unique positive root of (8.3.77) be $\eta > 0$, i.e. η denotes the final size of the epidemic at $t = \infty$. Then equation (8.3.67) integrates to give

$$t = \frac{1}{\gamma} \int_0^z \frac{w\, dw}{n - w - x_0 e^{-w/\rho}} \qquad (0 \leq z < \eta). \qquad (8.3.78)$$

This confirms that the whole time-axis $-\infty < t < \infty$ is involved. For not only does the integral (8.3.78) diverge as $z \to \eta = z(\infty)$, but it also diverges at the lower limit as $x_0 \uparrow n$, which means that an infinite time elapses before the epidemic gets going because the initial number of infectives $y_0 = n - x_0 \downarrow 0$. Any mathematical misgivings one may have on this lower limit are easily dispelled by changing the time-origin to the point where $x = \rho$, which may be called the *epidemic centre*. Since differentiating equations (8.3.63) yields

$$\frac{d^2z}{dt^2} = \gamma \frac{dy}{dt} = \gamma^2 y\left(\frac{x}{\rho} - 1\right), \tag{8.3.79}$$

not only does the peak of the epidemic (i.e. the maximum rate at which infectives are removed) occur at the centre, but also the maximum number of infectives present in the population occurs at the same time. If we place $x_0 = \rho$ we can still take $z_0 = 0$ without any loss of generality. For the value of $z(t)$ now corresponds to the number of removals in $(0, t)$ for $t > 0$, and in $(t, 0)$ if $t < 0$. The values $z(-\infty) = \psi_1$ and $z(\infty) = \psi_2$ are the number of removals occurring before and after the central point, with the total size of the epidemic being $\psi_1 + \psi_2$. Note that ψ_1 and ψ_2 will not in general be equal, so the (true) epidemic curve is asymmetric, in contrast to the Kermack and McKendrick result, which, being based on the quadratic approximation (8.3.68), is symmetric.

Stochastic behaviour

Admirable and pioneering though the Kermack and McKendrick Threshold Theorem is, a more important consideration is whether a deterministic threshold is appropriate at all. For the idea that an outbreak of infection *can or cannot* occur as the number of susceptibles switches from being just above to just below a threshold value is inherently unreasonable. Far more likely is that the *probability* that an outbreak occurs will change.

Let $X(t)$ and $Y(t)$ denote the stochastic number of susceptibles and infectives, respectively, at time t, with the number of removals $Z(t) = n - X(t) - Y(t)$. Then in the small time interval $(t, t + dt)$ we have the transition probabilities

$$\Pr\{(X, Y) \to (X - 1, Y + 1)\} = \beta XY\, dt \quad \text{(infection)}$$
$$\Pr\{(X, Y) \to (X, Y - 1)\} = \gamma Y\, dt \quad \text{(removal)}. \tag{8.3.80}$$

Whence the usual argument for constructing the equations for the probabilities $p_{ij}(t) = \Pr\{X(t) = i, Y(t) = j\}$ yields the forward Kolmogorov equation

$$dp_{ij}(t)/dt = \beta(i+1)(j-1)p_{i+1,j-1}(t) - (\beta ij + \gamma j)p_{ij}(t) + \gamma(j+1)p_{i,j+1}(t). \tag{8.3.81}$$

Even though this system is simpler than the analytically intractable competition and predator–prey systems, namely (8.1.29) with rates (8.1.25) and (8.2.17), respectively, since the total population size $X(t) + Y(t)$ is now bounded above (by n) the associated mathematical manipulations required to generate solutions can still only be described as heroic. The fact that the associated equation for the p.g.f. $G(z_1, z_2; t) \equiv \sum_{i,j=0}^{n} p_{ij}(t) z_1^i z_2^j$ takes the unfriendly form

$$\frac{\partial G}{\partial t} = \beta(z_2^2 - z_1 z_2)\frac{\partial^2 G}{\partial z_1 \partial z_2} + \gamma(1 - z_2)\frac{\partial G}{\partial z_2}, \tag{8.3.82}$$

that is a quadratic second-order p.d.e., provides a good indicator to this effect.

In principle, constructing a solution for the $\{p_{ij}(t)\}$ is certainly possible, since the transition rates (8.3.80) involve a natural ordering. For unlike the competition and predator–prey scenarios, in which each point (i,j) is connected to its four nearest-neighbours, equation (8.3.81) just involves the two-point connections $(i+1, j-1) \leftarrow (i,j) \to (i, j-1)$, which means that they can be solved sequentially in the order:

$$(i,j) = (n-1,1), \quad (n-1,0);$$
$$(n-2,2), \quad (n-2,1), \quad (n-2,0); \quad (8.3.83)$$
$$(n-3,3), \quad (n-3,2), \quad (n-3,1), \quad (n-3,0); \quad \text{etc.}$$

On noting that $p_{ij}(t) \equiv 0$ for $i+j > n$, we first have

$$dp_{n-1,1}(t)/dt = -[\beta(n-1) + \gamma]p_{n-1,1}(t) \quad (8.3.84)$$

which, for $p_{n-1,1}(0) = 1$, gives

$$p_{n-1,1}(t) = e^{-[\beta(n-1)+\gamma]t}. \quad (8.3.85)$$

This then feeds into

$$dp_{n-1,0}(t)/dt = \gamma p_{n-1,1}(t) \quad (8.3.86)$$

to yield

$$p_{n-1,0}(t) = \frac{\gamma}{\beta(n-1) + \gamma}\{1 - e^{-[\beta(n-1)+\gamma]t}\}. \quad (8.3.87)$$

We can now work through the second line of system (8.3.83), with

$$dp_{n-2,2}(t)/dt = \beta(n-1)p_{n-1,1}(t) - 2[\beta(n-2) + \gamma]p_{n-2,2}(t) \quad (8.3.88)$$

solving to give

$$p_{n-2,2}(t) = \frac{\beta(n-1)}{\beta(n-3) + \gamma}\{e^{-[\beta(n-1)+\gamma]t} - e^{-2[\beta(n-2)+\gamma]t}\}, \quad (8.3.89)$$

and so on. As the number of terms in each subsequent expression grows quite fast, with no obvious pattern, we see that the problem lies not in constructing sequential terms, but in determining a closed form expression for the recurrence integral

$$p_{ij}(t) = \int_0^t e^{(\beta i + \gamma)j(\tau - t)}[\beta(i+1)(j-1)p_{i+1,j-1}(\tau) + \gamma(j+1)p_{i,j+1}(\tau)]\,d\tau. \quad (8.3.90)$$

Since we are mainly interested in generating user-friendly results, we recommend that readers with a keen interest in series and matrix manipulation should investigate papers by (for example) Gani, Siskind, Sakino, Severo and Billard. Section 6.3 of Bailey (1975) provides an excellent introduction, and the *Journal* and *Advances in Applied Probability* provide a useful hunting ground. For example, the Laplace transform equivalent to equations (8.3.81) takes the form

$$p^*_{n-1,1}(s) = 1/\{s + [\beta(n-1) + \gamma]\}$$
$$p^*_{ij}(s) \quad = [\beta(i+1)(j-1)p^*_{i+1,j-1}(s) + \gamma(j+1)p^*_{i,j+1}(s)]/[s + (\beta ij + \gamma j)] \quad (8.3.91)$$

for $(i,j) \neq (n-1,1)$, which results in a sequence of partial fraction expansions which still need to be separated out and inverted. Whilst on writing the specific solutions (8.3.85), (8.3.87) and (8.3.89) in the general form

$$p_{ij}(t) = \sum_{u,v=0}^{\infty} c_{ij}(u,v) e^{-(\beta u + \gamma v)t}, \qquad (8.3.92)$$

we see that substituting (8.3.92) into (8.3.90) yields a set of recurrence equations for the coefficients $\{c_{ij}(u,v)\}$. Though both routes are mathematically tractable, the resulting solutions tell us little about the underlying dynamics of the process. Expressions for the final size of the epidemic, namely $\Pr(Z(\infty) = z) = p_{n-z,0}(\infty)$, are also mathematically opaque.

The Stochastic Threshold Theorem

The Deterministic Threshold Theorem (8.3.73) states that an epidemic can occur only when the initial number of susceptibles $x_0 > \rho = \gamma/\beta$, and that the number of removals $z(\infty) \simeq 2(x_0 - \rho)$. Whilst this result is a useful pointer to what happens in practice, it is clearly a very rough statement of reality. For if $x_0 < \rho$ there is still a positive probability that many infective events will occur. So a key question to ask is, *how likely* is it that a major epidemic will build up?

Starting from $(x_0, y_0) = (n-1, 1)$, the epidemic immediately dies out if the first event is a removal, and this has probability

$$p_{n-1,0}(\infty) = \frac{\gamma}{\beta(n-1) + \gamma} = \frac{\rho}{x_0 + \rho}. \qquad (8.3.93)$$

So the deterministic condition that $x_0 > \rho$ translates into the stochastic condition that the Pr(first event is an infection) $> 1/2$, i.e. the epidemic is likely 'to get going'. If we define a *major epidemic* as happening when a sizeable number of the x_0 susceptibles eventually become infected, say $z(\infty) \geq \xi$, then since the individual $p_{i0}(\infty)$ terms are not algebraically amenable, determining the probability that the final size of the epidemic is at least ξ, i.e. $\Pr(z(\infty) \geq \xi) = \sum_{i=1}^{n-\xi} p_{i0}(\infty)$, is even less so. We are therefore forced into using some form of stochastic approximation.

Now in the opening stages of the epidemic, if n is large then the infection rate $\beta xy = \beta(n - y - z)y \simeq (\beta n)y$, since $y, z \ll n$. So as the removal rate is γy, it follows that the population of infectives is approximately subject to a simple birth–death process with rates $\lambda = \beta n$ and $\mu = \gamma$. We have already seen through result (2.4.24) that the probability of ultimate extinction is $(\mu/\lambda)^{y_0} = \gamma/(n\beta) = \rho/n$ for $y_0 = 1$ if $\lambda > \mu$, i.e. if $\rho < n$; whilst it is one if $\lambda \leq \mu$, i.e. $\rho \geq n$. Thus in the latter case we might expect only a small outbreak to occur, but in the former either a major outbreak or a minor build-up with probabilities $1 - \rho/n$ and ρ/n, respectively.

Because the approximating birth rate $\beta n y$ is always greater than the true infection rate βxy we call this a "fast" approximating procedure. Conversely, if we are interested in the probability that not more than ξ removals occur, then because $\beta xy > \beta(n-\xi)y$ we may also construct the corresponding "slow" procedure based on the birth rate $\beta(n - \xi)$. So now only a small outbreak can occur if $\rho \geq (n - \xi)$, but if $\rho < (n - \xi)$

then either a major or a minor outbreak occurs with probabilities $1 - \rho/(n-\xi)$ and $\rho/(n-\xi)$, respectively. Combining these two bounding results gives rise to:

WHITTLE'S (1955) STOCHASTIC THRESHOLD THEOREM For $\pi(\xi) = \Pr(z(\infty) \leq \xi)$ and $y_0 = a$ initial infectives,

$$\pi(\xi) = 1 \qquad \text{if } n \leq \rho$$
$$(\rho/n)^a \leq \pi(\xi) < 1 \qquad \text{if } n - \xi \leq \rho < n \qquad (8.3.94)$$
$$(\rho/n)^a \leq \pi(\xi) < [\rho/(n-\xi)]^a \qquad \text{if } \rho < n - \xi.$$

This theorem is the stochastic analogue of the Deterministic Threshold Theorem (8.3.73), and states that if $n \leq \rho$ then there is a zero probability of an epidemic exceeding any (small) predetermined level ξ; whilst if $n > \rho$ then the probability of a major outbreak is roughly $(\rho/n)^a$.

If the final epidemic size $z(\infty)$ is small relative to the total population size n, then we may go further by combining the tractability of the simple birth–death process (Section 2.4) for infectives with the development of the counting process approach (Section 7.4). For infectives may be envisaged as giving birth at rate $\lambda = x\beta \simeq n\beta$, and dying (and being counted) at rate $\mu = \gamma$. This leads us to consider the bivariate probability $p_{jk}(t) = \Pr(\text{population contains } j \text{ infectives and } k \text{ removals at time } t)$. Whence a parallel argument to the construction of the simple birth–death equations (2.4.3) and immigration–count equations (7.4.19) yields

$$dp_{jk}(t)/dt = \lambda(j-1)p_{j-1,k}(t) - (\lambda + \mu)p_{jk}(t) + \mu(j+1)p_{j+1,k-1}(t). \qquad (8.3.95)$$

Thus in terms of the p.g.f. $G(z_1, z_2; t) \equiv \sum_{j,k=0}^{n} p_{jk}(t) z_1^j z_2^k$, we have

$$\frac{\partial G}{\partial t} - [\lambda z_1^2 - (\lambda + \mu)z_1 + \mu z_2] \frac{\partial G}{\partial z_1} = 0 \qquad (8.3.96)$$

which is almost identical to equation (2.4.7). Since z_2 does not play an active role in the p.d.e. (8.3.96), the analysis proceeds as before with the associated auxiliary equations, namely

$$\frac{dt}{1} = \frac{-dz_1}{\lambda z_1^2 - (\lambda + \mu)z_1 + \mu z_2} = \frac{dG}{0}, \qquad (8.3.97)$$

being virtually unchanged. Let ξ_1 and ξ_2 denote the two roots of the equation $\lambda z_1^2 - (\lambda + \mu)z_1 + \mu z_2 = 0$, i.e.

$$\xi_1, \xi_2 = (1/2\lambda)[(\lambda + \mu) \pm \sqrt{\{(\lambda + \mu)^2 - 4\lambda\mu z_2\}}]. \qquad (8.3.98)$$

Then the first equation in (8.3.97) integrates to give

$$\left(\frac{z_1 - \xi_1}{z_1 - \xi_2}\right) e^{\lambda(\xi_1 - \xi_2)t} = b \qquad (8.3.99)$$

for some arbitrary constant b. So (like before) for some arbitrary function f the general solution is

$$G(z_1, z_2; t) = f[\{(z_1 - \xi_1)/(z_1 - \xi_2)\} e^{\lambda(\xi_1 - \xi_2)t}]. \qquad (8.3.100)$$

Suppose that at $t = 0$ there are $j = y_0$ infectives with $k = 0$ removals. Then on placing $t = 0$ in (8.3.100) we see that

$$z_1^{y_0} = f[(z_1 - \xi_1)/(z_1 - \xi_2)]. \tag{8.3.101}$$

Whence writing $(z_1 - \xi_1)/(z_1 - \xi_2) = \phi$, i.e. $z_1 = (\xi_1 - \xi_2\phi)/(1 - \phi)$, gives

$$f(\phi) = [(\xi_1 - \xi_2\phi)/(1 - \phi)]^{y_0}. \tag{8.3.102}$$

Finally, on replacing ϕ in (8.3.102) by

$$\phi = \{(z_1 - \xi_1)/(z_1 - \xi_2)\}e^{\lambda(\xi_1 - \xi_2)t}, \tag{8.3.103}$$

the general result (8.3.100) takes the specific form

$$G(z_1, z_2; t) = \left\{ \frac{\xi_1(z_1 - \xi_2) - \xi_2(z_1 - \xi_1)e^{\lambda(\xi_1 - \xi_2)t}}{(z_1 - \xi_2) - (z_1 - \xi_1)e^{\lambda(\xi_1 - \xi_2)t}} \right\}^{y_0}. \tag{8.3.104}$$

On placing $z_2 = 1$ we see that the roots $\xi_1 = 1$ and $\xi_2 = \mu/\lambda$, whence (8.3.104) reduces to the (marginal) simple birth–death population size p.g.f. (2.4.11). Whilst placing $z_1 = 1$ yields the marginal removal p.g.f.

$$G(1, z_2; t) = \sum_{k=0}^{\infty} \Pr(z(t) = k) z_2^k. \tag{8.3.105}$$

In particular, letting $t \to \infty$ gives the total death p.g.f.

$$G(1, z_2; \infty) = \begin{cases} \xi_2^{y_0} & \text{if } \xi_1 > \xi_2 \\ \xi_1^{y_0} & \text{if } \xi_1 < \xi_2. \end{cases} \tag{8.3.106}$$

Since the probability that the ultimate number of removals is k is given by the coefficient of z_2^k in $G(1, z_2; \infty)$, (8.3.106) provides an approximation to the number of removals in a general epidemic with $\lambda = n\beta$ or $(n - \xi)\beta$, $\gamma = \mu$ and $y_0 = 1$. Namely, that for $n < \rho$

$$\Pr(z(\infty) = k) \simeq \left(\frac{\lambda + \mu}{2\lambda}\right) \left(\frac{2\lambda\mu}{(\lambda + \mu)^2}\right)^k \frac{1.3 \ldots |2k - 3|}{k!} \quad (k \geq 1). \tag{8.3.107}$$

If our main interest lies in the final configuration of the epidemic system, i.e. the number of removals $z(\infty)$ and susceptibles $x(\infty) = n - z(\infty)$ at $t = \infty$, then the time spent in each particular state (i, j, k) becomes irrelevant and only the specific *paths* taken to travel from $(x(0), y(0), 0)$ to $(x(\infty), 0, z(\infty))$ need to be examined. We have, of course, already explored this approach in Section 4.1 in the context of the simple random walk. However, whereas there the transition probabilities remain constant, now they change with i. For as $(i, j) \to (i - 1, j + 1)$ and $(i, j - 1)$ at rates βij and γj, respectively, it follows that

$$\Pr[(i, j) \to (i - 1, j + 1)] = \beta ij/(\beta ij + \gamma j) = \beta i/(\beta i + \gamma) = \alpha_i \quad \text{(say)}$$
$$\Pr[(i, j) \to (i, j - 1)] \quad = \gamma j/(\beta ij + \gamma j) = \gamma/(\beta i + \gamma) = \beta_i = 1 - \alpha_i. \tag{8.3.108}$$

So unlike the predator–prey and competition processes, which have four transition rates depending on i and j, not only are there now only two rates but these no longer involve j. In principle this means that the probabilities can be solved in order. For example, to evaluate $\Pr[z(\infty) = k | x(0) = n-1, y(0) = 1]$ we can evaluate all possible paths which end in $(n-k-1, 1) \to (n-k, 0)$. Moreover, since the $k-1$ infections have probability $\alpha_{n-1}\alpha_{n-2}\cdots\alpha_{n-k}$, regardless of the path taken, all we need to focus on are the probabilities for the states involving removal. Specifically, on denoting infection by I and removal by R, we have

k	events	event probability	
1	R	β_{n-1}	
2	IRR	$\alpha_{n-1}\beta_{n-2}^2$	(8.3.109)
3	$IIRRR + IRIRR$	$\alpha_{n-1}\alpha_{n-2}\beta_{n-3}^3 + \alpha_{n-1}\beta_{n-2}\alpha_{n-2}\beta_{n-3}^2$	
4	$IIIRRRR + \cdots$	$\alpha_{n-1}\alpha_{n-2}\alpha_{n-3}\beta_{n-4}^4 + \cdots,\quad$ etc.	

Thus for $z(\infty) = k > 1$, we have to consider all possible arrangements of $k-1$ I's and k R's starting with I and ending with R, such that: (a) in each arrangement the sequential number of I's - R's always exceeds -2; (b) the I's take probabilities $\alpha_{n-1}, \alpha_{n-2}, \ldots, \alpha_{n-k+1}$; and, (c) that each R to the immediate right of α_{n-s} takes probability β_{n-s-1}.

For small outbreaks of size k this is clearly a feasible operation. Suppose, for example, we wish to determine the chance of a very minor outbreak occurring, say $z(\infty) \leq 3$, when $n = 1000$ and $\beta = 0.001$. Then summing the first three sets of terms in (8.3.109) shows that for

γ:	10	1.5	1.0	0.5	0.1
$\rho = \gamma/\beta$:	$10000 \gg n$	$1500 > n$	$1000 = n$	$500 < n$	$100 \ll n$
$\Pr(z(\infty) \leq 3)$:	0.997	0.814	0.688	0.441	0.100

Hence the probability of such a minor outbreak occurring changes from near certainty when n lies well below the threshold value, to around 0.7 when n lies near the threshold value, and becomes small when n greatly exceeds it. Note that these values agree with those obtained from the approximate birth–death expression (8.3.107) to three decimal places, which highlights the usefulness of this neat approximation for small k.

Given that the state space has a natural ordering, since each event can only take one of two possible forms (infection or removal) and susceptibles cannot increase, early interest in combinatorial solutions blossomed. Although exact algebraic results are bound to be messy, as can be seen from result (8.3.109), elegantly simple asymptotic results have been obtained, and readers interested in this aspect of combinatorial analysis are advised to start with the arguments developed in Bailey (1975) and Williams (1971).

8.3.3 Recurrent epidemics

Of all the influences affecting human development, perhaps nothing has equalled the huge degree of concern generated by epidemic outbreaks. However, although recorded accounts go back to at least as far as the ancient Greeks, for example the *Epidemics* of Hippocrates (459-377 B.C.), scientific achievements in this field did not really begin to develop until the first half of the twentieth century. These were fueled by the spectacular rise of bacteriological science in the second half of the nineteenth century, primed by the researches of Pasteur (1822–1895) and Koch (1843–1910). It was Hamer (1906) who realized that the course of an epidemic was defined by the contact rate between susceptibles and infectives, and he used this feature to deduce the existence of periodic recurrences. As we shall soon see, this important idea was taken up and pushed forward by Soper (1929).

Historical study of epidemiological development shows that although useful deterministic advances were made in the 1920s and 1930s, stochastic progress essentially stems from the pioneering work of Bartlett (1946, 1949); for a concise historical outline see Chapter 2 of Bailey (1975). Since then, a great deal of progress has been made in using probabilistic arguments to increase our understanding of epidemic spread, and hence its control, in a wide variety of different situations. Specific examples include age-dependence, inclusion of latent and incubation periods, and household and carrier models. Whilst as well as application to mundane diseases such a chicken-pox, whooping cough and measles, crucially important developments (see, for example, Anderson, 1982) have taken place for malaria, tetanus, typhoid, diptheria, hookworm/roundworm/tapeworm, schistosomiasis, onchocerciasis, rabies and fascioliasis, etc. Moreover, huge media interest in AIDS/HIV and the potentially devastating cross-over of avian-flu from bird to human populations is constantly fueling intense research pressure to generate yet further understanding of epidemic processes. Running parallel to all of these are, of course, related studies involving plant, animal and bird populations. It is therefore no surprise that the associated mathematical literature is vast, and all we can usefully do here is to scratch the surface of awareness.

One particularly interesting feature ties in neatly with our earlier studies. For Soper's realization that many common diseases exhibit periodic flare-ups, with infection being sustained at a low level inbetween times by a gradual spread to new susceptibles, bears a striking similarity to the behaviour of the stochastic Volterra predator–prey process (see Figure 8.10). There the predator population is maintained through a logistic process, so suppose we maintain the corresponding susceptible population by the even simpler device of allowing fresh susceptibles to enter the population at rate α in order to replace the ones depleted by infection. Then the deterministic general epidemic equations (8.3.63) become

$$dx/dt = \alpha - \beta xy$$
$$dy/dt = \beta xy - \gamma y \qquad (8.3.110)$$
$$dz/dt = \gamma y.$$

Unlike the simple and general epidemics, $x(t)$ and $y(t)$ will now tend to non-trivial equilibrium values, namely $x^* = \gamma/\beta$ and $y^* = \alpha/\gamma$, respectively. Moreover, instead of being bounded above by n, the number of removals, $z(t)$, will rise unchecked, with $dz/dt \to \alpha$ as $t \to \infty$.

Deterministic paths

Showing that the process does indeed converge to these equilibrium values follows exactly the same route as that used for the deterministic Volterra predator–prey process (Section 8.2.2). For on writing $x(t) = x^*[1 + u(t)]$ and $y(t) = y^*[1 + v(t)]$ for small $u(t)$ and $v(t)$, the first two of equations (8.3.110) linearize to give

$$du/dt = -(\alpha\beta/\gamma)(u+v)$$
$$dv/dt = \gamma u. \qquad (8.3.111)$$

Whence on placing $\psi = \alpha\beta/\gamma$ with $k_{11} = k_{12} = -\psi$, $k_{21} = \gamma$ and $k_{22} = 0$, we see that these equations take the general form (8.2.21). So they have the solution (8.2.22), namely

$$u(t) = A_u e^{\lambda_1 t} + B_u e^{\lambda_2 t} \quad \text{and} \quad v(t) = A_v e^{\lambda_1 t} + B_v e^{\lambda_2 t} \qquad (8.3.112)$$

for constants $A_u, \ldots, B_v$ and eigenvalues (8.2.23), i.e.

$$\lambda_1, \lambda_2 = (\psi/2)[-1 \pm \sqrt{\{1 - 4\gamma/\psi\}}]. \qquad (8.3.113)$$

Thus provided the immigration rate α is sufficiently small to ensure that these two roots are complex, i.e. $\alpha < 4\gamma^2/\beta$, the trajectory of $(u(t), v(t))$ exhibits converging spirals at rate $e^{-\psi t/2}$ towards $(0,0)$ with period $T = 2\pi/\xi$ where $\xi = (1/2)\sqrt{\{4\psi\gamma - \psi^2\}}$. Whence on writing

$$u(t) = e^{-\psi t/2}[C\cos(\xi t) + D\sin(\xi t)] \qquad (8.3.114)$$

for constants C and D, direct substitution into equations (8.3.111) yields

$$v(t) = -e^{-\psi t/2}[\{(C/2) + (D\xi/\psi)\}\cos(\xi t) + \{(D/2) - (C\xi/\psi)\}\sin(\xi t)]. \qquad (8.3.115)$$

Placing $t = 0$ gives $C = u(0)$ and $D = -(\psi/2\xi)[u(0) + 2v(0)]$, where $u(0) = (x(0)/x^*) - 1$ and $v(0) = (y(0)/y^*) - 1$. Note that the phase difference is not $\pi/2$, unlike the earlier Lotka–Volterra process.

If $\alpha \simeq 0$, then $\psi = \alpha\beta/\gamma \simeq 0$ and so $\xi \simeq \sqrt{\psi\xi} = \sqrt{\alpha\beta} \simeq 0$. Thus a very small susceptible immigration rate leads (as one might expect) to a small decay rate combined with a long cycle period. Moreover, one of the coefficients in expression (8.3.115) for $v(t)$ is $-C\xi/\psi \simeq -C\gamma/\sqrt{\alpha\beta}$; so as α decreases the (localized) amplitude of the infectives increases relative to that of the susceptibles. As an example of real-life parameter values, Soper (1929) took γ^{-1} to be two weeks, since this is the approximate incubation period for measles, and estimated ψ^{-1} (from London data) to be roughly 68 weeks. So here $\xi = 0.0854$, giving a period (T) of $2\pi/\xi = 74$ weeks and a peak-to-peak damping factor of $e^{-\psi T/2} = 0.58$. Thus whilst the assumption of a constant influx of new susceptibles does give rise to epidemic waves with roughly the right period, the swift damping of the infective cycles towards a steady epidemic state clearly contradicts the epidemiological evidence.

Any misgivings one may have that this 'proof' of convergence to (x^*, y^*) relies on a linear approximation, and not on a globally exact argument, are easily dispelled by

Epidemic processes

recalling an elegant argument of G.E.H. Reuter (Bartlett, 1956; see also Bartlett, 1960 and Bailey, 1975). For consider the function

$$f(u,v) = \{(1+u) - \ln(1+u)\} + (\tau/\sigma)\{(1+v) - \ln(1+v)\} \qquad (8.3.116)$$

where $1+u = x/x^*$, $1+v = y/y^*$, $\sigma = \gamma/\alpha\beta$ and $\tau = 1/\gamma$. Observe the striking similarity between (8.3.116) and the closed Lotka–Volterra curves (8.2.9). Then on differentiating (8.3.116) to obtain

$$\frac{df}{dt} = \frac{\partial f}{\partial u}\frac{du}{dt} + \frac{\partial f}{\partial v}\frac{dv}{dt}, \qquad (8.3.117)$$

and noting that equations (8.3.110) take the form

$$\sigma du/dt = -(u+v+uv) \quad \text{and} \quad \tau dv/dt = u(1+v), \qquad (8.3.118)$$

we see that

$$\frac{df}{dt} = -\frac{u^2}{\sigma(1+u)} \leq 0. \qquad (8.3.119)$$

For we require $u \geq -1$ in order to ensure that $x \geq 0$. Thus $f(u,v)$ continually decreases along any path for which t increases. Since the function $g(z) = (1+z) - \ln(1+z)$ takes its minimum value at $z = 1$, it follows from (8.3.116) that $f \geq 1 + \tau/\sigma$. So f must tend to a finite limit $f_0 \geq 1 + \tau/\sigma$ as $t \to \infty$. Now the curves $f(u,v) = c$, for constant c, are closed, and contract inwards to the point $(u,v) = (0,0)$ corresponding to $f(0,0) = f_0$ as $c \to 1 + \tau/\sigma$ (note the paradigm with the Lotka–Volterra trajectories shown in Figure 8.7). Thus (x,y) must tend to (x^*, y^*), as is strongly suggested by the earlier linearization approach.

Stochastic paths

Although the damping of the deterministic infective cycles towards a steady state contradicts what happens in practice, our earlier experience with the associated Volterra predator–prey model, in which deterministic converging spirals switch into sustained stochastic cycles (Figure 8.10), suggests that a stochastic version of the model might well generate sustained epidemic cycles. Stochastic simulation follows the same route as for the competition and predator–prey processes, with only a slight change being required to parts (v) and (vi) of the general algorithm A8.1 (detailed in Section 8.1.3). For instead of working with B_1, D_1, B_2 and D_2, on denoting $X(t)$ and $Y(t)$ to be the random numbers of susceptibles and infectives at time t, we now have:

Recurrent epidemic simulation algorithm (A8.4)
    (i) read in parameters
    (ii) set $t = 0$, $X = x(0)$ and $Y = y(0)$
    (iii) print t, X and Y
    (iv) generate a new $U \sim U(0,1)$ random number
    (v)' evaluate $R = \alpha + \beta XY + \gamma Y$

(vi)' if $U \leq \beta XY/R$ then $(X,Y) \to (X-1, Y+1)$
if $U \leq (\gamma Y + \beta XY)/R$ then $(X,Y) \to (X, Y-1)$
else $(X,Y) \to (X+1, Y)$
(vi) generate a new $U \sim U(0,1)$ random number
(vii) update t to $t - [\ln(U)]/R$
(viii) return to (iii)

Let us retain Soper's values of $\gamma^{-1} = 2$ weeks and $\psi^{-1} = \gamma/\alpha\beta = 68$ weeks, and suppose (for illustration) that the equilibrium number of infectives in a small community is $y^* = \alpha/\gamma = 100$. Then $\alpha = \gamma y^* = 50$ with $\beta = \gamma/(68\alpha) = 0.000147$, and the equilibrium number of susceptibles $x^* = \gamma/\beta = 68\alpha = 3400$. Figure 8.15 shows a simulation run starting from $x(0) = x^*$ and $y(0) = 1$ which highlights the regular cyclic nature of the process. Moreover, it contrasts sharply with our earlier linearized deterministic analysis which suggests a peak-to-peak damping factor of 0.58. Though with 12 full cycles occurring between $t = 26.75$ and 957.79, giving a peak-to-peak wavelength of 77.6, there is quite close agreement with the deterministic wavelength 74.

Although Figure 8.15 creates the impression of a sustainable process, close examination shows that $Y(t)$ drops to as low as 3 at time $t = 310$, from which position infective extinction is non-negligible. Clearly the infective population must ultimately become extinct, followed by unlimited growth of the susceptible population as a Poisson immigration process at rate $\alpha = 50$. So the relevant question to ask is how long will cyclic behaviour continue? A rough idea of the probability of this event can be obtained by regarding $X(t)$ as being fixed equal to x^*, and then treating $Y(t)$ as a simple birth–death process with parameters $\lambda = \beta x^* = \gamma$, $\mu = \gamma$ and initial size y^*. Since $\lambda = \mu$, the probability of infective extinction by time t is given by (2.4.27), namely

$$p_0(t) = [\mu t/(1 + \mu t)]^{y^*} = [1 + (2/t)]^{-100}. \qquad (8.3.120)$$

So $p_0(t) = 0.14$, 0.82 and 0.98 at $t = 100$, 1000 and 10,000, respectively. This suggests that relatively few realizations will last less than 100 time units, the majority will die out by $t = 1000$ and virtually all will have ended by $t - 10,000$. Though these are only crude estimates, they nevertheless provide a feel for the degree of sustainability of the process. Note that the introduction of even a tiny rate of infective immigration would enable the process to persist indefinitely.

Whilst this simulation shows large cyclic fluctuations, set in motion by the single infective present at $t = 0$, realizations starting near the equilibrium point initially exhibit much smaller fluctuations and are more *endemic* than *epidemic* in nature. Conversely, if $(x(0), y(0))$ lies well above (x^*, y^*), then boom-and-bust dynamics may swiftly take over with the infective population dying out before the end of the first cycle. So the behaviour of the process is tightly controlled by the initial condition.

The behaviour of small fluctuations about the deterministic equilibrium values $x^* = \gamma/\beta$ and $y^* = \alpha/\gamma$ may be determined by examining the variance-covariances of the number of infectives and susceptibles under the assumption that oscillations are maintained indefinitely. Although we cannot parallel the derivation of the linearized second-order moments of the associated Lotka–Volterra process (see (8.2.30)–(8.2.36))

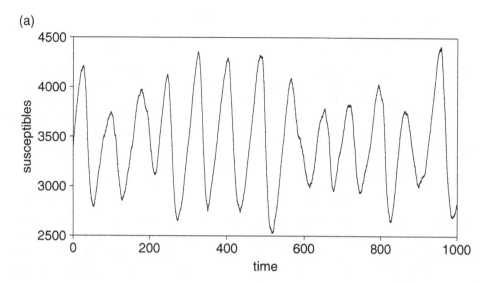

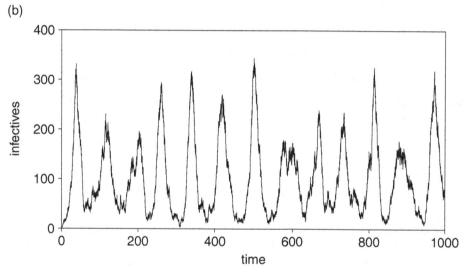

Figure 8.15 A simulated stochastic recurrent epidemic over $0 \leq t \leq 1000$ corresponding to the deterministic representation $dx/dt = 50 - 0.000147xy$ and $dy/dt = 0.000147xy - 0.5y$ showing the number of (a) susceptibles and (b) infectives.

directly – for whereas the instantaneous transitions of prey death and predator birth are independent, those of susceptible death and infective birth correspond to the same event – we can develop a closely related procedure. Now $X(t+dt) - X(t)$ comprises a gain $dZ_1(t)$ due to immigration and a loss $dZ_2(t)$ due to infection, and $Y(t+dt) - Y(t)$ comprises a gain $dZ_2(t)$ due to infection and a loss $dZ_3(t)$ due to infective removal. So the analogue to the Lotka–Voltera equations (8.2.30) is given by

$$X(t+dt) = X(t) + [\alpha - \beta X(t)Y(t)]dt + dZ_1(t) - dZ_2(t)$$
$$Y(t+dt) = Y(t) + [\beta X(t)Y(t) - \gamma Y(t)]dt + dZ_2(t) - dZ_3(t), \qquad (8.3.121)$$

where $dZ_1(t)$, $dZ_2(t)$ and $dZ_3(t)$ are independent Poisson variables, adjusted to have zero means, with parameters (i.e. variances) αdt, $\beta X(t)Y(t)dt$ and $\gamma Y(t)dt$, respectively. These $dZ_i(t)$ terms feature quite differently to those in equations (8.2.30). Write $X(t) = x^*[1 + u(t)]$ and $Y(t) = y^*[1 + v(t)]$, and assume that the variables $u(t)$ and $v(t)$ are sufficiently small for us to be able to neglect the quadratic terms $u^2(t)$, $u(t)v(t)$ and $v^2(t)$. Then on replacing α by γy^* and βx^* by γ, equations (8.3.121) linearize to

$$x^* u(t+dt) = x^* u(t) - \alpha[u(t) + v(t)]dt + dZ_1(t) - dZ_2(t)$$
$$y^* v(t+dt) = y^* v(t) + \alpha u(t)dt + dZ_2(t) - dZ_3(t). \qquad (8.3.122)$$

Next (as before) we square and cross-multiply equations (8.3.122), take expectations on both sides of the resulting expressions, and ignore terms in dt^2. We then presume that the process is in a state of quasi-equilibrium, for which $\text{Var}[u(t+dt)] = \text{Var}[u(t)]$, etc., and write $\sigma_u^2 = \text{Var}[u(t)] \simeq \text{E}[u^2(t)]$, etc., since $\text{E}[u(t)] \simeq 0$ and $\text{E}[v(t)] \simeq 0$. This procedure yields

$$x^{*2}\sigma_u^2 = x^{*2}\sigma_u^2 - 2\alpha x^*(\sigma_u^2 + \sigma_{uv})dt + \text{Var}(dZ_1) + \text{Var}(dZ_2)$$
$$y^{*2}\sigma_v^2 = y^{*2}\sigma_v^2 + 2\alpha y^* \sigma_{uv} dt + \text{Var}(dZ_2) + \text{Var}(dZ_3) \qquad (8.3.123)$$
$$x^* y^* \sigma_{uv} = x^* y^* \sigma_{uv} + \alpha x^* \sigma_u^{*2} dt - \alpha y^* \sigma_{uv} dt - \alpha y^* \sigma_v^2 dt - \text{Var}(dZ_2).$$

Hence as $\text{Var}[dZ_1(t)] = \alpha dt$, $\text{Var}[dZ_2(t)] \simeq \beta x^* y^* dt = \alpha dt$ and $\text{Var}[dZ_3(t)] \simeq \gamma y^* dt = \alpha dt$, we have

$$x^*(\sigma_u^2 + \sigma_{uv}) = 1, \quad y^* \sigma_{uv} = -1 \quad \text{and} \quad x^* \sigma_u^2 - y^*(\sigma_v^2 + \sigma_{uv}) = 1, \qquad (8.3.124)$$

which solve to give

$$\sigma_u^2 = 1/x^*, \quad \sigma_{uv} = -1/y^* \quad \text{and} \quad \sigma_v^2 = 1/y^* + x^*/y^{*2}. \qquad (8.3.125)$$

So for our example, we see that as $\sigma_v = 0.5916$ the linearized variable $v(t)$ only has to dip down by 1.7 standard deviations for the infective population to die out. If our aim is to achieve a high chance of long-term oscillations we might wish to reduce σ_v to say $< 1/3$. Since $x^* = \gamma/\beta$ and $y^* = \alpha/\gamma$, (8.3.125) would then yield the (rather severe) condition

$$(\gamma/\alpha)[1 + \gamma^2/(\alpha\beta)] \leq 1/9. \qquad (8.3.126)$$

Retaining the values $\gamma = 0.5$ and $\alpha = 50$ requires $\beta \geq 0.000495$ to achieve this, which is more than three times higher than the value 0.000147 used in the simulation shown in Figure 8.15.

Birth of susceptibles

Given that our success in simulating sustained stochastic epidemic cycles has been achieved through the use of an external agency, namely the immigration of fresh susceptibles from an external source, the question remains as to whether it is possible to obtain a similar effect by purely internal means? For example, one might consider increasing the complexity of the process by allowing for an incubation period or for seasonal variations in infection rate. However, such modifications also give rise to damped deterministic cycles. So let us return to the basic Lotka–Volterra predator–prey process, which exhibits neutrally stable deterministic cycles (Figure 8.7), and see whether the parallel stochastic epidemic process can avoid the inherent early extinction problem associated with population explosion (Figure 8.8). Since prey growth follows a simple birth process, let us retain this feature by replacing the susceptible immigration term α by the birth term λx. Then equations (8.3.110) become

$$dx/dt = \lambda x - \beta xy$$
$$dy/dt = \beta xy - \gamma y \qquad (8.3.127)$$
$$dz/dt = \gamma y.$$

On making the association $r_1 = \lambda$, $b_1 = b_2 = \beta$ and $r_2 = \gamma$, these equations are identical to the deterministic Lotka–Volterra equations (8.2.5), and so the conclusions derived earlier carry over to our new epidemic situation. In particular, we obtain a deterministic family of closed curves having the linearized solution (8.2.14), namely

$$x(t) = (\gamma/\beta)[1 + c\cos\{t\sqrt{\lambda\gamma} + d\}]$$
$$y(t) = (\lambda/\beta)[1 + c\sqrt{\gamma/\lambda}\sin\{t\sqrt{\lambda\gamma} + d\}], \qquad (8.3.128)$$

where the constants c and d are determined from the initial values $x(0) = (\gamma/\beta)[1 + c\cos(d)]$ and $y(0) = (\lambda/\beta)[1 + c\sqrt{\gamma/\lambda}\sin(d)]$.

The corresponding stochastic equations change slightly from (8.3.121) to

$$X(t + dt) = X(t) + [\lambda X(t) - \beta X(t)Y(t)]dt + dZ_1(t) - dZ_2(t)$$
$$Y(t + dt) = Y(t) + [\beta X(t)Y(t) - \gamma Y(t)]dt + dZ_2(t) - dZ_3(t). \qquad (8.3.129)$$

So as we now have $y^* = \lambda/\beta$ with $x^* = \gamma/\beta$ (as before), it follows that the linearized equations corresponding to (8.3.122) take the form

$$x^*u(t + dt) = x^*u(t) - \lambda x^*v(t)dt + dZ_1(t) - dZ_2(t)$$
$$y^*v(t + dt) = y^*v(t) + \lambda x^*u(t)dt + dZ_2(t) - dZ_3(t), \qquad (8.3.130)$$

where $dZ_1(t)$ has variance $\lambda X(t)dt \simeq \lambda x^*dt$. However, the parallel calculation to (8.3.123) fails, since it produces the contradiction $\sigma_{uv} = 1/x^* = -1/y^*$. The reason for this becomes apparent when we recall the paradigm with the Lotka–Volterra predator–prey process. For there we needed to include the logistic prey parameter c (see equations (8.2.20)) in order to obtain persistent stochastic fluctuations, and although the linearized second-order moments σ_2^2 and σ_{12} in (8.2.36) remain bounded as $c \to 0$, the prey variance $\sigma_1^2 \to \infty$.

This suggests that although stochastic simulations corresponding to (8.3.127) might produce a few cycles before extinction occurs, it will be necessary to change the susceptible birth rate from the simple form $\lambda X(t)$ to (for example) the logistic form

$X(t)[\lambda - cX(t)]$ if we are to secure long-term cyclic behaviour. This makes practical sense, for if $Y(t) = 0$ for $t > t_0$ then $X(t) \to \infty$ unless $c > 0$. In order to offer a good comparison with Figure 8.15, suppose we retain the equilibrium values $x^* = 3400$ and $y^* = 100$, together with the parameter values $\gamma = 0.5$ and $\beta = \gamma/x^* = 0.0001471$. Then $\lambda = \beta y^* = 0.01471$. Figure 8.16 shows a simulation run starting from $x(0) = x^* = 3400$ and $y(0) = y^* = 100$, and this clearly exhibits divergent oscillations leading to early extinction of the infectives after just three full cycles. So let us replace simple birth at rate λX by logistic birth with rate $\tilde{\lambda} X(1 - X/K)$. Then since in equilibrium we have $\tilde{\lambda} x^*(1 - x^*/K) = \beta x^* y^*$, to retain the same equilibrium values requires $\tilde{\lambda} = \beta y^* K/(K - x^*)$. Hence the carrying capacity K may be used as a fine-tuner. For the value of K determines the maximum possible size of the susceptible population and thereby the volatility of the process. When $K = 2x^* = 6800$ we have $\tilde{\lambda} = 2\lambda$, and although the infective population size moves down close to zero on its fourth, fifth and sixth troughs, we see that thereafter the population amplitudes become markedly less severe. Extinction is, of course, ultimately inevitable. Although both the susceptible immigration and logistic birth processes exhibit recurrent cycles, the periodic behaviour under the former (Figure 8.15) is far sharper than that under the latter (Figure 8.16) which produces small-scale endemic fluctuations intertwined with large-scale cycles.

As in the Lotka–Volterra predator-prey process (Section 8.2), trajectories for both the corresponding deterministic and stochastic update equations are easily obtained numerically by respectively computing:

$$x(t + dt) = x(t) + [\tilde{\lambda} x(t)(1 - x(t)/K) - \beta x(t)y(t)]dt$$
$$y(t + dt) = y(t) + [\beta x(t)y(t) - \gamma y(t)]dt \qquad (8.3.131)$$

over say $t = 0, 0.01, 0.02, \ldots$ (for $dt = 0.01$); and,

$$x(t_{k+1}) = x(t_k) + (B - I)/R$$
$$y(t_{k+1}) = y(t_k) + (I - D)/R \qquad (8.3.132)$$

for $t_{k+1} = t_k + 1/R$ where $B = \tilde{\lambda} x(1 - x/K)$, $I = \beta xy$, $D = \gamma y$ and $R = B + 2I + D$. The difference in nature between this latter form and that of the predator-prey stochastic update equations (8.2.18) and (8.2.19) is fairly minimal.

The number of possible variations on this epidemic theme is virtually endless, and many involve fascinating scenarios that are ripe for investigation. Marion, Renshaw and Gibson (1998), for example, study a complex (non-recurrent) stochastic extension to the three-variable system (8.3.127) that has considerable importance in farming. This models the gastrointestinal infection of ruminants by nematodes when the hosts maintain a fixed density. The incorporation of a feedback mechanism, which accounts for the immune response of the infected animals, results in a highly nonlinear process. So now x denotes the number of infected larvae, y the number of parasites infected by a host, and z the level of immunity to these parasites in the host population.

Epidemic processes 463

(a)
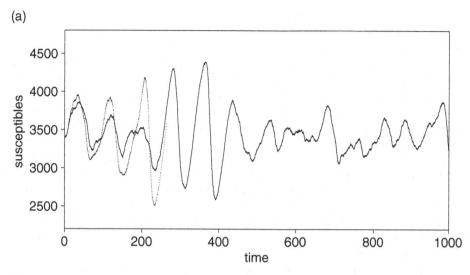

(b)
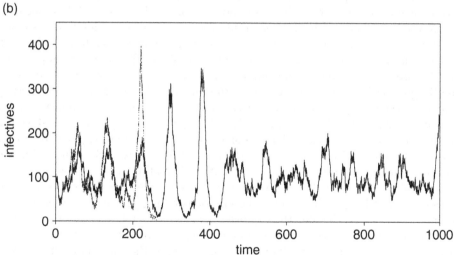

Figure 8.16 A simulated stochastic epidemic over $0 \leq t \leq 1000$ corresponding to the deterministic representation $dx/dt = \lambda_x - 0.0001471xy$ and $dy/dt = 0.0001471xy - 0.5y$, showing the number of (a) susceptibles and (b) infectives for simple ($\cdots$) and logistic (—) birth rates $\lambda_x = 0.01471x$ and $\lambda_x = 2 \times 0.01471x(1 - x/6800)$, respectively.

In deterministic terms, they parallel the model structure of Roberts and Grenfell (1991) via the representation

$$dx/dt = \alpha y \lambda(z) - (\beta + \rho)x$$
$$dy/dt = \beta x p(z) - y\mu(z) \qquad (8.3.133)$$
$$dz/dt = \beta x - \gamma z \qquad (8.3.134)$$

for suitably selected parameters α, β, ρ and γ, and functions $\lambda(z)$, $p(z)$ and $\mu(z)$ of the immunity level z. Given the absence of an exact analytic solution, a number of approximations are explored and compared to simulations of the full stochastic process. Because of the complexity of the process, several modes of behaviour may now occur. In the *endemic regime* the stochastic system fluctuates widely around the fixed points of the deterministic model. This contrasts with a *managed regime* where the system is subject to external periodic perturbation and stochastic effects are now negligible. Whilst in a regime in which the deterministic model predicts the long-term persistence of oscillations, the stochastic model gives rise to ultimate extinction. Note that although we have focussed our attention on the development of processes which develop over the non-negative integers, there is no reason why we cannot extend the arguments to cover continuous variables. Problems relating to moment closure, stochastic linearization, etc. carry through as before. As an illustration, Marion, Renshaw and Gibson (2000) extend this nemotode example by adopting the mean-reverting Ornstein–Uhlenbeck process (see Section 6.3) in order to induce continuous stochastic environmental noise.

Since the study of population processes subject to environmental noise is, potentially, a huge field in its own right, it has not been extensively covered in this current text. Nevertheless, although in some situations its presence has limited effect, it is important to appreciate that in others it can play a hugely significant role in the development of a process. Mao, Marion and Renshaw (2002), for example, show that the presence of even a tiny amount of environmental noise can suppress a potential population explosion. To prove this intrinsically interesting, and perhaps surprising, result they stochastically perturb the multivariate deterministic system $dx(t)/dt = f(x(t))$ into the Itô form $dx(t) = f(x(t))dt + g(x(t))dw(t)$, and show that although the solution to the original differential equation may explode to infinity in a finite time (recall the dishonest birth process discussed in Section 2.3.5), with probability one that of the associated stochastic differential equation does not.

On a final point, although much theoretical work has been developed for epidemic processes, relatively little has been done for competition, predator–prey, insect, etc. populations. So given that all these processes are closely related, it would be both interesting and highly worthwhile to carry techniques across from epidemic scenarios to their biological counterparts. Indeed, bearing in mind the continuing threat to the human population from animal, bird and plant viruses, a far more holistic combined approach to population and infection development would be extremely timely.

8.4 Cumulative size processes

Although the predator–prey, competition and epidemic processes discussed in this chapter play crucial roles in helping us to understand the behaviour of biological, ecological and human mechanisms, and hence improve our ability to predict likely outcomes, the potential for developing other important fields of application is clearly huge. To conclude, let us therefore illustrate a different kind of interaction effect to the ones shown previously that has so far received little attention in the literature.

Section 7.2 introduces a process for the population size, $N(t)$, of the black-margined pecan aphid, where the death rate of the aphid population is related not to $N(t)$ but to the cumulative population density

$$F(t) = \int_0^t N(s)\,ds. \tag{8.4.1}$$

Whence assuming that reproduction follows a simple birth process, and that the death rate per aphid is $\mu F(t)$, leads to the deterministic model (7.2.1), namely

$$dN(t)/dt = [\lambda - \mu F(t)]N(t). \tag{8.4.2}$$

Recall that the cumulative honey-dew excreted by the insects up to a given time forms a weak cover on the leaf surface which results in aphid starvation: the natural defence mechanism of plants to pecan aphid feeding, as well as the attractiveness of accumulated honey-dew to natural enemies of that aphid, is described in some detail in Bumroongsook and Harris (1991). This model clearly lies in sharp contrast to the logistic process, since the individual death rate now relates to the integrated number of individuals present up to time t (i.e. to the area $F(t)$ under the $N(t)$ curve), and not to $N(t)$ itself. It follows from (8.4.2) that since $dN/dt = 0$ when $F(t) = \lambda/\mu$, the maximum (deterministic) cumulative number of aphids $F_{max} = \lambda/\mu$. This parameter ratio relates to the total amount of nutrient (leaf area) available to sustain population growth. In the corresponding stochastic scenario, this biological interpretation appears to be more woolly, since values of $F(t)$ higher than λ/μ are possible; though the individual death rate ($\mu F(t) > \lambda$) relentlessly increases.

8.4.1 Deterministic models

The solution to (8.4.2) takes the form (7.2.5), namely

$$N(t) = ae^{-bt}/(1 + de^{-bt})^2, \tag{8.4.3}$$

where, for $N(0) = n_0$, the parameters $b = \sqrt{(\lambda^2 + 2\mu n_0)}$, $d = (b+\lambda)/(b-\lambda)$ and $a = 2d(\lambda^2/\mu + 2n_0)$. Thus

$$\mu = 2b^2 d/a, \quad \lambda = b(d-1)/(d+1) \quad \text{and} \quad n_0 = a/(1+d)^2. \tag{8.4.4}$$

Moreover, substituting (8.4.3) into (8.4.1) gives

$$F(t) = \left(\frac{a}{b(1+d)}\right)\left(\frac{1-e^{-bt}}{1+de^{-bt}}\right). \tag{8.4.5}$$

So $F(0) = 0$ (as required) and $F(\infty) = a/[b(1+d)]$.

Since $N(t)$ reaches its maximum value, N_{max}, when $dN(t)/dt = 0$, we see from (8.4.2) that this occurs at time $t = t_{max}$ where t_{max} is the solution to the equation $F(t) = \lambda/\mu$. Hence it follows from (8.4.5) that t_{max} is the solution to the equation

$$e^{-bt}[a\mu + \lambda bd(1+d)] = a\mu - \lambda b(1+d). \tag{8.4.6}$$

On substituting for λ and μ from (8.4.4) this reduces to $e^{-bt} = 1/d$, so $t_{max} = \ln(d)/b$. Whence substituting $e^{-bt} = 1/d$ into (8.4.3) gives $N_{max} = a/4d$.

Matis, Kiffe, Matis and Stevenson (2006) propose a (possibly) more realistic variable than $F(t)$, namely the cumulative insect count $C(t)$, which ties in better with the underlying biological rationale. For we may argue that the amount of leaf surface covered with honey-dew is proportional to the number of aphids that have ever lived, and since the population birth rate is $\lambda N(t)$ this implies that the (deterministic) number of births up to time t is given by

$$C(t) = \int_0^t \lambda N(s)\,ds = \lambda F(t). \tag{8.4.7}$$

So it follows from (8.4.5) that

$$C(t) = \left(\frac{\lambda a}{b(1+d)}\right)\left(\frac{1-e^{-bt}}{1+de^{-bt}}\right). \tag{8.4.8}$$

Substituting for λ from (8.4.4) and placing $a = 4dM_{max}$ then shows that, for large d, the eventual number of insects that have ever lived is given by

$$C(\infty) = \frac{\lambda a}{b(1+d)} = 4N_{max} \times \frac{d(d-1)}{(d+1)^2} = 4N_{max}. \tag{8.4.9}$$

Thus we have the rather intriguing result that the ultimate number of individuals born is four times the maximum population size. Moreover, replacing d and a in (8.4.3) by $d = e^{bt_{max}}$ and $a = 4dN_{max} = 4N_{max}e^{bt_{max}}$ yields the symmetric form

$$N(t) = \frac{4N_{max}e^{-b(t-t_{max})}}{(1+e^{-b(t-t_{max})})^2} = N_{max}\cosh^{-2}\{b(t-t_{max})/2\}. \tag{8.4.10}$$

Note that on using (8.4.7) the model (8.4.2) may be written in the equivalent form

$$dN(t)/dt = [\lambda - (\mu/\lambda)C(t)]N(t), \tag{8.4.11}$$

where under this new interpretation the death rate parameter $\phi = \mu/\lambda$. So from a purely deterministic perspective it is really irrelevant whether one works in terms of $F(t)$ or $C(t)$.

8.4.2 Stochastic simulation

However, this equivalence no longer holds when we consider the corresponding stochastic processes. For suppose we let $X(t)$, $Y(t)$ and $C(t)$ be *integer* random variables and $F(t)$ a *continuous* random variable, where $X(t)$ and $Y(t)$ denote the number of aphids at time t based on $C(t)$ and $F(t)$, respectively. Here $C(t)$ denotes the total number of aphids born by time t, and $F(t)$ the integrated population size. Then paralleling the extension of the two-parameter deterministic logistic model (3.2.29), namely $dN/dt = rN[1-(N/K)]$, into the four-parameter stochastic birth and death rates (3.2.35), namely $\lambda_N = N(a_1 - b_1 N)$ and $\mu_N = N(a_2 + b_2)N$, enables us to construct the transition rates for the two bivariate stochastic processes $\{X(t), C(t)\}$

and $\{Y(t), F(t)\}$. First, since the birth of a single new aphid leads to the simultaneous unit increase in both $X(t)$ and $C(t)$, in the small time interval $(t, t+dt)$

$$\Pr[(X(t), C(t)) \to (X(t)+1, C(t)+1)] = X(t)[a_1 - b_1 C(t)]dt \equiv V_1(t)dt$$
$$\Pr[(X(t), C(t)) \to (X(t)-1, C(t))] \quad = X(t)[a_2 + b_2 C(t)]dt \equiv W_1(t)dt, \quad (8.4.12)$$

else $(X(t), C(t))$ remains unchanged. Second, on presuming that $F(t) \to F(t) + Y(t)dt$ in $(t, t+dt)$, regardless of whether a birth or death occurs, we have

$$\Pr[(Y(t), F(t)) \to (Y(t)+1, F(t)+Y(t)dt)] = Y(t)[a_1 - c_1 F(t)]dt \equiv V_2(t)dt$$
$$\Pr[(Y(t), F(t)) \to (Y(t)-1, F(t)+Y(t)dt)] = Y(t)[a_2 + c_2 F(t)]dt \equiv W_2(t)dt, \quad (8.4.13)$$

else $Y(t)$ remains unchanged. The a_i denote the intrinsic growth rates, and b_i and c_i the crowding coefficients. As the corresponding deterministic models are based purely on their net value we have $\lambda = a_1 - a_2$ with $\mu/\lambda = b_1 - b_2$ (under (8.4.11)) and $\mu = c_1 - c_2$ (under (8.4.2)).

From a mathematical perspective the first set of rates is more appealing, since we can use them to construct the Kolmogorov forward and backward equations for $p_{ij}(t) = \Pr[X(t) = i, C(t) = j | X(t) = n_0, C(0) = 0]$. In contrast the second set comprises a mixture of integer and real variables, and so is harder to manipulate. However, simulation algorithms are easy to construct under both regimes. Though slight care needs to be taken with the initial conditions, since if we assume that the leaf will be free of honey-dew at (say) $t = 0$, then the interaction effects, namely $b_i n_0 C(0)$ and $c_i n_0 F(0)$, must be zero. Although the second condition follows directly from the definition (8.4.1), we can only take $C(0) = 0$ if $C(t)$ registers the cumulative number of births and not individuals.

Like the recurrent epidemic algorithm A8.4, the simulation procedure for constructing realizations of $\{X(t), C(t)\}$ involves only a slight modification to the general two-species algorithm A8.1, namely:

Cumulative sum simulation algorithm (A8.5)
    (i) read in parameters
    (ii) set $t = 0$, $X = n_0$ and $C(0) = 0$
    (iii) print t, X and C
    (iv) generate a new $U \sim U(0, 1)$ random number
    (v) evaluate V_1 and W_1
        if $U \leq V_1/(V_1 + W_1)$ then $X \to X + 1$ and $C \to C + 1$
        else $X \to X - 1$
    (vi) generate a new $U \sim U(0, 1)$ random number
    (vii) update t to $t - [\ln(U)]/(V_1 + W_1)$
    (viii) return to (iii)

Although the procedure for constructing realizations of $\{Y(t), F(t)\}$ is broadly similar, note that $F(t)$ must be updated *after* the time to the next event has been determined, viz:

Integrated population size simulation algorithm (A8.6)
  (i) read in parameters
  (ii) set $t = 0$, $Y = n_0$ and $F(0) = 0$
  (iii) print t, Y and F
  (iv) generate a new $U \sim U(0,1)$ random number
  (v) evaluate V_2 and W_2 and set $Y_{old} = N$
       if $U \leq V_2/(V_2 + W_2)$ then $Y \to Y + 1$
       else $Y \to Y - 1$
  (vi) generate a new $U \sim U(0,1)$ random number
  (vii) update t to $t + s$ where $s = -[\ln(U)]/(V_2 + W_2)$
  (viii) update F to $F + Y_{old}s$
  (ix) return to (iii)

To compare these two stochastic processes with their deterministic counterpart, let us consider the reduced model with $b_1 = c_1 = a_2 = 0$ by taking $\lambda = 2.5$, $\phi = 0.01$ and $\mu = \phi\lambda = 0.025$ for $n_0 = 2$. Figure 8.17a shows stochastic realizations of $X(t)$ and $Y(t)$ constructed from the same initial seed via (8.4.12) and (8.4.13), respectively, together with the deterministic curve (8.4.10) for $N(t)$. Not only do both $X(t)$ and $Y(t)$ lie well to the right of $N(t)$ (they take longer to get going), but $Y(t)$ starts faster than $X(t)$, and then proceeds to suffer a more dramatic and faster population bust once it passes its maximum value. Figure 8.17b shows the corresponding stochastic plots of $C(t)$ and $\lambda F(t)$, together with the associated deterministic curve (8.4.8). Whilst the asymptotic values are broadly similar, namely $X(\infty) = 528$, $\lambda Y(\infty) = 510$ and $N(\infty) = 500$, the ways in which the trajectories approach them clearly differ.

Not only is there a considerable difference between these stochastic and deterministic realizations, but $X(t)$ and $Y(t)$ exhibit substantially different structure even though they are underpinned by the same deterministic process. Indeed, such differences emerge right at the outset; for the death rate $\phi C(t) = 0$ until the first birth occurs, in contrast to $\mu F(t) > 0$ for all $t \geq 0$. Whilst in the extreme case of $\lambda = 0$, the individual death rates $\phi C(t) = 0$ and $\mu N(t) > 0$ for all $t > 0$. So $N(t) \equiv n_0$ in the first case, but $N(t) = 1/[(1/n_0) + \mu t] \to 0$ as $t \to \infty$ in the second. This example further highlights the danger is not comparing deterministic and stochastic realizations over a range of possible processes before deciding which particular model structure is to be employed.

8.4.3 Probability solutions

Let us now work with the easier discrete random variable process $\{X(t), C(t)\}$, rather than the mixed model $\{Y(t), F(t)\}$, and consider a simplified version of (8.4.12) in which $V_1(t) = \lambda X(t)$ and $W_1(t) = \phi X(t)C(t)$. Then since a birth corresponds to a unit increase in both $X(t)$ and $C(t)$, and a death to a decrease in $X(t)$ alone, using Bartlett's direct approach to write down the Kolmogorov equation for the m.g.f. $M(\theta_1, \theta_2; t) \equiv \sum_{i,j=0}^{\infty} p_{ij}(t)e^{i\theta_1}e^{j\theta_2}$ yields

$$\frac{\partial M}{\partial t} = \lambda(e^{\theta_1 + \theta_2} - 1)\frac{\partial M}{\partial \theta_1} + \phi(e^{-\theta_1} - 1)\frac{\partial^2 M}{\partial \theta_1 \partial \theta_2}. \qquad (8.4.14)$$

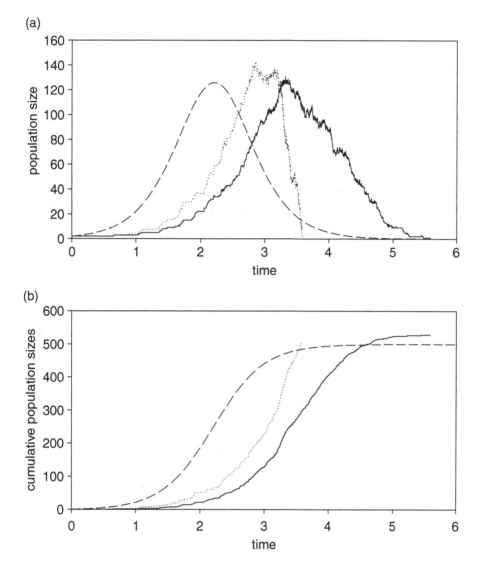

Figure 8.17 Realizations of $\{X(t), C(t)\}$ and $\{Y(t), F(t)\}$ for $\lambda = 2.5$, $\phi = 0.01$ and $\mu = \phi\lambda = 0.025$ with $n_0 = 2$, showing: (a) $X(t)$ (—), $Y(t)$ ($\cdots$) and $N(t)$ (- - -); and, (b) $C(t)$ (—), $F(t)$ ($\cdots$) and $\int_0^t N(s)\,ds$ (- - -).

Alternatively, we can employ the standard, albeit more laborious, approach of first constructing the forward probability equations

$$dp_{ij}(t)/dt = \lambda(i-1)p_{i-1,j-1}(t) + \phi(i+1)jp_{i+1,j}(t) - (\lambda i + \phi ij)p_{ij}(t), \quad (8.4.15)$$

and then using these to form the corresponding equation for the p.g.f. $G(z_1, z_2; t) \equiv \sum_{i,j=0}^{\infty} p_{ij}(t)z_1^i z_2^j$, namely

$$\frac{\partial G}{\partial t} = \lambda z_1(z_1 z_2 - 1)\frac{\partial G}{\partial z_1} + \phi z_2(1 - z_1)\frac{\partial^2 G}{\partial z_1 \partial z_2}, \qquad (8.4.16)$$

which is identical in structure to (7.2.10). Placing $z_i = e^{\theta_i}$ and noting that $\partial G/\partial \theta_1 = e^{-\theta_1}\partial M/\partial \theta_1$, etc., recovers the m.g.f. equation (8.4.14).

Now transforming (8.4.14) into the associated equation for the c.g.f. $K(\theta_1, \theta_2; t) \equiv \ln[M(\theta_1, \theta_2; t)]$ gives

$$\frac{\partial K}{\partial t} = \lambda(e^{\theta_1+\theta_2} - 1)\frac{\partial K}{\partial \theta_1} + \phi(e^{-\theta_1} - 1)[\frac{\partial^2 K}{\partial \theta_1 \partial \theta_2} + \frac{\partial K}{\partial \theta_1}\frac{\partial K}{\partial \theta_2}]. \qquad (8.4.17)$$

Whence differentiating (8.4.17) and placing $\theta_1 = \theta_2 = 0$ in the usual manner yields the equations for the bivariate cumulants $\kappa_{ij}(t) = \partial^{i+j} K(\theta_1,\theta_2;t)/\partial\theta_1^i \partial\theta_2^j \big|_{\theta_1=\theta_2=0}$, namely (e.g. Matis, Kiffe, Matis and Stevenson, 2006)

$$d\kappa_{10}/dt = \lambda\kappa_{10} - \phi(\kappa_{10}\kappa_{01} + \kappa_{11})$$
$$d\kappa_{01}/dt = \lambda\kappa_{10}$$
$$d\kappa_{20}/dt = \lambda(\kappa_{10} + 2\kappa_{20}) + \phi(\kappa_{11} - 2\kappa_{10}\kappa_{11} - 2\kappa_{21} + \kappa_{01}(\kappa_{10} - 2\kappa_{20}))$$
$$d\kappa_{11}/dt = \lambda(\kappa_{10} + \kappa_{20} + \kappa_{11}) - \phi(\kappa_{10}\kappa_{02} + \kappa_{01}\kappa_{11} + \kappa_{12}) \qquad (8.4.18)$$
$$d\kappa_{02}/dt = \lambda(\kappa_{10} + 2\kappa_{11}), \quad \text{etc.}$$

Note how the first two equations for the means compare with their deterministic counterparts. Although the equation for $E[C(t)] = \kappa_{01}(t)$ is identical to the differential of the cumulative density equation (8.4.7), namely $dC(t)/dt = \lambda N(t)$, that for $E[N(t)] = \kappa_{10}(t)$ is not the same as that implied by (8.4.2), namely $dN(t)/dt = N(t)[\lambda - \phi C(t)]$, since the former is diminished by $\phi\kappa_{11}(t)$. Because the covariance term, $\kappa_{11}(t)$, is initially positive, this difference implies that $N(t) > \kappa_{10}(t)$ at least in the opening stages of the process, as suggested by Figure 8.17a. Matis et al. (2006) remark that, in general, second-order cumulant truncation (i.e. placing $\kappa_{30}(t) = \cdots = \kappa_{03}(t) = 0$) appears to approximate the means well but the variances poorly; unfortunately, including third-order cumulants appears to offer little real improvement.

This means that direct application of the bivariate saddlepoint approximation developed in Section 7.3 is unlikely to offer good approximations for the probabilities $p_{ij}(t)$. Moreover, in spite of the saddlepoint approximation being the only realistic family-free technique available for constructing an associated probability distribution, and being optimal in the sense that it is based on the highly efficient numerical method of steepest descents, it also suffers from the problem of not always yielding full support. Although Wang's (1992) neat scaling approach provides a solution to this hurdle, it does lead to potentially inaccurate and aberrant results. Nevertheless, it is worthwhile noting that there are several ways of surmounting such difficulties. Gillespie and Renshaw (2007), for example: (i) extend the inversion of the cumulant generating function to second-order; (ii) select an appropriate probability structure for higher-order cumulants (instead of taking them to be zero); and, (iii) make subtle

8.4.4 Power-law processes

In Section 3.2.1 we extended the standard logistic process with quadratic birth and death rates (3.2.35) to encompass the power-law rates (3.2.38)–(3.2.39), namely $\lambda_N = a_1 N - b_1 N^{s+1}$ and $\mu_N = a_2 N + b_2 N^{s+1}$. A key feature is that the higher the value of the power coefficient, s, the nearer N is to the carrying capacity K before the nonlinear components take effect (see Figure 3.1). This structure is of considerable use when modelling a developing population, since s relates to the aggression of the individual population members. So by fine-tuning this parameter we can regulate the population size at which individual growth becomes seriously affected by competition. The underlying rationale is that the observed point of inflection in density-dependent situations often exceeds the $K/2$-value inherent in the ordinary logistic model. Now because the associated deterministic equation $dN/dt = \lambda_N - \mu_N$ takes the solution

$$N(t) = K/[1 + \{(K/n_0)^s - 1\}e^{-(a_1-a_2)st}]^{1/s}, \qquad (8.4.19)$$

where $K = [(a_1 - a_2)/(b_1 - b_2)]^{1/s}$ (see expression (3.2.42)), it follows that the point of inflexion $N_{infl} = K/(s+1)^{1/s} \to K$ as $s \to \infty$ (see result (3.2.43)). Hence we can choose s to ensure that the observed and theoretical points of inflection match.

This suggests that a natural extension to the aphid model is to replace the stochastic rates (8.4.12) by $V_1 = X[a_1 - b_1 C^s]$ and $W_1 = X[a_2 + b_2 C^s]$ (and similarly for (8.4.13)). Whence invoking Barlett's random variable technique for the m.g.f. $M(\theta_1, \theta_2; t)$ yields the p.d.e.

$$\frac{\partial M}{\partial t} = \{a_1(e^{\theta_1+\theta_2}-1) + a_2(e^{-\theta_1}-1)\}\frac{\partial M}{\partial \theta_1} + \{-b_1(e^{\theta_1+\theta_2}-1) + b_2(e^{-\theta_1}-1)\}\frac{\partial^{1+s} M}{\partial \theta_1 \partial \theta_2^s}. \qquad (8.4.20)$$

Note that if s is non-integer then this equation involves fractional differentiation. An alternative, and longer, construction is write down the forward equations

$$dp_{ij}(t)/dt = (i-1)[a_1 - b_1(j-1)^s]p_{i-1,j-1}(t) + (i+1)[a_2 + b_2 j^s]p_{i+1,j}(t)$$
$$- i[(a_1+a_2) - (b_1-b_2)j^s]p_{ij}(t), \qquad (8.4.21)$$

multiply each term by $e^{i\theta_1+j\theta_2}$, and then sum over $i, j = 0, 1, 2, \ldots$.

Replacing $M(\theta_1, \theta_2; t)$ in (8.4.20) by the c.g.f. $K(\theta_1, \theta_2; t) \equiv \ln[M(\theta_1, \theta_2; t)]$, and differentiating with respect to θ_1 and θ_2 in the usual manner, yields the cumulant equations. For example, on taking the $s = 2$ power-law version of the reduced process with $V_1(t) = \lambda X(t)$ and $W_1(t) = \mu X(t) C(t)^s$, the first- and second-order equations are given by

$$d\kappa_{10}/dt = \lambda\kappa_{10} - \mu(\kappa_{10}\kappa_{01}^2 + \kappa_{10}\kappa_{02} + 2\kappa_{01}\kappa_{11} + \kappa_{12})$$

$$d\kappa_{01}/dt = \lambda\kappa_{10}$$

$$d\kappa_{20}/dt = \lambda(2\kappa_{20} + \kappa_{10}) + \mu[\kappa_{10}(\kappa_{01}^2 + \kappa_{02} - 2\kappa_{12}) + 2\kappa_{01}(\kappa_{11} - 2\kappa_{21} - \kappa_{01}\kappa_{20})$$
$$+ (\kappa_{12} - 4\kappa_{11}^2 - 2\kappa_{20}\kappa_{02} - 4\kappa_{10}\kappa_{01}\kappa_{11})] \tag{8.4.22}$$

$$d\kappa_{11}/dt = \lambda(\kappa_{10} + \kappa_{11} + \kappa_{20}) - \mu[\kappa_{10}\kappa_{03} - \kappa_{01}(2\kappa_{12} + \kappa_{01}\kappa_{11}) - \kappa_{02}(3\kappa_{11} + 2\kappa_{10}\kappa_{01})]$$

$$d\kappa_{02}/dt = \lambda(\kappa_{10} + 2\kappa_{11})$$

(see Matis, Kiffe, Matis, and Stevenson, 2007). These equations are considerably more complex than the previous ones (8.4.18) for $s = 1$, and care is needed when developing higher-order moment closure approximations. A conservative approach would simply be to retain first- and second-order cumulants only, and just accept the restrictions imposed by the Normal approximation.

Figure 8.18 shows three simulations of $X(t)$ against t, using the same initial seed, for this process with $\lambda = 2.5$, $n_0 = 2$ and μ chosen to ensure that the deterministic value of $C(\infty)$ remains fixed at 500 as s varies; i.e. $\mu = 0.01$ at $s = 1$, $\mu = 0.01/335$ at $s = 2$ and $\mu = 0.01/126000$ at $s = 3$. Since $C(t)$ remains relatively small in the opening stage of each run, births predominate over deaths, and each simulation follows a similar, almost exponential, growth process until time $t = 2$. Thereafter deaths come into play, though the time at which they do so clearly increases with s. So the larger the value of s, the longer the trajectory follows an exponential path and hence the larger the peak. This must then be followed by a crash which steepens with s, since $C(\infty)$ has to lie around 500 (its deterministic value).

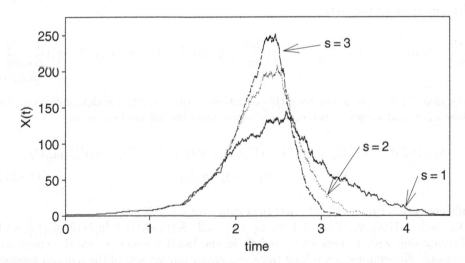

Figure 8.18 Realizations of $\{X(t)\}$ with $\lambda = 2.5$ and $n_0 = 2$, for: $s = 1$ and $\mu = 0.01$ (—); $s = 2$ and $\mu = 0.01/335$ ($\cdots$); and, $s = 3$ and $\mu = 0.01/126000$ (- - -).

The rates $V_1(t)$ and $W_1(t)$ may, of course, be made totally general. Though such constructs are easy to simulate, since the algorithm remains the same apart from changing the birth and death rates, mathematical analysis unfortunately becomes too opaque to be of much use. For the record, if for $-\infty < i, j < \infty$ and $0 \leq r, s < \infty$

$$\Pr[X(t) \text{ changes by } i \text{ and } C(t) \text{ by } j \text{ in } (t, t+dt)] = f_{ij} X^r C^s, \qquad (8.4.23)$$

then the associated m.g.f. $M(\theta_1, \theta_2; t)$ is the solution to the p.d.e.

$$\frac{\partial M}{\partial t} = \sum_{r,s} f_{ij}(e^{i\theta_1 + j\theta_2} - 1) \frac{\partial^{r+s} M}{\partial \theta_1^r \partial \theta_2^s}. \qquad (8.4.24)$$

Thus the m.g.f. for any bivariate process whose transition rates satisfy the form (8.4.23) be written down directly. Moreover, the extension to processes with three or more variables is straightforward.

9
Spatial processes

All our previous analyses are based on the assumption that populations develop at a single site where individuals mix homogeneously. Whilst this is mathematically ideal, in that it facilitates theoretical development, in reality there are many situations in which it may be violated. For not only may a population be spatially distributed across several interlinked sites, but even within a specific site the chance of two individuals meeting and interacting may well depend on the distance between them. Although this fact was realized early on in the development of theoretical population dynamics (see, for example, Bailey, 1968; Levin, 1974; Renshaw, 1991), the high degree of mathematical intractability which rides along with it has meant that little analytic progress has been made relative to non-spatial scenarios. In this chapter we shall expose the underlying theoretical difficulties, highlight directions in which some degree of progress can be made, and show that the introduction of space generates a whole new concept of a *stochastic dynamic*. In this latter construct, single-site processes, which on their own result in early extinction, can generate long-term persistence when linked together.

9.1 General results

Let $X_i(t)$ denote the number of individuals in colonies (i.e. sites) $i = 1, 2, \ldots, k$, and suppose that events may be of three types, namely the arrival of a new individual, the departure of an existing one, or the transfer of an individual from one colony to another. Denote the transition rates out of the states $\{X_i(t) = n_i\}$ by

$$q(\mathbf{n}, \mathbf{n} + \mathbf{e}_i) = \theta_i(\mathbf{n}) : \quad \text{arrival at } i \quad \text{(i.e. } X_i(t) \to n_i + 1\text{)}$$

$$q(\mathbf{n}, \mathbf{n} - \mathbf{e}_i) = \phi_i(\mathbf{n}) : \quad \text{departure from } i \quad \text{(i.e. } X_i(t) \to n_i - 1\text{)} \quad (9.1.1)$$

$$q(\mathbf{n}, \mathbf{n} - \mathbf{e}_i + \mathbf{e}_j) = \psi_{ij}(\mathbf{n}) : \quad \text{transfer from } i \text{ to } j \neq i \quad \text{(i.e. } X_i(t) \to n_i - 1 \text{ and}$$
$$X_j(t) \to n_j + 1\text{)}$$

where $\mathbf{n} = (n_1, \ldots, n_k)$ and $\mathbf{e}_i$ denotes the vector whose components are all zero except for 1 in the ith place. If these rates depend only on the numbers in the colonies affected by the transition then we may write

$$\theta_i(\mathbf{n}) = \theta_i(n_i) \, , \quad \phi_i(\mathbf{n}) = \phi_i(n_i) \quad \text{and} \quad \psi_{ij}(\mathbf{n}) = \psi_{ij}(n_i, n_j), \quad (9.1.2)$$

and describe the process as being *simple*. Thus both the linear immigration–birth–death–migration process with rates $\theta_i(\mathbf{n}) = \alpha_i + \lambda_i n_i$, $\phi_i(\mathbf{n}) = \mu_i n_i$ and $\psi_{ij}(\mathbf{n}) = \nu_{ij} n_i$, and the birth–epidemic process with rates $\theta_1(\mathbf{n}) = \lambda n_1$, $\phi_2(\mathbf{n}) = \gamma n_2$, $\psi_{12}(\mathbf{n}) = \beta n_1 n_2$ and $\theta_2(\mathbf{n}) = \phi_1(\mathbf{n}) = \psi_{21}(\mathbf{n}) = 0$, are simple. However, the Lotka–Volterra predator–prey process is not since its rates $\theta_1(\mathbf{n}) = n_1 r_1$, $\phi_1(\mathbf{n}) = b_1 n_1 n_2$, $\theta_2(\mathbf{n}) = b_2 n_1 n_2$ and $\phi_2(\mathbf{n}) = r_2 n_2$ with $\psi_{12}(\mathbf{n}) = \psi_{21}(\mathbf{n}) = 0$ do not conform to (9.1.2). Note the close paradigm between 'colony' and 'individual type'; with colony 1 relating to the status susceptible or prey, and colony 2 relating to infective or predator. So in this sense the term 'spatial' has a much wider interpretation than geographic location.

If a stationary distribution $\{\pi(\mathbf{n})\}$ exists, then it follows from (9.1.1) that

$$[\sum_{i=1}^{k}\theta_i(\mathbf{n}) + \sum_{i=1}^{k}\phi_i(\mathbf{n}) + \sum_{i,j=1}^{k}\psi_{ij}(\mathbf{n})]\pi(\mathbf{n}) = \sum_{i=1}^{k}\theta_i(\mathbf{n}-\mathbf{e}_i)\pi(\mathbf{n}-\mathbf{e}_i)$$

$$+ \sum_{i=1}^{k}\phi_i(\mathbf{n}+\mathbf{e}_i)\pi(\mathbf{n}+\mathbf{e}_i) + \sum_{i,j=1}^{k}\psi_{ij}(\mathbf{n}+\mathbf{e}_i-\mathbf{e}_j)\pi(\mathbf{n}+\mathbf{e}_i-\mathbf{e}_j). \quad (9.1.3)$$

As there is no hope of finding a general solution to equation (9.1.3) – even in simple cases the form of $\{\pi(\mathbf{n})\}$ may be complex – Kingman (1969) considers the restricted class of processes which satisfy the reversibility condition

$$\pi(\mathbf{n})q(\mathbf{n},\mathbf{m}) = \pi(\mathbf{m})q(\mathbf{m},\mathbf{n}); \quad (9.1.4)$$

for a general treatment of reversibility see Kelly (1979). For if (9.1.4) holds then it may be shown that

$$\pi(\mathbf{n}) = \pi(\mathbf{n}_0)\frac{q(\mathbf{n}_0,\mathbf{n}_1)q(\mathbf{n}_1,\mathbf{n}_2)\ldots q(\mathbf{n}_r,\mathbf{n})}{q(\mathbf{n}_1,\mathbf{n}_0)q(\mathbf{n}_2,\mathbf{n}_1)\ldots q(\mathbf{n},\mathbf{n}_r)}, \quad (9.1.5)$$

where $(\mathbf{n}_0,\mathbf{n}_1,\ldots,\mathbf{n}_r,\mathbf{n})$ is any path from the arbitrary fixed state $\mathbf{n}_0$ to $\mathbf{n}$, and reversibility guarantees that the expression is independent of the path chosen. Whilst in many situations the requirement of reversibility is perfectly reasonable, there are scenarios in which it is less so, and Kingman examines the consequences of 'partial balance'. Though we shall not progress this concept here. Whittle (1967, 1968) provides two interesting and highly useful generalizations. First, in 'closed systems' in which the total number of individuals is conserved, i.e. there are neither arrivals nor departures, it follows that $\theta_i(\mathbf{n}) = \phi_i(\mathbf{n}) = 0$. Whence if we also assume that $\psi_{ij}(\mathbf{n}) = \nu_{ij}\psi_i(n_i)$ for constants ν_{ij}, then $\{\pi(\mathbf{n})\}$ takes the multiplicative form

$$\pi(\mathbf{n}) = \pi_1(n_1)\pi_2(n_2)\ldots\pi_k(n_k). \quad (9.1.6)$$

Second, in 'open systems', where total population size is not conserved, suppose we also allow immigration at rate $\theta_i(\mathbf{n}) = \alpha_i$ and general death at rate $\phi_i(\mathbf{n}) = \mu_i \psi_i(n_i)$. Then the solution now takes the multiplicative form

$$\pi(\mathbf{n}) = C\prod_{i=1}^{k}\frac{\lambda_i^{n_i}}{\psi_i(1)\psi_i(2)\ldots\psi_i(n_i)}, \quad (9.1.7)$$

where the λ_i are determined from the linear equations

$$(\mu_i + \sum_{j=1}^{k} \nu_{ij})\lambda_i = \alpha_i + \sum_{j=1}^{k} \nu_{ij}\lambda_j \qquad (9.1.8)$$

and C is a normalizing constant. The linear immigration–death–migration system $\theta_i(\mathbf{n}) = \alpha_i$, $\phi_i(\mathbf{n}) = \mu_i n_i$ and $\psi_{ij}(\mathbf{n}) = \nu_{ij} n_i$ is clearly a special case of this open process and takes the equilibrium solution

$$\pi(\mathbf{n}) = \prod_{i=1}^{k} \frac{\lambda_i^{n_i} e^{-\lambda_i}}{n_i!} \qquad (9.1.9)$$

(Bartlett, 1949). Thus the numbers in the individual colonies have independent Poisson distributions.

Note that in these examples the migration rate $\psi_{ij}(\mathbf{n}) = \nu_{ij}\psi_i(n_i)$ from i to j depends only on the number of individuals in i and not j. We are therefore excluding, for example, flow models (e.g. Kingman, 1969) in which no colony can contain more than one individual at a time, i.e. an occupied site blocks migration to it (Beneš, 1965). Also excluded is the highly applicable and interesting class of research problems generated by interacting particle systems; see Durrett (1981) and Liggett (1985) for an early survey of results and open problems subsequent to the pioneering paper of Spitzer (1970).

Although the above approach involves the general migration function $\nu_{ij}\psi_i(n_i)$, the arrival function $\theta_i(\mathbf{n}) = \alpha_i$ is highly restrictive since it corresponds to pure immigration and so does not allow even a simple birth process to operate. In contrast, choice of the departure process $\phi_i(\mathbf{n}) = \phi_i(n_i)$ is totally unrestricted, since it can be rolled into the migration component. For on denoting $k+1$ to be the state from which there is no return, we can consider death as migration to an ethereal colony $k+1$ with $\phi_i(n_i) = \psi_{i,k+1}(n_i)$ and $\psi_{k+1,j}(n_{k+1}, n_j) = 0$ (for all $j = 1, \ldots, k$). Since it is the replacement of immigration by birth that causes severe mathematical difficulties, and *de facto* the recognition of spatial stochastic processes as a definitive field of study, it is worthwhile mentioning some key results without birth before we commence our journey down this route.

Bartlett's (1949) seminal paper analyses the conservative (i.e. closed) process with m individuals and general linear migration between N colonies (so $\theta_i(\mathbf{n}) = \phi_i(\mathbf{n}) = 0$ and $\psi_{ij}(\mathbf{n}) = \nu_{ij} n_i$). He derives the p.g.f. of population size at time t in the form

$$G(\mathbf{z};t) \equiv \sum_{\substack{n_1+\cdots+n_N=m \\ n_1,\ldots,n_N=0}} \Pr(\mathbf{X}(t) = \mathbf{n}(t)) z_1^{n_1} \ldots z_N^{n_N} = \left[\boldsymbol{\eta}_1^T \mathbf{z} + \sum_{s=2}^{r} c_s \boldsymbol{\eta}_s^T \mathbf{z} \exp(\omega_s t)\right]^r. \qquad (9.1.10)$$

Here ω_s and $\boldsymbol{\eta}_s$ are the eigenvalues and eigenvectors, respectively, of the transition matrix $\{\nu_{ij}\}$, where $\nu_{ii} = -\sum_{s \neq i} \nu_{is}$ and $\boldsymbol{\eta}^T$ denotes the transpose of $\boldsymbol{\eta}$. It is assumed, without loss of generality, that all colonies are initially empty apart from colony 1 which contains r individuals, i.e. $G(\mathbf{z};0) = z_1^r$. With this in mind, the c_s are defined through

$$z_1 = \sum_s c_s \boldsymbol{\eta}_s^T \mathbf{z}. \tag{9.1.11}$$

This is a truly pioneering result since it predates the mid-1950s which saw the start of an explosion of research into stochastic processes. Extensions took time to develop, with (for example) Radcliffe and Staff (1969) allowing for multiple eigenvalues, Ruben (1962) admitting immigration into the system, and Raman and Chiang (1973) considering time-dependent transition rates, albeit by restricting the matrix of migration rates to be of upper triangular form. Specific applications followed on swiftly behind, fuelled not only by the tremendous expansion in interest in stochastic processes during the 1960s and 1970s, but also by Joe Gani's formation of the *Journal* (and later the *Advances*) *of Applied Probability*. At that time few other journals published such papers, so current developments and applications were easily accessible (halycon days!). Early fields of application were surprisingly diverse, with Bartlett's result being used by Siegart (1949) to consider the approach to equilibrium of non-interacting gas molecules, and by Krieger and Gans (1960) and Gans (1960) to examine the effect of disturbing a multi-compartment system from its equilibrium distribution. Patil (1957) uses the model to obtain an approximate representation for the movement of bull spermatozoa, Bartholomew (1967) to study the recruitment and promotion of individuals in various grades of an organization, and Chiang (1968) derives explicit solutions for the probability structure of the illness–death model in which there are movements both between the various states (colonies) of illness and to the various types of death. Whilst Matis and Hartley (1971) present an early example which relates to our counting process discussion in Sections 7.4 and 7.5: they estimate migration and death parameters in situations where (as may well happen when studying living organisms) individual compartments (i.e. colonies) are inaccessible for observation, and time-series data are available only on the passage of material to the system exterior.

9.2   Two-site models

Bearing in mind that the simple stochastic birth–death–immigration model investigated in Chapter 2 is far more mathematically tractable than the (univariate) logistic models developed in Chapter 3, and especially the (bivariate) competition, predator–prey and epidemic extensions considered in Chapter 8, let us first concentrate our attention on the simple two-site birth–death–migration process in order to appreciate the problems which underlie this whole area.

Although expressions (9.1.10) and (9.1.11) provide a general solution for closed linear processes, as soon as birth is introduced no such representation is possible due to the intractability of the underlying Kolmogorov equations. This feature is best appreciated by restricting our attention to the simple immigration–birth–death–migration process with rates $\theta_i(\mathbf{n}) = \alpha_i + \lambda_i n_i$, $\phi_i(\mathbf{n}) = \mu_i n_i$ and $\psi_{ij}(\mathbf{n}) = \nu_{ij} n_i$. For in spite of the forward equations for the associated population size probabilities $p_{ij}(t) = \Pr(X_1(t) = i, X_2(t) = j | X_1(0) = a_1, X_2(0) = a_2)$ taking the relatively simple linear form

$$dp_{ij}(t)/dt = [\alpha_1 + \lambda_1(i-1)]p_{i-1,j}(t) + [\alpha_2 + \lambda_2(j-1)]p_{i,j-1}(t) + \mu_1(i+1)p_{i+1,j}(t)$$
$$+ \mu_2(j+1)p_{i,j+1}(t) + \nu_1(i+1)p_{i+1,j-1}(t) + \nu_2(j+1)p_{i-1,j+1}(t)$$
$$- [(\alpha_1 + \alpha_2) + i(\lambda_1 + \mu_1 + \nu_1) + j(\lambda_2 + \mu_2 + \nu_2)]p_{ij}(t), \quad (9.2.1)$$

multiplying both sides of (9.2.1) by $z_1^i z_2^j$ and summing over $i,j = 0,1,2,\ldots$ yields the p.d.e.

$$\frac{\partial G}{\partial t} = [(\lambda_1 z_1 - \mu_1)(z_1 - 1) + \nu_1(z_2 - z_1)]\frac{\partial G}{\partial z_1} + [(\lambda_2 z_2 - \mu_2)(z_2 - 1)$$
$$+ \nu_2(z_1 - z_2)]\frac{\partial G}{\partial z_2} + [\alpha_1(z_1 - 1) + \alpha_2(z_2 - 1)], \quad (9.2.2)$$

and the presence of the z_i^2 terms has so far prevented the construction of a closed-form solution.

9.2.1 Moments

Fortunately, the associated first- and second-order moment equations are far more amenable to solution. For example, differentiating (9.2.2) with respect to z_1 and then placing $z_1 = z_2 = 1$ yields the first-order linear differential equation

$$dm_i(t)/dt = (\lambda_i - \mu_i - \nu_i)m_i(t) + \nu_j m_j(t) + \alpha_i \quad (i,j = 1,2; i \neq j) \quad (9.2.3)$$

for the means $m_i(t) = \mathrm{E}[X_i(t)]$. Note that this is identical to the corresponding deterministic equation, in contrast to what happens in the nonlinear competition, predator–prey and epidemic situations. For recall that a deterministic equation is equivalent to first-order moment closure, so agreement with the mean equation will occur if and only if no higher-order moments are present, as is the case here. Equations (9.2.3) are straightforward to solve along standard lines, either by: (i) solving the two simultaneous equations for the Laplace transform

$$\mathcal{L}[m_i(t)] \equiv m_i^*(s) = \int_0^\infty e^{-st} m_i(t)\,dt \quad (s > 0)$$

and then inverting; (ii) eliminating say $m_2(t)$ to form a standard second-order differential equation for $m_1(t)$ alone; or, (iii) solving the associated vector-matrix equation for the column vector $\mathbf{m}(t) = (m_1(t), m_2(t))^T$. Denote the net individual growth rates by $\xi_i = \lambda_i - \mu_i - \nu_i$ ($i = 1,2$) and write

$$w_1, w_2 = (1/2)[(\xi_1 + \xi_2) \pm \sqrt{\{(\xi_1 - \xi_2)^2 + 4\nu_1\nu_2\}}]. \quad (9.2.4)$$

Then with the initial condition $m_i(0) = a_i$ the solution for the means is given (for $w_1 \neq w_2$) by

$$m_1(t) = (w_1 - w_2)^{-1}\{[a_1 w_1 + \alpha_1 - a_1\xi_2 + \nu_2 a_2 + w_1^{-1}(\alpha_2\nu_2 - \alpha_1\xi_2)]e^{w_1 t}$$
$$- [a_1 w_2 + \alpha_1 - a_1\xi_2 + \nu_2 a_2 + w_2^{-1}(\alpha_2\nu_2 - \alpha_1\xi_2)]e^{w_2 t}\} \quad (9.2.5)$$
$$+ (w_1 w_2)^{-1}(\alpha_2\nu_2 - \alpha_1\xi_2),$$

with a similar expression holding for $m_2(t)$ (Renshaw, 1972). If $\omega_1 = \omega_2$ then we see from (9.2.4) that $\xi_1 = \xi_2 = \xi$ (say) and $\nu_1 \nu_2 = 0$. Whence on assuming that $\nu_2 = 0$, equations (9.2.3) easily solve to give

$$m_1(t) = a_1 e^{\xi t} + (\alpha_1/\xi)(e^{\xi t} - 1) \tag{9.2.6}$$
$$m_2(t) = a_2 e^{\xi t} + (\alpha_2/\xi)(e^{\xi t} - 1) + \nu_1 a_1 t e^{\xi t} + (\nu_1 \alpha_1/\xi^2)[1 - (1 - \xi t)e^{\xi t}].$$

Since $\omega_1 \geq \omega_2$, the total population size either increases to infinity or approaches $m_i(\infty) = (\alpha_j \nu_j - \alpha_i \xi_j)/\omega_1 \omega_2$ (for $j \neq i$), i.e. zero if $\alpha_1 = \alpha_2 = 0$, according as $\omega_1 > 0$ or $\omega_1 < 0$. Indeed, it follows from (9.2.4) that for $\xi_1, \xi_2 < 0$ if $\xi_1 \xi_2 > \nu_1 \nu_2$ then the total size decreases, if $\xi_1 \xi_2 = \nu_1 \nu_2$ it remains constant, otherwise it increases. For $\lambda_1 - \mu_1 > 0$ and $\lambda_2 - \mu_2 < 0$ an increase or decrease will depend purely on the values of the migration parameters ν_i. Indeed, since the $m_i(t)$ remain constant when $\xi_1 \xi_2 = \nu_1 \nu_2$ with $\xi_1, \xi_2 < 0$, the (ν_1, ν_2)-boundary between these two regimes is defined by the *critical migration* condition

$$\lambda_1 - \mu_1 < \nu_1 = (\lambda_1 - \mu_1)[1 - \nu_2/(\lambda_2 - \mu_2)]. \tag{9.2.7}$$

Higher-order moments can be developed along similar lines, though as noted in earlier sections it is better to work in terms of the c.g.f. $K(\theta_1, \theta_2; t) \equiv G(e^{\theta_1 t}, e^{\theta_2 t}; t)$ rather than the p.g.f. $G(z_1, z_2; t)$. That is, we use the c.g.f. forward equation

$$\frac{\partial K}{\partial t} = [\lambda_1(e^{\theta_1} - 1) + \mu_1(e^{-\theta_1} - 1) + \nu_1(e^{-\theta_1 + \theta_2} - 1)]\frac{\partial K}{\partial \theta_1} + \alpha_1(e^{\theta_1} - 1)$$
$$+ [\lambda_2(e^{\theta_2} - 1) + \mu_2(e^{-\theta_2} - 1) + \nu_2(e^{-\theta_2 + \theta_1} - 1)]\frac{\partial K}{\partial \theta_2} + \alpha_2(e^{\theta_2} - 1) \tag{9.2.8}$$

with $K(\theta_1, \theta_2; 0) = a_1 \theta_1 + a_2 \theta_2$, rather than the p.g.f. equation (9.2.2). For example, differentiating both sides of equation (9.2.8) twice with respect to θ_1 and θ_2, and then placing $\theta_1 = \theta_2 = 0$, yields three first-order differential equations for the variances $V_{11}(t)$ and $V_{22}(t)$ and covariance $V_{12}(t)$ (i.e. the second-order cumulants), namely

$$dV_{11}(t)/dt = (\lambda_1 + \mu_1 + \nu_1)m_1(t) + \nu_2 m_2(t) + 2\xi_1 V_{11}(t) + 2\nu_2 V_{12}(t) + \alpha_1$$
$$dV_{12}(t)/dt = -\nu_1 m_1(t) - \nu_2 m_2(t) + \nu_1 V_{11}(t) + \nu_2 V_{22}(t) + (\xi_1 + \xi_2)V_{12}(t) \tag{9.2.9}$$
$$dV_{22}(t)/dt = (\lambda_2 + \mu_2 + \nu_2)m_2(t) + \nu_1 m_1(t) + 2\xi_2 V_{22}(t) + 2\nu_1 V_{12}(t) + \alpha_2.$$

These equations can be solved in ways parallel to those used to generate the means, namely: (i) first solve the three simultaneous equations for the Laplace transform

$$\mathcal{L}[V_{ij}(t)] \equiv V_{ij}^*(s) = \int_0^\infty e^{-st} V_{ij}(t)\, dt \quad (s > 0)$$

and then invert; (ii) eliminate say $V_{11}(t)$ and $V_{22}(t)$ to form a standard third-order o.d.e. for $V_{12}(t)$ alone; or, (iii) solve the associated vector-matrix equation for the column vector $\mathbf{V}(t) = (V_{11}(t), V_{12}(t), V_{22}(t))^T$. Since the first approach offers the least line of algebraic resistance, let us take Laplace transforms of equations (9.2.9), thereby obtaining

$$sV_{11}^*(s) = (\lambda_1 + \mu_1 + \nu_1)m_1^*(s) + \nu_2 m_2^*(s) + 2\xi_1 V_{11}^*(s) + 2\nu_2 V_{12}^*(s) + \alpha_1 s^{-1}$$
$$sV_{12}^*(s) = -\nu_1 m_1^*(s) - \nu_2 m_2^*(s) + \nu_1 V_{11}^*(s) + \nu_2 V_{22}^*(s) + (\xi_1 + \xi_2)V_{12}^*(s) \quad (9.2.10)$$
$$sV_{22}^*(s) = (\lambda_2 + \mu_2 + \nu_2)m_2^*(s) + \nu_1 m_1^*(s) + 2\xi_2 V_{22}^*(s) + 2\nu_1 V_{12}^*(s) + \alpha_2 s^{-1}.$$

Whence substituting for $V_{11}^*(s)$ and $V_{22}^*(s)$ from the first and third equations into the second yields

$$V_{12}^*(s) = \nu_1[2\lambda_1 m_1^*(s) - a_1](s - 2\xi_2)/(s - 2\omega_1)(s - \omega_1 - \omega_2)(s - 2\omega_2)$$
$$+ \nu_2[2\lambda_2 m_2^*(s) - a_2](s - 2\xi_1)/(s - 2\omega_1)(s - \omega_1 - \omega_2)(s - 2\omega_2). \quad (9.2.11)$$

The $m_i^*(s)$ are easy to derive, either by solving the transformed equations (9.2.3) or by transforming the solution (9.2.5), and are given (for $i, j = 1, 2; i \neq j$) by

$$m_i^*(s) = [a_i s^2 + (\alpha_i - a_i \xi_j + \nu_j a_j)s + (\alpha_j \nu_j - \alpha_i \xi_j)]/s(s - \omega_1)(s - \omega_2). \quad (9.2.12)$$

Hence as expression (9.2.11) can now be expanded into partial fractions, using the inverse transformation

$$(s - a)^{-1} \equiv \mathcal{L}[e^{at}] \quad (9.2.13)$$

yields (for $\omega_1 \neq \omega_2$)

$$V_{12}(t) = r_1 e^{2\omega_1 t} + r_2 e^{(\omega_1 + \omega_2)t} + r_3 e^{2\omega_2 t} + r_4 e^{\omega_1 t} + r_5 e^{\omega_2 t} + r_6 \quad (9.2.14)$$

where r_i ($i = 1, \ldots, 6$) denote the associated coefficients. For example, on placing $b_i = \alpha_i - a_i \xi_j + \nu_j a_j$ and $c_i = \alpha_i \nu_i - \alpha_j \xi_i$ we have

$$r_1 = -[\nu_1 a_1(\omega_1 - \xi_2) + \nu_2 a_2(\omega_1 - \xi_1)]/(\omega_1 - \omega_2)^2 + [\lambda_1 \nu_1(4a_1\omega_1^2 + 2b_1\omega_1 + c_2)(\omega_1 - \xi_2)$$
$$+ \lambda_2 \nu_2(4a_2\omega_1^2 + 2b_2\omega_1 + c_1)(\omega_1 - \xi_1)]/\omega_1^2(\omega_1 - \omega_2)^2(2\omega_1 - \omega_2)$$

and

$$r_6 = [\lambda_1 \nu_1 c_2 \xi_2 + \lambda_2 \nu_2 c_1 \xi_1]/\omega_1^2 \omega_2^2(\omega_1 + \omega_2).$$

To evaluate $V_{11}(t)$ and $V_{22}(t)$ we simply write the first and third of equations (9.2.9) in the integral form

$$V_{ii}(t) = \int_0^t e^{2\xi_i(t-s)}[(\lambda_i + \mu_i + \nu_i)m_i(s) + \nu_j m_j(s) + 2\nu_j V_{12}(s) + \alpha_i]\,ds, \quad (9.2.15)$$

and then replace $m_1(s)$, $m_2(s)$ and $V_{12}(s)$ by expressions (9.2.5) and (9.2.14), respectively. Thus for the non-equilibrium cases with $\omega_1 > 0$ and $\omega_1 > \omega_2$, for large t

$$(V_{11}(t), V_{12}(t), V_{22}(t)) \sim r_1 e^{2\omega_1 t}(\nu_2/(\omega_1 - \xi_1), 1, \nu_1/(\omega_1 - \xi_2)). \quad (9.2.16)$$

This suggests that asymptotic results for higher-order cumulants could be developed by presuming that $\kappa_{rs}(t) \propto e^{(r+s)\omega_1 t}$ and then determining the associated scaling coefficients by back-substituting into the cumulant equations. Note that in the equilibrium case the latter reduce to simultaneous equations in $\kappa_{rs}(\infty)$, which can then be solved via matrix manipulation. In the case of the second-order cumulants we have

$$V_{12}(\infty) = [\alpha_1\nu_1(\lambda_2\nu_2\xi_1 - \lambda_1\xi_2^2) + \alpha_2\nu_2(\lambda_1\nu_1\xi_2 - \lambda_2\xi_1^2)]/(\xi_1+\xi_2)(\xi_1\xi_2 - \nu_1\nu_2)^2$$
$$V_{11}(\infty) = (-\nu_2/\xi_1)V_{12}(\infty) + (\xi_1 - \lambda_1)(\alpha_2\nu_2 - \alpha_1\xi_2)/\xi_1(\xi_1\xi_2 - \nu_1\nu_2) \qquad (9.2.17)$$
$$V_{22}(\infty) = (-\nu_1/\xi_2)V_{12}(\infty) + (\xi_2 - \lambda_2)(\alpha_1\nu_1 - \alpha_2\xi_1)/\xi_2(\xi_1\xi_2 - \nu_1\nu_2).$$

For the homogeneous model $\lambda_i = \lambda$, $\mu_i = \mu$, $\nu_i = \nu$ without immigration (i.e. $\alpha_i = 0$), the exact results (9.2.5) and (9.2.14)–(9.2.15) for the mean and variance simplify considerably to give (for $i,j = 1,2; j \neq i$)

$$m_i(t) = (1/2)(a_i + a_j)e^{(\lambda-\mu)t} + (1/2)(a_i - a_j)e^{(\lambda-\mu-2\nu)t} \qquad (9.2.18)$$

with

$$\mathbf{V}(t) = (1/4)(a_1+a_2)(\lambda+\mu)(\lambda-\mu)^{-1}(e^{2(\lambda-\mu)t} - e^{(\lambda-\mu)t})(\mathbf{e}_1 + \mathbf{e}_2 + \mathbf{e}_3)$$
$$+ (1/2)(a_1 - a_2)(\lambda+\mu)(\lambda-\mu)^{-1}(e^{2(\lambda-\mu-\nu)t} - e^{(\lambda-\mu-2\nu)t})(\mathbf{e}_1 - \mathbf{e}_3) \qquad (9.2.19)$$
$$+ (1/4)(a_1+a_2)(\lambda+\mu+4\nu)(\lambda-\mu-4\nu)^{-1}(e^{2(\lambda-\mu-2\nu)t} - e^{(\lambda-\mu)t})(\mathbf{e}_1 - \mathbf{e}_2 + \mathbf{e}_3),$$

where the vectors $\mathbf{e}_1^T = (1,0,0)$, $\mathbf{e}_2^T = (0,1,0)$ and $\mathbf{e}_3^T = (0,0,1)$. Whence combining results (9.2.18) and (9.2.19) shows that for large t and $\nu > 0$ the correlation coefficient is given by $\rho \sim 1$ for $\lambda \geq \mu$ and $\rho \sim 4\nu\lambda/(\mu^2 + 4\nu\mu - \lambda^2)$ for $\lambda < \mu$.

9.2.2 Exact probabilities

Recall that we are developing this two-site migration process not only to investigate its dynamical structure but also to expose the difficulties which underlie the analytic development of multi-type processes in general. It is therefore worthwhile demonstrating how the degree of difficulty increases as we progress from the simple birth–death process of Chapter 2 to the analytically intractable probability structure defined by the forward equations (9.2.1) (see Renshaw, 1973b).

Model 1 ($\lambda_1 = \lambda_2 = 0$) The fact that Bartlett's result (9.1.10) applies to any closed process, and that immigration is relatively benign, suggests that the pure immigration–death–migration process should be fairly amenable to direct solution. Indeed, on placing $\lambda_1 = \lambda_2 = 0$ the auxiliary equations corresponding to the p.g.f. equation (9.2.2), namely

$$\frac{dt}{1} = \frac{-dz_1}{(\lambda_1 z_1 - \mu_1)(z_1 - 1) + \nu_1(z_2 - z_1)} = \frac{-dz_2}{(\lambda_2 z_2 - \mu_2)(z_2 - 1) + \nu_2(z_1 - z_2)}$$
$$= \frac{dG}{G[\alpha_1(z_1 - 1) + \alpha_2(z_2 - 1)]}, \qquad (9.2.20)$$

are all linear, and, after a little straightforward algebra, yield the solution

$$G(z_1, z_2; t) = [1 - (k_2 - k_1)^{-1}\{k_2(1 - z_1)(k_2 e^{-r_1 t} - k_1 e^{-r_2 t}) - (1 - z_2)k_1 k_2 (e^{-r_1 t} - e^{-r_2 t})\}]^{a_1}$$
$$\times [1 - (k_2 - k_1)^{-1}\{(1 - z_1)(e^{-r_1 t} - e^{-r_2 t}) - (1 - z_2)k_1 k_2(e^{-r_1 t} - e^{-r_2 t})\}]^{a_2}$$
$$\times \exp\{(k_2 - k_1)^{-1}[r_1^{-1}(\alpha_1 k_2 + \alpha_2)\{(1 - z_1) - k_1(1 - z_2)\}(e^{-r_1 t} - 1)$$
$$- r_2^{-1}(\alpha_1 k_1 + \alpha_2)\{(1 - z_1) - k_2(1 - z_2)\}(e^{-r_2 t} - 1)]\}. \tag{9.2.21}$$

Here $\theta_i = \mu_i + \nu_i$, with $r_1, r_2 = (1/2)[(\theta_1 + \theta_2) \pm \sqrt{\{(\theta_1 - \theta_2)^2 + 4\nu_1 \nu_2\}}]$ and $k_1, k_2 = (1/2\nu_2)[(\theta_2 - \theta_1) \pm \sqrt{\{(\theta_1 - \theta_2)^2 + 4\nu_1 \nu_2\}}]$. Note that if the population is initially empty, i.e. $a_1 = a_2 = 0$, then the process behaves as the product of two independent Poisson processes. In particular, the equilibrium p.g.f.

$$G(z_1, z_2; \infty) = \exp\{[(1 - z_1)(\alpha_2 \nu_2 + \alpha_1 \theta_2) + (1 - z_2)(\alpha_1 \nu_1 + \alpha_2 \theta_1)]/(\theta_1 \theta_2 - \nu_1 \nu_2)\}. \tag{9.2.22}$$

So as the corresponding p.g.f. of a single-colony immigration–death process is given by

$$G(z; \infty) = \exp\{(\alpha/\mu)(z - 1)\} \tag{9.2.23}$$

(see result (2.5.8)), it follows that in equilibrium the two colonies behave as though they are isolated from each other, the ratios of their immigration rates to death rates being as $(\alpha_2 \nu_2 + \alpha_1 \theta_2)$ to $(\theta_1 \theta_2 - \nu_1 \nu_2)$ and $(\alpha_1 \nu_1 + \alpha_2 \theta_1)$ to $(\theta_1 \theta_2 - \nu_1 \nu_2)$, for colonies 1 and 2, respectively.

Model 2 ($\lambda_1 = \nu_2 = 0$) Suppose we now assume that colony 1 behaves as a simple immigration-death process with parameters α_1 and μ_1, and that migration only takes place from colony 2 to 1. Then colony 2 develops as a birth–death process which possesses a time-dependent immigration component induced by colony 1 which is defined through the immigration–death p.g.f. (2.5.7). For simplicity assume that both colonies are initially empty, so that $a_1 = a_2 = 0$. Then since there is no feedback from colony 2 to 1, the p.g.f. $G(z_1, z_2; t)$ must factorize into the product

$$G(z_1, z_2; t) \equiv G_1(z_1; t)G_2(z_1, z_2; t),$$

where we see from solution (2.5.7) with $\alpha = \alpha_1$, $\mu = \mu_1 + \nu_1$ and $n_0 = 1$ that

$$G_1(z_1; t) = \exp\{[\alpha_1/(\mu_1 + \nu_1)](z_1 - 1)(1 - e^{-(\mu_1 + \nu_1)t})\} \tag{9.2.24}$$

is independent of z_2. Determination of $G_2(z_1, z_2; t)$ involves substantial detailed calculation (omitted here), but on writing $\psi = \mu_2 - \lambda_2$ and denoting the functions

$$P(z_2) = (z_2 - 1)/(\lambda_2 z_2 - \mu_2) \tag{9.2.25}$$

and

$$J(z_2; t) = \int_0^t P(z_2) e^{(\theta_1 - \psi)s}(1 - \lambda_2 P(z_2) e^{-\psi s})^{-1} ds, \tag{9.2.26}$$

it may be shown that the solution takes the compact form

$$G_2(z_1, z_2; t) \equiv G_2(z_2; t) = \left[\frac{1 - \lambda_2 P(z_2)}{1 - \lambda_2 P(z_2) e^{-\psi t}}\right]^{\left(\frac{\alpha_1 \nu_1}{\theta_1 \lambda_2} + \frac{\alpha_2}{\lambda_2}\right)} \times \exp\left\{\frac{\alpha_1 \nu_1 \psi}{\theta_1} e^{-\theta_1 t} J(z_2; t)\right\}. \quad (9.2.27)$$

Since this p.g.f. is independent of z_1, it follows that the probability structure of colony 2 behaves as a single-colony process, as also happens under Model 1. However, care must be taken not to place too much weight on this feature since it relates to an 'average' behaviour over the ensemble of all possible realizations. For in both models, if a particular realization exhibits a sudden upsurge in the number of individuals in colony 1, then these will immediately increase the rate of migration into colony 2 and thereby affect the process there.

If $\mu_2 > \lambda_2$ then an equilibrium distribution exists, with the p.g.f. now simplifying quite considerably to the product of two independent Poisson and negative binomial p.g.f.'s, namely

$$G(z_1, z_2; \infty) = \left[\frac{\lambda_2 z_2 - \mu_2}{\lambda_2 - \mu_2}\right]^{-\frac{1}{\lambda_2}\left(\frac{\alpha_1 \nu_1}{\mu_1 + \nu_1} + \alpha_2\right)} \exp\left\{\frac{\alpha_1(z_1 - 1)}{\mu_1 + \nu_1}\right\}. \quad (9.2.28)$$

Comparing the first one to the equilibrium p.g.f. (2.5.21) shows that colony 2 behaves as though it were a single-colony immigration–birth–death process with birth rate λ_2, death rate μ_2 and immigration rate $\alpha_2 + \alpha_1[\nu_1/(\mu_1 + \nu_1)]$. So the only (probabilistic) impact caused by migration from colony 1 to 2 is to increase the immigration rate α_2 by the proportion of colony 1 immigrants who eventually arrive there but do not die beforehand.

Model 3 ($\lambda_2 = \nu_2 = 0$) This third situation is mathematically more complicated than case 2, since colony 2 receives immigrants from an immigration–birth–death process rather than from the considerably simpler immigration–death process. This means that the solution of the auxiliary equations now revolves around a Riccati equation, and a fair amount of algebraic detail (again omitted here) is required in order to construct the p.g.f. $G(z_1, z_2; t)$. Though the resulting solution is of limited appeal in itself, it does serve the useful purpose of demonstrating how the degree of complexity quickly rises as parameters are introduced. Indeed, it strongly suggests that even if the full solution for $\lambda_i, \mu_i, \nu_i > 0$ could be found, it would be so opaque as to render it effectively meaningless. Even here, with $\lambda_2 = \nu_2 = 0$ and, for added simplicity, the assumption that both colonies are initially empty, we have

$$G(z_1, z_2; t) = \left[\frac{\psi(t, t; z_2)}{\psi(0, t; z_2)} \times \frac{H(z_1, z_2)}{H(z_1, z_2) + \lambda_1 L(t)}\right]^{\alpha_1/\lambda_1}$$

$$\times \exp\{-(\alpha_2/\mu_2)(1 - z_2)(1 - e^{-\mu_2 t}) - (\alpha_1 \xi_1/2\lambda_1)t\}; \quad (9.2.29)$$

where the functions

$$\psi(s, t; z_2) = I_{\xi_1/\mu_2}[w e^{\mu_2(s-t)/2}], \qquad L(t) = \int_0^t I_{\xi_1/\mu_2}^{-2}[w e^{-\mu_2 s/2}] \, ds \quad \text{and}$$

$$H^{-1}(z_1, z_2) = \lambda_1^{-1} \psi^2(t, t; z_2) \{\lambda_1(1 - z_1) - \xi_1/2 + (\mu_2/2) w \psi^{-1}(t, t; z_2) \frac{d}{dw}[I_{\xi_1/\mu_2}(w)]\}$$

for $w = (2/\mu_2)\sqrt{[\lambda_1\nu_1(1-z_2)]}$ and $I_{\xi_1/\mu_2}(w)$ a modified Bessel function of the first kind (see Abramowitz and Stegun, 1970). Not surprisingly, the equilibrium distribution does not possess a simple structure.

9.2.3 An approximate stochastic solution

Although equation (9.2.2) has so far evaded a solution in closed form, and is unlikely to yield one in the foreseeable future, we may derive an approximate solution by changing the probability of a birth in colony i in the time interval $(t, t + dt)$ from $\lambda_i dt$ to $\lambda m_i(t) dt$. For this effectively replaces birth by time-dependent immigration, which, as we have seen earlier, is likely to prove to be more algebraically amenable. However, care needs to be taken to ensure that the resulting change to the system dynamics is fully appreciated.

Since here we are essentially interested in the time-development of the process, let us further reduce algebraic awkwardness by placing $\alpha_1 = \alpha_2 = 0$ as the method of solution remains totally unchanged. Then the equation corresponding to (9.2.2) for the (new) p.g.f. $G(z_1, z_2; t)$ becomes

$$\frac{\partial G}{\partial t} = [\mu_1(1-z_1) + \nu_1(z_2 - z_1)]\frac{\partial G}{\partial z_1} + [\mu_2(1-z_2) + \nu_2(z_1 - z_2)]\frac{\partial G}{\partial z_2}$$
$$+ [\lambda_1 m_1(z_1 - 1) + \lambda_2 m_2(z_2 - 1)]G. \tag{9.2.30}$$

Solving (9.2.30) in the usual manner we first write down the auxiliary equations

$$\frac{dt}{1} = \frac{-dz_1}{\mu_1(1-z_1) + \nu_1(z_2 - z_1)} = \frac{-dz_2}{\mu_2(1-z_2) + \nu_2(z_1 - z_2)}$$
$$= \frac{dG}{G[\lambda_1 m_1(z_1 - 1) + \lambda_2 m_2(z_2 - 1)]}. \tag{9.2.31}$$

The first two equations integrate to yield

$$[d_i(z_1 - 1) + f_i(z_2 - 1)]e^{-\phi_i t} = \text{constant} \quad (i = 1, 2), \tag{9.2.32}$$

where, for $\theta_i = \mu_i + \nu_i$,

$$\phi_1, \phi_2 = (1/2)[(\theta_1 + \theta_2) \pm \sqrt{\{(\theta_1 - \theta_2)^2 + 4\nu_1\nu_2\}}] \tag{9.2.33}$$

and d_i, f_i satisfy the (equivalent) equations

$$d_i(\mu_1 + \nu_1 - \phi_i) = f_i\nu_2 \quad \text{and} \quad f_i(\mu_2 + \nu_2 - \phi_i) = d_i\nu_1.$$

We now put

$$\delta_1 = f_1/(d_1 f_2 - d_2 f_1), \quad \delta_2 = f_2/(d_1 f_2 - d_2 f_1),$$
$$\gamma_1 = -d_1/(d_1 f_2 - d_2 f_1), \quad \gamma_2 = -d_2/(d_1 f_2 - d_2 f_1),$$

and write solution (9.2.5) for the means (with $\alpha_1 = \alpha_2 = 0$) in the form

$$m_i(t) = h_i e^{\omega_1 t} + k_i e^{\omega_2 t} \tag{9.2.34}$$

for appropriate constants h_i and k_i. Then integrating the third part of equations (9.2.31) and combining the two solutions (9.2.32) yields

$$G(z_1, z_2; t) = [1 + \delta_2\{d_1(z_1 - 1) + f_1(z_2 - 1)\}e^{-\phi_1 t} - \delta_1\{d_2(z_1 - 1) + f_2(z_2 - 1)\}e^{-\phi_2 t}]^{a_1}$$
$$\times [1 + \gamma_2\{d_1(z_1 - 1) + f_1(z_2 - 1)\}e^{-\phi_1 t} - \gamma_1\{d_2(z_1 - 1) + f_2(z_2 - 1)\}e^{-\phi_2 t}]^{a_2}$$
$$\times \exp[(\omega_1 + \phi_1)^{-1}(\lambda_1 h_1 \delta_2 + \lambda_2 h_2 \gamma_2)\{d_1(z_1 - 1) + f_1(z_2 - 1)\}(e^{\omega_1 t} - e^{-\phi_1 t})$$
$$- (\omega_1 + \phi_2)^{-1}(\lambda_1 h_1 \delta_1 + \lambda_2 h_2 \gamma_1)\{d_2(z_1 - 1) + f_2(z_2 - 1)\}(e^{\omega_1 t} - e^{-\phi_2 t})$$
$$+ (\omega_2 + \phi_1)^{-1}(\lambda_1 k_1 \delta_2 + \lambda_2 k_2 \gamma_2)\{d_1(z_1 - 1) + f_1(z_2 - 1)\}(e^{\omega_2 t} - e^{-\phi_1 t})$$
$$- (\omega_2 + \phi_2)^{-1}(\lambda_1 k_1 \delta_1 + \lambda_2 k_2 \gamma_1)\{d_2(z_1 - 1) + f_2(z_2 - 1)\}(e^{\omega_2 t} - e^{-\phi_2 t})]. \tag{9.2.35}$$

Although at first glance this expression looks rather messy, all the component terms are linear in z_1 and z_2. So its underlying structure is essentially Poisson and hence readily amenable to direct interpretation. However, it is important to examine the extent to which changing the dependence of the two birth rates on the actual (i.e. varying) population sizes of colonies 1 and 2 to their respective mean values affects the overall probability structure. Suppose we consider the simplified homogeneous model in which $\lambda_i = \lambda$, $\mu_i = \mu$ and $\nu_i = \nu$. For result (9.2.35) then reduces to the much more transparent form

$$G(z_1, z_2; t) = [g_1(z_1, z_2; t)]^{a_1}[g_2(z_1, z_2; t)]^{a_2}, \tag{9.2.36}$$

where for $i, j = 1, 2$ and $j \neq i$

$$g_i(z_1, z_2; t) = [1 + (1/2)(z_i + z_j - 2)e^{-\mu t} + (1/2)(z_i - z_j)e^{-(\mu+2\nu)t}]$$
$$\times \exp\{(1/2)(e^{\lambda t} - 1)[(z_i + z_j - 2)e^{-\mu t} + (z_i - z_j)e^{-(\mu+2\nu)t}]\}. \tag{9.2.37}$$

Thus without loss of generality we need only consider the initial condition $a_1 = 1$, $a_2 = 0$. Whence extracting the coefficient of $z_1^i z_2^j$ in $g_1(z_1, z_2; t)$ and writing $\theta = e^{-\mu t}(1 + e^{-2\nu t})/2$, $\phi = e^{-\mu t}(1 - e^{-2\nu t})/2$ and $\psi = e^{\lambda t} - 1$ gives

$$p_{ij}(t) = \frac{\theta^i \phi^j \psi^{i+j-1}}{i! j!} [\psi(1 - e^{-\mu t}) + i + j] \exp\{-\psi e^{-\mu t}\}. \tag{9.2.38}$$

Hence for $\nu > 0$ and t large,

$$p_{ij}(t) \sim \frac{(e^{\lambda - \mu)t}/2)^{i+j}}{i! j!} \exp\{-e^{-(\lambda - \mu)t}\}. \tag{9.2.39}$$

So the two population sizes are asymptotically distributed as independent Poisson processes with parameter $e^{(\lambda-\mu)t}/2$. Thus replacing stochastic birth by deterministic birth has (for large t) the effect of destroying the inter-dependence between the two colonies. Though since birth has such a dominating influence this is clearly something to be anticipated. The marginal probabilities may be constructed in the same way, with

$$p_{i\cdot}(t) = \frac{\theta^i \psi^{i-1}}{i!}[\psi(1 - e^{-\mu t} + \phi) + i]\exp\{\psi(\phi - e^{-\mu t})\}; \qquad (9.2.40)$$

a similar expression holds for $p_{\cdot j}(t)$ (θ and ϕ are interchanged).

A further consequence of replacing birth by time-dependent immigration is that whereas in the original process the population becomes extinct immediately both colonies become empty, the modified process is subject to immigration at rate $\lambda(m_1(t) + m_2(t)) > 0$ and so cannot become extinct in a finite time. To examine the effect of this structural difference let $p_{00}(t)$ and $\bar{p}_{00}(t)$ respectively denote the probabilities associated with equations (9.2.2) (where $\alpha_1 = \alpha_2 = 0$) and (9.2.30) that both colonies are empty. Since we are assuming that the process is homogeneous the probability of extinction, $p_{00}(t)$, is identical to that of a single isolated colony. Hence it follows from (2.4.23) that for $a_1 = 1$ and $a_2 = 0$

$$p_{00}(t) = (\mu - \mu e^{-(\lambda - \mu)t})/(\lambda - \mu e^{-(\lambda - \mu)t}); \qquad (9.2.41)$$

whilst placing $z_1 = z_2 = 0$ in (9.2.38) yields the substantially different form

$$\bar{p}_{00}(t) = (1 - e^{-\mu t})\exp\{-e^{-\mu t}(e^{\lambda t} - 1)\}. \qquad (9.2.42)$$

If $\lambda > \mu$ then because the immigration rate $\lambda(m_1(t) + m_2(t))$ increases exponentially (see (2.5.2)) $\bar{p}_{00}(t)$ quickly tends to zero. So not only is state $(0,0)$ not absorbing in the modified process, but also the probability of being there swiftly decreases with increasing t. For recall that $\bar{p}_{00}(t)$ corresponds not to the probability of extinction, but to the probability of lying in state $(0,0)$ which it is free to move away from. Conversely, if $\lambda < \mu$, then the $m_i(t)$ decrease exponentially and so $\bar{p}_{00}(t)$ quickly tends to one since $\lambda(m_1(t) + m_2(t)) \to 0$ and hence death predominates. Finally, if $\lambda = \mu$ then $m_i(t) \sim \frac{1}{2}$, and the process reduces to a simple immigration–death process. Thus now an equilibrium situation develops, and with it a positive probability of lying in state $(0,0)$. So although the probability of ultimate extinction is zero, we expect $\bar{p}_{00}(t) \to \bar{p}$ for some $0 < \bar{p} < 1$. In fact it easily follows from (9.2.41) and (9.2.42) that

$$\begin{aligned}
\bar{p}_{00}(\infty) &= 0 & p_{00}(\infty) &= \mu/\lambda & (\lambda &> \mu), \\
\bar{p}_{00}(\infty) &= e^{-1} & p_{00}(\infty) &= 1 & (\lambda &= \mu), \\
\bar{p}_{00}(\infty) &= 1 & p_{00}(\infty) &= 1 & (\lambda &< \mu),
\end{aligned} \qquad (9.2.43)$$

together with

$$\bar{p}_{00}(t)/p_{00}(t) = \begin{cases} (\lambda/\mu)\exp\{-(\mu/\lambda)e^{-(\lambda-\mu)t}\} & (\lambda > \mu) \\ e^{-1}\exp\{(1+\lambda t)^{-1}\} & (\lambda = \mu) \\ \exp\{-(\lambda/\mu)e^{(\lambda-\mu)t}\} & (\lambda < \mu). \end{cases} \qquad (9.2.44)$$

Thus not only is the modified process conceptually different from the original one, in that being empty no longer corresponds to extinction, but the two probabilities of being in state $(0,0)$ at time t can also be markedly different.

To see whether the corresponding means and variances are similarly affected, let us first change the p.g.f. equation (9.2.30) into the c.g.f. equation

$$\frac{\partial K}{\partial t} = [\mu_1(e^{-\theta_1}-1) + \nu_1(e^{\theta_2-\theta_1}-1)]\frac{\partial K}{\partial \theta_1} + [\mu_2(e^{-\theta_2}-1) + \nu_2(e^{\theta_1-\theta_2}-1)]\frac{\partial K}{\partial \theta_2}$$
$$+ [\lambda_1 m_1(e^{\theta_1}-1) + \lambda_2 m_2(e^{\theta_2}-1)] \qquad (9.2.45)$$

by placing $z_1 = e^{\theta_1}$, $z_2 = e^{\theta_2}$ and $K = \log(G)$. Then on differentiating (9.2.45) with respect to θ_1 and θ_2, and setting $\theta_1 = \theta_2 = 0$, we obtain the differential equations

$$d\overline{m}_i(t)/dt = \lambda_i m_i(t) - (\mu_i + \nu_i)\overline{m}_i(t) + \nu_j \overline{m}_j(t) \qquad (i,j = 1,2; j \neq i) \qquad (9.2.46)$$

where $m_i(t)$ and $\overline{m}_i(t)$ denote the means of the ordinary and modified process, respectively. Direct comparison of (9.2.46) with the original mean equations (9.2.3) (for $\alpha_1 = \alpha_2 = 0$) shows that $\overline{m}_i(t) = m_i(t)$, and so modifying the process leaves the means unaltered.

The same is not true for the second-order moments, however. For differentiating (9.2.45) a second time yields

$$d\overline{V}_{11}(t)/dt = (\lambda_1 + \mu_1 + \nu_1)m_1(t) + \nu_2 m_2(t) - 2(\mu_1 + \nu_1)\overline{V}_{11}(t) + 2\nu_2 \overline{V}_{12}(t)$$
$$d\overline{V}_{12}(t)/dt = -\nu_1 m_1(t) - \nu_2 m_2(t) + \nu_1 \overline{V}_{11}(t) - (\mu_1 + \mu_2 + \nu_1 + \nu_2)\overline{V}_{12}(t) + \nu_2 \overline{V}_{22}(t)$$
$$d\overline{V}_{22}(t)/dt = (\lambda_2 + \mu_2 + \nu_2)m_2(t) + \nu_1 m_1(t) - 2(\mu_2 + \nu_2)\overline{V}_{22}(t) + 2\nu_1 \overline{V}_{12}(t),$$
$$(9.2.47)$$

where $\overline{V}_{ij}(t)$ denotes the variance-covariances of the modified process. The key difference between these three modified equations and the original equations (9.2.9) lies in the removal of the terms $2\lambda_1 V_{11}(t)$, $(\lambda_1 + \lambda_2)V_{12}(t)$ and $2\lambda_2 V_{22}(t)$: since $V_{ii}(t) > 0$ the variances of the modified process are smaller. Though given that (here) we have replaced stochastic birth by deterministic birth, this reduction is to be expected. The strong structural similarity between these two sets of equations means that the solution for the modified $\overline{V}_{ij}(t)$ can be developed in exactly the same way as that for the $V_{ij}(t)$. For example, with the homogeneous model we have

$$\overline{\mathbf{V}}(t) = (1/4)(a_1 + a_2)[2e^{(\lambda-\mu)t}(\mathbf{e}_1 + \mathbf{e}_3) - e^{-2\mu t}(\mathbf{e}_1 + \mathbf{e}_2 + \mathbf{e}_3) - e^{-(2\mu+4\nu)t}(\mathbf{e}_1 - \mathbf{e}_2 + \mathbf{e}_3)]$$
$$+ (1/2)(a_1 - a_2)[e^{(\lambda-\mu-2\nu)t}(\mathbf{e}_1 - \mathbf{e}_3) - e^{-(2\mu+2\nu)t}(\mathbf{e}_1 - \mathbf{e}_3)], \qquad (9.2.48)$$

which differs quite markedly from the original solution (9.2.19), especially in respect of the considerably reduced dependence on λ. Whilst the fast rise of the correlation coefficient $\rho(t) \sim -(1/2)\exp\{-(\lambda+\mu)t\}$ towards zero as t increases quantifies the rapid loss of dependence between the two colony sizes highlighted earlier through (9.2.39).

In summary, although this approximation procedure clearly has considerable shortcomings, it does enable us to develop at least some degree of analytic understanding of the stochastic behaviour of a process by converting a mathematically intractable model into a tractable one. Whether this is adequate for specific individual needs depends on both the problem and the properties under study. Conducting a comparative

488 Spatial processes

simulation exercise between the exact and modified processes can be a considerable help in this regard.

9.2.4 Slightly connected processes

Rather than modifying the process, an alternative strategy is to construct an approximating p.d.f. for the population size probabilities $p_{ij}(t)$ themselves. For example, we have already shown that for the simple univariate birth–death–immigration process the $p_i(t)$ follow the negative binomial distribution (2.5.15) provided that the colony is initially empty. This suggests that we might try to construct an approximating bivariate negative binomial distribution which provides reasonably accurate forms for say $m_i(t)$, $V_{ij}(t)$ and $p_{00}(t)$. Studies suggest that this is not an easy thing to do, which further strengthens the rationale for using the procedures developed earlier in Sections 7.3 and 8.1.4. However, rather than progressing these approaches here, let us now demonstrate an alternative procedure in which we presume that the colonies are only *slightly connected*. For if $\nu_i \simeq \nu \simeq 0$, then it seems likely that we might be able to force a solution which is correct to $O(\nu)$, since the individual population size probabilities are known exactly when $\nu = 0$.

Instead of working with the forward equations, in this exercise it is easier to use the backward equations. That is, rather than considering the *last* move in the small time interval $(t, t+dt)$, we now consider the *first* move in $(0, dt)$. Although these equations can easily be developed as a direct extension to those already constructed for the single colony situation (see equation (2.4.42)), it is instructive to consider an alternative, and considerably shorter, technique which enables us to write down the p.g.f. equations directly. Let $g_i(z_1, z_2; t)$ $(i = 1, 2)$ denote the p.g.f. of the process at time t given the start values $X_i(0) = 1$, $X_j(0) = 0$ $(j \neq i)$, and purely for algebraic simplicity suppose that $\nu_1 = \nu_2 = \nu$ and $\alpha_1 = \alpha_2 = 0$. Then since in the small time interval $(0, dt)$ colony 1 undergoes

a birth	with probability $\lambda_1 dt$	whence $g_1 \to g_1^2$
a death	with probability $\mu_1 dt$	whence $g_1 \to 1$
a migration to colony 2	with probability νdt	whence $g_1 \to g_2$
no event	with probability $1 - (\lambda_1 + \mu_1 + \nu)dt$	whence $g_1 \to g_1$,

we have

$$g_1(t+dt) = \lambda_1 g_1^2(t)dt + \mu_1 dt + \nu g_2(t)dt + [1 - (\lambda_1 + \mu_1 + \nu)dt]g_1(t),$$

i.e. on letting $dt \to 0$

$$dg_1(t)/dt = \lambda_1 g_1^2(t) - (\lambda_1 + \mu_1)g_1(t) + \mu_1 + \nu[g_2(t) - g_1(t)]. \qquad (9.2.49)$$

A similar expression follows for $dg_2(t)/dt$ (see Puri, 1968).

When $\nu = 0$ the solution of (9.2.49) is that of the single colony model, and so if ν is small in comparison to λ_i and μ_i it seems reasonable to expect the same solution apart from a slight perturbation. So let us express $g_1(t)$ and $g_2(t)$ as power series in ν, namely

$$g_1(t) \equiv \sum_{i=0}^{\infty} a_i(t)\nu^i \quad \text{and} \quad g_2(t) \equiv \sum_{i=0}^{\infty} b_i(t)\nu^i, \qquad (9.2.50)$$

where $a_i(t)$ and $b_i(t)$ are independent of the parameter ν. Since $g_i(z_1, z_2; 0) = z_i$ ($i = 1, 2$), the initial conditions are given by

$$a_0(0) = z_1, \quad b_0(0) = z_2, \quad a_i(0) = b_i(0) = 0 \quad (i > 0). \qquad (9.2.51)$$

Whence substituting (9.2.50) into (9.2.49), and equating like coefficients of ν^r for $r = 0, 1, 2, \ldots$, yields the differential equations

$$da_0/dt = \lambda_1 a_0^2 - (\lambda_1 + \mu_1)a_0 + \mu_1 \qquad (9.2.52)$$

$$da_n/dt = a_n[2a_0\lambda_1 - (\lambda_1 + \mu_1)] + \lambda_1 \sum_{r=1}^{n-1} a_r a_{n-r} + b_{n-1} - a_{n-1} \quad (n > 0), \qquad (9.2.53)$$

where we denote $\sum_{r=1}^{n-1} a_r a_{n-r} \equiv 0$ at $n = 1$. Given the obvious similarity between $a_n(t)$ and $b_n(t)$ all our results are stated for $a_n(t)$ alone.

Because $a_0(t)$ and $b_0(t)$ are the p.g.f.'s for their respective single colony cases when $\nu = 0$, it follows from (2.4.11) with $n_0 = 1$ that

$$a_0(t) = \frac{\mu_1(1 - z_1) - (\mu_1 - \lambda_1 z_1)e^{-(\lambda_1 - \mu_1)t}}{\lambda_1(1 - z_1) - (\mu_1 - \lambda_1 z_1)e^{-(\lambda_1 - \mu_1)t}}. \qquad (9.2.54)$$

Moreover, we can write (9.2.53) in the integral form

$$a_n(t) = \int_0^t [b_{n-1}(s) - a_{n-1}(s) + \lambda_1 \sum_{r=1}^{n-1} a_r(s) a_{n-r}(s)]$$

$$\times \exp\{\int_s^t [2\lambda_1 a_0(\tau) - (\lambda_1 + \mu_1)]\, d\tau\}\, ds, \qquad (9.2.55)$$

with a similar expression for $b_n(t)$. The solutions of the differential equations (9.2.49) for $g_1(t)$ and $g_2(t)$ are therefore given by the series expansions (9.2.50), where the coefficients $a_n(t)$ and $b_n(t)$ are evaluated recursively from (9.2.54) and (9.2.55).

In particular, on taking $n = 1$ we have

$$a_1(t) = \int_0^t [b_0(s) - a_0(s)] \exp\{\int_s^t [2\lambda_1 a_0(\tau) - (\lambda_1 + \mu_1)]\, d\tau\}\, ds. \qquad (9.2.56)$$

To simplify this integral write $\beta_i = \lambda_i(1 - z_i)$, $\gamma_i = \mu_i - \lambda_i z_i$, $\delta_i = \mu_i - \lambda_i$ and $\rho_i = \mu_i/\lambda_i$ for $i = 1, 2$, and define the function

$$R(x, y; a, t) = \int_0^t \frac{e^{xs}\, ds}{1 - ae^{ys}}. \qquad (9.2.57)$$

Then combining expression (9.2.54) with the integral (9.2.56) yields

$$g_1(z_1, z_2; t) = a_0(t) + \nu[I_1(t) + I_2(t)] + O(\nu^2), \qquad (9.2.58)$$

where

$$I_1(t) = e^{\delta_1 t}(1 - z_1)(\beta_1 - \gamma_1 e^{\delta_1 t})^{-2}[\beta_1(e^{-\delta_1 t} - 1) + \gamma_1 \delta_1 t] \qquad (9.2.59)$$

$$I_2(t) = e^{\delta_1 t}(\rho_2 - 1)(\beta_1 - \gamma_1 e^{\delta_1 t})^{-2}[\beta_1^2 R(-\delta_1, \delta_2; \gamma_2/\beta_2, t)$$
$$+ \gamma_1^2 R(\delta_1, \delta_2; \gamma_2/\beta_2, t) - 2(\beta_1 \gamma_1/\delta_2) \ln\{e^{\delta_2 t}(\beta_2 - \gamma_2)/(\beta_2 - \gamma_2 e^{\delta_2 t})\}]. \qquad (9.2.60)$$

The integral (9.2.57) is related to the Incomplete Beta function and so cannot generally be evaluated in closed form. However, it can in the case of the homogeneous model (i.e. $\lambda_i = \lambda$, $\mu_i = \mu$, $\nu_i = \nu$), which leads to a much simpler form for $I_2(t)$, namely

$$I_2(t) = e^{\delta t}(\rho_2 - 1)(\delta\beta_2)^{-1}(\beta_1 - \gamma_1 e^{\delta t})^{-2}[\delta t(\beta_1^2 \gamma_2 - 2\beta_1 \beta_2 \gamma_1) + \beta_1^2 \beta_2(1 - e^{-\delta t})$$
$$- \gamma_2^{-1}(\beta_2 \gamma_1 - \beta_1 \gamma_2)^2 \ln\{(\beta_2 - \gamma_2 e^{\delta t})/(\beta_2 - \gamma_2)\}]. \qquad (9.2.61)$$

The individual probabilities $p_{ij}(t)$ may now be retrieved from expressions (9.2.58), (9.2.59) and (9.2.61) by extracting the coefficients of $z_1^i z_2^j$. Indeed, the extinction probability $p_{00}(t)$ may be written down immediately by replacing z_i, β_i and γ_i by 0, λ and μ, respectively. The effect on the marginal population sizes $p_{i\cdot}(t)$ and $p_{\cdot j}(t)$ is (algebraically) more transparent, especially for the spatially homogeneous model. For on forming the marginal p.g.f.'s $g_1(z_1, 1; t)$ and $g_2(1, z_2; t)$ we obtain two relatively simple expressions from which the coefficients of z_1^i and z_2^j may be easily extracted. For example,

$$g_1(z_1, 1; t) = a_0(t) + \nu(1 - z_1)[\lambda(1 - z_1)(1 - e^{(\mu-\lambda)t})$$
$$+ (\mu - \lambda z_1)(\mu - \lambda)t e^{(\mu-\lambda)t}]/[\lambda(1 - z_1) - (\mu - \lambda z_1)e^{(\mu-\lambda)t}]^2 + O(\nu^2). \qquad (9.2.62)$$

Now the above results just involve the first two terms in the expansion of power-series in ν, and for certain values of λ_i, μ_i and ν_i it may well be necessary to consider higher-order terms which are likely to induce unfriendly algebra. An idea of when these might be needed may be gleaned by considering the mean population sizes (9.2.18), which with $a_1 = 1$, $a_2 = 0$ reduce to

$$m_1(t) = (1/2)e^{(\lambda-\mu)t}(1 + e^{-2\nu t}) \simeq e^{(\lambda-\mu)t}[1 - \nu t + (\nu t)^2 - \cdots]$$
$$m_2(t) = (1/2)e^{(\lambda-\mu)t}(1 - e^{-2\nu t}) \simeq e^{(\lambda-\mu)t}[\nu t - (\nu t)^2 + \cdots]. \qquad (9.2.63)$$

For to order $O(\nu)$ these expressions agree exactly with those formed by constructing $dg_1(z_1, 1; t)/dz_1|_{z_1=1}$ and $dg_2(z_2, 1; t)/dz_2|_{z_2=1}$. So the range of t over which the approximation applies corresponds to that for which $e^{\nu t}$ may be reasonably well approximated by $1 - \nu t$.

9.2.5 Sequences of integral equations

Although both of the above approximation procedures are certainly useful, they by no means provide a universal panacea, and their deficiencies highlight the problems to be faced when considering general interconnected processes. A third procedure is to construct bounding functions within which the true, but unknown, p.g.f. must lie. We shall therefore now develop two sequences of integral equations, and show that for suitable boundary conditions one converges monotonically upwards and the other monotonically downwards to the $g_i(z_1, z_2; t)$ (Renshaw, 1973a).

Let the vector $\mathbf{X}(t) = (X_1(t), X_2(t))$ represent the sizes of the two colonies at time t. Then using the terminology of Feller (1971) denote

$$Q_t^{(n)} = \Pr[\text{of a transition from } \mathbf{X}(0) = \mathbf{x} \text{ to } \mathbf{X}(t) = \mathbf{w} \text{ in at most } n \text{ jumps}],$$

and the vectors $\mathbf{e}_1 = (1, 0)$ and $\mathbf{e}_2 = (0, 1)$. Now a transition without jumps is only possible if $\mathbf{x} = \mathbf{w}$, and since the sojourn time at $\mathbf{x}$ is exponentially distributed with parameter $\theta_i = \lambda_i + \mu_i + \nu_i$ we have

$$Q_t^{(0)}(\mathbf{e}_i, \mathbf{w}) = e^{-\theta_i t} K^{(0)}(\mathbf{e}_i, \mathbf{w}) \qquad (9.2.64)$$

where

$$K^{(0)}(\mathbf{e}_i, \mathbf{w}) = \begin{cases} 1 & \text{if } \mathbf{w} = \mathbf{e}_i \\ 0 & \text{if } \mathbf{w} \neq \mathbf{e}_i. \end{cases} \qquad (9.2.65)$$

Suppose next that the first event occurs at epoch $\tau < t$ and leads from $\mathbf{e}_i$ to $\mathbf{y}$. Then summing over all possible τ and $\mathbf{y}$ leads to

$$Q_t^{(n+1)}(\mathbf{e}_i, \mathbf{w}) = e^{-\theta_i t} K^{(0)}(\mathbf{e}_i, \mathbf{w}) + \int_0^t e^{-\theta_i \tau} [\lambda_i Q_{t-\tau}^{(n)}(2\mathbf{e}_i, \mathbf{w}) + \mu_i Q_{t-\tau}^{(n)}(\mathbf{0}, \mathbf{w})$$

$$+ \nu_i Q_{t-\tau}^{(n)}(\mathbf{e}_j, \mathbf{w})] \, d\tau \qquad (9.2.66)$$

for $n = 0, 1, 2, \ldots$ and $i, j = 1, 2$. Define the p.g.f.

$$g_{it}^{(n)}(\mathbf{z}) = \sum_{w_1=0}^{\infty} \sum_{w_2=0}^{\infty} Q_t^{(n)}(\mathbf{e}_i, \mathbf{w}) z_1^{w_1} z_2^{w_2} \qquad (9.2.67)$$

where $\mathbf{w} = (w_1, w_2)$ and $\mathbf{z} = (z_1, z_2)$. Then on placing $s = t - \tau$ in (9.2.66) and multiplying both sides by $z_1^{w_1} z_2^{w_2}$, summing over all $w_i = 0, 1, 2, \ldots$ yields the recurrence relation

$$g_{it}^{(n+1)} = z_i e^{-\theta_i t} + \int_0^t e^{-\theta_i(t-s)} [\lambda_i (g_{is}^{(n)})^2 + \mu_i + \nu_i g_{js}^{(n)}] \, ds. \qquad (9.2.68)$$

By using two simple induction arguments on n we can easily show that the $g_{it}^{(n)}$ are monotonic increasing functions bounded above by 1. The limit $g_{it}^{(\infty)}$ therefore exists, whence letting $n \to \infty$ in (9.2.68) gives

$$g_{it}^{(\infty)} = z_i e^{-\theta_i t} + \int_0^t e^{-\theta_i(t-s)} [\lambda_i (g_{is}^{(\infty)})^2 + \mu_i + \nu_i g_{js}^{(\infty)}] \, ds. \qquad (9.2.69)$$

Multiplying both sides of (9.2.69) by $e^{\theta_i t}$, and then differentiating with respect to t, shows that this equation satisfies the backward equations (9.2.49). So the sequence of p.g.f.'s $\{g_{it}^{(n)}\}$ defined through equation (9.2.68) with

$$g_{it}^{(0)} = z_i e^{-\theta_i t} \qquad (9.2.70)$$

is monotonic increasing in n and converges to the solution $g_i(t)$ of the backward equations (9.2.49).

To develop a corresponding sequence of equations which converges monotonically downwards we shall use a technique similar to that developed by Puri (1968). First divide each colony $i = 1, 2$ into a countably infinite number of sub-colonies i_n ($n = 0, 1, 2, \ldots$). Next replace $X_i(t)$ by $X(i, n; j, m; t)$, which represents the size of the mth sub-colony j_m at time t given that there is initially only one individual at $t = 0$, that being in i_n. Now denote $f_{it}^{(n)}(\mathbf{Z})$ to be the p.g.f. of $X(i, n; j, m; t)$, where $\mathbf{Z}$ is the $2 \times \infty$ matrix of dummy variables with elements z_{hk} ($h = 1, 2; k = 0, 1, 2, \ldots$). Then on considering the first event for the initial individual in sub-colony i_n, we either allow it to:

(i) die with rate μ_i;
(ii) migrate to j_{n-1} ($j \neq i$) where it undergoes a similar birth–death migration process with rates λ_j, μ_j and ν_j; or,
(iii) give birth in such a way that as soon as the event of a birth takes place both it and its progeny instantly migrate to sub-colony i_{n-1}.

Thus following a migration or birth event the sub-colony 'level' reduces from n to $n - 1$. Whence on noting that $e^{\theta_i t} = \Pr[\text{there are no births, deaths or migrations in } (0, t)]$, and considering the first change which occurs in the period $(0, t)$, we can write down the integral equation

$$f_{it}^{(n)} = z_{in} e^{-\theta_i t} + \int_0^t e^{-\theta_i \tau} [\lambda_i (f_{i,t-\tau}^{(n-1)})^2 + \mu_i + \nu_i f_{j,t-\tau}^{(n-1)}] \, d\tau. \qquad (9.2.71)$$

Finally, we put $t - \tau = s$ and collapse the $2 \times \infty$ system back into the two-state process by replacing all the z_{hk} ($k = 0, 1, 2, \ldots$) by z_h, thereby reducing (9.2.71) to the form

$$f_{it}^{(n)} = z_i e^{-\theta_i t} + \int_0^t e^{-\theta_i (t-s)} [\lambda_i (f_{is}^{(n-1)})^2 + \mu_i + \nu_i f_{js}^{(n-1)}] \, d\tau. \qquad (9.2.72)$$

This is clearly identical to equation (9.2.68) for the $g_{it}^{(n)}$. The difference, of course, lies in the two boundary conditions, and to derive one for this second sequence let us assume that an individual dies as soon as it reaches either sub-colony 1_0 or 2_0. Then for all $t \geq 0$

$$f_{it}^{(0)} \equiv 1. \qquad (9.2.73)$$

Employing a parallel argument to that used above for the $\{g_{it}^{(n)}\}$ shows that the sequence of p.g.f.'s $\{f_{it}^{(n)}\}$ defined by (9.2.72) and (9.2.73) is monotonic decreasing in n and converges to $g_i(t)$. Thus as n increases, $g_i(t)$ becomes more and more tightly contained within these two bounding sequences.

However, this p.g.f. bounding approach does raise an issue of practical concern. For taking the simplest case $n = 1$, on integrating (9.2.68) with condition (9.2.70), and (9.2.72) with condition (9.2.73), we see that the relation $g_{it}^{(1)} \leq g_i(t) \leq f_{it}^{(1)}$ yields

$$z_i e^{-\theta_i t} + (\lambda_i/\theta_i) z_i^2 (e^{-\theta_i t} - e^{-2\theta_i t}) + (\mu_i/\theta_i)(1 - e^{-\theta_i t}) + (\nu_i/(\theta_i - \theta_j)) z_j (e^{-\theta_j t} - e^{-\theta_i t})$$
$$\leq g_i(t) \leq z_i e^{-\theta_i t} + (1 - e^{-\theta_i t}), \tag{9.2.74}$$

and direct inspection shows that the transition probabilities, which are given by the coefficients of the individual terms $z_1^r z_2^s$, do not converge monotonically. The one exception is the probability of extinction $q_i(t) = g_i(0; t)$. For placing $z_1 = z_2 = 0$ in (9.2.74) gives

$$(\mu_i/\theta_i)(1 - e^{-\theta_i t}) \leq q_i(t) \leq 1 - e^{-\theta_i t}, \tag{9.2.75}$$

and successively tighter bounds may be obtained by taking $n = 2, 3, \ldots$. When $\mathbf{z} \neq \mathbf{0}$ it is probably best to ignore the bounding considerations and regard $(f_{it}^{(n)} + g_{it}^{(n)})/2$ as an increasingly accurate approximation to $g_i(t)$ as $n \to \infty$.

9.2.6 Riccati representations

The difficulty of obtaining exact solutions for the extinction probabilities $\{q_i(t)\}$ is easily seen on placing $z_1 = z_2 = 0$ in the backward equations (9.2.49), thereby yielding for $i, j = 1, 2; j \neq i$

$$dq_i(t)/dt = \lambda_i q_i^2(t) - (\lambda_i + \mu_i + \nu_i) q_i(t) + \nu_i q_j(t) + \mu_i. \tag{9.2.76}$$

For these two simultaneous Riccati differential equations do not permit a direct solution. However, the probabilities of ultimate extinction, $q_i = \lim_{t \to \infty} q_i(t)$, may be determined. As $q_1 = q_2 = 1$ is always a solution of equations (9.2.76), the problem reduces to solving a cubic equation in q_1 or q_2. Whilst by considering the geometry of the two parabolas (9.2.76) at $t = \infty$, namely

$$\lambda_i q_i^2 - (\lambda_i + \mu_i + \nu_i) q_i + \nu_i q_j + \mu_i = 0 \quad (i, j = 1, 2; i \neq j), \tag{9.2.77}$$

Puri (1968) shows that for $q_1 = q_2 = 1$ to be the only admissible solution, i.e. ultimate extinction is inevitable regardless of whether the process starts off in state $(1, 0)$ or $(0, 1)$, it is necessary and sufficient that $A + B \geq C$ where $A = \nu_1 + \mu_1 - \lambda_1 = \xi_1$, $B = \nu_2 + \mu_2 - \lambda_2 = \xi_2$ and $C = \sqrt{\{(A - B)^2 + 4\nu_1 \nu_2\}}$. If say $\nu_2 = 0$, so that we have one-way migration, but all other rates are non-zero, then

$$q_1 = \begin{cases} (\lambda_1/2)[\theta_1 - \sqrt{\{\theta_1^2 - 4\lambda_1(\mu_1 + \nu_1 q)\}}] & (\lambda_2 > \mu_2) \\ (\mu_1 + \nu_1)/\lambda_1 & (\lambda_2 \leq \mu_2; \lambda_1 > \mu_1 + \nu_1) \\ 1 & (\lambda_2 \leq \mu_2; \lambda_1 \leq \mu_1 + \nu_1), \end{cases} \tag{9.2.78}$$

where $q = \min(1, \mu_2/\lambda_2)$ and $\theta_1 = \lambda_1 + \mu_1 + \nu_1$. Note that as $\nu_2 = 0$, the extinction probability q_2 is that of a single-colony process. That is, on using results (2.4.24) and (2.4.25) with $n_0 = 1$,

$$q_2 = \begin{cases} \mu_2/\lambda_2 & (\lambda_2 > \mu_2) \\ 1 & (\lambda_2 \leq \mu_2). \end{cases} \tag{9.2.79}$$

Aksland (1975) generalizes result (9.2.78) to cover the k-colony process, with Helland (1975) providing conditions for ultimate extinction which are given in terms of explicit functions of the rate parameters.

Although none of the above developments are totally satisfactory, they certainly highlight the reality that if exact mathematical analysis of the spatial extension to the single-colony birth–death process is difficult, then subsequent analysis of more complex spatial processes involving say predator–prey, competition or epidemic relations looks exceedingly bleak indeed. As we shall soon see, stochastic simulation provides the key to greater understanding in these scenarios. However, before we progress from our simple two-colony model to more challenging spatial situations, let us gain further insight into the nature of the underlying difficulties.

Reid (1959) extended early results on the scalar Riccati differential equation to matrix differential equations of the Riccati type. These have the form

$$d\mathbf{X}/dt = \mathbf{XCX} + \mathbf{DX} + \mathbf{XE} + \mathbf{K} \tag{9.2.80}$$

for the unknown matrix $\mathbf{X}$ and constant matrices $\mathbf{C}$, $\mathbf{D}$, $\mathbf{E}$ and $\mathbf{K}$. So if we wish to utilize this solution we first need to express the backward equations (9.2.49) for the $g_i(t)$ in matrix form. Since we have to transform the vector $(g_1(t), g_2(t))$ into a matrix to effect this, suppose we write

$$\mathbf{X} = \begin{pmatrix} \phi_1 & \phi_2 \\ \phi_3 & \phi_4 \end{pmatrix} \quad \text{for} \quad \phi_i = c_i g_1 + d_i g_2 + e_i \quad (i = 1, \ldots, 4), \tag{9.2.81}$$

where c_i, d_i and e_i are appropriately chosen constants. Now even after a fairly lengthy piece of algebraic manipulation, and in spite of this representation having a high degree of parametric flexibility, the nearest that equations (9.2.49) can be rearranged into the format (9.2.80) is

$$d\mathbf{X}/dt = \mathbf{XCX} + \mathbf{DX} + \mathbf{XE} + \mathbf{FXG} + \mathbf{K}, \tag{9.2.82}$$

where $\mathbf{F}$ and $\mathbf{G}$ are two more constant matrices. It is this extra term, $\mathbf{FXG}$, that seems to bar us from employing standard matrix theory results to construct solutions for $g_i(t)$. For the homogeneous model the constants may be chosen to satisfy $\max_{i,j}\{(\mathbf{FXG})_{ij}\} < \epsilon$ for any $\epsilon < 0$, which suggests regarding $\mathbf{FXG}$ as a perturbation. However, even for this simple case the algebra is awkward, and so the approach is probably not worth pursuing.

Severo (1967, 1969) partly overcame such difficulties by developing a technique (subsequently extended in several papers by Billard) for dealing with systems of ordinary difference-differential probability equations which can be written in the vector-matrix differential form

$$d\mathbf{x}(t)/dt = \mathbf{Bx}(t), \tag{9.2.83}$$

where $\mathbf{B}$ is a triangular matrix of constant coefficients. Although he illustrates the method of solution by discussing its application to bounded multi-dimensional pure birth and death processes, stochastic cross-infection among several otherwise isolated groups, and to generalized stochastic epidemics, it only applies to those processes which can be reduced to triangular form. Essentially, this means determining whether the elements of the ordered one-dimensional population size vector

$$\mathbf{p}(t) = (p_{00}(t); p_{10}(t), p_{01}(t); \ldots; p_{n0}(t), p_{n-1,1}(t), \ldots, p_{0n}(t); \ldots) \qquad (9.2.84)$$

can be rearranged to suit the form (9.2.83). Unfortunately, the penalty for imposing this structure on our birth–death–immigration–migration process is severe, since it requires $\mu_i = \alpha_i = 0$ with either $\nu_1 = 0$ or $\nu_2 = 0$. This is a pity, since the solution has the appearance of being eminently tractable. For example, with the one-way migration homogeneous-birth model (i.e. $\lambda_i = \lambda$, $\nu_1 > 0$, $\mu_i = \alpha_i = \nu_2 = 0$) the forward equations (9.2.1) reduce to

$$dp_{i0}(t)/dt = \lambda(i-1)p_{i-1,0}(t) - (\lambda + \nu_1)p_{i0}(t) \qquad (9.2.85)$$

$$dp_{ij}(t)/dt = \lambda(i-1)p_{i-1,j}(t) + \lambda(j-1)p_{i,j-1}(t) + (i+1)\nu_1 p_{i+1,j-1}(t)$$
$$- [(\lambda + \nu_1)i + \lambda j] p_{ij}(t) \qquad (j > 0). \qquad (9.2.86)$$

Equation (9.2.85) integrates sequentially to give $\{p_{i0}(t)\}$ ($i = 0, 1, 2, \ldots$), whence substitution into equations (9.2.86) yields $\{p_{i1}(t)\}$ when $j = 1$, and repeating this operation gives $\{p_{ij}(t)\}$ for $j = 2, 3, \ldots$. However, it is algebraically simpler to replace the associated ordering

$$\mathbf{p}(t) = (p_{00}(t), p_{10}(t), p_{20}(t), \ldots; p_{01}(t), p_{11}(t), p_{21}(t), \ldots; p_{02}(t), p_{12}(t), p_{22}(t), \ldots; \ldots) \qquad (9.2.87)$$

by (9.2.84). For then we have the lower triangular form (9.2.83), with the matrix $\mathbf{B}$ taking the diagonal elements $-\lambda - \nu_1, -\lambda; -2\lambda - 2\nu_1, -2\lambda - \nu_1, -2\lambda; -3\lambda - 3\nu_1, \ldots$ etc. Thus for some constants $c_{ij}(r, k)$ the solution to equation (9.2.83) may be expressed in the form

$$p_{ij}(t) = \sum_{r=0}^{i} e^{-r\lambda t} \sum_{k=0}^{i+j-r} c_{ij}(r,k) e^{-k(\lambda+\nu_1)t} \qquad (i, j = 0, 1, 2, \ldots). \qquad (9.2.88)$$

Substituting for $p_{ij}(t)$ from (9.2.88) into (9.2.86), and comparing coefficients of $e^{[-r\lambda - k(\lambda + \nu_1)]t}$, then yields a set of solvable equations for the $\{c_{ij}(r,k)\}$. For given the values of $c_{is}(r,k)$ for $s < j$ the technique generates $c_{ij}(r,k)$, and so the solution to equations (9.2.85) and (9.2.86) may be obtained recursively over the successive values $j = 1, 2, 3, \ldots$.

Even for this extremely basic model, which is essentially the simplest spatial birth process possible, the algebraic detail involved in constructing the coefficients $c_{ij}(r,k)$ (see Renshaw, 1976) is really too awkward to be of much practical use. So although the technique does provide a 'solution' for the $\{p_{ij}(t)\}$ in this simple case, it offers little hope of providing an acceptable way of generating the exact probability structure for more general spatial models. The only way forward would seem to be acceptance of approximate probability solutions instead, based on say the saddlepoint procedures developed in Section 7.3. However, it should be noted that such algebraic difficulties do not hinder the construction of more theoretically oriented results. For example, Renshaw and Dai (1997) consider two general birth–death processes that are linked by migrating individuals, and develop conditions for weak symmetry, regularity and recurrence.

9.3 Stepping-stone processes

Interest in the effect of migration between separate geographic regions has been widespread for many years, early examples being the spread of disease between cities in the USSR (Bailey, 1975), the spread of an advantageous mutant gene along a linear habitat such as a shore line (Fisher, 1937), and the spread of epidemics around the coastline of Iceland (Cliff et al., 1981). Whilst more recent examples include the spread of HIV/AIDS, and the deadly H5N1 strain of bird 'flu together with the likely effects of its potential crossover to the human population. Unfortunately, however, few mathematical tools have been developed in order to conduct useful stochastic analyses in such situations. The reason for this is perfectly clear when we recall the problems we have just encountered in the simple two-colony birth–death–migration system: extending the process to encompass many interacting colonies, never mind incorporating nonlinear interactions such as competition, predation and infection, is clearly a hope too far. All that can realistically be expected on the applied probability front is that equations for the first few moments might be solvable, and that these can then be incorporated into an appropriate saddlepoint-type approximation.

Returning to the USSR example, suppose we consider migration between N cities $(i = 1, \ldots, N)$ with individual migration rate ν_{ij} directly from city i to j, and that we exclude immigration (i.e. $\alpha_i = 0$). Then the deterministic two-site equations (9.2.3) extend to

$$dm_i(t)/dt = (\lambda_i - \mu_i - \sum_{j=1, j \neq i}^{N} \nu_{ij}) m_i(t) + \sum_{k=1, k \neq i}^{N} \nu_{ki} m_k(t) \qquad (i = 1, \ldots, N), \tag{9.3.1}$$

which, on defining $\nu_{ii} = \lambda_i - \mu_i - \sum_{j=1, j \neq i}^{N} \nu_{ij}$ and $\mathbf{m}(t) = (m_1(t), \ldots, m_N(t))$, reduce to the simple vector-matrix equation

$$d\mathbf{m}(t)/dt = \mathbf{m}(t)\mathbf{L} \qquad \text{where} \qquad \mathbf{L} = \begin{pmatrix} \nu_{11} & \cdots & \nu_{1N} \\ \vdots & \vdots & \vdots \\ \nu_{N1} & \cdots & \nu_{NN} \end{pmatrix}. \tag{9.3.2}$$

Now the general solution to (9.3.2) may be expressed in the form

$$\mathbf{m}(t) = \mathbf{m}(0) \sum_{i=1}^{N} e^{\omega_i t} \mathbf{q}_i \mathbf{r}_i, \tag{9.3.3}$$

where ω_i are the (presumed) distinct eigenvalues of $\mathbf{L}$, and $\mathbf{q}_i$ and $\mathbf{r}_i$ are the corresponding row and column eigenvectors normalized to ensure that

$$\mathbf{r}_i \mathbf{q}_i = 1 \qquad (i = 1, \ldots, N). \tag{9.3.4}$$

Thus determination of even the mean population vector $\mathbf{m}(t)$ depends on the migration matrix $\mathbf{L}$ having amenable closed form expressions for the eigenvalues and eigenvectors.

The same holds true for the variance-covariance matrix

$$\mathbf{V}(t) = \begin{pmatrix} V_{11}(t) & \cdots & V_{1N}(t) \\ \vdots & \vdots & \vdots \\ V_{N1}(t) & \cdots & V_{NN}(t) \end{pmatrix}. \tag{9.3.5}$$

For denote the diagonal matrices $\mathbf{M}(t) = \text{diag}\{m_i(t)\}$, $\mathbf{D} = \text{diag}\{\lambda_i\}$ and $\mathbf{H}(t) = \text{diag}\{\sum_{r=1}^{N} \nu_{ri} m_r(t)\}$. Then it is easy to show that $\mathbf{V}(t)$ satisfies the matrix differential equation

$$d\mathbf{V}(t)/dt = \mathbf{V}\mathbf{L} + \mathbf{L}^T \mathbf{V} + [\mathbf{H} + \mathbf{M}(\mathbf{D} - \mathbf{L}) + (\mathbf{D} - \mathbf{L}^T)\mathbf{M}], \tag{9.3.6}$$

which, on denoting $\mathbf{F} = \mathbf{H} + \mathbf{M}(\mathbf{D} - \mathbf{L}) + (\mathbf{D} - \mathbf{L}^T)\mathbf{M}$, takes the solution (e.g. Bellman, 1960)

$$\mathbf{V}(t) = \int_0^t e^{\mathbf{L}^T(t-s)} \mathbf{F}(s) e^{\mathbf{L}(t-s)} \, ds. \tag{9.3.7}$$

So as with $\mathbf{m}(t)$, determination of $\mathbf{V}(t)$ depends on the migration matrix $\mathbf{L}$ being easily decomposable.

A relatively straightforward way of effecting this is to let $\mathbf{L}$ be tri-diagonal. That is, only $\nu_{i,i+1}$ and $\nu_{i,i-1}$ are non-zero, so we have migration only between *nearest-neighbours*. This applies to any process in a linear habitat, in which an individual who moves between sites i and j has to visit all the intermediate sites between them; i.e. individuals are prohibited from jumping over an intermediate site. However, before we analyse this scenario, it is worth mentioning that the dimensionality of the process is crucial here. For in two or higher dimensions individuals can bypass directly intervening sites, and this vastly inflates the degree of difficulty of the associated mathematics. For example, whereas in one dimension movement from site 1 to 3 has to involve the two steps $1 \to 2 \to 3$, in two dimensions movement from $(1,1)$ to $(3,1)$ can easily bypass $(2,1)$ by taking the route $(1,1) \to (1,2) \to (2,2) \to (3,2) \to (3,1)$. So instead of having a single possible path between two sites we now have many, or even an infinite number.

This feature was recognized early on in statistical mechanics through the Ising model, in which atoms, animals, protein folds, biological membranes, etc. arranged on a lattice modify their own behaviour so as to conform to that of their immediate neighbours. Essentially, the spatial pattern produced switches from being nearly random when the association between neighbours is low, to having large clusters of individuals exhibiting the same behaviour when the association is high. A key feature is the existence of a *critical threshold value* as the process passes between these two modes of behaviour. Ising (1925) introduced it in an attempt to explain certain empirically observed facts about ferromagnetic materials based on a model proposed by Lenz (1920). More recent applications include phase separation in binary alloys, neural networks, flocks of birds and beating heart cells. It was referred to in Heisenberg (1928), who uses the exchange mechanism to describe ferromagnetism, and became well-established when Peierls (1936) gave a non-rigorous proof that spontaneous magnetization must exist. Theoretical development of this process has proved to be extremely difficult: although Onsager (1944) gave a complete analytic solution to the two-dimensional problem, the three-dimensional process is thought to be mathematically intractable. Nevertheless, Jones (1989) observes an amazing connection between this problem and knot theory, so perhaps an analytic breakthrough will eventually occur through successful developments in seemingly unrelated fields. As far as this text is concerned we shall therefore restrict theoretical attention to purely one-dimensional processes. Moreover, given our recent experience with a two-site multiplicative process, and that the birth–death–migration scenario is likely to be substantially more difficult to analyze than the Ising model since there the total number of individuals in the system is conserved, we shall mainly rely on simulation to gain insight into stochastic behaviour.

9.3.1 Birth–death–migration processes on the infinite line

The simplest way of ensuring that the migration matrix is readily decomposable is to assume that $\mathbf{L}$ is both homogeneous in λ_i and μ_i and tridiagonal, that is we take $\lambda_i = \lambda$, $\mu_i = \mu$, $\nu_{i,i+1} = \nu_1$, $\nu_{i,i-1} = \nu_2$ and $\nu_{ij} = 0$ for $j < i-1$ and $j > i+1$. Thus $\nu_{ii} = \lambda - \mu - \nu_1 - \nu_2 = \xi$ (say). Whilst we can avoid edge-effects by allowing the population to develop over the full integer axis $-\infty < i < \infty$. Let $X_i(t)$ denote the number of individuals in colony i at time t, with $X_i(0) = a_i$. Write $\mathbf{x} = (\ldots, x_i, \ldots)$ and $\mathbf{z} = (\ldots, z_i, \ldots)$, and let $\mathbf{e} = (\ldots, 0, 1, 0, \ldots)$ be the vector with 1 in the ith place and zeros elsewhere. Then for $p(\mathbf{x}; t) = \Pr\{X_i(t) = x_i | X_i(0) = a_i\}$, considering all possible events that can occur in the small time interval $(t, t + dt)$ yields the infinite-colony extension to the two-colony equations (9.2.1), namely

$$\frac{dp(\mathbf{x};t)}{dt} = \sum_{i=-\infty}^{\infty} \{\lambda(x_i - 1)p(\mathbf{x} - \mathbf{e}_i; t) + \mu(x_i + 1)p(\mathbf{x} + \mathbf{e}_i; t)$$

$$- (\lambda + \mu + \nu_1 + \nu_2)x_i p(\mathbf{x};t) + \nu_1(x_{i-1} + 1)p(\mathbf{x}_i + \mathbf{e}_{i-1} - \mathbf{e}_i; t)$$

$$+ \nu_2(x_{i+1} + 1)p(\mathbf{x}_i + \mathbf{e}_{i+1} - \mathbf{e}_i; t)\}. \tag{9.3.8}$$

Define the associated p.g.f.

$$G(\mathbf{z};t) \equiv \sum_{\mathbf{x}=-\infty}^{\infty} p(\mathbf{x};t) \prod_{i=-\infty}^{\infty} z_i^{x_i}. \tag{9.3.9}$$

Then multiplying both sides of equation (9.3.8) by $\prod_{i=-\infty}^{\infty} z_i^{x_i}$ and summing the resulting expression over $x_i = 0, 1, 2, \ldots$ ($-\infty < i < \infty$) yields the p.d.e.

$$\frac{\partial G}{\partial t} = \sum_{i=-\infty}^{\infty} \{\lambda(z_i^2 - z_i) + \mu(1 - z_i) + \nu_1(z_{i+1} - z_i) + \nu_2(z_{i-1} - z_i)\}\frac{\partial G}{\partial z_i}. \tag{9.3.10}$$

Such equations can be written down directly, without first developing the forward equations for $p(\mathbf{x};t)$, by following the general random variable technique described by Bailey (1964). For in essence, an individual in colony i that splits into $r = 0, 1, 2, \ldots$ individuals in i and $s = 0, 1, 2, \ldots$ in j corresponds to $z_i^r z_j^s$ in the p.d.e., whilst one that does not split corresponds to $-z_i$. So we can see straightaway that equation (9.3.10) relates to a simple birth–death–nearest-neighbour–migration process with rates λ, μ, ν_1 ($i \to i+1$) and ν_2 ($i \to i-1$), respectively.

Mean population size

Although the presence of the quadratic λz_i^2 terms renders this equation intractable to direct solution, the mean population sizes, $m_i(t)$, can still be obtained fairly readily (Bailey, 1968, examines the symmetric case $\nu_1 = \nu_2$). For differentiating (9.3.10) with respect to z_i gives

$$\frac{\partial^2 G}{\partial z_i \partial t} = [\lambda(2z_i - 1) - \mu - \nu_1 - \nu_2]\frac{\partial G}{\partial z_i} + \nu_1 \frac{\partial G}{\partial z_{i-1}} + \nu_2 \frac{\partial G}{\partial z_{i+1}}$$

$$+ \sum_{j=-\infty}^{\infty} [\lambda(z_j^2 - z_j) + \mu(1 - z_j) + \nu_1(z_{j+1} - z_j) + \nu_2(z_{j-1} - z_j)]\frac{\partial^2 G}{\partial z_i \partial z_j}. \quad (9.3.11)$$

Whence placing $z_i = 1$ ($-\infty < i < \infty$) yields the set of first-order linear equations

$$dm_i(t)/dt = (\lambda - \mu - \nu_1 - \nu_2)m_i(t) + \nu_1 m_{i-1}(t) + \nu_2 m_{i-1}(t). \quad (9.3.12)$$

Although these can be solved via the decomposition (9.3.3), or by employing Laplace transforms, it is instructive to use a generating function approach that exploits the properties of known functions.

Define $H(z;t) \equiv \sum_{i=-\infty}^{\infty} m_i(t) z^i$. Then on multiplying equations (9.3.12) by z^i and summing over $-\infty < i < \infty$ we obtain

$$dH(z;t)/dt = [(\lambda - \mu - \nu_1 - \nu_2) + (\nu_1 z + \nu_2 z^{-1})]H(z;t), \quad (9.3.13)$$

which integrates to give

$$H(z;t) = H(z;0)\exp\{[(\lambda - \mu - \nu_1 - \nu_2) + (\nu_1 z + \nu_2 z^{-1})]t\}. \quad (9.3.14)$$

To extract the coefficient of z^i from this expression we simply note its close resemblance to the standard series expansion

$$\exp\{x(y + y^{-1})\} = \sum_{k=-\infty}^{\infty} I_k(2x) y^k \quad (y \neq 0), \quad (9.3.15)$$

where

$$I_i(2x) \equiv x^i \sum_{k=0}^{\infty} \frac{x^{2k}}{k!(i+k)!} \quad (9.3.16)$$

denotes the modified Bessel function of the first kind (Abramowitz and Stegun, 1970). Specifically, on writing $x = t\sqrt{(\nu_1 \nu_2)}$ and $y = z\sqrt{(\nu_1/\nu_2)}$, expression (9.3.14) takes the form

$$H(z;t) = e^{(\lambda-\mu-\nu_1-\nu_2)t} \sum_{j=-\infty}^{\infty} a_j z^j \sum_{k=0}^{\infty} I_k(2t\sqrt{(\nu_1\nu_2)})[z\sqrt{(\nu_1/\nu_2)}]^k, \quad (9.3.17)$$

and extracting the coefficient of z^i gives

$$m_i(t) = e^{(\lambda-\mu-\nu_1-\nu_2)t} \sum_{j=-\infty}^{\infty} a_j(\nu_1/\nu_2)^{i/2} I_{i-j}(2t\sqrt{(\nu_1\nu_2)}) \qquad (-\infty < i < \infty).$$
(9.3.18)

If the initial population consist of a single individual at site 0, so that $a_0 = 1$ and $a_j = 0$ $(j \neq 0)$, then this result condenses to

$$m_i(t) = e^{(\lambda-\mu-\nu_1-\nu_2)t}(\nu_1/\nu_2)^{i/2} I_i(2t\sqrt{(\nu_1\nu_2)}).$$
(9.3.19)

Whence on using the asymptotic result that when i is fixed and $|x|$ is large

$$I_i(x) \sim \frac{e^x}{\sqrt{2\pi x}}\{1 - \frac{4i^2-1}{8x} + \cdots\} \qquad (|\arg x| < \pi/2)$$
(9.3.20)

(Abramowitz and Stegun, 1970), it follows that for large t

$$m_i(t) \sim \frac{e^{(\lambda-\mu-(\sqrt{\nu_1}-\sqrt{\nu_2})^2)t}}{2\sqrt{\pi t \sqrt{(\nu_1\nu_2)}}}(\nu_1/\nu_2)^{i/2}$$
(9.3.21)

provided $|i| \ll 2\sqrt{t\sqrt{(\nu_1\nu_2)}}$. So for $|i|$ small relative to $\sqrt{t}$, we see that $m_i(t)$ changes geometrically in i at rate $\sqrt{(\nu_1/\nu_2)}$.

A better feel for the shape of the mean curve over the full range of i can be gained by allowing immigration at rate α into site 0. For then an equilibrium situation develops if $\lambda < \mu$, since the total population size follows a non-spatial simple immigration–birth–death process (Section 2.5). The forward p.g.f. equation now has the added term $\alpha(z_0 - 1)G$ on its right hand side, whilst the mean equation (9.3.12) is augmented to

$$dm_i(t)/dt = (\lambda - \mu - \nu_1 - \nu_2)m_i(t) + \nu_1 m_{i-1}(t) + \nu_2 m_{i+1}(t) + \alpha\delta(i)$$
(9.3.22)

where $\delta(0) = 1$ and $\delta(i) = 0$ for $i \neq 0$. Since $m_i(\infty)$ is independent of t, letting $t \to \infty$ in the modified equation (9.3.13) leads to

$$dH(z;\infty)/dt = 0 = [(\lambda - \mu - \nu_1 - \nu_2) + (\nu_1 z + \nu_2 z^{-1})]H(z;\infty) + \alpha,$$
(9.3.23)

which, on writing $\phi = \mu + \nu_1 + \nu_2 - \lambda$, gives

$$H(z;\infty) = (\alpha/\phi)\sum_{n=0}^{\infty}[(\nu_1 z + \nu_2 z^{-1})/\phi]^n.$$
(9.3.24)

Extracting the coefficient of z^i then yields (for $i \geq 0$)

$$m_i(\infty) = (\alpha/\phi)(\nu_1/\phi)^i \sum_{r=0}^{\infty} \binom{i+2r}{r}(\nu_1\nu_2/\phi^2)^r,$$
(9.3.25)

together with a parallel expression for $i < 0$.

Second-order moments

Not only is this approach easily extended to multi-dimensional models, but higher-order moments can also be constructed using a similar procedure. For placing $z_i = e^{\theta_i}$ in (9.3.10) to form the c.g.f. $K(\theta;t) \equiv \ln G(e^{\theta};t)$ gives

$$\frac{\partial K}{\partial t} = \sum_{i=-\infty}^{\infty} \{\lambda(e^{\theta_i} - 1) + \mu(e^{-\theta_i} - 1) + \nu_1(e^{\theta_{i+1}-\theta_i} - 1) + \nu_2(e^{\theta_{i-1}-\theta_i} - 1)\}\frac{\partial K}{\partial \theta_i}.$$

(9.3.26)

Equations for the variances $V_{ii}(t)$ and covariances $V_{ij}(t)$ may now be derived directly by twice differentiating (9.3.26) with respect to θ_i and θ_j and then placing $\theta_i = 0$ ($-\infty < i < \infty$); though care must be taken since each e^{θ_i} occurs in three successive terms of the summation. The resulting first-order differential equations can be solved using a similar p.g.f. approach to that used above for the $m_i(t)$, or by employing Laplace transforms. Although both methods of solution are straightforward, the resulting algebra is slightly messy, and so here we shall simply state the result for the symmetric migration case $\nu_1 = \nu_2 = \nu$, namely

$$V_{ij}(t) = \int_0^t e^{2\xi(t-s)} \sum_{r=-\infty}^{\infty} [(\lambda + \mu + 2\nu)I_{i-r}(2\nu(t-s))I_{j-r}(2\nu(t-s))$$

$$- \nu I_{i-1-r}(2\nu(t-s))I_{j-r}(2\nu(t-s)) - \nu I_{i+1-r}(2\nu(t-s))I_{j-r}(2\nu(t-s))$$

$$- \nu I_{i-r}(2\nu(t-s))I_{j-1-r}(2\nu(t-s)) - \nu I_{i-r}(2\nu(t-s))I_{j+1-r}(2\nu(t-s))$$

$$+ \nu I_{i-1-r}(2\nu(t-s))I_{j-1-r}(2\nu(t-s)) + \nu I_{i+1-r}(2\nu(t-s))I_{j+1-r}(2\nu(t-s))]$$

$$\times e^{\xi s} \sum_{n=-\infty}^{\infty} a_n I_{r-n}(2\nu s) \, ds \qquad (9.3.27)$$

where $\xi = \lambda - \mu - 2\nu$ and $a_n = m_n(0)$. Comparison with expression (9.3.18) shows that whereas the $m_i(t)$ have a fairly simple form involving linear sums of Bessel functions, the corresponding result for the second-order moments is substantially more complicated since it involves the Laplace transform of pair-products of Bessel functions. Indeed, if we also allow immigration at rate α into site $i = 0$, then this form becomes even more convoluted, with the last term in (9.3.27) being augmented to

$$\cdots \times \{e^{\xi s} \sum_{n=-\infty}^{\infty} a_n I_{r-n}(2\nu s) + \alpha \int_0^s e^{\xi \tau} I_r(2\nu \tau) \, d\tau\} \, ds$$

$$+ \alpha \int_0^t e^{2\xi(t-s)} I_i(2\nu(t-s)) I_j(2\nu(t-s)) \, ds \qquad (9.3.28)$$

(Renshaw, 1986). So whilst the development of higher-order moments is certainly mathematically tractable, its value in helping to provide a deeper understanding of the underlying process is rather limited.

Probability solution ($\lambda = 0$)

Although this level of complexity suggests that attempts to find exact solutions for the probability vector $\mathbf{p}(\mathbf{x}; t)$ are likely to founder, it is worthwhile mentioning two specific situations in which progress can be made. First, suppose that $\lambda = 0$ and consider death as being transfer into a new site from which 'there is no escape'. Then if $\alpha = 0$ the process is closed, and so the p.g.f. $G(\mathbf{z}; t)$ is a special case of the general result (9.1.10).

502 Spatial processes

So bearing in mind that immigration is a fairly anodyn process, it seems likely that a solution for $G(\mathbf{z};t)$ can be constructed when $\alpha > 0$. The forward p.g.f. equation corresponding to (9.3.10) for the symmetric case $\nu_1 = \nu_2 = \nu$ is now

$$\frac{\partial G}{\partial t} = \sum_{i=-\infty}^{\infty} \{\mu(1-z_i) + \nu(z_{i+1}-z_i) + \nu(z_{i-1}-z_i)\}\frac{\partial G}{\partial z_i} + \alpha(z_0-1)G, \quad (9.3.29)$$

which takes the associated auxiliary equations

$$\frac{dt}{1} = \frac{-dz_i}{\mu(1-z_i) + \nu(z_{i+1}-z_i) + \nu(z_{i-1}-z_i)} = \frac{dG}{\alpha(z_0-1)G} \quad (-\infty < i < \infty). \tag{9.3.30}$$

The solution of the first pair runs very close to that for the mean equations (9.3.12). For on writing $r_i = 1 - z_i$, so that

$$dr_i/dt = \mu r_i + \nu(r_i - r_{i+1}) + \nu(r_i - r_{i-1}), \tag{9.3.31}$$

with $H(y;t) \equiv \sum_{i=-\infty}^{\infty} r_i y^i$ replacing $H(z;t)$, we obtain

$$dH(y;t)/dt = [(\mu+2\nu) - \nu(y+y^{-1})]H(y;t). \tag{9.3.32}$$

Whence on writing the constant of integration in the general form $\sum_{k=-\infty}^{\infty} A_k y^k$, where A_k ($-\infty < k < \infty$) are arbitrary constants, and using the Bessel representation (9.3.15), equation (9.3.32) integrates to give

$$H(y;t) = e^{(\mu+2\nu)t} \sum_{k=-\infty}^{\infty} A_k y^k \sum_{j=-\infty}^{\infty} I_j(-2\nu t) y^j. \tag{9.3.33}$$

Extracting the coefficient of y^i yields

$$r_i = e^{(\mu+2\nu)t} \sum_{j=-\infty}^{\infty} A_{i-j} I_j(-2\nu t), \tag{9.3.34}$$

and so

$$z_i = 1 - e^{(\mu+2\nu)t} \sum_{j=-\infty}^{\infty} A_{i-j} I_j(-2\nu t). \tag{9.3.35}$$

Whilst on writing the solution to equation (9.3.32) in the form

$$H(y;t)e^{\nu t(y+y^{-1})} = e^{(\mu+2\nu)t} \sum_{k=-\infty}^{\infty} A_y y^k, \tag{9.3.36}$$

and then re-using the Bessel representation (9.3.15), we obtain

$$\sum_{k=-\infty}^{\infty} A_k y^k = e^{-(\mu+2\nu)t} \sum_{j=-\infty}^{\infty} I_j(2\nu t) y^j \sum_{i=-\infty}^{\infty} r_i y^i. \tag{9.3.37}$$

Whence extracting the coefficient of y^k yields

$$A_k = e^{-\mu t} - e^{-(\mu+2\nu)t} \sum_{j=-\infty}^{\infty} z_{k-j} I_j(2\nu t). \tag{9.3.38}$$

It now remains to integrate the remaining auxiliary equation in (9.3.30). First note that (9.3.35) with $i = 0$ gives

$$z_0 - 1 = -e^{(\mu+2\nu)t} \sum_{j=-\infty}^{\infty} A_j I_j(-2\nu t). \tag{9.3.39}$$

So this last equation may be written as

$$\frac{d(\log G)}{dt} = \alpha(z_0 - 1) = -\alpha e^{(\mu+2\nu)t} \sum_{j=-\infty}^{\infty} A_j I_j(-2\nu t), \tag{9.3.40}$$

which integrates to give

$$G = \text{constant} \times \exp\{-\alpha \sum_{j=-\infty}^{\infty} A_j P_j(t)\} \tag{9.3.41}$$

where

$$P_j(t) = \int_0^t e^{(\mu+2\nu)s} I_j(-2\nu s)\, ds. \tag{9.3.42}$$

Thus the general solution to equation (9.3.29) is given by

$$G(\mathbf{z}; t) = f(\ldots, A_i, \ldots) \exp\{-\alpha \sum_{j=-\infty}^{\infty} A_j P_j(t)\} \tag{9.3.43}$$

where $f(\cdot)$ is an arbitrary function.

In order to find the specific solution which satisfies the initial condition

$$G(\mathbf{z}; 0) = \prod_{k=-\infty}^{\infty} z_k^{a_k}, \tag{9.3.44}$$

we now have to consider the 'constants' A_i as functions of z and t. Since $I_0(0) = 1$ and $I_j(0) = 0$ ($j \neq 0$), at $t = 0$ expression (9.3.35) reduces to $z_i = 1 - A_i$; whilst (9.3.42) gives $P_j(0) = 0$. Hence on combining (9.3.43) and (9.3.44) we obtain

$$f(\ldots, A_i, \ldots) = \prod_{i=-\infty}^{\infty} (1 - A_i)^{a_i}. \tag{9.3.45}$$

So on writing (9.3.38) as the (z, t)-dependent function

$$Q_i(2\nu t) = -A_i = e^{-(\mu+2\nu)t} \sum_{j=-\infty}^{\infty} z_{i-j} I_j(2\nu t) - e^{-\mu t}, \tag{9.3.46}$$

it follows that the general solution (9.3.43) takes the particular form

$$G(\mathbf{z};t) = \exp\{\alpha \sum_{j=-\infty}^{\infty} Q_j(2\nu t) P_j(t)\} \prod_{i=-\infty}^{\infty} [1 + Q_i(2\nu t)]^{a_i}. \tag{9.3.47}$$

Note that setting $x = \nu t$ and $y = 1$ in (9.3.15) yields the relation

$$\sum_{k=-\infty}^{\infty} I_k(2\nu t) = e^{2\nu t}, \tag{9.3.48}$$

whence placing $z_i = z$ in (9.3.46) gives

$$Q_i(2\nu t) = e^{-\mu t}(z-1) \qquad (-\infty < i < \infty).$$

Thus the p.g.f. of total population size is

$$G(z;t) = \exp\{\alpha e^{-\mu t}(z-1) \int_0^t e^{(\mu+2\nu)s} \sum_{j=-\infty}^{\infty} I_j(-2\nu s)\, ds\} \prod_{i=-\infty}^{\infty} [1 + e^{-\mu t}(z-1)]^{a_i}$$

$$= \exp\{(\alpha/\mu)(z-1)(1 - e^{-\mu t})\}[1 + e^{-\mu t}(z-1)]^{\sum a_i}, \tag{9.3.49}$$

which is the well-known result (2.5.7) for the simple immigration–death process.

It is possible to simplify the solution (9.3.47) by replacing $Q_j(2\nu t) P_j(t)$ by a more transparent expression. The algebraic detail (shown in Renshaw, 1974) centres around the lemma

$$[1 - \theta(z + z^{-1})]^{-1} \equiv (1 - 4\theta^2)^{-1/2} \sum_{n=-\infty}^{\infty} \beta^{|n|} z^n \qquad (0 < \theta < 1/2) \tag{9.3.50}$$

where $\beta = [1 - (1 - 4\theta^2)^{1/2}]/(2\theta)$. For on defining $\theta = \nu/(\mu + 2\nu)$, and noting that $0 < \theta < 1/2$ for $\mu > 0$, this leads to

$$G(\mathbf{z};t) = \exp\{-(\alpha/\mu)(1 - e^{-\mu t})\} \exp\{(\alpha\theta/\nu)(1 - 4\theta^2)^{-1/2} \tag{9.3.51}$$

$$\times \sum_{h=-\infty}^{\infty} z_h[\beta^{|h|} - e^{-(\mu+2\nu)t} \sum_{i=-\infty}^{\infty} \beta^{|h-i|} I_i(2\nu t)]\} \prod_{i=-\infty}^{\infty} [1 + Q_i(2\nu t)]^{a_i}.$$

If the population is initially empty, so that $a_i = 0$ $(-\infty < i < \infty)$, then this expression simplifies still further, since $G(\mathbf{z};t)$ may be factored into the form

$$G(\mathbf{z};t) = \exp\{-(\alpha/\mu)(1 - e^{-\mu t})\} \prod_{i=-\infty}^{\infty} \exp\{(\alpha\theta/\nu)(1 - 4\theta^2)^{-1/2} z_i f_i(t)\} \tag{9.3.52}$$

where

$$f_i(t) = \beta^{|i|} - e^{-(\mu+2\nu)t} \sum_{n=-\infty}^{\infty} \beta^{|i-n|} I_n(2\nu t). \tag{9.3.53}$$

Whence the coefficients of $\prod_{i=-\infty}^{\infty} z_i^{x_i}$ are now easily extracted, yielding

$$p(\mathbf{x};t) = \exp\{-(\alpha/\mu)(1-e^{-\mu t})\} \prod_{i=-\infty}^{\infty} \frac{1}{x_i!}\{(\alpha\theta/\nu)(1-4\theta^2)^{-1/2}f_i(t)\}^{x_i}. \quad (9.3.54)$$

Thus in probability terms, each colony behaves as though it is independent of all other colonies, with the size of colony i being distributed as a Poisson variable with parameter

$$(\alpha\theta/\nu)(1-4\theta^2)^{-1/2}f_i(t). \quad (9.3.55)$$

This exposes a trap which needs to be avoided when interpreting results of this nature. For although the colonies are independent *in probability* they are clearly not independent *in realization*. The former relates to an 'average property' taken over the uncountable infinity of all possible realizations, and lies in total contrast to the behaviour of a single realization. In this latter situation, if $X_i(t)$ is excessively high then migrants will escape in large numbers from colony i to $i-1$ and $i+1$ and hence raise the values of $X_{i-1}(t)$ and $X_{i+1}(t)$; so neighbouring large colonies will be positively correlated. So care must clearly be taken when jumping from probability 'solutions' to inferring the behavioural characteristics of single realizations.

Moments are easily evaluated by differentiating (9.3.52) with respect to z_i and then placing $\mathbf{z} = 1$. In particular, the means are given by

$$m_i(t) = (\alpha\theta/\nu)(1-4\theta^2)^{-1/2}f_i(t) + e^{-(\mu+2\nu)t}\sum_{n=-\infty}^{\infty} a_n I_{n-i}(2\nu t), \quad (9.3.56)$$

whilst the variance-covariances take the form

$$V_{ij}(t) = m_i(t)\delta(i-j) - e^{-2(\mu+2\nu)t}\sum_{n=-\infty}^{\infty} a_n I_{n-i}(2\nu t)I_{n-j}(2\nu t). \quad (9.3.57)$$

Here the Kronecker delta function $\delta(i-j) = 1$ if $i = j$ and zero otherwise. Note that for large t, applying the asymptotic result (9.3.20) with $x = 2\nu t$ yields

$$m_i(t) \sim \beta^{|i|}(\alpha\theta/\nu)(1-4\theta^2)^{-1/2} + e^{-\mu t}(4\pi\nu t)^{-1/2}(\sum_{j=-\infty}^{\infty} a_j - \alpha/\mu). \quad (9.3.58)$$

Whence on placing $\theta = \nu/(\mu+2\nu)$ we see that as $t \to \infty$

$$m_i(t) \to \alpha\beta^{|i|}(\mu^2+4\mu\nu)^{-1/2} \quad (9.3.59)$$

at rate $t^{-1/2}e^{-\mu t}$, where $\beta = [(\mu+2\nu) - \sqrt{\mu^2+4\mu\nu}]/(2\nu)$.

Moreover, on letting $t \to \infty$ in (9.3.53) and (9.3.55) it follows that in equilibrium the $X_i(\infty)$ are distributed as independent Poisson variables with parameter $\alpha(\mu^2+4\mu\nu)^{-1/2}\beta^{|i|}$. So $m_i(\infty) = V_{ii}(\infty) = \alpha(\mu^2+4\mu\nu)^{-1/2}\beta^{|i|}$ and $V_{ij}(\infty) = 0$ ($j \neq i$). Thus since $0 < \beta < 1$, the equilibrium means and variances decrease geometrically away from the origin at rate β.

506 *Spatial processes*

Approximate probability solution ($\lambda \geq 0$)

Although only slight extensions of the arguments leading to results (9.3.18) for $m_i(t)$ and (9.3.27) for $V_{ij}(t)$ are required when $\lambda > 0$ and $\alpha > 0$ (see Renshaw, 1974 for details), developing the probability structure akin to (9.3.51) for $G(\mathbf{z};t)$ and (9.3.54) for $p(\mathbf{x};t)$ is a different matter. In Section 9.2.3 we developed an approximate stochastic solution for the 2-colony process by changing the probability of a birth in colony i in $(t, t+dt)$ from $\lambda i\, dt$ to $\lambda m_i(t)dt$, and whilst we showed that this was not ideal, in that the properties of this modified process did not completely align with those of the original one, it did yield some useful information on the underlying probability structure.

A stochastic approximation for our process on the infinite line can be developed in a similar manner. For example, when $\alpha = 0$ (taking $\alpha > 0$ does not add much extra algebraic complexity), on paralleling the construction of the approximate two-colony p.g.f. equation (9.2.30) the forward equations (9.3.10) and (9.3.29) become

$$\frac{\partial G}{\partial t} = \sum_{i=-\infty}^{\infty} \lambda m_i(z_i - 1)G + \sum_{i=-\infty}^{\infty} \{\mu(1 - z_i) + \nu(z_{i+1} - z_i) + \nu(z_{i-1} - z_i)\}\frac{\partial G}{\partial z_i}. \tag{9.3.60}$$

The associated auxiliary equations now take the form

$$\frac{dt}{1} = \frac{-dz_i}{\mu(1-z_i) + \nu(z_{i+1}-z_i) + \nu(z_{i-1}-z_i)} = \frac{dG}{\sum_{k=-\infty}^{\infty} \lambda m_k(z_k - 1)G}, \tag{9.3.61}$$

and as the first pair are identical to those in (9.3.30) it follows that their solution is given by (9.3.35). Whence on substituting for $m_k(t)$ from (9.3.18) and z_k from (9.3.35), the last equation in (9.3.61) takes the form

$$\frac{d(\ln G)}{dt} = -\lambda e^{\lambda t} \sum_{k=-\infty}^{\infty} \sum_{j=-\infty}^{\infty} a_j I_{k-j}(2\nu t) \sum_{l=-\infty}^{\infty} A_{k-l} I_l(-2\nu t)$$

$$= -\lambda e^{\lambda t} \sum_{j=-\infty}^{\infty} \sum_{i=-\infty}^{\infty} a_j A_i \sum_{k=-\infty}^{\infty} I_{k-j}(2\nu t)I_{k-i}(-2\nu t) \tag{9.3.62}$$

(on replacing $k - l$ by i). Now replacing x in (9.3.15) by $2\nu t$ and $-2\nu t$, and multiplying the two resulting expressions, yields

$$1 = \sum_{p=-\infty}^{\infty} \sum_{r=-\infty}^{\infty} I_p(2\nu t)I_{r-p}(-2\nu t)y^r. \tag{9.3.63}$$

Comparing coefficients on both sides of this expression shows that when $p = k - j$ and $r - p = k - i$

$$\sum_{k=-\infty}^{\infty} I_{k-j}(2\nu t)I_{k-i}(-2\nu t) = \begin{cases} 1 : j = i \\ 0 : j \neq i. \end{cases} \tag{9.3.64}$$

Thus (9.3.62) integrates to give

$$G = \text{constant} \times \exp\{-e^{\lambda t} \sum_{i=-\infty}^{\infty} a_i A_i\}.$$

So on paralleling result (9.3.43), it follows that the general solution to equation (9.3.60) is given by

$$G(\mathbf{z};t) = f(\ldots, A_i, \ldots) \exp\{-e^{\lambda t} \sum_{i=-\infty}^{\infty} a_i A_i\} \qquad (9.3.65)$$

for some new arbitrary function $f(\cdot)$. As before, at $t=0$ expression (9.3.38) reduces to $A_i = 1 - z_i$, whence

$$G(\mathbf{z};0) = \prod_{i=-\infty}^{\infty} z_i^{a_i} = \prod_{i=-\infty}^{\infty} (1 - A_i)^{a_i} = f(\ldots, A_i, \ldots) \exp\{\sum_{j=-\infty}^{\infty} -a_j A_j\}. \qquad (9.3.66)$$

Hence on replacing A_i by $-Q_i(2\nu t)$ (see (9.3.46)), we have

$$G(\mathbf{z};t) = f(\ldots, Q_i(2\nu t), \ldots) \exp\{e^{\lambda t} \sum_{j=-\infty}^{\infty} a_j Q_j(2\nu t)\}$$

$$= \prod_{i=-\infty}^{\infty} [1 + Q_i(2\nu t)]^{a_i} \exp\{(e^{\lambda t} - 1) \sum_{j=-\infty}^{\infty} a_j Q_j(2\nu t)\}. \qquad (9.3.67)$$

Now the p.g.f. solution (9.3.67) may be written in the much neater form

$$G(\mathbf{z};t) = \prod_{i=-\infty}^{\infty} [g_i(\mathbf{z};t)]^{a_i}, \qquad (9.3.68)$$

where

$$g_i(\mathbf{z};t) = [1 + Q_i(2\nu t)]^{a_i} \exp\{(e^{\lambda t} - 1) Q_i(2\nu t)\} \qquad (9.3.69)$$

is the probability generating function conditional on the initial population consisting solely of a single member in colony i. Without any loss of generality it is therefore sufficient just to consider $g_0(\mathbf{z};t)$. Moreover, differentiating $G(\mathbf{z};t)$ with respect to z_i and then placing $\mathbf{z} = 1$ yields the first moment

$$\overline{m}_i(t) = e^{(\lambda - \mu - 2\nu)t} \sum_{j=-\infty}^{\infty} a_j I_{i-j}(2\nu t), \qquad (9.3.70)$$

which agrees with the exact result (9.3.19) for $m_i(t)$ (with $\nu_1 = \nu_2 = \nu$). So the approximation leaves the mean values unaltered. However, it does substantially change the variance-covariances, since differentiating $G(\mathbf{z};t)$ a second time (or the c.g.f. $K(\boldsymbol{\theta};t)$ twice) leads to

508 *Spatial processes*

$$\overline{V}_{ii}(t) = e^{(\lambda-\mu-2\nu)t}I_i(2\nu t) - e^{-2(\mu+2\nu)t}I_i^2(2\nu t)$$

$$\overline{V}_{ij}(t) = -e^{-2(\mu+2\nu)t}I_i(2\nu t)I_j(2\nu t) \qquad (i \neq j), \qquad (9.3.71)$$

which are substantially different from the far more involved exact expressions (9.3.27). For large t and $i \neq j$, $\overline{V}_{ij}(t) = o(e^{-2\mu t})$ and $\overline{V}_{ii}(t) = m_i(t) + o(e^{-2\mu t})$. So asymptotically the covariances between the colonies approach zero, with the ratio $\overline{m}_i(t)$ to $\overline{V}_{ii}(t)$ approaching unity.

That the approximate and exact models give rise to different probability structures is to be expected, since replacing birth by time-dependent immigration not only prevents the chance of extinction but it also reduces the inter-dependence between neighbouring colonies. Nevertheless, it is surely better to have at least some kind of probability solution than none at all, especially when it has such a transparent form as (9.3.68)–(9.3.69) and the behavioural effect of the approximation used is fully understood.

9.3.2 Birth–death–migration processes on the finite line

Whilst allowing the birth–death–migration process to develop on the infinite line is certainly mathematically convenient, in practice the number of separate colony locations will be finite. So it is natural to ask what happens when boundary conditions are placed on the process at say $i = M$ and $i = N > M$. Let us first suppose that migration is wholly contained within the region $M \leq i \leq N$; so in the interior $\nu_{i,i+1} = \nu_{i,i-1} = \nu$ ($M < i < N$), at the boundary $\nu_{M,M+1} = \nu_{N,N-1} = \nu$, and $\nu_{ij} = 0$ otherwise. Then the mean equations (9.3.12) for the infinite-colony case now have to be replaced by

(i) $dm_i(t)/dt = (\lambda - \mu - 2\nu)m_i(t) + \nu(m_{i+1}(t) + m_{i-1}(t)) \qquad (M < i < N)$

(ii) $dm_M(t)/dt = (\lambda - \mu - \nu)m_M(t) + \nu m_{M+1}(t)$ \hfill (9.3.72)

(iii) $dm_N(t)/dt = (\lambda - \mu - \nu)m_N(t) + \nu m_{N-1}(t).$

As the boundary equations (ii) and (iii) are anomalous to the interior equations (i), we solve this system by employing the Method of Images. In essence, we regard equation (i) as applying over the full range $-\infty < i < \infty$, and then manipulate the resulting general solution in order to ensure that it satisfies the boundary equations (ii) and (iii). More explicitly, consider the boundary equations

$$m_M(t) = m_{M-1}(t) \qquad \text{and} \qquad m_N(t) = m_{N+1}(t), \qquad (9.3.73)$$

and let the initial conditions be

$$m_i(0) \equiv b_i = \begin{cases} a_i & (M \leq i \leq N) \\ q_i & (i < M, i > N) \end{cases} \qquad (9.3.74)$$

where the q_i are arbitrary and provide the flexibility which enables us to ensure that the solution to (i) for $-\infty < i < \infty$ also satisfies (ii) and (iii). Now define the generating function $H(z;t) \equiv \sum_{i=-\infty}^{\infty} m_i(t)z^i$ over an appropriately chosen domain of z. Then multiplying both sides of (9.3.72i) by z^i and summing over $-\infty < i < \infty$ leads to the equation

$$dH(z;t)/dt = H(z;t)[\xi + \nu(z + z^{-1})] \tag{9.3.75}$$

for $\xi = \lambda - \mu - 2\nu$. This integrates to give

$$H(z;t) = A\exp\{t[\xi + \nu(z + z^{-1})]\} \tag{9.3.76}$$

where A is a constant of integration. Using (9.3.74) we see that $A = H(z;0) = \sum_{i=-\infty}^{\infty} b_i z^i$, and so

$$H(z;t) = (\sum_{i=-\infty}^{\infty} b_i z^i) \exp\{t[\xi + \nu(z + z^{-1})]\}. \tag{9.3.77}$$

Whence on recalling from (9.3.15) that

$$\exp\{\nu t(z + z^{-1})\} = \sum_{k=-\infty}^{\infty} I_k(2\nu t) z^k, \tag{9.3.78}$$

it follows that

$$H(z;t) = e^{\xi t} \sum_{i=-\infty}^{\infty} b_i z^i \sum_{k=-\infty}^{\infty} I_k(2\nu t) z^k. \tag{9.3.79}$$

It now remains to determine the $\{b_i\}$. Extracting the coefficients of z^{M-1}, z^M, z^N and z^{N+1} from expression (9.3.79) gives

$$m_{M-1}(t) = e^{\xi t} \sum_{i=-\infty}^{\infty} b_{i-1} I_{M-i}(2\nu t), \quad m_M(t) = e^{\xi t} \sum_{i=-\infty}^{\infty} b_i I_{M-i}(2\nu t)$$

$$m_{N+1}(t) = e^{\xi t} \sum_{i=-\infty}^{\infty} b_{i+1} I_{N-i}(2\nu t), \quad m_N(t) = e^{\xi t} \sum_{i=-\infty}^{\infty} b_i I_{N-i}(2\nu t). \tag{9.3.80}$$

So on noting from the symmetry of (9.3.78) that $I_{-k}(2\nu t) = I_k(2\nu t)$, equations (9.3.73) become

$$b_M I_0(2\nu t) + \sum_{i=1}^{\infty} (b_{M+i} + b_{M-i}) I_i(2\nu t) = b_{M-1} I_0(2\nu t) + \sum_{i=1}^{\infty} (b_{M+i-1} + b_{M-i-1}) I_r(2\nu t)$$

$$b_N I_0(2\nu t) + \sum_{i=1}^{\infty} (b_{N+i} + b_{N-i}) I_i(2\nu t) = b_{N+1} I_0(2\nu t) + \sum_{i=1}^{\infty} (b_{N+i+1} + b_{N-i+1}) I_i(2\nu t). \tag{9.3.81}$$

Equating coefficients of $I_i(2\nu t)$ on both sides of these equations shows that the b_i are determined by the simple recursive relations $b_{M-i} = b_{M+i+1}$ and $b_{N+i} = b_{N-i+1}$ ($i \geq 0$). For the sequence $\{b_i\}$ to be unique, these two equations must be equivalent, and on recalling (9.3.74) it is easy to show that this implies $q_{2N+1-i} = q_{L+i} = a_i$ with $b_{M-Lp-q} = b_{M-q}$ and $b_{N+LP+q} = b_{N+q}$ ($p, q = 1, 2, \ldots$) where $L = 2(N - M + 1)$. Thus the $\{b_i\}$ are periodic with period L. Now extracting the coefficient of z^i from (9.3.79) gives

510 *Spatial processes*

$$m_i(t) = e^{\xi t} \sum_{j=0}^{\infty} b_j I_{i-j}(2\nu t). \tag{9.3.82}$$

Whence using the above relations in order to write $\{b_i\}$, and hence $\{q_i\}$, in terms of the initial values $\{a_i\}$ yields the solution

$$m_i(t) = e^{(\lambda-\mu-2\nu)t} \sum_{j=M}^{N} a_j \sum_{k=-\infty}^{\infty} [I_{i+kL+j-2N-1}(2\nu t) + I_{i+kL-j}(2\nu t)]. \tag{9.3.83}$$

The unrestricted result (9.3.18) with $\nu_1 = \nu_2 = \nu$ follows as a limiting case of (9.3.83) with $M = -\infty$ and $N = \infty$ (see Renshaw, 1972, for details).

Although the above algebra might look slightly opaque, it becomes transparently clear when we envisage the initial values $\{a_i\}$ as being reflected indefinitely between two parallel mirrors placed at $i = M$ and $i = N$. It is this analogy with the physics of reflected light that gives rise to the procedure being called the Method of Images.

Let us now use the following lemma to obtain a more manageable form for solution (9.3.83).

LEMMA 9.1   *The modified Bessel function summation*

$$2N \sum_{r=-\infty}^{\infty} I_{s+2rN}(x) = e^x + (-1)^s e^{-x} + 2 \sum_{r=1}^{N-1} \cos(\pi rs/N) \exp\{x \cos(\pi r/N)\}. \tag{9.3.84}$$

Proof   Define

$$\theta_s(v) \equiv \sum_{n=-\infty}^{\infty} I_{s+n}(x) v^n, \tag{9.3.85}$$

and let $1, \omega, \ldots, \omega^{2N-1}$ denote the $2N$ roots of unity. Consider

$$\Phi(s) \equiv \sum_{r=0}^{2N-1} \theta_s(\omega^r) = \sum_{r=0}^{2N-1} \sum_{p=-\infty}^{\infty} I_{s+p}(x) \omega^{rp} \equiv \sum_{p=-\infty}^{\infty} c_{s+p}(x) I_{s+p}(x). \tag{9.3.86}$$

Now $c_{s+p} = 1 + \omega^p + \omega^{2p} + \cdots + \omega^{p(2N-1)}$. Hence if $p = 0 \pmod{2N}$, then $c_{s+p} = 2N$ since $\omega^{2qN} = 1$ for every integer q; whilst if $p \neq 0 \pmod{2N}$, then $c_{s+p} = (1 - \omega^{2pN})(1 - \omega^p)^{-1} = 0$ since $\omega^{2pN} = 1$ but $\omega^p \neq 1$. Thus

$$\Phi(s) = 2N \sum_{r=-\infty}^{\infty} I_{s+2rN}(x). \tag{9.3.87}$$

Comparing (9.3.78) and (9.3.85) shows that

$$\theta_s(v) = v^{-s} \exp\{(x/2)(v + v^{-1})\}. \tag{9.3.88}$$

So

$$\Phi(s) = \sum_{r=0}^{2N-1} \theta_s(\omega^r) = \sum_{r=0}^{2N-1} \omega^{-rs} \exp\{(x/2)(\omega^r + \omega^{-r})\}$$

$$= \sum_{r=0}^{2N-1} \omega^{-rs} \exp\{x\cos(\pi r/N)\}$$

$$= e^x + (-1)^s e^{-x} + \sum_{r=1}^{N-1} (\omega^{-rs} + \omega^{rs}) \exp\{x\cos(\pi r/N)\}$$

$$= e^x + (-1)^s e^{-x} + 2\sum_{r=1}^{N-1} \cos(\pi rs/N) \exp\{x\cos(\pi r/N)\}. \qquad (9.3.89)$$

Result (9.3.84) then follows by comparing expressions (9.3.87) and (9.3.89). □

It now remains to substitute result (9.3.84) into the general solution (9.3.83). Without any loss of generality we can take $M = 1$, whence $L = 2N$ and so

$$m_i(t) = e^{(\lambda-\mu-2\nu)t} \sum_{j=1}^{N} a_j \sum_{k=-\infty}^{\infty} [I_{(i+j-1)+2(k-1)N}(2\nu t) + I_{(i-j)+2kN}(2\nu t)]. \qquad (9.3.90)$$

Place $s = i+j-1$ and $r = k-1$ in the first summation, and $s = i-j$ and $r = k$ in the second. Then on denoting $\beta = \pi/N$ we have

$$m_i(t) = (1/2N)e^{(\lambda-\mu-2\nu)t} \times \sum_{j=1}^{N} a_j\{[e^{2\nu t} + (-1)^{i+j-1}e^{-2\nu t}$$

$$+ 2\sum_{r=1}^{N-1} \cos[\beta r(i+j-1)] \exp\{2\nu t \cos(\beta r)\}] \qquad (9.3.91)$$

$$+ [e^{2\nu t} + (-1)^{i-j}e^{-2\nu t} + 2\sum_{r=1}^{N-1} \cos[\beta r(i-j)] \exp\{2\nu t \cos(\beta r)\}]\}.$$

On using the cosine sum rule $\cos(A) + \cos(B) = 2\cos[(A+B)/2]\cos[(A-B)/2]$, for mean initial colony population size $\bar{a} = \sum_{j=1}^{N} a_j$ this expression simplifies to give

$$m_i(t) = \bar{a}e^{(\lambda-\mu)t} + (2/N)e^{(\lambda-\mu-2\nu)t} \sum_{j=1}^{N} a_j \sum_{r=1}^{N-1} \cos[\beta r(i-1/2)] \cos[\beta r(j-1/2)]$$

$$\times \exp\{2\nu t \cos(\beta r)\} \qquad (i = 1, 2, \ldots, N). \qquad (9.3.92)$$

Because $\cos(\beta r) = \cos(\pi r/N)$ decreases as r increases over $r = 1, 2, \ldots, N$, it therefore follows that

$$m_i(t) = \bar{a}e^{(\lambda-\mu)t}[1 + O(\exp\{-2\nu t[1 - \cos(\pi/N)]\})]. \qquad (9.3.93)$$

Thus as t increases, $m_i(t) \sim \bar{a} e^{(\lambda-\mu)t}$ for all i: the speed of approach increases with increasing ν due to individuals migrating across colonies more quickly, and decreases with increasing N due to there being more colonies to spread themselves over.

The effect of end emigration

Since the boundary conditions at the two end colonies may well play a major role in determining how the $m_i(t)$ develop, let us now examine the effect of allowing emigration to occur at rate ν from colonies $i = 1$ and N. Moreover, we shall also demonstrate how the matrix approach (9.3.3) may be applied, since its *modus operandi* is quite different from the generating technique used above. On noting that an emigration out of the system is equivalent to a death, the matrix $\mathbf{L}$ in (9.3.2) takes the form

$$\mathbf{L} = \begin{pmatrix} \xi & \nu & 0 & \cdots & 0 \\ \nu & \xi & \nu & \cdots & 0 \\ 0 & \nu & \xi & \cdots & 0 \\ \vdots & \vdots & \vdots & \cdots & \vdots \\ 0 & 0 & 0 & \cdots & \xi \end{pmatrix} \tag{9.3.94}$$

where, as before, $\xi = \lambda - \mu - 2\nu$. On exploiting the tri-diagonal nature of this matrix by expanding about the top row, it is simple to show that the corresponding characteristic equation $|\mathbf{L} - \omega \mathbf{I}| = 0$ is given by

$$\sin\{(N+1)\theta\}/\sin(\theta) = 0 \quad \text{where} \quad \theta = \cos^{-1}\{(\xi - \omega)/2\nu\}. \tag{9.3.95}$$

Thus for $\sin(\theta) \neq 0$,

$$\sin[(N+1)\cos^{-1}\{(\xi - \omega_m)/2\nu\}] = 0,$$

i.e.

$$(N+1)\cos^{-1}\{(\xi - \omega_m)/2\nu\} = m\pi$$

for $m = 1, \ldots, N$, and so the eigenvalues of $\mathbf{L}$ are

$$\omega_m = \xi - 2\nu \cos\{m\pi/(N+1)\}. \tag{9.3.96}$$

The corresponding column eigenvectors $\mathbf{q}_m = (q_{1m}, \ldots, q_{Nm})^T$ are given by the solutions to the equations

$$\mathbf{L}\mathbf{q}_m = \omega_m \mathbf{q}_m \quad (m = 1, \ldots, N), \tag{9.3.97}$$

i.e. to

$$\xi q_{1m} + \nu q_{2m} = \omega_m q_{1m}$$
$$\nu q_{i-1,m} + \xi q_{im} + \nu q_{i+1,m} = \omega_m q_{im} \quad (i = 2, \ldots, N-1)$$
$$\nu q_{N-1,m} + \xi q_{Nm} = \omega_m q_{Nm}. \tag{9.3.98}$$

This system of linear equations is easily solved sequentially, and, after a little algebra, leads to

$$q_{im} = b_m(-1)^i \sin(im\gamma) \quad (i, m = 1, \ldots, N), \quad (9.3.99)$$

where b_m denotes the normalizing constant and $\gamma = \pi/(N+1)$.

We can avoid having to determine the corresponding row eigenvectors, which are required by the general solution (9.3.3)–(9.3.4), by working instead in terms of the equivalent form

$$\mathbf{m}(t) = \mathbf{m}(0) \sum_{i=1}^{N} e^{\omega_i t} \mathbf{q}_i \mathbf{q}_i^T \quad \text{for} \quad \mathbf{q}_i^T \mathbf{q}_i = 1. \quad (9.3.100)$$

To determine b_m we combine (9.3.99) and the normalizing part of (9.3.100), obtaining

$$b_m^2[\sin^2(m\gamma) + \sin^2(2m\gamma) + \cdots + \sin^2(Nm\gamma)] = 1,$$

which sums to give

$$b_m^2 = 4[(2N+1) - \sin\{(2N+1)m\gamma\}/\sin(m\gamma)]^{-1}. \quad (9.3.101)$$

Whilst it follows from (9.3.99) that the (i,j)th element of the matrix $\mathbf{A}_m = \mathbf{q}_m\mathbf{q}_m^T$ is

$$(\mathbf{A}_m)_{ij} = b_m^2(-1)^{i+j}\sin(im\gamma)\sin(jm\gamma). \quad (9.3.102)$$

Since the process is linear, without any loss of generality we can let the initial population vector $\mathbf{m}(0) = \mathbf{a}$ comprise a single member at colony s, i.e. $a_s = 1$ and $a_j = 0$ ($j \neq s$). In which case

$$m_i(t) = \sum_{m=1}^{N} e^{\omega_m t} (\mathbf{A}_m)_{si} = 4(-1)^{i+s} e^{(\lambda-\mu)t} \sum_{m=1}^{N} \frac{e^{-2\nu t[1+\cos(m\gamma)]}\sin(im\gamma)\sin(sm\gamma)}{(2N+1) - \sin[(2N+1)m\gamma]/\sin(m\gamma)}. \quad (9.3.103)$$

Inspection of the eigenvalues ω_m in (9.3.96) shows that $m = N$ gives the term of maximum order in t, and $m = N - 1$ the second largest. So for large t

$$m_i(t) = e^{(\lambda-\mu-2\nu)t}[(2/N)\sin(i\pi/(N+1))\sin(s\pi/(N+1))\exp\{2\nu t\cos(\pi/(N+1))\} + O(\exp\{2\nu t\cos(2\pi/(N+1))\})] \quad (9.3.104)$$

(see also Crump, 1970). Whence on neglecting the error term we see that:

(i) if i is fixed: then $m_i(t)$ is a minimum when $s = 1$ or N, rising to a maximum with rate $\sin(s\pi/(N+1))$ as s moves towards the centre value $(N+1)/2$; whilst
(ii) if s is fixed: $m_i(t)$ is a minimum when $i = 1$ or N, rising to a maximum with rate $\sin(i\pi/(N+1))$ as i moves towards the centre value $(N+1)/2$.

Thus in direct contrast to the reflecting barrier solution (9.3.93), the effect of including end emigration is to cause each $m_i(t)$ to vary almost sinusoidally both with the position of the initial particle and the state i being considered. This behavioural difference neatly demonstrates that edge effects may permeate right through a large dynamical system.

9.3.3 Basic simulation algorithms

Given that the probability equations for even the simple two-colony birth–death–migration process are intractable to direct solution, the most pragmatic way of gaining an understanding of general multi-colony stochastic stepping-stone dynamics is through simulation. Consider the finite set of colonies $i = M, M+1, \ldots, N$ with general birth, death, right migration and left migration rates $B_i[\mathbf{x}(t)]$, $D_i[\mathbf{x}(t)]$, $U_i[\mathbf{x}(t)]$ and $V_i[\mathbf{x}(t)]$, where $\mathbf{x}(t) = (x_M(t), x_{M+1}(t), \ldots, x_N(t))$ denotes the population size vector at time t. So for the linear birth–death–migration process with no end emigration we have $B_i = \lambda x_i$ ($M \leq i \leq N$), $D_i = \mu x_i$ ($M \leq i \leq N$), $U_i = \nu_1 x_i$ ($M \leq i < N$) and $V_i = \nu_2 x_i$ ($M < i \leq N$); whilst with end emigration we also have the 'death' rates $U_N = \nu_1 x_N$ and $V_M = \nu_2 x_M$.

Simulation of this multi-colony process proceeds in exactly the same way as for the multi-species processes considered earlier; that is, we compare each of the $4(N - M + 1)$ possible birth, death and migration rates in turn against a $U(0,1)$ random number. Thus on writing $R = \sum_{i=M}^{N}(B_i + D_i + U_i + V_i)$, the simplest procedure is as follows.

Exact multi-colony single-species simulation algorithm (A9.1)
- (i) read in parameters
- (ii) set $t = 0$ and $x_i = x_i(0)$ over $M \leq i \leq N$
- (iii) print t and $x_M, \ldots, x_N$
- (iv) generate a new $U \sim U(0,1)$ random number
- (v) evaluate B_i, D_i, U_i, V_i and R and set $test = 0$
- (vi) cycle over $i = M, \ldots, N$
  set $test = test + B_i/R$
  if $U \leq test$ then $x_i = x_i + 1$ and go to (vii)
  set $test = test + D_i/R$
  if $U \leq test$ then $x_i = x_i - 1$ and go to (vii)
  set $test = test + U_i/R$
  if $U \leq test$ then $x_i = x_i - 1$, $x_{i+1} = x_{i+1} + 1$ and go to (vii)
  set $test = test + V_i/R$
  if $U \leq test$ then $x_i = x_i - 1$, $x_{i-1} = x_{i-1} + 1$ and go to (vii)
- (vii) generate a new $U \sim U(0,1)$ random number
- (viii) update t to $t - [\ln(U)]/R$
- (ix) if t is (say) integer, print t and $x_M, \ldots, x_N$
- (x) return to (iv)

This algorithm is surprisingly speedy provided that neither the number of colonies nor the number of cycles is too large. If required, efficiency gains can be made by cycling through colonies in decreasing order of population size, though care must be taken to ensure that the cost of the associated 'housekeeping' does not negate any savings made. Whilst if all colonies to the left of M_0 and right of N_0 are initially empty, then by cycling over $M_0, \ldots, N_0$ and changing $M_0 \to M_0 - 1$ and $N_0 \to N_0 + 1$ when left and right migrations occur at $i = M_0$ and $i = N_0$, respectively, we can avoid testing for events in empty tail colonies. The same procedure allows us to simulate efficiently over the infinite line.

To illustrate this algorithm consider the homogeneous process on the infinite line with $\lambda = 0.2$, $\mu = 0.1$, $\nu_1 = 0.5$ and $\nu_2 = 0.2$. First note that since the process is homogeneous, the total population size follows a simple birth–death process with rates $\lambda = 0.2$ and $\mu = 1.0$. Thus it follows from (2.4.24) that the probability of ultimate extinction $p_0(\infty) = (\mu/\lambda)^{n_0} = 2^{-n_0}$. So if the initial population is of size $n_0 = 2$, then a quarter of simulation runs will eventually become extinct. Whilst from (2.4.2), the deterministic number of individuals alive at time t is given by $m_{total}(t) = n_0 e^{(\lambda - \mu)t} = n_0 e^{0.1t}$. So for $n_0 = 2$ this increases to only 2.72 by $t = 10$, but reaches 22026 by $t = 100$. Moreover, result (9.3.18), i.e.

$$m_i(t) = 2 \times e^{-0.6t}(5/2)^i I_i(2\sqrt{0.1}t), \qquad (9.3.105)$$

shows how the population is expected to spread over the line. In particular, $m_i(t) > 0.5$ over

$i = 1, \ldots, 6$ at $t = 10$; $i = 1, \ldots, 11$ at $t = 20$
$i = 0, \ldots, 31$ at $t = 50$; $i = -3, \ldots, 65$ at $t = 100$.

Thus the population spreads much faster to the right than to the left, in accord with $\nu_1 > \nu_2$. The asymptotic expression (9.3.21) gives similar lower values of i, namely 0, 0, 0 and -3. To simulate the infinite line process up to say $t = 100$, it therefore seems reasonable (as a first try) to let $M = -20$ and $N = 80$. Alternatively, we could initially set $M = -1$ and $N = 1$ and then let $M \to M - 1$ and $N \to N + 1$ whenever $x_M > 0$ and $x_N > 0$, respectively (as outlined above). Figure 9.1 shows a simulation run starting from $x_0 = 2$ that did not lead to extinction, and it clearly takes some time to get going. For by $t = 10$ the population has only shifted slightly to $(x_1, x_2) = (1, 1)$, and even by $t = 20$ it has only moved to $(x_3, x_4) = (1, 2)$. Although the total population has grown to size 103 by $t = 50$, comparison with the deterministic spatial distribution (9.3.105) (scaled down by a factor of 0.5) shows considerable spatial volatility (Figure 9.1a). At $t = 100$, 13363 individuals are now present, and the population is distributed far more smoothly. However, since this particular simulation run takes some time to develop, the total population size at $t = 100$ is considerably smaller than $m_{total}(100) = 22026$, and so the spread over the line is also smaller, being over $5 \leq i \leq 51$ rather than $-3 < i < 65$ expected. Nevertheless, the stochastic and deterministic shapes are clearly compatible even though their overall magnitudes are substantially different.

If the number of colonies, total population size or time interval $(0, T)$ are too large then this exact algorithm may take too long to compute, and in such cases alternative strategies need to be employed. A general approximate approach, that works well if

$$\max_{M \leq i \leq N; 0 \leq t \leq T}\{B_i(t) + D_i(t) + U_i(t) + V_i(t)\}dt \equiv \phi dt \ll 1, \qquad (9.3.106)$$

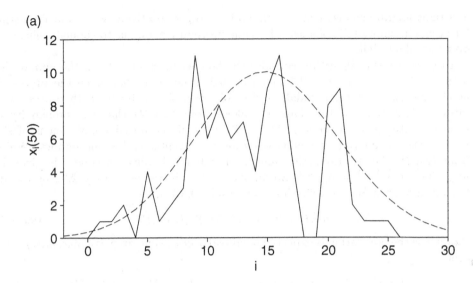
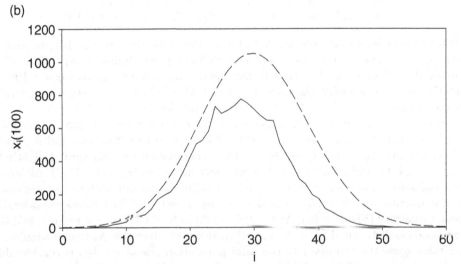

Figure 9.1 A birth–death–migration realization (—) on the infinite line starting at $x_0 = 2$ for $\lambda = 0.2$, $\mu = 0.1$, $\nu_1 = 0.5$ and $\nu_2 = 0.2$ at (a) $t = 50$ and (b) $t = 100$; together with the scaled deterministic values $0.5 \times m_i(t)$ (- - -).

is to extend the approximate algorithm A2.2 for the simple birth–death process by assuming that all the colonies develop independently during each appropriately small time interval $(t, t + dt)$. For then we can cycle through $i = M, \ldots, N$ and check for an event in each colony in turn. Moreover, there is now no need to construct the inter-event times. First choose dt to ensure that $\phi dt < 0.1$ (say) over the time period of interest, and then apply the following procedure.

Approximate multi-colony single-species simulation algorithm (A9.2)
  (i) read in parameters
  (ii) set $t = 0$ and $x_i = x_i(0)$ over $M \leq i \leq N$
  (iii) cycle over $i = M, \ldots, N$
       generate a new $U \sim U(0,1)$ random number
       evaluate B_i, D_i, U_i and V_i using $\{x_i(t)\}$
       if $U \leq B_i dt$ then $x_i = x_i + 1$
       else if $U \leq (B_i + D_i)dt$ then $x_i = x_i - 1$
       else if $U \leq (B_i + D_i + U_i)dt$ then $x_i = x_i - 1$, $x_{i+1} = x_{i+1} + 1$
       else if $U \leq (B_i + D_i + U_i + V_i)dt$ then $x_i = x_i - 1$, $x_{i-1} = x_{i-1} + 1$
       else x_i remains unchanged
  (iv) update $\{x_i\}$ and t to $t + dt$
  (v) if t is (say) integer, print t and $x_M, \ldots, x_N$
  (vi) return to (iii)

The downside is that if most of the $B_1(t) + D_i(t) + U_i(t) + V_i(t) \ll \phi$ then the process can be highly inefficient since many of the $(t, t+dt)$ intervals may not witness a change in state. Although this problem can be partially overcome by making the algorithm more sophisticated, any savings may be negated by the additional cost of the associated computational housekeeping. In practice, pilot runs could be made for both algorithms, and the fastest one chosen.

In some situations the range of colonies occupied, namely $N - M + 1$, may substantially exceed the total number of individuals, and in such cases an individual-based approach may be more computationally efficient than the above colony-based procedures. Typical scenarios include processes with:

(a) $\nu_1, \nu_2 \gg \lambda - \mu$;
(b) non-nearest migration rates $\nu(r)$ from i to $i + r$ for $r = \ldots, -2, -1, 1, 2, \ldots$;
(c) diffusion (see Chapter 6) where for some small $dx \ll 1$ colonies are sited at $x = (\ldots, -dx, 0, dx, 2dx, \ldots)$ rather than at $x = (\ldots, -1, 0, 1, 2, \ldots)$;
(d) marked point processes (discussed later in Section 10.2) on the real line, where x now takes continuous values over $-\infty < x < \infty$.

So let $x(i)$ denote the position of individual i, where we assign labels $i = 1, \ldots, k_0$ at random to the k_0 individuals initially present at $t = 0$, and then label subsequent individuals born thereafter in order of appearance. Now choose an appropriately small $dt \ll 1$ (e.g. $dt = 0.01$). Then to illustrate this procedure, consider nearest-neighbour migration with individual birth, death and migration rates λ, μ and ν_1, ν_2, respectively.

Approximate individual-based simulation algorithm (A9.3)
  (i) read in parameters
  (ii) set $t = 0$, $k = k_0$ and record positions $\{x(i)\}$ for labels $i = 1, \ldots, k_0$
  (iii) set $ind(i) = 1$ for $i = 1, \ldots, k_0$
  (iv) cycle over individuals $i = 1, k$ for $ind(i) = 1$
       generate a new $U \sim U(0,1)$ random number
       if $U \leq \lambda dt$ then $k \to k + 1$ with $x(k) = x(i)$ and $ind(k) = 1$

else if $U \le (\lambda + \mu)dt$ then $ind(i) = 0$
　　　else if $U \le (\lambda + \mu + \nu_1)dt$ then $x(i) \to x(i) + 1$
　　　else if $U \le (\lambda + \mu + \nu_1 + \nu_2)dt$ then $x(i) \to x(i) - 1$
(vi) update t to $t + dt$
(vii) if t is (say) integer, print t and $x(1), \ldots, x(k)$ for $ind(k) = 1$
(viii) return to (iv)

Here a dead individual i is denoted by $ind(i) = 0$, and for long simulation runs it is advisable to sweep through all the individuals $i = 1, \ldots, k$ from time to time, deleting dead ones and then relabelling the alive individuals in order.

These three algorithms may be easily extended to cover more general scenarios, such as processes involving non-homogenous parameters and general migration rates $\nu(r)$, diffusion processes, marked point processes, etc. Moreover, multi-species situations may also be treated in exactly the same way: the underlying simulation principles remain unchanged. For example, suppose we replace the (Bernoulli) nearest-neighbour migration rates ν_1 and ν_2 from i to $i+1$ and $i-1$ by the geometric rates $\nu_1 b_1 (1-b_1)^{r-1}$ and $\nu_2 b_2 (1-b_2)^{s-1}$ from i to $i+r$ and $i-s$, respectively, for $r,s = 1, 2, \ldots$ and $0 < b_1, b_2 < 1$. Values of r and s may be simulated via the inversion technique developed in Section 2.1.4. Specifically, on paralleling result (2.1.40), which shows how to generate Poisson random variables efficiently, first place

$$S_k = \sum_{r=1}^{k} b_1(1-b_1)^{r-1} = 1 - (1-b_1)^k \qquad (k \ge 1).$$

Then for $U \sim U(0,1)$, set $r = k$ if $S_{k-1} \le U < S_k$, i.e. if $(1-b_1)^{k-1} \ge 1 - U > (1-b_1)^k$. On replacing $1-U$ by U this becomes

$$(k-1)\ln(1-b_1) \ge \ln(U) > k\ln(1-b_1)$$

i.e.

$$k - 1 \le \frac{\ln(U)}{\ln(1-b_1)} < k.$$

Thus

$$r = 1 + \left[\frac{\ln(U)}{\ln(1-b_1)}\right], \qquad (9.3.107)$$

where $[y]$ denotes the integer part of y; a similar expression holds for s. So to incorporate this particular model into the exact algorithm A9.1, with $B_i = \lambda x_i$, $D_i = \mu x_i$, $U_i = \nu_1 x_i$ and $V_i = \nu_2 x_i$, we could change parts (v) and (vi) as follows.

Exact algorithm A9.1 with geometric migration rates
(v) evaluate $R = \sum_{i=M}^{N}(\lambda + \mu + \nu_1 + \nu_2)x_i$ and set $test = 0$
(vi) cycle over $i = M, \ldots, N$
　　set $test = test + \lambda x_i / R$
　　if $U \le test$ then $x_i = x_i + 1$ and go to (vii)
　　set $test = test + \mu x_i / R$

if $U \leq test$ then $x_i = x_i - 1$ and go to (vii)
set $test = test + \nu_1 x_i / R$
if $U \leq test$ then generate a new $U = U'$, determine r via (9.3.107) with $U = U'$,
set $x_i = x_i - 1$, $x_{i+r} = x_{i+r} + 1$ and go to (vii)
set $test = test + \nu_2 x_i / R$
if $U \leq test$ then generate a new $U = U''$, determine s via (9.3.107) with b_1 replaced by b_2 and $U = U''$, set $x_i = x_i - 1$, $x_{i-s} = x_{i-s} + 1$ and go to (vii)

Note that after each migration M and N must be updated to ensure that they contain all the non-empty sites.

Figure 9.2 shows the development of a simulated realization of a homogeneous birth–migration process with geometric right migration over $i = 0, 1, 2, \ldots$ starting at $x_0 = 2$. Here $\lambda = 0.1$, $\nu_1 = 1$ and $b_1 = 0.2$, so the migration rates take the form $\nu_1(r) = 0.2(0.8)^{r-1}$ over $r = 1, 2, \ldots$; for ease of illustration we have placed $\mu = \nu_2 = 0$. We see that no births have occurred by $t = 20$ in spite of the total number of deterministic individuals $m_{total}(20) = 2e^2 = 14.58$, though the two initial individuals have migrated from $i = 0$ to $i = 95$ and 120 in rough accord with their 'expected' position $\nu_1 t \times (1 - b_1)/b_1 = 80$. Births have, however, occurred by $t = 40$ and the population is spread out quite thinly over $i = 135$ to 270 with at most two individuals in each location. Thereafter the spatial distribution takes on a slightly right-skew form that becomes increasingly smooth as t increases, apart from the tails which remain sparsely populated.

9.3.4 Tau-leaping and other extensions

The exact algorithm A9.1 has a simple and elegant beauty in that a straightforward generalization enables us to simulate the development of *any* Markov universe in terms of the ordered event-time pairs $\{E_i, t_i\}$. Moreover, even in its most general form the computation involves nothing more than a direct extension of the basic birth–death results (2.4.58) and (2.4.59) used in algorithm A2.1. For on denoting $\{A_{ij}; j = 0, 1, 2, \ldots\}$ to be the set of all possible events $\{E_i\}$ at time t_i, with associated rates r_{ij} and $R_i = \sum_{j=0}^{\infty} r_{ij}$, we see that

$$\Pr(A_{ij}|\text{event in } (t, t+dt)) = r_{ij}/R_i \quad \text{with} \quad t_{i+1} - t_i = -[\ln(U)]/R_i. \quad (9.3.108)$$

Although these results follow as simple consequences of the law of conditional probability and the cumulative exponential distribution, respectively, and so were not specifically named *per se* in early research on stochastic population dynamics, their relevance to complex processes such as chemical reaction systems (in particular) means that they have attracted considerable attention in the applied literature. Indeed, a series of papers evolving from the pioneering work of Gillespie (1976, 1977), together with the later expository treatments of Gillespie and Petzold (2006) and Samad, Khammash, Petzold and Gillespie (2005), has earned the technique the title of the "Gillespie Method". On a historical note, the basic procedure of using a cumulative function to select an event, and a time-scale calculation of the form $1/R$, was published by Young and Elcock (1966). The residence-time algorithm appeared at about the same time in Cox and Miller (1965).

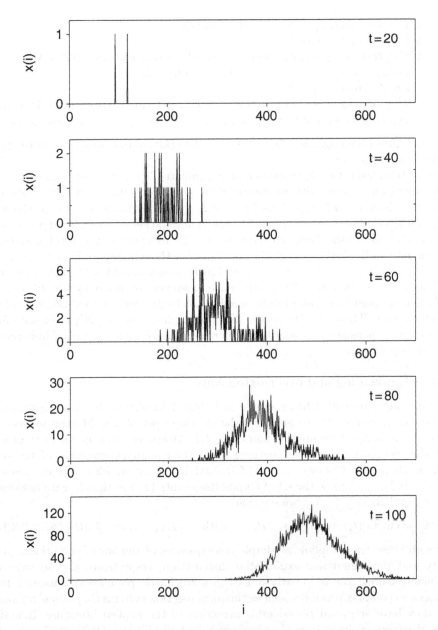

Figure 9.2 A birth-geometric migration realization on $i = 0, 1, 2, \ldots$ starting at $x_0 = 2$, for $\lambda = 0.1$ and migration rates $\nu_1(r) = 0.2(0.8)^r$ from i to $i + r$, shown at times $t = 20, 40, \ldots, 100$.

Although the exact approach A9.1 (and its generalization via (9.3.108)) is both trivial to explain and easy to compute, and hence ideal for constructing realizations of fairly simple population dynamics processes, it is typically too computationally expensive to be of use in more complex systems. It is this aspect that has provided the motivation for generating multi-scale algorithms that maximize speed and efficiency by operating at the most coarse-grained level possible. Before we proceed to discuss these, it is worth returning to the two approximate algorithms A9.2 and A9.3. For in both of these procedures we assume that the time increment dt is small enough to ensure that the probability of more than one event occurring in $(t, t+dt)$ is negligible. Whilst this is perfectly feasible for the individual-based algorithm A9.3, since it is straightforward to ensure that $(\lambda + \mu + \nu_1 + \nu_2)dt < 0.01$ (say), the equivalent condition for the multi-colony algorithm A9.2, namely (9.3.106), requires that $max_{i,t}\{(B_i(t) + D_i(t) + U_i(t) + V_i(t))dt\} = \phi(t)dt \ll 1$ over the time period of interest. Since this function changes value with the population sizes x_i, this means pre-judging the likely range of $\phi(t)$. Unfortunately, if $\phi(t)dt$ is too small then the process is grossly inefficient since most events will simply be 'no change'; whilst too large a value produces error since $Pr(> 1$ event at i in $(t, t+dt))$ is no longer negligible. One way round this would be to change dt at each time step by writing (say) $dt = 0.01/ max_{i,t}\{B_i(t) + D_i(t) + U_i(t) + V_i(t)\}$. Though a better approach is to work with a reasonably small, but fixed, time-increment of size τ, and to admit the possibility of multiple events. For suppose that τ is small enough to ensure that the rates $B_i, \ldots, V_i$ do not change by a significant amount over $(t, t+\tau)$. Then over this time period we may assume that the number of births $b_i \sim$ Poisson$(B_i\tau)$, deaths $d_i \sim$ Poisson$(D_i\tau)$, etc., and that all such events are independent of each other. Whence for each site i, $x_i \to x_i + b_i - d_i - u_i - v_i$ with $x_{i+1} \to x_{i+1} + u_i$ and $x_{i-1} \to x_{i-1} + v_i$. Now recall that Poisson$(\gamma\tau)$ random variables, Y, are easily generated over time $(t, t+\tau)$ through the routine (2.1.40): that is, if $\{Z_0, Z_1, \ldots\}$ denotes a series of $U(0,1)$ pseudo-random variables, then Y is the largest integer n such that $\prod_{j=0}^{n} Z_j \leq e^{-\gamma\tau}$. So for appropriately small τ we have the following Poisson version of A9.2.

Tau–leaping multi-colony single-species simulation algorithm (A9.4)
    (i) read in parameters
   (ii) set $t = 0$ and $x_i = x_i(0)$ over $M \leq i \leq N$
  (iii) cycle over $i = M, \ldots, N$
        evaluate B_i, D_i, U_i and V_i
        generate $Z_0 \sim U(0,1)$: if $Z_0 \leq e^{-B_i\tau}$ then $b_i = 0$
        else generate $Z_1 \sim U(0,1)$: if $Z_0Z_1 \leq e^{-B_i\tau}$ then $b_i = 1$
        else generate $Z_2 \sim U(0,1)$: if $Z_0Z_1Z_2 \leq e^{-B_i\tau}$ then $b_i = 2$,   etc.
        generate $Z_0 \sim U(0,1)$: if $Z_0 \leq e^{-D_i\tau}$ then $d_i = 0$, etc.
        generate $Z_0 \sim U(0,1)$: if $Z_0 \leq e^{-U_i\tau}$ then $u_i = 0$, etc.
        generate $Z_0 \sim U(0,1)$: if $Z_0 \leq e^{-V_i\tau}$ then $v_i = 0$, etc.
  (iv) update x_i to $x_i + b_i - d_i - u_i - v_i$, x_{i+1} to $x_{i+1} + u_i$ and x_{i-1} to $x_{i-1} + v_i$
      over $i = M, \ldots, N$
   (v) update t to $t + \tau$
  (vi) if t is (say) integer, print t and $x_M, \ldots, x_N$
 (vii) return to (iii)

This technique is called *tau-leaping*, and simply involves modifying algorithm A9.2 by replacing the routine used to generate uniform random numbers by an efficient Poisson routine such as (2.1.40). The approach can work well in situations where the transition rates do not change substantially in relative terms over the time intervals $(t, t + \tau)$. Moreover, accuracy can be enhanced by changing the leap time, τ, adaptively, based on the current state of the system and the transition rates.

The exact and approximate approaches, A9.1 and A9.2, were used almost exclusively in early simulation studies in population dynamics, though a parallel (albeit virtually independent) interest in developing computationally accurate and efficient simulation algorithms in the chemical reaction literature (and elsewhere) followed the tau-leaping route. Unfortunately this split between modellers in stochastic processes and applied science was not helped by the tau-leaping approach spawning its own terminology. It is therefore worthwhile noting the correspondence between the chemical reaction and the corresponding applied probability definitions. For example, the Kolmogorov forward differential equation (9.3.8), which describes the evolution of the probability $p(\mathbf{x}; t)$ for each state $\mathbf{x}$ of the system at time t, is called the *chemical master equation*. Whilst the exact simulation algorithm (e.g. our A9.1) is called the *stochastic simulation*, or *Gillespie's, algorithm*. Remember though that there is a substantial paradigm shift between the population dynamics and chemical reaction scenarios. For in the former we have N separate colonies (taking $M = 1$) with birth and death within each site, and migration, infection, etc. across them. Whilst in the latter we have changes to N different types of *molecules* or *chemical species* which take place through one or more types of chemical reactions. For example, suppose we have $N = 4$ types, labelled A, B, C and D, undergoing two types of reaction, namely:

(i) an A and a B molecule combine to create a molecule of C and a molecule of D, i.e. $A + B \rightarrow C + D$; and,
(ii) the reverse reaction in which $C + D \rightarrow A + B$.

Then in population dynamics terms these state changes correspond to $\mathbf{x} = (x_1, x_2, x_3, x_4) \rightarrow$ (i) $(x_1 - 1, x_2 - 1, x_3 + 1, x_4 + 1)$ and (ii) $(x_1 + 1, x_2 + 1, x_3 - 1, x_4 - 1)$. So we have a kind of 'symbiotic migration' from states $(1, 2)$ to $(3, 4)$ and from $(3, 4)$ to $(1, 2)$, respectively, over the state space $\{(N, N, 0, 0), (N - 1, N - 1, 1, 1), \ldots, (0, 0, N, N)\}$.

In spite of their mathematical equivalence (provided we equate 'colony' and 'molecule type') the conceptual distance between these two quite distinct fields of study means that desires and objectives differ quite markedly between them. For example, in molecular systems it may well be the case that the number of molecules of each type i always remains high, and so the probability that $X_i(t) = 0$ is negligible over all $i = 1, 2, \ldots, N$ and $t \geq 0$. Conversely, in population dynamics a key feature of interest is often precisely the opposite, namely that a given site i becomes empty at some time t. So whereas in the former the transition rates should remain relatively constant, in the latter they are likely to fluctuate quite substantially. This degree of rate variability over $(t, t + \tau)$ defines the *stiffness* of the system, and over the years extensive theory has been developed for efficiently solving *stiff* ordinary differential equations. That is, systems of o.d.e.'s which involve processes that occur on vastly different time scales, the fastest of which is stable. However, the practical consequence of stiffness is that, even though a system may itself be stable, naïve simulation techniques may well be unstable unless they progress in extremely small time steps. So applied probabilists wishing to use the tau-leaping method are advised to make extensive study of recent developments in this o.d.e. field before using A9.4 in earnest.

These four algorithms should be sufficient for modelling population dynamics. Nevertheless, it is worthwhile presenting a brief summary of other approaches that have been inspired through the study of chemical reaction systems. If $\{E_1(\mathbf{X}(t)), E_2(\mathbf{X}(t)), \ldots\}$ denotes the set of all possible event-types based on $\mathbf{X}(t)$, with associated transition rates $\{r_1(\mathbf{X}(t)), r_2(\mathbf{X}(t)), \ldots\}$, then the tau-leaping method involves sequentially computing

$$\mathbf{X}(t+\tau) = \mathbf{X}(t) + \sum_j \text{Poisson}(r_j(\mathbf{X}(t))\tau). \qquad (9.3.109)$$

Now if τ can be chosen large enough to ensure that every possible event is expected to occur many times during $(t, t+\tau)$, i.e. $E[r_j(\mathbf{X}(t))\tau]$ is large for all $j = 1, 2, \ldots$, without violating the condition that none of the $r_j(\mathbf{X}(t))$ change substantially over $(t, t+\tau)$, then we may replace the Poisson variables by Normal ones with the same mean and variance. So on denoting $\{Z_j\}$ to be a series of independent $N(0, 1)$ random variables, we may replace the recurrence relations (9.3.109) by the *Langevin-leaping formula* (algorithm A9.5)

$$\mathbf{X}(t+\tau) = \mathbf{X}(t) + \tau \sum_j r_j(\mathbf{X}(t)) + \sqrt{\tau} \sum_j \sqrt{r_j(\mathbf{X}(t))} Z_j. \qquad (9.3.110)$$

Provided a computationally efficient routine is used to generate the $\{Z_j\}$, this switch in distribution may speed up the tau-leaping algorithm A9.4 quite substantially. Though note that because the integer-valued Poisson valued random variables have been replaced by real-valued Normal ones, the state space has been fundamentally changed.

The next stage is to take the reverse limit $\tau \to dt$, whence (9.3.110) represents the discretization of the basic stochastic differential equation (see Section 6.2.5)

$$d\mathbf{X}(t) = \sum_j r_j(\mathbf{X(t)})dt + \sum_j \sqrt{r_j(\mathbf{X}(t))} dW_j(t) \qquad (9.3.111)$$

where $\{dW_j\}$ are independent Brownian motions. The system (9.3.111) is called the *chemical Langevin equation* (algorithm A9.6). Finally, ignoring the stochastic part of (9.3.111) yields the set of ordinary differential *reaction rate equations* (algorithm A9.7)

$$d\mathbf{X}(t)/dt = \sum_j r_j(\mathbf{X}(t)), \qquad (9.3.112)$$

which is, of course, the classic deterministic representation. Though these later stages, from tau-leaping onwards, do not usually feature in studies of population dynamics (apart from the purely deterministic approach A9.7), it would be interesting to

9.4 Velocities of propagation

The spatial propagation demonstrated in Figure 9.2 can be summarized in a variety of different ways. Obvious choices include: (a) measures of central location, such as the mean, mode or median values of $\{X_i(t)\}$; and, (b) the total range of spread, defined by the positions of the left-most (i.e. trailing) and right-most (i.e. leading) individuals. For example, we have

t:	0	20	40	60	80	100
left:	0	95	135	186	250	316
mean:	0	106.5	192.7	291.6	389.1	490.7
right:	0	120	270	426	555	687

and it is apparent that the velocities of propagation of these three features differ quite markedly. Crude estimates at $t=100$ are 3.2, 4.9 and 6.9, respectively, and as t rises the wave clearly increases both in length and right-skewness. Since we have already noted that the presence of the birth rate $\lambda > 0$ renders the Kolmogorov equations for the probabilities $p(\mathbf{x}; t) = \Pr\{X_i(t) = x_i | X_i(0) = a_i\}$ intractable to direct solution, there is negligible hope of developing theoretical results for the leading and trailing stochastic wavefronts. Though one way forward might be to progress the development of limit theorems for the position of the furthermost particle in an exponentially growing number of *independent* random walks, by relating these to known results for branching random walks (e.g. Durrett, 1979). However, as this route is likely to be extremely challenging, here we shall simply use deterministic results for the $\{m_i(t)\}$ to develop theoretical velocities, and simulation to examine stochastic velocities.

The key issues which underpin the former are easily illustrated by considering the one-way nearest-neighbour case with $\nu_2 = 0$. For with $a_0 = 1$ and $a_i = 0$ for $i > 0$, the deterministic equations (9.3.12) reduce to

$$dm_0(t)/dt = (\lambda - \mu - \nu_1)m_0(t)$$
$$dm_i(t)/dt = (\lambda - \mu - \nu_1)m_i(t) + \nu_1 m_{i-1}(t) \quad (i > 0). \quad (9.4.1)$$

These are trivial to solve sequentially and have the solution

$$m_i(t) = e^{(\lambda-\mu-\nu_1)t}(\nu_1 t)^i/i! \quad (i = 0, 1, 2, \ldots). \quad (9.4.2)$$

Thus

$$m_i(t) = (\text{total mean pop. size}) \times \text{Poisson}(\nu_1 t). \quad (9.4.3)$$

So for $\lambda > \mu$ this exponentially growing population spreads out over $i = 0, 1, 2, \ldots$ with the expected number of individuals at each colony being proportional to a Poisson distribution. This process may therefore be envisaged as generating a population wave which expands over the integer axis. We may define the position, i^*, of the resulting *wavefront* at time t through the condition

$$m_{i^*}(t) = K, \qquad (9.4.4)$$

where K is some pre-assigned constant (e.g. $K = 1$). Two issues to resolve therefore concern the way in which i^* depends on t, and the associated velocity of spread.

First note that since $m_i(t)$ reaches its maximum value for given t at $i = \tilde{i}$ when $m_{\tilde{i}}(t) \simeq m_{\tilde{i}+1}(t)$, it follows from (9.4.2) that for large t the (asymptotic) velocity of the mode is

$$c_{mode} = \tilde{i}/t \simeq \nu_1. \qquad (9.4.5)$$

Second, suppose we define the positions of the left and right wavefronts, R^- and R^+, at time $t > 0$ to be the two solutions of the equation $m_R(t) = K = 1$; that is, where we expect to find exactly one individual at time t. Then on replacing $R!$ in (9.4.2) by Stirling's formula, we have

$$1 = m_R(t) \simeq e^{(\lambda-\mu-\nu_1)t} \frac{(\nu_1 t)^R}{(R)!} \sim e^{(\lambda-\mu-\nu_1)t} \frac{(\nu_1 t)^R}{\sqrt{2\pi} e^{-R} R^{R+1/2}}. \qquad (9.4.6)$$

Whence taking logarithms yields

$$0 \simeq (\lambda - \mu - \nu_1)t - R\ln(R/\nu_1 t) + R - \ln(\sqrt{2\pi}) - (1/2)\ln(R). \qquad (9.4.7)$$

Now suppose that R increases with constant velocity c, so that $R = ct$. Then on dividing both sides of (9.4.7) by t and letting t become large, we have, to $O(\ln(t)/t)$,

$$\mu + \nu_1 - \lambda = c - c\ln(c/\nu_1). \qquad (9.4.8)$$

The required *asymptotic* velocities c^- and c^+, corresponding to the wavefront positions R^- and R^+, are therefore the roots of equation (9.4.8).

To illustrate this approach consider simulating a one-sided pure birth version of Figure 9.1 with $\lambda = 0.1$, $\mu = \nu_2 = 0$ and $\nu_1 = 1$ starting from $X_0 = 0$. Then for one run the stochastic wavefront positions $R^-(t)$ and $R^+(t)$ at time t, and their associated velocities $c^-(t) = R^-(t)/t$ and $c^+(t) = R^+(t)/t$, took the values

t	$R^-(t)$	$R^+(t)$	$c^-(t)$	$c^+(t)$
0	0	0		
50	35	66	0.70	1.32
100	66	134	0.66	1.34
150	97	208	0.65	1.39

Thus here $c^-(t) \downarrow c^- = 0.588$ and $c^+(t) \uparrow c^+ = 1.479$. Strictly speaking, as the deterministic velocity $c(t)$ depends on t, it is given by $dR(t)/dt$ which is formed by differentiating (9.4.7). However, since the stochastic wavefronts are likely to exhibit substantial variability around the deterministic ones there seems to be little point in progressing this more precise theoretical route. Moreover, numerical values of $c(t) = R(t)/t$ can be easily determined by solving (9.4.7) (via *Maple*, for example), so it is fairly simple to assess the speed of convergence of $c^-(t)$ and $c^+(t)$ to their asymptotic values c^- and c^+. Note that the *instantaneous* stochastic velocities, based on the right-most and left-most individuals present, will not converge as $t \to \infty$ since they depend only on local migration events at the two wavefronts, and hence on the behaviour of a small number of individuals.

9.4.1 Wave profiles for two-way migration

With one-way migration (here $\nu_2 = 0$) the process is not only constrained to lie on the non-negative integers $i = 0, 1, 2, \ldots$, but the existence of a proper left wavefront depends on the relative size of ν_1. For (9.4.2) shows that $m_0(t) = e^{(\lambda - \mu - \nu_1)t} \to 0$ if and only if $\lambda < \mu + \nu_1$, in which case colonies $i = 0, 1, 2, \ldots$ successively empty (i.e. have $m_i(t) < K$) and so $c^-(t) > 0$.

In contrast, if $\nu_2 > 0$ then the wave develops over $i = 0, \pm 1, \pm 2, \ldots$, and the velocities $c^- < c^+$ can be either positive or negative. To investigate their properties (Renshaw, 1977, 1979), we first place $i = ct$ in (9.3.19) to obtain

$$m_{ct}(t) = e^{(\lambda - \mu - \nu_1 - \nu_2)t}(\nu_1/\nu_2)^{ct/2} I_{ct}(2t\sqrt{(\nu_1 \nu_2)}). \tag{9.4.9}$$

Now for large x

$$I_x(xy) \sim (2\pi x)^{-1/2}(1 + y^2)^{-1/4} e^{\eta x} \tag{9.4.10}$$

(result (9.7.7)–(9.7.11) of Abramowitz and Stegun, 1970) where

$$\eta = (1 + y^2)^{1/2} + \ln\{y/[1 + (1 + y^2)^{1/2}]\}. \tag{9.4.11}$$

Whence on writing $x = ct$ and $y = 2\sqrt{(\nu_1 \nu_2)}/c$, expression (9.4.9) takes the asymptotic form

$$m_{ct}(t) \sim (2\pi ct)^{-1/2}(1 + 4\nu_1\nu_2/c^2)^{-1/4}(\nu_1/\nu_2)^{ct/2} e^{\eta ct} e^{(\lambda - \mu - \nu_1 - \nu_2)t} \tag{9.4.12}$$

where

$$\eta = (1 + 4\nu_1\nu_2/c^2)^{1/2} + \ln\{2(\nu_1\nu_2)^{1/2}/[c + (c^2 + 4\nu_1\nu_2)^{1/2}]\}. \tag{9.4.13}$$

Putting $m_{ct} = K$ and taking logarithms on both sides of (9.4.12) then yields for large t

$$c\eta + (\lambda - \mu - \nu_1 - \nu_2) + (c/2)\ln(\nu_1/\nu_2) \sim (1/2t)\ln(t). \tag{9.4.14}$$

Hence on letting $t \to \infty$ it follows that the required asymptotic velocities c^- and c^+ are the roots of the equation

$$\nu_1 + \nu_2 + \mu - \lambda = (c^2 + 4\nu_1\nu_2)^{1/2} - c\ln\{[c + (c^2 + 4\nu_1\nu_2)^{1/2}]/(2\nu_1)\}. \tag{9.4.15}$$

Equation (9.4.15) may also be derived directly from the saddlepoint approximation developed earlier in Section 3.7.

Suppose $\lambda > \mu$ so that the total mean population size is increasing. Then the left-hand side of (9.4.15) gives $\nu_1 + \nu_2 + \mu - \lambda < \nu_1 + \nu_2$. In order to examine the right-

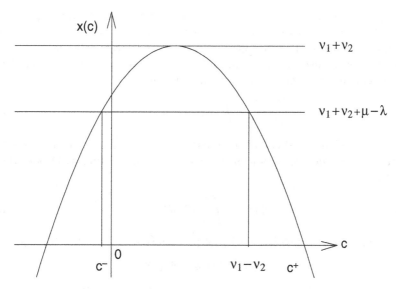

Figure 9.3 Sketch of the function $x(c)$ against c.

hand side put

$$x(c) = (c^2 + 4\nu_1\nu_2)^{1/2} - c\ln\{[c + (c^2 + 4\nu_1\nu_2)^{1/2}]/(2\nu_1)\}. \tag{9.4.16}$$

Now as c tends towards either $+\infty$ or $-\infty$ the function $x(c)$ tends to $-\infty$, whilst $x(0) = 2(\nu_1\nu_2)^{1/2} > 0$. Also, $x(c)$ reaches a maximum value at $c = \nu_1 - \nu_2$, namely $x(\nu_1 - \nu_2) = \nu_1 + \nu_2 > \nu_1 + \nu_2 + \mu - \lambda$. Hence we have the situation shown in Figure 9.3. First, equation (9.4.15) clearly has two real roots c^- and c^+ which correspond to the velocities of the left and right wavefronts, respectively. Second, if $\nu_1 > \nu_2$, so that there is a drift to the right, then $c^- = 0$ when $x(0) = 2(\nu_1\nu_2)^{1/2} = \nu_1 + \nu_2 + \mu - \lambda$, i.e. when

$$\lambda - \mu = (\nu_1^{1/2} - \nu_2^{1/2})^2. \tag{9.4.17}$$

Third, if $\nu_1 > \nu_2$ then the left-hand boundary moves asymptotically towards $+\infty$ or $-\infty$ depending on whether $\lambda - \mu < (\nu_1^{1/2} - \nu_2^{1/2})^2$ or $\lambda - \mu > (\nu_1^{1/2} - \nu_2^{1/2})^2$, respectively. If $\nu_1 < \nu_2$ then expression (9.4.17) still holds, but these inequalities, which now refer to the right-hand boundary, are reversed.

A possible objection to this approach is that combining (9.4.15) and (9.4.12) leads to

$$m_{ct}(t) \sim (2\pi ct)^{-1/2}(1 + 4\nu_1\nu_2/c^2)^{-1/4}, \tag{9.4.18}$$

which appears to contradict condition (9.4.4). However, it turns out that (9.4.4) is far more stringent than is really necessary. For provided we employ a condition of the form

$$m_{i*}(t) = Kt^\alpha \tag{9.4.19}$$

for some constants $K > 0$ and α in (9.4.12), then we obtain the same asymptotic velocities c^- and c^+.

Daniels (1975) presents a more general viewpoint by considering a spatial version of the simple epidemic process discussed in Section 8.3.1, in which a population of susceptibles is distributed with constant density along the real line $(-\infty, \infty)$. For when the number of susceptibles is unlimited, the equation for $Y(s;t)$, the density of infected individuals at point s at time t, takes the form

$$\frac{\partial Y(s;t)}{\partial t} = \int_{-\infty}^{\infty} Y(s-u;t)\,dF(u). \tag{9.4.20}$$

The function $dF(u)$ is called the *contact distribution*, and relates to the probability of a susceptible being infected by an infective at a distance u from it. Denote the m.g.f. of $dF(u)$ by

$$\Psi(\theta) = \int_{-\infty}^{\infty} e^{\theta u}\,dF(u). \tag{9.4.21}$$

Then the minimum asymptotic velocity of propagation, c_0, satisfies the equations

$$c_0 = \Psi(\theta)/\theta = \Psi'(\theta). \tag{9.4.22}$$

Now on comparing the discrete state-space equation (9.3.12) with the continuous state-space analogue (9.4.20), we see that these two equations are equivalent provided that

$$dF(u) = (\lambda - \mu - \nu_1 - \nu_2)\delta(u) + \nu_1\delta(u-1) + \nu_2\delta(u+1), \tag{9.4.23}$$

where $\delta(x)$ denotes the Dirac delta function. Substituting (9.4.23) into (9.4.21) yields

$$\Psi(\theta) = \lambda - \mu - \nu_1 - \nu_2 + \nu_1 e^{\theta} + \nu_2 e^{-\theta}, \tag{9.4.24}$$

whence equations (9.4.22) become

$$c_0 = \nu_1 e^{\theta} - \nu_2 e^{-\theta} \tag{9.4.25}$$

$$c_0\theta = \lambda + \mu + \nu_1(e^{\theta} - 1) + \nu_2(e^{-\theta} - 1). \tag{9.4.26}$$

Equation (9.4.25) solves to give

$$e^{\theta} = (1/2\nu_1)[c_0 + (c_0^2 + 4\nu_1\nu_2)^{1/2}], \tag{9.4.27}$$

and substituting (9.4.27) into (9.4.26) recovers equation (9.4.15). Thus c and c_0 are identical.

Using the general solution (9.4.22) clearly carries a huge advantage over the development of specific solutions such as (9.4.15). For in this case we have had to use the modified Bessel function results (9.3.15) and (9.4.10); in general such forms would not be available to us.

Continuous-state space solution

At this point it is worth developing the equivalent diffusion approximation. First note that the spatially homogeneous version of the general discrete state space model (9.3.1) has migration rates $\nu_{i,i+j} = h_j$ (say) for all $-\infty < i, j < \infty$. Denote $\nu = \sum_{j=-\infty}^{\infty} h_j$. Thus our specific nearest-neighbour migration process has $h_1 = \nu_1$, $h_{-1} = \nu_2$ with

$h_j = 0$ otherwise, and $\nu = \nu_1 + \nu_2$. The analogous continuous state space process has migration rate $\nu(y)$ from x to $x + y$, with the corresponding equation to (9.3.1) for continuous population density $f(x;t)$ being

$$\frac{\partial f(x;t)}{\partial t} = (\lambda - \mu - \nu) f(x;t) + \int_{-\infty}^{\infty} \nu(y) f(x - y; t)\, dy, \qquad (9.4.28)$$

where

$$\nu = \int_{-\infty}^{\infty} \nu(y)\, dy$$

is the overall rate of migration. On denoting the Fourier transforms of $f(x;t)$ and $\nu(y)$ by

$$f^*(\theta; t) = \int_{-\infty}^{\infty} e^{-\theta x} f(x; t)\, dx \quad \text{and} \quad \nu^*(\theta) = \int_{-\infty}^{\infty} e^{-\theta y} \nu(y)\, dy, \qquad (9.4.29)$$

equation (9.4.28) takes the form

$$\partial f^*(\theta; t)/\partial t = (\lambda - \mu - \nu) f^*(\theta; t) + \nu^*(\theta) f^*(\theta; t) \qquad (9.4.30)$$

which integrates to give

$$f^*(\theta; t) = e^{(\lambda - \mu - \nu)t} f^*(\theta; 0) \exp\{\nu^*(\theta) t\}. \qquad (9.4.31)$$

So provided we can invert $\exp\{\nu^*(\theta)t\}$ we can derive the population density $f(x;t)$.

In general this will not, of course, be possible, but fortunately the solution is amenable to analysis when the migration rates take the Normal form

$$\nu(y) = \nu e^{-(y-\rho)^2/2\sigma^2} / \sqrt{2\pi\sigma^2}, \qquad (9.4.32)$$

which has Fourier transform

$$\nu^*(\theta) = \nu e^{-\rho\theta + \theta^2\sigma^2/2}. \qquad (9.4.33)$$

Moreover, if we assume that the initial population at $t = 0$ comprises a single individual at $x = 0$, i.e. $f(x; 0) = \delta(x)$, then $f^*(\theta; 0) = 1$. Whence (9.4.31) becomes

$$f^*(\theta; t) = e^{(\lambda - \mu - \nu)t} \exp\{\nu t e^{-\rho\theta + \theta^2\sigma^2/2}\} = e^{(\lambda - \mu - \nu)t} \sum_{j=0}^{\infty} \frac{(\nu t)^j}{j!} e^{-j[\rho\theta - \theta^2\sigma^2/2]},$$

$$(9.4.34)$$

which inverts to give

$$f(x; t) = e^{(\lambda - \mu - \nu)t} \left[\delta(x) + \sum_{j=1}^{\infty} \frac{(\nu t)^j}{j!} \times \frac{e^{-(x - \rho j)^2/2j\sigma^2}}{\sqrt{2\pi j\sigma^2}} \right] \quad (-\infty < x < \infty). \quad (9.4.35)$$

Thus the population density is an infinite sum of Poisson-weighted Normal densities.

A simple approximation to this solution can be easily obtained by considering the first-order expansion of expression (9.4.34), namely

$$f^*(\theta; t) = e^{(\lambda-\mu-\nu)t} \exp\{\nu t e^{-(\rho\theta+\theta^2\sigma^2/2)}\}$$
$$\simeq e^{(\lambda-\mu)t} \exp\{\nu t(-\rho\theta + \theta^2\sigma^2/2)\}. \tag{9.4.36}$$

For this inverts to give

$$f(x; t) \simeq e^{(\lambda-\mu)t} \frac{e^{-(x-\rho\nu t)^2/(2\nu t \sigma^2)}}{\sqrt{2\pi\nu t \sigma^2}}, \tag{9.4.37}$$

which being of Normal form constitutes a basic diffusion approximation. Whence placing $f(ct; t) \equiv K$, denoting the associated velocities by c_d^+ and c_d^-, and letting $t \to \infty$, shows that

$$c_d^+, c_d^- = \rho\nu \pm \sqrt{2\nu(\lambda-\mu)\sigma^2}. \tag{9.4.38}$$

So for $\lambda - \mu > 0$ the left wavefront (for example) moves towards $-\infty$ if $\lambda - \mu > (\nu\rho^2)/(2\sigma^2)$, and towards $+\infty$ otherwise.

To see whether the stepping-stone velocities c^+, c^- are compatible with the diffusion velocities c_d^+, c_d^-, let us equate ρdt and $\sigma^2 dt$ to the mean and variance of the distance travelled by an individual having the migration probabilities

$$\Pr(\text{at } j \text{ at } t+dt| \text{ at } i \text{ at } t) = \begin{cases} (\nu_1/\nu)dt & \text{if } j = 1 \\ (\nu_2/\nu)dt & \text{if } j = -1, \end{cases} \tag{9.4.39}$$

where $\nu = \nu_1 + \nu_2$. Then $\rho dt = (\nu_1 - \nu_2)dt/\nu$ and $\sigma^2 dt = (\nu_1 + \nu_2)dt/\nu = dt$, whence the velocities (9.4.38) become

$$c_d^+, c_d^- = (\nu_1 - \nu_2) \pm \sqrt{2(\nu_1 + \nu_2)(\lambda - \mu)}. \tag{9.4.40}$$

Whilst on denoting

$$\lambda - \mu = \epsilon\nu \quad \text{and} \quad \delta = 1 - [c + (c^2 + 4\nu_1\nu_2)^{1/2}]/(2\nu_1), \tag{9.4.41}$$

we may write (9.4.15) as

$$\nu(1 - \epsilon) = (c^2 + 4\nu_1\nu_2)^{1/2} - c\ln(1 - \delta) \tag{9.4.42}$$

(note that $1 - \delta > 0$). Since inverting the expression for δ in (9.4.41) gives $c = \nu_1(1 - \delta) - \nu_2(1 - \delta)^{-1}$, expression (9.4.42) takes the form

$$\nu(1 - \epsilon)(1 - \delta) = [\nu_1(1 - \delta)^2 + \nu_2] - [\nu_1(1 - \delta)^2 - \nu_2]\ln(1 - \delta). \tag{9.4.43}$$

If we now consider ϵ and δ to be 'small' and disregard terms with power greater than ϵ and δ^2, then (9.4.43) reduces to

$$\delta = \pm(2\epsilon)^{1/2}. \tag{9.4.44}$$

Whence a little manipulation recovers (9.4.40), i.e. $c \simeq \nu_1 - \nu_2 \pm \{2\nu(\lambda - \mu)\}^{1/2}$.

Thus provided δ and ϵ are sufficiently small, the velocities derived from the stepping-stone and diffusion models are in close agreement. However, we see from (9.4.41) and

(9.4.44) that this implies $0 < \lambda - \mu \ll \nu$. Thus unless the net growth rate is small in comparison with the overall migration rate, 'similar' migration structures can give rise to substantially different velocities in the discrete and continuous cases. This reinforces what we have seen earlier, namely that care needs to be exercised when taking continuous approximations to spatial-temporal processes in order to obtain simpler forms of solution.

An example with boundary effects

A good illustration of a successful modelling scenario is provided by the development, and subsequent application, of a stepping-stone process with boundary effects. Neyman, Park and Scott (1956) describe an experiment to determine the spatial distribution of 2257 adult flour beetles (*Tribolium confusum*) placed in a one-to-one sex ratio within a $10 \times 10 \times 10$-inch cubic container filled with fresh flour; this was kept in a dark incubator at a constant temperature and humidity, and periodically rotated. After four months, at which time the population totalled 73009 individuals excluding eggs, the contents of each of the 1000 individual $1 \times 1 \times 1$-inch cubes were lifted and examined. Both sexes showed a gradual increase in density towards the edges and along the edges towards the corners (Table 9.1 of Renshaw, 1991, shows the 10×10 array from the second layer). Renshaw (1980) develops a discrete space–time stepping-stone approach through a multi-dimensional discrete-time inhomogeneous version of Bailey's deterministic equations (9.3.12), namely (in the two-dimensional case)

$$m_{ij}(t+1) = \begin{cases} (1+\lambda-\mu-4\nu)m_{ij}(t) + \nu H_{ij}(t) & \text{(interior)} \\ (1+\lambda'-\mu'-3\nu)m_{ij}(t) + \nu H_{ij}(t) & \text{(sides)} \\ (1+\lambda''-\mu''-2\nu)m_{ij}(t) + \nu H_{ij}(t) & \text{(corners)}. \end{cases} \quad (9.4.45)$$

Here $m_{ij}(t)$ denotes the population size at sites $(i,j = 0, \ldots, n)$ at the discrete times $t = 0, 1, 2, \ldots$; the birth and death rates are λ and μ in the interior, λ' and μ' on the sides, and λ'' and μ'' in the corners. Whilst

$$H_{ij}(t) = m_{i-1,j}(t) + m_{i,j-1}(t) + m_{i+1,j}(t) + m_{i,j+1}(t),$$

where $m_{ij}(t) \equiv 0$ for any sites (i,j) lying outside the square lattice $i, j = 0, 1, \ldots, n$. On assuming that the $\{m_{ij}(t)\}$ eventually all grow at the same rate ω, so that for large t

$$m_{ij}(t+1) = \omega m_{ij}(t), \quad (9.4.46)$$

the interior solution (Renshaw, 1980) takes the form

$$m_{ij}(t) \sim K(t) \cosh[(n-2i)(\theta/2)] \cosh[(n-2j)(\theta/2)] \quad (9.4.47)$$

for some scaling factor $K(t)$, where

$$\cosh(\theta) = (\omega - 1 - \lambda + \mu + 4\nu)/(4\nu). \quad (9.4.48)$$

The side and corner equations in (9.4.45) are equivalent to the conditions

$$(\lambda'' - \mu'') - (\lambda - \mu) = 2[(\lambda' - \mu') - (\lambda - \mu)] \quad (9.4.49)$$

$$= 2\nu\{2\cosh(\theta) - 1 - \cosh[(n-2)(\theta/2)]/\cosh(n\theta/2)\}, \quad (9.4.50)$$

i.e. the excess growth activity at the corners is twice that at the sides. Note that the two-dimensional catenary solution (9.4.47) extends directly to three dimensions. Not only is this representation highly amenable to parameter estimation, but both the resulting deterministic and simulated stochastic solutions provide a reasonably good fit to the data and highlight the striking difference in spatial behaviour between the two sexes.

A possible point of contention is that since the data relate to 10×10 one-inch squares we have chosen $i, j = 0, 1, \ldots, 9$ even though the distance travelled by a beetle at each migration event will almost certainly be much less than one inch. Fortunately, it may be shown that dividing each square into $h \times h$ sub-squares, and allowing migration to occur only between adjacent sub-squares, does not affect the total (deterministic) number in each one-inch square. Thus the distribution (9.4.47) is independent of the chosen scale of migration. One important consequence of this equivalence is that it holds true even when $h \to \infty$ and the process turns into Brownian motion. So in the steady state, the stepping-stone and Brownian motion models yield the same population sizes. However, this equivalence is a purely deterministic one and it does not follow that the corresponding stochastic motions are compatible. Simulation experiments would have to be performed to select the best empirical values of h and ν in any given situation.

Nevertheless, the basic diffusion approach can provide a useful way of gaining a qualitative feel for the density and associated velocity structure of specific processes. Skellam (1951), for instance, uses a bivariate Normal distribution in pioneering research into the dispersal of oaks, wingless ground beetles and muskrats. The latter study provides an excellent illustration of his classic diffusion approach and relates to data from Elton (1958) on the invasion of muskrats through central Europe from 1905 to 1927 (see Ulbrich, 1930; Matis and Kiffe, 2000). Study has also been made of their spread through western and northern Europe between 1930 and 1960 (van den Bosch, Metz and Dickmann, 1990; van den Bosch, Hengeveld and Metz, 1992). In contrast, Broadbent and Kendall (1953) consider a fixed population size, but let each individual cease moving after an exponentially distributed time. The resulting convolution of this random stop time with the diffusion approximation produces a distribution that is closely related to the modified Bessel function of the second kind, $K_\nu(z)$ (Watson, 1952), a form that repeatedly appears in situations in physics.

The attraction of the stepping-stone approach is that it generates a model that lies fairly close to reality. Although other continuous space approaches can be developed which mimic the broad spatial characteristics of the *Tribolium* data, they rather stretch the imagination. Sherman (1956), for example, investigates a one-dimensional random walk scenario, and shows that distributions roughly like the one observed can be obtained by presuming that once a beetle strikes the boundary it remains there a randomly distributed time and then instantaneously jumps a finite distance within the region of motion. Whilst Cox and Smith (1957) adopt a completely different approach for motion within a circular region by assuming that: (i) within the region a beetle follows a straight line path; (ii) at the boundary it returns along the original path with probability p; or, (iii) with probability $1 - p$ it moves along the boundary before choosing a new direction of motion. They tried to extend their results to the case of

The saddlepoint profile

Use of the saddlepoint approximation enables us to construct a general form for the wave profile. For on recalling results (3.7.14) and (3.7.15) (with $n = 1$), we see that in terms of our current notation the population density

$$Y(s,t) \sim \frac{\exp\{t\Psi(\theta) - s\theta\}}{\{2\pi t\Psi''(\theta)\}^{1/2}}[1 + O(t^{-1})] \qquad (9.4.51)$$

where

$$t\Psi'(\theta) = s. \qquad (9.4.52)$$

Thus for the nearest-neighbour migration process, substituting for $\Psi(\theta)$ from (9.4.24) into (9.4.52) yields

$$t[\nu_1 e^\theta - \nu_2 e^{-\theta}] = s,$$

which solves to give

$$e^\theta = [s \pm \sqrt{(s^2 + 4\nu_1\nu_2 t^2)}]/(2\nu_1 t). \qquad (9.4.53)$$

Since $e^\theta > 0$ (for real θ) we discard the negative root, whence (9.4.51) becomes

$$Y(s,t) \sim e^{(\lambda-\mu-\nu)t}\exp\{\sqrt{(s^2 + 4\nu_1\nu_2 t^2)}\}\frac{\{[\sqrt{(s^2 + 4\nu_1\nu_2 t^2)} - s]/(2\nu_2 t)\}^s}{\{2\pi\sqrt{(s^2 + 4\nu_1\nu_2 t^2)}\}^{1/2}}. \qquad (9.4.54)$$

Not only is this representation more transparent than the exact solution (9.3.19), since the latter relies on structural knowledge of the modified Bessel function, but it is also asymptotically equivalent to it. To see this we simply replace $I_i(2t\sqrt{(\nu_1\nu_2)})$ by the asymptotic result (9.4.10) with $x = s$ and $y = 2t\sqrt{(\nu_1\nu_2)}/s$. For large s and t this procedure leads directly to $m_s(t) \sim Y(s,t)$. In particular, if we choose y sufficiently close to zero, so that $s \gg c^+ t$, then $m_s(t)$ and $Y(s,t)$ match even in the extreme tails of the wave.

Daniels (1975) also shows that $Y(s,t)$ becomes proportional to $\exp\{-\theta(s-ct)\}$ for s near to ct as t becomes large. Thus in the vicinity of the wavefront the density $Y(s,t)$ asymptotically decreases at rate $e^{-\theta} = [\sqrt{(c^2 + 4\nu_1\nu_2)} - c]/(2\nu_2)$ as s increases (for fixed t), which provides a useful and concise description of the wavefront profile. Moreover, he goes on to develop the spatial covariance density at a wavefront (Daniels, 1977), and shows that there are three ranges of the velocity c within each of which the covariance density has quite different properties. His results clearly place a limit on the applicability of deterministic velocities to individual realizations of the process.

9.4.2 Wave profiles for non-nearest-neighbour migration

In spite of the 'health warning' which deterministic approaches should generally carry, given the considerable difficulties faced when analysing spatial-temporal stochastic processes deterministic results can be justified on the grounds that some knowledge is usually worth more than none. Moreover, the reliability, or even appropriateness, of deterministic solutions can always be checked through stochastic simulation. So let us first consider the deterministic extension from nearest-neighbour to general migration on the infinite line, and then construct stochastic realizations.

As before, each colony i is considered to be subject to a simple birth–death process with common birth and death rates λ and μ, respectively. However, the probability that an individual in colony i migrates to $i+j$ ($-\infty < i,j < \infty$) in $(t, t+dt)$ is now taken to be $\nu h_j dt$, where $\sum_{j=-\infty}^{\infty} h_j = 1$ and $h_0 = 0$. Thus the total rate of migration out of a colony is equal to ν. So for nearest-neighbour migration, $h_1 = \nu_1/\nu$, $h_{-1} = \nu_2/\nu$ and $h_j = 0$ otherwise. The probability distribution $\{h_j\}$ is often called the *contact distribution* (Mollison, 1972a). The nearest-neighbour equation (9.3.10) extends to give

$$\frac{\partial G}{\partial t} = \sum_{i=-\infty}^{\infty} \{\lambda(z_i^2 - z_i) + \mu(1 - z_i) + \nu \sum_{j=-\infty}^{\infty} h_j(z_{i+j} - z_i)\} \frac{\partial G}{\partial z_i}. \qquad (9.4.55)$$

Whence differentiating (9.4.55) with respect to z_i, and then placing $z_i = 1$ ($-\infty < z_i < \infty$), yields

$$dm_i/dt = (\lambda - \mu - \nu)m_i(t) + \nu \sum_{j=-\infty}^{\infty} h_j m_{i-j}(t) \qquad (-\infty < i < \infty), \qquad (9.4.56)$$

which is a direct extension of (9.3.12). Thus equation (9.3.13) for the generating function $H(z;t) \equiv \sum_{i=-\infty}^{\infty} m_i(t)z^i$ now takes the more general form

$$\frac{\partial H(z;t)}{\partial t} = [(\lambda - \mu - \nu) + \nu \sum_{j=-\infty}^{\infty} h_j z^j] H(z;t), \qquad (9.4.57)$$

which integrates to yield

$$\sum_{i=-\infty}^{\infty} m_i(t)z^i = e^{(\lambda-\mu-\nu)t} \exp\{\nu t \sum_{j=-\infty}^{\infty} h_j z^j\} \times \sum_{k=-\infty}^{\infty} a_k z^k \qquad (9.4.58)$$

where $m_i(0) = a_k$. Though since individuals behave independently from each other we may take $a_0 = 1$ and $a_k = 0$ ($k \neq 0$) without any loss of generality. Even so, extracting the coefficient of z^i will, in general, produce totally opaque results, so progress relies on being able to identify user-friendly special cases.

Mollison (1972b) simulated the spatial propagation of infection for several simple nonlinear epidemic models and found that when the contact distribution corresponding to $\{h_j\}$ was exponentially bounded the wavefront was observed to advance at a steady rate. However, when this condition was not satisfied then the advance was unsteady, and for a contact distribution of 'just infinite variance' (equivalent to $\sum j^{2-\epsilon} h_j$ converging for arbitrarily small $\epsilon > 0$ but diverging for $\epsilon \leq 0$) the wavefront progressed in 'wilder and wilder leaps forward'. We shall now parallel these two markedly different situations by first taking $\{h_j\}$ to be a geometric distribution and then a generalized Cauchy distribution.

Geometric contact distribution

Consider the one-sided migration distribution

$$h_j = \begin{cases} \beta^{j-1}(1-\beta) & (j = 1, 2, 3, \ldots) \\ 0 & (j = 0, -1, -2, \ldots), \end{cases} \quad (9.4.59)$$

where $0 < \beta < 1$. Thus individuals may move only to the right, and the probability of migration from colony i to colony $i+j$ decreases geometrically at rate β as j increases. If the initial population at time $t = 0$ consists of a single individual in colony 0, so that $a_0 = 1$ and $a_k = 0$ ($k \neq 0$), then expression (9.4.58) simplifies to

$$\sum_{i=-\infty}^{\infty} m_i(t) z^i = e^{(\lambda-\mu-\nu)t} \exp\{\nu t z (1-\beta)(1-\beta z)^{-1}\}$$

$$= e^{(\lambda-\mu-\nu)t}[1 + \sum_{j=1}^{\infty}\{\nu(1-\beta)tz\}^j \sum_{k=0}^{\infty} \binom{j+k-1}{k}(\beta z)^k/j!].$$

Whence extracting the coefficient of z^i gives

$$m_0(t) = e^{(\lambda-\mu-\nu)t} \quad (9.4.60)$$

$$m_i(t) = e^{(\lambda-\mu-\nu)t}\beta^i \sum_{j=1}^{i} \binom{i-1}{i-j}[\nu t(1-\beta)/\beta]^j/j! \quad (i=1,2,\ldots). \quad (9.4.61)$$

On assuming that the wavefront i^* possesses an asymptotic expectation velocity c for large t, we could, in theory, set $m_{ct}(t) = K$ (as before) in order to determine c. Unfortunately, placing $i = ct$ in expression (9.4.61) does not appear to yield an amenable expression for c. However, this difficulty can be overcome by considering the associated saddlepoint approximation (9.4.22). For the contact distribution is now given by

$$dF(u) = (\lambda - \mu - \nu)\delta(u) + \nu \sum_{j=1}^{\infty} h_j \delta(u-j) \quad (9.4.62)$$

where $\delta(x)$ denotes the Dirac delta function, and substituting for h_j from (9.4.59) into (9.4.62) and then (9.4.21) gives

$$\Psi(\theta) = (\lambda - \mu - \nu) + \nu e^{\theta}(1-\beta)(1-\beta e^{\theta})^{-1} \quad (9.4.63)$$

for $|\beta e^{\theta}| < 1$. Thus equations (9.4.22) become

$$c\theta = (\lambda - \mu - \nu) + \nu e^{\theta}(1-\beta)(1-\beta e^{\theta})^{-1}$$

$$0 = e^{2\theta}(c\beta^2) - e^{\theta}[2\beta c + \nu(1-\beta)] + c.$$

So on eliminating θ it follows that c is a root of the equation

$$\lambda - \mu - \nu = [\nu(1-\beta)/2\beta][1 - \sqrt{\{1 + 4\beta c/(\nu(1-\beta))\}}]$$
$$+ c\ln[(1/\beta)\{D - \sqrt{(D^2-1)}\}] \quad (9.4.64)$$

where $D = 1 + \nu(1-\beta)/(2\beta c)$.

On denoting the right-hand side of expression (9.4.64) by $R(c)$, a little algebra shows that:

(i) $R(0) = 0$ and $R(\infty) = \infty$;
(ii) $R(c)$ has only one turning point for $c \geq 0$ and this corresponds to $R_{min} = -\nu$.

Thus we have the following two situations:

(a) if $\lambda - \mu - \nu > 0$ then equation (9.4.64) has only one real positive root, c^+;
(b) if $0 < \lambda - \mu - \nu < 0$ then it has two positive real roots $0 < c^- < c^+$.

For we see from (9.4.60) that $m_0(t) \to \infty$ and 0, respectively under cases (a) and (b), whence the corresponding velocities of the left-hand wavefront are 0 and $c^- > 0$.

Now in the example illustrated in Figure 9.2 we have $\lambda = 0.1$, $\mu = 0$, $\nu = 1$ and $\beta = 0.2$, whence equation (9.4.64) yields the two roots $c^- = 2.448$ and $c^+ = 8.448$. Inspection shows that the process is still burning-in at time $t = 100$, which explains why the average stochastic velocities $i^*/100 = 3.16$ and 6.87 are respectively larger and smaller than these asymptotic deterministic values. Note that the left wavefront exists since $\lambda - \mu - \nu = -0.9 < 0$, so populations in the sites $i = 0, 1, 2, \ldots$ successively die out.

The associated wave profile may be determined in a similar manner. For substituting for $\Psi(\theta)$ from (9.4.63) into the saddlepoint profile (9.4.51) gives

$$Y(s,t) \sim e^{(\lambda-\mu-\nu)t} \exp\{s[(1-E) + \sqrt{(E^2-1)}]\} \times \frac{\{\beta[E + \sqrt{(E^2-1)}]\}^s}{\sqrt{\{2\pi s \sqrt{[1 + 4\beta s/(\nu t(1-\beta))]}\}}} \quad (9.4.65)$$

where $E = 1 + \nu t(1-\beta)/(2\beta s)$.

An extreme contact distribution

Our second example parallels a simple spatial epidemic process considered by Mollison (1972b) which propagates via the contact distribution

$$dV(x) = k_r \left[\prod_{u=0}^{r}(|x|+u)\right]^{-1} \quad (-\infty < x < \infty). \quad (9.4.66)$$

Here $r = 3$ and 4, and k_r is an appropriate normalizing constant. The function $dV(x)$ is clearly not exponentially bounded since $\int_{-\infty}^{\infty} e^{\theta x}\, dV(x)$ diverges for all θ. It therefore follows from Mollison (1972b) that the asymptotic velocity of the corresponding right-hand wavefront is infinite. When $r = 3$ the mean exists but the variance does not, so the stochastic wavefronts progress in wilder and wilder leaps forward. Whilst when $r = 4$ both the mean and variance exist, which gives rise to an intermediate case in which the wavefronts develop as a mixture of steady progress and great leaps forward, and this behaviour is unlikely to be mirrored through any local deterministic approximation. Mollison remarks that if one could show that the distribution of light windborn objects (such as germs and plant seeds) are of this type, then fresh insight might be gained on a number of problems which involve geographical spread. For example, it might explain why outbreaks of epidemics of mutant species sometimes appear to have several origins.

To develop a parallel migration process suppose we write the discrete analogue to (9.4.66) in the one-sided form

$$h_j = k_r \left[\prod_{u=0}^{r}(j+u) \right]^{-1} \qquad (j=1,2,\ldots; r \geq 1). \qquad (9.4.67)$$

Then the associated p.g.f.

$$S_r(z) \equiv \sum_{j=1}^{\infty} h_j z^j = k_r \sum_{j=1}^{\infty} \frac{z^j}{j(j+1)\ldots(j+r)} = k_r \sum_{j=0}^{\infty} \frac{z^{j+1}\Gamma^2(1+j)}{j!\Gamma(r+2+j)}, \qquad (9.4.68)$$

where $\Gamma(\cdot)$ denotes the Gamma function. Now the Gauss hypergeometric series is defined by

$$_2F_1(a,b;c;z) = \frac{\Gamma(c)}{\Gamma(a)\Gamma(b)} \sum_{j=0}^{\infty} \frac{z^j \Gamma(a+j)\Gamma(b+j)}{j!\Gamma(c+j)} \qquad (|z| \leq 1) \qquad (9.4.69)$$

(Abramowitz and Stegun, 1970, result (15.1.1)). Whence comparing expressions (9.4.68) and (9.4.69) yields

$$S_r(z) = [zk_r/(r+1)!] {}_2F_1(1,1;r+2;z). \qquad (9.4.70)$$

To determine the normalizing constant k_r, we set $z=1$ and use the results $S_r(1) = 1$ and

$$_2F_1(a,b;c;1) = \frac{\Gamma(c)\Gamma(c-a-b)}{\Gamma(c-a)\Gamma(c-b)} \qquad (c \neq 0, -1, -2, \ldots; \Re(c-a-b) > 0) \qquad (9.4.71)$$

(Abramowitz and Stegun, 1970, result (15.1.20)). This yields $k_r = r.r!$, whence for the initial distribution $a_0 = 1$, $a_i = 0$ ($i \neq 0$) the solution (9.4.58) takes the form

$$\sum_{i=-\infty}^{\infty} m_i(t) z^i = e^{(\lambda-\mu-\nu)t} \exp\{[\nu t r z/(r+1)]{}_2F_1(1,1;r+2;z)\}. \qquad (9.4.72)$$

Since this result involves the exponent of a power series in z, it is difficult to derive an expression for $m_i(t)$. However, an alternative representation may be constructed which appears to have a more promising form. For on splitting $S_r(x)$ in (9.4.68) into partial fractions, we obtain (Renshaw, 1979)

$$S_r(z) = r.r! \sum_{n=1}^{\infty} z^n \left[\frac{\{r!\}^{-1}}{n} + \cdots + \frac{(-1)^i \{i!(r-i)!\}^{-1}}{n+i} + \cdots + \frac{(-1)^r \{r!\}^{-1}}{n+r} \right]$$

$$= -r(1-z^{-1})^r \ln(1-z) - r \sum_{i=1}^{r} \sum_{s=1}^{i} s^{-1} \binom{r}{i}(-1)^i z^{s-i}. \qquad (9.4.73)$$

Whence combining (9.4.72) and (9.4.73) yields

$$\sum_{i=0}^{\infty} m_i(t) z^i = e^{(\lambda-\mu-\nu)t}(1-z)^{-r\nu t(1-z^{-1})^r} \exp\{-r\nu t \sum_{i=1}^{r} \sum_{s=1}^{i} s^{-1}\binom{r}{i}(-1)^i z^{s-i}\}. \qquad (9.4.74)$$

This result provides analytically tractable solutions for $m_i(t)$ when r is small, since unlike result (9.4.72), whose exponent involves an infinite power series expansion, the exponent in (9.4.74) just contains powers of $z^0, \ldots, z^{-r}$. In particular, when $r = 1$ we have

$$\sum_{i=0}^{\infty} m_i(t) z^i = e^{(\lambda-\mu)t}(1-z)^{-\nu t(1-z^{-1})}. \qquad (9.4.75)$$

Now

$$(x)_n \equiv x(x-1)\ldots(x-n+1) = \sum_{k=0}^{n} s(n,k) x^k, \qquad (9.4.76)$$

where $s(n,k)$ denotes the Stirling numbers of the first kind (Riordan, 1958, p.33). Thus since

$$(1-z)^{\nu t z^{-1}} = 1 + \sum_{n=1}^{\infty} \frac{(-z)^n}{n!} (\nu t z^{-1})(\nu t z^{-1} - 1)\ldots(\nu t z^{-1} - n + 1)$$

$$= 1 + \sum_{n=1}^{\infty} \frac{(-z)^n}{n!} \sum_{k=0}^{n} s(n,k)(\nu t z^{-1})^k,$$

we have

$$\text{coeff. of } z^r \equiv Q_r = (-1)^r \sum_{k=0}^{\infty} \frac{(-\nu t)^k}{(r+k)!} s(r+k, k). \qquad (9.4.77)$$

Whence on expanding

$$(1-z)^{-\nu t} = 1 + \sum_{k=1}^{\infty} \frac{z^k}{k!} (\nu t)(\nu t + 1)\ldots(\nu t + k - 1),$$

we see that the coefficient of z^i in (9.4.75) takes the form

$$m_i(t) = e^{(\lambda-\mu)t} \sum_{r=0}^{i} \frac{Q_r}{(i-r)!} (-1)^{i-r} (-\nu t)_{i-r} \quad (i = 0, 1, 2, \ldots), \qquad (9.4.78)$$

where the $\{Q_r\}$ are defined by expression (9.4.77).

Ideally, we would like to obtain analytically amenable expressions for $m_i(t)$ from the coefficients of the z^i in the expansion of either solutions (9.4.72) or (9.4.74), and use them to develop a family of velocities (as functions of time) corresponding to $r = 1, 2, \ldots$. For moments of the distribution (9.4.67) for $\{h_j\}$ are finite up to order $r - 1$, whilst all moments of higher order are infinite, and it would be interesting to compare the velocities for different values of r. If the same qualitative results hold for our spatial migration process as for Mollison's epidemic process then the velocities of greatest interest should correspond to $r = 2$ and $r = 3$. Unfortunately, the construction of user-friendly expressions currently remains an open problem, and even reverting back to the saddlepoint approximation exposes non-trivial issues.

For example, when $r = 1$ the distribution (9.4.67) reduces to the simple form

$$h_j = 1/[j(j+1)] \qquad (j = 1, 2, \ldots),$$

and expression (9.4.62) becomes

$$dF(u) = (\lambda - \mu - \nu)\delta(u) + \nu \sum_{j=1}^{\infty} \delta(u-j)/[j(j+1)]. \qquad (9.4.79)$$

Thus (9.4.21) integrates to

$$\Psi(\theta) = (\lambda - \mu - \nu) + \nu \sum_{j=1}^{\infty} e^{\theta j}/[j(j+1)] = (\lambda - \mu) + \nu(e^{-\theta} - 1)\ln(1 - e^{\theta}). \qquad (9.4.80)$$

Clearly, $\Psi(\theta)$ does not exist for $\theta > 0$, and so there is no saddlepoint on the real positive axis of θ (it is replaced by two conjugate complex ones). Hence there is no solution to equations (9.4.22) for the minimum asymptotic velocity of propagation c_0. This comes as no surprise, for expression (9.4.79) is exponentially unbounded and we have already remarked that the wavefront is likely to progress in increasingly wilder surges. Note, however, that there is a real negative saddlepoint, provided $\lambda - \mu - \nu < 0$, and it may be shown that this gives rise to a positive value of c which corresponds to the finite velocity c^- of the (less interesting) left-hand wavefront.

In order to determine how the velocity $c^+(t)$ increases with t, we therefore have to revert back to the wave profile. However, even this is not straightforward. For unlike the derivation of (9.4.65) for the geometric wave profile, the fact that (9.4.79) is not exponentially bounded implies that as $t \to \infty$ then $s/t \to \infty$ also, where here s denotes the position of the right-hand wavefront. This means that the asymptotic expression (9.4.51) for $Y(s,t)$ is no longer valid because the $O(t^{-1})$ term cannot be ignored. The situation becomes clearer if we include the second-order term in the saddlepoint expansion (3.7.13), namely

$$Y(s,t) \sim \frac{\exp\{t\Psi(\phi) + s\phi\}}{\{2\pi t \Psi''(\phi)\}^{1/2}}[1 + (1/24t)(3\alpha_4 - 5\alpha_3^2) + \cdots] \qquad (9.4.81)$$

where $\phi = -\theta$ and $\alpha_j(\phi) = \Psi^{(j)}(\phi)/[\Psi''(\phi)]^{1/2}$ $(j \geq 3)$. Now for large t and s/t the second-order term

$$(1/24t)(3\alpha_4 - 5\alpha_3^2) \sim (1/24\nu t)e^{s/\nu t}$$

becomes exponentially large as $s/t \to \infty$. So the asymptotic behaviour of the right-hand wavefront cannot be determined from the first-order approximation for $Y(s,t)$. Indeed, there is no guarantee that the general expansion (3.7.13) can be successfully truncated after a finite number of terms, and this whole question of analysing wavefronts and profiles for non-exponentially bounded migration/contact distributions presents an interesting challenge.

Simulated examples

Although the above deterministic velocity and wave profile results are of intrinsic theoretical interest, in practice we have to concern ourselves with stochastic developments. Moreover, for small values of r theoretical and observed velocities (and their associated wave profiles) will probably have little in common with each other because of the leaping behaviour at the wavefront. This highlights the importance of using stochastic simulation studies in order to discover likely modes of behaviour. To demonstrate the main comparisons between nearest-neighbour, exponentially bounded and extreme migration distributions suppose we: (i) ignore death by placing $\mu = 0$; (ii) label the individuals $i = 1, \ldots, k$ in order of arrival; and, (iii) let the positions of individual i be x_i, y_i and z_i under the three migration distributions

(a) $h_1 = 1$ right nearest-neighbour migration
(b) $h_j = \beta^{j-1}(1-\beta)$ $(j = 1, 2, \ldots)$ geometric migration
(c) $h_j = 1/[j(j+1)]$ $(j = 1, 2, \ldots)$ Cauchy migration.

Then conditional on the next event being a migration, given a $U \sim U(0,1)$ random variable the associated jump lengths $j = 1, 2, \ldots$ are

(a) $j = 1$
(b) $j = 1 + $ integer part$\{\ln(U)/\ln(\beta)\}$ (placing the c.d.f. $1 - \beta^j = 1 - U$)
(c) $j = $ integer part$\{1/U\}$ (placing the c.d.f. $\sum_{s=1}^{j}\left(\frac{1}{s} - \frac{1}{s+1}\right) = 1 - \frac{1}{1+j} = U$).

So as U decreases, the jump size j (for $\beta = 0.8$) changes as

U	0.9	0.5	0.1	0.01	0.001
(a)	1	1	1	1	1
(b)	1	4	11	21	31
(c)	1	2	10	100	1000

Thus the difference between the geometric (b) and Cauchy (c) regimes becomes marked only for values of $U < 0.1$.

To compare the effect of using these three very different distributions, let us develop simulations based on the same quadruples $\{T_j, U_j, V_j, W_j\}$ of $U(0,1)$ pseudo-random variables through the following exact comparative procedure.

Exact individual-based simulation algorithm (A9.8)

(i) read in parameters λ, ν and β
(ii) set $t = 0$, $i = k = j = 1$ and $x_i = y_i = z_i = 0$
(iii) update t to $t - \ln(T_j)/[k(\lambda + \nu)]$
    determine affected individual
    $i = 1 + $ integer part$\{kU_j\}$
    determine type of event
    if $V_j \leq \lambda/(\lambda + \nu)$ then $x_{k+1} = x_i$, $y_{k+1} = y_i$, $z_{k+1} = z_i$ and $k \to k+1$
    else $x_i \to x_i + 1$
      $y_i \to y_i + 1 + $ integer part$\{\ln(W_j)/\ln(\beta)\}$
      $z_i \to z_i + $ integer part$\{1/W_j\}$
(iv) if t is (say) integer print t and $(x_1, y_1, z_1), \ldots, (x_k, y_k, z_k)$
(v) return to (iii)

Figure 9.4 shows the spatial distribution at time $t = 40$ under each of these migration regimes for $\lambda = 0.2$, $\mu = 0$, $\nu = 1$ and $\beta = 0.8$ starting with a single individual in colony 0 at time $t = 0$. By this time the population has grown to 1161 individuals, and the choice of migration distribution clearly has a major impact. For under (a) all the individuals are tightly contained with a Poisson-shaped bunch between 26 and 60, whilst (b) exhibits a similar, albeit more volatile, distribution between 127 and 315 with three pioneers at 340, 341 and 383. Whilst the non-exponentially bounded case (c) exhibits a very different structure showing a main, extremely skewed, clump between 98 and 1287 followed by widely dispersed and isolated small groups from 2689 to 7318. Visually, Figure 9.4a is well-described by the deterministic Poisson result (9.4.2), namely $m_i(40) = e^{-32} 40^i / i!$, since not only do the associated wavefronts at positions 21 $(m_{21}(40) = 1.09)$ and 62 $(m_{62}(40) = 0.86)$ closely match the observed simulated wavefronts, but the simulated mode at 40–43 concurs with $m_i(40)$ reaching its maximum at $i \sim 40$. Even Figure 9.4b broadly agrees with the deterministic velocities $c^- = 2.45$ and $c^+ = 8.45$ evaluated via equation (9.4.64), since the range $(c^-t, c^+t) = (98, 338)$ matches the simulated range quite well. However, the deterministic profile (9.4.78) for the extreme Cauchy regime exhibited in Figure 9.4c provides only a very rough outline of the stochastic development, and further stochastic analysis needs to be undertaken if we are to gain greater insight into such spatial-temporal development. A possible way forward might be to try and construct analytic results for loosely-connected random walks with non-exponentially bounded jumps.

9.4.3 Travelling waves

For the birth–death–migration process all individuals develop independently of each other, and so the total population size is determined solely by the simple (non-spatial) birth–death process developed much earlier in Section 2.4. The migration component merely affects how the population is spread over the sites available to it. However, with nonlinear systems such as predator–prey and infection processes this is no longer the case, since population size and spatial distribution are closely intertwined.

Nonlinear spatial–temporal situations can exhibit quite complex behaviour, and for illustration let us return to the classic Lotka–Volterra model of Section 8.2.1. Dubois (1975) proposed a spatial extension in the framework of marine populations. He considered the spatial diffusion of both phytoplankton (prey) and herbivorous zooplankton (predator), and a stochastic simulation of his model (Dubois and Monfort, 1978) shows that it leads to the spontaneous emergence of very strong spatial heterogeneities similar to those observed in the sea. Now whilst a diffusion approach (e.g. Okubo, 1980) may be adequate for describing predator–prey interactions over a continuous medium such as sea or a large homogeneous area of land (Mimura and Murray, 1978), in other situations potential sites for colonization may possess well-defined local boundaries, such as ponds interconnected by small waterways, or a collection of oceanic islands. MacArthur and Wilson (1967) present an early, and interesting, account of the importance of stepping-stone islands to the rate and

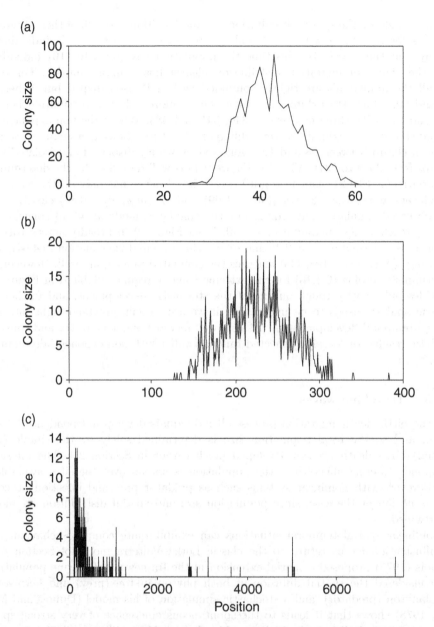

Figure 9.4 Comparative simulations at time $t = 40$ for the birth–(right)migration process with $\lambda = 0.2$ and $\nu = 1$ under (a) nearest-neighbour, (b) geometric ($\beta = 0.8$) and (c) extreme Cauchy ($r = 2$) migration regimes.

success of population dispersal. So suppose we now replace this diffusion setting by a stepping-stone process. Though note that there are fundamental issues concerning the extent to which the behaviour of a spatial process changes when diffusion on the real axis (e.g. Nisbet and Gurney, 1982) is replaced by migration over a multi-dimensional stepping-stone model, and the development of conditions (if any) under which spatial instabilities can occur (e.g. Levin, 1974). These are such huge topics that all we can do here is to touch on the velocity of propagation and its associated wave profile.

As before, consider a linear community of individuals spread over colonies $i = \ldots, -1, 0, 1, \ldots$, but now let each colony i contain $m_i(t)$ prey and $n_i(t)$ predators at time t which interact according to a simple Lotka–Volterra process. Suppose that both predators and prey may migrate to neighbouring colonies, with migration rates from i to $i-1, i+1$ being η_-, η_+ for prey and ν_-, ν_+ for predators. Moreover, to avoid local population explosions let us introduce a prey carrying capacity so that (deterministically) $m_i(t)$ may not exceed some value K. Now whilst we may parallel the single-species birth–death–migration process by introducing (say) a single predator into colony $i = 0$ at time $t = 0$ (so $n_0(0) = 1$ but $n_i(0) = 0$ for $i \neq 0$), we can either (i) introduce prey just into colony 0 at $t = 0$, or else (ii) allow prey to have become established by that time, whence $m_i(0) = K$ for all $-\infty < i < \infty$. Note, however, that (ii) involves conceptual difficulties, since it implies that the total prey population is infinite. In both cases a predator–prey cycle initially commences within colony 0, and since the system of colonies is spatially connected the predators will then spread through it in a travelling wave. Though in scenario (i) prey must advance ahead of the predators in order for population cycles to be generated over all $|i| = 1, 2, \ldots$.

Given that this process is considerably more complex than our previous birth–death–migration model, there is clearly little hope of being able to construct exact stochastic results for it. Let us therefore restrict our (analytic) attention to the deterministic equations, namely

$$dm_i(t)/dt = -\alpha m_i(t)n_i(t) + \lambda m_i(t)[1 - m_i(t)/K] - (\eta_+ + \eta_-)m_i(t)$$
$$+ \eta_+ m_{i-1}(t) + \eta_- m_{i+1}(t) \tag{9.4.82}$$
$$dn_i(t)/dt = \beta m_i(t)n_i(t) - \mu n_i(t) - (\nu_+ + \nu_-)n_i(t) + \nu_+ n_{i-1}(t) + \nu_- n_{i+1}(t).$$

So in terms of the Volterra system (8.2.20) we have placed $r_1 = \lambda$, $c = \lambda/K$, $b_1 = \alpha$, $r_2 = \mu$ and $b_2 = \beta$. As these equations are difficult to handle, instead of trying to find their most general solution let us look instead for *travelling waves* by writing $s = i - ct$ and denoting $x(s) = m_{s+ct}$ with $y(s) = n_{s+ct}$, where c denotes the velocity with which the wavefront travels under the assumption of a constant waveform. This is eminently reasonable since nearest-neighbour migration is exponentially bounded. Indeed, it should also apply to geometric migration, though not to non-exponentially bounded extreme migration distributions such as the Cauchy form discussed in the previous section. Thus our frame of reference is no longer determined by the origin $i = 0$, but by the position, ct, of the wavefront.

544 Spatial processes

Equations (9.4.82) now reduce to the more algebraically amenable form

$$cdx(s)/ds = \alpha x(s)y(s) - \lambda x(s)[1 - x(s)/K] + (\eta_+ + \eta_-)x(s)$$
$$- \eta_+ x(s-1) + \eta_- x(s+1) \qquad (9.4.83)$$
$$cdy(s)/dt = -\beta x(s)y(s) + (\mu + \nu_+ + \nu_-)y(s) - \nu_+ y(s-1) - \nu_- y(s+1),$$

and to solve them suppose we write

$$x(s) = \sum_{r=0}^{\infty} a_r e^{-r\theta s} \quad \text{and} \quad y(s) = \sum_{r=0}^{\infty} b_r e^{-r\theta s}, \qquad (9.4.84)$$

for real $\theta > 0$ and $\{a_r\}$, $\{b_r\}$ unknown constants. Then (9.4.83) takes the form

$$c\sum_{r=0}^{\infty} a_r(r\theta)e^{-r\theta s} = -\alpha \sum_{r=0}^{\infty} a_r e^{-r\theta s} \sum_{r=0}^{\infty} b_r e^{-r\theta s} + \lambda \sum_{r=0}^{\infty} a_r e^{-r\theta s}[1 - K^{-1}\sum_{r=0}^{\infty} a_r e^{-r\theta s}]$$
$$- (\eta_+ + \eta_-)\sum_{r=0}^{\infty} a_r e^{-r\theta s} + \eta_+ \sum_{r=0}^{\infty} a_r e^{-r\theta(s-1)} + \eta_- \sum_{r=0}^{\infty} a_r e^{-r\theta(s+1)}$$

$$c\sum_{r=0}^{\infty} b_r(r\theta)e^{-r\theta s} = \beta \sum_{r=0}^{\infty} a_r e^{-r\theta s} \sum_{r=0}^{\infty} b_r e^{-r\theta s} - (\mu + \nu_+ + \nu_-)\sum_{r=0}^{\infty} b_r e^{-r\theta s} \qquad (9.4.85)$$
$$+ \nu_+ \sum_{r=0}^{\infty} b_r e^{-r\theta(s-1)} + \nu_- \sum_{r=0}^{\infty} b_r e^{-r\theta(s+1)}.$$

Whence on extracting the coefficients of $e^{-r\theta s}$ we see that the resulting expressions for $r = 0$, namely

$$-\alpha a_0 b_0 + \lambda a_0(1 - a_0/K) = 0 = \beta a_0 b_0 - \mu b_0, \qquad (9.4.86)$$

are satisfied provided one of the following three conditions holds.

(a) $a_0 = b_0 = 0$: there are neither prey nor predators a long way ahead of the wavefront;
(b) $a_0 = K$ and $b_0 = 0$: a wave of predators advances through an established field of prey;
(c) $a_0 = \mu/\beta$ and $b_0 = (\lambda/\alpha)[1 - \mu/(K\beta)]$: since these represent the equilibrium points both prey and predators are already established at all sites.

Let us first consider scenario (b), that is we need to examine the velocity and profile of the wave of advancing predators. On taking $a_0 = K$ and $b_0 = 0$, extracting the coefficients of $e^{-r\theta s}$ for $r = 1$ in equations (9.4.85) leads to

$$ca_1[c\theta + \lambda + \eta_+(1 - e^\theta) + \eta_-(1 - e^{-\theta})] = -\alpha K b_1 \qquad (9.4.87)$$
$$c\theta + \nu_+(1 - e^\theta) + \nu_-(1 - e^{-\theta}) = \beta K - \mu. \qquad (9.4.88)$$

Whilst for $r > 1$ we have

$$cr\theta a_r = -\alpha(a_0 b_r + \cdots + a_r b_0) + \lambda a_r - (\lambda/K)(a_0 a_r + \cdots + a_r a_0)$$
$$- a_r[\eta_+(1 - e^{r\theta}) + \eta_-(1 - e^{-r\theta})] \qquad (9.4.89)$$
$$cr\theta b_r = \beta(a_0 b_r + \cdots + a_r b_0) - b_r[\mu + \nu_+(1 - e^{-r\theta}) + \nu_-(1 - e^{-r\theta})].$$

Thus once c and θ have been determined, relations (9.4.89) may be used iteratively to generate the coefficients $\{a_r\}$ and $\{b_r\}$ for $r > 1$ in terms of a_1 and b_1. The remaining coefficients a_1 and b_1 may then be obtained from (9.4.87) and (9.4.84) with $y(0) = \sum_{r=0}^{\infty} b_r$ placed equal to some specified value (e.g. $y(0) = 1$).

To determine c and θ we first note from Mollison (1977) that in theory all velocities (c) greater than or equal to some minimum value (c_{min}) are possible. Since in practice the conditions necessary to maintain a velocity in excess of c_{min} are always violated, it follows that c_{min} is the unique velocity of predator spread. Now on differentiating (9.4.88) we see that $dc/d\theta = 0$ at

$$e^\theta = (1/2\nu_+)[c + (c^2 + 4\nu_+\nu_-)^{1/2}]. \qquad (9.4.90)$$

Whence substituting this value back into (9.4.88) shows that c_{min} is the solution of the equation

$$c \ln[(1/2\nu_+)\{c + (c^2 + 4\nu_+\nu_-)^{1/2}\}] - (c^2 + 4\nu_+\nu_-)^{1/2} = \beta K - \mu - \nu_+ - \nu_-. \qquad (9.4.91)$$

Whilst the determination of an exact expression for the waveform itself requires the derivation of all the coefficients $\{a_r\}$ and $\{b_r\}$ for $r \geq 0$, provided we remain ahead of the wavefront (i.e. $s > 0$) then we may derive approximations to $x(s)$ and $y(s)$ by truncating expressions (9.4.84). For example, including just the first two terms gives

$$x(s) \simeq K + a_1 e^{-\theta s} \quad \text{and} \quad y(s) \simeq b_1 e^{-\theta s}. \qquad (9.4.92)$$

So if we define the position of the wavefront to be at $y(0) = 1$, then $b_1 = 1$ whilst equation (9.4.87) yields

$$a_1 = -\alpha K/[c\theta + \lambda + \eta_+(1 - e^\theta) + \eta_-(1 - e^{-\theta})]. \qquad (9.4.93)$$

More accurate approximations may be derived by successively evaluating $(a_2, b_2), (a_3, b_3), \ldots$ To illustrate this technique Renshaw (1982) takes $\alpha = 0.05$, $\beta = 0.2$, $\lambda = 1.3$, $\mu = 0.5$ and $K = 100$ together with $\nu_+ = 0.8$ and $\nu_- = \eta_- = \eta_+ = 0$, so migration only involves predator movement to the right. Then equation (9.4.91) solves to give $c = 11.32952$, whence (9.4.90) yields $\theta = 2.65056$. On taking $b_1 = 1$, expression (9.4.93) then gives $a_1 = -0.15960$. Now we may write equations (9.4.89) in the recurrence form

$$b_r = \beta \sum_{i=1}^{r-1} a_i b_{r-i}/[cr\theta - \beta K + \mu + \nu_+ - \nu_+ e^{r\theta}]$$

$$a_r = -[\alpha K b_r + \sum_{i=1}^{r-1} a_i \{\alpha b_{r-i} + (\lambda/K) a_{r-i}\}]/(\lambda + cr\theta), \qquad (9.4.94)$$

which is easily evaluated sequentially, yielding

$$a_0 = 100 \qquad\qquad b_0 = 1$$
$$a_1 = -1.60 \times 10^{-1} \qquad\qquad b_1 = 2.68 \times 10^{-4}$$
$$a_2 = 1.03 \times 10^{-4} \qquad\qquad b_2 = -5.46 \times 10^{-9}$$
$$\vdots \qquad\qquad \vdots$$
$$a_{17} = 2.51 \times 10^{-58} \qquad\qquad b_{17} = 1.06 \times 10^{-74}$$
$$\vdots \qquad\qquad \vdots$$

These computed values converge rapidly to zero, and so we only need the first few terms in order to construct an accurate solution. Moreover, as θ is real and positive the power series representations (9.4.84) hold for $s \geq 0$ and hence for $y(s) \leq 1.000268\ldots$. To evaluate larger values of $y(s)$ we simply choose a larger value for $y(0) \simeq b_1$, i.e. place the wavefront further back in the leading edge of the wave.

Not only is this wavefront technique totally general, but it also applies to the slightly more complex case (a) in which an initially empty population is suddenly perturbed by (say) an influx of prey and predators into colony $i = 0$ at time $t = 0$. For now there are two wavefronts to consider, namely one each for predators and prey. If the migration rates for prey (η_+, η_-) are greater than those for predators (ν_+, ν_-), then the prey will spread faster than predators and so a situation akin to the above case (b) will develop. Conversely, since any predators in colonies ahead of the advancing wave of prey will be unable to give birth, the resulting numbers of pioneering predators will remain small, and so the two wavefronts will effectively coalesce. Note that it is far easier to envisage process development from a stochastic perspective rather than a deterministic one. For in the former both prey and predators spread out over $i = 0, \pm 1, \pm 2, \ldots$ as t increases; whilst in the latter $m_i(t)$ and $n_i(t)$ instantaneously change for all $-\infty < i < \infty$ as t increases from 0 to 0+, and so it is less easy to speculate what the resulting deterministic behaviour might be.

Within the body of the wave

Although the above approach yields information on behaviour near the wavefront, within the body of the wave predators and prey interact to form a system of linked cycles around the equilibrium values, and so a different analytic approach is required. To simplify the analysis let us suppose that the carrying capacity is large in relation to the equilibrium values. Then on placing $dm_i(t)/dt = dn_i(t)/dt = 0$ in equations (9.4.82) with $1/K$ replaced by zero, we see that the equilibrium prey and predator values are given by $m_i^* = \mu/\beta$ and $n_i^* = \lambda/\alpha$ for all $-\infty < i < \infty$. For $t \geq 0$ write $m_i(t) = m_i^* + h_i(t)$ and $n_i(t) = n_i^* + p_i(t)$. Then for cycles of low amplitude we may take $h_i(t)$ and $p_i(t)$ small enough to neglect the product $h_i(t)p_i(t)$, whence equations (9.4.82) become

$$dh_i(t)/dt = -(\alpha\mu/\beta)p_i(t) - \eta_+[h_i(t) - h_{i-1}(t)] - \eta_-[h_i(t) - h_{i+1}(t)]$$
$$dp_i(t)/dt = (\beta\lambda/\alpha)h_i(t) - \nu_+[p_i(t) - p_{i-1}(t)] - \nu_-[p_i(t) - p_{i+1}(t)]. \qquad (9.4.95)$$

Being both homogeneous and linear, these equations may be solved by the standard procedure of switching into a generating function representation with

$$H(z;t) \equiv \sum_{i=-\infty}^{\infty} h_i(t)z^i \quad \text{and} \quad P(z;t) \equiv \sum_{i=-\infty}^{\infty} p_i(t)z^i. \qquad (9.4.96)$$

For multiplying equations (9.4.95) by z^i, and then summing over $-\infty < i < \infty$ in the usual manner, yields the two simultaneous equations

$$dH(z;t)/dt = -(\alpha\mu/\beta)P(z;t) - [\eta_+(1-z) + \eta_-(1-z^{-1})]H(z;t)$$
$$dP(z;t)/dt = (\beta\lambda/\alpha)H(z;t) - [\nu_+(1-z) + \nu_-(1-z^{-1})]P(z;t), \qquad (9.4.97)$$

and these are easily solved to yield the solution

$$H(z;t) = A(z)e^{\omega_1 t} + B(z)e^{\omega_2 t}$$
$$P(z;t) = -(\beta/\alpha\mu)\{A(z)\omega_1 e^{\omega_1 t} + B(z)\omega_2 e^{\omega_2 t} + [\eta_+(1-z) + \eta_-(1-z^{-1})]$$
$$\times [A(z)e^{\omega_1 t} + B(z)e^{\omega_2 t}]\}. \qquad (9.4.98)$$

Here the eigenvalues

$$\omega_1, \omega_2 = (1/2)[-(\eta_+ + \nu_+)(1-z) - (\eta_- + \nu_-)(1-z^{-1}) \pm \Delta] \qquad (9.4.99)$$

where

$$\Delta^2 = [(\nu_+ - \eta_+)(1-z) + (\nu_- - \eta_-)(1-z^{-1})]^2 - 4\lambda\mu. \qquad (9.4.100)$$

Whilst the functions of integration, $A(z)$ and $B(z)$, are easily determined by solving equations (9.4.98) at $t = 0$ in terms of the initial conditions $H(z;0) = \sum_{i=-\infty}^{\infty} h_i(0)z^i$ and $P(z;0) = \sum_{i=-\infty}^{\infty} p_i(0)z^i$.

Extracting $h_i(t)$ and $p_i(t)$ from the coefficients of z^i in $H(z;t)$ and $P(z;t)$ is algebraically straightforward, but the resulting expressions are generally rather messy and hence hide the underlying properties of the process. However, for some parameter values their solution simplifies quite considerably. For example, suppose that at time $t = 0$ all colonies are in equilibrium except for colony 0 which is perturbed by a sudden influx of predators. That is, $h_i(0) = p_i(0) = 0$ ($-\infty < i < \infty$) apart from $p_0(0) = p$, and prey and predators have the same migration rates, i.e. $\eta_- = \nu_-$ and $\eta_+ = \nu_+$. Then the solution (9.4.98) leads to

$$h_i(t) = -(p\alpha/\beta)\sqrt{\mu/\lambda}\sin(t\sqrt{\lambda\mu})e^{-(\nu_+ + \nu_-)t}(\nu_+/\nu_-)^{i/2}I_i(2t\sqrt{\nu_+\nu_-})$$
$$p_i(t) = p\cos(t\sqrt{\lambda\mu})e^{-(\nu_+ + \nu_-)t}(\nu_+/\nu_-)^{i/2}I_i(2t\sqrt{\nu_+\nu_-}), \qquad (9.4.101)$$

where $I_i(x)$ denotes the modified Bessel function of the first kind (see (9.3.16)). Whence on noting from (9.3.20) that $I_i(x) \sim e^x/\sqrt{2\pi x}$ when x is large relative to i^2, it follows that for large t and appropriately small i both $h_i(t)$ and $p_i(t)$ change geometrically with i at rate $\sqrt{\nu_+/\nu_-}$. Moreover, all colonies are in phase and attenuate at the same rate $t^{-1/2}\exp\{-t(\sqrt{\nu_+} - \sqrt{\nu_-})^2\}$.

Less restrictive conditions may be invoked by supposing, for example, that $\lambda\mu$ is large in comparison with $(\nu_+ - \eta_+)$ and $(\nu_- - \eta_-)$. For then expression (9.4.100) reduces to $\Delta^2 \simeq -4\lambda\mu$ which is independent of z. This simplification allows us obtain a similar, albeit slightly more complex, solution to (9.4.101) in which prey and predators no longer cycle exactly $\pi/2$ out of phase.

Although in both of these special cases cycles decay to zero as t increases, stable cycles may also be generated by modifying the spatial configuration. For example, if colonies are sited at $i = 0, 1, 2, \ldots$, and migration occurs only to the right, then not only will any deterministic cycle present in colony 0 be maintained indefinitely, but it will also affect behaviour in all the other colonies. For illustration consider the simple case with $\eta_+ = \eta_- = \nu_- = 0$, so that migration just involves predator movement to the right. Then the equilibrium values are $m_0^* = (\mu + \nu_+)/\beta$, $m_i^* = \mu/\beta$ $(i > 0)$ and $n_i^* = \lambda/\alpha$ $(i \geq 0)$. Whence on taking $p_0(0) = p$ and $h_0(0) = 0$, equations (9.4.95) reduce to the linear non-spatial Lotka–Volterra form (8.2.13), namely

$$dh_0(t)/dt = -(\alpha/\beta)(\mu + \nu_+)p_0(t)$$
$$dp_0(t)/dt = (\lambda\beta/\alpha)h_0(t), \qquad (9.4.102)$$

and hence yield the standard solution

$$h_0(t) = -(\alpha p/\beta)\sqrt{(\mu + \nu_+)/\lambda}\sin(\xi t)$$
$$p_0(t) = p\cos(\xi t) \qquad (9.4.103)$$

where $\xi = \sqrt{\lambda(\mu + \nu_+)}$. For $i > 0$ equations (9.4.95) become

$$dh_i(t)/dt = -(\alpha\mu/\beta)p_i(t) \qquad (9.4.104)$$
$$dp_i(t)/dt = (\beta\lambda/\alpha)h_i(t) - \nu_+(p_i(t) - p_{i-1}(t)), \qquad (9.4.105)$$

and apart from the anomaly at colony 0, where there is no immigration from the left, each colony has the same parametric structure. Let us therefore suppose that all colonies have the same period $2\pi/\xi$, forced by the cyclic behaviour of colony 0, and consider the potential solution

$$p_i(t) = a_i\cos(\xi t) + b_i\sin(\xi t) \qquad (i \geq 0) \qquad (9.4.106)$$

for some constants $\{a_i\}$, $\{b_i\}$. Then equation (9.4.104) yields

$$h_i(t) = -(\alpha\mu/\beta\xi)[a_i\sin(\xi t) - b_i\cos(\xi t)] + k \qquad (9.4.107)$$

where k is a constant of integration. On substituting expressions (9.4.106) and (9.4.107) back into equations (9.4.104) and (9.4.105), and equating coefficients, we have $k = 0$ together with

$$a_i + (\lambda/\xi)b_i = a_{i-1} \quad \text{and} \quad (\lambda/\xi)a_i - b_i = -b_{i-1}.$$

Whence solving these two simultaneous difference equations yields for $i > 0$

$$h_i(t) = -(\alpha\mu/\xi\beta)pr^i\sin(\xi t - i\delta) \quad \text{and} \quad p_i(t) = pr^i\cos(\xi t - i\delta), \qquad (9.4.108)$$

where $r = \sqrt{\{(\mu+\nu_+)/(\mu+\nu_++\lambda)\}}$ and $\delta = \tan^{-1}[\sqrt{\{\lambda/(\mu+\nu_+)\}}]$. Thus in contrast to the two previous special cases the behaviour of the system is now governed solely by the initial number of predators and prey in colony 0; the result being a train of linked elliptical cycles around the equilibrium values, in which the amplitude decreases geometrically with i at rate r and the phase-lag increases linearly at rate δ.

However, although this deterministic result has an appealing simplicity, it must not be forgotten that its stochastic counterpart may well be radically different, and further simulation studies are needed if we are to expose the potential difference between them. For example, the stochastic results of Section 8.2.1 have already shown that the non-spatial stochastic Lotka–Volterra trajectory is highly likely to collapse during the first cycle, so there is a fundamental issue here of whether the imposition of spatial structure will stabilize the system in the sense that it enables population cycles to persist over fairly long periods of time. This key issue is addressed in the following sections. Though before doing so, it should be stressed that the above techniques for determining velocities of propagation and population behaviour, both at the wavefront and within the main body of the population itself, may be applied to a whole host of spatial scenarios across a wide range of different fields of application. Renshaw (1981), for example, develops a parallel study for the spread of infection, and not only allows a given *infective* in colony i to migrate to $i+j$ at rate νh_j (so nearest-neighbour migration is replaced by general migration) but also considers a different mechanism of spread in which the phenomenon of infection moves rather than the infectives themselves. So now a given susceptible in colony i may be infected at rate γh_j by a given infective in colony $i-j$.

9.5 Turing's model for morphogenesis

The question of whether an inherently unstable stochastic system can be made stable by inducing spatial structure is an old one, and was first examined on a serious experimental basis by Huffaker (1958). Specific details of his heavily contrived experiment are contained in Renshaw (1991), but the essence is that the six-spotted mite *Eotetranychus sexmaculatus* acted as prey for the predatory mite *Typhlodromus occidentalis* over increasingly large rectangular arrays of oranges and a range of migration options. Although a 252-orange system did produce four full oscillations in population numbers before the process collapsed, the biological process was so convoluted that all it really illustrated was the extreme difficulty in sustaining such oscillations in that particular situation. Fortunately, elegant mathematical solutions were developed in a pioneering paper by Turing (1952). His model specifically relates to a system of chemical substances, called morphogens, which react together and diffuse through tissue. This classic work is not only of general applicability, but it was at least 20 years ahead of its time, and it is interesting to speculate where the development of spatial processes might have reached had he not met an untimely death shortly after this paper was published. Rather than following Huffaker's later route of increasing spatial complexity, Turing took the reverse approach by allowing only nearest-neighbour interaction, and avoiding potential problems of edge-effects by considering a ring of N cells. Let $X_r(t)$ and $Y_r(t)$ denote the concentrations of two morphogens in cell $r = 1, \ldots, N$ at time t, and for convenience regard cells 0 and $N+1$ as being synonymous with cells N and 1, respectively. Then on taking the rates of chemical reaction to be respectively $f(X_r, Y_r)$ and $g(X_r, Y_r)$ for some specified functions f and g, it follows that the deterministic predator–prey equations (9.4.82) over $-\infty < i < \infty$ switch into the general form

$$dX_r(t)/dt = f(X_r(t), Y_r(t)) + \eta(X_{r+1}(t) - 2X_r(t) + X_{r-1}(t))$$
$$dY_r(t)/dt = g(X_r(t), Y_r(t)) + \nu(Y_{r+1}(t) - 2Y_r(t) + Y_{r-1}(t)) \quad (9.5.1)$$

over $r = 1, \ldots, N$. For ease of illustration we have taken the left and right migration rates to be equal, i.e. $\eta_- = \eta_+ = \eta$ and $\nu_- = \nu_+ = \nu$, though this restriction can be easily relaxed. Thus for the Volterra predator–prey process $f(X, Y) = -\alpha XY - (\lambda/K)X^2 + \lambda X$ and $g(X, Y) = \beta XY - \mu Y$; whilst for the SIS epidemic process $f(X, Y) = -\gamma XY + \delta Y$ and $g(X, Y) = \gamma XY - \delta Y$.

Although Turing developed his ideas in a continuous (i.e. reaction-diffusion) setting (see Murray, 1989; Okubo, 1980) since he was primarily interested in the dispersion of chemical concentrations, the discrete-space representation is important in two respects. First, suppose we write X_r and Y_r as $X(\theta)$ and $Y(\theta)$ where $\theta = r/N$. Then $X_{r+1} = X(\theta + h)$, etc. where $h = 1/N$, and on allowing η and ν to be dependent on the small space increment $h > 0$ we see that equations (9.5.1) take the form

$$\frac{\partial X}{\partial t} \simeq f(X, Y) + h^2 \eta(h) \frac{\partial^2 X}{\partial \theta^2} \quad \text{and} \quad \frac{\partial Y}{\partial t} \simeq g(X, Y) + h^2 \nu(h) \frac{\partial^2 Y}{\partial \theta^2}. \quad (9.5.2)$$

Whence letting $h^2 \eta(h) \to \eta'$ and $h^2 \nu(h) \to \nu'$ as $h \to 0$ (i.e. $N \to \infty$) shows that the reaction-stepping-stone equations contain the classic reaction-diffusion equations

$$\frac{\partial X}{\partial t} = f(X, Y) + \eta' \frac{\partial^2 X}{\partial \theta^2} \quad \text{and} \quad \frac{\partial Y}{\partial t} = g(X, Y) + \nu' \frac{\partial^2 Y}{\partial \theta^2} \quad (9.5.3)$$

as a special limiting case. Though note that since (9.5.3) is effectively the spatial Taylor series expansion of (9.5.1), there is no reason why for moderate values of N these two systems should generate structurally similar patterns unless the discarded terms in the expansion are small. So unless N is large, discrete- and continuous-space systems will generally need to be analysed separately. Second, since numerical solution of the continuous space–time equations (9.5.3) is often performed over a finite grid of (θ, t)-values, it is worthwhile examining the effects of replacing (9.5.3) by the discrete-space equations (9.5.1) in order to assess the associated error structure.

The potential class of function-pairs (f, g) is clearly huge, but three simple forms (discussed in Murray, 1989) that provide a useful flavour are (for constants $k_i > 0$):

(a) the Schakenburg (1979) reaction in which X is created autocatalytically at rate $k_3 X^2 Y$, i.e.

$$f(X, Y) = k_1 - k_2 X + k_3 X^2 Y \ , \ g(X, Y) = k_4 - k_3 X^2 Y;$$

(b) the Gierer and Meinhardt (1972) activator–inhibitor mechanism in which X is created autocatalytically at rate $k_3 X^2/Y$ (so $X^2 Y$ in (a) $\to X^2/Y$), i.e.

$$f(X,Y) = k_1 - k_2 X + k_3 X^2/Y \ , \ g(X,Y) = k_4 X^2 - k_5 Y;$$

(c) an empirical substrate-inhibition system studied experimentally by Thomas (1975) in which f and g involve the fractional quadratic form $h(X,Y) = (k_5 XY)/(k_6 + k_7 X + k_8 X^2)$ via

$$f(X,Y) = k_1 - k_2 X - h(X,Y) \ , \ g(X,Y) = k_3 - k_4 Y - h(X,Y).$$

9.5.1 Solution of the linearized equations

Since this general stepping-stone process develops over a finite number of sites, we shall not consider problems associated with velocities of propagation and behaviour at the wavefront (though we could), but shall instead concentrate on the overall stability and structure of the system. Suppose there exist positive values, X^* and Y^*, such that $f(X^*, Y^*) = g(X^*, Y^*) = 0$. Then this ring system of equations possesses an equilibrium, which may be either stable or unstable, in which each $X_i(t) = X^*$ and $Y_i(t) = Y^*$. Paralleling the predator–prey linearization approach based on equations (9.4.95), consider small departures from equilibrium by writing $X_r(t) = X^* + x_r(t)$ and $Y_r(t) = Y^* + y_r(t)$ ($r = 1, \ldots, N$). Then the functions f and g may be approximated by

$$f(X_r, Y_r) \simeq a x_r + b y_r \quad \text{and} \quad g(X_r, Y_r) \simeq c x_r + d y_r, \qquad (9.5.4)$$

where the constant *local* reaction rates $a = \partial f(X^*, Y^*)/\partial x_r$, $b = \partial f(X^*, Y^*)/\partial y_r$, $c = \partial g(X^*, Y^*)/\partial x_r$ and $d = \partial g(X^*, Y^*)/\partial y_r$. Thus under this linear approximation equations (9.5.1) take the much simpler form

$$dx_r/dt = a x_r + b y_r + \eta(x_{r+1} - 2x_r + x_{r-1})$$
$$dy_r/dt = c x_r + d y_r + \nu(y_{r+1} - 2y_r + y_{r-1}). \qquad (9.5.5)$$

Here we denote $x_0 = x_N$ and $y_0 = y_N$ with $x_{N+1} = x_1$ and $y_{N+1} = y_1$. Instead of using the generating functions (9.4.96) to develop a comparable solution to (9.4.98), let us switch into Fourier transform mode by defining

$$u_r = (1/N) \sum_{s=1}^{N} x_s \exp\{-2\pi i r s/N\} \quad \text{and} \quad v_r = (1/N) \sum_{s=1}^{N} y_s \exp\{-2\pi i r s/N\}$$
$$(9.5.6)$$

where i denotes $\sqrt{-1}$. Now a routine inversion exercise shows that

$$x_r = (1/N) \sum_{s=1}^{N} u_s \exp\{2\pi i r s/N\} \quad \text{and} \quad y_r = (1/N) \sum_{s=1}^{N} v_s \exp\{2\pi i r s/N\}. \qquad (9.5.7)$$

Whence substituting for x_r and y_r from (9.5.7) into equations (9.5.5), extracting terms involving $\exp(2\pi irs/N)$, and then using $e^{iz} = \cos(z) + i\sin(z)$, yields

$$du_s/dt = [a - 4\eta \sin^2(\pi s/N)]u_s + bv_s$$
$$dv_s/dt = [d - 4\nu \sin^2(\pi s/N)]v_s + cu_s. \tag{9.5.8}$$

Equations (9.5.8) are considerably more manageable than (9.5.5) since each pair contains just two variables, namely u_s and v_s. Their solution takes the standard form

$$u_s = A_s \exp(p_s t) + B_s \exp(q_s t)$$
$$v_s = C_s \exp(p_s t) + D_s \exp(q_s t), \tag{9.5.9}$$

where the constants $A_s, \ldots, D_s$ satisfy the relations

$$bC_s = A_s[p_s - a + 4\eta \sin^2(\pi s/N)]$$
$$bD_s = B_s[q_s - a + 4\eta \sin^2(\pi s/N)] \tag{9.5.10}$$

and p_s, q_s are the roots of the equation

$$[p - a + 4\eta \sin^2(\pi s/N)][p - d + 4\nu \sin^2(\pi s/N)] = bc. \tag{9.5.11}$$

Substituting these results back into (9.5.7), and replacing the local variables x_r and y_r by X_r and Y_r, then yields

$$X_r = X^* + \sum_{s=1}^{N} [A_s \exp(p_s t) + B_s \exp(q_s t)] \exp(2\pi irs/N)$$

$$Y_r = Y^* + \sum_{s=1}^{N} [C_s \exp(p_s t) + D_s \exp(q_s t)] \exp(2\pi irs/N). \tag{9.5.12}$$

Although expression (9.5.12) is the required solution to the linearized deterministic equations (9.5.5), from a practical perspective it is too opaque to be of much use as it stands. So Turing (1952) split the process into four sub-cases in order to appreciate its underlying properties. In essence, as t increases, (9.5.12) becomes dominated by the eigenvalue, p_{dom}, which has the largest real part. Two distinct situations arise depending on whether p_{dom} is real or complex.

(a) If p_{dom} is real then a series of *morphologically stable* waves develops as t increases. That is, their relative amplitudes become fixed, though their true amplitudes become exponentially large or small depending on whether $\Re(p_{dom})$ is positive or negative. Note, however, that only does the initial structure affect the position of the wave around the ring, but it can also control the type of wave structure produced by the original nonlinear process (9.5.1) (Renshaw, 1994a).

(b) If p_{dom} is complex, then these waves oscillate through time.

Of the four sub-cases, the one that gives rise to patterns of greatest interest generates stationary waves of finite length (for examples of the other three sub-cases see Renshaw, 1991). The general condition for this to occur is that

$$bc < 0 \quad \text{and} \quad \frac{4\sqrt{\eta\nu}}{\eta+\nu} < \frac{d-a}{\sqrt{-bc}} < \frac{\eta+\nu}{\sqrt{\eta\nu}}, \qquad (9.5.13)$$

with exponential growth occurring if, in addition (for $\eta < \nu$),

$$d\sqrt{\eta/\nu} - a\sqrt{\nu/\eta} > 2\sqrt{-bc}. \qquad (9.5.14)$$

When $\nu > \eta$ we simply interchange the morphogen parameters.

To illustrate this phenomenon let us return to Turing's specific example in which

$$a = I - 2, \quad b = 2.5, \quad c = -1.25, \quad d = I + 1.5 \text{ and } \eta = 2\nu. \qquad (9.5.15)$$

Write

$$U = 4\eta \sin^2(\pi s/N). \qquad (9.5.16)$$

Then with these special values the eigenvalue equation (9.5.11) reduces to

$$2(p-I)^2 + (1+3U)(p-I) + (U-1/2)^2 = 0. \qquad (9.5.17)$$

So if $U = 1/2$ then the solution is simply $p = I$ (with $q = I - 1 < p$), which is particularly useful since the condition (9.5.14) for exponential growth reduces to $I > 0$. Suppose we now choose the migration parameter η to ensure that for some integer $\tilde{s}$

$$2U = 8\eta \sin^2(\pi \tilde{s}/N) = 1. \qquad (9.5.18)$$

Then there will be $\tilde{s}$ stationary waves around the ring since all the other roots p_s and q_s must have real part smaller than $p_{\tilde{s}} = I$. If $\tilde{s}$ is not an integer, then the actual number of waves is one of the two nearest integers to $\tilde{s}$. This deterministic linearized system is therefore controlled by the eigenvalue I and the wave-number $\tilde{s}$.

9.5.2 An example of wave formation

The restrictive nature of condition (9.5.13) suggests that by no means all processes will lend themselves to the generation of interesting wave patterns. For example, we see from the linearized spatial Lotka–Volterra equations (9.4.95) that $a = d = 0$, $b = -\alpha\mu/\beta$ and $c = \beta\lambda/\alpha$. Hence although the requirement that $bc = -\mu\lambda < 0$ is automatically satisfied, the second inequality in (9.5.13) is bound to fail since $4\sqrt{\eta\nu}/(\eta+\nu) > 0 = d - a$. So to produce spectacular effects, such as the *Hydra's* ability to regenerate, we need to consider a more general model. This fresh water creature, which resembles a sea-anemone, has between five to ten tentacles and when part of it is cut-off it rearranges itself to form a new complete organism (Child, 1941, demonstrates this phenomena through an interesting staining experiment). Applying Turing's theory, we may presume that these new tentacles occur at the lobes of a developing wave system.

On assuming that the *individual* rate functions f and g are linear in X and Y, suppose we take $f(X,Y) = X(r_1 + a_{11}X + a_{12}Y)$ and $g(X,Y) = Y(r_2 + a_{21}X + a_{22}Y)$. Then the corresponding equilibrium values X^*, Y^* are the solutions of the equations

$$r_1 + a_{11}X^* + a_{12}Y^* = 0 = r_2 + a_{21}X^* + a_{22}Y^*. \qquad (9.5.19)$$

So the linear coefficients (9.5.4) are given by
$$a = (r_1 + a_{11}X^* + a_{12}Y^*) + a_{11}X^* = a_{11}X^* \ , \ b = a_{12}X^*$$
$$d = (r_2 + a_{21}X^* + a_{22}Y^*) + a_{22}Y^* = a_{22}Y^* \ , \ c = a_{21}Y^*. \tag{9.5.20}$$

Whence using the Turing values (9.5.15) gives
$$a_{11} = (I-2)/X^* \ , \ a_{12} = 2.5/X^* \ , \ a_{21} = -1.25/Y^* \ , \ a_{22} = (I+1.5)/Y^* \tag{9.5.21}$$

together with
$$r_1 = 2 - I - 2.5(Y^*/X^*) \ , \ r_2 = -1.5 - I + 1.25(X^*/Y^*). \tag{9.5.22}$$

Thus if $I < 2$ then
$$\begin{pmatrix} a_{11} & a_{12} \\ a_{21} & a_{22} \end{pmatrix} = \begin{pmatrix} - & + \\ - & + \end{pmatrix},$$

and although this system clearly makes sense when applied to a chemical reaction process, in biological terms it corresponds to the rather strange situation in which the X-species eats both itself and the Y-species, and the Y-species cultivates both itself and the X-species.

Suppose, for example, that we wish to generate $\tilde{s} = 5$ stationary waves around a ring of $N = 50$ cells. Then it follows from result (9.5.18) that
$$\eta = 1/[8\sin^2(\pi/10)] = 1.309017 \quad \text{with} \quad \nu = \eta/2 = 0.654509. \tag{9.5.23}$$

Let us retain the parameter values (9.5.15), and choose $I = 0.005 > 0$ so that the pattern grows gently as t increases. Then on applying the perturbation $X_1(0) = 200$ at $t = 0$ to the equilibrium spatial configuration $X_r(0) = X^* = 100$ and $Y_r(0) = Y^* = 50$ over $r = 1,\ldots,N$, the linearized equations (9.5.5) lead to the symmetric five-wave structure at $t = 500$ shown in Figure 9.5a. Comparison with Figure 9.5b, which shows the solution of the original equations (9.5.1) with parameter values (9.5.21) and (9.5.22), namely
$$dX_r/dt = X_r(2 - I - 2.5Y^*/X^*) + X_r^2(I-2)/X^* + 2.5X_rY_r/X^*$$
$$+ 2\nu(X_{r+1} - 2X_r + X_{r-1})$$
$$dY_r/dt = Y_r(-1.5 - I + 1.25X^*/Y^*) - 1.25X_rY_r/Y^* + Y_r^2(I+1.5)/Y^*$$
$$+ \nu(Y_{r+1} - 2Y_r + Y_{r-1}), \tag{9.5.24}$$

exhibits a remarkable similarity in structure. For the linearized solution is technically valid only near the equilibrium points, i.e. $X(t) \simeq 100$ and $Y(t) \simeq 50$. Moreover, there are an infinite number of nonlinear processes that give rise to it, all of which generate different numerical solutions. Observe that although the pattern grows steadily up to $t = 940$, thereafter the wave amplitudes swiftly explode; Figure 9.5b shows the pattern at $t = 935$ shortly before this occurs.

Given that the linearized solution appears to be a good indicator of the general qualitative features of the full nonlinear deterministic system, a key question is

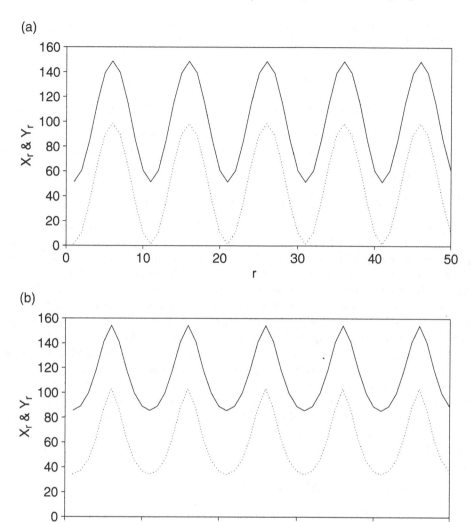

Figure 9.5 Solution of (a) the linearized equations (9.5.5) at $t = 500$, and (b) the full equations (9.5.24) at $t = 935$, showing X_r (—) and Y_r (···).

whether such strong similarities also carry across to the corresponding stochastic system. Provided the number of cells N is not too large we can easily apply the exact simulation algorithm A9.1 of Section 9.3 by allowing migration between cells 1 and N. First, however, we need to address a serious issue concerning the construction of the birth and death rates. For associated with each deterministic system there exists an infinite number of possible stochastic processes that correspond to it. Denote B_r^X, D_r^X, B_r^Y and D_r^Y to be the (non-negative) population birth and death rates in cell r, for species X and Y. Then for *any* functions $h_r^X, h_r^Y \geq 0$, the augmented birth and death rates $B_r^X + h_r^X$ and $D_r^X + h_r^X$, $B_r^Y + h_r^Y$ and $D_r^Y + h_r^Y$, yield the *same* deterministic equations

$$dX_r/dt = (B_r^X + h_r^X) - (D_r^X + h_r^X) = B_r^X - D_r^X$$
$$dY_r/dt = (B_r^Y + h_r^Y) - (D_r^Y + h_r^Y) = B_r^Y - D_r^Y. \qquad (9.5.25)$$

For pragmatic reasons suppose we take $h_r^X = h_r^Y \equiv 0$. Then one way of decomposing the deterministic equations (9.5.24) into stochastic transition rates would be to choose

$$B_r^X = 2.5 X_r Y_y / X^*, \qquad D_r^X = X_r(I - 2 + 2.5 Y^*/X^*) + X_r^2(2 - I)/X^*$$
$$B_r^Y = Y_r^2(I + 1.5)/Y^*, \qquad D_r^Y = Y_r(1.5 + I - 1.25 X^*/Y^*) + 1.25 X_r Y_r / Y^*, \qquad (9.5.26)$$

provided I, X^* and Y^* are such that each of these four component terms are non-negative. Renshaw (1994a), for example, takes $I = -0.2$ and (for computational speed) the small equilibrium values $X^* = Y^* = 10$ together with the migration rates (9.5.23). This configuration produces stochastic realizations that exhibit a marked build-up from the deterministic equilibrium values $X_r(0) = X^*$ and $Y_r(0) = Y^*$ into a wave structure with strongly oscillating amplitudes. Figures 9.6a&b illustrate this behaviour at times 10 and 40, showing around 5 to 6 X- and Y-waves of irregular cycle length and varying amplitude. The contrast between this and the highly regular deterministic process illustrated in Figure 9.5 highlights the substantial difference between deterministic and stochastic space-time behaviour. Note that: (i) the degree of synchrony between $X_r(t)$ and $Y_r(t)$ is far higher at $t = 40$ than at $t = 10$; (ii) the peaks quickly wax and wane as t increases; (iii) the maximum peak sizes greatly exceed the deterministic equilibrium values; and, (iv) the minimum $X_r(t)$ and $Y_r(t)$ values lie well below them (in proportional terms).

Such behaviour is easily explained by examining the birth, death and migration rates

$$B_r^X = 0.25 X_r Y_r, \qquad B_r^Y = 0.13 Y_r^2 \qquad (X_r, Y_r \text{ birth})$$
$$D_r^X = X_r(0.3 + 0.22 X_r), \quad D_r^Y = Y_r(0.05 + 0.125 X_r) \quad (X_r, Y_r \text{ death}) \qquad (9.5.27)$$
$$M_{r+}^X = M_{r-}^X = 1.309017 X_r, \quad M_{r+}^Y = M_{r-}^Y = 0.654509 Y_r \quad (X_r, Y_r \text{ right/left migration}).$$

For $B_r^X > D_r^X$ if $Y_r > 1.2 + 0.88 X_r$; whilst $B_r^Y > D_r^Y$ if $Y_r > 0.3846 + 0.9615 X_r$. Thus if $Y_r > X_r$, and X_r and Y_r are both large, then the population may boom (i.e. produce a large local maximum) before the joint influences of migration and the net growth rate of X_r, namely $(B_r^X - D_r^X)$, exceeding that of Y_r, namely $(B_r^Y - D_r^Y)$, can drag it back down. Conversely, when X_r and Y_r are small, it is only fresh migrants arriving from neighbouring colonies $r - 1$ and $r + 1$ that prevents the local (X_r, Y_r)-population from dying out: when $X_r = Y_r = 1$, for example, $B_r^X = 0.25 < D_r^X = 0.52$ and $B_r^Y = 0.13 < D_r^Y = 0.175$.

The rate decomposition (9.5.27) works over $X_r, Y_r > 0$ because the transition rates all remain positive. However, the same does not hold true for the earlier rates (9.5.26) when $X^* = 100$ and $Y^* = 50$ (we previously took $X^* = Y^* = 10$). For with

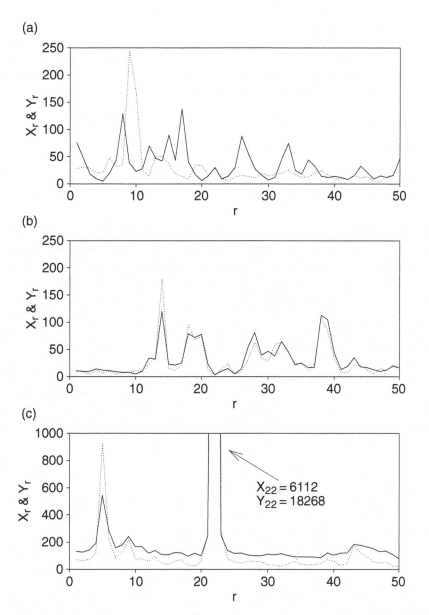

Figure 9.6 A stochastic realization showing X_r (—) and Y_r ($\cdots$) at times (a) $t = 10$ and (b) $t = 40$ corresponding to the deterministic five-wave Turing ring system (9.5.24) over 50 cells with transition rates (9.5.26), $I = -0.2$ and start position $X_r(0) = X^* = 10$ and $Y_r(0) = Y^* = 10$; (c) a realization for rates (9.5.28) with $X^* = 100$, $Y^* = 50$ and $I = 0.001$.

558 Spatial processes

$I = 0.001$ the two death rates may now be negative. Specifically, $D_r^X = -0.749X_r + 0.01999X_r^2 < 0$ when $X \le 38$ and $D_r^Y = -0.999Y_r + 0.025X_rY_r < 0$ when $X_r \le 39$. So we need to reallocate the deterministic rate components to prevent this. When $-1.5 \le I \le 2$ the obvious approach is to transfer the two negative terms in (9.5.26) from the death rates to the birth rates, thereby yielding

$$B_r^X = 2X_r + 2.5X_rY_r/X^* \qquad\qquad = 2X_r + 0.025X_rY_r$$
$$D_r^X = X_r(I + 2.5Y^*/X^*) + X_r^2(2-I)/X^* = 1.251X_r + 0.01999X_r^2$$
$$B_r^Y = 1.25Y_r(X^*/Y^*) + Y_r^2(I+1.5)/Y^* \quad = 2.5Y_r + 0.03002Y_r^2 \qquad (9.5.28)$$
$$D_r^Y = Y_r(1.5+I) + 1.25X_rY_r/Y^* \qquad = 1.501Y_r + 0.025X_rY_r.$$

Unlike the earlier rates (9.5.27), when $X_r = Y_r = 1$ both $B_r^X > D_r^X$ and $B_r^Y > D_r^Y$ which reduces the chance of a local population temporarily dying out. Figure 9.6c shows a stochastic realization of this process starting from $X_r(0) = X^*$ and $Y_r(0) = Y^*$, and even by time $t = 1.24$ the process has produced a huge tentacle whose size greatly exceeds the population values at all other sites. Using different start seeds produced the same effect at each attempt; only the location of the tentacle and the exact time of its maximum size changed. As $X_{22} = 6112$ and $Y_{22} = 18268$ are still rapidly increasing at this time, the two-species extension to the exact simulation algorithm A9.1 clearly requires huge computational effort. Indeed, since at $t = 1.24$ the rates (9.5.28) are $B_{22}^X = 2.80 \times 10^6 \gg D_{22}^X = 0.75 \times 10^6$ and $B_{22}^Y = 10.06 \times 10^6 \gg D_{22}^Y = 2.82 \times 10^6$, this tentacle exhibits violent growth; very soon afterwards, at $t = 1.2407$, the large values $X_{22} = 9800$ and $Y_{22} = 31887$ effectively cause the algorithm to collapse. Fortunately such computational difficulties can be circumvented by making appropriate modifications to each of the subsequent approximate algorithms in turn; namely the cell-based A9.2, the individual-based A9.3 and the tau-leaping approach A9.4. Conducting a detailed comparative study of compute time versus error effects for these approximate approaches would be well worthwhile.

The considerable discrepancy between the deterministic and stochastic versions of the Turing process offers great scope for future research. For although many deterministic studies have been made of pattern generation (see, for example, Othmer, Maini and Murray, 1993; Maini and Othmer, 2000; Sekimura, Noji, Ueno and Maini, 2003), comparable stochastic studies are few and far between. Not only do issues involving the decomposition of deterministic equations into stochastic rates, and the effect of initial start values, need to be addressed, but we have seen that fundamental computational questions arise when X_r and Y_r, and also N and t, become large. For unless high-powered parallel computers are employed the exact algorithm 9.1 may well be too slow to be of practical use, and analysis needs to be undertaken to determine optimal ways of using the approximate algorithms A9.2 to A9.6 instead. These issues are particularly relevant to the diffusion of chemicals, since our stepping-stone procedure is then merely a convenient discrete-space approximation to a continuous-space problem. Serious questions therefore arise as to how close realizations of these two structures lie to each other, and what conditions are required for the stepping-stone approximation

9.5.3 Stability and the 'Stochastic Dynamic'

At the start of Section 9.5.2 we noted that the Lotka–Volterra predator-prey system does not satisfy the Turing conditions for the development of an established wave system. In order to visualize what happens in such situations let us consider two extreme situations. First, if the migration rates $\eta = \nu = 0$ then each of the N colonies behaves as an independent non-spatial Lotka–Volterra process and so is highly likely to exhibit boom-and-bust behaviour with early extinction (usually prey) during the first cycle (see Figure 8.8). Second, allowing $\eta, \nu \to \infty$ ensures increasingly fast mixing through every colony which effectively reduces the system to a single-colony since the resulting population synchrony removes the spatial structure, and thereby also gives rise to boom-and-bust dynamics. Simulation experiments show that if $\eta \simeq \nu$ or $\eta > \nu$, then the time gap between the arrival of the first prey into an empty colony and the first predator there is likely to be high enough for the prey population to explode before the predator population can start to control it. This suggests that to construct a persistent system we need to keep η fairly low, with $\nu > \eta$ sufficiently high to enable predators to start killing the relatively inert prey before their population becomes too large. The aim is to have appropriately small numbers of prey and predators migrating from r to empty colonies $r-1$ and $r+1$, and hence starting new population cycles there, before X_r and Y_r decay to zero. The balance between η and ν appears to be crucial. This gives rise to the concept of a *Stochastic Dynamic*, in which a non-spatial transient process can be made persistent, at least in the medium term, by placing it within a suitable spatial environment. For then local short-term population cycles within the system can spread to empty sites and start fresh cycles there before the original ones have decayed.

For example, suppose we combine the non-spatial parameters of Figure 8.8, namely $B_r^X = 1.5X_r$, $D_r^X = 0.1X_rY_r$, $B_r^Y = 0.01X_rY_r$ and $D_r^Y = 0.25Y_r$, with the migration parameters $\eta = 0.05$ and $\nu = 0.1$, i.e. predators are twice as mobile as prey. Then on placing $X_1(0) = 1$ and $Y_1(0) = 10$ with $X_r(0) = Y_r(0) = 0$ $(r = 2, \ldots, N)$, we have an initial population spike in colony 0 which (hopefully) develops and spreads through the system. One series of experiments with varying N produced the population extinction times (T_{ext})

$$N: \quad 1 \; 3 \; 5 \; 6 \quad 7 \quad 8 \quad\;\; 9 \quad\;\; 10 \quad\;\; 11 \quad\;\; 12 \quad\;\; 13$$
$$T_{ext}: 8\;12\;15\;277\;990\;1525\;1414\;11373\;22153\;32187\;45825 \tag{9.5.29}$$

Here increasing N from 1 to 5 gives only a rough doubling of the time to extinction, but thereafter the increase becomes much stronger. Figure 9.7 shows a realization for a ten-colony system at times $t = 20, 40, 60$ and 100, and we see that although the initial prey member has developed only locally by $t = 20$, the more mobile predator population has already spread through the system ready to take advantage of new migrating prey. By $t = 40$ the initial prey population in cell 1 has virtually died out, and two new population cycles have developed elsewhere, as happens yet again by

560 *Spatial processes*

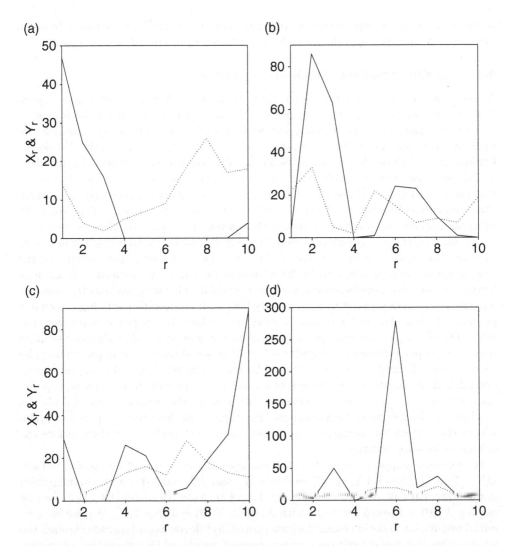

Figure 9.7 A stochastic realization of X_r (—) and Y_r (···) at times (a) $t = 20$ (b) 40, (c) 60 and (d) 100 for a 10-cell Lotka–Volterra ring system with transition rates $B_r^X = 1.5X_r$, $D_r^X = 0.1X_rY_r$, $B_r^Y = 0.01X_rY_r$, $D_r^Y = 0.25Y_r$ and $\eta = 0.05$, $\nu = 0.1$, for an initial population spike $X_1(0) = 1$, $Y_1(0) = 10$.

$t = 60$. Whilst by $t = 100$ a global decline in the predator population has allowed a sudden large surge in prey. Note that unlike Figures 9.6a&b for our earlier chemical-reaction system, $X_r(t)$ and $Y_r(t)$ are not in synchrony over $r = 1,\ldots,10$. For in the non-spatial predator-prey process predators lag prey by a quarter cycle, which exerts a considerable impact on the spatial ring dynamics.

To date, no theory exists either for the conditions for long-term persistence or the time to extinction itself. Intuitively, the problem is that systems may persist even if almost all sites are empty. All that is needed is for there to be a few non-empty ones present that can, with high probability, restart the others before they themselves temporarily die out. For large systems there are clearly a huge number of configurations that can lead to this. Construction of the exact probability of extinction, $\Pr(\mathbf{X}(t) = \mathbf{0}$ for all $t \geq T_{ext})$, through the Kolmogorov forward (or backward) equations requires determining the p.g.f. $G(\mathbf{0}; t)$ (or $g_i(\mathbf{0}; t)$) which is likely to prove intractable in all but the most trivial of cases. Issues are further compounded on noting that the Turing ring system is just one of many types of possible spatial structure. Other configurations include:

1. *Bicycle wheel with hub*: Each colony $r = 1, \ldots, N$ is connected to a central hub ($r = 0$). So instead of an individual having to migrate from cell i to j via all the intervening clockwise or anti-clockwise sites $i + 1, \ldots, j - 1$ or $i - 1, \ldots j + 1$, it can now also migrate from i to j in just two steps via $r = 0$.
2. *Bicycle spokes with hub*: Here nearest-neighbour migration is prohibited, and movement from i to j occurs only through the hub at $r = 0$.
3. *General migration*: All cells now have the potential to interconnect, with general prey and predator migration rates $\eta_{ij} \geq 0$ and $\nu_{ij} \geq 0$, respectively, from cell i to j.
4. *Distance dependent migration*: Generalizing (3), we allow the migration rates $\eta_{ij} = \eta_{ij}(||i - j||)$ and $\nu_{ij} = \nu_{ij}(||i - j||)$ to depend on the distance $||i - j||$ between sites i and j.
5. *Distance dependence connection*: Now the probability of connection itself varies with $||i - j||$. That is, $\eta_{ij} = \eta$ and $\nu_{ij} = \nu$ with probability $d(||i - j||)$, and $\eta_{ij} = \nu_{ij} = 0$ otherwise, for some appropriate decreasing function $d(\cdot)$.
6. *Small World migration*: In addition to nearest-neighbour migration (1), sites i and j are connected with small probability ϵ independently of separation distance and of all other sites. Thus we now have the Small World structure previously introduced in the Metropolis MSW setting of Section 7.5.4. So even when N is large and i and j are well separated the length of the connected path between i and j is likely to be small.

Although simulation can provide much detailed information on a specific process, it would be extremely useful if general mathematical results could be developed which provide deeper insight into how the behaviour of stochastic dynamical systems varies between spatial configuration, number of sites, parameter selection, etc. Key questions relating to long-term persistence include: is there a range of parameters over which a given structure can persist virtually indefinitely; and, are there critical threshold values below which a process is unable to exploit fully the entire system available to it?

9.6 Markov chain approach

Some progress can be made provided we are prepared to sacrifice a fully quantitative solution in favour of a purely qualitative one. For if our primary interest is the probability of extinction, then in a two-species scenario we may consider a given colony to be in one of four states:

E empty (devoid of both species)
H contains only species 1
P contains only species 2
M contains both species. (9.6.1)

Thus a system of N connected sites has 4^N total states, which, for small N, may well be both algebraically and computationally manageable.

Suppose that the time (i.e. number of events) to extinction, T_{ext}, is reasonably large, and that we concentrate our attention on one specific site, say i. Then development of the population size, $X_i(t)$, at the discrete times $t = 0, 1, 2, \ldots$ may be crudely represented by the transition probability matrix

$$\mathbf{R} = \begin{pmatrix} p_{EE} & p_{EH} & p_{EP} & p_{EM} \\ p_{HE} & p_{HH} & p_{HP} & p_{HM} \\ p_{PE} & p_{PH} & p_{PP} & p_{PM} \\ p_{ME} & p_{MH} & p_{MP} & p_{MM} \end{pmatrix}. \tag{9.6.2}$$

So the state population vector for colony i at time t is given by

$$\mathbf{p}(t) = \mathbf{p}(0)\mathbf{R}^t, \tag{9.6.3}$$

to which all our earlier Markov chain results apply (see Chapter 5). This huge reduction in state space now allows us to undertake a basic study of extinction times. For example, a simulation run with the $N = 5$ version of the prey-predator Turing ring system shown in Figure 9.7 led to prey extinction by $t = 882$ and total extinction by $t = 898$. Colony $i = 1$ experienced the state changes

$$25(P \to P), \quad 4(P \to M), \quad 5(M \to P), \quad 66(M \to M)$$

over $t = 0, 1, 2, \ldots, 100$, and as states E and H remained empty during this time period (i.e. prey were always present), it follows that crude estimates for the transition rates are

$$\hat{p}_{PP} = 25/29, \quad \hat{p}_{PM} = 4/29, \quad \hat{p}_{MP} = 5/71, \quad \hat{p}_{MM} = 66/71, \tag{9.6.4}$$

with the remaining twelve estimated rates $\hat{p}_{EE}, \ldots, \hat{p}_{MH}$ being zero. Thus the non-zero quasi-equilibrium probabilities are (roughly)

$$\pi_P \simeq \hat{p}_{MP}/(\hat{p}_{PM} + \hat{p}_{MP}) = 0.338 \quad \text{and} \quad \pi_M \simeq \hat{p}_{PM}/(\hat{p}_{PM} + \hat{p}_{MP}) = 0.662.$$

Although this highly approximate analysis relates to just a single colony, an assessment of the whole five-site system can be made by assuming that each colony behaves independently of all the others. The argument is that spatial connectivity is already enshrined in the estimates (9.6.4), since the transition rates have been determined from the full spatial process and not from a realization of an isolated colony. The equilibrium probability that the process is devoid of prey at a particular time is then given by $\pi_P^5 = 0.0044$. Suppose we make the further assumption that the probability of prey extinction occurring exactly at time t is given by $\Pr(\text{prey somewhere at times } 1, \ldots, t-1) \times$

Pr(no prey anywhere at time t) = $(1 - 0.0044)^{t-1} \times 0.0044$. Then the time to prey extinction has a geometric distribution with parameter $h = 0.0044$. Whence a crude ball-park estimate for the mean time to prey extinction is $T_{prey\ ext} = (1-h)/h = 226$, which is of the right order of magnitude. Moreover, this simple approach produces an indication of how extinction times change with increasing N.

9.6.1 A more refined approximating process

The derivation of this estimate involves grossly simplifying assumptions both for parameter estimation and spatial interaction, and in order to make the approach more credible we need to improve the way in which spatial structure is woven into the approximating system. Moreover, since this reduction to simple presence-absence for each species is highly restrictive, let us broaden our predator–prey illustration by expanding the state space to cover the eight situations (Maynard Smith, 1974):

E empty
HA few prey
HB increasing prey
HC many prey
MA many prey, few predators
MB many prey, increasing predators
MC many prey, many predators
MD few prey, many predators.

Thus H denotes prey but no predators, whilst M denotes both prey and predators. This list is clearly not exhaustive, and precludes any possibility of there being predators but no prey. For in the absence of prey, predators soon die out, and to enlarge the system further would defeat the whole point of having a simple representation. Bearing in mind the shape of the non-spatial stochastic realizations illustrated in Figure 8.8, let us assume that for a given site:

(i) the transitions $MA \to MB \to MC \to MD$ must occur;
(ii) the predator death rate is high enough, relative to the inwards predator migration from connected colonies, for $MD \to E$;
(iii) a migrant predator arriving at an E colony soon starves;
(iv) a migrant predator arriving at an HA, HB or HC colony converts it to an MA colony;
(v) a migrant prey arriving at an E colony converts it to an HA colony.

There is nothing sacrosanct about these rules, but as well making biological sense they also enable stochastic realisations to be computed extremely quickly.

For a given colony i, let x_i denote the number of colonies $j \neq i$ directly connected to it that contain many prey (i.e. HC, MA, MB and MC). Then if α denotes the probability that in one time unit prey will migrate from such a connected colony to i, we have Pr(no prey migrate from a connecting colony j to i) = $(1-\alpha)^{x_i}$, since colonies send out migrants independently from each other. Thus Pr($E \to HA$) is given by

$$a = \Pr(\text{at least 1 prey reaches } i \text{ from any connected colony}) = 1 - (1-\alpha)^{x_i}. \quad (9.6.5)$$

Similarly, on denoting y_i to be the number of connecting colonies which contain migratory predators, and β the predator migration rate, we have

$$b = \Pr(\text{at least 1 predator reaches } i \text{ from any connected colony}) = 1 - (1-\beta)^{y_i}. \tag{9.6.6}$$

Note the big change in migration strategy here from the original process where prey and predators migrate as independent individuals. Moreover, whilst it is reasonable to presume that prey may migrate only from colonies containing many prey (i.e. migration out of states HA, HB and MD is precluded), the situation for migrating predators is less obvious. We might argue, for example, that predators may only migrate when they are hungry, in which case y_i is the number of connected MD colonies. Conversely, if all predators may migrate then we could claim that y_i is the number of connected MB, MC and MD colonies; here we have just excluded MA (few predators), though we could also have excluded MB (increasing predators) as well.

As we have to assume that the arrival of a migrating predator turns HA, HB and HC into MA (a price paid for just having eight states), it follows that

$$\Pr(HA \to MA) = \Pr(HB \to MA) = \Pr(HC \to MA) = b.$$

Whence

$$\Pr(HA \to HB) = \Pr(HB \to HC) = \Pr(HC \to HC) = 1 - b.$$

Whilst the only effect induced by an arriving prey is to turn E to HA, i.e.

$$\Pr(E \to E) = 1 - a \quad \text{and} \quad \Pr(E \to HA) = a.$$

Finally, we parallel the behaviour of the non-spatial Lotka–Volterra process by taking

$$\Pr(MA \to MB) = \Pr(MB \to MC) = \Pr(MC \to MD) = \Pr(MD \to E) = 1.$$

Since the associated transition probability matrix

$$\mathbf{P} = \begin{pmatrix} 1-a & a & 0 & 0 & 0 & 0 & 0 & 0 \\ 0 & 0 & 1-b & 0 & b & 0 & 0 & 0 \\ 0 & 0 & 0 & 1-b & b & 0 & 0 & 0 \\ 0 & 0 & 0 & 1-b & b & 0 & 0 & 0 \\ 0 & 0 & 0 & 0 & 0 & 1 & 0 & 0 \\ 0 & 0 & 0 & 0 & 0 & 0 & 1 & 0 \\ 0 & 0 & 0 & 0 & 0 & 0 & 0 & 1 \\ 1 & 0 & 0 & 0 & 0 & 0 & 0 & 0 \end{pmatrix} \begin{matrix} E \\ HA \\ HB \\ HC \\ MA \\ MB \\ MC \\ MD \end{matrix} \tag{9.6.7}$$

is virtually cyclic, the equilibrium equations

$$\pi_E = (1-a)\pi_E + \pi_{MD} \qquad \pi_{MA} = b(\pi_{HA} + \pi_{HB} + \pi_{HC})$$
$$\pi_{HA} = a\pi_E \qquad \pi_{MB} = \pi_{MA}$$
$$\pi_{HB} = (1-b)\pi_{HA} \qquad \pi_{MC} = \pi_{MB} \qquad (9.6.8)$$
$$\pi_{HC} = (1-b)(\pi_{HB} + \pi_{HC}) \qquad \pi_{MD} = \pi_{MC}$$

are particularly easy to solve. Specifically, on denoting $D = a + b + 4ab$, we have $\pi_{MA} = \pi_{MB} = \pi_{MC} = \pi_{MD} = ab/D$ with $\pi_{HA} = ab/D$, $\pi_{HB} = ab(1-b)/D$, $\pi_{HC} = a(1-b)^2/D$ and $\pi_E = b/D$. Whence writing

$$\pi_H = \pi_{HA} + \pi_{HB} + \pi_{HC} \qquad \text{(prey, but no predators)}$$
$$\pi_M = \pi_{MA} + \pi_{MB} + \pi_{MC} + \pi_{MD} \qquad \text{(prey and predators)},$$

and using $\pi_E + \pi_H + \pi_M = 1$, leads to

$$(\pi_E, \pi_H, \pi_M) = (b, a, 4ab)/D. \qquad (9.6.9)$$

Now $\pi_M = 4/[(1/a) + (1/b) + 4]$ takes its maximum value of $2/3$ when $a, b \simeq 1$, at which point $\pi_E \simeq \pi_H \simeq 1/3$. Hence it follows from (9.6.5) and (9.6.6) that the probability of long-term persistence is maximized when $(1-\alpha)^{x_i} \simeq 0$ and $(1-\beta)^{y_i} \simeq 0$. On replacing x_i and y_i by their expected values $\bar{x}$ and $\bar{y}$, this result implies that either $\alpha, \beta \simeq 1$, in which case both prey and predators are highly mobile, or else $\bar{x}, \bar{y}$ are fairly large, which means that each colony is connected to many others. Though care should be taken when invoking these two conditions, since if both hold true then we are effectively removing the spatial structure from the system, and thereby reducing it to a non-spatial process which does not exhibit long-term persistence.

The high computational speed of this matrix representation means that it is ideal for studying how different types of network, i.e. the way in which colonies are connected, affects the quasi-equilibrium structure. Two extreme types are 'island' models in which migration may occur between many, if not all, of the N colonies ($\bar{x}, \bar{y} \simeq N - 1$), and 'stepping-stone' models where migration takes place between nearest-neighbours only ($\bar{x}, \bar{y} \simeq 2$). So if we wish to compare various network strategies, it seems sensible to keep a in (9.6.5) and b in (9.6.6) fixed across models by using the migration probabilities

$$\alpha \simeq 1 - (1-a)^{1/\bar{x}} \qquad \text{and} \qquad \beta \simeq 1 - (1-b)^{1/\bar{y}}. \qquad (9.6.10)$$

Simulations by Maynard Smith (1974) suggest that although the island model is more persistent than the stepping-stone model, the difference is not particularly great. The inference is that in nature persistence would be favoured if individuals could occasionally migrate long distances, so the Small World scenario (configuration (6) in Section 9.5.3) possibly offers the best scope for long-term survival.

9.6.2 Simulating the Markov chain representation

Although the above approach takes account of the way in which colonies are connected, it suffers from the defect that the equilibrium probabilities relate to a single colony, and so can only provide a flavour of likely outcomes. Moreover, its qualitative nature means that we cannot use it to develop precise statements relating to occupation and extinction probabilities. Fortunately, it is relatively straightforward to construct a general simulation algorithm which parallels the exact multi-colony algorithm A9.1.

Continuing our eight-state predator–prey illustration, first label the states $E, HA, \ldots, MC, MD$ by $1, 2, \ldots, 8$. Second, label the colonies $i = 1, 2, \ldots, N$ and construct the connection matrix $\mathbf{V} = \{\nu_{ij}\}$; so $\nu_{ij} = 1$ if colonies i and j are connected and $\nu_{ij} = 0$ if they are not. Third, construct the transition probability matrix $\mathbf{P} = \{p_{ij}\}$ corresponding to example (9.6.7). Finally, decide on the initial start configuration, e.g. $x_1(0) = 5$ and $x_i(0) = 1$ for $i = 2, \ldots, N$. Then cycle through each colony over the discrete times $t = 1, 2, \ldots$, testing both for potential change within each colony and for migration events between colonies. So whereas algorithm A9.1 is based on a unique set of event–time pairs, here there is no exact event–time ordering. Hence we have to select whether the examination of $\mathbf{P}$ precedes, is made in concurrence with, or follows that for $\mathbf{V}$. Although the following construct is based on the first of these, it is completely general; the specific transition rules employed below are used purely for illustration.

Exact Markov chain simulation algorithm (A9.9)

(i) read in parameters, transition rules for $\mathbf{P}$, and connection matrix $\mathbf{V}$
(ii) set $t = 0$ and $x_i = x_i(0)$ over $1 \leq i \leq N$
(iii) print t and $x_1, \ldots, x_N$
(iv) cycle over $i = 1, \ldots, N$ (under $\mathbf{P}$ with $a = b = 0$)
generate a new $U \sim U(0, 1)$ random number
set $x_i = k$ where $p_{x_i,1} + \cdots + p_{x_i,k-1} \leq U < p_{x_i,1} + \cdots + p_{x_i,k}$
(so here if $x_i = 5, \ldots, 8$ set $x_i = 6, 7, 8$ or 1, respectively)
(v) cycle over $i = 1, \ldots, N$ (under $\mathbf{V}|\mathbf{P}$)
cycle over $j = 1, \ldots, N$ ($j \neq i$)
generate a new $U \sim U(0, 1)$ random number
if $x_i = 1$, $\nu_{ji} = 1$, $x_j = 4, 5, 6$ or 7 and $U \leq \alpha$ then $y_i = 2$
generate a new $U \sim U(0, 1)$ random number
if $x_i = 2, 3$ or 4, $\nu_{ji} = 1$, $x_j = 6, 7$ or 8 and $U \leq \beta$ then $y_j = 5$
(vi) cycle over $i = 1, \ldots, N$
set $x_i = y_i$
(vii) update t to $t + 1$ and return to (iii)

This algorithm takes a particularly simple form under the (almost cyclic) transition scheme (9.6.7), since stochasticity is induced purely by the relative states of the nearest-neighbours. This is especially so if we simplify the migration rules even further by taking $MA \to MC \to MD \to E$ together with

$\Pr(E \to HA) = \alpha$ if a neighbour is HA, MA, MB or MC
$\Pr(HA \to MA) = \beta$ if a neighbour is MB, MC or MD, else $HA \to HB$
$\Pr(HB \to MA) = \beta$ if a neighbour is MB, MC or MD, else $HB \to HC$
$\Pr(HC \to MA) = \beta$ if a neighbour is MB, MC or MD, else $HC \to HC$.

The removal of the requirement to test each neighbouring colony separately to see whether migration occurs from it greatly reduces compute time for long runs, though it does, of course, negate the interpretation (9.6.5) for a and (9.6.6) for b.

Consider, for example, a simple spatial configuration comprising a small 3×3 lattice where the random variable $X_{ij}(t) = 1, 2, \ldots, 8$ (corresponding to $E, HA, \ldots, MD$) denotes the state occupied by cell (i, j) at time t. A single simulation run over the parameter values $\alpha, \beta = 0, 0.05, 0.1, \ldots, 1$ for $t = 1, 2, \ldots, 1000$, starting from $X_{ij}(0) = 1$ (i.e. E) with $X_{11}(0) = 5$ (i.e. MA), gave the following minimum values of β for both prey and predators to be present in at least one of the nine colonies at $t = 1000$:

α	1.00	0.95	0.90	0.85	0.80	0.75	0.70	0.65	0.60	0.55	0.50	0.45	0.40	0.35
$\beta_{min}\|\alpha$	0.45	0.45	0.60	0.60	0.55	0.50	0.65	0.75	0.80	0.80	0.75	0.95	0.95	*

So if prey are highly mobile ($\alpha \simeq 1$) then the populations persist provided the predators are reasonable mobile ($\beta \simeq 0.45$). However, as prey mobility decreases, predator mobility must gradually increase if both populations are to persist; though once α drops to 0.35 even perfect predator mobility is not sufficient to sustain the process. Figure 9.8 shows a simulated time-series for $\alpha = 0.85$ and $\beta = 0.5$ over $t = 0, \ldots, 100$, and the highly cyclic wave structure is clearly evident. However, decreasing the predator mobility rate slightly, from $\beta = 0.5$ to 0.43, causes a critical number of sites to 'jam' in state 4 (i.e. HC) (Figure 9.9). For then there are too few MB, MC and MD sites available to ensure a sufficiently high probability of predators migrating to them before the predator population becomes extinct.

Although this example is even simpler than the already highly simplified construct (9.6.7), it nevertheless highlights how extremely quick simulation experiments can be performed which provide a good feel for how the within- and between-colony rules interact with spatial structure to produce long-term persistence and interesting spatial patterns of population size. Though the discrete-time Markov chain approach can clearly never match the full continuous-time Markov process representation in terms of accuracy, it is computationally far faster, and hence offers considerable potential in the initial investigation of a spatial–temporal process.

9.6.3 Stochastic cellular automata

Whilst the use of a lattice might seem to be overly simple in terms of establishing interesting spatial structures, it actually yields a surprisingly rich source of results. Moreover, lattices are directly relevant to real-life, since in practice many situations involve sites that are located on a rectangular grid of coordinates. Though this was appreciated by Ising as early as 1925 in a Physics context, it was not until 1970 that the power of the lattice construct became widely recognised when Conway's *Game of Life* took the popular scientific press by storm.

The game is purely deterministic, and hence its evolution is totally determined by the initial state. The basic idea is that each cell (i, j) on a rectangular lattice is in one of two possible states, alive or dead. Every cell interacts with its eight direct neighbours, and at each time step $t = 1, 2, \ldots$ the following transitions occur:

568 *Spatial processes*

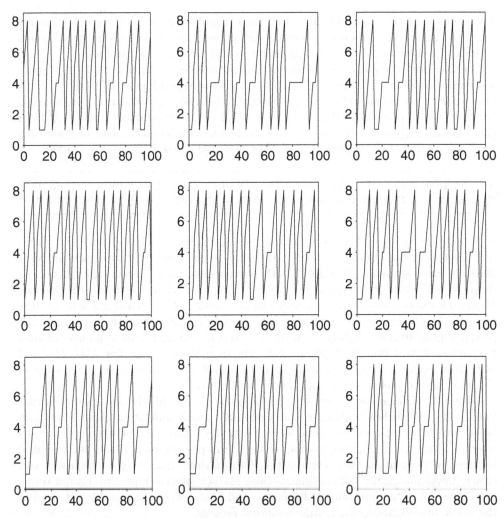

Figure 9.8 Development of a predator–prey Markov chain on a 3×3 lattice with $\alpha = 0.85$ and $\beta = 0.5$ starting with $X_{11}(0) = MA$ and $X_{ij}(0) = E$ otherwise.

(i) a live cell with $0, 1$ or $4, 5, \ldots, 8$ live neighbours dies;
(ii) a live cell with $2, 3$ live neighbours survives to the next step; and,
(iii) a dead cell with exactly 3 live neighbours comes to life.

These rules were carefully chosen, following a substantial period of experimentation, in order to meet the following three desired criteria. There should be:

(a) no initial pattern for which we can construct a simple proof that the population can grow without limit;
(b) initial patterns that appear (on the basis of long-run simulations) to grow without limit;

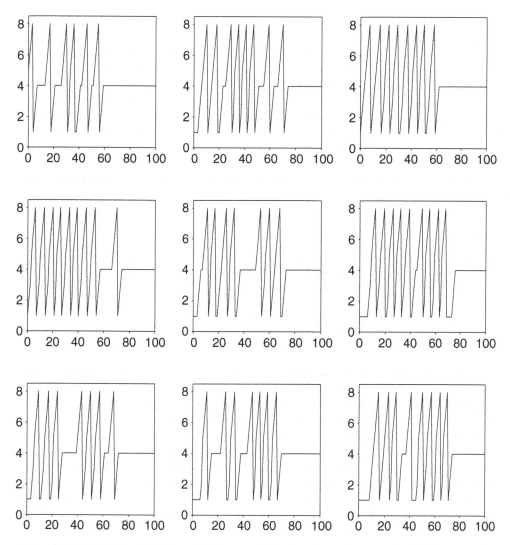

Figure 9.9 As Figure 9.8, but with β now taking the sub-persistent value 0.43.

(c) simple initial patterns that wax and wane for some time before ending by either (i) fading away completely (from overcrowding or becoming too sparse), (ii) settling into a stable configuration, or (iii) entering an oscillating phase in which cycles of length two or more are repeated indefinitely.

Conway's motivation lay in a problem posed in the 1940s by Von Neumann, who tried to construct a 'machine' that could build copies of itself. For although Von Neumann succeeded in developing such a process on a Cartesian grid, the fact that his rules were quite complicated inspired Conway to develop a far simpler algorithm. The resulting Game of Life opened up the whole new field of *cellular automata*.

570 Spatial processes

A vast array of fascinating patterns generated both by this algorithm, and countless variations of it, have been produced since his original discovery, many of which may be found on the web. Their ramifications go well beyond the production of pretty mathematical pictures, and raise important questions across a wide range of mathematics, including number theory and the study of fractals. Moreover, there are thousands of 'Life' programs online, so interested readers can easily begin trial runs of their own. The question here, however, is what happens when deterministic rules are replaced by stochastic ones? We have already seen a 3×3 predator–prey illustration of this (in the previous section) with migration probabilities α and β, though there we were primarily interested in long-term persistence. So to illustrate how the patterns themselves develop, we shall now consider two simple examples over much larger grids. The first relates to the spread of forest fires, and the second to the spread and control of foot-and-mouth disease.

Stochastic fire

Consider an $N \times N$ array of trees over the torus $i, j = 1, \ldots, N$, so there are no edge-effects, and let each tree interact independently with its four nearest-neighbours over the time-steps $t = 1, 2, \ldots$ according to the following rules:

(i) a dead tree is reborn at the next step;
(ii) a burning tree is dead at the next step; and,
(iii) a burning tree sets a neighbouring tree on fire with probability p.

Moreover, let us take the simple initial configuration of a single burning tree. Then when $p = 1$ we have a true (i.e. deterministic) cellular automata, and at each time-step fire spreads out remorsely in a diamond shape until every tree has died (and then subsequently reborn at the next step) by time $t = N$. Figure 9.10a shows this diamond configuration at $t = 20$ with $N = 50$, for a fire starting at $i = j = 26$. However, reducing p very slightly to 0.99 enables a single susceptible tree to remain unburnt, which is sufficient to enable the fire to burn back on itself (Figure 9.10b; $t = 30$) and hence establish a pattern of travelling nested diamonds (Figure 9.10c; $t = 100$). This structure is still very much in evidence at $t = 500$ (Figure 9.10d), and indeed persists for a long time thereafter. Thus introducing a tiny amount of random variation not only results in a completely different pattern, but unlike the pure automata (i.e. $p = 1$) which ceases movement at $t = N$, the fire continues to burn virtually indefinitely with a long-term underlying (diagonal) pattern. Successively reducing p reduces the strength of this diagonal structure, until a point is reached where local fires come and go virtually independently of each other. A plot of the total number of trees on fire, $n_f(t)$, against t for $p = 0.4$ (Figure 9.11) exhibits four full cycles of development before the fire eventually becomes extinct at $t = 993$. The process clearly takes some time to get going, and the spatial pattern at $t = 100$ (Figure 9.12a) displays five localized fires in the 'lower' part of the torus, with no fire occurring at all in the 'upper' part. After reaching a local maximum of 57 at $t = 211$, $n_f(t)$ then decays back to 1 at $t = 344$ and 363, at which point the fire could die out completely. For the probability that none of the four neighbouring trees catches fire, and hence continues the process, is $(1 - p)^4 = 0.13$; with just three and two neighbouring alive trees this probability

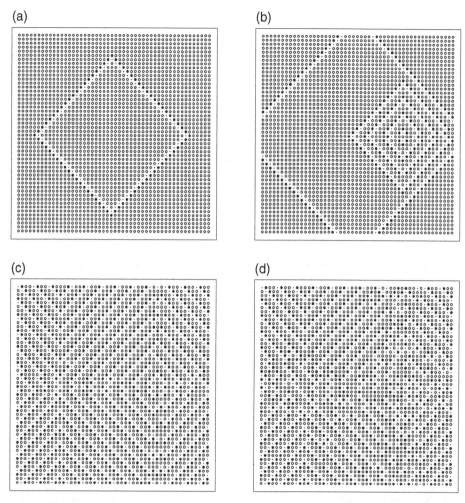

Figure 9.10 Patterns of stochastic fire for $p = 1$ at (a) $t = 20$, and $p = 0.99$ at (b) $t = 30$, (c) $t = 100$ and (d) $t = 500$; showing trees that are dead ($\cdot$), alive (o) and on fire ($\bullet$).

increases to 0.22 and 0.36, respectively. So random variation ensures that one of these population drops will eventually result in extinction, which does indeed occur at $t = 993$. Figure 9.12b shows the highest-density pattern which takes place shortly beforehand, at $t = 792$ when $\max(n_f) = 99$. Note that the fire population consistently comprises localised outbreaks, with substantial gaps existing between them, that wax and wane and spawn new small outbreaks as time progresses.

The extent to which natural variation causes large fluctuations both in pattern structure and population size poses interesting questions such as: over what range of p and associated time $0 \leq t \leq T_p$ does some form of diagonal structure exist; and, for a given initial configuration how does the p.d.f. of the time to extinction, T_{ext}, vary with

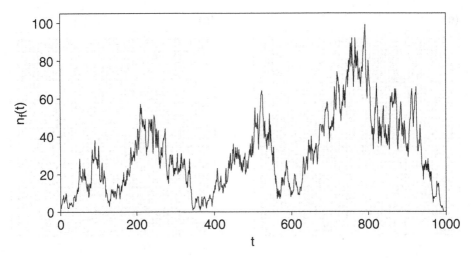

Figure 9.11 Number of trees, $n_f(t)$, on fire at $t = 0, 1, \ldots$ until extinction at $t = T_{ext} = 993$, for $p = 0.4$.

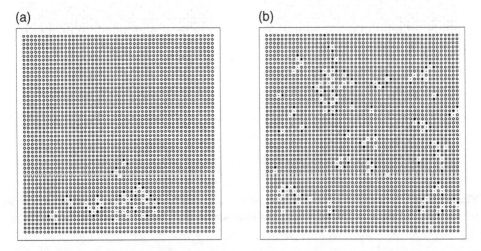

Figure 9.12 Patterns of stochastic fire for $p = 0.4$ at (a) $t = 100$ (low intensity) and (b) $t = 792$ (maximum intensity); showing trees that are dead ($\cdot$), alive (o) and on fire ($\bullet$).

p and N? Note that for this simulation run $p = 0.4$ was the smallest value that did not lead to short-term extinction. As the lattice size N increases, much lower values of p will lead to sustained fire, which raises the issue of determining the associated limiting value $p_{min} = \lim_{N \to \infty} p$.

Although such models can clearly be nothing more than highly simplified versions of reality, their simplicity ensures that they can be understood and interpreted by non-mathematicians, and therein lies their power. For if different types of fire control strategy are proposed, such as regular versus random introduction of firebreaks, or water-bombing, then they can be swiftly tested in a computer simulation. Indeed, use of *QBasic* is to be highly recommended in such situations, since it is trivial to program, produces acceptably good real-time graphics, and is eminently transferable between PCs.

Foot-and-mouth disease

The 2001 UK-wide outbreak of foot-and-mouth disease (e.g. http://www.defra.gov.uk/footandmouth) provides a good illustration of using this approach to gain an initial insight into various control strategies. Whilst a full-scale analysis would

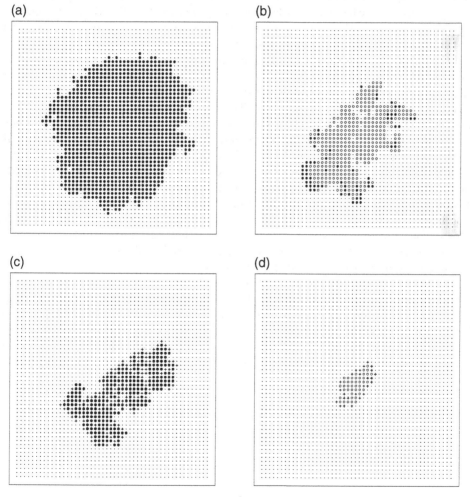

Figure 9.13 Patterns of foot-and-mouth disease at time $t = 50$ for infection rate $p = 0.2$: (a) $q = r = 0$, (b) $q = 0.2$, $r = 0$, (c) $q = 0$, $r = 0.1$, and (d) $q = r = 0.1$ (at termination time $t = 44$) showing farms that are susceptible (·), infected (•), burnt (o) and culled (+).

have to be built around complete spatial knowledge of farm locations, boundaries, roads, wildlife abundance, etc., an initial study of possible control measures can be undertaken extremely quickly just by making minor modifications to the stochastic fire algorithm. For simplicity, suppose that farms are located on a rectangular lattice, and that each day there is a probability p that the disease may be transmitted from an infected farm to each one of its four nearest-neighbours, independently. Moreover, let us presume that control takes place either by burning cattle on an infected farm with probability q, or by culling all susceptible cattle on its four nearest-neighbouring farms with probability r.

Suppose, for example, that the daily infection probability $p = 0.2$. Then in the absence of any control measures (i.e. $q = r = 0$) infection spreads rapidly outwards in a roughly circular path (Figure 9.13a shows the pattern at $t = 50$). Although employing a burning strategy with $q = 0.2$ does stop the infection spreading in some directions, it still develops in others (Figure 9.13b), and we need to increase q to 0.3 in order to end the outbreak; even then it can take around 40 days for this to happen. Similarly, using a pure culling strategy with $r = 0.1$ only slows down the relentless spread of the infection (Figure 9.13c); though increasing r to 0.2 does cause the outbreak to end relatively quickly in about 30 days. Whilst this implies that burning is less effective than culling, it should be noted that burning may well require substantially less labour effort; for up to four neighbouring farms may need to be culled rather than burning cattle at the single infected farm. Nevertheless, the environmental implications of using a high burn rate, and the labour costs involved in using a high cull rate, suggests that a combination of these two strategies might prove to be optimal. Figure 9.13d demonstrates that such a mixed control procedure works well even when $q = r = 0.1$ (the infection terminates at $t = 44$). Indeed, by exploring a range of values for q and r for various infection rates p, it is easy to develop ball-park estimates for appropriate cull and burn rates. These can then be used as initial trial values in more detailed, and hopefully more accurate, GIS-based simulations.

The beauty of this technique is that by distilling the essential ingredients of a complex process into a few discrete states, with movement described through simple transition rates over a lattice, we can glean a considerable amount of qualitative behavioral information on its spatial–temporal development. Once armed with a quiver of likely properties to investigate, we can then target specific quantitative questions via either a theoretical lattice-based study (Section 10.1) or a simulation-based marked point process approach (Section 10.2).

10
Spatial–temporal extensions

Although the Turing model (Section 9.5) and the Markov chain process (Section 9.6) are ideal for studying systems which involve spatial interaction between adjacent sites, in practice interaction may occur across much larger spatial levels. Moreover, there is no reason why locations have to lie on a lattice structure, which has been our presumption so far in order to enable some degree of mathematical tractability. To conclude we shall therefore present two extensions to our earlier spatial analyses. The first introduces the concept of long-range dependence, whilst the second examines processes which develop over real, rather than discrete, space.

10.1   Power-law lattice processes

Examples of six types of migration structure have already been noted in Section 9.5.3, whilst a major component in the initial spread of the 2001 UK foot-and-mouth epidemic (discussed in Section 9.6.3) is the unwitting long-distance transport of infected cattle. However, since we have already shown that even the simple two-colony birth–death–migration process is intractable to a full stochastic analysis (Section 9.2), there is clearly no possibility of developing complete theoretical investigations of multiplicative models involving large-scale spatial interaction. This forces us into adopting pragmatic approximations if we are to develop reasonably transparent, and hence meaningful, probabilistic results.

Following the growth-competition route of Renshaw (1984), let us retain our two-dimensional grid of points $-\infty < i, j < \infty$, thereby avoiding problems of edge-effects, and develop a linear interaction model comprising a deterministic component that is subject to random variation. Denote $\{X_{ij}(t)\}$ to be the set of random variables at time $t \geq 0$ over the lattice $-\infty < i, j < \infty$. Let each $X_{ij}(t)$ grow deterministically at rate λ, with two variables a distance (r, s) apart interacting deterministically at rate a_{rs} times their difference. Whilst in each time increment $(t, t+dt)$ suppose that $X_{ij}(t)$ is subject to a white noise perturbation $dZ_{ij}(t)$, where the $\{dZ_{ij}(t)\}$ are i.i.d. random variables with mean 0 and variance $\sigma^2 dt$. Then we have the linear equation

$$X_{ij}(t+dt) = (1+\lambda dt)X_{ij}(t) + \sum_{r,s=-\infty}^{\infty} a_{rs}[X_{ij}(t) - X_{i+r,j+s}(t)]dt + dZ_{ij}(t) + \mathrm{o}(dt),$$

(10.1.1)

where $\mathrm{E}[dZ_{ij}(t)] = 0$ and

$$\mathrm{E}[dZ_{ij}(t)dZ_{kl}(s)] = \begin{cases} \sigma^2 dt & \text{if } i=k, j=l, t=s \\ 0 & \text{otherwise.} \end{cases}$$

It follows from (10.1.1) that if $a_{rs} > 0$ then not only will an $X_{i+r,j+s} > X_{ij}$ *suppress* X_{ij}, but also an $X_{i+r,j+s} < X_{ij}$ will *enhance* X_{ij}. So for situations in which suppression is realistic but enhancement is not (e.g. plant growth), λ must be interpreted as being the average growth rate under spatial interaction and not as the growth rate in the absence of such interaction. In Section 10.2 we shall not only consider more general interaction regimes which circumvent this issue, but we shall also remove the lattice assumption and thereby develop processes over the real plane.

If $a_{rs} < 0$ ($-\infty < r, s < \infty$) then spatial interaction tries to regress large and small X_{ij} (situated a distance (r, s) apart) back to the mean mark. So we may anticipate the existence of an equilibrium structure based around the overall population mean. Also, if the interaction weights are symmetric, then in two dimensions the one-dimensional second-order difference term at lag r, namely $a_r\{X_{i+r} - 2X_r + X_{i-r}\}$ (since $a_{-r} \equiv a_r$), becomes $a_r \Delta^2_{(r)} X$; Whittle (1962) develops this analogous diffusion model within the context of agricultural field trials, with $\Delta^2 X$ representing the spread of fertility through the medium.

10.1.1   First- and second-order moments

To develop the basic underlying ideas let us first construct the associated means and variances. Denote $m_{ij}(t) = \mathrm{E}[X_{ij}(t)]$. Then taking expectations on both sides of (10.1.1) and letting $dt \to 0$ gives

$$dm_{ij}(t)/dt = (\lambda + \sum_{r,s=-\infty}^{\infty} a_{rs}) m_{ij}(t) - \sum_{r,s=-\infty}^{\infty} m_{i+r,j+s}(t). \qquad (10.1.2)$$

Let the generating function

$$G(z_1, z_2; t) \equiv \sum_{i,j=-\infty}^{\infty} m_{ij}(t) z_1^i z_2^j \qquad (10.1.3)$$

be defined within some suitable domain of convergence. Then multiplying equation (10.1.2) by $z_1^i z_2^j$ and summing over $-\infty < i, j < \infty$ yields

$$\frac{\partial G(z_1, z_2; t)}{\partial t} = [\lambda + \sum_{r,s=-\infty}^{\infty} a_{rs}] G(z_1, z_2; t) - [\sum_{r,s=-\infty}^{\infty} a_{rs} z_1^{-r} z_2^{-s}] G(z_1, z_2; t), \qquad (10.1.4)$$

which integrates to give

$$G(z_1, z_2; t) = [\sum_{i,j=-\infty}^{\infty} m_{ij}(0) z_1^i z_2^j] \exp\{t[\lambda + \sum_{r,s=-\infty}^{\infty} a_{rs}(1 - z_1^{-r} z_2^{-j})]\}. \qquad (10.1.5)$$

Thus if the process starts off with a single individual at $i = j = 0$ at time $t = 0$, then $m_{ij}(t)$ is given by the rather awkward coefficient of $z_1^i z_2^j$ in

$$G^0(z_1, z_2; t) = \exp\{t[\lambda + \sum_{r,s=-\infty}^{\infty} a_{rs}]\} \times \exp\{-t \sum_{r,s=-\infty}^{\infty} a_{rs} z_1^{-r} z_2^{-s}\}. \quad (10.1.6)$$

Though if $m_{ij}(0) = \mu$ for $-\infty < i, j < \infty$, then (10.1.5) collapses to give

$$m_{ij}(t) = \mu e^{\lambda t} \quad (-\infty < i, j < \infty), \quad (10.1.7)$$

which is the expected size of an isolated, i.e. non-spatially interactive, population.

To generate the variance-covariances

$$\sigma_{uv}(t) \equiv \mathrm{E}[\{X_{ij}(t) - m_{ij}(t)\}\{X_{i+u,j+v}(t) - m_{i+u,j+v}(t)\}],$$

let us assume that the $\{X_{ij}(0)\}$ are i.i.d. random variables with mean μ and variance τ^2. Then using (10.1.7) gives

$$\mathrm{E}[X_{ij}(t) X_{i+u,j+v}(t)] = \sigma_{uv}(t) + \mu^2 e^{2\lambda t}.$$

Whence forming $\mathrm{E}[X_{ij}(t+dt) X_{i+u,j+v}(t+dt)]$ from (10.1.1) and letting $dt \to 0$ leads to

$$\frac{d\sigma_{uv}(t)}{dt} = 2\lambda \sigma_{uv}(t) + \sum_{r,s=-\infty}^{\infty} a_{rs}[2\sigma_{uv}(t) - \sigma_{u+r,v+s}(t) - \sigma_{u-r,v-s}(t)] + \sigma^2 \delta(u,v;0,0),$$

$$(10.1.8)$$

where $\delta(u,v;0,0)$ denotes the Kronecker delta function. On defining the generating function

$$H(z_1, z_2; t) \equiv \sum_{u,v=-\infty}^{\infty} \sigma_{uv}(t) z_1^u z_2^v, \quad (10.1.9)$$

this equation takes the form

$$\frac{\partial H(z_1, z_2; t)}{\partial t} = \{2\lambda + \sum_{r,s=-\infty}^{\infty} a_{rs}(2 - z_1^{-r} z_2^{-s} - z_1^r z_2^s)\} H(z_1, z_2; t) + \sigma^2, \quad (10.1.10)$$

which integrates to yield the solution

$$H(z_1, z_2; t) = H(z_1, z_2; 0) e^{\psi t} + (\sigma^2/\psi)(e^{\psi t} - 1) \quad (10.1.11)$$

where

$$\psi(z_1, z_2) = 2\lambda + \sum_{r,s=-\infty}^{\infty} a_{rs}(2 - z_1^{-r} z_2^{-s} - z_1^r z_2^s). \quad (10.1.12)$$

In the absence of any initial spatial structure $H(z_1, z_2; 0) = \tau^2$, whence (10.1.11) reduces to

$$H(z_1, z_2; t) = \tau^2 e^{\psi t} + (\sigma^2/\psi)(e^{\psi t} - 1). \quad (10.1.13)$$

Even so, determining $\sigma_{uv}(t)$ from the coefficient of $z_1^u z_2^v$ in (10.1.13) does not produce a user-friendly expression, since it involves the exponential series expansion of a doubly infinite series. This reflects the fact that the autocovariances of a spatial–temporal process are highly-interdependent. Indeed, the autocovariance structure $\{\sigma_{uv}(t)\}$ may be substantially more involved than the process $\{X_{ij}(t)\}$ under study. Fortunately, we can avoid such difficulties by working with the associated spectral density function instead.

10.1.2 The spectrum

Spectral analysis is an especially important technique in the analysis of spatial–temporal stochastic processes, since the relative lack of dependence between values in the spectrum at different frequencies enables it to provide a comprehensive description of both the structure and scales of pattern in a spatial data set. Moreover, it assumes no structural characteristics in the data prior to analysis, which is of great advantage in many practical situations where pattern may exist over a range of scales, be anisotropic (i.e. have directional components) and non-stationary. All of these may be key features in real-life pattern generating mechanisms, and, ideally, should be explained prior to model building. For an introduction to the basic spectral techniques involved in the interrogation and subsequent interpretation of spatial pattern see Renshaw and Ford (1983) and Ford and Renshaw (1984).

The spectrum is the Fourier transform of the autocovariances, namely

$$f(\omega_1, \omega_2; t) = \sum_{u,v=-\infty}^{\infty} \sigma_{uv}(t) e^{i(\omega_1 u + \omega_2 v)} \equiv H(e^{i\omega_1 u}, e^{i\omega_2 v}; t). \quad (10.1.14)$$

So we see from (10.1.11) that

$$f(\omega_1, \omega_2; t) = f(\omega; 0) e^{\psi t} + (\sigma^2/\psi)(e^{\psi t} - 1), \quad (10.1.15)$$

where, from (10.1.12),

$$\psi \equiv \psi(e^{i\omega_1}, e^{i\omega_2}) = 2\{\lambda + \sum_{r,s=-\infty}^{\infty} a_{rs}[1 - \cos(r\omega_1 + s\omega_2)]\}. \quad (10.1.16)$$

Since results for the one-dimensional process $\{X_i(t)\}$ easily translate into higher dimensions, for simplicity let us just consider the one-dimensional weights $\{a_r\}$. Expression (10.1.15) then becomes

$$f(\omega; t) = f(\omega; 0) e^{\psi t} + (\sigma^2/\psi)(e^{\psi t} - 1), \quad (10.1.17)$$

with (10.1.16) changing to

$$\psi(\omega) = 2\{\lambda + \sum_{r=-\infty}^{\infty} a_r[1 - \cos(r\omega)]\}. \quad (10.1.18)$$

If the $\{X_i(0)\}$ are i.i.d. with mean μ and variance τ^2, as in (10.1.13) for the autocovariances, then $f(\omega; 0) = \tau^2$. Whilst in the absence of added noise, i.e. $\sigma^2 = 0$, expression (10.1.17) reduces to straight exponential growth with

$$\ln[f(\omega; t)] = \ln[f(\omega; 0)] + 2t\{\lambda + \sum_{r=-\infty}^{\infty} a_r[1 - \cos(r\omega)]\}. \quad (10.1.19)$$

As t increases, the asymptotic behaviour of $f(\omega;t)$ clearly depends on the sign of ψ. First, if $a_r \geq 0$ for all r (i.e. straight competition) and $\lambda > 0$, then we see from (10.1.18) that $\psi(\omega) > 0$ for all ω. So as $t \to \infty$

$$f(\omega;t) \simeq [f(\omega;0) + (\sigma^2/\psi)]e^{\psi t}, \qquad (10.1.20)$$

and hence any initial structure does not lose its influence as $t \to \infty$. This is because the additive noise $\{dZ_i(t)\}$ becomes swamped by the exponential growth of the $\{X_i(t)\}$. Second, if some $a_r < 0$, then we may have a situation in which $\psi(\omega) < 0$ for $\omega \in \Omega$ for some set $\Omega \subset [0,\pi]$. Thus as $t \to \infty$

$$f(\omega;t) \sim \begin{cases} -\sigma^2/\psi & : \text{ for } \omega \in \Omega \\ [f(\omega;0) + (\sigma^2/\psi)]e^{\psi t} & : \text{ otherwise.} \end{cases} \qquad (10.1.21)$$

Since $-\sigma^2/\psi$ is independent of both the initial spatial configuration and time, we denote a process which confirms to the structure (10.1.21) as being Ω-*stationary*. The *threshold frequencies*, $\{\omega_c\}$, between these two types of asymptotic behaviour are the solutions to the equation $\psi(\omega) = 0$, i.e.

$$\lambda + \sum_{r=-\infty}^{\infty} a_r[1 - \cos(r\omega)] = 0. \qquad (10.1.22)$$

Hence by making a suitable choice of the interaction weights $\{a_r\}$ we can construct a process which is stationary only within chosen scales of pattern, i.e. frequencies $\omega \in \Omega$. Note that this scenario does not give rise to an equilibrium distribution unless $\Omega = [0,\pi]$.

Although the spectral result (10.1.15) has been derived by first determining the autocovariance structure (10.1.11), it may also be derived directly. For denote the (scaled) Fourier transform

$$L(\omega;t) = \lim_{N \to \infty} (1/\sqrt{2N}) \sum_{k=-\infty}^{\infty} e^{i\omega k}\{X_k(t) - m_k(t)\}, \qquad (10.1.23)$$

with complex conjugate $L^*(\omega;t) = L(-\omega;t)$. Then on retaining the implicit underlying assumption of spatial stationarity, writing $k = l + u$ shows that

$$E[L(\omega;t)L^*(\omega;t)] = \lim_{N \to \infty} (1/2N) E\left[\sum_{k=-N}^{N} e^{ik\omega}\{X_k(t) - m_k(t)\} \sum_{l=-N}^{N} e^{-il\omega}\{X_l(t) - m_l(t)\}\right]$$

$$\simeq \lim_{N \to \infty} (1/2N) E\left[\sum_{u=-N}^{N} \sum_{l=-N}^{N} e^{i(l+u)\omega}\{X_{l+u}(t) - m_{l+u}(t)\}e^{-il\omega}\{X_l(t) - m_l(t)\}\right]$$

$$\simeq \lim_{N \to \infty} (1/2N) \sum_{u=-N}^{N} \sum_{l=-N}^{N} e^{iu\omega} E[\{X_l(t) - m_l(t)\}\{X_{l+u}(t) - m_{l+u}(t)\}]$$

$$\simeq \lim_{N \to \infty} (1/2N) \sum_{u=-N}^{N} (2N) e^{iu\omega} \sigma_u(t) = \sum_{u=-\infty}^{\infty} e^{iu\omega} \sigma_u(t) = f(\omega;t) \qquad (10.1.24)$$

(using (10.1.14)). This illustrative limiting argument can be made exact through proper choice of the summation limits. Now taking the Fourier transform of equation (10.1.1) leads to

$$L(\omega; t+dt) = (1+\lambda dt)L(\omega; t) + \sum_{r=-\infty}^{\infty} a_r\{L(\omega; t) - e^{-ir\omega}L(\omega; t)\} + \sum_{r=-\infty}^{\infty} e^{ir\omega}dZ_r(t) + o(dt).$$

Whence forming (10.1.24) by multiplying this equation by its conjugate complex, taking expectations of both sides, and recalling (10.1.18), leads to the simple differential equation

$$df(\omega; t)/dt = [2\lambda + \sum_{r=-\infty}^{\infty} a_r(2 - e^{ir\omega} - e^{-ir\omega})]f(\omega; t) + \sigma^2 = \psi(\omega)f(\omega; t) + \sigma^2$$
(10.1.25)

which integrates to yield the spectrum (10.1.17), as required. Note that this is the purely *spatial* spectrum conditional on a fixed time t; one could also construct the full *spatial–temporal* spectrum as well.

On a general point, in spite of the excellent early attempt by Bartlett (1971, 1974) to promote the spectral approach, statisticians and probabilists have generally placed much greater emphasis on the space domain. This contrasts with the viewpoint taken by physicists and electrical engineers who have fully embraced the spectral domain for many decades, presumably because Fourier coefficients are both a natural choice for the transmission of data and also suffer far less interdependence than moments. So whilst space domain analyses certainly have a lot to offer, conducting frequency domain analyses in concert with them is likely to lead to a greater diversity of results, and hence a deeper understanding of the process under study.

Nearest-neighbour interaction For our first example let us consider symmetric nearest-neighbour interaction with $a_1 = a_{-1} = \alpha$ and $a_r = 0$ otherwise. Then expression (10.1.18) reduces to

$$\psi(\omega) = 2[\lambda + 2\alpha - 2\alpha\cos(\omega)] = 2\lambda + 8\alpha\sin^2(\omega/2). \quad (10.1.26)$$

Thus if $\lambda > 0$ and $\alpha > -\lambda/4$, then $\psi(\omega) > 0$ for all $0 \le \omega \le \pi$. However, if $\alpha < -\lambda/4$ then both $\psi(\omega) > 0$ and $\psi(\omega) < 0$ are possible, with the threshold frequency (from equation (10.1.22)) between these two regimes being

$$\omega_c = \cos^{-1}(1 + \lambda/2\alpha). \quad (10.1.27)$$

Combining results (10.1.21) and (10.1.26) therefore gives rise to two contrasting situations.

(a) $\alpha > -\lambda/4$: here $\psi(\omega) > 0$ for all $0 \le \omega \le \pi$. So for large t the spectrum grows exponentially with

$$f(\omega; t) \sim [f(\omega; 0) + \sigma^2\{2\lambda + 8\alpha\sin^2(\omega/2)\}^{-1}]\exp\{t[2\lambda + 8\alpha\sin^2(\omega/2)]\}.$$
(10.1.28)

(b) $\alpha < -\lambda/4$: now $\psi(\omega) > 0$ for $0 \leq \omega \leq \omega_c$, so expression (10.1.28) holds true only within this lower frequency range. Since $\psi(\omega) < 0$ for $\omega_c < \omega \leq \pi$, within this higher, Ω-*stationary*, range

$$f(\omega;t) \sim -\sigma^2/\psi(\omega) = -\sigma^2/[2\lambda + 8\alpha \sin^2(\omega/2)]. \quad (10.1.29)$$

In (a) some of the $\{X_i(t)\}$ explode exponentially fast, since a large $|X_i(t)|$ (relative to its two neighbours) retains its dominance because the stochastic perturbations $\{dZ_i(t)\}$ have fixed variance as t increases. The same holds true in (b) over large scales of distance; though over small scales the inhibition factor α is sufficiently strong to render the process *locally stationary*. So the wavelength $2\pi/\omega_c$ corresponds to an *outer scale of pattern*. Features of smaller wavelength than this will appear stationary; those of longer wavelength continually grow in amplitude.

If $\lambda = 0$ and $\alpha > 0$ then expression (10.1.29) reduces to

$$f(\omega;t) \sim -(\sigma^2/2\alpha)\omega^{-2} \quad \text{(for } \omega \simeq 0\text{)}, \quad (10.1.30)$$

which is a genuine power law reflecting correlation effects right across the spatial process. This inverse square phenomenon is encountered throughout the natural and man-made physical universe. Akaike (1960), for example, uses it to study the roughness of roads and runways; whilst Sayles and Thomas (1978) cite examples with wavelengths from less than 10^{-5} m to 10^2 m, including the Fort Knox tank-proving ground, lava-flows, hip-joints and ships' hulls. The parameter $k = -\pi\sigma^2/\alpha$ is called the 'topothesy' of the random surface.

More specifically, it follows from (10.1.21) and (10.1.26) that

$$\ln[f(\omega;\infty)] = \ln(-\sigma^2/8\alpha) - 2\ln[\sin(\omega/2)]. \quad (10.1.31)$$

Now

$$\ln[\sin(\theta)] = \ln(\theta) - \sum_{n=1}^{\infty} \frac{2^{2n-1}}{n(2n)!} B_n \theta^{2n} \quad (\theta^2 < \pi^2) \quad (10.1.32)$$

where the Bernoulli numbers $B_0 = -1$, $B_1 = 1/6$, $B_2 = 1/30$, $B_3 = 1/42$, $B_4 = 1/30$, $B_5 = 5/66$, ... (e.g. Jolley, 1961). Thus

$$\ln[f(\omega;\infty)] = \ln(-\sigma^2/8\alpha) - 2\ln(\omega/2) + \frac{\omega^2}{12} + \frac{\omega^4}{1440} + \frac{\omega^6}{90720} + \frac{\omega^8}{4838400} + \cdots, \quad (10.1.33)$$

and so $f(\omega;\infty)$ decays as ω^{-2} even when ω is quite close to, albeit less than, 1.

At this point it is worth reflecting on two forms of *non-temporal* spatial process. The first of these is the *simultaneous* model (Whittle, 1954) for which the nearest-neighbour case is

$$X_i = \beta(X_{i-1} + X_{i+1}) + Z_i, \quad (10.1.34)$$

where the $\{Z_i\}$ are i.i.d. random variables with $E(Z_i) = 0$. The corresponding spectral function is proportional to

$$[1 - 2\beta \cos(\omega)]^{-2}, \quad (10.1.35)$$

in direct contrast to expression (10.1.29), namely (with $\lambda = 0$)

$$f(\omega; t) \sim (\sigma^2/4\alpha)[1 - \cos(\omega)]^{-1} \qquad (10.1.36)$$

which has negative index -1. Let us therefore follow Bartlett (1974, 1975) and expand (10.1.35) in the form

$$[1 + 2\beta^2 - 4\beta \cos(\omega) + 2\beta^2 \cos(2\omega)]^{-1}. \qquad (10.1.37)$$

For this corresponds to the second-order *conditional* scheme (Besag, 1972)

$$E[X_i | \text{all other values}] = \gamma_1(X_{i-1} + X_{i+1}) + \gamma_2(X_{i-2} + X_{i+2}) \qquad (10.1.38)$$

with $\gamma_1 = 2\beta/(1 + 2\beta^2)$ and $\gamma_2 = -\beta^2/(1 + 2\beta^2)$. Now at first sight one might suppose that the equivalent spatial–temporal scheme may be found by comparing expression (10.1.37) with that derived from the general result (10.1.18) and (10.1.20). A suitable second-nearest-neighbour model would then have

$$\psi(\omega) = 2\{\lambda + 2a_1(1 - \cos(\omega)) + 2a_2(1 - \cos(2\omega))\}$$
$$= 2k\{(1 + 2\beta^2) - 4\beta \cos(\omega) + 2\beta^2 \cos(2\omega)\} \qquad (10.1.39)$$

for some constant k. Whence equating coefficients would give $a_1 = 2k\beta$ and $a_2 = -k\beta^2$ with $\lambda = k(1 + 2\beta^2) - 4k\beta + 2k\beta^2 = k(1 - 2\beta)^2$. However, we see from (10.1.39) that $\psi(\omega) = 2k(1 - 2\beta \cos(\omega))^2$, which is negative (for all ω) if and only if $k < 0$. This implies that $\lambda < 0$, so the simultaneous scheme (10.1.34) is compatible only with the stationary spatial–temporal *decay* model (10.1.1). For then negative λ prevents individual $X_i(t)$-values from becoming too large.

An obvious question to consider is whether the spatial structure changes substantially when interaction is allowed over distances $r > 1$, i.e. $a_r > 0$ for some $|r| > 1$? An indication that it might well do so comes from early work on spatial epidemic models (Mollison, 1977) which introduces a *contact distribution* of infection that is analogous to our $\{a_r\}$-distribution of interaction. For as we have already noted in Section 9.4.2, simulations based on several simple epidemic models show that when the contact distribution is exponentially bounded then the wavefront advances at a steady rate. However, when this condition is not satisfied then the advance is unsteady, and for a contact distribution possessing just infinite variance the "wavefront progresses in wilder and wilder leaps forward" (Mollison, 1972). Although this comparison is not perfect, it does suggest that as $\{a_r\}$ ranges through increasingly tail-heavy distributions the underlying process may well exhibit marked changes in behaviour.

Geometric interaction Taking the exponentially bounded case first, consider the geometric interaction regime

$$a_0 = 0 \quad \text{and} \quad a_r = k\beta^{|r|} \quad (r = \pm 1, \pm 2, \ldots). \qquad (10.1.40)$$

Expression (10.1.18) then reduces to

$$\psi(\omega) = 2\lambda + \{2d(1 + \beta)(1 - \cos(\omega))\}/(1 - 2\beta \cos(\omega) + \beta^2), \qquad (10.1.41)$$

where $d = \sum_{r=-\infty}^{\infty} a_r = 2k\beta/(1-\beta)$. Thus the associated threshold frequency (determined from $\psi(\omega_c) = 0$) is given by

$$\omega_c = \cos^{-1}[\{d(1+\beta) + \lambda(1+\beta^2)\}/\{d(1+\beta) + 2\beta\lambda\}], \qquad (10.1.42)$$

provided $-2 < \lambda(1-\beta)^2/\{d(1+\beta) + 2\beta\lambda\} < 0$. On letting $\beta \to 0$ whilst keeping d fixed, we see from (10.1.41) that $\psi(\omega) \to 2\lambda + 2d(1 - \cos(\omega))$, which agrees with the nearest-neighbour result (10.1.26) for $d = 2\alpha$. Moreover, if $\omega \simeq 0$ then $\psi(\omega) \simeq 2\lambda + d(1+\beta)\omega^2/(1-\beta)^2$, which has the same form as that obtained under the nearest-neighbour regime, namely $\psi(\omega) \simeq 2\lambda + 2\alpha\omega^2$. So for small frequencies the spectra are essentially equal provided $\alpha = d(1+\beta)/[2(1-\beta)^2]$. Similarly, equality at high frequencies (i.e. $\omega \simeq \pi$) requires $\alpha = d/[2(1+\beta)]$.

Logarithmic interaction To examine the effect of introducing non-exponentially bounded interaction weights, suppose we now consider the much more slowly decaying interaction regime

$$a_0 = 0 \quad \text{and} \quad a_r = k(-1)^{|r|+1}/|r| \quad (r = \pm 1, \pm 2, \ldots). \qquad (10.1.43)$$

Here the $\{a_r\}$ have to oscillate in sign to enable $\sum a_r$ to converge. Since $\sum_{r=-\infty}^{\infty} a_r = 2k\ln(2)$, expression (10.1.18) reduces to

$$\psi(\omega) = 2\lambda - 4k\ln[\cos(\omega/2)]. \qquad (10.1.44)$$

Note that the process can be Ω-stationary (i.e. $\psi(\omega) < 0$ for some $\omega \in (0, \pi)$) only if $k < 0$, i.e. if $a_{\pm 1} < 0$, in line with our earlier nearest-neighbour result; the threshold frequency $\omega_c = 2\cos^{-1}[\exp(\lambda/2k)]$. Result (10.1.44) can be made more transparent by applying the Bernoulli number expansion (e.g. Jolley, 1961)

$$\ln[\cos(\theta)] = -\sum_{n=1}^{\infty} \frac{2^{2n-1}(2^n - 1)}{n(2n)!} B_n \theta^{2n} \qquad (\theta^2 < \pi^2/4) \qquad (10.1.45)$$

to obtain

$$\psi(\omega) = 2\lambda + k\left(\frac{\omega^2}{6} + \frac{\omega^4}{240} + \frac{\omega^6}{6480} + \cdots\right) \qquad (\omega < \pi). \qquad (10.1.46)$$

So, for appropriately small $\omega > \omega_c$, the Ω-stationary spectrum $f(\omega; \infty) = -\sigma^2/\psi(\omega) = -\sigma^2/(2\lambda + k\omega^2/6)$. Being an ω^{-2} law for $\lambda = 0$ and $\omega \simeq 0$, this therefore tells us nothing fundamentally new at low frequencies. At high frequencies, however, putting $\omega = \pi - \epsilon$ gives $f(\omega; \infty) \simeq \sigma^2/[4k\ln(\epsilon)]$ as $\epsilon \to 0$, so here the heavy tail of the logarithmic interaction distribution has left its mark.

Cauchy interaction This retention of the ω^{-2} law for small ω is not really surprising, since the weights (10.1.43) alternate in sign and therefore relate to a high-frequency structure. To construct a process which has a fundamentally different structure we require variables $X_i(t)$ that are far apart to be highly correlated with each other. That is, we need to construct a process with *long-range memory dependence*. Suppose, therefore, that we take $\{a_r\}$ to be positive and slowly decaying. Consider the rather extreme example

$$a_0 = 0 \quad \text{and} \quad a_r = k/[|r|(|r|+1)] \quad (r = \pm 1, \pm 2, \ldots) \qquad (10.1.47)$$

since none of the moments of $\{a_r\}$ exist. Expression (10.1.18) reduces to

$$\psi(\omega) = 2\lambda + 2k[1 - \cos(\omega)]\ln[2 - 2\cos(\omega)] - 2k(\pi + \omega)\sin(\omega). \qquad (10.1.48)$$

Whence on writing $\ln[2 - 2\cos(\omega)] = \ln(4) + 2\ln[\sin(\omega/2)]$, applying the Bernoulli number expansion (10.1.32) yields

$$\psi(\omega) = 2\lambda + 2k[1 - \cos(\omega)]\left[\ln(4) + 2\ln(\omega/2) - \sum_{n=1}^{\infty} \frac{2^{2n}}{n(2n)!} B_n (\omega/2)^{2n}\right] - 2k(\pi + \omega)\sin(\omega). \qquad (10.1.49)$$

Thus if ω is small, then $\psi(\omega) \simeq 2\lambda + 2k\omega^2 \ln(\omega/2) - 2k\omega$. Now $\omega^\beta \ln(\omega) \to 0$ as $\omega \to 0$ for all $\beta > 0$. Hence $\omega^2 \ln(\omega/2) = o(\omega)$, and so $\psi(\omega) \simeq 2\lambda - 2k\pi\omega$. This is an especially interesting case, for if $\lambda = 0$ then, for small ω, $f(\omega; \infty) \simeq (\sigma^2/(2k\pi))\omega^{-1}$ which corresponds to *pure $1/\omega$-noise*. A feature of this process is that it is 'just non-stationary' (though it is invertible).

10.1.3 General power-law spectra

This switch from $f(\omega; \infty)$ behaving like ω^{-2} to ω^{-1} raises a second question, namely whether our spatial interaction structure can be extended to produce $f(\omega; \infty) \propto \omega^{-d}$ for general $d < 0$ as $\omega \to 0$? Fortunately, the existence of spectral densities with this long-term persistence property has been well known in time-series analysis for some considerable time (e.g. Cox, 1977; Hosking, 1981), especially in applications relating to economics and hydrology. Such time-series appear to exhibit cycles and changes of level at all orders of magnitude, with their spectral densities increasing indefinitely as the frequency decreases to zero (Granger, 1966). It is worthwhile noting that the $d = 2$ case corresponds to (ordinary) Brownian motion, and the generalization to $1 < d < 3$ to fractional Brownian motion (Mandelbrot, 1965). The latter has covariance structure proportional to $|k|^{-3}$ (for distance k): for a general discussion of power-law covariance functions see Whittle (1962).

A discrete-time analogue of continuous-time fractional noise was proposed by Mandelbrot and Wallis (1969) to enable the development of hydrological simulation studies of long-term persistence; Lawrance and Kottegoda (1977) provide many useful references. To generate fractionally differenced white noise with parameter $H = (d-1)/2$ we simply take the $(\frac{1}{2} - H)$th fractional difference of discrete-time white noise $\{\epsilon(t)\}$ (Mandelbrot and van Ness, 1968); the fractional difference operator Δ^d (here $d = H - 1/2$) is defined through the binomial expansion

$$\Delta^d \equiv (1 - B)^d = \sum_{k=0}^{\infty} \binom{d}{k}(-B)^k, \qquad (10.1.50)$$

where B denotes the backwards difference operator $B[X(t)] = X(t-1)$. Hosking (1981) exploits this by considering the basic autoregressive integrated moving average ARIMA$(0, d, 0)$ process

$$(1 - B)^d X(t) = \epsilon(t) \qquad (10.1.51)$$

for non-integer d. When $-1/2 < d < 1/2$ the process $X(t)$ is both stationary and invertible, and has a long memory for $0 < d < 1/2$; when $-1/2 < d < 0$ it has a short memory and is antipersistent in the terminology of Mandelbrot (1977). A stationary process is defined as having a long or short memory according to whether its correlations have an infinite or a finite sum (McLeod and Hipel, 1978). On inverting (10.1.51) we see that $X(t)$ has the infinite moving average representation

$$X(t) = (1-B)^{-d}\epsilon(t) = \sum_{k=0}^{\infty}\binom{k+d-1}{k}\epsilon(t-k) \equiv \phi(B)\epsilon(t) \quad \text{(say)}, \quad (10.1.52)$$

with spectral density

$$s(\omega) \propto \phi(e^{i\omega})\phi(e^{-i\omega}) = [2\sin(\omega/2)]^{-2d} \simeq \omega^{-2d} \quad \text{(for } \omega \simeq 0\text{)}. \quad (10.1.53)$$

In particular, when $d = 1/2$ we have $s(\omega) \sim \omega^{-1}$ for $\omega \simeq 0$. Thus the ARIMA$(0, 1/2, 0)$ process corresponds to discrete-time $1/\omega$-noise (Mandelbrot, 1967).

Although this result relates to a purely temporal process, it suggests that we should consider using the same binomial structure (10.1.50) in our spatial–temporal interaction process. So suppose the spatial interaction weights $\{a_r\}$ now take the form

$$a_0 = 0 \quad \text{and} \quad a_r = c\binom{d}{|r|}(-1)^r \quad (r \neq 0) \quad (10.1.54)$$

for appropriate values of $d > 0$. Since $\sum_{r=-\infty}^{\infty} a_r = -2c$, expression (10.1.18) gives

$$\psi(\omega) = 2\lambda - 4c - 4c\sum_{r=1}^{\infty} a_r \cos(r\omega) = 2\lambda - 4c - 2c\sum_{r=1}^{\infty}\binom{d}{r}(-1)^r(e^{ir\omega} + e^{-ir\omega})$$

$$= 2\lambda - 2c\{2\sin(\omega/2)\}^d([\exp\{(i(-\pi+\omega)\}]^d + [\exp\{i(\pi-\omega)\}]^d)$$

$$= 2\lambda - 4c[2\sin(\omega/2)]^d \cos\{(\pi-\omega)d/2\} \quad (d > 0). \quad (10.1.55)$$

Thus if $\lambda = 0$ and $c > 0$ then $\psi(\omega) < 0$ for all $0 < \omega < \pi$. Whence for small ω and non-integer d, expression (10.1.17) yields the general power-law spectrum

$$f(\omega;\infty) = -\sigma^2/\psi(\omega) \simeq \{\sigma^2/[4c\cos(\pi d/2)]\}\omega^{-d} \quad (d > 0). \quad (10.1.56)$$

If d is integer, then for integer h

$$f(\omega;\infty) \simeq \begin{cases} (\sigma^2/2cd)(-1)^{h+1}\omega^{-2h} & \text{for } c(-1)^{h+1} > 0, \ d = 2h-1 \\ (\sigma^2/4c)(-1)^h\omega^{-2h} & \text{for } c(-1)^h > 0, \ d = 2h, \end{cases} \quad (10.1.57)$$

and so the same power-law applies regardless of whether d is even or odd. Thus the binomial weights can generate all ω^{-d} spectral powers *except* for odd d; as d approaches an odd integer the ω-region within which the power-law holds shrinks to zero.

10.1.4 The inverse problem

Whilst the above examples enable us to construct processes that possess asymptotic ω^{-d} spectra, it would be even more useful if we could develop a general method for constructing interaction weights that lead to any specified target spectrum $f(\omega;\infty)$

(Renshaw, Phayre and Jakeman, 2000). Let us first consider a purely spatial, i.e. non-temporal, scenario.

The fractional integration process For a stationary zero-mean stochastic process $\{y(x)\}$ the autocovariance function is $\gamma(k) = \mathrm{E}[y(x)y(x+k)]$ $(k = 0, 1, 2, \ldots)$, with the spectrum, $f(\omega)$, being the Fourier transform of $\gamma(k)$. Since fractals are characterized (Mandelbrot, 1977) by the inverse power-law form of their spectrum, Jefferson and Anderson (1987) use this characterization to describe a fractal in terms of $\{y(x)\}$ through

$$f(\omega) = \int_{-\infty}^{\infty} e^{-i\omega k} \mathrm{E}[y(x)y(x+k)]\,dk = |\omega|^{-2\nu} \quad (1/2 < \nu < 3/2). \quad (10.1.58)$$

The restriction that $1/2 < \nu < 3/2$ may be relaxed, though it should be noted that outside this range the notion of a fractal no longer holds; for $\nu > 3/2$ the process becomes sub-fractal (i.e. once-differentiable). Let $\{\epsilon(x)\}$ denote a zero-mean, unit-variance, white-noise process with Fourier transform

$$\overline{\epsilon}(\omega) = \int_{-\infty}^{\infty} e^{i\omega x} \epsilon(x)\,dx. \quad (10.1.59)$$

Then $\{y(x)\}$ may be written in the form

$$y(x) = \frac{1}{2\pi} \int_{-\infty}^{\infty} \overline{g}(\omega)\overline{\epsilon}(\omega) e^{-i\omega x}\,d\omega, \quad (10.1.60)$$

where $g(\omega)$, with Fourier transform $\overline{g}(\omega)$, is called the kernel of the process. The key point is that by making a careful choice of $g(\omega)$, we may ensure that $\{y(x)\}$ satisfies the differential equation

$$\frac{d^{\nu} y(x)}{dx^{\nu}} = \epsilon(x), \quad (10.1.61)$$

i.e. $\{y(x)\}$ is ν-times integrated white noise.

To obtain a purely spatial process $\{y(x)\}$ which possesses the specific spectrum $f(\omega)$ write

$$\mathrm{E}[y(x)y(x+k)] = (1/2\pi)\int_{-\infty}^{\infty} e^{i\omega k} |\overline{g}(\omega)|^2\,d\omega = (1/2\pi)\int_{-\infty}^{\infty} e^{-i\omega k} f(\omega)\,d\omega. \quad (10.1.62)$$

For then a direct comparison of these two integrands yields a simple relationship between $f(\omega)$ and $\overline{g}(\omega)$, namely

$$|\overline{g}(\omega)|^2 = f(\omega). \quad (10.1.63)$$

Note that $\overline{g}(\omega)$ may take several different forms. For example, in the case of the pure power-law spectrum $f(\omega) = |\omega|^{-2\nu}$ Jefferson and Anderson (1987) highlight the three kernels $\overline{g}(\omega) = |\omega|^{-\nu}$, $(i/\omega)^{\nu}$ and $i\omega|\omega|^{-\nu}/|\omega|$. Using of any of these in (10.1.60) yields a process which possesses the power-law spectrum (10.1.58).

The spatial–temporal interaction process Having seen how to construct a kernel $g(\omega)$ appropriate to a purely spatial process $\{y(x)\}$, let us now develop a parallel temporal approach. The key lies in replacing the spatial interaction term $\sum_{r=-\infty}^{\infty} a_r\{X_n(t) - X_{n+r}(t)\}$ by an equivalent integration term $Y_n(t)$. For this enables us to write the linear equation (10.1.1), namely (in one dimension)

$$X_n(t+dt) = X_n(t) + \sum_{r=-\infty}^{\infty} a_r[X_n(t) - X_{n+r}(t)]dt + dZ_n(t) + o(dt), \qquad (10.1.64)$$

as

$$X_n(t+dt) = X_n(t) + Y_n(t)dt + dZ_n(t) + o(dt). \qquad (10.1.65)$$

Here we have taken $\lambda = 0$ in order to avoid the process developing on an exponentially growing carrier wave. Since the (above) purely spatial approach consists of taking the inverse Fourier transform of the product of $\bar{\epsilon}(\omega)$ with some kernel $\bar{g}(\omega)$, we shall use a similar expression for the integration term, but replace $\bar{\epsilon}(\omega)$ by $\bar{L}(\omega;t)$, the Fourier transform of $\{X_n(t)\}$. We then define $\bar{g}(\omega)$ by presupposing that we can write $\bar{L}(\omega;t)\bar{g}(\omega)$ as the Fourier transform of $\{Y_n(t)\}$, i.e.

$$\mathcal{F}\{X_n(t)\} = \bar{L}(\omega;t) \quad \text{and} \quad \mathcal{F}\{Y_n(t)\} = \bar{L}(\omega;t)\bar{g}(\omega). \qquad (10.1.66)$$

The rationale behind the second transform in (10.1.66) can be seen by writing

$$\mathcal{F}\{Y_n(t)\} = \sum_{n=-\infty}^{\infty} \sum_{r=-\infty}^{\infty} a_r\{X_n(t) - X_{n+r}(t)\}e^{in\omega}$$

$$= \mathcal{F}\{X_n(t)\} \sum_{r=-\infty}^{\infty} a_r[1 - \cos(r\omega) + i\sin(r\omega)] = \bar{L}(\omega;t)\bar{g}(\omega).$$

So for $\{a_r\}$ an even function, i.e. $a_{-r} = a_r$, comparison with (10.1.18) (for $\lambda = 0$) suggests that the real part of $\bar{g}(\omega)$ should be $\psi(\omega)/2$, i.e.

$$\psi(\omega) = \bar{g}(\omega) + \bar{g}^*(\omega). \qquad (10.1.67)$$

To prove that (10.1.67) does indeed produce a surface with the required spectral properties, consider the relationship

$$f(\omega;t) \equiv \mathrm{E}[\bar{L}(\omega;t)\bar{L}^*(\omega;t)]. \qquad (10.1.68)$$

For on taking the Fourier transform of equation (10.1.65) we obtain

$$\bar{L}(\omega;t+dt) = \bar{L}(\omega;t) + \bar{L}(\omega;t)\bar{g}(\omega)dt + \bar{L}_Z(\omega;t) + o(dt), \qquad (10.1.69)$$

where $\bar{L}_Z(\omega;t)$ denotes the Fourier transform of $\{dZ_n(t)\}$. Whence multiplying (10.1.69) by its complex conjugate, taking expectations of both sides of the resulting expression, and letting $dt \to 0$, produces the differential equation

$$df(\omega;t)/dt = f(\omega;t)[\bar{g}(\omega) + \bar{g}^*(\omega)] + \sigma^2. \qquad (10.1.70)$$

This solves to give

$$f(\omega;t) = f(\omega;0)\exp\{[\bar{g}(\omega)+\bar{g}^*(\omega)]t\} + \frac{\sigma^2}{[\bar{g}(\omega)+\bar{g}^*(\omega)]}(\exp\{[\bar{g}(\omega)+\bar{g}^*(\omega)]t\}-1), \quad (10.1.71)$$

which is directly equivalent to (10.1.17) provided (10.1.67) holds true. This equivalence between (10.1.17) and (10.1.71) not only validates our presumptions (10.1.66) and (10.1.67), but it also highlights the duality between space–time interaction and spatial integration. Indeed, the close spectral relationship that exists between these two different representations provides considerable potential for future exploitation.

If we wish to generate a static surface with given spectral properties then we follow the route of Jefferson and Anderson (1987); whilst if our objective is to grow the associated space–time process then we first use (10.1.67) to determine the appropriate weight structure $\{a_r\}$ (outlined below), and then employ (10.1.64). Either way involves a (theoretically) simple construction for the kernel $\bar{g}(\omega)$ or the weights $\{a_r\}$. For example, to develop a static process with spectrum $f(\omega;\infty) = (\sigma^2/2)\omega^{-\nu}$, i.e. a pure power-law spectrum valid over all frequencies $0 < \omega < \pi$, we just place the kernel $\bar{g}(\omega) = -|\omega|^\nu$ in (10.1.60). Whilst to determine the associated space-time weights $\{a_r\}$, here with $\psi(\omega) = -2\omega^\nu$, we return to (10.1.18) with $\lambda = 0$, namely

$$\sum_{s=1}^{\infty} a_s[1 - \cos(s\omega)] = \psi(\omega)/4.$$

Taking the inverse Fourier cosine transform yields

$$\int_0^\pi \sum_{s=1}^\infty a_s[1-\cos(s\omega)]\cos(r\omega)\,d\omega = \frac{1}{4}\int_0^\pi \psi(\omega)\cos(r\omega)\,d\omega,$$

which gives the general solution

$$a_r = \frac{-1}{2\pi}\int_0^\pi \psi(\omega)\cos(r\omega)\,d\omega \quad (r > 0) \quad \text{with} \quad \sum_{r=1}^\infty a_r = \frac{1}{4\pi}\int_0^\pi \psi(\omega)\,d\omega. \quad (10.1.72)$$

The last expression provides a useful way of checking that the $\{a_r\}$ have been correctly evaluated.

To illustrate result (10.1.72), suppose we wish to construct a space-time surface having a pure power-law spectrum $f(\omega;\infty) = \sigma^2/\omega^4$ valid over all frequencies $0 < \omega \leq \pi$. Then we require $\psi(\omega) = -\omega^4$, whence applying (10.1.72) yields

$$a_r = \frac{1}{2\pi}\int_0^\pi \omega^4 \cos(r\omega)\,d\omega = \left(\frac{2\pi^2}{r^2} - \frac{12}{r^4}\right)(-1)^r. \quad (10.1.73)$$

Using standard results for $\sum r^{-n}$ (e.g. Spiegel, 1990) yields $\sum_{r=1}^\infty a_r = -\pi^4/20$, which agrees with the checking condition in (10.1.72). Similarly, for the inverse-square spectrum $f(\omega;\infty) = \sigma^2/\omega^2$, we take $\psi(\omega) = -\omega^2$ and recover the alternating-sign weights

$$a_r = (-1)^r/r^2 \quad (r = \pm 1, \pm 2, \ldots). \quad (10.1.74)$$

Whilst to develop a pure $1/\omega$-noise process, we take $\psi(\omega) = -\omega$ and obtain

$$a_{2r} = -1/\pi r^2 \quad \text{with} \quad a_{2r+1} = 0 \quad (r = \pm 1, \pm 2, \ldots). \tag{10.1.75}$$

From an algebraic viewpoint we may regard 'success' as being able to obtain a closed form for the a_r via the integral (10.1.72). For example, if $\psi(\omega) = \omega^m$ for integer m, then use of standard Fourier cosine results leads to

$$a_r = (1/2\pi)r^{-m-1}\{(-1)^r \sum_{d=0}^{[(m-1)/2]} \frac{(-1)^d (r\pi)^{m-2d-1} m!}{(m-2d-1)!}$$

$$+ (-1)^{[(m+1)/2]}(2[(m+1)/2] - m)m!\}, \tag{10.1.76}$$

where $[s]$ denotes the integer part of s. Whilst to generate a surface with spectrum $f(\omega; \infty) = \sigma^2/\sin^{2m}(\omega)$ we require

$$a_{2r} = (-1)^r 2^{-2m-2} \binom{2m}{m-r} \quad \text{with} \quad a_{2r+1} = 0. \tag{10.1.77}$$

Note that unlike (10.1.76), the non-zero weights in (10.1.77) are assigned exclusively to even values of r, as happens in (10.1.75) for the pure $1/\omega$-noise process.

10.1.5 Simulated realizations

Any attempt at simulating the process $\{X_n(t)\}$ through (10.1.64) hits the immediate problem that n takes all integer values $-\infty < n < \infty$. Although we could restrict n to lie within the finite range $n = 1, \ldots, N$ (say), and just observe values only within the central part of this interval, unless N is chosen to be appropriately large the resulting edge effects are likely to distort the output. Alternatively, we can retain the essential characteristics of the process by arranging $\{X_n(t)\}$ around a ring of N cells, and regard as synonymous cells 0 and N, likewise cells 1 and $N+1$. Thus $X_{n+pN} \equiv X_n(t)$ for all $n = 1, \ldots, N$ and $p = \ldots, -1, 0, 1, \ldots$, which gives rise to a Turing ring process (Renshaw, 1994b; Smith and Renshaw, 1994). So now (10.1.18) is replaced by the finite sum

$$\psi(\omega) = 2\lambda + 2\sum_{r=0}^{N-1} b_r(1 - \cos(2\pi rs/N)). \tag{10.1.78}$$

Here the continuous frequency range $0 < \omega < \infty$ has been replaced by the finite set of discrete Fourier frequencies $\omega = 2\pi/N, \ldots, 2\pi(N-1)/N$, and the interaction weights $\{a_r\}$ by the finite set

$$b_r \equiv \sum_{p=-\infty}^{\infty} a_{r+pN} \quad (0 \le r \le N-1). \tag{10.1.79}$$

Clearly N must be chosen large enough to ensure that all frequencies of potential interest are included. Note that unlike a_0, b_0 is not necessarily zero. The relation (10.1.79) means that we have to make a choice when transforming the $\{a_r\}$ to the ring process. For we can either retain the full set $\{a_r\}$ and use (10.1.79) to generate the $\{b_r\}$; or else we can place $b_r = a_r$ over the restricted range $r = \pm 1, \ldots, \pm N/2$ and take $b_r \equiv 0$ elsewhere.

At first sight the strong association between the line process spectrum $f(\omega; t)$ and the corresponding ring spectrum generated through (10.1.78) and (10.1.79) suggests that the two processes are fundamentally similar. Unfortunately this is not entirely true, since finite N not only curtails long-range behaviour but it also imposes a maximum wavelength of size $N/2$. Since $f(\omega; \infty) = \sigma^2/\omega^d \to \infty$ as $\omega \to 0$ (for $d > 0$), the variance of the line process is exponentially unbounded. So as the ring process has minimum frequency $2\pi/N$ it cannot therefore mimic the spatial transience of the line process. What it can do perfectly well, however, is to demonstrate spatial behaviour for wavelengths up to $N/2$.

Figure 10.1 shows a simulation of the process with power-law spectrum $f(\omega, \infty) = \sigma^2/\omega^4$ for restricted weights (10.1.73) at times $t = 100$ and 1500 over a ring of $N = 1024$ sites (so $r = -512, \ldots, 512$), starting from $X_1(0) = \cdots = X_N(0) = 0$. At $t = 100$ the process is (visually) dominated both by high-frequency oscillations and medium-range components. However, as t increases, the impact of the former declines, and by $t = 1500$ there is evidence of long-range dependence. The increasing influence of medium- and long-term components is reflected in the log-periodograms and theoretical log-spectra (Figure 10.2); the former are smoothed by applying successive Daniell windows of length 11, 7 and 11. Both figures exhibit a straight line with gradient -4 over the whole mid- to high-frequency range. The sharp cut-off point between this power-law part of the process (gradient -4) and low-frequency white noise (gradient 0), i.e. the outer scale of pattern, moves slowly back towards zero as long-range dependence develops in strength. It is the presence of this flat log-spectrum across low-frequencies, which disappears only as $t \to \infty$, that illustrates the fundamental difference between the space-time process $\{X_n(t)\}$ and the spatial integration process $\{y(x)\}$. For the latter construction generates the required target spectrum over the full frequency range in a single pass.

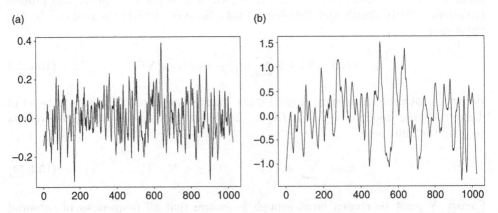

Figure 10.1 Simulated realization of the space–time interaction process $\{X_n(t)\}$ with power-law spectrum $f(\omega; \infty) = \sigma^2/\omega^4$ for weights a_r given by (10.1.73): (a) $t = 100$ and (b) $t = 1500$ over a ring of $N = 1024$ sites; note the change of scale (Renshaw, Phayre and Jakeman, 2000).

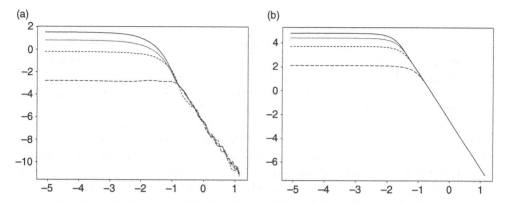

Figure 10.2 (a) Smoothed log-periodograms and (b) theoretical log-spectra corresponding to Figure 10.1: $t = 100$ (———), 500 (- - -), 1000 ($\cdots$) and 1500 (———) (Renshaw, Phayre and Jakeman, 2000).

It is easy to construct simulated realizations which enable us to compare the spatial structures generated by the fractional integration and space–time interaction approaches. Consider, for example, Bartlett's (1964) suggested form for the spectrum of a spatial inhibition process, namely

$$f(\omega) = \phi[1 - v\gamma^2\omega^2 \exp(-\gamma^2\omega^2/2)]. \tag{10.1.80}$$

Here ϕ reflects the intensity of the inhibition process, and v and γ^2 the strength and scale of inhibition, respectively. Then result (10.1.63) shows that we may take

$$\overline{g}(\omega) = +\sqrt{[\phi\{1 - v\gamma^2\omega^2 \exp(-\gamma^2\omega^2/2)\}]}. \tag{10.1.81}$$

This generates a continuous process $\{y(x)\}$ which has strong low- and high-frequency components, but only weak middle-frequency components. To simulate it we first need to replace the continuous variable x over $-\infty < x < \infty$ in (10.1.60) by the discrete variable $x_j = \{-L\Delta x, -(L-1)\Delta x, \ldots, L\Delta x\}$ for appropriately chosen small increment step $0 < \Delta x \ll 1$ and large integer L. We then generate a sequence of i.i.d. $N(0,1)$ random variables $\{\epsilon_j\}$ over $j = -L, \ldots, L$; so for given frequency ω the Fourier transform (10.1.59) is replaced by

$$\overline{\epsilon}(\omega) \simeq \sum_{j=-L}^{L} e^{ij\omega\Delta x}\epsilon_j \Delta x. \tag{10.1.82}$$

Whilst to discretize the frequencies $-\infty < \omega < \infty$ we choose an appropriately large bound M and small increment $0 < \Delta\omega \ll 1$, so that now $\omega = \{-M\Delta\omega, -(M-1)\Delta\omega, \ldots, M\Delta\omega\}$. Whence simulated values $\{y_j\}$ may be generated by replacing the continuous solution (10.1.60) by its discrete approximation

$$y_j = \left(\frac{\Delta x \Delta\omega}{2\pi}\right) \sum_{k=-M}^{M} e^{-ijk\Delta x\Delta\omega}\overline{g}(k\Delta\omega) \sum_{s=-L}^{L} e^{isk\Delta x\Delta\omega}\epsilon_s \quad (j = -L, \ldots, L). \tag{10.1.83}$$

Since the associated computation involves a single cycle over $j = -L, \ldots, L$; $k = -M, \ldots, M$, appropriately large values of L and M may be coupled with suitably small values of Δx and ΔM without involving a high compute time. This contrasts markedly with the space–time approach which may well involve an extremely high compute time if the equilibrium configuration is to be approached at all closely (note the slow rate of convergence implied by Figure 10.2). To simulate the corresponding space–time process we place $\psi(\omega) = -\sigma^2/f(\omega)$ and then employ the weights (10.1.72), either over $-\infty < r < \infty$ via (10.1.79) or by truncating the range to $-N/2 \leq r \leq N/2$.

The success of the space–time algorithm (10.1.64) in constructing processes which possess precise target spectra over the full frequency range has already been demonstrated above. For example, interaction weights $\{a_r\}$ which generate pure power-law spectra with powers -1, -2 and -4 are given by (10.1.75), (10.1.74) and (10.1.73), respectively. However, all of these schemes are of infinite extent; yet if we are to simulate such processes it would be preferable to work with finite schemes in order to avoid having to select between the wrap-round mapping (10.1.79) and the truncated mapping $b_r = a_r$ for $|r| \leq N/2$ and $b_r = 0$ elsewhere. Result (10.1.77) shows that this is certainly possible when $\psi(\omega) \propto \sin^\nu(\omega)$, instead of $\psi(\omega) \propto \omega^\nu$, since the former has $a_r = 0$ for all $|r| > m$. Moreover, this emulates a pure power-law when $\omega \simeq 0$. So continuing our ω^{-4} example, suppose we take

$$\psi(\omega) = k\sin^4(\omega) \sim k\omega^4 \qquad (\omega \simeq 0).$$

Since

$$\sin^4(\omega) = (3/8) - (1/4)[\cos(-2\omega) + \cos(2\omega)] + (1/16)[\cos(-4\omega) + \cos(4\omega)],$$

comparison with (10.1.18) shows that the required non-zero weights are $a_{\pm 2} = -1/8$ and $a_{\pm 4} = 1/32$, which are much simpler than the infinite weight scheme (10.1.73). Note that $\sum_{r \neq 0} a_r < 0$, which ensures that the asymptotic form $f(\omega; \infty) = \sigma^2/\sin^4(\omega)$ is valid over all $0 < \omega < \pi$. Clearly, we may approximate any power-law of the form $\psi(\omega) = k\omega^{2s}$ ($k < 0$, s integer) by $\psi(\omega) = k\sin^{2s}(\omega)$, and thereby evaluate the finite set of weights (10.1.77) which gives rise to a spectrum of $O(\omega^{-2s})$ for $\omega \sim 0$. This approach is easily extended to construct more complex finite weight schemes $\{a_r\}$. Renshaw, Phayre and Jakeman (2000), for example, develop non-zero weights for $r = \pm m, \pm 2m, \ldots, \pm nm$ that give rise to a general power-sine-law of the form $\psi(\omega) = \sin^{2n}(m\omega/2)$.

10.1.6 An application to sea waves

To conclude, we present a two-dimensional example of considerable practical consequence in the study of the loading of sea-waves on offshore structures. Here the Neumann (1953) spectrum

$$f_D(\omega) = (Kg^2/\omega^6)\exp\{-B/\omega^2\} \tag{10.1.84}$$

relates to fully developed sea conditions, whilst the Bretschneider (1959) spectrum

$$f_B(\omega) = (5H_s^2/16\omega_0)(\omega/\omega_0)^{-5}\exp\{-(5/4)(\omega/\omega_0)^{-4}\} \tag{10.1.85}$$

relates to developing seas; see Sarpkaya and Isaacson (1981) for a full description of parameter values and engineering considerations. For appropriate ranges of ω these are power-law spectra of degree -6 and -5, respectively, in sharp contrast to the ω^{-2} and ω^{-1} spectra developed previously in Section 10.1.3. The ability to simulate the growth of sea waves possessing this spectral structure has clear relevance to the offshore engineering industry.

For a given two-dimensional spectrum $f(\omega_x, \omega_y)$, the required interaction weights $\{a_{rs}\}$ which correspond to the one-dimensional values (10.1.72) are

$$a_{rs} = \frac{-1}{2\pi^2} \int_0^\pi \int_0^\pi \psi(\omega_x, \omega_y) \cos(r\omega_x) \cos(s\omega_y) \, d\omega_x \, d\omega_y. \qquad (10.1.86)$$

Whence paralleling our one-dimensional sine-approximation, with $\psi(\omega_x, \omega_y) = k(\sin^2(\omega_x) + \sin^2(\omega_y))^3$ $(k < 0)$, generates the finite set of weights $a_{\pm 1,0} = a_{0,\pm 1} = 57k/128$, $a_{\pm 2,0} = a_{0,\pm 2} = -12k/128$, $a_{\pm 3,0} = a_{0,\pm 3} = k/128$, $a_{\pm 2,\pm 2} = -24k/128$ and $a_{\pm 4,\pm 2} = a_{\pm 2,\pm 4} = 3k/128$. Figures 10.3bc&d show a realization of this process at $t = 100, 500$ and 1500, starting from $X_{ij}(0) \equiv 0$. It is visually apparent that all three realizations possess a substantially greater degree of low-frequency structure than that

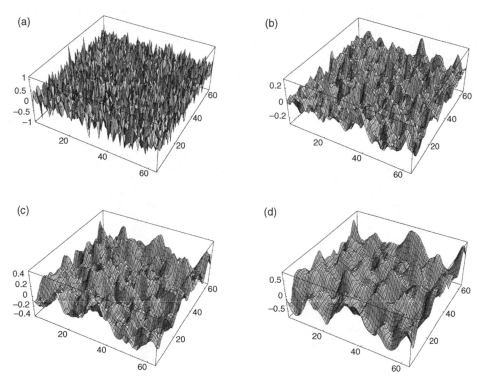

Figure 10.3 Simulated spatial realizations: (a) an $O(\omega^{-2})$-process at $t = 100$; an $O(\omega^{-6})$-process at (b) $t = 100$, (c) $t = 500$ and (d) $t = 1500$ (Renshaw, Phayre and Jakeman, 2000).

generated by using just second-nearest-neighbour weights (Figure 10.3a); the latter yields a power-law process of degree -2 for $|\omega_x, \omega_y| \simeq 0$. Observe that the realizations become smoother as t increases, and that the process becomes dominated by medium- and long-range components. This is exactly what occurs during the development of a calm sea building up into large waves under the influence of wind.

The fact that fractals may be characterized by the inverse power-law form of their spectrum (see expression (10.1.58)) allows us to study optical phenomena such as the scattering of light from developing sea surfaces. Jefferson and Anderson (1987), for example, highlight the propagation of light rays in random media by drawing rays perpendicular to the wavefront in order to show the direction of travel. In particular, on selecting the kernel $\bar{g}(\omega) = |\omega|^{-\nu}$, and superimposing rays perpendicular to simulated $y(x)$-values generated from (10.1.60) with $\nu = 1, \ldots, 4$, there is an absence of 'focussing' for $\nu = 1$ (Brownian fractal) and $\nu = 2$ (Brownian sub-fractal); see Figures 10.4a&b (from Phayre, 1995). In the former case the rays propagate from a slope represented by $\epsilon(x)$, whereas in the latter the slope is $(1/2\pi) \int_{-\infty}^{\infty} \bar{\epsilon}(\omega) e^{-i\omega x} |\omega|^{-1} d\omega$. However, when $\nu = 3$ the rays begin to focus, and a caustic forms which becomes sharper for the smoother curve $\nu = 4$; see Figures 10.4c&d. Although geometrical optics are not really applicable when dealing with a true fractal wavefront since this has detail on infinitely fine scales (fractals are stochastically non-differentiable), this problem can be circumvented by considering the geometrical optics limit as the growth rate $\lambda \to \infty$ in the generating process (10.1.1). This is an important limit for a large class of scatterers, where interference and diffraction effects are of secondary concern.

Figure 10.5 (from Phayre, 1995) provides a two-dimensional illustration of this approach, showing plots of cross-sections through the rays propagating from the

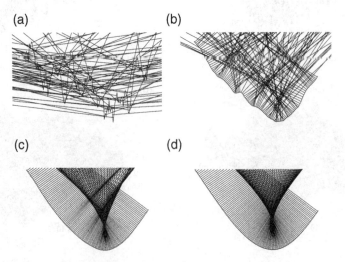

Figure 10.4 Formation of caustics as the degree of integration is increased from 1 to 4 in steps of 1: (a) $\nu = 1$, (b) $\nu = 2$, (c) $\nu = 3$ and (d) $\nu = 4$ (taken from Phayre, 1995).

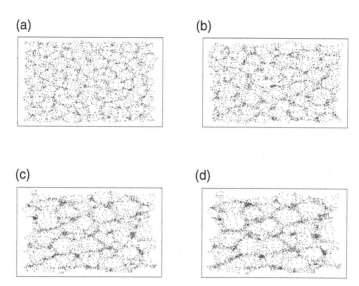

Figure 10.5 Contrast plots of rays (intersecting the fixed plane $z = 20$) propagated from surfaces at times: (a) $t = 100$, (b) $t = 500$, (c) $t = 1500$ and (d) $t = 2500$ (taken from Phayre, 1995).

surfaces in Figures 10.3b,c&d corresponding to times $t = 100$, 500 and 1500, together with one corresponding to the later time $t = 2500$. Each dot represents the intersection of a ray with a horizontal plane at height $z = 20$. As t increases, and low-frequency components have a greater influence on the surface topography, we see that there is a marked increase in the definition of the pattern in the contrast plots. Moreover, Figure 10.6 shows that this property is confined to a small range of values of z. For although some of the surface characteristics are defined at $z = 10$, a large proportion of the rays have not yet deviated far from their start coordinates. Hence the lattice structure is still evident in the resultant pattern. Figure 10.6b is a repeat of Figure 10.5d, and of the given values of z this one ($z = 20$) yields a contrast plot which best characterizes the surface topography. The final two plots show that as z increases still further, this clarity is lost, resulting first in a blurred representation of the surface ($z = 50$), and finally in the loss of all definition by $z = 100$. This is precisely the fractal structure an observer sees when looking, for example, onto the surface of a choppy illuminated swimming pool.

Although the study of scattering geometries is considered in detail within the physics domain (see, for example, Jakeman, 1984; Jefferson, Jakeman and Beale, 1986), within the stochastic process domain the fields of spatial and spatial–temporal ray propagation lie virtually untouched. There are clearly a large number of challenging problems, both in theory and application, that surround the space-time structural analysis of caustics and cross-sectional spatial pattern, and their interaction with the underlying stochastic surfaces that generate them. So this potentially rewarding field is clearly ripe for development.

596 *Spatial–temporal extensions*

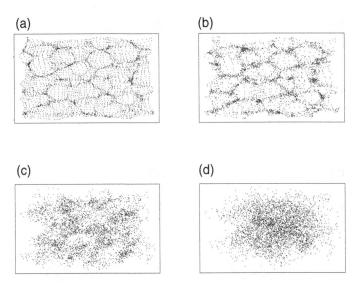

Figure 10.6 Contrast plots of rays (for fixed time $t = 2500$) generated as for Figure 10.5, shown intersecting the planes (a) $z = 10$, (b) $z = 20$, (c) $z = 50$ and (d) $z = 100$ (taken from Phayre, 1995).

10.2 Space–time marked point processes

Until now we have presumed that colonies of individuals are arranged over integer arrays, and have considered properties such as developing wavefronts (Section 9.3), velocities of propagation (Section 9.4), Turing profiles (Section 9.5), stochastic cellular automata (Section 9.6), and spectral and ray propagation issues (Section 10.1). In all of these studies the assumption of integer space has been crucial to the development of theoretical results. For example, the frequency domain constructions of Section 10.1 are heavily dependent on the orthogonality of the Fourier sine and cosine transformations which underpin them. Given a process $\{X_i(t)\}$ over the integer sites $i = 1, 2, \ldots, N$, the associated frequencies ω take the (Fourier) values $1, 2, \ldots, N/2$; the maximal frequency, $N/2$, is called the Nyquist frequency. Mapping these sites onto the fractional locations $\Omega_N = \{1/N, 2/N, \ldots, 1\}$ produces exactly the same frequencies; whilst the associated wavelengths simply reduce from $N, N/2, \ldots, 2$ to $1, 1/2, \ldots, 2/N$. However, on making the fundamental switch from the discrete space Ω_N to the continuous range $[0, 1]$ the underlying orthogonality breaks down and mathematical theory becomes far harder to construct. This means that the avoidance of relatively heavy analytical constructions necessitates reliance on simulation-based approaches.

The essential difference between the lattice and continuous space domains is that in the former the minimum inter-site distance is fixed, whilst in the latter sites can be arbitrarily close together. So although the (lattice) Nyquist frequency takes the upper bound $N/2$, in continuous space the maximum possible frequency is unbounded.

Moreover, there are now two fundamentally different ways of interpreting 'colony size', namely:

(a) In a *pure point process* scenario an individual can live anywhere in $[0,1]^2$ (say), so here each colony comprises a single individual. Locations may change either according to (for example) an immigration–death process with fresh immigrants being placed at random within the study region, or else individuals may move.
(b) In a *marked point process* scenario each 'alive' point location carries with it an attribute which is a random variable. Although this might be the number of individuals resident there, it could equally well be a characteristic, or *mark*, of a single individual developing at that location. For example, in a forestry context the ith labelled tree has location x_i and mark $m_i(t)$ at time t, where $m_i(t)$ denotes say height, d.b.h. (diameter at breast height) or even commercial value.

Whereas in lattice space the separation of neighbouring individuals is governed by the known inter-lattice point distance, which means that the population size variables $\{X_{ij}(t)\}$ play a far more important role than the associated locations $\{(i,j)\}$, in the continuous paradigm both the mark $m_i(t)$ and its location x_i can be equally significant. Moreover, since marks and points may be highly interdependent, disentangling them presents a challenging problem. Now although Monte Carlo frequency-domain analyses (e.g. Renshaw, 2002) can separate mark and point structure, theoretical advances for marks have so far been based on the conditional mark spectrum for a given point structure. Renshaw, Mateu and Saura (2007) therefore develop a *discrepancy function* which isolates the spatial structure of the marks alone, and involves a harmonic decomposition of the mark frequencies. By admitting a much greater degree of 'disentanglement' between marks and points, this approach enables us to make a far more definitive, and quantifiably precise, assessment of spatial structure than that provided by standard space- or frequency-domain techniques. Nevertheless, much further work needs to be undertaken if we are to develop a holistic approach to mark and point interaction.

The huge upsurge of interest in spatio-temporal modelling is due in no small measure to the large amount of data generated by environmental scientists studying pollution and global climate modelling through geographical information systems, remote sensing platforms, monitoring networks and computer simulation models. Now whilst such studies are often motivated by the desire to make predictions or inferences about space–time structure based on sampling at fixed locations (see Särkkä and Renshaw, 2006 for a wide variety of references), in many ecological situations the locations of the measurements also play a crucial role in the process generating mechanism. So employing fixed location sampling strategies can result in a considerable loss of information, and a far better approach is to base estimation and inference procedures on spatial–temporal mark-point process models, as outlined in case (b) above.

Such concerns raise several important issues. First, we have to recognize that in many situations there is a genuine dependence between point locations and their associated mark variables. For example, in forestry the relative positions of trees directly affects competition for light or nutrient; so any analysis of tree size which does not allow for possible dependencies between points and marks must be questioned

(Schlather, Ribeiro and Diggle, 2004). Second, many spatial structures develop through time, and so point/mark development has to reflect this, rather than presuming some kind of purely spatial structure. For although a spatial pattern can be expressed mathematically in terms of a non-temporal simultaneous or conditional scheme, the actual process generating mechanism will often be temporal. Third, only rarely will we know the exact process generating mechanism, so we need to construct a general and flexible stochastic model that can act as a realistic surrogate, yet have the potential for generating simulated events quickly. Fourth, as time progresses, new locations may be created and old ones destroyed, as happens, for example, in forest growth. A key objective is therefore to develop a general space–time process that satisfies each of these four attributes. Since all marks may (potentially) interact with all others, we need to ensure that the process is kept as simple as possible in order to keep the level of spatial complexity, and hence the speed of computation, under reasonable control. Moreover, given the inherent level of mathematical intractability associated with marked point structures, we are forced into using a simulation based approach. The development of an underpinning mathematical theory remains a serious challenge for the future.

10.2.1 The general model

Interest in the construction and analysis of spatial-temporal marked point patterns has grown substantially in the past few years fuelled by two separate, but essentially linked, fields of study. In the biological sciences many plants grow in a competing environment, with the development of individual plants being affected by others that compete with them for nutrient and natural resources. Whilst in the study of porous and granular materials, fundamental to the fields of spatial statistics and stochastic geometry is the modelling and statistical analysis of random systems of hard particles. Here randomly placed balls of material are moved until the system 'jams'. Renshaw and Särkkä (2001) and Särkkä and Renshaw (2006) construct a generic packing algorithm that covers both of these general situations in order to infer properties and generating mechanisms of space–time stochastic processes. For simplicity they induce variability by allowing immigrants to enter a unit torus according to a Poisson process with rate α and uniformly distributed locations; though this is easily changed to cover development within bounded regions and any other type of stochastic event such as birth and batch immigration.

In order to introduce this process let us recall that the linear growth-interaction lattice model (10.1.1) can generate a process with *any* specific target spectrum $f(\omega_x, \omega_y)$, simply by using the inverse relation (10.1.86) to select the appropriate interaction weights $\{a_{rs}\}$. So an obvious paradigm is to replace the variable $X_{ij}(t)$ at the lattice location (i, j) by the variable $m_i(t)$ for the ith point located at x_i. Whence the deterministic part of (10.1.1) becomes

$$m_i(t+dt) = m_i(t) + f(m_i(t))dt + \sum_{j \neq i} h(m_i(t), m_j(t); \|x_i - x_j\|)dt. \qquad (10.2.1)$$

There are four main differences between these two processes. First, the city-block metric distance (r, s) has been replaced by the Cartesian distance $\|x_i - x_j\|$ between individuals (i.e. points) i and $j \neq i$. Second, the linear growth rate, $\lambda X_{ij}(t)$, has been replaced by the general form $f(\cdot)$ which denotes the individual growth function in the absence of spatial interaction. Third, the linear interaction function, $a_{rs}[X_{ij}(t) - X_{i+r,j+s}(t)]$, has been replaced by the general spatial interaction function $h(\cdot)$ taken over all points $j \neq i$. Fourth, in order to minimize compute time the added error term, $dZ_{ij}(t)$, has been removed, with stochasticity now being induced solely through the random arrival and location mechanisms of new individuals entering the system, and the death of old ones. Although this last feature works well in some situations, it is not universally appropriate and in Section 10.2.8 we examine a range of alternative stochastic strategies.

For now we shall remain with the deterministic growth-interaction process (10.2.1) and couple it with an appropriate stochastic arrival–departure mechanism. Given that we have already highlighted the ubiquitous nature of the simple immigration–death process, this seems a natural first choice, though there are clearly many other possibilities. For example, parent marks (e.g. oaks) might produce offspring (acorns) which are displaced around it according to some spatial distribution. Fortunately, the ideas developed in this section may be easily extended to cover such more complex scenarios. So we shall use the immigration-death model for the purpose of illustration; it is relatively trivial to replace it with any other mechanism deemed to be appropriate. Moreover, although we shall assume that population development takes place on the unit torus in order to avoid the added complexity of edge-effects, the algorithm works equally well for any domain having a well-defined border. Here two main options are open to us. If the study area lies well within a bounded region and is large relative to the maximum interaction distance between individuals, then we can retain the torus as a realistic and useful approximation. For the study area may be envisaged as developing as though it were a sub-region of a much larger torus. However, if this is not the case, and the boundaries play a crucial role in process development, then we can easily modify the algorithm by including appropriate edge-effects.

Let us therefore assume that: (i) the process is simple, i.e. there is at most one individual at each location; and, (ii) new immigrants arrive randomly in time according to a Poisson process with rate α, have uniformly distributed locations on $U(0,1)^2$, and are assigned marks, $\tilde{m}$, from some distribution, e.g. $\tilde{m} \sim U(0,\epsilon)$. In the successive small time intervals $(t, t+dt)$ each individual either dies 'naturally' according to a simple death process with probability μdt, or, if $m_i(t+dt) \leq 0$ and $m_i(t) > 0$, then the individual is deemed to have died 'interactively' at time $t+dt$; in both cases the point i is deleted. This coupling of stochastic arrival and death mechanisms with deterministic growth-interaction results in a powerful, yet highly flexible and computationally fast, generator of space–time structure. Moreover, it is especially suitable for analysing dynamic mark point processes where the underlying point structure is constantly changing.

10.2.2 Choosing growth and interaction functions

A key question in pattern analysis is to determine whether it is possible to construct summary statistics which not only highlight the spatial properties of interest but which are also sensitive to change in the underlying process. Renshaw and Särkkä (2001) base their analyses around marked Gibbs processes (see Stoyan, Kendall and Mecke, 1995), and study whether the interaction parameters of such purely spatial constructions provide a useful and sensible characterization for our space-time process. The reason for considering Gibbs processes is that they possess some similarities with the linear lattice system (10.1.1) and its spectral representation (10.1.15)–(10.1.16). For these are defined through the probability density function

$$f(\varphi) = (1/V) \exp\{-U(\varphi)\} \qquad (10.2.2)$$

where $\varphi = \{[x_1; m_1], [x_2; m_2], \ldots, [x_n; m_n]\}$ is a realization over the points $i = 1, \ldots, n$ having location x_i and mark m_i; V is a normalizing constant and U denotes an energy (reaction) function. Now this representation has two features in common with our original space–time process. First, U can take exactly the same form as the right hand side of expression (10.1.1). Second, both the numerators in (10.2.2) and (10.1.15) have a similar exponential form; the main structural difference lies in the denominator, with the normalizing constant V replacing the function ψ. For inhibitive interaction (competition) it is usually sufficient to write the energy function U in (10.2.2) in terms of pairwise interactions through

$$U(\varphi) = \sum_i \beta(x_i, m_i) + \sum_{i<j} \phi(x_i, x_j, m_i, m_j), \qquad (10.2.3)$$

where $\beta : T \times M \to (-\infty, \infty)$, M being the mark space, is the so-called *chemical activity function* and $\phi : T \times T \times M \times M \to (-\infty, \infty)$ is an appropriate *pair potential function*. This chemical activity function can be interpreted as the ability of the system to receive a point at location x_i with mark m_i. Whilst the pair potential function describes the interaction between two points: in broad terms $\phi > 0$ indicates inhibition, $\phi < 0$ clustering, and $\phi = 0$ corresponds to the Poisson case with no interaction between points. In principle, pairwise interaction processes are appropriate models both for clustered and regular point patterns, but problems can occur in the clustered case since the models are not always well-defined. Even if they are well-defined their use is still not to be recommended (e.g. Møller, 1999).

The simplest type of growth function with $f(m_i(t)) = K$ for some constant $K > 0$ and $m_i(t) > 0$ is mildly related to the 'lilypond', or 'germ-grain', model conceived by Häggeström and Meester (1996), and later generalized by (for example) Daley, Mallows and Shepp (2000) and Cotar and Volkov (2004). There the circle centres are *initially* distributed according to a Poisson point process with intensity α, rather than *arriving* as a simple immigration process. At time $t = 0$ all the circles start growing at unit rate, with each one ceasing to grow immediately it touches a neighbour. This is in marked contrast to the growth-interaction regime (10.2.1) where each $m_i(t) > 0$ either continually increases or decreases depending on whether the growth term exceeds, or is exceeded by, the interaction term. Moreover, in our model individuals are also subject to natural and interactive death.

If the constant lilypond growth rate is applied to (10.2.1) under spatial inhibition (i.e. $h(\cdot) < 0$), then in the absence of a 'cease-growing' variable the marks, $m_i(t)$, grow without bound if $K + \sum_{j \neq i} h(\cdot) < 0$. However, such explosive growth is easily avoided if we ensure that $f(m_i(t)) \downarrow 0$ as $m_i(t)$ increases. The growth function $f(\cdot)$ can exert a considerable influence on the generated pattern structure, and two simple stable forms that remain bounded are

$$f_1(m_i(t)) = \lambda m_i(t)(1 - m_i(t)/K) \quad \text{and} \quad f_2(m_i(t)) = \lambda(1 - m_i(t)/K), \quad (10.2.4)$$

for intrinsic rate of growth λ and (non-spatial) carrying capacity K. These deterministic forms correspond to the classic logistic growth and immigration–death processes, respectively (see Sections 3.2 and 2.5). We shall use them purely for the purpose of illustration, since both the growth and interaction functions can be easily modified to suit specific situations. For example, suppose we wish to develop a biological interpretation of a packing problem in physics, chemistry or engineering, in which all particles have marks which lie in the range (K_1, K_2). Then by modifying the logistic growth function to

$$f_3(m_i(t)) = \lambda(m_i(t) - K_1)(1 - m_i(t)/K_2), \quad (10.2.5)$$

we ensure that any mark $m_i(t) < K_1$ tends to zero irrespective of whether or not it is being inhibited, whilst any $m_i(t) > K_2$ reduces to K_2 in the absence of spatial inhibition. So once the process has 'burned-in', all established marks will lie in the range (K_1, K_2).

Not only is the choice of plausible growth functions clearly large, but so is the number of potential spatial interaction functions. Hence their combination gives rise to a huge range of models. All we shall do here is to highlight two or three, and stress that the simulation algorithm developed in the subsequent section is totally general in its applicability. Following Renshaw and Särkkä (2001), let us first consider the symmetric *hard-core* interaction function

$$h_1(m_i(t), m_j(t); \|x_i - x_j\|) = -bI(\|x_i - x_j\| < r(m_i(t) + m_j(t)), \quad (10.2.6)$$

where the indicator function $I(x) = 1$ if x is true and $I(x) = 0$ otherwise. We assume that the area, in horizontal projection, over which the mark $m_i(t)$ competes for resources can be represented by a disk of radius $rm_i(t)$. Thus in the tree analogy, r is the scaling parameter that relates say tree height or diameter to canopy radius. As soon as two disks overlap, i.e. $\|x_i - x_j\| < r(m_i(t) + m_j(t))$, then competitive interaction takes place with force b. Since this function is symmetric, the larger and smaller of an interactive-pair are affected equally, and although this represents a reasonable approximation for interaction between marks of similar size it may be inappropriate if $m_i(t)$ and $m_j(t)$ are radically different. For then the smaller mark can exert an appreciable influence on the growth of the larger mark irrespective of how tiny it may be.

To construct a *soft-core* form that takes account of the relative sizes of two interacting marks, suppose that the amount of competition experienced by a given mark is proportional to the extent to which its zone of influence is overlapped by those of its neighbours (Staebler, 1951; Gerrard, 1969). Let $D(x_i, s)$ denote the disk with centre x_i and radius s, and place

$$h_2(m_i(t), m_j(t); \|x_i - x_j\|) = -b \text{ area}\{D(x_i, rm_i(t)) \cap D(x_j, rm_j(t))\}/\pi r^2 m_i^2(t). \tag{10.2.7}$$

This function clearly promotes asymmetric effects, since larger marks have a competitive advantage over smaller ones, so the smaller of two interacting marks is affected substantially more than the larger. Whilst marks with similar size experience similar spatial inhibition.

Here we use the terms 'disk' and 'area' in a loose sense, since the models apply irrespective of the dimensionality of the process. Extensions are clearly endless. For example, to take account of one-sided interaction, i.e. a large mark influencing a small mark but not vice versa, we could replace $h_1(\cdot)$ by

$$h_3(m_i(t), m_j(t); \|x_i - x_j\|) = -bI\{\|x_i - x_j\| < r(m_i(t) + m_j(t))I(m_j(t) > m_i(t))\}. \tag{10.2.8}$$

So locally dominant marks will not now decline as smaller fresh immigrants pack in around them. Whilst a balance, $h_4(\cdot)$, between $h_1(\cdot)$ and $h_3(\cdot)$ may be constructed by replacing $I(\cdot)$ by, for example, $J(\cdot) = 2m_i^\gamma/(m_i^\gamma + m_j^\gamma)$ for some parameter γ. If $\gamma = 0$ then $J(\cdot) = 1$ which recovers $h_1(\cdot)$; letting $\gamma = \to \infty$ yields $h_5(\cdot) = 2h_3(\cdot)$; whilst the reverse limit $\gamma \to -\infty$ generates

$$h_6(m_i(t), m_j(t); \|x_i - x_j\|) = -bI\{\|x_i - x_j\| < r(m_i(t) + m_j(t))I(m_i(t) > m_j(t))\}. \tag{10.2.9}$$

So now a small mark affects a large mark but a large mark does not affect a small one.

To appreciate the interplay between growth and interaction functions, suppose that point i lies within the zone of influence of θ neighbouring disks under the hardcore regime $h_1(\cdot)$. Then on letting $dt \to 0$, the growth-interaction equation (10.2.1) reduces to

$$dm_i/dt = \begin{cases} \lambda m_i(1 - m_i/K) - b\theta & \text{(under } f_1\text{)} \\ \lambda(1 - m_i/K) - b\theta & \text{(under } f_2\text{)}. \end{cases} \tag{10.2.10}$$

Taking f_1 first, we see that m_i survives provided

$$dm_i/dt = \lambda m_i(1 - m_i/K) - b\theta > 0. \tag{10.2.11}$$

Hence as $dm_i/dt = 0$ at $m_i = (m', m'') = (K/2)[1 \pm \sqrt{1 - 4b\theta/\lambda K}]$, it follows that if $m_i < m'$ then m_i decays to zero, if $m' < m_i < m''$ then m_i grows to m'', whilst if $m_i > m''$ then m_i decreases to m''. This presupposes that $\sqrt{1 - 4b\theta/\lambda K}$ is real, i.e. that $\theta \leq \lambda K/4b$. So for m_i to be able to survive within even a single zone of influence (i.e. $\theta = 1$) we need the interaction force b to be less than $\lambda K/4$.

For example, suppose that $\lambda = 1$, $K = 20$ and $b = 1$, and assume that a new immigrant k has initial mark $\tilde{m}_k$. Then since $(m', m'') = (1.06, 18.94)$ when $\theta = 1$, k can survive under the shadow of a single covering disk only if $\tilde{m}_k > 1.06$; if it does, then m_k has the potential for reaching 18.94 if no other disc further impedes its growth. Similarly, if $\theta = 4$, then since $(m', m'') = (5.53, 14.47)$, k's survival depends on $\tilde{m}_k > 5.53$. Clearly, the larger the number of covering disks, θ, the larger the new immigrant mark must be in order for it to survive. Here, if $\theta > \lambda K/4b = 5$ then $dm_k/dt < 0$ for all values of m_k, and so survival is impossible. In direct contrast, under f_2 we see that $dm_k/dt \uparrow \lambda - b\theta$ as $m_k \to 0$. So now k's survival depends purely on whether θ is less than λ/b, and not on the size of $\tilde{m}_k$. Thus if $\lambda = b = 1$ then a new immigrant cannot survive under any zone of influence since $\lambda/b = 1$; whilst when $b = 0.2$ it can survive under as many as four influence zones since now $\lambda/b = 5$.

In reality, if a small m_k is affected by a larger m_i, then that, in turn, will be affected by m_k and so reduce in size. Continuing our two-point example, suppose that under the model (f_2, h_1) we have $\lambda < b$, so that no point can survive under a covering disk. Then if marks $m_i > m_k$ interact at time $t = 0$, but $rm_i < \|i - k\| < r(m_i + m_k)$, both m_i and m_k will decrease until their two disks just touch, at which point m_k will start to increase, and m_i decrease, until $m_i = m_k$. If, however, $rm_i > \|i - k\|$, then k's survival depends on whether it is uncovered by m_i's shrinking zone of influence before it dies. Since the second of equations (10.2.10) solves to give

$$m_k(t) = m_k(0)e^{-\lambda t/K} - (K/\lambda)(b - \lambda)[1 - e^{-\lambda t/K}], \qquad (10.2.12)$$

on writing $\xi = (K/\lambda)(b - \lambda)$ we see that $m_k(t)$ reaches zero at time $\tau_k = (K/\lambda)\ln[1 + m_k(0)/\xi]$. Thus particle k dies under the influence of i if $rm_i(\tau_k) > \|i - k\|$, i.e. if the inter-point distance

$$\|i - k\| < r\phi[m_i(0) - m_k(0)]/[m_k(0) + \xi]. \qquad (10.2.13)$$

This feature of interactive death means that each of the n marks $m_i(t)$ in the deterministic dynamical system (10.2.1) has to be replaced by $m_i(t)H(m_i(t))$, where $H(x) = 1$ if $x > 0$ and $H(x) = 0$ if $x \leq 0$. It is the all-invading presence of this Heaviside function, as well as nonlinearity in $f(\cdot)$ and $h(\cdot)$, that renders theoretical progress well nigh impossible unless n is very small. Thus in general, simulation provides the only practical route forward.

10.2.3 Parameter selection

Once the growth and interaction functions, $f(\cdot)$ and $h(\cdot)$, have been selected we still have to choose the associated parameters. In many ecological scenarios, for example, a considerable amount of field and laboratory information will usually be available on the growth parameters; though far greater uncertainty often surrounds the force of interaction b and, albeit to a lesser extent, the scaling parameter r. So if we are studying the growth of trees in a forest and the mark zone of influence is directly observable, e.g. tree canopy radius, and we record say radius at breast height (r.b.h.), then r is simply the ratio of the two radii. Conversely, if hidden rooting systems play a major role in interaction then the likely value of r is less clear cut. The wisest policy is to relate any prior knowledge to estimated parameter values; for a substantive difference between them suggests a wrong choice of model selection.

An arguably more difficult problem than 'pragmatically correct' model selection is acquiring data in the first place. For the acquisition of good quality spatial, and especially spatial–temporal, data is often expensive both in financial and in human resource terms. Ideally, the same sample area should be measured on several different occasions so that the estimation procedure can utilize information on the change in pattern structure through time. For a multitude of reasons this may not be possible, whence data are only available from a single snap-shot. Fortunately, both scenarios lend themselves to statistical analysis.

Taking the *purely spatial* case first, Renshaw and Särkkä (2001) use marked Gibbs processes to develop a maximum pseudo-likelihood approach. For in addition to being commonly used in the analysis of point patterns, we have already seen through (10.2.3) that they possess similar interaction properties to the model structure (10.2.1). Moreover, only a few statistical indicators are needed to determine the nature of the generating process. This contrasts markedly with the relatively large number of estimators usually required to determine spectral, autocorrelation or K-function representations. The procedure is also robust, in the sense that it is not crucial to know the exact process-generating mechanism (the norm in ecological studies); if we do possess additional information about the true mechanism, then it becomes even more effective. Note that maximum likelihood could be used in place of maximum pseudo-likelihood, but it is much more computationlly demanding (see, for example, Møller and Waagepetersen, 2003).

Though in theory the pseudo-likelihood approach could be extended to analyse spatial–temporal patterns, the computational expenditure involved may well be overwhelming. Särkkä and Renshaw (2006) therefore construct a least-squares estimation procedure for estimating the growth and interaction parameters *through time*. Their paper not only provides a comprehensive overview of a wide variety of spatial–temporal approaches currently on offer, and the relationships which exist between them, but it also shows through several simple examples that little, if anything, is likely to be lost by moving to a least squares procedure. As before, a robustness study confirms that the approach works well even when an incorrectly presumed model is employed. Hence both techniques are likely to prove to be very useful in general practical applications where the precise nature of the underlying process generating mechanism is almost certain to be unknown. Note that a possible 'middle-ground' approach for generating space-time likelihoods might be to extend the importance sampling ideas developed in Marion, Gibson and Renshaw (2003) for lattice-based spatial–temporal epidemics; though this is currently uncharted territory.

Key issues that are currently being addressed include: (a) how to select the 'best' growth and interaction functions to apply in a given situation; and, (b) how to construct appropriate statistical measures that will enable us to test whether the chosen model represents a realistic and useful approximation to reality. To date, both questions have been answered via a purely subjective assessment based on comparing real and simulated data plots by eye.

10.2.4 Simulation algorithm

We have already noted that the stochastic mechanism that determines the arrival, location, initial mark size and death rates can be as complex as deemed necessary. For example, little extra effort is required to augment it in order to study a general nonlinear birth–immigration–death process for mark size coupled with a density dependent and anisotropic (i.e. direction dependent) location variable for new arrivals. In the case of births these would be centred around parent marks, in contrast to immigrants whose location distribution is determined solely by the (potentially non-homogeneous) properties of the study region itself. Moreover, any initial spatial configuration of points and mark size could be employed. However, since the linear immigration–death process has already been used successfully as a basic portrayer of stochastic population development, there is no loss in retaining it here for the purpose of illustration.

For simplicity let us therefore assume that the population is initially empty, and that new immigrants, $k = 1, 2, \ldots$, arrive according to a Poisson process with rate α, are located within the unit torus at $x_k = U[0,1)^2$, and either take variable marks $\tilde{m}_k = U(0, \epsilon)$ for some small $\epsilon > 0$, or fixed marks $\tilde{m}_k = \epsilon$. Now suppose that our basic stochastic immigration-death process is coupled with the deterministic linear growth function $f_2(\cdot)$ and hard-core spatial interaction function $h_1(\cdot)$ given by (10.2.4) and (10.2.6), respectively. Then the parallel construction to our earlier first-order approximation algorithm A9.2 in Section 9.3.3 for multi-colony populations takes the following form.

Space–time marked point process algorithm (A10.1)
  (i) read in parameters
  (ii) set $t = 0$ and population size $n = 0$
  (iii) cycle over individuals $i = 1, \ldots, n$
       generate a new $U_1 \sim U(0,1)$ random number
       if $U_1 \leq \mu dt$ then $m_i \to 0$; if not, apply deterministic growth and spatial interaction, i.e.

$$m_i \to m_i + \lambda(1 - m_i/K)dt - b \sum_{j=1, j \neq i}^{n} I\{\|x_i - x_j\| < r(m_i + m_j)\}dt; \tag{10.2.14}$$

  (iv) generate new $U_2, U_3, U_4 \sim U(0,1)$ random numbers
       test for new immigrants: if $U_2 \leq \alpha dt$ then $n \to n+1$ with new mark $m_n = \tilde{m}$ having location $x_n = (U_3, U_4)$
  (v) delete zero sites: if $m_i \leq 0$ then delete $(x_i; m_i)$ from list, relabelling all remaining marks with $j > i$ to $j - 1$, and set $n \to n - 1$.

Without loss of generality the growth and interaction parameters, λ and b, can be replaced by $\lambda = 1$ and b/λ, respectively, simply by changing the time increment dt to λdt. Suppose that $\alpha = 5$ and $\mu = 0.02$, so that the equilibrium population size follows a Poisson(250) distribution in the absence of spatial interaction. Since under spatial interaction we also have interactive, as well as natural, death, each marked point therefore has an expected lifetime of less than $1/\mu = 50$ time units. So the equilibrium mean population size is less than 250. Let $b = 0.4$, immigrant marks have size $\tilde{m} = 1$, and take $dt = 0.01$. Then on returning to the argument following (10.2.11), we see that because $\lambda/b = 2.5$ a mark can survive under the influence of at most $\theta = 2$ covering disks. Finally, placing the scaling factor $r = 0.005$ means that the unit torus can just accommodate 25 touching disks with maximum mark size $K = 20$ placed on a 5×5 square lattice with spacing equal to the disk diameter, namely $2rK = 0.2$. So a structure comprising around 200 points is likely to exhibit interacting disks with size

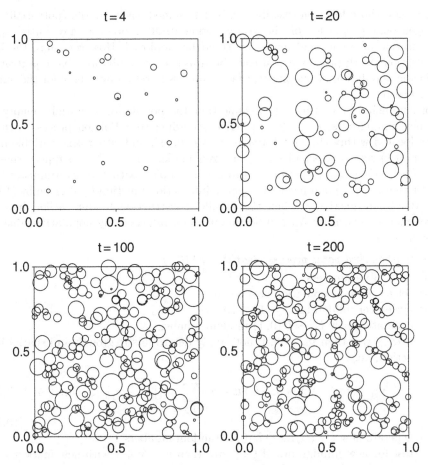

Figure 10.7 Simulation based on the growth-interaction algorithm A10.1 shown at times $t = 4$, 20, 100 and 200, with $\alpha = 5$, $\mu = 0.02$, $\lambda = 1$, $K = 20$, $b = 0.4$, $r = 0.005$ and $\bar{m} = 1$; disk radius is proportional to rm_i.

substantially less than K. Figure 10.7 shows a realization of this process at times $t = 4$, 20, 100 and 200. Transient development is clearly visible through the first three frames, with empty space being exploited by new immigrants. The third and fourth frames have similar statistical properties, since by now the process has burned-in, though as more than two expected lifetimes have elapsed between them, few points are common to both. Note that $\max\{m_i(200)\} = 12.56 < K = 20$, as anticipated; whilst some of the small marks with $m_i(200) \simeq 2$ or less correspond to later immigrants that have had insufficient time to develop against the surrounding interaction pressure. Reducing b would enable more disks to intersect, whilst increasing b to $b > \lambda = 1$ would prevent established marks intersecting each other since now $\theta = \lambda/b < 1$.

We have taken $\mu = 0.02$ in order to ensure a reasonably fast turn-around of points. However, when $\mu = 0$ a point can die only through spatial inhibition (i.e. it constantly has more than θ neighbouring interactions), and the resulting increase in population size means that burn-in may take far longer to occur. If we wish to be certain that equilibrium has indeed been reached, then applying the perfect simulation algorithm of Propp and Wilson (1996) to this spatial scenario could well prove useful (see Kendall and Møller, 2000). Though whether the computation could be performed in an acceptably short compute time, for all but very small values of equilibrium population size, is questionable.

Possibilities for modifying this process are virtually endless. For example, Figure 10.8a shows an asymmetric soft-core simulation at time $t = 200$. Here the symmetric hard-core interaction function h_1 (expression (10.2.6)) used for Figure 10.7 has been replaced by the soft-core area-interaction form h_2 (expression (10.2.7)). Now it follows from (10.2.7) that a mark m_i can survive provided that the proportion of area covering it does not exceed $\theta = (1/b)(1 - m_i/K)$. Thus a tiny mark of size $m_i \simeq 0$ can exist if $\theta \leq 2.5$, which explains why the smallest mark with $\min[m_i(200)] \simeq 0.071$ is able to survive even though it is is subject to the large relative area coverage value $\theta = (1 - 0.071/20)/0.4 = 2.49$. In contrast, a large mark of size $m_i \simeq K = 20$ can avoid decreasing only if its canopy is virtually devoid of competing marks, which explains why $\max[m_i(200)]$ is only $\simeq 15$. Note that the choice of growth function $f(\cdot)$ is crucial to the type of spatial structure generated. If, for example, $f_2(\cdot)$ is replaced by the logistic form in (10.2.4), i.e. $f_1(\cdot)$, then an $m_i \simeq 0$ has critical coverage parameter $\theta \simeq \lambda m_i/b \simeq 0$. So the survival of small marks is in much greater jeopardy.

Figure 10.8b shows an extreme form of asymmetry in which the larger of two competing marks is totally unaffected by the smaller. The interaction function $h_3(\cdot) = h_1(\cdot) = b$ if the competing mark $m_j > m_i$, but $h_3(\cdot) = 0$ if not (see (10.2.8)). Increasing the interaction parameter from $b = 0.4$ to 1 ensures that any mark touching a larger

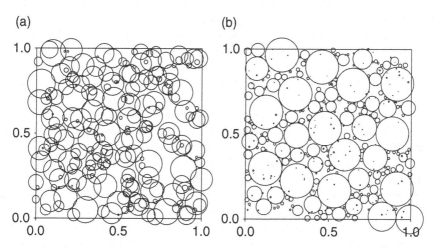

Figure 10.8 Two simulations shown at $t = 200$ based on: (a) area-interaction (f_2, h_2) with $\mu = 0.02$ and $b = 0.4$; (b) one-sided hard-core interaction (f_2, h_3) with $\mu = 0$ and $b = 1$. In both cases $\alpha = 5$, $\lambda = 1$, $K = 20$, $r = 0.005$, $\tilde{m} = 1$ and $dt = 0.01$; disk radius is proportional to rm_i.

one must decrease, since the 'critical coverage' value $\theta = (1 - m_i/K)/b < 1$. So as time develops, the spatial structure involving marks $m_i > \tilde{m} = 1$ becomes increasingly static, with change being restricted to the addition of small disks which can pack into the remaining empty space. In contrast, replacing the model (f_2, h_3) by (f_2, h_6) (see (10.2.9)) leads to an entirely different pattern structure, since now it is the larger of two competing marks that is affected by the smaller. Whilst employing the more general model (f_2, h_4) with $-\infty < \gamma < \infty$ enables a gradual progression to be made between these two extreme cases.

10.2.5 Convergence issues

Consider the purely one-sided interaction regime (f_2, h_3), illustrated in Figure 10.8b, with $b > \lambda$. If points i and j interact with $m_j < m_i$, then m_i will remain unchanged but m_j will decrease until either i and j no longer interact or else j dies interactively. Thus in the absence of further immigration the system should soon settle down to a fixed state. If immigrants with initial mark size $\tilde{m} = \epsilon \ll 1$ then enter the system, only those arriving in empty space will survive since $\theta = b/\lambda < 1$. Eventually, when all the remaining empty space is too small to be able to accept any more ϵ-radius disks, the developing spatial pattern finally 'jams'. So in order to add yet more points suppose we now cool the system by letting ϵ slowly tend to zero. Then as a new point j arriving in empty space at time t has mark size $m_j(t + dt) \simeq \epsilon + \lambda dt$ after the next time increment, this suggests that we should also cool dt in unison by taking $dt = O(\epsilon)$. Unfortunately, whilst this idea looks fine in theory, in practice there are severe compute time issues at stake; though these can be partially resolved by letting $\alpha = O(1/\epsilon)$.

As $t \to \infty$ and $\epsilon \to 0$ the pattern becomes fractal with $m_i(\infty)$ taking values between 0 and K. However, serious convergence questions arise, such as: "how does the expected number of established points, n, and occupied area, $A(t)$, increase with t"; and, "what happens when the extreme interaction function h_3 is replaced by say h_1"? For then a large mark will experience downwards pressure from small marks that interact with it. A simplistic viewpoint is to argue that if i and j interact only with each other, then since $dm_i/dt = \lambda(1 - m_i/K) - b < \lambda(1 - m_j/K) - b = dm_j/dt$, it follows that if $||i - j||$ does not satisfy (10.2.13) then both marks will decrease until their disks just touch. At this point m_j will start to increase until such time as both marks are of equal size. Extending this argument to densely populated patterns is risky, but it does suggest that, under some conditions, for fixed ϵ every established mark will gradually reduce to ϵ as $t \to \infty$. At this point new immigrants will no longer be at a competitive disadvantage to established marks, and the pattern will gradually 'turn-over' in a stationary state with mean mark size $\sim \epsilon$ due to randomly sited deaths.

In spite of the inherent complexity of this interacting particle system, an understanding of the underlying dynamics is easily gained through a simple illustration. Consider an empty torus at time $t = 0$ and let the time step $dt = 0.001$; this is small enough to ensure accurate numerical integration of the $\{m_i(t)\}$, yet large enough to allow the process to develop reasonably quickly over fairly large time intervals. Each new point k has initial size $\tilde{m}_k = 1$, since this is large enough to allow it time to try and dominate neighbouring disks before it might die. Finally, take the carrying capacity $K = 20$ along with (the relatively high) interaction force $b = 2 > \lambda = 1$. Since $\alpha = 2$ is fairly small, for large r it is likely that m_1 will become totally dominant since rm_1 may well exceed $1/\sqrt{2}$ (the maximum inter-point distance) before a successful immigrant can arrive outwith its covering shadow. In one series of simulations m_1 swiftly approached K, whereupon it killed every subsequent immigrant m_k ($k > 1$) for all $r \geq 1.59$. In marked contrast, for $r \leq 1.58$, i.e. just below this critical threshold, some m_k avoided falling foul of condition (10.2.13), and so were able to establish a 'successful' spatial pattern. Figure 10.9a illustrates three such developments over $t = 0, 1, 2, \ldots, 800$. First, when both $r = 1$ and $\tilde{m}_k = 1$, since each zone of influence (with radius $r\tilde{m}_k = 1$) covers the torus it exerts an inhibitory effect on all other marks present. In the run shown, the total number of marks at any one time never exceeded 13, since the combined influence of existing large marks was sufficiently dominant (the average mark size at $t = 800$ is 0.3150) to prohibit the development of a larger number of smaller marks. Second, slightly reducing $\tilde{m}_k$ (and hence $r\tilde{m}_k$) to 0.9 dramatically changes this effect, since far more individuals are able to survive. Now the population grows fairly steadily until $t = 671$, when the number of alive individuals suddenly 'flatlines' at 451; the average mark size at $t = 800$ shows a correspondingly large reduction to 0.0178. Third, and in direct contrast, on taking $\tilde{m}_k = 0.001 \ll 0.0178$, so that $m_k(t_k + dt) = 0.001 + (1 - 2\theta)0.001 \leq 0$ for arrival time t_k and all $\theta \geq 1$, we see that new immigrants arriving under the cover of even a single zone of influence are immediately killed. Nevertheless, fresh immigrants can still exploit the large number

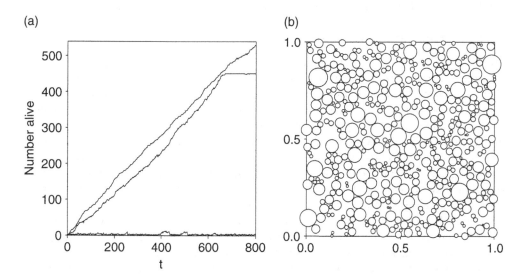

Figure 10.9 Convergence to high-intensity packing under the linear-growth hard-core model (f_2, h_1) with $\alpha = 2$, $\lambda = 1 < b = 2$, $K = 20$, $r = 1$ and $dt = 0.001$: (a) $\tilde{m}_k = 1$ (*lower*), $\tilde{m}_k = 0.9$ (*middle*) and $\tilde{m}_k = 0.001$ (*top*); (b) packing structure at $t = 800$ for $\tilde{m}_k = 0.9$ (Renshaw and Comas, 2009).

of small empty spaces, filling them in until such time as we have a parallel pattern to the second (i.e. $\tilde{m}_k = 0.9$) case above (shown in Figure 10.9b for $t = 800$). This example and related issues are considered in detail in Renshaw and Comas (2009).

A considerable merit of this modelling approach is that the same generating process can give rise to different types of pattern structure simply by making appropriate changes to the parameter values. For example, we have already seen that under (f_2, h_1), i.e. linear growth and hard-core interaction, a point can survive within $\theta < \lambda/b$ zones of influence. Taking $b = 0.3$ and 2 for fixed $\lambda = 1$ gives $\theta = 3$ and 0, respectively, and whilst the former suggests clumps of overlapping disks the latter implies regularity. Not only do statistical tests based on nearest-neighbour distances and the pair-correlation function confirm this, but also the choice of the immigrant size, $\tilde{m}$, has a considerable effect both on the strength of the resulting pattern and the shape of the corresponding mark histograms. In particular, under strong interaction ($b = 2$) small immigrants can explore the available empty space more efficiently than large ones, and thereby find it far easier to establish a stable population of small marks than when $\tilde{m}$ is large. Whilst under weak interaction ($b = 0.3$) the converse applies. Not only would further investigation of the critical value, b_{crit}, between the clumping and regular regimes, in terms of the choice of functions $f(\cdot)$ and $h(\cdot)$, be useful, but study of the associated convergence criteria is also of interest. For example, under clumping with $b = 0.3$ the number of alive points rises less quickly when $\tilde{m} \simeq 0$ than happens when $\tilde{m} = K/2$ or K, since increasing the size of the immigrants increases their chance of successfully forcing their way into a group of established marks. Whilst under regularity with $b = 2$, success depends far more on an immigrant being small enough to exploit available empty space without enduring competition in its early stage of development.

10.2.6 Application to forestry

To illustrate how this modelling approach may be applied in practise, let us briefly consider an application to forest management. This is done on the understanding that the general principles and procedures employed are equally appropriate to any spatial-temporal scenario, whether it be in biometry, chemistry, engineering, mining or physics. Thinning strategies are a prime factor in generating spatial patterns in managed forests, and have a dramatic effect on stand development, and hence product yields. Moreover, because trees generally have long life spans relative to the length of typical research projects, the design and analysis of complex long-term spatial-temporal experiments in forest stands is clearly difficult. This means that forest modelling is a key tool in the formulation and development of optimal management strategies. Fortunately, our highly flexible growth-interaction process (GIP) is easily adapted to enable a comparative study to be made of different thinning regimes, and it not only provides a powerful descriptor of forest stand growth but there is also considerable evidence that it is particularly robust to the accuracy of model choice.

The first task is to choose the growth and interaction functions, $f(\cdot)$ and $h(\cdot)$. Both $f_1(\cdot)$ and $f_2(\cdot)$ in (10.2.4) are special cases of the logistic power-law process (3.2.41), namely

$$dm_i(t)/dt = \lambda m_i(t) - d[m_i(t)]^{s+1}, \qquad (10.2.15)$$

which has the solution (3.2.42). Now as well as providing an extremely useful descriptor for modelling the dynamics of insect populations (e.g. Matis, Kiffe and Parthasarathy, 1998) in which the more aggressive the insect the higher the value of $s > 1$, the general representation (10.2.15) has also played a major role in the modelling of tree growth. For it specifically relates to the Von Bertalanffy–Chapman–Richards (VBCR) growth function

$$f_4(m_i(t)) = (\beta m_i(t)/\nu)[(K/m_i(t))^\nu - 1] \qquad (10.2.16)$$

(von Bertalanffy, 1949; Richards, 1959; Chapman, 1961), which is used extensively to model both diameter and height growth. Experience has shown that this pragmatically simple form is sufficiently flexible to be the natural first choice in all the scenarios so far considered. Here K denotes the tree-size carrying capacity, β scales the time axis, and ν defines the curve shape. Note the implicit sign change between (10.2.15) and (10.2.16); in the former the power term $dm_i(t)^{s+1}$ 'controls' the linear growth term $\lambda m_i(t)$, whilst the converse holds in the latter. Moreover, since the Gerrard-based asymmetric area-interaction function (10.2.7) has been used as a measure of tree competition since the late 1960's, we shall retain it here.

It is vitally important to ensure that as much knowledge of forest development as possible is enshrined in this chosen model structure, namely (f_4, h_2), by relating the (theoretical) mark variables $m_i(t)$ to key features of tree growth. Given that d.b.h. is highly affected by tree density, and so is sensitive to forest management, it seems eminently reasonable to work with the tree radius at breast height (r.b.h.), i.e. $m_i(t) = \text{d.b.h.}/2$. Whence the stand basal area at time t is $G(t) = \pi \sum_{i=1}^{n} m_i^2(t)$, where n denotes the number of trees. Note that $G(t)$ is closely related to the quadratic mean diameter (q.m.d.), namely $D(t) = 2\sqrt{G(t)/(\pi n)}$, which is essential for estimating total biomass production. The influence zone of a tree is taken to be the horizontal projection of tree canopy, and hence has radius $\delta m_i(t)$ where δ denotes the constant of proportionality between r.b.h. and crown radius. Renshaw, Comas and Mateu (2009) provide detailed information on how the associated parameter values, namely K, β, ν, δ, b and maximum tree age, can be determined from selected forest databases.

Four thinning procedures are generally used in the tree selection process. *Thinning from below* favours the upper crown classes by removing suppressed trees from the lower crown. So it mimics natural selection by favouring vigorous trees and accelerating the mortality of suppressed ones. In contrast, *thinning from above* favours the growth of 'promising' target trees from the upper crown by removing trees there. It therefore regulates tree competition by eliminating strong competitors rather than by removing weak ones, as happens when thinning from below. *Selection thinning* enhances the growth of target trees in the lower class; whilst *systematic thinning* involves predetermining the tree-tree spacing configuration. Combining these methods yields a wide variety of possible strategies.

Any chosen management strategy is easily incorporated into algorithm A10.1 by including an extra step in the computation. For example, to include a thinning procedure we simply insert:

(vi) apply the selected thinning strategy if either: t corresponds to a 'thinning time'; or, the spatial structure satisfies the 'thinning criteria'.

612 Spatial–temporal extensions

Two specific thinning regimes, chosen from the huge number available, neatly illustrate how this modelling procedure provides considerable insight into the way in which the choice of strategy affects overall forest output. These are: (1) a single-thinning from below with 100% and 50% of trees being cut from small and medium diameter classes, respectively; and, (2) a two-thinning regime in which (1) is followed by a thinning from above which cuts 50% of trees from both the remaining medium and also the large diameter classes. Thinning from below is triggered when the basal area in the 25×25 m^2 stand reaches 3 m^2 (around year seven). Thinning from above is applied when the basal area reaches 6 m^2 (around year fifteen).

Figure 10.10 (from Renshaw, Comas and Mateu, 2009) shows the associated change in population size $n(t)$, quadratic mean diameter $D(t)$ and stand basal area $G(t)$ over 100 years of simulated forest growth. Under linear growth ($\nu = 1$) the population size is

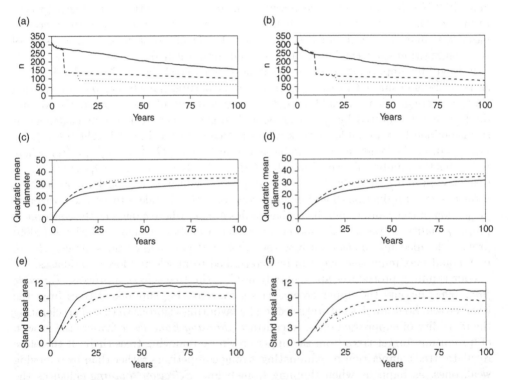

Figure 10.10 Time-series plots for the growth-interaction model (10.2.1) based on (f_4, h_2), i.e. the VBCR growth function (10.2.16) and area-interaction function (10.2.7), with $\beta = 0.052$, $K = 25$, $b = 0.6$, $r = 0.01$ and $\delta = 9$. The initial population comprises $n = 312$ trees having $U(0.25, 0.75)$ radii at $t = 0$, $\mu = 0.05$ and $\alpha = 0$. Cases (a), (c) and (e) relate to $\nu = 1$, and (b), (d) and (f) to $\nu = 0.7$. Simulations relate to a non-treated forest (—), a single-thinning from below (- - -), and a two-thinning programme from below and above ($\cdots$). Quadratic mean diameter is given in cm and basal area in m^2 for a forest region of 0.0625 ha (Renshaw, Comas and Mateu, 2009).

larger than under sigmoid growth ($\nu = 0.7$) independently of the thinning programme. This is in direct contrast to the development of quadratic mean diameter, since linear growth results in slightly smaller individual tree size, though larger basal area. Nevertheless, the choice of tree growth function has only a small effect on the result of applying the two thinning strategies. When thinning from below, individual tree size is larger than that for a non-treated forest, suggesting that an early thinning contributes to a reduction in between-tree competition by allowing timber to be produced more efficiently. The second thinning from above applied several years later contributes to increased tree size when compared to both the single-thinning from below and the untreated regimes. Thus this simulation exercise demonstrates that whilst the first thinning procedure causes a substantial increase in stand basal area, and hence likely commercial profit, application of the second thinning increases this profitability still further. Continued investigation of other thinning strategies for a given forest stand scenario would then lead to an approach which maximizes expected forest output. Though caution should be exercised to guard against adopting a 'black-box' mentality in which the resulting spatial patterns are not critically examined. For both the single- and double-thinning strategies result in empty space in which additional trees could exist without interacting with those already present. This highlights the danger in taking a purely formulaic approach, and demonstrates the need for a symbiotic relationship between modeller and practitioner.

Other important management features that could be similarly examined include planting regimes, construction of extraction routes, consideration of stand shape and size, single- versus multi-species stands, etc. Indeed, the number of potential spatial–temporal fields in which this general algorithmic procedure may be applied is clearly huge, and further development across a broad range of different areas and disciplines is to be strongly encouraged.

10.2.7 Application to tightly packed particle systems

One such field concerns the study of properties of porous and granular materials. For this is of great practical importance since the way in which individual elements pack together, and their overall packing intensity, are crucial to defining material structure and strength. Such systems have very interesting characteristics, complicated topological structures, and potential phase transitions (i.e. from fluid to crystalline). Indeed, their analysis exposes deep mathematical issues in spatial statistics and stochastic geometry which, until recently, have been avoided in most popular texts. Stoyan, Kendall and Mecke (1995) provide an early outline; whilst Stoyan (2003) presents an overview of the various modelling opportunities on offer. Since simulation studies of random hard particle systems, involving the development of powerful simulation techniques and ingenious statistical methods for the evaluation and quality control of the ensuing results, have been progressed almost entirely by the physics community, this field is wide open for development by mathematicians, statisticians and probabilists.

A popular approach involves simple sequential inhibition (SSI) in which particles are placed in random locations until the system jams, i.e. there is no available space left for any more to be added. This technique is particularly useful when generating patterns with a relatively low packing intensity. The resulting stationary distribution of point locations is discussed in Stoyan and Schlather (2000); closely related models are the second Matern hard-core process and the Matheron dead leaves model. Readers interested in classical models for random systems of hard particles of constant diameter should consult Torquato (2002) (see also Hansen and McDonald, 1986; Löwen, 2000). Whilst examples of novel, more statistically oriented, approaches are contained in Döge, Mecke, Møller, Stoyan and Waagepetersen (2004), Hermann, Elsner, Hecker and Stoyan (2005) and Kadashevich, Schneider and Stoyan (2005). Here, however, we shall promote our own radically different packing strategy by allowing particles to overlap with, and even contain, each other. For this approach easily relates to systems involving hard particles in engineering, chemistry and physics applications on allowing the process to 'burn-out' at the end of its development period by preventing the arrival of any more fresh points. Moreover, increasing the spatial interaction parameter recovers classical hard-packing as a limiting case. Note that in this chapter the algorithm is presented in two dimensions purely for illustrative convenience; it extends easily and naturally to three (and higher) dimensions.

The advantage, and indeed novelty, of the GIP approach lies in the ability of points to exploit the available space local to them by changing the size of their associated marks in response to the changing status, and hence interaction pressure, of their neighbours. The disadvantage from a tightly-packed perspective is that the algorithm seldom generates packing intensities in excess of 50%, which is substantially less than is often observed in practice. For as more marks arrive, the existing ones get smaller, and so the proportion of empty space remaining stays more or less the same. Only extreme cases like the (f_2, h_3)-regime shown in Figure 10.8b avoid this. One way of exploiting empty space would be to allow points to give birth, with their progeny following some spatial distribution (e.g. isotropic Normal, exponential or Cauchy) around the parents, as this just involves trivial modifications to the basic algorithm. However, a more straightforward, and non-multiplicative, approach is simply to allow individual points to move. This creates a whole new suite of spatial structures that are directly relevant to the study of materials science. For since marks can reduce in size they are now able to squeeze (under sufficiently high local interaction pressure) through small gaps from high-density to low-density space and thereby provide a totally fresh extension to the maximum packing scenarios described in Torquato (2002).

Although the simplest mechanism for movement is to let each point location follow Brownian motion, it is more reasonable to assume that the movement of a point depends on the degree and direction of the interaction pressure being placed upon it. Moreover, when a large mark interacts with a small one, the relative interaction pressure is likely to be greater on the smaller than the larger. So following Renshaw and Comas (2009) let us assume, for example, that the vector force on i from j is $v\min(1, m_j(t)/m_i(t))$ along the line connecting points i and j (as illustrated in Figure 10.11). Denote the components of the position vector $x_i(t)$ by (x_i, y_i), write $r_{ij} = \sqrt{\{(x_i - x_j)^2 + (y_i - y_j)^2\}}$, and let $\sum_{j\setminus i}$ denote the sum over all points j

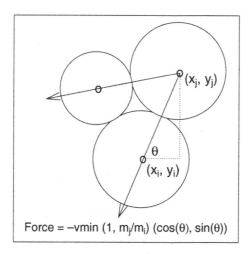

Figure 10.11 Forces exerted on interacting disks.

that interact with i. Then during the small time interval $(t, t + dt)$, i's position is perturbed to

$$x_i(t + dt) = x_i(t) + vdt \sum_{j \backslash i} \min(1, m_j(t)/m_i(t))(x_i - x_j)/r_{ij}$$

$$y_i(t + dt) = y_i(t) + vdt \sum_{j \backslash i} \min(1, m_j(t)/m_i(t))(y_i - y_j)/r_{ij}. \quad (10.2.17)$$

This change in position from $x_i(t) \to x_i(t + dt)$ is straightforward to incorporate into the simulation algorithm, and may be achieved either deterministically or stochastically. However, if there is no real practical gain in allowing for stochastic movement then the best option is simply to use the mechanism (10.2.17) in order to update $x_i(t) \to x_i(t + \eta_i)$ where η_i denotes either the fixed time increment dt or (as we shall soon see in Section 10.2.8) either the tau-leaping time increment τ or the variable inter-event time s_k. Here η_i must be sufficiently small to prevent points jumping far enough in one step to escape from highly constrained to empty space in a single movement. For if they are not then the whole nature of the process changes: the difference between the resulting spatial structures provides a useful line for future investigation.

Figure 10.12 illustrates the effect of allowing particles to move, and shows the initial stages of development under the two velocity regimes $v = 0$ and $v = 0.01$ over $0 \leq t \leq 200$. Here we retain the linear growth (f_2) and hard-core interaction (h_1) functions used in Figure 10.9, and let the interaction parameter $b = 2 > \lambda = 1$, scaling parameter $r = 0.005$, carrying capacity $K = 20$, immigration rate $\alpha = 10$, immigrant size $\tilde{m}_k = 0.01$ and time increment $dt = 0.01$. If points are not allowed to move (i.e. $v = 0$) then the mean mark, $\overline{m}(t)$, swiftly rises to a maximum value of $\overline{m}(20) = 4.22$, at which time there are $n = 198$ alive individuals; after $t = 20$ we see that $\overline{m}(t)$ slowly decreases to 2.37 at $t = 200$ when $n = 1010$. In contrast, allowing points to move with small velocity v causes a considerable change in spatial structure. For $\overline{m}(t)$ reaches a higher

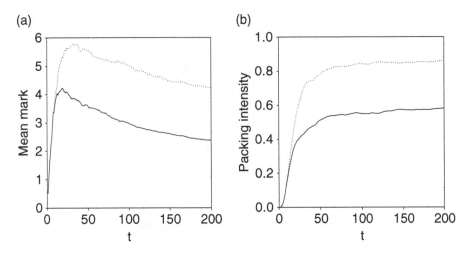

Figure 10.12 Convergence to high-intensity packing under the linear-growth hard-core model (f_2, h_1) with $\alpha = 10$, $\lambda = 1 < b = 2$, $K = 20$, $r = 0.005$, $\tilde{m}_k = 0.01$ and $dt = 0.01$: (a) average mark size and (b) proportion of filled space for $v = 0$ (—) and $v = 0.01$ ($\cdots$).

maximum value of 5.79 based on fewer individuals ($n = 235$) at the later time $t = 36$; with $\overline{m}(200) = 4.23$ now being based on only 477 individuals.

The overall packing intensity is of key interest in chemistry, physics and engineering applications, and Figure 10.12b shows that allowing points to move with velocity $v = 0.01$ can dramatically increase the proportion of filled space (here from 58.3% to 86.2% at $t = 200$), even though the resulting spatial structure may involve fewer marks than that under $v = 0$. Figure 10.13 shows the corresponding patterns at $t = 200$,

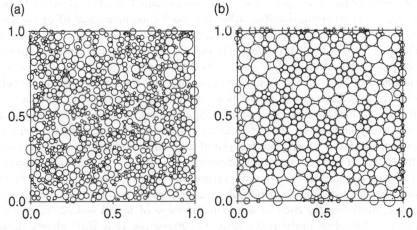

Figure 10.13 Packing pattern at $t = 200$ for the model of Figure 10.12: (a) $v = 0$ and (b) $v = 0.01$.

and highlights the way in which allowing movement generates patterns with greater internal structure. Some feel for how 'hard-packed' they are can gleaned by making a comparison with the packing intensity for regularly packed disks of equal radius 1. For the area between four neighbouring centres has size 4, and contains four quarter disks each of area $\pi/4$, so the packing intensity is $4(\pi/4)/4 = 0.7854$. Whilst on a triangular lattice the area between three neighbouring centres is $\sqrt{3}$ and contains 3 1/6th size segments each of area $\pi/6$, so now the packing intensity is $3(\pi/6)/\sqrt{3} = 0.9069$, which is only slightly higher than the above packing intensity of 86%. Though since our marks are of unequal size and so can be packed more efficiently, yet are not regularly distributed, these benchmarks provide only a rough guide.

To demonstrate the considerable flexibility of the GIP approach let us consider an example, similar to one analysed by Renshaw (2009), which was inspired by the simulations developed in Bezrukov, Bargiel and Stoyan (2002) and Lautensack, Schladitz and Särkkä (2006). Suppose we wish to simulate a system which contains both large and small particles. Then two different modelling approaches (here illustrated in two dimensions) are as follows.

1. Let n_1 large and n_2 small disks have carrying capacities $K = k_1$ and $K = k_2$, respectively, with the $n = n_1 + n_2$ initial marks having size $m_i(0) = k_1/2$ ($i = 1, \ldots, n_1$) and $m_i(0) = k_2/2$ ($i = n_1 + 1, \ldots, n$) and randomly distributed locations over $U(0,1)^2$. Choose, for example, the simple linear growth function $f(m_i(t)) = \lambda(1 - m_i(t)/K)$, and with no loss of generality place $\lambda = 1$. Since n_1 and n_2 are fixed we take the natural death rate $\mu = 0$; but if any mark suffers interactive death it is immediately replaced by a new randomly located mark of size $m_i(0)$. We use the hard-core inhibition function (10.2.6) and the displacement function (10.2.17).

2. A second possibility is to use the same growth function for all marks, but to replace the linear form by a quartic polynomial. For if we wish the mark sizes m_i to lie within the region $\theta_1 < m_i < \theta_2$ for small discs, and within the non-overlapping region $\theta_3 < m_i < \theta_4$ for large discs, then we can simply take $f(m_i) = -\lambda(m_i - \theta_1)(m_i - \theta_2)(m_i - \theta_3)(m_i - \theta_4)$. It is clear from the way in which the sign of $f(m_i)$ changes with m_i, that in the absence of spatial interaction: $m_i \to \theta_4$ if $m_i > \theta_3$; $m_i \to \theta_2$ if $\theta_3 > m_i > \theta_1$; and $m_i \to 0$ if $m_i < \theta_1$. Thus θ_2 and θ_4 correspond to stable equilibrium values, and θ_1 and θ_3 to unstable ones. Note that this is a fairly basic construct, and other forms involving fractional polynomials (for example) also have considerable promise.

Figure 10.14 shows the spatial patterns and histograms of mark sizes under both growth regimes for $n_1 = n_2 = 50$ large and small disks. In each case the intrinsic growth rate $\lambda = 1$, the velocity $v = 0.01$ and the scale parameter $r = 1$. In (1) we place $c = 2 > \lambda = 1$ to ensure hard-core packing, choose $k_2 = 0.025$ and $k_1 = 0.1$ so that in the absence of spatial interaction the radii (i.e. marks) of large disks are four times larger than the radii of small ones, and let the time increment $dt = 0.001$. Although the packing intensity doubled from 38% to 76% by $t = 87$, it took until $t = 884$ to exceed 85%: the discrete nature of the algorithm then causes the packing intensity to hover around 85%–86%. The rate of convergence and the degree of 'hover'

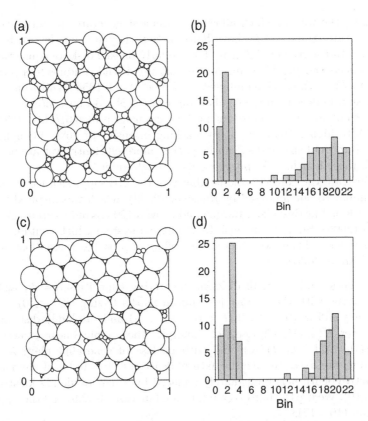

Figure 10.14 Simulated spatial patterns and associated mark distributions for two GIP regimes which couple hard-core interaction with: (a&b) (1) dual linear growth at $t = 5000$; (c&d) (2) single quartic growth at $t = 40000$.

following burn-in can be controlled by tweaking dt and v. Figure 10.14a shows the simulated pattern at time $t = 5000$, and demonstrates a tendency for large disks to group into (maximal) hexagonal packing, with small disks tending to form small groups within the remaining available space since the displacement function (10.2.17) exerts a greater force on them relative to large ones. The associated mark size histogram (Figure 10.14b), which has 22 bins from $0 - 0.004, \ldots, 0.084 - 0.088$, shows that spatial inhibition has induced a fairly wide distribution of values for both large and small disks, but not so large that the clear separation between them is removed. In (2) we choose $\theta_1 = 0.01$, $\theta_2 = 0.02$, $\theta_3 = 0.05$ and $\theta_4 = 0.09$, with $m_i(0) = 0.015$ and 0.07 for small and large disks, respectively. Since $\max_{0.01<m<0.02}[f(m)] = 0.657 \times 10^{-7}$ at $m = 0.014$, and $\max_{0.05<m<0.09}[f(m)] = 0.135 \times 10^{-5}$ at $m = 0.076$, any value of $c \geq 0.14 \times 10^{-5}$ will guarantee hard-core interaction, i.e. non-overlapping disks. We therefore took $c = 10^{-5}$. Moreover, because the resulting values for $f(\cdot)$ and $h(\cdot)$ are much smaller than under (1), we increased the time increment to $dt = 0.01$ in order to speed convergence.

Because vdt is now 10 times larger, the point location displacements $\|x_i(t+dt) - x_i(t)\|$ are substantially larger than the mark size displacements $|m_i(t+dt) - m_i(t)|$. This causes the initial overlapping disks to move swiftly away from each other, with the result that the packing intensity of 80.43% at $t = 1$ drops only very slightly to 79.34% at $t = 56$ before it starts to rise again. However, since $dm_i(t)/dt \ll 1$, convergence for this second regime is far slower than under the first regime, with the times taken to reach specific levels of packing intensity being

80%	82%	84%	86%	88%	90%
367	1582	3587	9611	18227	29728

Thereafter, the packing intensity effectively ceased rising at $t = 37000$ when it hovered around 90.5%. Figure 10.14c shows the pattern at $t = 40000$ which has a packing intensity of 90.54%. As this value drops slightly to 90.52% at $t = 60000$, we may assume that burn-in has occurred. The pattern structure is visually similar to that generated under (1), though the associated distribution of mark size (Figure 10.14d) shows less variability. Note that algorithm A10.1 is now virtually deterministic since $\alpha = \mu = 0$; stochasticity occurs only if a particle suffers interactive death and is then replaced by a new mark at a random location. Thus the final spatial configuration is almost wholly dependent on the start configuration. If this is an issue then we may replace A10.1 by one of the five alternative stochastic approaches A10.2–A10.6 developed in the following section.

10.2.8 Other stochastic strategies

At this point it is worth recalling that the original raison d'être which underpinned the construction of the growth-interaction algorithm A10.1 was to provide a fast and computationally efficient way of simulating complex spatial–temporal dynamical systems. The deterministic growth-interaction component (f, h) is fundamental to computational speed, whilst the immigration and natural death rates, α and μ, inject stochasticity into the system. However, whilst such random arrival and departure events may well be appropriate in many ecological scenarios, physics and chemistry situations often concern a fixed, and hence non-random, number of interacting particles. Thus both α and μ in A10.1 now equal zero, whence virtually all stochasticity is lost from the temporal development of the system. In some practical situations this may not be perceived as being an issue, since different final spatial configurations may still be generated by allowing the initial spatial configuration to be random. However, if we wish to work with a genuine stochastic process then the form of the algorithm clearly needs to be amended. Fortunately, we have already explored a number of alternative constructions in Sections 9.3.3 and 9.3.4. So all we need do here is to restructure the procedures underlying algorithms A9.1 to A9.6, i.e. exact event–time pairs through to the chemical Langevin equation, in terms of marked point processes.

First, we have to decompose the growth and interaction functions in (10.2.1) into general stochastic birth and death components. So let us split $f_i = f(m_i(t)) \to f_i^+ - f_i^-$ and $h_{ij} = h(m_i(t), m_j(t); \|i - j\|) \to h_{ij}^+ - h_{ij}^-$; where f_i^+ and f_i^- denote pure birth and death, whilst h_{ij}^+ and h_{ij}^- denote spatial enhancement and inhibition. Second, we have to appreciate that each marked point is affected differently, since it is subject to its own specific growth and spatial interaction forces. Thus we cannot follow the 'usual' Kolmogorov type argument by pooling individuals together to form population-based transition rates, but instead we have to use an individual-based approach.

Exact marked point process algorithm (A10.2)

Since our process is Markov we may generate exact stochastic realizations by constructing a complete set of events together with their times of occurrence. Label the n points by $i = 1, \ldots, n$, and the successive event–time pairs by $\{e_k, t_k\}$ over $k = 1, 2, \ldots$. For each mark construct the individual birth and death rates $\lambda_i = f_i^+ + \sum_{j \neq i} h_{ij}^+$ and $\mu_i = f_i^- + \sum_{j \neq i} h_{ij}^-$, and compute the overall event rate $R = \sum_{i=1}^{n}(\lambda_i + \mu_i)$. Now under the GIP algorithm marks generated via (10.2.1) are not allowed to take negative values, so point i is immediately removed if $m_i(t) \leq 0$. Although this feature can easily be incorporated into a corresponding stochastic event-time algorithm, it is mathematically neater, and indeed more appealing, to make the paradigm with population processes in which population size is automatically non-negative. Let us therefore assume that a 'birth' or 'death' at point i now corresponds to $m_i(t)$ increasing or decreasing by some suitably small amount $\delta > 0$. Thus $m_i(t) = \delta l_i(t)$ for integer $l_i(t) = 0, 1, 2, \ldots$. Whilst if any point i suffers interactive death (i.e. $l_i(t) = 0$) it is immediately replaced by a new point randomly located in the study region with a mark size $\tilde{m}$ that is an integer multiple of δ. So for a sequence $\{U\}$ of i.i.d. $U(0,1)$ pseudo-random numbers, if $\sum_{j=1}^{i-1}(\lambda_j + \mu_j) \leq U \leq \lambda_i + \sum_{j=1}^{i-1}(\lambda_j + \mu_j)$ then event e_k corresponds to $m_i(t) \to m_i(t) + \delta$, whilst if $\lambda_i + \sum_{j=1}^{i-1}(\lambda_j + \mu_j) \leq U < \sum_{j=1}^{i}(\lambda_j + \mu_j)$ it corresponds to $m_i(t) \to m_i(t) - \delta$. Finally, since the inter-event times $s_k = t_k - t_{k-1}$ ($k = 1, 2, \ldots$) are exponentially distributed with parameter R it follows that $s_k = -\ln(U)/R$. Whence on retaining the deterministic point displacement function (10.2.17) the exact algorithm A9.1 becomes:

(i) set $t = 0$ and $m_i = \tilde{m}$ for $i = 1, \ldots, n$
(ii) generate new $U(0,1)$ random numbers U_1, U_2, U_3, U_4
(iii) cycle over individuals $i = 1, 2, \ldots, n$
   compute $\lambda_i = f_i^+ + \sum_{j \neq i} h_{ij}^+$, $\mu_i = f_i^- + \sum_{j \neq i} h_{ij}^-$ and $R = \sum_{i=1}^{n}(\lambda_i + \mu_i)$
   if $U_1 \times R \leq \lambda_1$ then $m_1 = m_1 + \delta$
   else if $U_1 \times R \leq \lambda_1 + \mu_1$ then $m_1 = m_1 - \delta$
   $\vdots$
   else if $U_1 \times R \leq \sum_{i=1}^{n-1}(\lambda_i + \mu_i) + \lambda_n$ then $m_n = m_n + \delta$
   else $m_n = m_n - \delta$
   if $i = r$ is the altered mark then recalculate h_{rj} and h_{jr} ($j \neq r$)
   if $m_r = 0$ then set $m_i = \tilde{m}$ at new location (U_2, U_3)
(iv) update time t to $t + s$ for $s = -\ln(U_4)/R$
(v) cycle over all individuals $i = 1, \ldots, n$ alive at time $t - s$
   use (10.2.17) with $dt = s$ to update locations $(x_i(t), y_i(t))$
(vi) return to (ii)

Note that only a simple extension to this procedure is required in order to include immigration, birth, natural death, etc.

Time-increment algorithm (A10.3)

Whilst this exact algorithm is computationally fast when the number of particles is fairly small, e.g. $n \sim 100$, for larger values of n it may be far too slow. For not only is the expected number of "if-checks" in (iii) of order $O(n)$, but the resulting time increment (iv) is $O(n^{-1})$. So if say $\lambda_i + \mu_i \simeq 1$ over $i = 1, \ldots, n = 10^4$, then $R \simeq 10^4$ and roughly 10^4 steps in (iii) are required in order to increase t by a mere $-10^4 \ln(U)$. Now we have already seen through the construction of the multi-colony population algorithm A9.2 that a general approximation procedure, which often works well in such situations, is to replace calculation of the time to the next event, s_k, by a fixed time increment of size dt that is small enough to ensure that for all practical purposes the probability of a multiple event during the time interval $(t, t+dt)$ is negligible. Thus in our marked point process scenario we presume that each mark acts independently of all others during each of the small time intervals $(t, t+dt)$. A key advantage of this approach is that instead of step (iii) leading to the update of just a single mark, all the n marks are now updated in turn. Thus steps (ii)-(iv) become:

(iiia) cycle over individuals $i = 1, 2, \ldots, n$
generate new $U(0,1)$ random numbers U_1, U_2, U_3
compute $\lambda_i = f_i^+ + \sum_{j \neq i} h_{ij}^+$, $\mu_i = f_i^- + \sum_{j \neq i} h_{ij}^-$
if $U_1 \leq \lambda_i dt$ then $m_i(t+dt) = m_i(t) + \delta$
else if $U_1 \leq (\lambda_i + \mu_i)dt$ then $m_i(t+dt) = m_i(t) - \delta$
else $m_i(t+dt) = m_i(t)$
if $m_i(t+dt) = 0$ then set $m_i = \tilde{m}$ at new location (U_2, U_3)
(iva) update t to $t + dt$

The velocity step (v) has been omitted (here and later) purely for brevity.

Tau–leaping algorithm (A10.4)

One of the problems with this approach is that if dt is sufficiently small to ensure that $\max(\lambda_i + \mu_i)dt \ll 1$ then many of the steps in (iiia) will result in 'no change', which induces considerable computational inefficiency into the procedure. Fortunately, we have already examined ways of circumventing this problem when discussing the simulation of multi-colony population processes in Section 9.3.4. For recall that although the exact event-time approach, here displayed in algorithm A10.2, stems from the mid-1960s, interest in the development of large-scale chemical reaction systems led to the generation of more pragmatically oriented approximations in the mid-1970s, with the pioneering papers of Gillespie (1976, 1977) being particularly noteworthy. Sadly, this work was developed completely independently of the earlier results, and it is unfortunate that population and molecular research groups have remained virtually isolated from each other. In essence, we eschew the tiny time increment dt in favour of the much larger increment $\tau \gg dt$, under the presumption that the number of positive (B_i) and negative (D_i) changes in size undertaken by mark i in $(t, t+\tau)$ follow independent Poisson($\lambda_i \tau$) and Poisson($\mu_i \tau$) distributions, respectively. Though since our marked process system needs to remain stiff, $\tau \gg dt$ must be small enough to ensure that with high probability the n marks remain relatively unchanged during $(t, t+\tau)$. Poisson random variables are particularly easy to simulate via result (2.1.40), for if $\{U_0, U_1, \ldots\}$ denotes a series of $U(0,1)$ pseudo-random variables then B_i (for example) is the largest integer j such that $\prod_{j=0}^{\infty} U_j \leq e^{-\lambda_i \tau}$. Whence on replacing dt by τ, and the independent Bernoulli events in A10.3 by Poisson variables, we have the tau-leaping algorithm:

(i) set $t = 0$ and $m_i = \tilde{m}$ for $i = 1, \ldots, n$
(ii) cycle over individuals $i = 1, 2, \ldots, n$
  compute $\lambda_i = f_i^+ + \sum_{j \neq i} h_{ij}^+$, $\mu_i = f_i^- + \sum_{j \neq i} h_{ij}^-$
  generate new $U(0,1)$ random numbers $U_0, U_1, \ldots$
  determine $B_i \sim \text{Poisson}(\lambda_i \tau)$
  generate new $U(0,1)$ random numbers $U_0, U_1, \ldots$
  determine $D_i \sim \text{Poisson}(\mu_i \tau)$
  set $m_i(t+\tau) = m_i(t) + (B_i - D_i)\delta$
  if $m_i(t+\tau) \leq 0$ then set $m_i = \tilde{m}$ at location (U_0, U_1) for new U_0, U_1
(iii) update time t to $t + \tau$ and return to (ii)

Accuracy can be enhanced by changing the leap time, τ, adaptively, based on the current state of the system and the transition rates. Moreover, if one or more marks are likely to lie near zero during parts of a realization, then we can switch back onto algorithm A10.3, with tau-leaping being resumed once the system has regained the required level of stiffness.

Langevin–leaping algorithm (A10.5)

If δ is small relative to $\max(m_i(t))$ then an appreciable number of points may have to undergo a large number of events before their associated marks show any real change. This is especially so when interest centres on the diffusion limit for mark size, i.e. $\delta \ll 1$. For then the scheme (2.1.40) for generating Poisson variables becomes grossly time-consuming, and it is more sensible to parallel the multi-colony algorithm A9.5 by replacing the Poisson variables by Normal variables with the same mean and variance, since this is likely to speed up the tau-leaping algorithm quite substantially. Thus we now take $B_i \sim N(\lambda_i \tau, \lambda_i \tau)$ and $D_i \sim N(\mu_i \tau, \mu_i \tau)$. Though note that if either $\lambda_i \tau$ or $\mu_i \tau$ become small enough for B_i or D_i to be negative with non-negligible probability, e.g. $\min(\sqrt{\lambda_i \tau}, \sqrt{\mu_i \tau}) < 2.5$, then the approach clearly loses its population-based rationale. There are a variety of different ways of simulating Normal random variables, some more efficient than others, but one of the easiest is to employ the Box–Muller method (Box and Muller, 1958). Here independent pairs of Normal variates, (Z_1, Z_2), are generated from pairs of uniform variates, (U_1, U_2), through the polar transformation $(Z_1, Z_2) = (R \cos \Theta, R \sin \Theta)$ where $R = \sqrt{-2 \ln(U_1)}$ and $\Theta = 2\pi U_2$. Whence the above tau-leaping procedure A10.4 becomes the Langevin-leaping algorithm:

(i) set $t = 0$ and $m_i = \tilde{m}$ for $i = 1, \ldots, n$
(ii) cycle over individuals $i = 1, 2, \ldots, n$
  compute $\lambda_i = f_i^+ + \sum_{j \neq i} h_{ij}^+$, $\mu_i = f_i^- + \sum_{j \neq i} h_{ij}^-$
  generate new $U(0,1)$ random numbers U_1, U_2

determine $Z = \sqrt{-2\ln(U_1)}\cos(2\pi U_2)$
set $m_i(t + \tau) = m_i(t) + [(\lambda_i - \mu_i)\tau + Z\sqrt{(\lambda_i + \mu_i)\tau}]\delta$
if $m_i(t + \tau) \leq 0$ then set $m_i = \tilde{m}$ at location (U_3, U_4) for new U_3, U_4
(iii) update time t to $t + \tau$ and return to (ii)

Observe that the mark state space has been fundamentally changed, since the integer-valued Poisson random variables have been replaced by real-valued Normal ones.

Chemical Langevin algorithm (A10.6)

Reverting τ back to dt gives rise to the chemical Langevin equation, namely the discretized version of the stochastic differential equation representation previously discussed in Section 6.2.5. As commented there, this process is fundamentally different to the time-increment algorithm A10.3 since the distribution for the change in mark size, $\Delta_i = m_i(t + \tau) - m_i(t)$, switches from the trinomial form $\Delta_i = -1$, 0, 1 with probability $\mu_i(t)dt$, $1 - (\mu_i(t) + \lambda_i(t))dt$, $\lambda_i(t)dt$, to the Normal form $\Delta \sim N[(\lambda_i(t) - \mu_i(t))dt, (\lambda_i(t) + \mu_i(t))dt]$. Whilst this may be appropriate if all the n mark sizes lie well way from zero, if they do not then the two processes may exhibit different characteristics. So if algorithm A10.6 is adopted as a mainstream approach in a given situation, then comparative simulation studies should be undertaken between it and A10.3 beforehand.

Reaction rate algorithm (A10.7)

Algorithms A10.2 to A10.6 all involve the generation of random mark size increments. In contrast, the purely deterministic, i.e. reaction rate, approach runs totally counter to this in that the mark size increments are determined solely through expression (10.2.1). Thus in terms of the above notation, we now simply have $\Delta_i = (\lambda_i - \mu_i)dt$. Nevertheless, in addition to the initial random placement of the n marks at time $t = 0$ we can also inject a further degree of stochasticity by replacing a mark that dies interactively at time t (i.e. $m_i(t - dt) > 0$ and $m_i(t) \leq 0$) with a new one of size $m_i(t + dt) = \tilde{m}$ at a fresh random location. Moreover, we could also allow marks to die naturally at rate μ provided they are likewise replaced by new ones. Conversely, if we require the process to be wholly deterministic for $t > 0$, then if mark i dies interactively at location $(x_i(t), y_i(t))$ at time t we could (for example) replace it with a new one of size $\tilde{m}$ at a location that is displaced a fixed distance from the old one. Thus in terms of the above (λ_i, μ_i)-notation we have:

(i) set $t = 0$ and $m_i = \tilde{m}$ at random locations for $i = 1, \ldots, n$
(ii) cycle over individuals $i = 1, 2, \ldots, n$
compute $\lambda_i = f_i^+ + \sum_{j \neq i} h_{ij}^+$, $\mu_i = f_i^- + \sum_{j \neq i} h_{ij}^-$
set $m_i(t + dt) = m_i(t) + (\lambda_i - \mu_i)dt$
if $m_i(t + dt) \leq 0$ then set $m_i = \tilde{m}$ at a new random or deterministically displaced location
(iii) update time t to $t + \tau$ and return to (ii)

Not only do the algorithms A10.2–A10.7 highlight the many and varied ways that stochastic realizations may be constructed, but it must be emphasised that within this text we have restricted attention purely to random events and times of occurrence.

There are, of course, many other mechanisms for inducing stochasticity, such as letting parameters be random, or allowing for feedback mechanisms and random external inputs to the system. Thus the number of possible model structures, and hence the potential for future development in this ever-widening arena of applied stochastic processes, is truly enormous. As a preliminary taster, Higham (2008) points to current challenges, and presents short downloadable MATLAB codes that can be used to generate exploratory studies.

In summary, our algorithms cover a wide stochastic spectrum, from the exact procedure A10.2, down through successive levels of approximation, to the almost deterministic form A10.7. However, whilst A10.2 and A10.3 are in current use, A10.4–A10.6, i.e. tau-leaping onwards, have not so far featured in simulation studies of spatial population dynamics. Since all six techniques may be easily generalized to cover a large variety of situations across a wide range of disciplines, a future task is to examine whether the exact algorithm A10.2 produces significantly different results to the Renshaw and Särkkä (2001) approach illustrated in A10.1. If it does, but incurs too large a compute-time penalty, then the other algorithms should be analysed in sequence in order to assess the trade-off between pattern structure and computational efficiency. This is likely to be of prime importance when developing models of granular structures under high-intensity packing, in which, unlike our ecologically based scenarios, particles may move under spatial interaction pressure. Such studies are currently being undertaken in order to generate models that accurately replicate mixed-size particle systems which previously could only be simulated by using 'sequential packing under gravity' and 'collective rearrangement' strategies.

10.2.9 Final comments

The general GIP process (10.2.1), with the inclusion of movement as a natural extension, clearly provides a powerful and highly flexible technique for constructing a large family of spatial–temporal marked point processes which exhibit widely varying properties and characteristics. Moreover, it is fast to compute in one, two and three dimensions, and may be easily generalized to cover specific situations across a wide range of applied disciplines, and thereby provide increased understanding of many packing structures in the real world. In addition, randomness can be induced in a variety of different ways, from the exact algorithm A10.2 right through to the original deterministic algorithm A10.1, which comprises A10.7 with added stochastic arrival and departure components. This flexibility allows quite complex processes to be investigated virtually independent of their size. To date only a few growth and interaction functions have been used, and there are many other functional forms that could also be explored. Indeed, it would be well worthwhile constructing an index of (f, h)-pairs that lists their key properties so that researchers studying a particular type of pattern could have a useful starting point at their fingertips.

An obvious challenge is to find optimal ways of selecting the growth and interaction functions f and h, and initial mark size $\tilde{m}$, in order to be able to 'fit' our general model structure to observed patterns. For there are a huge number of possible variations, and here we have just presented a few examples that highlight salient points of interest.

Indeed, not only will changing f and h result in fundamental differences in the patterns produced, but for given f and h altering the parameter and initial mark values can also lead to fairly profound changes. For example, Figure 10.8b is based on the hard-core asymmetric (i.e. one-sided) interaction function h_3 and shows a mixture of large and small marks which parallel the simulated dense-packing structures of Bezrukov, Bargiel and Stoyan (2001). So can similar patterns be produced from two-sided interaction functions? Possible approaches include allowing for both small and large growth rates, $0 < \lambda_1 < \lambda_2 < b$, or small and large interaction rates with $0 < \lambda < b_1 < b_2$. The ratio of small to large marks could be manipulated through judicious use of the immigration rules. For instead of considering the space to be empty at time $t = 0$, we could start with the observed number of small and large marks, and replace them by ones of similar type at a random location immediately they suffer interactive death.

In order to be able to estimate the parameters of a chosen function-pair (f, h), we first need to develop a sufficient set of statistics that summarizes marked point patterns with sufficient precision. For example, Tscheschel and Stoyan (2006) describe a general reconstruction method for simulating point patterns which possess prescribed summary statistics that are free of explicit model conditions. The characteristics considered include the packing intensity, the L-function, the spherical contact distribution function and the kth nearest-neighbour distance function. Whilst in addition to the spectral approach of Renshaw (2002), Capobianco and Renshaw (1998) present two alternative autocovariance estimators to the mark correlation function; these apply to both planar and lattice situations, and to anisotropic as well as isotropic spatial patterns. For marked point processes, constructs based on $\{m_i; m_j, \|x_i - x_j\|\}$ for all particles j that interact with i offer a promising way forward. Initial parameter estimates may be obtained by using the pseudo-likelihood or least squares approaches described in Renshaw and Särkkä (2001) and Särkkä and Renshaw (2006). These may be fine-tuned by employing established iterative techniques such as hill-climbing. Though a simple and computationally expedient method is to use a random walk approach by randomizing one, some or all of the parameters at each step in the iteration. For example, we might rotate through the parameters, either in sequence or randomly, using Bernoulli increments such as $\hat{r}^{p+1} = \hat{r}^p + Z_r$ where the $\{Z_r\}$ are i.i.d. variables with say $\Pr(Z_r = -0.01) = \Pr(Z_r = 0.01) = 0.5$ in order to minimize an appropriate sum of squares based on the difference between the observed and simulated patterns. This could be with respect to appropriate summary spatial statistics involving the mark distribution, point distribution, or some combination thereof.

We applied the GIP approach to two quite different scenarios. The first is ecologically motivated with the aim being to simulate the spatial-temporal development of trees or plants. So unless either the number of points n, or the elapsed time t, are huge compute time should not be an issue. However, in the second we are interested in constructing densely packed patterns that have (ideally) burnt-in, and in this context the compute time to convergence may be a problem. For example, the two-dimensional pattern of 100 points shown in Figure 10.14c required 4×10^6 time iterations to reach burn-in, and whilst this may be swiftly attained on a desktop computer a similar three-dimensional exercise involving say 10^4 points would most likely run too slowly. One way forward would be to investigate the use of annealing, in which the time increment dt starts off quite large and is then gradually reduced to zero as t increases in order to reduce the level of hover around the final packing intensity once burn-in has been achieved.

Not only are the opportunities and potential for exploring the range of patterns that may be generated using the general GIP procedure clearly immense, but it would be highly worthwhile comparing these with those produced under other algorithms. Current approaches ripe for comparison include: the variable radius technique of Lochmann, Oger and Stoyan (2006), who employ mono, binary, power-law and Gaussian size distributions; and non-overlapping germ-grain models developed (for example) by Andersson, Häggström and Månsson (2006). Other algorithms include sequential packing under gravity (the Jodrey–Tory algorithm) and collective rearrangement and force-biased strategies (Mościński and Bargiel, 1991). The latter start with a dense system of overlapping particles which are then stepwise moved and shrunk in order to reduce overlappings (e.g. Bezrukov, Bargiel and Stoyan, 2001).

The study of porous and granular material occupies a large field and has been explored using various physical techniques involving particles that move under pressure or random rearrangements. So let us briefly examine the extent to which the GIP procedure is related to these. Specifically, let us consider its relation to the force-biased (FB) algorithm (Bezrukov, Bargiel and Stoyan, 2001; Mościński and Bargiel, 1991; Mościński, Bargiel, Ryerz and Jacobs, 1989) since this has been one of the most commonly used ways of generating random close packings. For example, two particular applications involve the modelling of the microstructure of concrete (Ballani, Daley and Stoyan, 2006) and sintered copper (Lautensack, Schladitz and Särkkä, 2006). In both of these cases FB generates very dense isotropic packings with patterns that are too regular compared to the data even when random shifts of balls are introduced in order to disturb the regularity. In FB each ball is given an inner and an outer diameter: the inner one is chosen to ensure that there are no overlaps in the system and exactly two balls are in contact; whilst the outer diameter is based on the desired volume fraction and ball diameters. Each step in the algorithm attempts to reduce overlaps between balls by: (i) moving them in order to push them apart; and, (ii) gradually shrinking them by reducing the outer diameter. The algorithm stops when the inner and outer diameters merge. So in the final packing, although there are still only two balls in contact, many nearly touch. Thus we have the following similarities and dissimilarities. (a) In both FB and GIP randomness stems from the initial configuration, though GIP does have additional stochasticity since balls which suffer interactive death are reborn at random locations. Moreover, on replacing the reaction rate algorithm A10.7 by one of the stochastic versions A10.2–A10.6 we can induce as much event–time randomness as we wish. Though the simplest, and computationally most expedient, approach is to allow each ball to die with a suitably small natural death rate $\mu > 0$, and immediately a death occurs to replace it with a similar ball at a fresh location. (b) At each step changes are made globally in FB, but more locally in GIP since two balls interact only if their respective zones of influence overlap. Though the local GIP h-function can easily be replaced by a global one. (c) During FB the relative sizes of the balls do not change, but in GIP they wax and wane depending on the interaction pressure they are subjected to. (d) In both cases the displacement rate of ball i is proportional to a repulsive force $\sum_{j \sim i} F_{ij}$ where $j \sim i$ denotes all balls j that interact with i. Figure 10.11, for example, illustrates the GIP case with a force along the x-direction given by

$$F_{ij}(t) = v \min(1, m_j(t)/m_i(t))[x_i(t) - x_j(t)]/\|x_i(t) - x_j(t)\|.$$

Whilst in FB we have

$$F_{ij}(t) = (\rho/2d_i)\phi_{ij}[x_i(t) - x_j(t)]/\|x_i(t) - x_j(t)\|,$$

where ϕ_{ij} is a potential function, d_i denotes the diameter of ball i, and ρ is a scaling function. So by replacing $v \min(1, m_j(t)/m_i(t))$ by $(\rho/2d_i)\phi_{ij}$ we recover the FB function. Taking (a)–(d) as a whole suggests that to a fairly large extent FB may indeed be considered as being a special case of GIP. So intuitively GIP might be capable of producing a greater range of spatial structures, and hence have the potential for producing simulated patterns that lie closer to the observed ones. Särkkä, Redenbach and Renshaw (2010) are currently undertaking a study in order to examine this.

Finally, it must be stressed that virtually no theoretical analyses have yet been made on the underpinning mathematical and statistical properties of these spatial structures and their approach to convergence. Nor, in the case of $\alpha > 0$, have necessary and sufficient conditions for ergodicity been determined. The simulation route has been chosen since the mathematical analysis of nonlinear dynamical systems is notoriously difficult. However, it should be noted that the underlying stochastic population-based structures of algorithms A10.2 and A10.3 both lend themselves to the formation of the associated Kolmogorov forward and backward equations. So could useful, approximate analytic results be derived from them by employing moment closure techniques? Moreover, the model structure expands far beyond the single-species process considered here. For not only may we have several types of different species, or marks, involving competition, attraction, predator–prey, infection, chemical reaction, etc., but severe problems of edge-effects may arise when these processes are placed in a finite bounded domain. The development of such stochastic spatial–temporal systems within a marked point process environment generates a whole new field of study which provides great mathematical, statistical and computational challenges for the future.

References

Abramowitz, M. & Stegun, I.A. (1970) *Handbook of Mathematical Functions* (Dover Edition). AMS 55, US Department of Commerce.

Achelis, S.B. (2001) *Technical Analysis from A to Z* (2nd edition). McGraw-Hill: New York.

Akaike, H. (1960) On a limiting process which asymptotically produces f^{-2} spectral density. *Annals of the Institute of Statistical Mathematics, Tokyo*, **12**, 7–11.

Aksland, M. (1975) A birth, death and migration process with immigration. *Advances in Applied Probability*, **7**, 44–60.

Albert, R. & Barabási, A.-L. (2000) Topology of evolving networks: local events and universality. *Physics Review Letters*, **24**, 5234–5237.

Allaart, P. (2004a) Optimal stopping rules for correlated random walks with a discount. *Journal of Applied Probability*, **41**, 483–496.

Allaart, P. (2004b) Stopping the maximum of a correlated random walk with cost for observation. *Journal of Applied Probability*, **41**, 998–1007.

Allaart, P. & Monticino, M. (2001) Optimal stopping rules for directionally reinforced processes. *Advances in Applied Probability*, **33**, 483–504.

Anderson, R.M. (1982) *The Population Dynamics of Infectious Diseases: Theory and Applications*. Chapman & Hall: London.

Anderson, W.J. (1991) *Continuous Time Markov Chains*. Springer: Berlin.

Andersson, J., Häggström, O. & Månsson, M. (2006) The volume fraction of a non-overlapping germ-grain model. *Electronic Communications in Probability*, **11**, 78–88.

Asmussen, S. & Hering, H. (1983) *Branching Processes*. Birkhäuser: Boston.

Athreya, K.B. & Jagers, P. (1996) *Classical and Modern Branching Processes*. Springer: Berlin.

Athreya, K.B. & Ney, P.E. (1972) *Branching Processes*. Springer: Berlin.

Bachelier, L. (1900) Théorie de las speculation. *Annales Scientifiques de l'Ecole Normale Superiere*, **3**, 21–86.

Bailey, N.T.J. (1954) A continuous time treatment of a simple queue using generating functions. *Journal of the Royal Statistical Society*, **B 16**, 288–291.

Bailey, N.T.J. (1957) *The Mathematical Theory of Epidemics* (1st edition). Griffin: London.

Bailey, N.T.J. (1964) *The Elements of Stochastic Processes*. Wiley: New York.

Bailey, N.T.J. (1968) Stochastic birth, death and migration processes for spatially distributed populations. *Biometrika*, **55**, 189–198.

Bailey, N.T.J. (1975) *The Mathematical Theory of Infectious Diseases* (2nd edition). Griffin: High Wycombe.

Ballani, F., Daley, D.J. & Stoyan, D. (2006) Modelling the microstructure of concrete with spherical grains. *Computational Materials Science*, **35**, 399–407.

Banks, R.B. (1994) *Growth and Diffusion Phenomena*. Springer: Berlin.

Barabási, A.-L., & Albert, R. (1999) Emergence of scaling in random networks. *Science*, **286**, 509–512.

Barndorff-Nielson, O. & Cox, D.R. (1979) Edgeworth and saddle-point approximations with statistical applications. *Journal of the Royal Statistical Society*, **B 41**, 279–312.

Bartholomew, D.J. (1967) *Stochastic Models for Social Processes* (2nd edition). Wiley: London.

Bartlett, M.S. (1946) *Stochastic Processes*. Notes of a course given at the University of North Carolina in 1946.

Bartlett, M.S. (1949) Some evolutionary stochastic processes. *Journal of the Royal Statistical Society*, **B 11**, 211–229.

Bartlett, M.S. (1956) Deterministic and stochastic models for recurrent epidemics. *Proceedings of the 3rd Berkeley Symposium on Mathematical Statistics and Probability*, **4**, 81–109. University of California Press: Berkeley and Los Angeles.

Bartlett, M.S. (1960) *Stochastic Population Models in Ecology and Epidemiology*. Methuen: London.

Bartlett, M.S. (1964) The spectral analysis of two-dimensional point processes. *Biometrika*, **51**, 299–311.

Bartlett, M.S. (1971) Physical nearest-neighbour models and non-linear time series. *Journal of Applied Probability*, **8**, 222–232.

Bartlett, M.S. (1974) The statistical analysis of spatial pattern. *Advances in Applied Probability*, **6**, 336–358.

Bartlett, M.S. (1975) *The Statistical Analysis of Spatial Pattern*. Chapman and Hall: London.

Bartlett, M.S., Gower, J.C. & Leslie, P.H. (1960) A comparison of theoretical and empirical results for some stochastic population models. *Biometrika*, **47**, 1–11.

Bellman, R. (1953) *Stability Theory of Differential Equations*. McGraw-Hill: New York.

Bellman, R. (1960) *Introduction to Matrix Analysis*. McGraw-Hill: New York.

Bellman, R. (1964) *Perturbation Techniques in Mathematics, Physics and Engineering*. Dover: New York.

Beneš, V.E. (1965) *Mathematical Theory of Connecting Networks and Telephone Traffic*. Academic Press: London.

Berlolotti, M. (1974) Photon statistics. In *Photon Correlation and Light Beating Spectroscopy* (eds. H.Z. Cummins & E.R. Pike), 41–74. Plenum: New York.

Besag, J. (1972) Nearest-neighbour systems and the auto-logistic model for binary data. *Journal of the Royal Statistical Society*, **B 34**, 75–83.

Bezrukov, A., Bargiel, M. & Stoyan, D. (2002) Statistical analysis of simulated random packings of spheres. *Particle and Particle Systems Characterization*, **19**, 111–118.

Bhat, U.N. & Miller, G.K. (2002) *Elements of Applied Stochastic Processes* (3rd edition). Wiley: New York.

Birch, L.C. (1953) Experimental background to the study of the distribution and abundance of insects. III. The relations between innate capacity for increase and survival of different species of beetles living together on the same food. *Evolution*, **7**, 136–144.

Black, F. & Scholes, M. (1973) The pricing of options and corporate liabilities. *Journal of Political Economy*, **81**, 637–654.

Box, G.E.P. & Muller, M.E. (1958) A note on the generation of random Normal deviates. *Annals of Mathematical Statistics*, **29**, 610–611.

Bretschneider, C.L. (1959) Wave variability and wave spectra for wind-generated gravity waves. *Technical Memorandum*, **118**, Beach Erosion Board: US Army Corps of Engineers.

Broadbent, S.R. & Kendall, D.G. (1953) The random walk of *Trichostrongylus retortaeformis*. *Biometrics*, **9**, 460–466.

Brown, R. (1828) A brief account of microscopical observations made in the months of June, July and August, 1827, on the particles contained in the pollen of plants; and on the general existence of active molecules in organic and inorganic bodies. *Philosophical Magazine*, **4**, 161–173.

Bumroongsook, S. & Harris, M.K. (1991) Nature of the conditioning effect on pecan by black margined aphid. *Southwestern Entomology*, **16**, 267–275.

Burnham, D.C. & Weinberg, D.L. (1970) Observation of simultaneity in parametric production of optical photon pairs. *Physics Review Letters*, **25**, 84–87.

Cappé, O., Robert, C.P. & Rydén, T. (2003) Reversible jump, birth-and-death and more general continuous time Markov chain Monte Carlo samplers. *Journal of the Royal Statistical Society*, **B 65**, 679–700.

Capobianco, R. & Renshaw, E. (1998) The autocovariance function for marked point processes: a comparison between two different approaches. *Biometrical Journal*, **40**, 1–16.

Chandrasekhar, S. (1943) Stochastic problems in physics and astronomy. *Reviews of Modern Physics*, **15**, 1–89.

Chandrasekhar, S. (1955) A theory of turbulence. *Proceedings of the Royal Society of London*, **A 229**, 1–19.

Chapman, D.G. (1961) Statistical problems in dynamics of exploited fisheries populations. *Proceedings of the 4th Berkeley Symposium on Mathematics, Statistics and Probability*. University of California Press: Berkeley.

Chatfield, C. (1991) *The Analysis of Time Series: An Introduction*. Chapman and Hall: London.

Chen, A. & Renshaw, E. (1990) Markov branching processes with instantaneous immigration. *Probability Theory and Related Fields*, **87**, 209–240.

Chen, A. & Renshaw, E. (1992) The Gillis–Domb–Fisher correlated random walk. *Journal of Applied Probability*, **29**, 792–813.

Chen, A. & Renshaw, E. (1993a) Existence and uniqueness criteria for conservative uni-instantaneous denumerable Markov processes. *Probability Theory and Related Fields*, **94**, 427–456.

Chen, A. & Renshaw, E. (1993b) Recurrence of Markov branching processes with immigration. *Stochastic Processes and Their Applications*, **45**, 231–242.

Chen, A. & Renshaw, E. (1994) The general correlated random walk. *Journal of Applied Probability*, **31**, 869–884.

Chen, A. & Renshaw, E. (1995) Markov branching processes regulated by emigration and large immigration. *Stochastic Processes and Their Applications*, **57**, 339–359.

Chen, A. & Renshaw, E. (1997) The M/M/1 queue with mass exodus and mass arrivals when empty. *Journal of Applied Probability*, **34**, 192–207.

Chen, A. & Renshaw, E. (2000) Existence, recurrence and equilibrium properties of Markov branching processes with instantaneous immigration. *Stochastic Processes and Their Applications*, **88**, 177–193.

Chen, A. & Renshaw, E. (2004) Markovian bulk-arriving queues with state-dependent control at idle time. *Advances in Applied Probability*, **36**, 1–27.

Chiang, C.L. (1968) *Introduction to Stochastic Processes in Biostatistics*. Wiley: New York.

Child, C.M. (1941) *Patterns and Problems of Development*. Chicago University Press.

Chung, K.L. (1956) Foundations of the theory of continuous parameter Markoff chains. *Proceedings of the 3rd Berkeley Symposium on Mathematical Statistics and Probability*, **2**, 29–40. University of California Press: Berkeley and Los Angeles.

Chung, K.L. (1960) *Markov Chains with Stationary Transition Probabilities*. Springer: Berlin.

Cliff, A.D., Haggett, P., Ord, J.K. & Versey, G.R. (1981) *Spatial Diffusion: An Historical Geography of Epidemics in an Island Community*. Cambridge University Press.

Conway, J.H. (1970) In Martin Gardner's 'Mathematical Games' column. *Scientific American*, **223**, 120–123.

Cotar, C. & Volkov, S. (2004) A note on the lilypond model. *Advances in Applied Probability*, **36**, 325–339.

Cowles, M.K. & Carlin, B.P. (1996) Markov chain Monte Carlo convergence diagnostics: a comparative review. *Journal of the American Statistical Society*, **91**, 883–904.

Cox, D.R. (1948) A note on the asymptotic distribution of the range. *Biometrika*, **35**, 311–315.

Cox, D.R. (1962) *Renewal Theory*. Methuen: London.

Cox, D.R. (1977) Discussion on stochastic modelling of riverflow time series (by A.J. Lawrance and N.T. Besag). *Journal of the Royal Statistical Society*, **A 140**, 34.

Cox, D.R. & Lewis, P.A.W. (1966) *The Statistical Analysis of Series of Events*. Methuen: London.

Cox, D.R. & Miller, H.D. (1965) *The Theory of Stochastic Processes*. Methuen: London.

Cox, D.R. & Smith, W.L. (1957) On the distribution of *Tribolium confusum* in a container. *Biometrika*, **44**, 328–335.

Cramér, H.R. (1928) On the composition of elementary errors. *Skandinavisk Aktuarietids*, **11**, 13–74, 141–180.

Cramér, H.R. (1937) *Random Variables and Probability Distributions*. Cambridge University Press.

Cramér, H.R. (1938) Sur un nouveau théorème-limite de la théorie des probabilités. *Actualités Scientifiques et Industrielles*, **736**, Hermann & Cie: Paris.

Crump, K.S. (1970) Migratory populations in branching processes. *Journal of Applied Probability*, **7**, 565–572.

Dai, Y. & Renshaw, E. (2000) The Markov oscillation problem in discrete time. *Journal of the London Mathematical Society*, **61**, 301–314.

Daley, D.J., Mallows, C.L. & Shepp, L.A. (2000) A one-dimensional Poisson growth model with non-overlapping intervals. *Stochastic Processes and Their Applications*, **90**, 223–241.

Daniels, H.E. (1952) The statistical theory of stiff chains. *Proceedings of the Royal Society of Edinburgh*, **53**, 290–311.

Daniels, H.E. (1954) Saddlepoint approximations in statistics. *Annals of Mathematical Statistics*, **25**, 631–650.

Daniels, H.E. (1960) Approximate solutions of Green's type for univariate stochastic processes. *Journal of the Royal Statistical Society*, **B 22**, 376–401.

Daniels, H.E. (1975) The deterministic spread of a simple epidemic. In *Perspectives in Probability and Statistics: Papers in Honour of M.S. Bartlett* (ed. J. Gani), Applied Probability Trust, Sheffield; Academic Press, London, 373–386.

Daniels, H.E. (1977) The advancing wave in a spatial birth process. *Journal of Applied Probability*, **14**, 689–701.

Daniels, H.E. (1983) Saddlepoint approximations for estimating equations. *Biometrika*, **70**, 89–96.

Daniels, H.E. (1987) Tail probability approximations. *International Statistical Review*, **55**, 37–48.

Daniels, H.E. (1995) An alternative to the saddlepoint approximation. Conference at the University of Bath (manuscript).

Davison, A.C. (2001) Biometrika centenary: theory and general methodology. *Biometrika*, **88**, 13–52.

Dimakos, X.K. (2001) A guide to exact simulation. *International Statistical Review*, **69**, 27–48.

Döge, G., Mecke, K., Møller, J., Stoyan, D. & Waagepetersen, R. (2004) Grand canonical simulations of hard-disk systems by simulated tempering. *International Journal of Modern Physics*, **15**, 129–147.

Domb, C. & Fisher, M.E. (1958) On random walks with restricted reversals. *Proceedings of the Cambridge Philosophical Society*, **54**, 48–59.

Dubois, D.M. (1975) A model of patchiness for prey–predator plankton populations. *Ecological Modelling*, **1**, 67–80.

Dubois, D.M. & Monfort, G. (1978) Stochastic simulation of a space–time dependent predator–prey model. *Compstat*, **78**, 384–390.

Durrett, R. (1979) Maxima of branching random walks versus independent random walks. *Stochastic Processes and Their Applications*, **9**, 117–136.

Durrett, R. (1981) An introduction to infinite particle systems. *Stochastic Processes and Their Applications*, **11**, 109–150.

Dvoretzky, A. & Motzkin, T. (1947) A problem of arrangements. *Duke Mathematical Journal*, **14**, 305–313.

Einstein, A. (1905) Über die von der molekularkinetischen Theorie der Wärme geforderte Bewegung von in ruhenden Flüssigkeiten suspendierten Teilchen. *Annalen der Physik*, **17**, 549–560.

Erdélyi, A. (1953) *Higher Transcendental Functions*. Vol. II. McGraw-Hill: New York.

Erdös, P., Feller, W. & Pollard, H. (1949) A property of power series with positive coefficients. *Bulletin of the American Mathematical Society*, **55**, 201–204.

Ethier, S.N. & Kurtz, T.G. (1986) *Markov Processes: Characterization and Convergence*. Wiley: New York.

Feller, W. (1952) The parabolic differential equations and the associated semigroups of transformations. *Annals of Mathematics*, **55**, 468–519.

Feller, W. (1968) *An Introduction to Probability Theory and its Applications*. Vol. 1. Wiley: New York.

Feller, W. (1971) *An Introduction to Probability Theory and its Applications*. Vol. 2. Wiley: New York.

Fisher, R.A. (1937) The wave of advance of advantageous genes. *Annals of Eugenics*, **7**, 355–369.

Ford, E.D. & Renshaw, E. (1984) The interpretation of process from pattern using two-dimensional spectral analysis: modelling single species patterns in vegetation. *Vegetatio*, **56**, 113–123.

Forsyth, A.R. (1928) *A Treatise on Differential Equations*. Macmillan: London.

Fowler, R.H. (1936) *Statistical Mechanics*. Cambridge University Press.

Gans, P.J. (1960) Open first order stochastic processes. *Journal of Chemical Physics*, **33**, 691–694.

Gause, G.F. (1932) Experimental studies on the struggle for existence. I. Mixed population of two species of yeast. *Journal of Experimental Biology*, **9**, 389–402.

Gause, G.F. (1934) *The Struggle for Existence*. Williams & Wilkins: Baltimore.

Gerrard, D.J. (1969) Competition quotient - a new measure of competition affecting individual forest trees. *Michigan State University Agricultural Experimental Station Research Bulletin*, No. **20**.

Gibson, G.J. (1997a) Markov chain Monte Carlo methods for fitting spatio-temporal stochastic models in plant epidemiology. *Applied Statistics*, **46**, 215–233.

Gibson, G.J. (1997b) Investigating mechanisms of spatio-temporal spread using stochastic models. *Phytopathology*, **87**, 139–146.

Gibson, G.J. & Renshaw, E. (1998) Estimating parameters in stochastic compartmental models using Markov chain methods. *IMA Journal of Mathematics Applied in Medicine and Biology*, **15**, 19–40.

Gibson, G.J. & Renshaw, E. (2001a) Inference for immigration–death processes with single and paired immigrants. *Inverse Problems*, **17**, 455–466.

Gibson, G.J. & Renshaw, E. (2001b) Likelihood estimation for stochastic compartmental models using Markov chain methods. *Statistics and Computing*, **11**, 347–358.

Gierer, A. & Meinhardt, H. (1972) A theory of biological pattern formation. *Kybernetik*, **12**, 30–39.

Gilks, W.R., Richardson, S. & Spiegelhalter, D.J. (1996) *Markov Chain Monte Carlo in Practice*. Chapman & Hall: London.

Gillespie, C.S. & Renshaw, E. (2005) The evolution of a batch-immigration death process subject to counts. *Proceedings of the Royal Society of London*, **A 461**, 1563–1581.

Gillespie, C.S. & Renshaw, E. (2007) An improved saddlepoint approximation. *Mathematical Biosciences*, **208**, 359–374.

Gillespie, C.S. & Renshaw, E. (2008) The evolution of a single-paired immigration death process. *Journal of Physics A: Mathematical and Theoretical*, **41**, 35002+20.

Gillespie, D.T. (1976) A general method for numerically simulating the stochastic time evolution of coupled chemical reactions. *Journal of Computational Physics*, **22**, 403–434.

Gillespie, D.T. (1977) Exact stochastic simulation of coupled reaction equations. *Journal of Physical Chemistry*, **81**, 2340–2361.

Gillespie, D.T. & Petzold, L. (2006) Numerical simulation for biochemical kinetics. In *System Modelling in Cellular Biology* (eds. Z. Szallasi, J. Stelling & V. Periwal). MIT Press: Cambridge MA.

Gillis, J. (1955) Correlated random walk. *Proceedings of the Cambridge Philosophical Society*, **51**, 639–651.

Gradshteyn, I.S. & Ryzhik, M. (1994) *Tables of Integrals, Series and Products* (5th edition). Academic Press: London.

Granger, C.W.J. (1966) The typical spectral shape of an economic variable. *Econometrica*, **34**, 150–161.

Green, P.J. (1995) Reversible jump Markov chain Monte Carlo computation and Bayesian model determination. *Biometrika*, **82**, 711–732.

Grimmett, G.R. & Stirzaker, D.R. (1992) *Probability and Random Processes* (2nd edition). Clarendon Press: Oxford.

Guillemin, F. & Pinchon, D. (1999) Excursions of birth and death processes, orthogonal polynomials, and continued fractions. *Journal of Applied Probability*, **36**, 752–770.

Häggeström, O. & Meester, R. (1996) Nearest neighbour and hard sphere models in continuum percolation. *Random Structures Algorithms*, **9**, 295–315.

Hallam, T.G. (1986) Community dynamics in a homogeneous environment. In Biomathematics, vol. 17, *Mathematical Ecology* (eds. T.G. Hallam & S.A. Levin), 61–94. Springer-Verlag: Berlin.

Hamer, W.H. (1906) Epidemic disease in England. *Lancet*, **1**, 733–739.

Hansen, J.-P. & McDonald, I.R. (1986) *Theory of Simple Liquids*. Academic Press: London.

Harris, T.E. (1963) *The Theory of Branching Processes*. Berlin: Springer.

Haskey, H.W. (1954) A general expression for the mean in a simple stochastic epidemic. *Biometrika*, **41**, 272–275.

Hastings, W.K. (1970) Monte Carlo sampling methods using Markov chains and their applications. *Biometrika*, **57**, 97–109.

Heisenberg, W. (1928) Zur Theorie des Ferromagnetismus. *Zeitschrift für Physik*, **49**, 619–636.

Helland, I.S. (1975) The condition for extinction with probability one in a birth, death and migration process. *Advances in Applied Probability*, **7**, 61–65.

Henderson, R., Renshaw, E. & Ford, E.D. (1983) A note on the recurrence of a correlated random walk. *Journal of Applied Probability*, **20**, 696–699.

Henderson, R., Renshaw, E. & Ford, E.D. (1984) A correlated random walk model for two-dimensional diffusion. *Journal of Applied Probability*, **21**, 233–246.

Hermann, H., Elsner, A., Hecker, M. & Stoyan, D. (2005) Computer simulated denserandom packing models as approach to the structure of porous low-k dielectrics. *Microelectronic Engineering*, **81**, 535–543.

Higham, D.J. (2007) A matrix perturbation view of the small world phenomenon. *SIAM Review*, **49**, 91–108.

Higham, D.J. (2008) Modelling and simulating chemical reactions. *SIAM Review*, **50**, 347–368.

Holling, C.S. (1965) The functional response of predators to prey density and its role in mimicry and population regulation. *Memoirs of the Entomological Society of Canada*, **45**, 3–60.

Hopcraft, K.I., Jakeman, E. & Matthews, J.O. (2004) Discrete scale-free distributions and associated limit theorems. *Journal of Physics*, **A 37**, 635–642.

Hopcraft, K.I., Jakeman, E. & Tanner, R.M. (2001) Characterization of structural reorganization in rice piles. *Physics Review*, **E 64**, 016116.

Hopcraft, K.I., Jakeman, E. & Tanner, R.M. (2002) *Fractal Geometry: Mathematical Methods, Algorithms & Applications* (eds. J.M. Blackledge, A.K. Evans & M.J. Turner). Horwood: Chichester.

Hosking, J.R.M. (1981) Fractional differencing. *Biometrika*, **68**, 165–176.

Huffaker, C.B. (1958) Experimental studies on predation: dispersion factors and predator–prey interactions. *Hilgardia*, **27**, 343–383.

Hutton, G. (2007) The development and application of correlated random walk theory to the construction of share trading strategies. *PhD Thesis*. University of Strathclyde (unpublished).

Ising, E. (1925) Beitrag zur Theorie des Ferromagnetismus. *Zeitschriften Physik*, **31**, 253–258.

Jacquez, J.A. & Simon, C.P. (1993) The stochastic SI model with recruitment and deaths I. Comparison with the closed SIS model. *Mathematical Biosciences*, **117**, 77–125.

Jakeman, E. (1984) Scattering by fractal objects. *Nature*, **307**, 110.

Jakeman, E., Hopcraft, K.I. & Matthews, J.O. (2003) Distinguishing population processes by external monitoring. *Proceedings of the Royal Society of London*, **A 459**, 623–639.

Jakeman, E., Phayre, S. & Renshaw, E. (1995) The evolution and measurement of a population of pairs. *Journal of Applied Probability*, **32**, 1048–1062.

Jakeman, E. & Renshaw, E. (1987) Correlated random-walk model for scattering. *Journal of the Optical Society of America*, **A 4**, 1206–1212.

Jakeman, E. & Shepherd, T.J. (1984) Population statistics and the counting process. *Journal of Physics*, **A 17**, L745–L750.

Jefferson, J.H. & Anderson, J.D. (1987) Generation and properties of self-similar stochastic processes with application to ray propagation in random media. *Conference Proceedings of the Advisory Group for Aerospace Research and Development*, **419**, 14.1–14.19.

Jefferson, J.H., Jakeman, E. & Beale, J.E.P. (1986) Computation of ray statistics beyond a multi-scale diffuser. *Proceedings of the Institute of Acoustics*, **8**, Part 5, 30–35.

Jeffreys, H. (1948) *Theory of Probability*. Oxford University Press.

Jeffreys, H. & Jeffreys, B.S. (1950) *Methods of Mathematical Physics*. Cambridge University Press.

Jensen, J.L. (1995) *Saddlepoint Approximations*. Oxford University Press.

Jeong, H., Tombor, B., Albert, R., Oltvai, Z.N. & Barabasi, A.-L. (2000) The large-scale organization of metabolic networks. *Nature*, **407**, 651–654.

Jolley, L.B.W. (1961) *Summation of Series*. Dover: New York.

Jones, V.F.R. (1989) On knot invariants related to some statistical mechanical models. *Pacific Journal of Mathematics*, **137**, 311–334.

Kac, M. (1947) Random walk and the theory of Brownian motion. *American Mathematical Monthly*, **54**, 369–391.

Kadashevich, I., Schneider, H.-J. & Stoyan, D. (2005) Statistical modeling of the geometrical structure of the system of artificial air pores in autoclaved aerated concrete. *Cement and Concrete Research*, **35**, 1495–1502.

Karlin, S. & McGregor, J. (1955) Representation of a class of stochastic processes. *Proceedings of the National Academy of Sciences U.S.A.*, **41**, 387–391.

Karlin, S. & McGregor, J. (1957a) The differential equations of birth and death processes and the Stieltjes moment problem. *Transactions of the American Mathematical Society*, **85**, 489–546.

Karlin, S. & McGregor, J. (1957b) The classification of birth and death processes. *Transactions of the American Mathematical Society*, **86**, 366–400.

Karlin, S. & McGregor, J. (1958) Linear growth, birth and death processes. *Journal of Mathematics and Mechanics*, **7**, 643–662.

Kelly, F.P. (1979) *Reversibility and Stochastic Networks*. Wiley: Chichester.

Kemp, C.D. & Kemp, A.W. (1965) Some properties of the 'Hermite' distribution. *Biometrika*, **52**, 381–394.

Kendall, D.G. (1952) Les processus stochastiques de croissance en biologie. *Annales des L'institute Henri Poincaré*, **13**, 43–108.

Kendall, D.G. (1956) Deterministic and stochastic epidemics in closed populations. *Proceedings of the 3rd Berkeley Symposium on Mathematical Statistics and Probability*, **4**, 149–165. University of California Press: Berkeley and Los Angeles.

Kendall, W.S. & Møller, J. (2000) Perfect simulation using dominating processes on ordered state spaces, with application to locally stable point processes. *Advances in Applied Probability*, **32**, 844–865.

Kendall, M. & Stuart, A. (1977) *The Advanced Theory of Statistics: Vol. 1. Distribution Theory* (4th edition). Griffin: London.

Kermack, W.O. & McKendrick, A.G. (1927) Contributions to the mathematical theory of epidemics. *Proceedings of the Royal Society of London*, **A 115**, 700–721.

Keyfitz, N. (1977) *Introduction to the Mathematics of Population (with Revisions)*. Addison-Wesley: Reading, Massachusetts.

Khinchin, A.I. (1949) *Mathematical Foundations of Statistical Mechanics*. Dover: New York.

Khintchine, A. (1929) Sur la loi des grands nombres. *Comptes rendus de l'Académie des Sciences*, **129**, 477–479.

Kingman, J.F.C. (1969) Markov population processes. *Journal of Applied Probability*, **6**, 1–18.

Kingman, J.F.C. (1972) *Regenerative Phenomena*. Wiley: New York.

Kingman, J.F.C. (1982) The thrown string. *Journal of the Royal Statistical Society*, **B 44**, 109–138.

Krebs, C.J. (1985) *Ecology: The Experimental Analysis of Distribution and Abundance* (3rd edition). Harper & Row: New York.

Krieger, I.M. & Gans, P.J. (1960) First order stochastic processes. *Journal of Chemical Physics*, **32**, 247–250.

Kryscio, R.J. & Lefèvre, C. (1989) On the extinction of the S-I-S stochastic logistic epidemic. *Journal of Applied Probability*, **26**, 685–694.

Lautensack, C., Schladitz, K. & Särkkä, A. (2006) Modelling the microstructure of sintered copper. In *Proceedings of the 6th International Conference on Stereology, Spatial Statistics and Stochastic Geometry*, Prague.

Lawrance, A.J. & Kottegoda, N.T. (1977) Stochastic modelling of riverflow time series (with Discussion). *Journal of the Royal Statistical Society*, **A 140**, 1–47.

Lenz, W. (1920) Beitrag zum Verständnis der magnetischen Eigenschaften in festen Körpern. *Zeitschriften Physik*, **21**, 613–615.

Leslie, P.H. & Gower, J.C. (1960) The properties of a stochastic model for the predator-prey type of interaction between two species. *Biometrika*, **47**, 219–234.

Levin, S.A. (1974) Dispersion and population interactions. *American Naturalist*, **108**, 207–228.

Lévy, P. (1937) *Theorie de l'Addition des Variables Aléatoires*. Gaulthier-Villars: Paris.

Liggett, T.M. (1985) *Interacting Particle Systems*. Springer-Verlag: New York.

Lindley, D.V. (1959) Discussion of C.B. Wisten's paper on 'Geometric distributions in the theory of queues'. *Journal of the Royal Statistical Society*, **B 21**, 22–23.

Lochmann, K., Oger, L. & Stoyan, D. (2006) Statistical analysis of random sphere packings with variable radius distribution. *Solid State Sciences*, **8**, 1397–1413.

Lotka, A.J. (1925) *Elements of Physical Biology*. Williams and Wilkins: Baltimore.

Lotka, A.J. (1931) The extinction of families. *Journal of the Washington Academy of Sciences*, **21**, 377–380 & 453–459.

Löwen, H. (2000) Fun with hard spheres. In: *Statistical Physics and Spatial Statistics* (eds. K.R. Mecke & D. Stoyan). Springer Lecture Notes in Physics, **554**, 215–331.

Lugannani, R. & Rice, S. (1980) Saddlepoint approximations for the distribution of the sum of independent random variables. *Advances in Applied Probability*, **12**, 475–490.

MacArthur, R. & Wilson, E.O. (1967) *Theory of Island Biogeography*. Princeton University Press.

Maini, P.K. & Othmer, H.G. (eds) (2000) *Mathematical Models for Biological Pattern Formation*. In *IMA Volumes in Mathematics and its Applications*, **121**. Berlin: Springer-Verlag.

Mandel, L. (1959) Fluctuations of photon beams: the distribution of the photo-electrons. *Proceedings of the Physics Society*, **74**, 233–243.

Mandelbrot, B.B. (1965) Une classe de processus stochastiques homothétiques à soi; application à la loi climatologique de H.E. Hurst. *Comptes Rendus del Academie des Sciences de Paris*, **260**, 3274–3277.

Mandelbrot, B.B. (1967) Some noises with $1/f$ spectrum, a bridge between direct current and white noise. *IEEE Transactions on Information Theory*, **13**, 289–298.

Mandelbrot, B.B. (1977) *Fractals: Form, Chance and Dimension.* Freeman: San Francisco.

Mandelbrot, B.B. & van Ness, J.W. (1968) Fractional Brownian motions, fractional noises and applications. *SIAM Review,* **10**, 422–437.

Mandelbrot, B.B. & Wallis, J.R. (1969) Computer experiments with fractional Gaussian noises. *Water Resources Research,* **5**, 228–267.

Mao, X., Marion, G. & Renshaw, E. (2002) Environmental Brownian noise suppresses explosions in population dynamics. *Stochastic Processes and Their Applications,* **97**, 95–110.

Mao, X., Sabanis, S. & Renshaw, E. (2003) Asymptotic behaviour of the stochastic Lotka–Volterra model. *Journal of Mathematical Analysis and Applications,* **287**, 141–156.

Marion, G., Gibson, G.J. & Renshaw, E. (2003) Estimating likelihoods for spatio-temporal models using importance sampling. *Statistics and Computing,* **13**, 111–119.

Marion, G., Mao, X. & Renshaw, E. (2002) Convergence of the Euler scheme for a class of stochastic differential equation. *International Journal of Mathematics,* **1**, 9–22.

Marion, G., Renshaw, E. & Gibson, G. (1998) Stochastic effects in a model of nematode infection in ruminants. *IMA Journal of Mathematics Applied in Medicine and Biology,* **15**, 97–116.

Marion, G., Renshaw, E. & Gibson, G. (2000) Stochastic modelling of environmental variation for biological populations. *Theoretical Population Biology,* **57**, 197–217.

Marsaglia, G. (1968) Random numbers fall mainly in the planes. *Proceedings of the National Academy of Sciences,* **61**, 25–28.

Matis, J.H. & Hartley, H.O. (1971) Stochastic compartmental analysis: model and least squares estimation from time series data. *Biometrics,* **27**, 77–102.

Matis, J.H. & Kiffe, T.R. (1996) On approximating the moments of the equilibrium distribution of a stochastic logistic model. *Biometrics,* **52**, 980–991.

Matis, J.H. & Kiffe, T.R. (2000) *Stochastic Population Models: A Compartmental Perspective.* Lecture Notes in Statistics **145**. Springer: New York.

Matis, J.H., Kiffe, T.R., Matis, T.I. & Stevenson, D.E. (2006) Application of population growth models based on cumulative size to pecan aphids. *Journal of Agricultural, Biological and Environmental Statistics,* **11**, 425–449.

Matis, J.H., Kiffe, T.R., Matis, T.I. & Stevenson, D.E. (2007) Stochastic modeling of aphid population growth with nonlinear, power-law dynamics. *Mathematical Biosciences,* **208**, 469–494.

Matis, J.H., Kiffe, T.R. & Parthasarathy, P.R. (1998) On the cumulants of population size for the stochastic power law logistic model. *Theoretical Population Biology,* **53**, 16–29.

Matis, J.H., Rubink, W.L. & Makela, M. (1992) Use of the gamma distribution for predicting arrival times of invading insect populations. *Environmental Entomology,* **21**, 436–440.

Matis, J.H., Zheng, Q. & Kiffe, T.R. (1995) Describing the spread of biological populations using stochastic compartmental models with births. *Mathematical Biosciences*, **126**, 215–247.

Matthews, J.O., Hopcraft, K.I. & Jakeman, E. (2003) Generation and monitoring of discrete stable random processes using multiple immigration population models. *Journal of Physics*, **36**, 585–603.

May, R.M. (1971a) Stability in multispecies community models. *Mathematical Biosciences*, **12**, 59–79.

May, R.M. (1971b) Stability in model ecosystems. *Proceedings of the Ecological Society of Australia*, **6**, 18–56.

May, R.M. (1974) *Stability and Complexity in Model Ecosystems* (2nd edition). University Press: Princeton.

Maynard Smith, J. (1974) *Models in Ecology*. Cambridge University Press.

McLeod, A.I. & Hipel, K.W. (1978) Preservation of the rescaled adjusted range: 1. A reassessment of the Hurst phenomenon. *Water Resources Research*, **14**, 491–508.

Metropolis, N., Rosenbluth, A.W., Teller, A.H. & Teller, E. (1953) Equations of state calculations by fast computing machines. *Journal of Chemical Physics*, **21**, 1087–1092.

Mimura, M. & Murray, J.D. (1978) On a diffusive prey–predator model which exhibits patchiness. *Journal of Theoretical Biology*, **75**, 249–262.

Møller, J. (1999). Markov chain Monte Carlo and spatial point processes. In *Stochastic Geometry. Likelihood and Computation* (eds. O.E. Barndorff-Nielsen, W.S. Kendall & N.M.N. van Lieshout). Chapman and Hall: London.

Møller, J. (2000) A review on perfect simulation in stochastic geometry. *Technical Report R-00-2016*, Aalborg University.

Møller, J. & Waagepetersen, R. (2003). *Statistical Inference and Simulation for Spatial Point Processes*. Chapman and Hall/CRC: Boca Raton, Florida.

Mollison, D. (1972a) Possible velocities for a simple epidemic. *Advances in Applied Probability*, **4**, 233–258.

Mollison, D. (1972b) The rate of spatial propagation of simple epidemics. *Proceedings of the 6th Berkeley Symposium on Mathematical Statistics and Probability*, **3**, 283–326. University of California Press: Berkeley and Los Angeles.

Mollison, D. (1977) Spatial contact models for ecological and epidemic spread (with Discussion). *Journal of the Royal Statistical Society*, **B 39**, 283–326.

Mollison, D. (1991) Dependence of epidemic and population velocities on basic parameters. *Mathematical Biosciences*, **107**, 255–287.

Moran, P.A.P. (1948) The statistical distribution of the length of a rubber molecule. *Proceedings of the Cambridge Philosophical Society*, **44**, 342–344.

Morgan, B.J.T. (1984) *Elements of Simulation*. Chapman and Hall: London.

Morgan, B.J.T. & Watts, S.A. (1980) On modelling microbial infections. *Biometrics*, **36**, 317–321.

Mościński, J. & Bargieł, M. (1991) C. Language program for the irregular close packing of hard spheres. *Computer Physics Communications*, **64**, 183–192.

Mościński, J., Bargiel, M., Ryerz, Z.A. & Jacobs, P.W.M. (1989) The force-biased algorithm for the irregular packing of equal hard spheres. *Molecular Simulation*, **3**, 201–212.

Moyal, J.E. (1950) The momentum and sign of fast cosmic ray particles. *Philosophical Magazine*, **56**, 1058–1077.

Murray, J.D. (1989) *Mathematical Biology*. Springer-Verlag: Berlin.

Nåsell, I. (1996) The quasi-stationary distribution of the closed endemic SIS model. *Advances in Applied Probability*, **28**, 895–932.

Nåsell, I. (1999) On the quasi-stationary distribution of the stochastic logistic epidemic. *Mathematical Biosciences*, **156**, 21–40.

Nåsell, I. (2001) Extinction and quasi-stationarity in the Verhulst logistic model. *Journal of Theoretical Biology*, **211**, 11–27.

Neyman, J., Park, T. & Scott, E.L. (1956) Struggle for existence. The *Tribolium* model: biological and statistical aspects. *Proceedings of the Third Berkeley Symposium on Mathematics, Statistics and Probability*, **4**, 41–79. University of California Press: Berkeley and Los Angeles.

Neumann, G. (1953) On ocean wave spectra and a new method of forecasting wind-generated sea. *Technical Memorandum*, **43**, Beach Erosion Board: US Army Corps of Engineers.

Newman, M.E.J., Moore, C. & Watts, D.J. (2000) Mean-field solution of the small-world network model. *Physical Review Letters*, **84**, 3201–3204.

Nisbet, R.M. & Gurney, W.S.C. (1982) *Modelling Fluctuating Populations*. Wiley: New York.

Okubo, A. (1980) *Diffusion and Ecological Problems: Mathematical Models*. Springer-Verlag: Berlin.

O'Neill, P.D. & Roberts, G.O. (1997) Bayesian inference for partially observed epidemics. Report SOR97–80, Bradford University.

Onsager, L. (1944) Crystal statistics. I. A two-dimensional model with an order-disorder transition. *Physics Review*, **65**, 117–149.

Osbourne, M. (1959) Brownian motion in the stock market. *Operations Research*, **11**, 145–173.

Othmer, H.G., Maini, P.K. & Murray, J.D. (eds) (1993) *Experimental and Theoretical Advances in Biological Pattern Formation*. Plenum Press.

Otis, G.W. (1991) Population biology of the Africanized honey bee. In *The "African" Honey Bee* (eds. M. Spivak, D.J.C. Fletcher & M.D. Breed), 213–234. Westview: Boulder, Colorado.

Park, T. (1954) Experimental studies on interspecies competition. II. Temperature, humidity and competition in two species of *Tribolium*. *Physiological Zoology*, **27**, 177–238.

Patil, V.T. (1957) The consistency and adequacy of the Poisson-Markoff model for density fluctuations. *Biometrika*, **44**, 43–56.

Pearl, R. (1927) The growth of populations. *Quarterly Review of Biology*, **2**, 532–548.

Pearl, R. (1930) *Introduction of Medical Biometry and Statistics*. Saunders: Philadelphia.

Pearl, R. & Reed, L.J. (1920) On the rate of growth of the population of the United States since 1790 and its mathematical representation. *Proceedings of the National Academy of Sciences*, **6**, 275–288.

Peierls, R. (1936) On Ising's model of ferromagnetism. *Proceedings of the Cambridge Philosophical Society*, **32**, 477–481.

Phayre, S. (1995) Physical applications of stochastic processes. *PhD Thesis*. University of Strathclyde (unpublished).

Piaggio, H.T.H. (1962) *An Elementary Treatise on Differential Equations and Their Applications*. Bell: London.

Pielou, E.C. (1977) *Mathematical Ecology*. Wiley: New York.

Prajneshu (1998) A nonlinear statistical model for aphid population growth. *Journal of the Indian Society of Agricultural Statistics*, **51**, 73–80.

Propp, J.G. & Wilson, D.B. (1996) Exact sampling with coupled Markov chains and applications to statistical mechanics. *Random Structures and Algorithms*, **9**, 223–252.

Puri, P.S. (1968) Interconnected birth and death processes. *Journal of Applied Probability*, **5**, 334–349.

Radcliffe, J. & Staff, P.J. (1969) First-order conservative processes with multiple latent roots. *Journal of Applied Probability*, **6**, 186–194.

Raman, S. & Chiang, C.L. (1973) On a solution of the migration process and the application to a problem in epidemiology. *Journal of Applied Probability*, **10**, 718–727.

Rayleigh, Lord (1919) On the problem of random vibrations and of random flights in one, two and three dimensions. *Philosophical Magazine*, **37**, 321–347.

Reid, N. (1988) Saddlepoint methods and statistical inference. *Statistical Science*, **3**, 213–238.

Reid, W.T. (1959) Solutions of a Riccati matrix differential equation as functions of initial values. *Journal of Mathematics and Mechanics*, **8**, 221–230.

Renshaw, E. (1972) Birth, death and migration processes. *Biometrika*, **59**, 49–60.

Renshaw, E. (1973a) Interconnected population processes. *Journal of Applied Probability*, **10**, 1–14.

Renshaw, E. (1973b) The effect of migration between two developing populations. *Proceedings of the 39th Session of the International Statistical Institute*, **2**, 294–298.

Renshaw, E. (1974) Stepping stone models for population growth. *Journal of Applied Probability*, **11**, 16–31.

Renshaw, E. (1976) Spatial Population Processes. *PhD Thesis*. University of Edinburgh (unpublished).

Renshaw, E. (1977) Velocities of propagation for stepping-stone models of population growth. *Journal of Applied Probability*, **14**, 591–597.

Renshaw, E. (1979) Waveforms and velocities for non-nearest-neighbour contact distributions. *Journal of Applied Probability*, **16**, 1–11.

Renshaw, E. (1980) The spatial distribution of *Tribolium confusum*. *Journal of Applied Probability*, **17**, 895–911.

Renshaw, E. (1981) Waveforms and velocities for models of spatial infection. *Journal of Applied Probability*, **18**, 715–720.

Renshaw, E. (1982) The development of a spatial predator-prey process on interconnected sites. *Journal of Theoretical Biology*, **94**, 355–365.

Renshaw, E. (1984) Competition experiments for light in a plant monoculture: an analysis based on two-dimensional spectra. *Biometrics*, **40**, 717–728.

Renshaw, E. (1986) A survey of stepping-stone models in population dynamics. *Advances in Applied Probability*, **18**, 581–627.

Renshaw, E. (1987) The discrete Uhlenbech-Ornstein process. *Journal of Applied Probability*, **24**, 908–917.

Renshaw, E. (1988) The high-order autocovariance structure of the Telegraph Wave. *Journal of Applied Probability*, **25**, 744–751.

Renshaw, E. (1991) *Modelling Biological Populations in Space and Time*. Cambridge University Press.

Renshaw, E. (1994a) Non-linear waves on the Turing ring. *Mathematical Scientist*, **19**, 22–46.

Renshaw, E. (1994b) The linear spatial-temporal interaction process and its relation to $1/\omega$-noise. *Journal of the Royal Statistical Society*, **B 56**, 75–91.

Renshaw, E. (1994c) Chaos in Biometry. *IMA Journal of Mathematics Applied in Medicine and Biology*, **11**, 17–44.

Renshaw, E. (1998) Saddlepoint approximations for stochastic processes with truncated cumulant generating functions. *IMA Journal of Mathematics Applied in Medicine and Biology*, **15**, 41–52.

Renshaw, E. (2000) Applying the saddlepoint approximation to bivariate stochastic processes. *Mathematical Biosciences*, **168**, 57–75.

Renshaw, E. (2002) Two-dimensional spectral analysis for marked point processes. *Biometrical Journal*, **44**, 718–745.

Renshaw, E. (2004) Metropolis-Hastings from a stochastic population dynamics perspective. *Computational Statistics and Data Analysis*, **45**, 765–786.

Renshaw, E. (2009) Spatial-temporal marked point processes: a spectrum of stochastic models. *Environmetrics*, **20**, 1–17.

Renshaw, E. & Chen, A. (1997) Birth-death processes with mass annihilation and state-dependent immigration. *Stochastic Models*, **13**, 239–254.

Renshaw, E. & Comas, C. (2009) Space-time generation of high intensity patterns using growth-interaction processes. *Statistics and Computing*, **19**, 423–437.

Renshaw, E., Comas, C. & Mateu, J. (2009) Analysis of forest thinning strategies through the development of space-time growth-interaction simulation models. *Stochastic Environmental and Research Risk Assessment*, **23**, 275–288.

Renshaw, E. & Dai, Y. (1997) Regularity and reversibility results for birth-death-migration processes. *Journal of Applied Probability*, **34**, 192–207.

Renshaw, E. & Ford, E.D. (1983) The interpretation of process from pattern using two-dimensional spectral analysis: methods and problems of interpretation. *Applied Statistics*, **32**, 51–63.

Renshaw, E. & Gibson, G.J. (1998) Can Markov chain Monte Carlo be usefully applied to stochastic processes with hidden birth times? *Inverse Problems*, **14**, 1581–1606.

Renshaw, E. & Henderson, R. (1981) The correlated random walk. *Journal of Applied Probability*, **18**, 403–418.

Renshaw, E., Mateu, J. & Saura, F. (2007) Disentangling mark/point interaction in marked point processes. *Computational Statistics and Data Analysis*, **51**, 3123–3144.

Renshaw, E., Phayre, S. & Jakeman, E. (2000) The development of space-time interaction processes with given spectral structure. *Inverse Problems*, **16**, 877–890.

Renshaw, E. & Särkkä, A. (2001) Gibbs point processes for studying the development of spatial-temporal stochastic processes. *Computational Statistics and Data Analysis*, **36**, 85–105.

Reuter, G.E.H. & Lederman, W. (1953) On the differential equations for the transition probabilities of Markov processes with enumerably many states. *Proceedings of the Cambridge Philosophical Society*, **49**, 247–262.

Richards, F.J. (1959) A flexible growth function for empirical use. *Journal of Experimental Botany*, **10**, 290–300.

Riordan, J. (1958) *An Introduction to Combinatorial Analysis*. Wiley: New York.

Ripley, B.D. (1987) *Stochastic Simulation*. Wiley: New York.

Roberts, M.G. & Grenfell, B.T. (1991) The population dynamics of nematode infections of ruminants: periodic perturbations as a model for management. *IMA Journal of Mathematics Applied in Medicine and Biology*, **8**, 83–93.

Roberts, G.O. & Stramer, O. (2001) On inference for partially observed nonlinear diffusion models using the Metropolis-Hastings algorithm. *Biometrika*, **88**, 603–621.

Rothschild, Lord (1953) A new method of measuring the activity of spermatozoa. *Journal of Experimental Biology*, **30**, 178–199.

Ruben, H. (1962) Some aspects of the emigration-immigration process. *Annals of Mathematical Statistics*, **33**, 119–129.

Saaty, T.L. (1961) Some stochastic processes with absorbing barriers. *Journal of the Royal Statistical Society*, **B 23**, 319–334.

Saleh, B.E.A. (1978) *Photoelectron Statistics*. Springer-Verlag: Berlin.

Samad, H.E., Khammash, M., Petzold, L. & Gillespie, D.T. (2005) Stochastic modeling of gene regulatory networks. *International Journal of Robust and Nonlinear Control*, **15**, 691–711.

Sang, J.H. (1950) Population growth in *Drosophila* cultures. *Biological Reviews*, **25**, 188–219.

Särkkä, A., Redenbach, C. & Renshaw, E. (2010) Space-time growth-interaction processes with moving objects versus the force biased algorithm (in preparation).

Särkkä, A. & Renshaw, E. (2006) The analysis of marked point patterns evolving through space and time. *Computational Statistics and Data Analysis*, **51**, 1698–1718.

Sarpkaya, T. & Isaacson, M. (1981) *Mechanics of Wave Forces on Offshore Structures*. Van Nostrand Reinhold: New York.

Sayles, R.S. & Thomas, T.R. (1978) Surface topography as a nonstationary random process. *Nature*, **271**, 431–434.

Schakenburg, J. (1979) Simple chemical reaction systems with limit cycle behaviour. *Journal of Theoretical Biology*, **81**, 389–400.

Schlather, M., Ribeiro, P. & Diggle, P.J. (2004) Detecting dependence between marks and locations of marked point processes. *Journal of the Royal Statistical Society*, **B 66**, 79–93.

Schoenberg, I.J. (1983) Problem 650. *Nieuw Archiff Vor Wiskunde, Series 4*, **1**, 377–378.

Sekimura, T., Noji, S., Ueno, N. & Maini, P.K. (eds.) (2003) *Morphogenesis and Pattern Formation in Biological Systems: Experiments and Models*. Proceedings of the Chubu 2002 Conference. Berlin: Springer-Verlag.

Severo, N.C. (1967) Two theorems on solutions of differential-difference equations and applications to epidemic theory. *Journal of Applied Probability*, **4**, 271–280.

Severo, N.C. (1969) A recursion theorem on solving differential-difference equations and applications to some stochastic processes. *Journal of Applied Probability*, **6**, 673–681.

Shepherd, T.J. (1984) Photoelectron counting – semiclassical and population monitoring approaches. *Optica Acta*, **31**, 1399–1407.

Sherman, B. (1956) The limiting distribution of Brownian motion on a finite interval with instantaneous return. Westinghouse Research Laboratory Scientific Paper 60-94698-3-P3.

Shimoda, K., Takahasi, H. & Townes, C.H. (1957) Fluctuations in amplification of quanta with application to maser amplifiers. *Journal of the Physical Society of Japan*, **12**, 686–700.

Siegart, A.J.F. (1949) On the approach to statistical equilibrium. *Physics Review*, **76**, 1708–1714.

Simmonds, J.G. & Mann, J.E. (1986) *A First Look at Perturbation Theory*. Robert E. Krieger Publishing Company: Malabar, Florida.

Skellam, J.G. (1951) Random dispersal in theoretical populations. *Biometrika*, **38**, 196–218.

Smith, M. & Renshaw, E. (1994) Parallel-prefix remapping for efficient data-parallel implementation of unbalanced simulations. *Advances in Parallel Computing*, **9**, 215–222.

Smith, A.F.M. & Roberts, G.O. (1993) Bayesian computation via the Gibbs sampler and related Markov chain Monte Carlo methods. *Journal of the Royal Statistical Society*, **B 55**, 2–23.

Sommerfield, A. (1949) *Partial Differential Equations in Physics*. Academic Press: New York.

Soper, H.E. (1929) Interpretation of periodicity in disease-prevalence. *Journal of the Royal Statistical Society*, **92**, 34–73.

Spiegel, M.R. (1990) *Mathematical Handbook of Formulae and Tables*. McGraw-Hill: New York.

Spitzer, F. (1970) Interaction of Markov processes. *Advances in Mathematics*, **5**, 246–290.

Spivak, M., Fletcher, D.J.C. & Breed, M.D. (1991) *The "African" Honey Bee*. Westview: Boulder, Colorado.

Srinivasan, S.K. (1988) *Point Process Models of Cavity Radiation and Detection*. Oxford University Press.

Staebler, G.R. (1951) Growth and spacing in an even-aged stand of Douglas fir. Master's Thesis: University of Michigan.

Stark, J., Ianelli, P. & Baigent, S. (2001) A nonlinear dynamics perspective of moment closure for stochastic processes. *Nonlinear Analysis*, **47**, 753–764.

Stoyan, D. (2003) Hard problems with random systems of hard particles. *Proceedings of the 54th Session of the International Statistical Institute*, Berlin.

Stoyan, D., Kendall, W.S. & Mecke, J. (1995) *Stochastic Geometry and Its Applications* (2nd edition). Chichester: Wiley.

Stoyan, D. & Schlather, M. (2000) Random sequential adsorption: relationship to dead leaves and characterization of variability. *Journal of Statistical Physics*, **100**, 969–979.

Stoyanov, J.M. (1988) *Counterexamples in Probability*. Wiley: New York.

"Student" (1908) The probable error of a mean. *Biometrika*, **6**, 1–25.

Tanner, J.T. (1975) The stability and the intrinsic growth rates of prey and predator populations. *Ecology*, **56**, 855–867.

Thomas, D. (1975) Artificial enzyme membranes, transport, memory, and oscillatory phenomena. In *Analysis and Control of Immobilised Enzyme Systems* (eds. D. Thomas & J.-P. Kernevez), 115–150. Springer-Verlag: Berlin.

Thomas, M.U. & Barr, D.R. (1977). An approximate test for Markov chain lumpability. *Journal of the American Statistical Association*, **72**, 175–179.

Tierney, L. (1994) Markov chains for exploring posterior distributions. *The Annals of Statistics*, **22**, 1701–1762.

Titchmarsh, E.C. (1939) *Theory of Functions* (2nd edition). Oxford University Press.

Torquato, S. (2002) *Random Heterogeneous Materials: Microstructure and Macroscopic Properties*. Springer-Verlag: New York.

Tscheschel, A. & Stoyan, D. (2006) Statistical reconstruction of random point patterns. *Computational Statistics and Data Analysis*, **51**, 859–871.

Turing, A.M. (1952) The chemical basis of morphogenesis. *Philosophical Transactions of the Royal Society of London*, **B 237**, 37–72.

Uhlenbeck, G.E. & Ornstein, L.S. (1930) On the theory of Brownian motion. *Physics Review*, **36**, 823–841. (Also reproduced in Wax, 1954).

Ulbrich, J. (1930) *Die Bisamratte*. Heinrich: Dresden.

van den Bosch, F., Metz, J.A.J. & Dickmann, O. (1990) The velocity of spatial population expansion. *Journal of Mathematical Biology*, **28**, 529–565.

van den Bosch, F., Hengeveld, R. & Metz, J.A.J. (1992) Analyzing the velocity of animal range expansion. *Journal of Biogeography*, **19**, 135–150.

Verhulst, P.F. (1838) Notice sur la loi que la population suit dans son accroissement. *Correspondence Mathematique et Physique*, **10**, 113–121.

Volterra, V. (1926) Fluctuations in the abundance of a species considered mathematically. *Nature*, **118**, 558–560.

Volterra, V. (1931) *Leçons sur la Théorie Mathématique de la Lutte pour la Vie*. Paris: Gauthier-Villars.

von Bertalanffy, L. (1949). Problems of organic growth. *Nature*, **163**, 156–158.

Wang, S. (1992) General saddlepoint approximations in the bootstrap. *Statistics and Probability Letters*, **27**, 61–66.

Watson, G.N. (1952) *Theory of Bessel Functions*. Cambridge University Press.

Watts, D.J. & Strogatz, S.H. (1998) Collective dynamics of 'small-world' networks. *Nature*, **393**, 440–442.

Wax, N. (ed.) (1954) *Selected Papers on Noise and Stochastic Processes*. New York: Dover.

Weiss, G.H. & Dishon, M. (1971) On the asymptotic behaviour of the stochastic and deterministic models of an epidemic. *Mathematical Biosciences*, **11**, 261–265.

Whittle, P. (1954) On stationary processes in the plane. *Biometrika*, **41**, 434–449.

Whittle, P. (1955) The outcome of a stochastic epidemic – a note on Bailey's paper. *Biometrika*, **42**, 116–122.

Whittle, P. (1957) On the use of the Normal approximation in the treatment of stochastic processes. *Journal of the Royal Statistical Society*, **B 19**, 268–281.

Whittle, P. (1962) Topographic correlation, power-law covariance functions, and diffusion. *Biometrika*, **49**, 305–314.

Whittle, P. (1967) Non-linear migration processes. *Proceedings of the 36th Session of the International Statistical Institute*, **2**, 642–646.

Whittle, P. (1968) Equilibrium distributions for an open migration process. *Journal of Applied Probability*, **5**, 567–571.

Williams, D. (1991) *Probability and Martingales*. Cambridge University Press.

Williams, T. (1971) An algebraic proof of the threshold theorem for the general stochastic epidemic. *Advances in Applied Probability*, **3**, 223.

Wilson, D.B. (2000) How to couple from the past using a read-once source of randomness. *Random Structures and Algorithms*, **16**, 85–113.

Wintner, A. (1947) *The Fourier Transforms of Probability Distributions*. Johns Hopkins University: Baltimore.

Wolfram, S. (1999) *The Mathematica Book* (4th edition). Wolfram Media/Cambridge University Press.

Young, W.M. & Elcock, E.W. (1966) Monte Carlo studies of vacancy migration in binary ordered alloys. *Proceedings of the Physical Society*, **89**, 735–746.

Yule, G.U. (1925) A mathematical theory of evolution, based on the conclusions of Dr. J.C. Willis, F.R.S. *Philosophical Transactions of the Royal Society of London*, **B 213**, 21–87.

Subject Index

Africanized honey bees
  description 9, 159
  equilibrium solutions 159–161
  moment approximation 162–164
annealing 626
anomalous equation 6, 95
aphids 337

ballot theorems 10, 203–205
birth–death–migration, 2-site
  approximate probabilities 22, 484–488
  critical migration 479
  equations for 477–478
  exact probabilities 481–484
  moments 478–481, 487
  Riccati representations 22, 493–496
  sequences of integral equations 22, 490–493
  slightly connected processes 488–490
  with immigration 347–349
birth–death–migration, n-site
  boundary effects 23, 531–533
  description 22, 496–498, 508, 532
  end emigration 512–513
  moments 496–497, 508–512
  simulation of 23, 514
birth–death–migration, ∞-site
  approximate probabilities ($\lambda \geq 0$) 506–508
  description 498
  diffusion approximation 528–531
  equations for 498
  exact probabilities ($\lambda = 0$) 501–505
  long-range jumps 533–536, 540–541
  moments 499–501, 505, 507–508
  saddlepoint profile 533, 535, 539
  simulation of 25, 515–519, 520–524, 539–541
  velocity of propagation 24, 524–528, 530, 535–536
  wavefront 24
  wave profile 24, 526–527, 536, 539–541
  wild jumps 24, 534, 536–541, 582
  with immigration 500
branching processes
  criticality 286
  extinction 283, 286–290
  martingales for 291–294
  moments 285–286

p.g.f. representation 13, 284–285
transition probabilities 282–283

carrying capacity 107, 123, 126
catastrophe 101, 385
Cauchy form 24, 383
chemical Langevin equation 619
chemical reaction systems 522, 550–551
clipping procedure 264
competition process
  biological experiments 392, 398
  description 389–390, 414
  deterministic 17, 389–392
  extinction probabilities 18, 403–404
  moments 18, 408–413
  quasi-equilibrium probabilities 406–408
  population size probabilities 402
  simulation of 18, 399–402
  stability general 397–398
  stability global 18, 393–397
  stability local 17, 394–395
  steps to extinction 404–406
  time to extinction 406
contact distribution 24, 528, 534
correlated random walk
  description 238–239, 254–255, 325
  occupation probabilities 239–243
  share trading 11–12, 246–251
  simulation of 243–246
count-dependent growth
  deterministic 337
  moments 338–339
  stochastic 337–338
counting processes
  batch-immigration–death 357–364
  birth–death 364–367
  external monitoring 353–354
  paired-immigration–death 350–353
  Schoenberg–Poisson immigration 359–362
  simulation of 360–361
  single-paired-immigration–death 353–357
  triple-immigration–death 358
cumulative size process
  cumulant equations 470, 472
  deterministic 16, 21, 464–466
  moments 16, 21, 470
  power-law rates 471–473

648 Subject Index

cumulative size process (*cont.*)
   simulation of 467–469, 472
   stochastic 466–477

diffusion processes
   absorbing barriers 304–309
   batch immigration 175–177
   boundary conditions 301–303, 305, 321–322
   Brownian motion 14, 295–297, 532, 584
   diffusion approximation 164–168
   diffusion equations 296–300
   equilibrium probabilities 301, 317–321
   Fokker–Planck equation 14, 309–323
   immigration–death 312–317
   immigration–emigration 177–178
   Ornstein–Uhlenbeck process 15, 246, 325–330
   rapid oscillations 329–330
   reflecting barriers 300–304
   simulation of 15, 297–298, 302, 312–317, 328
   stochastic differential equation 15, 323–325, 415
   Wiener process 14, 296–309, 311, 329–330
dominant leader process 5, 86

embedded process 139
epidemic process, general
   deterministic 445–446
   epidemic curve 446–447
   stochastic 449–454
   threshold theorem, deterministic 20, 447–449
   threshold theorem, stochastic 20, 451–454
epidemic process, recurrent
   cyclic behaviour 20–21, 456
   description 454–455
   deterministic 454–457, 461
   endemic 458, 462, 464
   extensions 462
   extinction 458
   moments 460
   simulation of 457–459, 462–463
   stochastic 459–460
   susceptible birth 460–464
epidemic process, simple
   deterministic 19, 414, 434–435
   duration time 443–444
   epidemic curve 435, 441–442
   ϵ-modified 437–440
   moments 20, 439–443
   SIS model 444–445
   stochastic 19, 435–439, 441, 443, 445
Euler's constant 51

Euler's product form 62
expected update equation 151, 410, 419–420, 462

Fano factor 350
First Arc Sine Law 11, 211–212
FORTRAN
   NAG library 128
   sample program 128–129
fractal 298, 586, 594–595, 608
fractional
   Brownian motion 584
   difference 584
   integration 586, 591
   noise 584
functions
   Bernoulli numbers 581, 583
   Bessel 99, 499–500, 506, 510, 526
   Gauss hypergeometric 537
   generalized Riemann zeta 135
   Lerch transcendent 135
   Pareto 138
   Riemann zeta 68
   Stirling numbers 538
   theta 67

Galerkin approximation 8, 150
general birth process
   dishonest 65–70, 115
   general birth rate 4, 64–65
   simulation of 69
   stochastic solution 64–65, 115
   time-dependent 64
general birth–death process
   adding noise 9, 153–156
   approximate solutions 141–143
   approximate time-dependent probabilities 9, 141–143
   equilibrium solutions 117–118
   mean time to ultimate extinction 145–148
   multiple equilibria 8, 129–133
   power-law rates 8
   probability of ultimate extinction 143–145
   stability index 145, 148
generating functions, definition of 33, 36–37, 39, 192, 412

Hougaard's principle of pragmatism 191–192

Ising model 408, 497, 567

Laws of Large Numbers 202–203
linear birth process

Subject Index   649

assumptions in 55
deterministic 55
Laplace transform solution 57–59
moments 57
p.g.f. solution 59–60
sequential solution 56–57
simulation of 63–64
time to a given state 60–63
linear birth–death process
    backward solution 77–79
    deterministic 70
    extinction 5, 74–76
    mixture representation 76
    moments 73–74
    reverse transition probabilities 80–81
    simulation of 81–85
    stochastic 70–72
    time to a given state 79–80
linear birth–mass annihilation–immigration process
    equilibrium distribution 102
    moments 102–103
    occupation probability 101–102
linear death process
    assumptions in 50
    bridge probabilities 54–55
    deterministic 50
    moments 51
    reverse transition probabilities 53–54
    simulation of 52–53
    stochastic 50–54
    time to extinction 50–51
linear immigration–birth–death process
    deterministic 86–87
    equilibrium probabilities 90–91, 118
    perfect simulation of 91–95
    stochastic 88–89, 264
linear immigration–death process
    equilibrium solution 88, 119, 263–264
    likelihood 369–378
    stochastic 87–88
linear immigration–death–switch process
    bivariate 16
    moments 334–336
    p.g.f. solution 331–333
linear immigration–emigration process
    equilibrium solution 96–97
    extended state-space approach 99–101
    perturbation approximation 177–178
    queueing process 95–96, 114, 230
    time-dependent solutions 97–101
    traffic intensity 96

linear mass immigration–death process
    cumulants 362–363
    equilibrium solution 103–104, 135
    occupation probabilities 362–364
    simulation of 135–138
logistic process
    assumptions in 8, 123–124
    deterministic 123–124
    diffusion approximation 166–167
    moment closure 195
    quasi-equilibrium 124–125
    simulation of 126–129

marked point processes
    alternative simulation strategies 30, 619–624, 626
    convergence 608–610, 616–619, 625–626
    description 596–598
    discrepancy function 597
    disentangling 28, 597
    fitting parameters 30, 604
    force-biased 626–627
    forest growth 29, 597, 603, 610–613
    general growth-interaction 28, 598–599
    germ–grain 600
    Gibbs processes 600, 604
    growth functions 29, 601, 611, 619–620, 624–625
    hard-core 601
    interaction functions 29, 601–602, 619–620, 624–625
    jammed systems 598, 608
    lilypond model 600
    maximum packing 29
    moving particles 29, 614–619
    packing intensity 614, 616–617
    parameter estimation 625
    parameter selection 603–604
    pseudo-likelihood 604
    pure point process 28, 597
    sequential packing 613–614
    simulation of 29, 604–609, 612–613, 615–624
    soft-core 601–602, 607
    thinning procedures 29–30, 610–613
    tightly packed systems 613–619, 624
    two-point example 602–603
Markov chain Monte Carlo
    basic principles 17, 367–369
    comparison of strategies 381–388
    immigration–death 369–378
    simulation of 370–371
    single/paired immigration–death 378–381

650 Subject Index

Markov chains
  aperiodic 273, 276
  classification of states 13, 274–278
  closed states 280–281
  decomposition theorem 280–281
  definition of 252
  Ehrenfest model 13, 265–271
  ephemeral state 274
  equilibrium solution 257–258, 266–269, 281
  ergodic 276–277, 282
  examples 252–255, 273
  first passage probability 274, 276–280
  first passage time 275
  first return probability 274, 276–280
  irreducible chains 272, 281–282
  limit theorem 278
  lumpable chains 264–265
  matrix solution 12–13, 257–261, 265, 267
  null recurrence 274
  occupation probabilities 255–257, 273
  parameter estimation 256
  periodic 275
  Perron–Frobenius theorem 272, 281
  positive recurrence 274
  recurrence time 261–262, 274
  relation to continuous process 262–265
  stationary distribution 258
  transience 274
  transition probabilities 257
  two-state chain 256–265, 275–276, 279
Markov oscillation problem 105
Markov processes
  classification of states 108–117
  definition of 1, 56
  dishonest 65–70, 109, 112, 116–117
  first passage time distribution 112–113
  honest 66, 115
  integral representation 7
  Kolmogorov equations, forward 5, 108
  Kolmogorov equations, backward 5, 77, 108
  local approximations 156–159
  matrix solution 78, 108
  null recurrent 112, 114
  orthogonal polynomial solution 7, 109–111
  periodic 119
  positive recurrent 112, 114
  quasi-equilibrium distribution, exact 7–8, 119–122
  quasi-equilibrium distribution, pragmatic 7–8, 121–123, 125
  recurrence time distribution 112–113
  semi-group property 109
  stationary distribution 119
  transient 112, 114–116
mass annihilation–immigration process 6, 384–388
Method of Images 6, 100, 235, 303, 305, 308, 508, 510
Method of Separation of Variables 308–309
Metropolis–Hastings algorithm 368, 381–382, 384
moment closure 16, 153–156
moments
  definitions 37–38
  relations between 37–38
  relation with distributions 47–50

network systems 254

odd-even effect 17
optical phenomena
  caustics 594–596
  classical light 350
  examples 349–350
  non-classical light 16, 350, 353
  photon counts 16, 349, 354, 365

perfect simulation 6, 29, 91–95, 607
perturbation techniques
  additive mass-immigration process 175–177
  ϵ-power series 9, 174–175
  examples 168–173
  tilting 173
  time-dependent generalization 178–181
Poisson process
  cumulants 39, 47
  definition of 3, 31
  doubly stochastic 350
  generating functions for 36, 46
  moments 33, 36–40
  multi-rate 44–45
  number of events 34–37
  power of 2 state space 4, 47, 359
  Schoenberg 47, 359
  simulation of 40–41
  simultaneous occurrences 45–47
  super-Poissonness 350, 358
  time-dependent 43–44
  time to events 31–34
polynomials
  Hermite 189, 192, 351
  Laguerre 351
  terminating hypergeometric 440

power-law logistic process
  cumulant equations 149, 151
  deterministic solution 125–126
  m.g.f. equation 148–149
  moment closure 149–153
  quasi-equilibrium 125, 167
  rates 125
  simulation of 126–129
power-law lattice process
  Bartlett's form 591–592
  Cauchy 583–584
  general power-law spectra 584–585
  geometric 582–583
  inverse problem 28, 585–589
  local stationarity 581
  logarithmic 583
  long-range memory dependence 583, 585, 590
  long-term persistence 584
  moments 576–578
  nearest-neighbour 580–582
  Ω-stationarity 27
  Pólya process 89
  pure $1/\omega$-noise 28, 584, 588–589
  scales of pattern 581
  sea waves 592–596
  simulation of 589–594
  spatial-temporal interaction 587–589
  spectrum 578–584
  stochastic equation 27, 575–576
  threshold frequency 580–581
power-law process
  flight length 134
  Lévy-stable 134
  Pareto form 138–140
  sandpile paradigm 134
  scale-free 8, 132–134
  stability 134
  stable densities 134
predator–prey process
  description 18, 413–414
  global stability index 19
  Holling–Tanner model 428–434
  Leslie–Gower model 429
  limit cycles 19, 416, 426, 428–434
  local linearization 417–418, 421–425
  Lotka–Volterra model 414–419
  moments 423–424
  prey cover 426–428
  simulation of 18, 418, 421, 424, 427–428
  stochastic behaviour 418–420, 423, 427–428, 432–434
  time to extinction 425
  Volterra model 419–425

random number generators
  exponential 40
  geometric 316, 518
  inversion 40
  mixed congruential 42–43
  NAG 43
  Normal 622
  Poisson 41
  RANDU 42–43
  uniform 41, 43
Reflection of Images 308
reflection principle 10, 204
regenerative phenomena 104–106
renewal theory 105–106, 231
reverse transition probabilities 53–55, 94–95
reversibility 91

saddlepoint approximation, bivariate
  algorithm for 344
  cumulant truncation 342–345
  examples 340–342, 349, 343–345
  general form 339
  need for caution 346–347, 470–471
saddlepoint approximation, univariate
  basic derivation 9, 178, 181–185
  examples 185–188
  open questions 197–198
  Poisson truncation example 193–195
  relation to Method of Steepest Descents 10, 182, 188–190
  tail approximations 195–196
  truncated 10, 190–197
saw-tooth distributions 49, 187–188, 352, 355–356, 364
simple random walk
  absorbing barriers 11, 215
  definition of 10, 199–201, 252–253
  equilibrium probabilities 228–229, 231
  first passage probability 206–208
  first return probability 206–208
  infimum and supremum 226
  long leads 208–211
  Normal approximation 201–202
  occupation probabilities 201, 209–210, 214–215, 234–238
  path analysis 10, 205–206, 208–211
  probability of absorption 215–223
  reflecting barriers 11, 228–229, 234–238, 280–281
  relations between barriers 232–234
  return to the origin 206–208, 212, 226–228
  simulation of 212–214
  time to absorption 221, 223–225

simulation algorithms
  chemical Langevin 23, 523, 623
  cumulative sum 467
  discrete-time approximation 82, 84, 517, 621
  exact event–time pairs 81, 84–85, 128, 136,
    514, 518–519, 620
  general two-species 399–400
  Gillespie's method 82, 519, 522, 621
  individual based 517–518, 540
  integrated population size 468
  Langevin-leaping 24, 523, 622
  marked point process 605, 620–624
  Markov chain 381–382
  Markov chain Monte Carlo 373
  Metropolis 382
  multi-colony 514, 517–521
  perfect 92
  reaction rate 523, 623
  recurrent epidemic 457–458
  separating births and deaths 84–85
  spatial Markov chain 566
  tau-leaping 23, 519, 521–522, 621–622
skeleton process 229, 256, 282
spatial networks
  bicycle wheel 561
  distance dependent 26, 561
  general connectivity 26, 561
  island 541, 565
  Small World 384, 561, 565
spatial process
  approximation technique 563–565
  conditional model 582
  general results 474–477, 541
  open and closed systems 475–477
  predator–prey 541, 543–549, 559, 562,
    566–567
  Markov chain representation 26, 561–567
  reversibility 475
  simple 475
  simulation of 565–569
  simultaneous model 581–582
  stationary distribution 475
  stochastic population dynamic 21, 26, 474,
    559–561
  travelling waves 25, 526–527, 541–549
  Tribolium confusum application 24, 531–533
stiff systems 23, 522–523
stochastic cellular automata
  Conway's Game of Life 26, 567–570
  foot-and-mouth disease 27, 573–574
  stochastic fire 27, 570–573
symbiosis 389, 414

Telegraph wave 197, 261–264
Turing process
  deterministic systems 26, 549–551, 589
  linearized equations 25, 551–554
  morphogenesis 549–550, 552
  simulation of 555–560
  stable waves 25, 552–555